Multi-Energy Systems

Multi-Energy Systems

Editors

Zbigniew Leonowicz
Arsalan Najafi
Michał Jasiński

MDPI

Basel • Beijing • Wuhan • Barcelona • Belgrade • Novi Sad • Cluj • Manchester

Editors
Zbigniew Leonowicz
Wroclaw University of Science and Technology
Wroclaw
Poland

Arsalan Najafi
Wroclaw University of Science and Technology
Wroclaw
Poland

Michał Jasiński
Wroclaw University of Science and Technology
Wroclaw
Poland

Editorial Office
MDPI
St. Alban-Anlage 66
4052 Basel, Switzerland

This is a reprint of articles from the Topic published online in the open access journals *Energies* (ISSN 1996-1073), *Mathematics* (ISSN 2227-7390), *Smart Cities* (ISSN 2624-6511), *Designs* (ISSN 2411-9660), and *Clean Technologies* (ISSN 2571-8797) (available at: https://www.mdpi.com/topics/multi_energy).

For citation purposes, cite each article independently as indicated on the article page online and as indicated below:

Lastname, A.A.; Lastname, B.B. Article Title. *Journal Name* **Year**, *Volume Number*, Page Range.

ISBN 978-3-0365-9426-2 (Hbk)
ISBN 978-3-0365-9427-9 (PDF)
doi.org/10.3390/books978-3-0365-9427-9

Contents

About the Editors

Zbigniew Leonowicz

Zbigniew Leonowicz (Senior Member, IEEE) received his M.Sc. and Ph.D. degrees in electrical engineering from Wroclaw University of Science and Technology, Wroclaw, Poland, in 1997 and 2001, respectively, as well as a habilitation degree from Bialystok University of Technology, Bialystok, Poland, in 2012. Since 1997, he has worked for the Electrical Engineering Faculty, Wroclaw University of Technology. He was also awarded the title of full professor by the President of Poland and the President of the Czech Republic in 2019. Since 2019, he has worked as a Professor at the Department of Electrical Engineering, where he is currently the Chair of Electrical Engineering Fundamentals.

Arsalan Najafi

Arsalan Najafi received his B.Sc. degree in electrical engineering from the University of Kurdistan, Sanandaj, Iran, in 2009, as well as his M.Sc. and Ph.D. degrees in electrical engineering from the University of Birjand, Birjand, Iran, in 2011 and 2016, respectively. He won a grant from the Polish National Agency for Academic Exchange (NAWA) under the Ulam Program hosted by Wroclaw University of Science and Technology, Wroclaw, Poland. He is currently an associate professor with the Wroclaw University of Science and Technology. His research interests include the operation and planning of multi-energy systems, electric vehicles, electricity markets, and optimization theory and its application in power systems.

Michał Jasiński

Michał Jasiński received his Ph.D. and D.Sc. degrees in electrical engineering from Wrocław University of Science and Technology in 2019 and 2022, respectively. Since 2018, he has worked for the Electrical Engineering Faculty, Wroclaw University of Science and Technology, where he is currently an assistant professor. He is the author and coauthor of more than 100 scientific publications. Currently, he is an Editorial Board Member at *Clean Energy* (Oxford University Press) and the Associate Editor of *Network: Computation in Neural Systems* (Taylor and Francis).

energies

MDPI

Article

Multi-Objective Optimization of Autonomous Microgrids with Reliability Consideration

Maël Riou [1,*], Florian Dupriez-Robin [2], Dominique Grondin [3], Christophe Le Loup [1], Michel Benne [3] and Quoc T. Tran [4]

1 Entech Smart Energies, 29000 Quimper, France; christophe.leloup@entech-se.com
2 France Energies Marines, 29280 Plouzané, France; florian.dupriez.robin@france-energies-marines.org
3 ENERGY Lab—LE2P (FRH2 CNRS), 97744 Saint-Denis, France; dominique.grondin@univ-reunion.fr (D.G.); michel.benne@univ-reunion.fr (M.B.)
4 CEA Tech, 44340 Nantes, France; quoctuan.tran@cea.fr
* Correspondence: mael.riou@entech-se.com

Abstract: Microgrids operating on renewable energy resources have potential for powering rural areas located far from existing grid infrastructures. These small power systems typically host a hybrid energy system of diverse architecture and size. An effective integration of renewable energies resources requires careful design. Sizing methodologies often lack the consideration for reliability and this aspect is limited to power adequacy. There exists an inherent trade-off between renewable integration, cost, and reliability. To bridge this gap, a sizing methodology has been developed to perform multi-objective optimization, considering the three design objectives mentioned above. This method is based on the non-dominated sorting genetic algorithm (NSGA-II) that returns the set of optimal solutions under all objectives. This method aims to identify the trade-offs between renewable integration, reliability, and cost allowing to choose the adequate architecture and sizing accordingly. As a case study, we consider an autonomous microgrid, currently being installed in a rural area in Mali. The results show that increasing system reliability can be done at the least cost if carried out in the initial design stage.

Keywords: microgrid; off-grid; reliability; sizing; genetic algorithm

Citation: Riou, M.; Dupriez-Robin, F.; Grondin, D.; Le Loup, C.; Benne, M.; Tran, Q.T. Multi-Objective Optimization of Autonomous Microgrids with Reliability Consideration. *Energies* **2021**, *14*, 4466. https://doi.org/10.3390/en14154466

Academic Editor: Zbigniew Leonowicz

Received: 29 June 2021
Accepted: 13 July 2021
Published: 23 July 2021

1. Introduction

Electricity is at the heart of modern economies, and its share in the global energy demand continues to increase [1]. Global electricity demand is expected to grow by 30% by 2040, this growth is largely dominated by developing countries. Most modern economies have robust electricity grids, which guarantee a high degree of reliability to end-users. There is a direct link between access to reliable electricity and economic and social development. However, in many places in the world, electricity access is still lacking. Around 759 million people had no access to electricity worldwide in 2019 [2]. Most of the concerned population lives in Sub-Saharan Africa and Asia. For the regions where the electricity grid is not present, different solutions are available. Grid extension appears to be the logical option, however, this solution becomes less viable as the distance from existing grid infrastructure increases, and as the density, load demand, and revenues of the concerned population decrease [3]. One promising alternative is to build small electricity grids known as microgrids which mutualize production assets to consumers as opposed to standalone systems [4]. It is estimated that at least 34 million people had access to electricity from standalone systems (71%) and microgrids (29%) between 2010 and 2017 [3]. Microgrids integrate more and more renewable resources as the prices of these technologies get more competitive. In terms of the type of renewable resources used, we can cite solar photovoltaics, wind, biomass, micro-hydro, and tidal energy [5]. Autonomous microgrids, which have no connection to the national electricity grid, are the topic of interest in this paper. In Section 2, a review is given including important aspects related to reliability and existing methods for designing these types of power systems. In Section 3, the method developed to size autonomous microgrids taking into consideration reliability aspects is introduced. In Section 4, a microgrid project used as a case study for this article is described. In Section 5, the results obtained after applying the proposed method to the case study are presented.

2. Literature Review on Autonomous Microgrid Design

This paper focuses on microgrids that have no ability to connect to the grid and are therefore referred to as autonomous microgrids, also known as mini-grids in the context of rural energy access [6–8]. These systems have been used for a long time as a solution to bring access to electricity to remote locations where grid extension is unaffordable. The cut in renewable energy prices has introduced new types of autonomous microgrids, based on renewable energy resources and energy storage. The power ratings of these systems can range from as little as 50 kVA to a few MVA. Only PV systems and diesel generators (*Gensets*) are considered as the potential sources of generation. PV arrays can be either DC coupled (*PVdc*), AC coupled (*PVac*), or integrated into a hybrid architecture where one part of the solar system is connected to a DC bus and another part is connected to an AC bus. Figure 1 shows the microgrid architecture that is considered for the paper. Energy storage systems (*ESSs*) are used to store excess renewable energy, allowing for a further decrease in the use of fossil-based generation and can be in the form of electrochemical, kinetic, compressed air, or gravitational; however, only battery storage is considered in this paper. A power conversion system (*PCS*) is required to interface the DC sources (*ESS* and *PVdc*) to the AC bus.

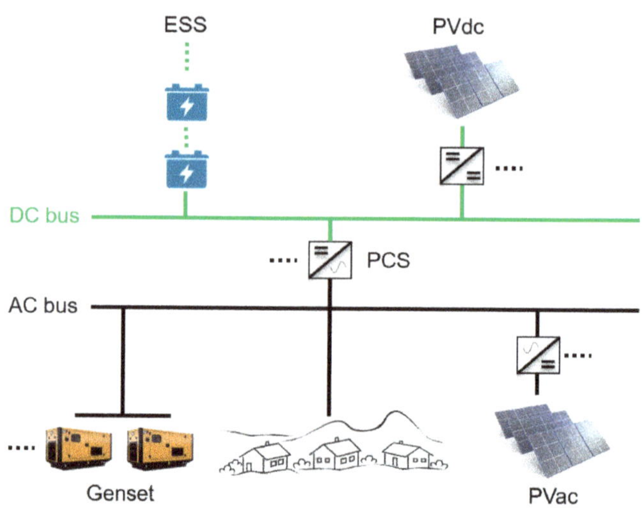

Figure 1. Architecture of the considered microgrids.

Several articles have reviewed the methodologies proposed for the sizing and optimization of autonomous hybrid systems [9–14]. Al-falahi et al. [10] listed various indicators used as design objectives. Most of them are economic, reliability, and environmental indicators. Social indicators can also be found in some papers. The authors have observed that single-objective articles were focusing on the optimization of a cost indicator, whereas multi-objective articles were often focusing on the objectives of cost and reliability. The cost thus represents the first optimization objective and includes investment, operation, maintenance, and replacement costs. A recent review on the sizing methodologies of hybrid renewable energy systems from Lian et al. [15] shows that a large proportion of the reviewed sizing methodologies were focusing on off-grid/autonomous applications

(79%). To classify the different methods available, Tezer et al. [11] distinguish classical optimization approaches to meta-heuristic approaches. Classical methods require limiting mathematical properties linked to the objective function and include for example the iterative optimization method and linear programming. Meta-heuristic methods include higher-level algorithms to control the whole process of search to explore the solution space efficiently and avoid local optima. These methods can be applied to a wide range of optimization problems. We can distinguish "neighborhood" meta-heuristics developing a single solution at a time to "distributed" or "population-based" meta-heuristics that process a whole population at a time, such as particle swarm optimization (PSO) and genetic algorithms (GA). Various articles have used genetic algorithms to size hybrid energy systems. Katsigiannis et al. [16] use the NSGA-II to design a small autonomous hybrid power system comprising of both renewable and conventional power sources with the objectives of minimizing the energy cost of the system and total greenhouse gas emission during the system lifetime. Reliability was however not considered. Kamjoo et al. [17] have used the NSGA-II algorithm to obtain the trade-offs between cost and reliability in order to size a wind/solar/battery system. Roy et al. [18] used the NSGA-II algorithm to size a multi-energy system solely on renewable energy under the objectives of cost (*LCOE*) and reliability. Refs. [19–23] have considered long-term sizing with multi-step investments using optimization techniques.

Some reviews also focus on available tools for the design and planning of hybrid renewable energy systems [24–26]. Various articles use the software Homer to size hybrid energy systems [27]. Homer is an optimization software that is used to design hybrid systems for microgrid/stand-alone applications. It performs simulations of all possible configurations, calculates energy flows, and lists results according to their relative cost of energy (*COE*). iHOGA is another hybrid system optimization tool that can be used similarly to model and simulate various components [28]. The authors in [29] make a comparative assessment of Homer and iHoga using a case study, with the motivation that the latter has not been explored as much in the literature. An interesting feature of iHOGA is its ability to perform multi-objective optimization, using up to three objectives (Net Present Cost, CO_2 emissions, and Unmet Load), it offers also more flexibility in the control strategies used in the simulations.

Reliability can have different meanings and can account for different aspects depending on the application and field. It can be summarized as the ability of a system to perform as intended without any failure and within the desired performance limits for a specified time, in its lifetime conditions [30]. In power systems, reliability deals with power interruptions, whereas power quality concerns the quality of the sine wave when power is available. Therefore, phenomena of interest in power quality studies, such as swells, swags, impulses, and harmonics are not explored in reliability studies. The reliability of power systems can be separated into adequacy and security [31]. Adequacy relates to the ability of power systems to supply the demand with adequate generation and transmission facilities with a desired level of reserve and can be evaluated in long-term planning studies [30]. Security relates to the ability of the power system to withstand sudden contingencies and outages and is more often integrated into short-term reliability assessment.

Reviews have investigated the use of reliability objectives in designing hybrid systems. Several studies involve reliability assessment in the design of microgrids [32–36]. Most of the reliability indicators used relate to adequacy assessment and account for the risk that generation is lower than consumption [11]. The main adequacy indicators used in the literature for sizing hybrid systems are loss of power supply probability (*LPSP*), loss of load probability (*LOLP*), expected energy not supplied (*EENS*), deficiency of power supply probability (*DPSP*), loss of load expected (*LOLE*), and loss of energy expected (*LOEE*) [15]. The software tools for hybrid system optimization mentioned above also account solely for system adequacy. In the Homer software, reliability can be used as a constraint, specifying a maximum capacity shortage fraction allowed. This capacity shortage accounts for the shortage of generation to power the load, as well as insufficient reserves from the reserve requirement set up by the user. iHOGA also includes a reliability constraint using the indicator of unmet load; however, this indicator does not account for operating reserves.

Some papers in the literature have investigated the security assessment of microgrids. In [37], Paliwal et al. use a Particle Swarm Optimization method to determine optimal autonomous microgrid component sizing with the incorporation of reliability constraints. The reliability analysis of the microgrid is carried out using a multi-state availability model (MSAM) of different generators to calculate the percentage of risk state probability (generation is inadequate to supply load) and the percentage of healthy state probability (system has adequate reserves). Xu et al. [38–40] have integrated the consideration of protection and operation into the reliability evaluation of microgrids. However, the reliability analysis developed is not focusing on purely autonomous microgrids with centralized generation and is not integrated into a design method. Escalera et al. [41] suggest that security aspects could be incorporated into reliability analysis in the design phases of microgrids, as the size of the considered systems is small enough to limit the computation time. Security assessment, which in conventional power systems is performed with a short-term horizon, could thus be implemented in long-term planning. Peyghami et al. [30] introduce a new framework for the reliability evaluation of modern power systems. According to the author, security assessment would focus on static phenomena, dynamic and transient, and cybersecurity. In rural autonomous microgrids, security issues are mainly concerned with the stability and thermal limitations of the power electronic interfaces. These limitations impact considerably the protection scheme of the microgrids, as those are typically based on conventional overcurrent devices.

There is thus a research gap in the literature related to the consideration of reliability in the design of isolated microgrid systems, often focusing solely on adequacy. There is a need to model how the design can influence reliability, considering other aspects, such as component failure and protection. There is also a need to explore further the trade-off between design objectives such as cost, reliability, and renewable integration. Therefore, this paper proposes a novel method to size individual components as well as redundancy by exploring the trade-offs between the objectives mentioned above and considering the impact of component failure and protection malfunction on reliability. It aims not to return the optimal sizing of the system, but rather to give the designer the means to carefully select the preferred option according to the observed trade-offs. The method is described in Section 3. A case study is presented in Section 4, and the results obtained from applying the method to this case study are discussed in Section 5 before a conclusion is drawn.

3. Method for Sizing Autonomous Microgrids

This section describes the methodology developed for the sizing and design of autonomous microgrids. The methodology aims to give the user the means to select the optimal component sizing, architecture, as well as control strategy, regarding the objectives of cost, renewable integration, and reliability. There are two general approaches to solve multi-objective optimization problems. The first approach consists of collecting all objectives into a single objective function, using a weighted sum, or treating some objectives as constraints [11]. The second approach is Pareto-based optimization, which uses the Pareto-dominance concept. The Pareto-front is the set of all solutions for which the corresponding objective vectors cannot be improved in any dimension without degradation in another [20]. When considering three objectives, the Pareto-front becomes a three-dimensional surface. The Pareto-based approach was preferred because it does not require fixing a priority or a limit on one of the objectives and it gives the ability to observe trade-offs between optimization objectives.

3.1. Global Multi-Objective Optimisation Method

A genetic algorithm was selected for its ability to implement various control algorithms and component models without the need to adapt the optimization formulation. A schematic of the global method developed is presented in Figure 2. This method is presented in [42]. In the evaluation of each microgrid configuration, the simulation gives indicators for the objectives of cost and renewable energy integration and the reliability analysis gives the indicator of unavailability. Both evaluations are performed in a python environment [43]. The planning horizon is 15 years.

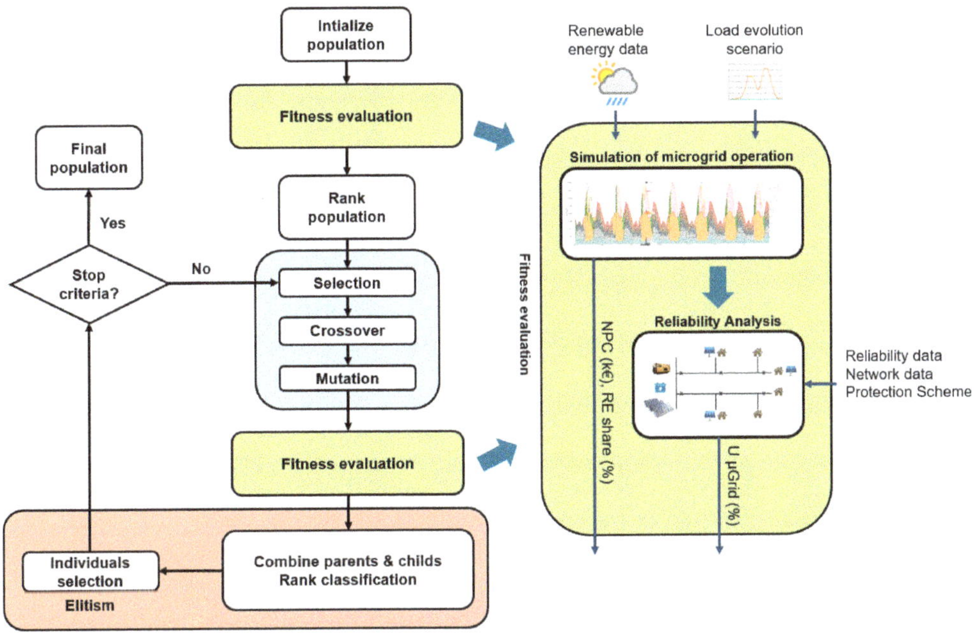

Figure 2. Global multi-objective optimization method developed using the NSGA-II algorithm.

The genetic algorithm NSGA-II, developed by K. Deb et al. [44], is used to obtain the non-dominant pareto frontier of the objectives. The advantage of this type of algorithm is that it can efficiently explore the search space. It starts with creating an initial population of a predefined size. Each individual from the population is then evaluated with the simulation and reliability analysis. The population is then ranked based on three indicators (cost, renewable integration, reliability). The algorithm applies a selection, crossover, and mutation to create a new child population. The parent population and children are then combined and ranked to select individuals for the new generation. This process is replicated until the stop criteria are reached. The selection is based on elitism, ensuring the non-dominated individuals from the combined parent and child populations are passed to the next generation.

The variables to optimize are shown in Table 1. Redundancy of diesel generators and PCS inverters is also considered. Different dispatch strategies are considered including "load following" and "cycle charging" as defined in [45]. The NSGA-II algorithm is implemented with the package Pymoo [46], whose settings are given in Table 2. The computation requirement for a population size of 65 is around 11 min per generation.

Table 1. Sizing variables of the genetic algorithm.

Sizing Variables	Redundancy Variables	Control Variables
$Pnom_{PVac}$		*Dispatch strategy*
$Pnom_{PVdc}$		
$Pnom_{Genset}$	N_{Genset}	
$Pnom_{PCS}$	N_{PCS}	
$Cnom_{ESS}$		

Table 2. Tuning of the genetic algorithm.

Parameter	Value
Population size	65
Number of generations	100
Selection type	Elitism
Crossover probability	1.0
Mutation probability	1.0
Stop criteria	Max number of generations

The first optimization objective is the net present cost (*NPC*) expressed in k€, and is calculated through Equation (1), where NPC_i is the net present cost of component *i* and includes investment cost $C_{inv,i}$, yearly operation cost $C_{O\&M,i,t}$, and replacement cost $C_{rep,i,t}$, as calculated in Equation (2). *r* is the discount rate.

$$NPC = NPC_{PV_{ac}} + NPC_{PV_{dc}} + NPC_{ESS} + NPC_{Genset} + NPC_{AFE} \tag{1}$$

$$NPC_i = C_{inv,i} + \sum_{y=1}^{Y} \frac{C_{O\&M,i,t} + C_{rep,i,t}}{(1+r)^t} \tag{2}$$

The second optimization objective is the renewable integration as is calculated with Equation (3). The objective is calculated through Equation (3), with $P_{Genset}(t)$ being the power produced by all gensets at time *t*, $P_{Load}(t)$ being the load consumption at time *t*, Δt being the simulation time step, and *T* the total number of time steps in the project duration considered.

$$Share_{R.E.} = 1 - \frac{\sum_{t=1}^{T} P_{Genset}(t) \times \Delta t}{\sum_{t=1}^{T} P_{Load}(t) \times \Delta t} \tag{3}$$

The third optimization objective concerns reliability. The unavailability $U_{\mu Grid}$ is to be minimized and is given as the ratio of the expected energy not served (*EENS*) to the yearly load demand. The *EENS* indicator is a sum of three components which are detailed in Section 3.3.

$$U_{\mu Grid} = 100 \cdot \frac{EENS_{\mu Grid}}{E_{Load}} = 100 \cdot \frac{EENS_{Adequ.} + EENS_{Cont.} + EENS_{Prot.}}{E_{Load}} \tag{4}$$

3.2. Simulation Platform Developed

The simulation is made in Python 3.6 (Python Software Foundation, https://www.python.org/ (accessed on 22 July 2021)) and is based on various models describing the behavior of the different microgrid components [43]. The simulation time step is taken as 10 min to account for variability in the load and renewable energy production as well as to model the control of microgrid components with sufficient time resolution. The input data is available as 1-year irradiation and temperature data as well as 15-years load consumption data. Only active power flows are considered in the simulation. The same model is used to calculate the power at the Maximum Power Point for the PV array connected to the AC bus ($P_{PVac,mppt}$) and the one connected to the DC bus ($P_{PVdc,mppt}$). Equation (5) describes the model where $P_{PV,nom}$ is the nominal power of the installed PV array (kWp), $G_{tot,\ \beta}(t)$ is the global horizontal irradiance in the plane of array (W/m^2), and $\eta_{PV,\ glob}(t)$ is the efficiency of the global PV array.

$$P_{PV,mppt}(t) = P_{PV,nom} \times \frac{G_{tot,\ \beta}(t)}{1000} \times \eta_{PV,\ glob}(t) \tag{5}$$

$\eta_{PV,\ glob}(t)$ (p.u.) includes temperature losses, inverter losses, and other miscellaneous losses as calculated in Equation (6), where $\eta_{inv}(t)$ is the inverter efficiency at time t (p.u.), $Losses_{const}$ are constant losses and account for cable losses, mismatch, and dirt (p.u.), α_{temp} is the temperature derating coefficient according to the datasheet of the PV module (%/°C),

$T_c(t)$ is the module cell temperature (°C), and $T_{c,\ ref}$ is the reference cell temperature at Standard Test Conditions (°C).

$$\eta_{PV,\ glob}(t) = \eta_{inv}(t) \times (1 - Losses_{const} \times \left[1 - \frac{\alpha_{temp}}{100} \times \left(T_c(t) - T_{c,\ ref}\right)\right] \quad (6)$$

The module cell temperature $T_c(t)$ is calculated as per Equation (7), where $T_a(t)$ is the ambient temperature at time t (°C), $T_{c\ NOCT}$ is the nominal operating cell temperature [°C], $T_{a\ NOCT}$ is the nominal operating ambient temperature (°C), G_{NOCT} is the nominal operating irradiance (W/m^2).

$$T_c(t) = T_a(t) + (T_{c\ NOCT} - T_{a\ NOCT}) \times \frac{G_{tot,\ \beta}(t)}{G_{NOCT}} \quad (7)$$

The EMS model calculates active power setpoints for each microgrid component. Only the battery system does not receive a setpoint as its power output is the difference between the power from the bi-directional inverter and the power produced by the PV array connected to the DC bus, both controlled by the energy management system (*EMS*).

The genset controller model decides to start/stop individual gensets and dispatches the global genset power setpoint to each available unit. The PV converter model applies a saturation of the active power setpoint to the nominal power rating of the converter as well as an efficiency based on an efficiency versus operating power curve. The *PCS* also applies a saturation and an efficiency to the setpoint but allows for bi-directional power flow.

3.3. *Reliability Analysis Method*

As discussed in Section 2, reliability can address both adequacy and security aspects. In the sizing method developed, security aspects of component failure and protection failure are considered in addition to generation adequacy. These two aspects considered are described in this section. The system size is sufficiently small to be able to integrate these aspects in a genetic algorithm with acceptable computation time. The methodology is described in this section.

3.3.1. Adequacy Assessment

Adequacy relates to the ability of power systems to supply the demand with adequate generation and transmission facilities with a desired level of reserve and can be evaluated in long-term planning studies [30]. The indicator used in this paper for assessing adequacy is the expected energy not supplied (*EENS*), which can be calculated from the simulation results. At each time-step, the load power not supplied due to insufficient generation capability $P_{N.S.}(t)$ is obtained from Equation (8), $P_{prod,\ total}(t)$ being the sum of active powers from all generating sources. The *EENS* indicator is then calculated from Equation (9).

$$P_{N.S.}(t) = \begin{cases} P_{Load}(t) & if\ P_{prod,\ total}(t) < P_{Load}(t) \\ 0 & otherwise \end{cases} \quad (8)$$

$$EENS_{Adequ.} = \sum_{t}^{T} P_{N.S.}(t) \times \Delta t \quad (9)$$

3.3.2. Contingency Enumeration Method

The first security aspect considered is the static response to component failure. Different methods exist to obtain reliability indices in this regard. An enumerative contingency analysis, often used for reliability analysis on conventional power systems, can be easily applied for this application as a small number of components are present in the type of microgrids considered. The following component failures are considered:

- Failure of an AC-coupled PV array;
- Failure of a DC-coupled PV array;

- Failure of a diesel generator;
- Failure of the battery system;
- Failure of a *PCS* inverter.

Each component is modeled with a short-term failure rate $\lambda_{n,t}$ corresponding to the failure probability of component n at time step t. Failure rates are assumed constant throughout the component life. A blackout state is obtained when there is not enough reserve power to counteract the contingency or when no backup master unit is available to take over the role of grid-forming. For each of the considered failures, the steps illustrated in Figure 3 are followed.

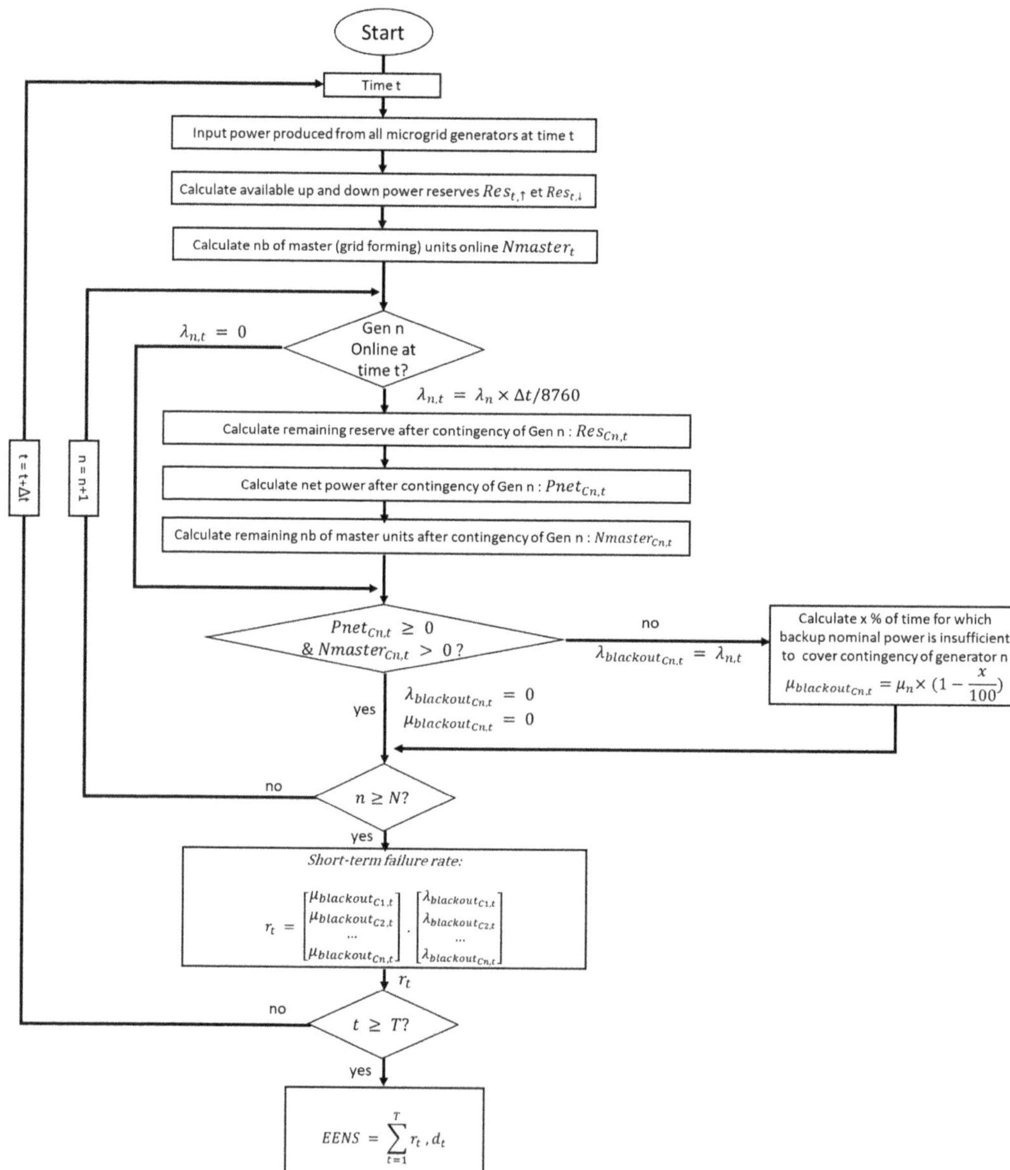

Figure 3. Schematic of the contingency enumeration method.

First, the available up and down reserves before contingency are calculated for each time step from the simulation results. The reserve of the storage system and the reserve of the diesel generators are assumed to be effective in counterbalancing generator failures and to be independent of the grid-forming configuration. The storage system reserve (up and down) is the minimum between the power reserve available on the inverters and the power reserve available in the batteries that could be released during the time required to turn on/off an additional generator (if available). The reserve on the diesel generators is calculated based on their nominal power rating for the up-regulating reserve and based on their minimum acceptable operating power for the down-regulating reserve. The number of master units (operated in grid-forming) depends on the selected grid-forming configuration. In this paper, we consider a single-master configuration, where the grid-forming unit(s) is switched between a diesel generator(s) and the *PCS* inverter(s).

For each considered contingency, the available up and down reserves after the failure of element n, $Pres_{Cn,t}$, are calculated by subtracting the reserve provided by the failed unit from the available reserve before contingency. The net power after contingency is then calculated by subtracting the power produced by the failed unit from the available reserve after failure. It is used to estimate whether the available reserve at time t is sufficient to counterbalance the loss of element n. If at t, component n is generating power, then the up-regulating reserve is used. However, if it is absorbing power, the down-regulating reserve is used. The loss of element n at time t induces a blackout of the microgrid if one of the following conditions are met:

- The net power after contingency is negative: $Pnet_{Cn,t} < 0$;
- The number of remaining master units after contingency is zero: $Nmaster_{Cn,t} = 0$.

If a blackout state is predicted, then the blackout rate $\lambda_{blackoutCn,t}$ due to contingency n is equal to the short-term failure rate of element n $\lambda_{n,t}$ and a repair time $\mu_{blackoutCn,t}$ is allocated. This repair time depends on the remaining nominal power available in the microgrid after contingency. If sufficient nominal power is available to power the load, the repair time is only the time taken to restart the microgrid. Otherwise, the repair time is calculated according to the time the microgrid can be maintained online with the remaining nominal power. The short-term reliability index at each time step t is calculated by summing each product of failure rate and repair time corresponding to all considered contingencies:

$$r_t = \begin{bmatrix} \lambda_{blackoutC1,t} \\ \lambda_{blackoutC1,t} \\ \cdots \\ \lambda_{blackoutCn,t} \end{bmatrix} \cdot \begin{bmatrix} \mu_{blackoutC1,t} \\ \mu_{blackoutC2,t} \\ \cdots \\ \mu_{blackoutCn,t} \end{bmatrix} \tag{10}$$

The chosen index to evaluate the reliability related to component failures is the expected energy not supplied (*EENS*) and is calculated with Equation (11).

$$EENS_{cont.} = \Delta t \cdot \sum_{t=1}^{T} d_t . r_t \tag{11}$$

3.3.3. Protection Reliability Assessment

Protection selectivity is another important issue to address in autonomous microgrids, especially as the microgrids of interest can operate in various modes with different short-circuit levels available. There is a need to design a protection scheme that is operating well in all configurations of the microgrid. The protection scheme that is used in our case is based on conventional overcurrent relays and fuses. These devices require a sufficient level of short-circuit current to operate in case of a fault.

Reliability analysis of the protection scheme aims at assessing how well the protection will perform for a particular architecture and sizing, regarding coordination and selectivity, considering two possible causes of protection malfunction:

- The insensitivity of protections in the case of insufficient short-circuit contribution to trigger the right protection;
- Overtripping in the case of protection triggering in healthy operation of the microgrid.

The different steps toward the protection reliability assessment method are described in Figure 4. The microgrid is modeled with the package Pandapower [47] in the Python environment, which is used for static network analysis. All buses, lines, converters, and loads are modeled. The first step consists of calculating short-circuit currents on each microgrid configuration observed from the simulation. These configurations correspond to all possible on/off combinations of the different short-circuit current contributors, including gensets, *PCS* inverters, and *PVac* inverters. Next, load flow simulations are made for each simulation time-step to calculate the current flowing through each protection. Reliability indicators are then calculated to assess the protection scheme. Three probability distributions must be obtained to calculate these reliability indicators:

- The probability distribution of short-circuit current in each protection *Isc* (blue line in Figure 4). This distribution is obtained using the minimum and maximum short-circuit current calculated for each microgrid configuration and the frequency of occurrence of each configuration;
- The probability distribution of the load flow current *In* (red line in Figure 4). This distribution is obtained directly from the load flow calculation made for each simulation time step;
- The probability distribution of the pick-up current *Ir* (orange line in Figure 4). This distribution is modeled with a normal distribution with a mean equal to the pick-up setting.

The probability of insensitivity of protection i is the probability that the pick-up current is higher than the short-circuit current available at the protection. This probability is calculated by Equation (12) using a convolution of the probability distributions of *Ir* and *Isc*:

$$P_{insensitivity,i} = p(Ir_i > Isc_i) \tag{12}$$

The probability of the over-tripping of protection i is the probability that the load flow current *In* is higher than the pick-up current *Ir*. This is illustrated in Equation (13) and also calculated by convolution:

$$P_{overtripping,i} = p(In_i > Ir_i) \tag{13}$$

These indicators are then combined into a single indicator for reliability assessment of the protection scheme, which is the expected energy not supplied (*EENS*), calculated with Equations (14)–(17), λ_i being the short-circuit rate, $|P_i|$ the mean power flowing through protection i (obtained from simulation results), r_{sc} the short-term repair time of faults, and r_{blc} the repair time following a blackout.

$$EENS_{protection} = \sum_{i}^{I} EENS_{insensitivity,i} + EENS_{overtripping,i} + EENS_{normal,i} \tag{14}$$

$$EENS_{insensitivity,i} = \lambda_i \cdot P_{insensitivity,i} \cdot |P_i| \cdot (r_{sc} + r_{blc}) \tag{15}$$

$$EENS_{overtripping,i} = (1 - \lambda_i) \cdot P_{overtripping,i} \cdot |P_i| \cdot (r_{blc}) \tag{16}$$

$$EENS_{normal,i} = \lambda_i \cdot (1 - P_{insensitivity,i}) \cdot |P_i| \cdot (r_{sc}) \tag{17}$$

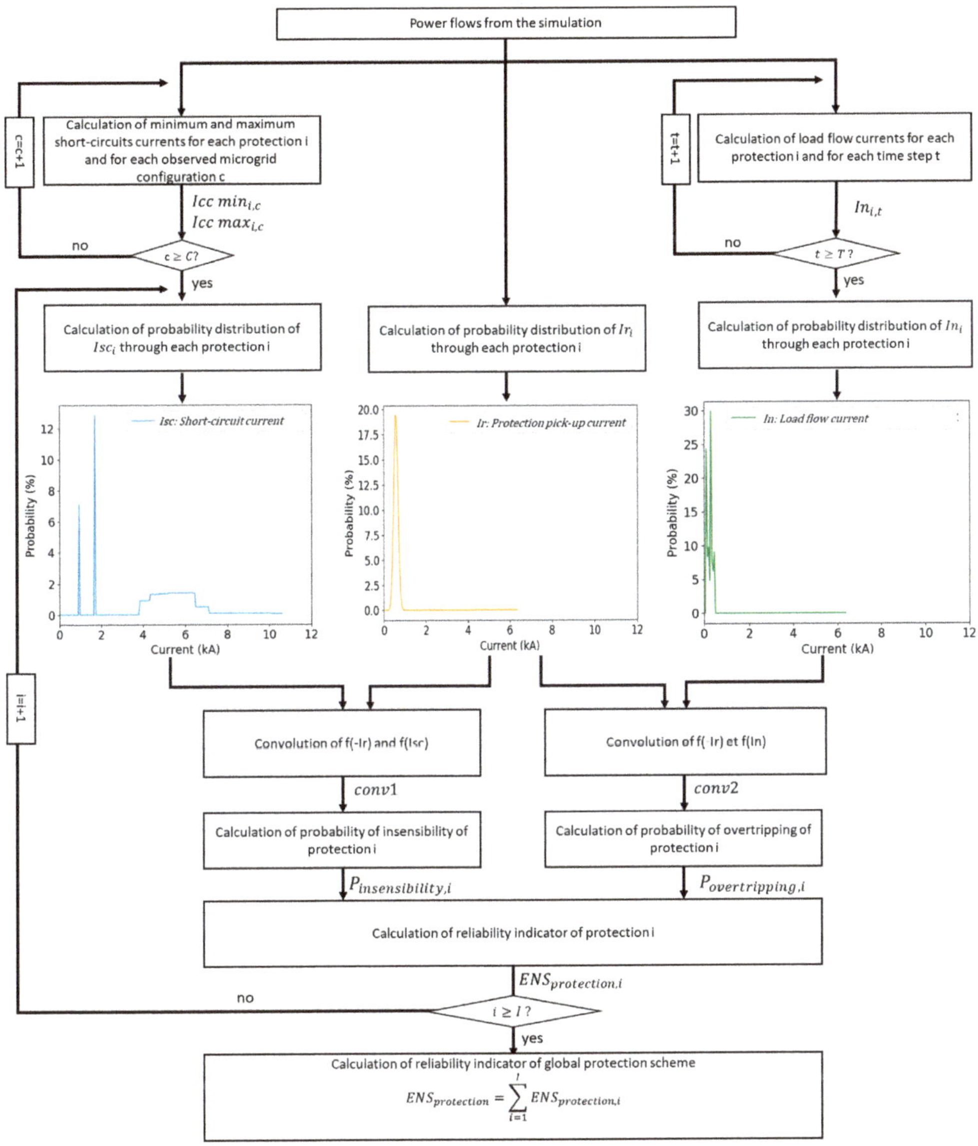

Figure 4. Overview of the method developed to assess protection reliability.

4. Case Study Description

The methodology introduced in the previous section was applied to a case study of an autonomous microgrid currently in installation in the rural localities of Sanando and Tissala in Mali, shown on a map in Figure 5. This microgrid project was enabled by the Energizing Development Program (EnDev) and coordinated by GIZ together with AMADER and the municipality of Sanando. This project aims to build a hybrid power station including a solar PV array, a diesel generator (*Genset*), and a battery storage system (*ESS* + *PCS*) connecting both villages. The operation and maintenance of the microgrid will then be carried out by a consortium including Entech Smart Energies and Sinergie SA. There was initially no electricity grid available to inhabitants, some of them relying on individual solutions (gensets or small solar systems).

Figure 5. Localization of the microgrid case study.

The objectives of the operation of the hybrid system to be installed in the villages of Sanando-Tissala are to minimize fuel consumption on-site, to limit the aging of the equipment, and minimize the risk of blackout. To optimize the performance of the system, the following functions will be implemented in the Energy Management System:

- Energy shifting—this function allows excess solar energy to be stored during the day for later redistribution.
- Genset off capability—this function enables the microgrid to operate on solar + storage only without any diesel generator online. It requires the grid-forming capability on the *PCS* inverters to be able to stabilize frequency and voltage.
- Spinning reserve—this function enables the monitoring of available reserves on the different generators and control the *PCS* inverters to ensure a certain level of reserve (both upregulation and downregulation).

Figure 6 shows the layout of the case study with the variables to optimize using the method. A wide range of values were considered for the optimization variables as shown in Table 3.

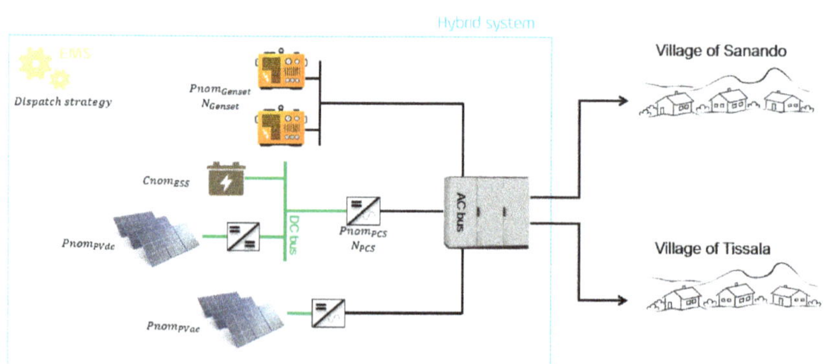

Figure 6. Schematic of the case study.

Table 3. Variable range of the genetic algorithm.

Sizing Variables (Unit)	Minimum Value	Maximum Value	Step Size
P_{Genset} (kW)	0	100	20
P_{PVac} (kWp)	0	1000	20
P_{PVdc} (kWp)	0	1000	20
P_{PCS} (kVA)	0	150	20
C_{ESS} (kWh)	0	1000	20
N_{Genset} (nb)	0	4	1
N_{PCS} (nb)	0	4	1

The simulation of the system operation requires various technical parameters whose values are shown in Table A1 in Appendix A. To calculate the net present cost of each sizing configuration, cost parameters regarding investment, operation, and replacement are also required for each component and are shown in Table A2. Investment costs are modeled with two coefficients, as proposed in [48]. Coefficient b accounts for the decreasing unit cost of the equipment with increasing size. The resulting investment cost is given by Equation (18).

$$C_{inv,i} = P_i \cdot \left(a \cdot P_i^{\ -b} \right) \tag{18}$$

The reliability parameters are given in Appendix A in Table A3 for the contingency enumeration method and in Table A4 for the protection reliability assessment.

5. Results and Discussion

The multi-objectives optimization method presented in Section 3 was applied to the case study. The NSGA-II has led to the 3D Pareto surface shown in Figure 7. There is a strong relationship between all three objectives. To increase the renewable energy integration, an increase in net present cost is required. Configurations without gensets (in blue) lead to an increased unavailability and increased net present cost compared to configurations with gensets (in orange). Configurations with renewable energy integration less than 93% are not included in the Pareto frontier. They are thus not leading to an improvement in either net present cost or reliability. Considering the control strategy, only load following dispatch was found in the Pareto surface, which proves this type of control more interesting for this level of renewable integration.

Figure 8 shows the same Pareto points in a 2D graph with the third objective of reliability shown in a color scale. It can be observed that reliability can be improved with a small increase in net present cost for a similar renewable energy integration. In this figure, six configurations of interest have been selected for more detailed analysis:

1. Least cost: the configuration with the least net present cost;
2. Cost/reliability trade-off: the configuration with the least net present cost that has an unavailability less than 0.1%;
3. Most reliable: the configuration with the lowest unavailability;
4. Cost/reliability/RE trade-off: the configuration with the least net present cost that has an unavailability less than 0.1% and a renewable energy integration above 95%;
5. Most renewable: the cheapest configuration with 100% renewable energy integration;
6. Reliability/RE trade-off: The configuration with the highest renewable integration and unavailability less than 0.1%.

The least-cost configuration (config 1) can be obtained at a net present cost of 1.050 M€ over the 15-year period considered. An improvement in reliability (config 2) can be obtained for a net present cost of 1.074 M€. A solution with 0% unavailability (config 3) can be obtained at a cost of 1.089 M€. A compromise between all three objectives (config 4) can be found for a net present cost of 1.091 M€, having a renewable energy integration above 95% and an unavailability under 0.1%. A 100% renewable energy solution can be obtained for a net present cost of 1.130 M€ but with high unavailability of 1.2% (config 5). Reaching a high level of reliability with 100% renewable energy integration (config 6) would require oversizing considerably the components and, therefore, adds significant costs to the design (1.76 M€).

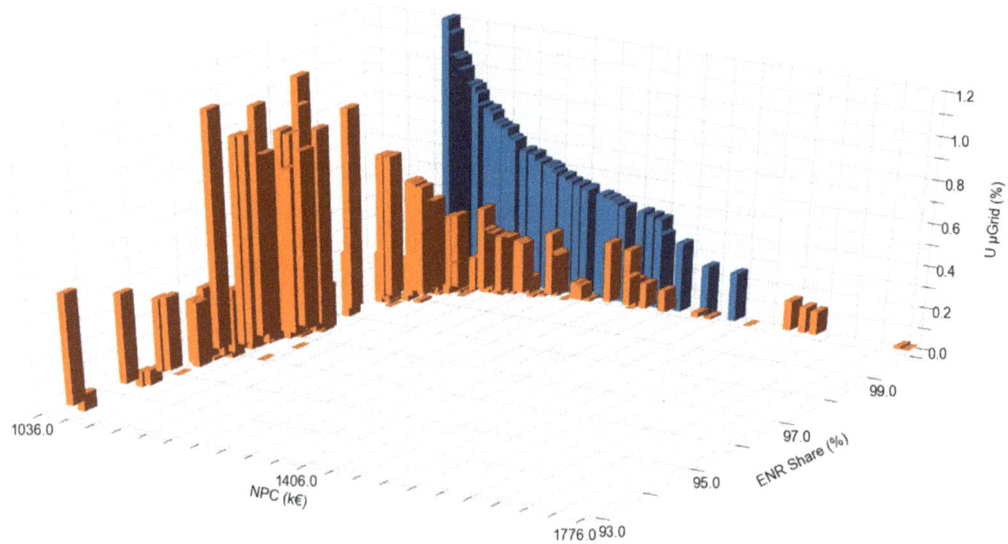

Figure 7. 3D Pareto surface of non-dominated solutions (in orange: solutions with gensets, in blue: solutions with no gensets).

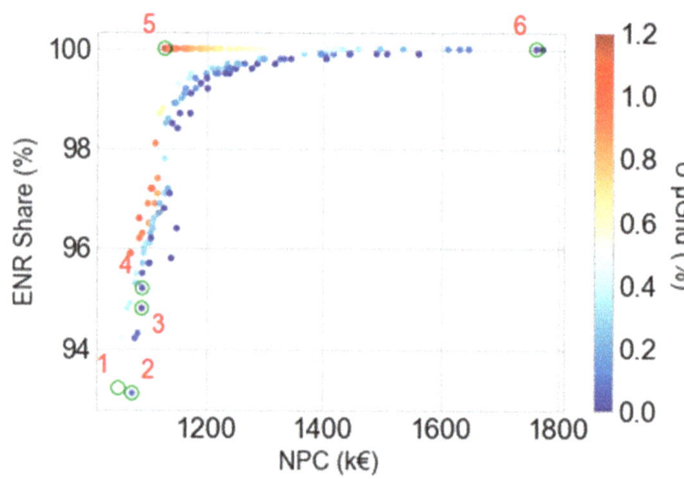

Figure 8. 2D Pareto front with the third objective of reliability shown in color and six selected configurations of interest.

These six configurations are further explored in the following figures. In Figure 9, the reliability indicator is decomposed into the different aspects considered. The least-cost configuration has a significant lack of generation capacity ($U_{Adequacy}$ of 0.4%). The most renewable configuration also has a significant unavailability related to adequacy and to contingencies. For other configurations, unavailability is essentially related to contingencies. The aspect of protection is well managed in these six configurations with a sufficient short-circuit capacity and the configurations of the Pareto surface have an unavailability related to protection malfunction that is null.

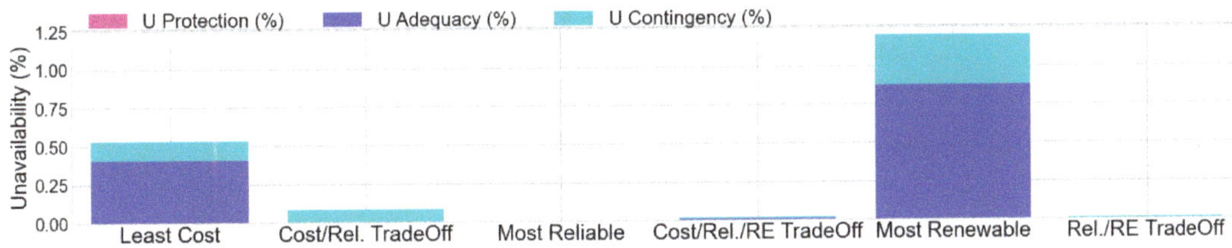

Figure 9. Sources of unavailability for the six selected configurations.

Figure 10 shows the installed PV power, battery capacity, *PCS* power, and the genset nominal power. In terms of architecture, AC-coupled PV power was preferred, except for the reliability/RE trade-off which is a hybrid AC/DC configuration. The least-cost configuration has a small renewable energy capacity in terms of PV power and battery capacity installed (180 kWp/400 kWh). To increase the reliability (cost/reliability trade-off), an increase in the genset capacity is required (60 kW). The most reliable configuration is similar to the "cost/reliability trade-off" with an increase in PV power (200 kWp) and an additional genset unit (2 × 40 kW). The fourth configuration, being a compromise on the three objectives, requires a small increase in PV power and battery capacity as compared to the least cost option as well as three genset units installed (3 × 20 kW). The most renewable configuration has a significant amount of PV power installed (260 kWp) and *ESS* (620 kWh). The configuration with the most renewable integration and constrained unavailability (reliability/RE trade-off) leads to a further increase in PV and battery capacity, without reaching 100% renewable integration. This configuration has also three 13 kW gensets installed, as well as three *PCS* units of 27 kVA each. Apart from this configuration, an optimum size of one unit of 80 kVA for the *PCS* inverter is found.

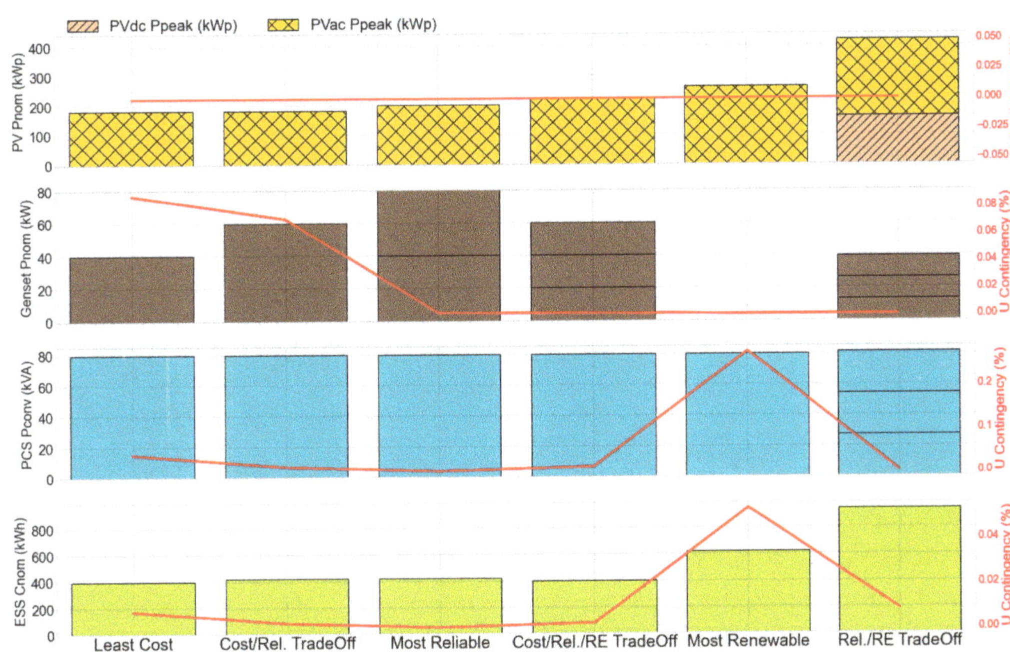

Figure 10. Component sizes for the six selected solutions with unavailability contribution.

Figure 11 shows the energy flows in the 15-year period for each of the six selected configurations. A small part of the energy production in all configurations is from the gensets. When looking at how this energy is consumed, an important share of the PV production is curtailed, from 31% for the least cost configuration up to 68% for the reliability/RE trade-off.

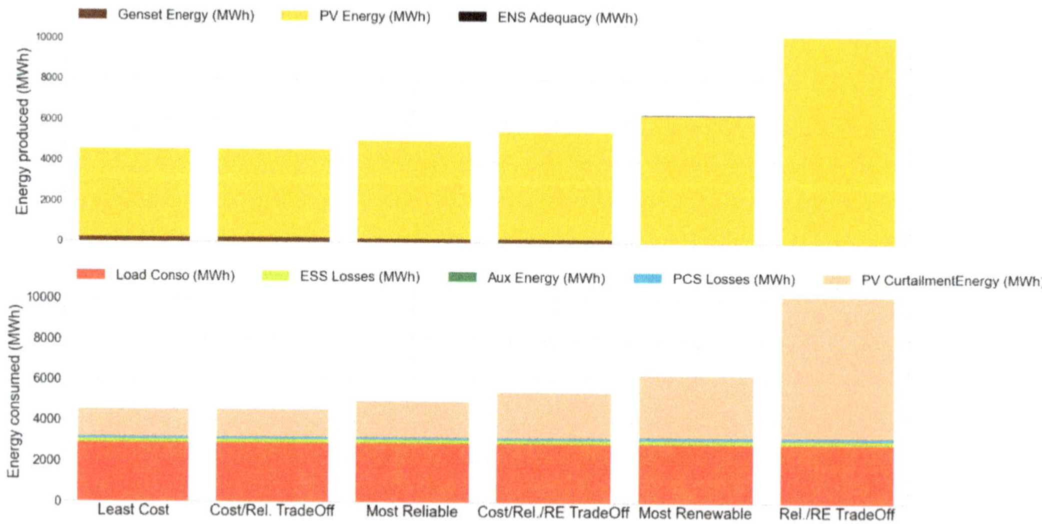

Figure 11. Energy flows for the six selected solutions.

Figure 12 shows the cash flows involved in these six configurations. The investment costs are dominated by the *ESS* and PV systems. BOS corresponds to the balance of system costs to integrate the storage system. Regarding O&M costs, fuel and genset maintenance costs are a significant part of the two first configurations but are less dominant as renewable integration is increased. The replacement costs are dominated by battery costs. Although, the "most renewable" and "reliability/RE trade-off" configurations have a larger battery capacity installed, the cycling is expected to be less and, therefore, the battery replacement cost over the 15 years is less important.

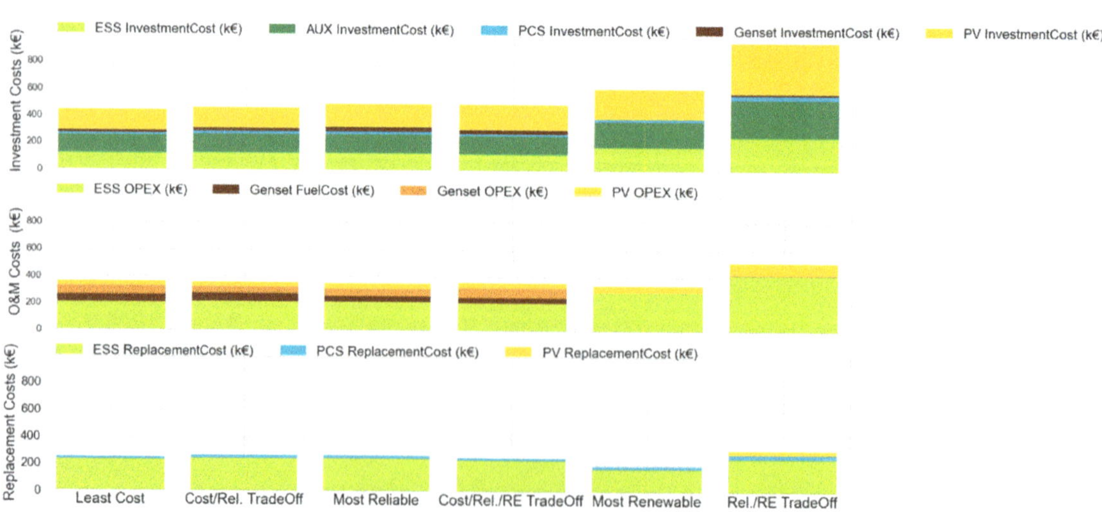

Figure 12. Cashflows for the six selected solutions.

6. Conclusions

This paper presents a method to optimize an autonomous microgrid considering the three design objectives of cost, renewable integration, and reliability. The multi-objective optimization is implemented with a genetic algorithm and involves a simulation of the system operation as well as a reliability analysis for each configuration evaluated. By accounting for reliability related to component failure and protection, the method gives an additional investigation on the impact of microgrid design on power availability. Additionally, rather than finding a single optimal configuration, it helps to understand the trade-offs between all objectives and estimating the cost of improving either renewable integration, reliability, or both. This method was applied to a case study of a rural microgrid in Mali to size the different microgrid components. The Pareto surface obtained shows all non-dominated solutions over the three design objectives. It was first observed that high renewable integration could be obtained without impacting the long-term cost and reliability. The cost of having high reliability was found to be low in this typical case study. Six different solutions were illustrated, each representing a trade-off between the three design objectives. With the proposed method, the user can decide to select a sizing according to the chosen trade-off. Reaching a high level of reliability with 100% renewable energy integration would require oversizing considerably the components and, therefore, add significant costs to the design. Moreover, it leads to an important curtailment of surplus renewable energy. This energy could however be used for other applications such as long-term energy storage, water heating, or water pumping.

Author Contributions: Writing and editing, M.R.; supervision, F.D.-R., D.G., C.L.L., Q.T.T. and M.B. All authors have read and agreed to the published version of the manuscript.

Funding: This research was supported by ANRT.

Data Availability Statement: Not applicable.

Acknowledgments: The authors would like to thank the GIZ for its contribution to this paper and for giving us the opportunity to apply our method to a real case study.

Conflicts of Interest: The authors declare no conflict of interest.

Appendix A. Parameters of the Optimization Method

Table A1. Technical parameters of the optimization method.

Component	Variable	Unit	Value
Genset	P_{min}	%	30
	P_{max}	%	120
	$Cons._{fuel}$ (at 10%, 25%, 50%, 75%, 100% P_{nom})	L/kWh	(0.466, 0.304, 0.305, 0.325, 0.375)
	$NbHours_{ToOverhaul}$	h	15,000
	$Threshold_{Start}$	%	90
	$Threshold_{Stop}$	%	40
Pvac	η_{conv}	%	96
	$Losses_{const}$	%	10
	α_{temp}	%/°C	−0.35
PVdc	η_{conv}	%	97
	$Losses_{const}$	%	10
	α_{temp}	%/°C	−0.35
PCS	η_{conv} (at 20%, 30%, 50%, 75%, 90%, 100% P_{nom})	%	(0.952, 0.962, 0.97, 0.973, 0.974, 0.975)
	$Losses_{const}$	%	5

Table A1. *Cont.*

Component	Variable	Unit	Value
Batteries	SOC_{ini}	%	50
	C-rate	p.u.	1
	η_{ch}	%	93
	η_{dch}	%	93
Aux	$P_{aux\ ,\ const}$	kW	3
	Air conditioning use	On/Off	Off
EMS	Genset Off fun	On/Off	On
	Spinning reserve fun	On/Off	On
	SOC_{min}	%	20
	SOC_{max}	%	100

Table A2. Economical parameters of the optimization method.

Component	C_{inv}—Coef a	C_{inv}—Coef b	$C_{O\&M}$	$C_{rep.}$	$T_{rep.}$
GE	1821 €/kW	0.51	5 €/OpHr	-	-
PVac	730 €/kW	0	1.5% C_{inv} /y	100% C_{inv}	fixed year 15
PVac conv	130 €/kW	0	1.5% C_{inv} /y		fixed year 15
PVdc	730 €/kW	0	1.5% C_{inv} /y		fixed year 15
PVdc conv	200 €/kW	0	1.5% C_{inv} /y		fixed year 10
PCS	1816 €/kW	0.45	1.5% C_{inv} /y		fixed year 10
ESS	593 €/kWh	0.12	5% C_{inv} /y		at EOL
BOS	50% overall ESS cost	0	5% C_{inv} /y	-	-

Table A3. Parameters for the contingency enumeration method [49].

Component Type	Failure Rate (f/Year)	Repair Time (h)
Genset	0.20	438
PVac	0.04	480
PVdc	0.04	480
PCS	0.14	168
ESS	0.03	168

Table A4. Parameters for the protection reliability assessment.

Parameter	Unit	Value
λ_i	occ/an	0.2
r_{cc}	h	4
r_{blc}	h	4
r_{trip}	h	2
δ_{trip}	%	10

References

1. IEA. *World Energy Outlook 2019*; IEA: Paris, France, 2019.
2. Iea, I.; UNSD, W. *WHO Tracking SDG 7: The Energy Progress Report*; World Bank: Washington, DC, USA, 2021.
3. Bhatia, M.; Angelou, N. *Beyond Connections: Energy Access Redefined*; World Bank: Washington, DC, USA, 2015.
4. Aresti, M.; Barclay, A.; Cherubini, P. *RE-Thinking Access to Energy Business Models: Ways to Walk the Water-Energy-Food Nexus Talk in Sub-Saharan Africa*; Res4Africa Foundation: Rome, Italy, 2019.
5. IRENA. *Africa 2030: Roadmap for a Renewable Energy Future*; IRENA: Abu Dhabi, United Arab Emirates, 2015; p. 72.
6. Amanda Kahunzire Off-Grid, Mini-Grid and On-Grid Solar PV Solutions in Africa: Opportunities and Challenges. *Int. Support Netw. Afr. Dev. ISNAD Afr.* **2018**. Available online: https://isnad-africa.org/2018/09/30/off-grid-mini-grid-and-on-grid-solar-pv-solutions-in-africa-opportunities-and-challenges-renewable-energy-green-energy-solar-energy-amanda-kahunzire/ (accessed on 20 June 2021).

7. Asian Development Bank. *Deployment of Hybrid Renewable Energy Systems in Minigrids*; Asian Development Bank: Manila, Philippines, 2017.
8. Bhattacharyya, S.C.; Palit, D. (Eds.) *Mini-Grids for Rural Electrification of Developing Countries*; Green Energy and Technology; Springer International Publishing: Cham, Switzerland, 2014; ISBN 978-3-319-04815-4.
9. Siddaiah, R.; Saini, R.P. A Review on Planning, Configurations, Modeling and Optimization Techniques of Hybrid Renewable Energy Systems for off Grid Applications. *Renew. Sustain. Energy Rev.* **2016**, *58*, 376–396. [CrossRef]
10. Al-falahi, M.D.A.; Jayasinghe, S.D.G.; Enshaei, H. A Review on Recent Size Optimization Methodologies for Standalone Solar and Wind Hybrid Renewable Energy System. *Energy Convers. Manag.* **2017**, *143*, 252–274. [CrossRef]
11. Tezer, T.; Yaman, R.; Yaman, G. Evaluation of Approaches Used for Optimization of Stand-Alone Hybrid Renewable Energy Systems. *Renew. Sustain. Energy Rev.* **2017**, *73*, 840–853. [CrossRef]
12. Sinha, S.; Chandel, S.S. Review of Recent Trends in Optimization Techniques for Solar Photovoltaic–Wind Based Hybrid Energy Systems. *Renew. Sustain. Energy Rev.* **2015**, *50*, 755–769. [CrossRef]
13. Rey, J.M.; Vergara, P.P.; Solano, J.; Ordóñez, G. Design and Optimal Sizing of Microgrids. In *Microgrids Design and Implementation*; Zambroni de Souza, A.C., Castilla, M., Eds.; Springer International Publishing: Cham, Switzerland, 2019; pp. 337–367. ISBN 978-3-319-98686-9.
14. Bourennani, F.; Rahnamayan, S.; Naterer, G.F. Optimal Design Methods for Hybrid Renewable Energy Systems. *Int. J. Green Energy* **2015**, *12*, 148–159. [CrossRef]
15. Lian, J.; Zhang, Y.; Ma, C.; Yang, Y.; Chaima, E. A Review on Recent Sizing Methodologies of Hybrid Renewable Energy Systems. *Energy Convers. Manag.* **2019**, *199*, 112027. [CrossRef]
16. Katsigiannis, Y.A.; Georgilakis, P.S.; Karapidakis, E.S. Multiobjective Genetic Algorithm Solution to the Optimum Economic and Environmental Performance Problem of Small Autonomous Hybrid Power Systems with Renewables. *IET Renew. Power Gener.* **2010**, *4*, 404. [CrossRef]
17. Kamjoo, A.; Maheri, A.; Dizqah, A.M.; Putrus, G.A. Multi-Objective Design under Uncertainties of Hybrid Renewable Energy System Using NSGA-II and Chance Constrained Programming. *Int. J. Electr. Power Energy Syst.* **2016**, *74*, 187–194. [CrossRef]
18. Roy, A. Optimal Energy Management of a Multi-Source System for a Remote Maritime Area. Ph.D.'s Thesis, Université de Nantes (UNAM), Nantes, France, 2019.
19. Fioriti, D.; Pintus, S.; Lutzemberger, G.; Poli, D. Economic Multi-Objective Approach to Design off-Grid Microgrids: A Support for Business Decision Making. *Renew. Energy* **2020**, *159*, 693–704. [CrossRef]
20. Fioriti, D.; Poli, D.; Duenas-Martinez, P.; Perez-Arriaga, I. Multi-Year Stochastic Planning of off-Grid Microgrids Subject to Significant Load Growth Uncertainty: Overcoming Single-Year Methodologies. *Electr. Power Syst. Res.* **2021**, *194*, 107053. [CrossRef]
21. Petrelli, M.; Fioriti, D.; Berizzi, A.; Poli, D. Multi-Year Planning of a Rural Microgrid Considering Storage Degradation. *IEEE Trans. Power Syst.* **2021**, *36*, 1459–1469. [CrossRef]
22. Stevanato, N.; Lombardi, F.; Balderrama, S.; Guidicini, G.; Rinaldi, L.; Pavičević, M.; Quoilin, S.; Colombo, E. Long-Term Optimal Sizing of Rural Microgrids: Accounting for Load Evolution through Multi-Step Investment Plan. 2. In Proceedings of the 14th Conference on Sustainable Development of Energy, Water and Environment Systems, Dubrovnik, Croatia, 1–6 October 2019; SDEWES Centre: Dubrovnik, Croatia, 2019.
23. Pecenak, Z.K.; Stadler, M.; Fahy, K. Efficient Multi-Year Economic Energy Planning in Microgrids. *Appl. Energy* **2019**, *255*, 113771. [CrossRef]
24. Kumar, A.P. Analysis of Hybrid Systems: Software Tools. In Proceedings of the 2016 2nd International Conference on Advances in Electrical, Electronics, Information, Communication and Bio-Informatics (AEEICB), Chennai, India, 27–28 February 2016; pp. 327–330.
25. Sinha, S.; Chandel, S.S. Review of Software Tools for Hybrid Renewable Energy Systems. *Renew. Sustain. Energy Rev.* **2014**, *32*, 192–205. [CrossRef]
26. Erdinc, O.; Uzunoglu, M. Optimum Design of Hybrid Renewable Energy Systems: Overview of Different Approaches. *Renew. Sustain. Energy Rev.* **2012**, *16*, 1412–1425. [CrossRef]
27. HOMER—Hybrid Renewable and Distributed Generation System Design Software. Available online: https://www.homerenergy.com/ (accessed on 6 April 2020).
28. IHOGA—Simulation and Optimization of Stand-Alone and Grid-Connected Hybrid Renewable Systems. Available online: https://ihoga.unizar.es/en/ (accessed on 12 March 2021).
29. Saiprasad, N.; Kalam, A.; Zayegh, A. *Comparative Study of Optimization of HRES Using HOMER and IHOGA Software*; NISCAIR-CSIR: New Delhi, India, 2018; pp. 677–683.
30. Peyghami, S.; Palensky, P.; Blaabjerg, F. An Overview on the Reliability of Modern Power Electronic Based Power Systems. *IEEE Open J. Power Electron.* **2020**, *1*, 34–50. [CrossRef]
31. Vrana, T.K.; Johansson, E. Overview of Analytical Power System Reliability Assessment Techniques. In Proceedings of the CIGRE Symposion, Recife, Brasil, 3–6 April 2011; p. 14.
32. Mitra, J.; Singh, C. Optimal Deployment of Distributed Generation Using a Reliability Criterion. *IEEE Trans. Ind. Appl.* **2016**, *52*, 1989–1997. [CrossRef]

33. Arefifar, S.A.; Mohamed, Y.A.-I.; EL-Fouly, T.H.M. Optimum Microgrid Design for Enhancing Reliability and Supply-Security. *IEEE Trans. Smart Grid* **2013**, *4*, 1567–1575. [CrossRef]
34. Baghaee, H.R.; Mirsalim, M.; Gharehpetian, G.B.; Talebi, H.A. Reliability/Cost-Based Multi-Objective Pareto Optimal Design of Stand-Alone Wind/PV/FC Generation Microgrid System. *Energy* **2016**, *115*, 1022–1041. [CrossRef]
35. Vallem, M.R.; Mitra, J. Siting and Sizing of Distributed Generation for Optimal Microgrid Architecture. In Proceedings of the 37th Annual North American Power Symposium, Ames, IA, USA, 25 October 2005; pp. 611–616.
36. Mitra, J.; Vallem, M.R. Determination of Storage Required to Meet Reliability Guarantees on Island-Capable Microgrids With Intermittent Sources. *IEEE Trans. Power Syst.* **2012**, *27*, 2360–2367. [CrossRef]
37. Paliwal, P.; Patidar, N.P.; Nema, R.K. Determination of Reliability Constrained Optimal Resource Mix for an Autonomous Hybrid Power System Using Particle Swarm Optimization. *Renew. Energy* **2014**, *63*, 194–204. [CrossRef]
38. Xu, X.; Mitra, J.; Wang, T.; Mu, L. Evaluation of Operational Reliability of a Microgrid Using a Short-Term Outage Model. *IEEE Trans. Power Syst.* **2014**, *29*, 2238–2247. [CrossRef]
39. Xu, X.; Mitra, J.; Wang, T.; Mu, L. Reliability Evaluation of a Microgrid Considering Its Operating Condition. *J. Electr. Eng. Technol.* **2016**, *11*, 47–54. [CrossRef]
40. Xu, X.; Mitra, J.; Wang, T.; Mu, L. An Evaluation Strategy for Microgrid Reliability Considering the Effects of Protection System. *IEEE Trans. Power Deliv.* **2016**, *31*, 1989–1997. [CrossRef]
41. Escalera, A.; Hayes, B.; Prodanović, M. A Survey of Reliability Assessment Techniques for Modern Distribution Networks. *Renew. Sustain. Energy Rev.* **2018**, *91*, 344–357. [CrossRef]
42. Riou, M.; Le Loup, C.; Dupriez-Robin, F.; Tran, Q.T.; Grondin, D. Michel Benne A Method for Planning Evolutive Design of Isolated Renewable Microgrids with Multi-Objective Optimisation. Unpublished work. 2021.
43. Welcome to Python.Org. Available online: https://www.python.org/ (accessed on 3 September 2020).
44. Deb, K.; Pratap, A.; Agarwal, S.; Meyarivan, T. A Fast and Elitist Multiobjective Genetic Algorithm: NSGA-II. *IEEE Trans. Evol. Comput.* **2002**, *6*, 182–197. [CrossRef]
45. Aziz, A.; Tajuddin, M.; Adzman, M.; Ramli, M.; Mekhilef, S. Energy Management and Optimization of a PV/Diesel/Battery Hybrid Energy System Using a Combined Dispatch Strategy. *Sustainability* **2019**, *11*, 683. [CrossRef]
46. Pymoo: Multi-Objective Optimization in Python. Available online: https://pymoo.org/index.html (accessed on 3 September 2020).
47. Pandapower. Available online: http://www.pandapower.org/ (accessed on 14 May 2020).
48. Tsuanyo, D.; Azoumah, Y.; Aussel, D.; Neveu, P. Modeling and Optimization of Batteryless Hybrid PV (Photovoltaic)/Diesel Systems for off-Grid Applications. *Energy* **2015**, *86*, 152–163. [CrossRef]
49. Adefarati, T. Reliability Assessment of Distribution System with the Integration of Renewable Distributed Generation. *Appl. Energy* **2017**, *185*, 14. [CrossRef]

energies

MDPI

Article

Thermodynamic Analysis and Systematic Comparison of Solar-Heated Trigeneration Systems Based on ORC and Absorption Heat Pump

Jesús García-Domínguez * and J. Daniel Marcos

Department of Energy Engineering, National Distance Education University, UNED, 28040 Madrid, Spain; jdmarcos@ind.uned.es

* Correspondence: jgarcia5088@alumno.uned.es

Abstract: Modular and scalable distributed generation solutions as combined cooling, heating and power (CCHP) systems are currently a promising solution for the simultaneous generation of electricity and useful heating and cooling for large buildings or industries. In the present work, a solar-heated trigeneration approach based on different organic Rankine cycle (ORC) layouts and a single-effect H_2O/LiBr absorption heat pump integrated as a bottoming cycle is analysed from the thermodynamic viewpoint. The main objective of the study is to provide a comprehensive guide for selecting the most suitable CCHP configuration for a solar-heated CCHP system, following a systematic investigation approach. Six alternative CCHP configurations based on single-pressure and dual-pressure ORC layouts, such as simple, recuperated and superheated cycles, and their combinations, and seven organic fluids as working medium are proposed and compared systematically. A field of solar parabolic trough collectors (SPTCs) used as a heat source of the ORC layouts and the absorption heat pump are kept invariant. A comprehensive parametric analysis of the different proposed configurations is carried out for different design operating conditions. Several output parameters, such as energy and exergy efficiency, net electrical power and electrical to heating and cooling ratios are examined. The study reveals that the most efficient CCHP configuration is the single-pressure ORC regenerative recuperated superheated cycle with toluene as a working fluid, which is on average 25% and 8% more efficient than the variants with single-pressure simple cycle and the dual-pressure recuperated superheated cycle, respectively. At nominal design conditions, the best performing CCHP variant presents 163.7% energy efficiency and 12.3% exergy efficiency, while the electricity, cooling and heating productions are 56.2 kW, 223.0 kW and 530.1 kW, respectively.

Keywords: trigeneration (CCHP); organic rankine cycle (ORC); solar thermal energy; parametric optimisation; performance comparison

Citation: García-Domínguez, J.; Marcos, J.D. Thermodynamic Analysis and Systematic Comparison of Solar-Heated Trigeneration Systems Based on ORC and Absorption Heat Pump. *Energies* **2021**, *14*, 4770. https://doi.org/10.3390/en14164770

Academic Editor: Zbigniew Leonowicz

Received: 15 July 2021
Accepted: 1 August 2021
Published: 5 August 2021

1. Introduction

One of the potential applications that combine the use of low or medium temperature solar energy and organic Rankine cycle (ORC) is a trigeneration thermal system, which can be defined as combined cooling, heating and power (CCHP) production simultaneously from the same energy source [1]. In this regard, the thermodynamic analysis to optimise the performance of this system is an important area of research to improve energy efficiency.

In particular for ORC technology, in the last few years, different investigations have been carried out aimed at evaluating its technical, economic and market penetration differentiating its wide range of application according to the driven energy source [2–7]. In order to compare different configurations of the ORC system and different working fluids, Branchini et al. [8] carried out a parametric analysis through different performance indexes, concluding that both the evaporation pressure and the maximum temperature of the heat source are determining parameters in the performance of the power cycle. Delgado-Torres et al. [9] carried out an analysis and optimisation of a low temperature solar driven ORC system considering different solar collector technologies as well as different cycle configurations and organic working fluids. The results obtained indicate that a recovery stage downstream of the turbine implies higher average temperatures in the cycle, and therefore, higher cycle efficiency.

Likewise, for CCHP systems based on ORC power cycle, several studies were done in recent years to determine the thermal and economic performance for different system configurations [10–15]. Al-Sulaiman et al. [16] analysed and compared three CCHP systems with different prime mover approaches: a solid oxide fuel cell (SOFC), a biomass boiler and SPTCs. The results indicated that the maximum electrical efficiency is achieved for the SOFC system with a value of 19%, being 15% for the biomass system, and 15% for the solar energy system. Al-Sulaiman et al. [17] designed and assessed a trigeneration system driven by solar parabolic trough collectors (SPTCs) to produce 500 kW of electricity through an ORC system. The results show that the maximum electrical efficiency is 15%, while the overall efficiency of the CCHP is 94%. Suleman et al. [18] proposed a new system combining solar and geothermal energy as prime movers for multigeneration applications. The overall energy and exergy efficiencies of the system are found to be 54.7% and 76.4%, respectively. Bellos and Tzivanidis [19] analysed a solar-driven CCHP system through a parametric optimisation for different working fluids and design parameters. In the optimum case, the electric exergy and energy efficiency found are 27.9% and 22.5%, respectively, while the energetic performance varied from 130% to 180%.

The use of SPTCs in combination with different ORC layouts and absorption heat pumps for trigeneration systems have been already examined to date. However, there are no known studies aimed at optimising solar-powered trigeneration systems by means of systematic comparison of multiple ORC configuration and the correspondent parametric analysis. Therefore, the current investigation has a significant contribution by analysing and optimising the use of concentrated solar energy and ORC technology as a prime mover for a trigeneration plant. In this paper, the performance of six alternative CCHP configurations based on single-pressure and dual-pressure ORC layouts, such as simple, recuperated, and superheated cycles, and their combinations, is analysed and compared, considering seven working fluids. All the analysed CCHP configurations are fed with thermal input from SPTCs through a close loop that constrains the minimum temperature of the heat source at the evaporator outlet. A single-effect H_2O/LiBr absorption heat pump is integrated as a bottoming cycle to meet heating and cooling demands simultaneously.

The objective of this work is twofold: on one hand, to provide a comprehensive guide for selecting the most suitable solar-heated CCHP configuration in terms of system energy and exergy efficiency by means of a fair systematic comparison between the six layouts and the seven working fluids; on the other, to evaluate parametrically all the CCHP alternatives for a wide range of solar field outlet temperature and ORC condensation temperature aiming for the design of the most efficient system that may be coupled with buildings or industries for combined generation, or as a back-up, of electricity, cooling and heating.

2. Thermodynamic Analysis of CCHP Solutions

The CCHP system assessed in this study is mainly composed of an ORC as a power generator, which is driven by a field of SPTCs. Six alternative ORC layouts are compared under steady-state conditions and seven organic fluids are considered as working medium. A single-effect H_2O/LiBr absorption heat pump is integrated as a bottoming cycle to meet heating and cooling demands simultaneously.

2.1. Investigated Thermodynamic CCHP Configurations

In order to determine the most suitable solar-heated CCHP configuration, a thermodynamic analysis is conducted for the six configurations represented in Figures 1–6. The six ORC layouts are: (i) single-pressure simple cycle (1P SC), (ii) single-pressure superheated cycle (1P SH), (iii) single-pressure recuperated cycle (1P REC), (iv) single-pressure recuperated superheated cycle (1P REC + SH), (v) single-pressure regenerative recuperated superheated cycle (1P REG + REC + SH) and (vi) dual-pressure recuperated superheated cycle (2P REC + SH).

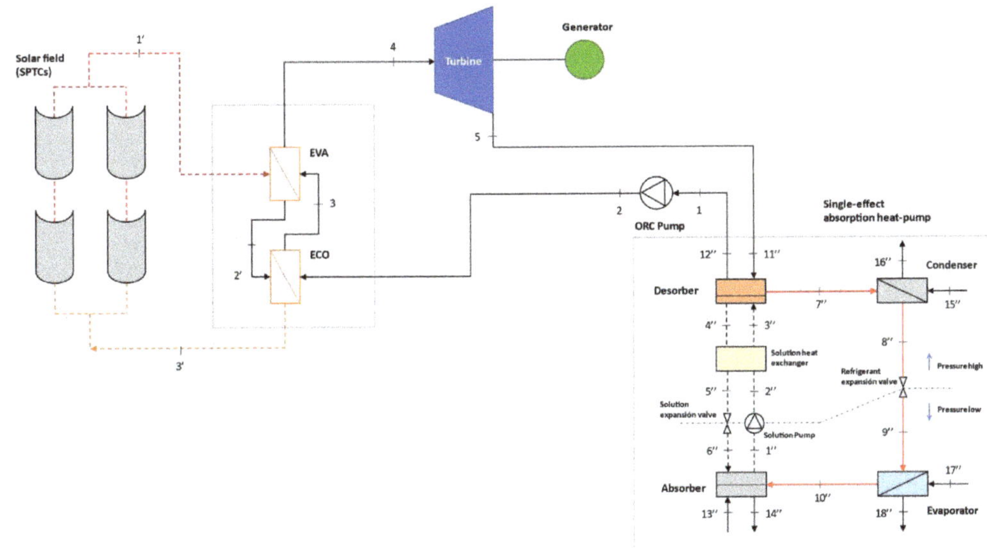

Figure 1. Case 1: CCHP with single-pressure ORC simple cycle (1P SC).

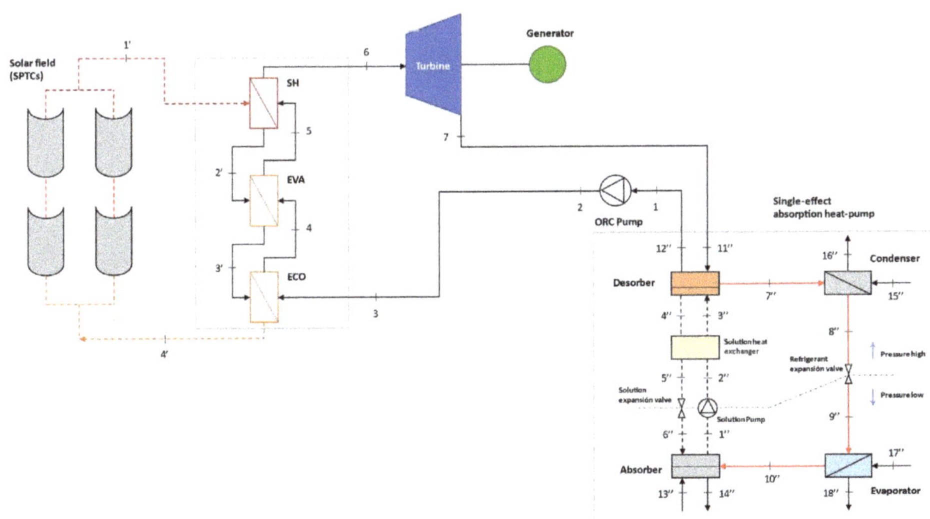

Figure 2. Case 2: CCHP with single-pressure ORC superheated cycle (1P SH).

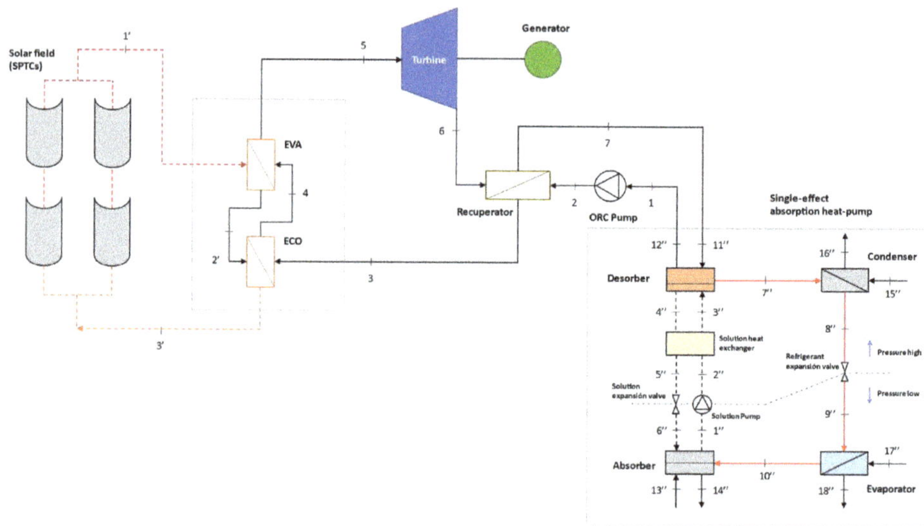

Figure 3. Case 3: CCHP with single-pressure ORC recuperated cycle (1P REC).

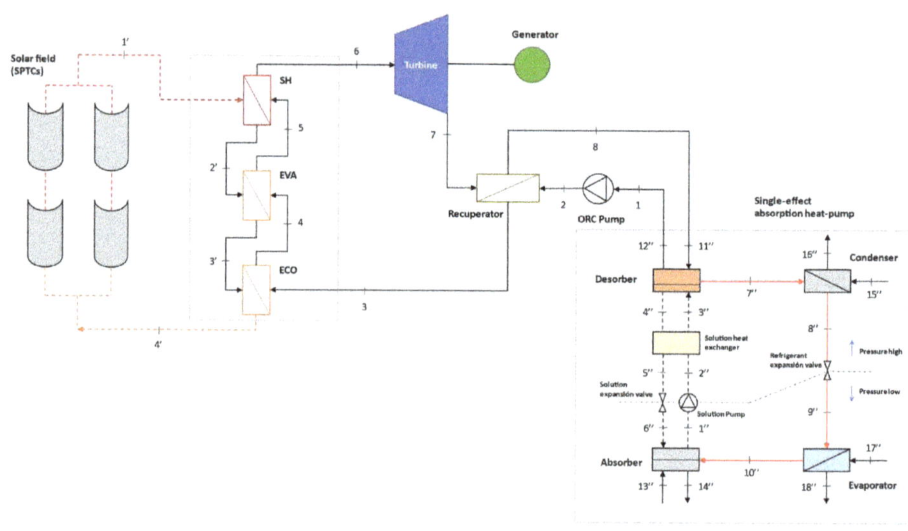

Figure 4. Case 4: CCHP with single-pressure ORC recuperated superheated cycle (1P REC + SH).

The selection of the appropriate working fluid plays a highly important role in the system design as the ORC energy and exergy efficiency must be as high as possible, and the fluid must be chemically stable in the selected working temperature range. Environmental and safety issues must also be considered. For the present work, seven organic working fluids have been selected in order to deal with solar field outlet temperature values between 180 °C and 260 °C, typical values for a field of SPTCs used in existing ORC systems.

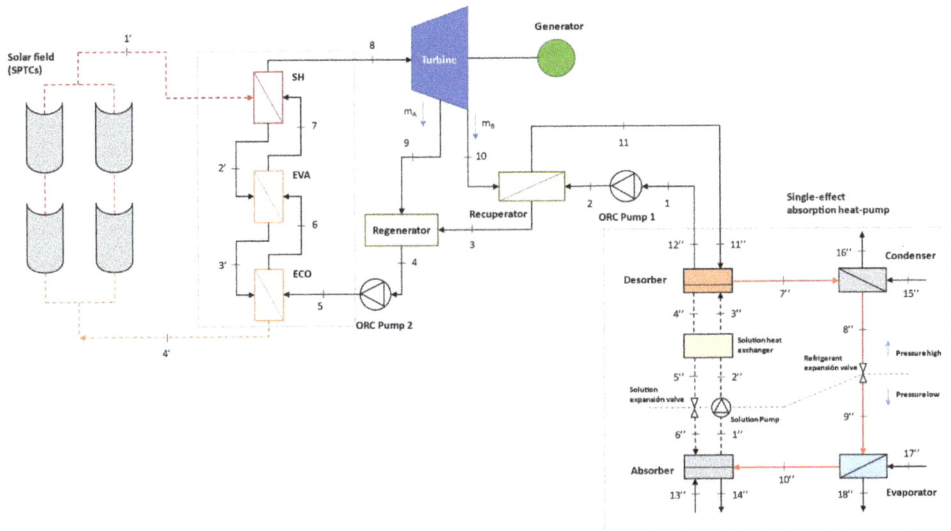

Figure 5. Case 5: CCHP with single-pressure ORC regenerative recuperated superheated cycle (1P REG + REC + SH).

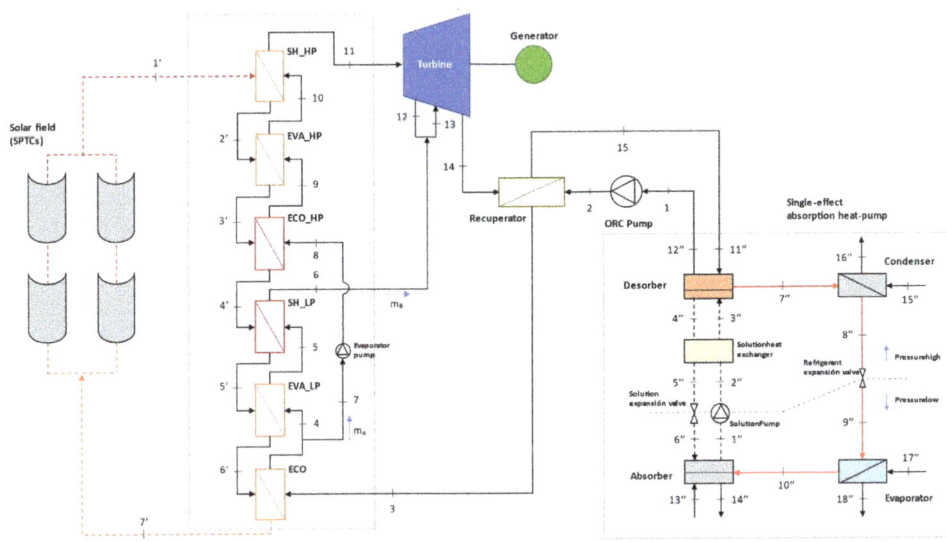

Figure 6. Case 6: CCHP with dual-pressure ORC recuperated superheated cycle (2P REC + SH).

2.2. CCHP Performance Indexes

The overall performance assessment equations of the CCHP considered are presented in this section. The energy and exergy efficiency of the ORC are calculated, taking into account the efficiency of SPTC. The Petela model [20] is used for the exergy flow of the solar irradiation presented in Equation (7).

$$\begin{aligned}
\eta_{en,ORC} &= \frac{W_{turb} - W_{ORC,pump}}{Q_{sol}}; && \text{for Cases 1–4} \\
\eta_{en,ORC} &= \frac{W_{turb} - W_{ORC,pump1} - W_{ORC,pump2}}{Q_{sol}}; && \text{for Case 5} \\
\eta_{en,ORC} &= \frac{W_{turb} - W_{ORC,pump} - W_{Evap,pump}}{Q_{sol}}; && \text{for Case 6}
\end{aligned} \quad (1)$$

$$\eta_{ex,ORC} = \frac{W_{turb} - W_{ORC,pump}}{Q_{sol}\cdot\left(1-\frac{4}{3}\cdot\frac{T_0}{T_{sun}}+\frac{1}{3}\cdot\left(\frac{T_0}{T_{sun}}\right)^4\right)}; \quad for\ Cases\ 1\text{–}4$$
$$\eta_{ex,ORC} = \frac{W_{turb} - W_{ORC,pump1} - W_{ORC,pump2}}{Q_{sol}\cdot\left(1-\frac{4}{3}\cdot\frac{T_0}{T_{sun}}+\frac{1}{3}\cdot\left(\frac{T_0}{T_{sun}}\right)^4\right)}; \quad for\ Case\ 5 \quad (2)$$
$$\eta_{ex,ORC} = \frac{W_{turb} - W_{ORC,pump} - W_{Evap,pump}}{Q_{sol}\cdot\left(1-\frac{4}{3}\cdot\frac{T_0}{T_{sun}}+\frac{1}{3}\cdot\left(\frac{T_0}{T_{sun}}\right)^4\right)}; \quad for\ Case\ 6$$

where:

$$W_{turb} = \dot{m}_{ORC}\cdot(h_4 - h_5); \quad for\ Case\ 1$$
$$W_{turb} = \dot{m}_{ORC}\cdot(h_6 - h_7); \quad for\ Cases\ 2,\ 4$$
$$W_{turb} = \dot{m}_{ORC}\cdot(h_5 - h_6); \quad for\ Case\ 3 \quad (3)$$
$$W_{turb} = (\dot{m}_{ORC,A} + \dot{m}_{ORC,B})\cdot h_8 - \dot{m}_{ORC,A}\cdot h_9 - \dot{m}_{ORC,B}\cdot h_{10}; \quad for\ Case\ 5$$
$$W_{turb} = \dot{m}_{ORC,A}\cdot(h_{11} - h_{12}) + (\dot{m}_{ORC,A} + \dot{m}_{ORC,B})\cdot(h_{13} - h_{14}); \quad for\ Case\ 6$$

$$W_{ORC,pump} = \dot{m}_{ORC}\cdot(h_2 - h_1); \quad for\ Cases\ 1\text{–}4$$
$$W_{ORC,pump1} = \dot{m}_{ORC}\cdot(h_2 - h_1); \quad for\ Case\ 5 \quad (4)$$
$$W_{ORC,pump2} = \dot{m}_{ORC}\cdot(h_5 - h_4); \quad for\ Case\ 5$$
$$W_{Evap,pump} = \dot{m}_{ORC}\cdot(h_8 - h_7); \quad for\ Case\ 6$$

$$Q_{sol} = DNI\cdot w_{ap}\cdot L_{SPTC}\cdot N_{SPTC} \quad (5)$$

The energy and exergy efficiency of the trigeneration system are defined as

$$\eta_{en,tri} = \left(\frac{W_{turb} - W_{ORC,pump} + Q_e + Q_a + Q_c}{Q_{sol}}\right); \quad for\ Cases\ 1\text{–}4$$
$$\eta_{en,tri} = \left(\frac{W_{turb} - W_{ORC,pump1} - W_{ORC,pump2} + Q_e + Q_a + Q_c}{Q_{sol}}\right); \quad for\ Case\ 5 \quad (6)$$
$$\eta_{en,tri} = \left(\frac{W_{turb} - W_{ORC,pump} - W_{Evap,pump} + Q_e + Q_a + Q_c}{Q_{sol}}\right); \quad for\ Case\ 6$$

$$\eta_{ex,tri} = \left(\frac{W_{turb} - W_{ORC,pump} + Q_e\cdot(t_0/t_{17}'' - 1) + (Q_a + Q_c)\cdot(1 - t_0/t_{13}'')}{Q_{sol}\cdot\left(1-\frac{4}{3}\cdot\frac{T_0}{T_{sun}}+\frac{1}{3}\cdot\left(\frac{T_0}{T_{sun}}\right)^4\right)}\right); \quad for\ Cases\ 1\text{–}4$$
$$\eta_{ex,tri} = \left(\frac{W_{turb} - W_{ORC,pump1} - W_{ORC,pump2} + Q_e\cdot(t_0/t_{17}'' - 1) + (Q_a + Q_c)\cdot(1 - t_0/t_{13}'')}{Q_{sol}\cdot\left(1-\frac{4}{3}\cdot\frac{T_0}{T_{sun}}+\frac{1}{3}\cdot\left(\frac{T_0}{T_{sun}}\right)^4\right)}\right); \quad for\ Case\ 5 \quad (7)$$
$$\eta_{ex,tri} = \left(\frac{W_{turb} - W_{ORC,pump} - W_{Evap,pump} + Q_e\cdot(t_0/t_{17}'' - 1) + (Q_a + Q_c)\cdot(1 - t_0/t_{13}'')}{Q_{sol}\cdot\left(1-\frac{4}{3}\cdot\frac{T_0}{T_{sun}}+\frac{1}{3}\cdot\left(\frac{T_0}{T_{sun}}\right)^4\right)}\right); \quad for\ Case\ 6$$

The coefficient of performance (COP) of the heat pump for cooling and heating mode is defined as

$$COP_{cool} = \frac{Q_e}{Q_d + W_{S.pump}} \quad (8)$$

$$COP_{heat} = \frac{Q_c + Q_a}{Q_d + W_{S.pump}} \quad (9)$$

2.3. CCHP Thermodynamic Calculation Procedure and Numerical Assumptions

The mathematical modelling of the proposed trigeneration system with all its variants is based on mass and energy balances applied to each component of the system under steady-state conditions. For a given configuration and a given working fluid, the inlet and outlet thermodynamic states of each component are calculated on the basis of the same given input data and assumptions using Engineering Equations Solver (EES) software.

The energy formulations of the SPTC model (Equations (10)–(14)) are based on the equations presented in [21] for an absorber pipe with a glass envelope, as shown in Figure 7. The energy balance in a section of the absorber pipe depends mainly on: (i) radiation losses from the glass envelope to the open sky ($\dot{q}'_{57rad}$); (ii) convection losses from the glass envelope to the environment ($\dot{q}'_{56conv}$); (iii) radiation losses from the selective coating of the metal tube to the glass envelope ($\dot{q}'_{34rad}$); (iv) conduction losses through metal pipe supports ($\dot{q}'_{cond,bracket}$).

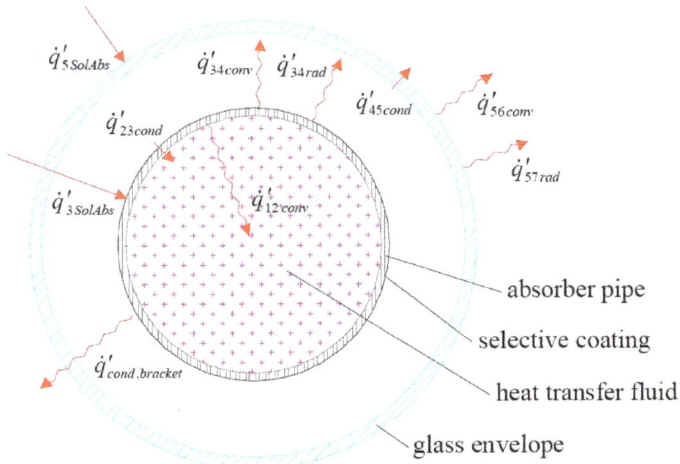

Figure 7. One-dimensional steady-state energy balance of SPTC [21].

All heat losses described in this section are evaluated in an analytical manner using the thermodynamic and fluid-mechanical equations and correlations governing heat transfers by conduction, convection and radiation. A stationary energy balance for the cross-section of the absorber pipe is then proposed, applying the principle of energy conservation to each of the surfaces of the section. Due to the complexity involved in this type of development, numerous simplifying hypotheses have been made. Most of these assumptions are made considering that temperatures, heat fluxes and thermodynamic properties are uniform around the perimeter of the absorber pipe.

Absorber inner surface. The useful heat that the solar thermal oil receives is the result of transfer by conduction through the absorber tube.

$$\dot{q}'_{12conv} = \dot{q}'_{23cond} \tag{10}$$

Absorber outer surface. The heat that the surface of the absorber receives from the sun, after taking into account both the optical and geometric effects of the collector, is the result of the sum of the heat fluxes due to the absorber–glass radiation, internal convection, heat loss through the absorber pipe support brackets and the fraction of energy that is finally conducted through the thickness of the absorber pipe into the fluid.

$$\dot{q}'_{3SolAbs} = \dot{q}'_{23cond} + \dot{q}'_{34rad} + \dot{q}'_{34conv} + \dot{q}'_{cond,bracket} \tag{11}$$

Glass envelope inner surface. The heat that is evacuated from the absorber outer surface through the space between the absorber and the glass envelope (regardless of whether there is a vacuum or not) is the same as that is transferred by conduction through the thickness of the glass.

$$\dot{q}'_{34rad} + \dot{q}'_{34conv} = \dot{q}'_{45cond} \tag{12}$$

Glass envelope outer surface. The heat that falls upon the external surface is in balance with the heat that the system releases to the outside from the external surface of the glass envelope.

$$\dot{q}'_{5SolAbs} + \dot{q}'_{45cond} = \dot{q}'_{56conv} + \dot{q}'_{57rad} \tag{13}$$

Considering that the region between the absorber pipe and the glass envelope has been vacuumed, the convective heat transfer between the two surfaces ($\dot{q}'_{34conv}$) can be considered negligible. Hence, under these assumptions, the useful thermal power ($\dot{q}'_{12conv}$) can be reformulated as follows:

$$\dot{q}'_{util} = \dot{q}'_{3SolAbs} + \dot{q}'_{5SolAbs} - (\dot{q}'_{56conv} + \dot{q}'_{57rad} + \dot{q}'_{cond,bracket}) \tag{14}$$

The overall efficiency of the SPTC considers all types of losses [21,22]: optical, geometric and thermal, and it can be defined as the ratio between the useful thermal power delivered to the solar thermal oil, and the solar resource available based on the direct normal irradiance (DNI).

$$\eta_{SPTC} = \frac{\dot{q}'_{u}}{\dot{q}'_{sol}} = \frac{\dot{q}'_{u}}{DNI \cdot w_{ap}} \tag{15}$$

where $\dot{q}'_{u}$ is defined as

$$\dot{q}'_{u} = \frac{\dot{m}_{sol} Cp_{sol} (T_{sol.out} - T_{sol.in})}{L_{SPTC}} \tag{16}$$

The solar field includes SPTCs (PTMx-24 from the company Soltigua) with a total collecting area of 617.4 m^2, consisting of five rows with two collectors per row. The specifications of the collector and the parameters of the solar system that have been selected in this analysis are defined in Table 1. The selected values are reasonable, and they were taken from Refs [9,17,19,21].

Table 1. Input data for SPTC model.

Parameter	Value
Collector aperture width, –w_{ap}	2.36 m
Collector length–L_{SPTC}	26.16 m
Collector nominal mass flow rate	1 kg/s
Absorber outer diameter	20.5 mm
Absorber inner diameter	22 mm
Glass envelope outer diameter	37.5 mm
Glass envelope inner diameter	40 mm
Number of collectors-N_{SPTC}	10
Solar field outlet temperature–$T_{1'}$	200 °C
Ambient temperature	30 °C
Reference temperature	298.15 K
Sun Temperature	5770 K
Direct Normal Irradiance–DNI	800 W/m^2
Solar incident angle	0°
Wind velocity	3 m/s

The ORC modelling is performed for the six CCHP configuration variants represented in Figures 1–6. Apart from the inputs coming from the solar field model, which are the solar field outlet temperature and mass flow rate, the key input thermodynamic variables required for the calculations are:

- The turbine isentropic efficiency;
- The ORC pump isentropic efficiency;
- For SH cycles, the superheating temperature;
- For REC cycles, the recuperator effectiveness;

- The condensation temperature.

For the ORC layouts corresponding to Case 5 and Case 6, the extraction pressure is selected strategically between condensation and evaporation pressures with the aim to obtain the maximum thermodynamic efficiency of each cycle.

The evaporator, or so-called heat recovery system, is the element that serves as the link between the heat source, provided by the SPTCs, and the steam cycle. In the evaporator, the fluid passes through different stages depending on the ORC layout considered. Initially, in the economiser the fluid is heated to the fluid evaporation temperature minus a Delta T, called the approach point (AP); in the evaporator, heat is added to the saturated liquid to produce saturated vapor at constant temperature and pressure. In case a superheater is considered, the saturated vapor is heated above the evaporation temperature until design conditions are reached. The evaporator design parameters used in the study are the pinch point (PP)—difference between the solar field mass flow and the organic fluid—the approach point (AP)—difference between the organic fluid temperature leaving the evaporator and the saturation temperature—and the live steam outlet temperature T_{LS}. All these values are given in Table 2.

Table 2. Input data for ORC model.

Parameter	Value
Condensation temperature–T_1	90 °C
Turbine efficiency–η_{turb}	85%
Pump isentropic efficiency-$\eta_{ORC,pump}$, $\eta_{Evap,pump}$,	70%
Recuperator efficiency *–η_{REC}	70%
Superheating **–ΔT_{SH}	10 °C
Live steam outlet temperature ***–T_{LS}	$T_{1'}$-25 °C
Live steam outlet temperature **–T_{LS}	$T_{1'}$-25 °C-ΔT_{SH}
Pinch Point–PP	8 K
Approach Point–AP	5 K

* For recuperated cycles (Cases 3–6). ** For superheated cycles (Cases 2, 4–6). *** For non-superheated cycles (Cases 1, 3)

Figure 8 represents the correspondent heat transfer–temperature diagrams for a single-pressure evaporator with superheater (a) that applies to Cases 2, 4–5, and for a dual-pressure evaporator with low-pressure and high-pressure superheaters, (b) that applies to Case 6.

With regard to the absorption heat pump, several modelling studies with experimental validation for specific and generic absorption machines can be found in the literature reviewed [23–26]. In the proposed absorption heat pump model, there is a total of 18 states, each of which is determined by its temperature, pressure, enthalpy, flow, H_2O/LiBr concentration, etc. The assumptions used in the single-effect absorption chiller are:

- Saturated liquid solution at states 1 and 4;
- Subcooled liquid solution at states 2, 3 and 5;
- Vapor-liquid mixture solution at state 6;
- Superheated water vapor at state 7;
- Saturated water liquid at high pressure at state 8;
- Vapor-liquid mixture (water) at state 9;
- Saturated water vapor at low pressure at state 10.

The input data used in the absorption heat pump model is given in Table 3. The selected values are reasonable and conservative to avoid the formation of crystals from the H_2O/LiBr solution.

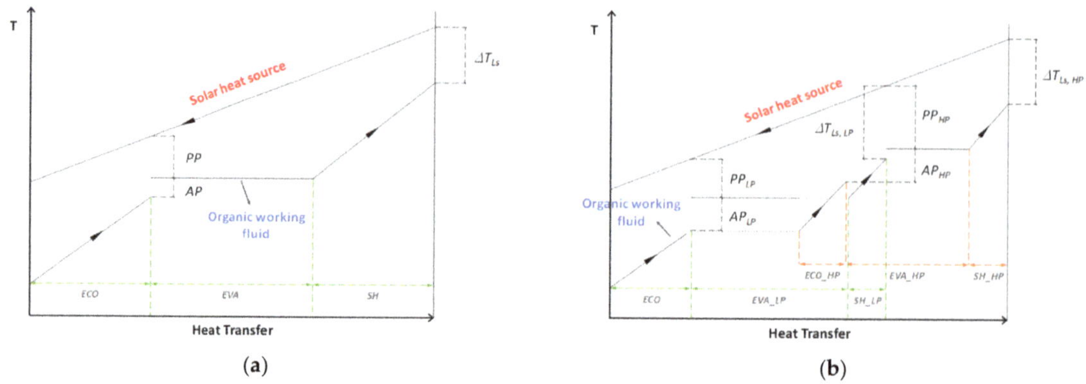

Figure 8. Scheme and heat transfer-temperature diagram for two variants of evaporators: (**a**) Single-pressure with superheater; (**b**) Dual-pressure with low-pressure and high-pressure superheaters.

Table 3. Input data for absorption heat pump model.

Parameter	Value
Maximum solution concentration,–$x_{4''}$	65%
Condensation temperature–$T_{13''}$; $T_{15''}$	20 °C
Condensation mass flow rate	12 kg/s
Evaporation temperature–$T_{17''}$	12 °C
Evaporation mass flow rate	15 kg/s
Solution heat exchanger efficiency–$\eta_{sol.he}$	70%
UA desorber	30 kW/K
UA condenser	70 kW/K
UA absorber	20 kW/K
UA evaporator	30 kW/K

3. Results and Discussion

In the framework of the above constraints and assumptions, the methodology pursued to analyse the CCHP configuration variants from the thermodynamic viewpoint is organised as follow. First of all, for a given configuration and a given working fluid, an analysis of each pair is performed according to the nominal conditions indicated in Tables 1–3. Then, a systematic comparison of each combination is carried out by means of the evaluation of the performance indexes indicated in Section 2.1. Thereafter, a parametric approach is conducted for the best pair (configuration variant and working fluid) to evaluate the effect of different system operating parameters on the energy and exergy efficiency of the ORC and on the overall CCHP system performance. Finally, for each of the identified best pair, a muti-objective optimisation study is performed based on the same operating parameters following the criteria of system energy and exergy.

With such methodology, it is possible to determine the best performing CCHP variant in terms of system energy and exergy efficiency within the six analysed alternatives and for the seven organic working fluids, and on the other hand, to assess how the variation of some design operating parameters can affect the performance of the system and what the optimum values for such parameters are for each variant in terms of system performance.

3.1. Analysis of CCHP Variants

Tables 4–6 represent the energy and exergy efficiency of the ORC and the overall CCHP system performance for each of the proposed CCHP configurations and organic working fluids at nominal conditions, indicated in Tables 1–3.

Table 4. Results for Case 1: CCHP with single-pressure ORC Simple cycle (1P SC).

Fluid	$\eta_{en,ORC}$ [%]	$\eta_{ex,ORC}$ [%]	$\eta_{en,tri}$ [%]	$\eta_{ex,tri}$ [%]	W_{turb} [kW]	Q_e [kW]	Q_a [kW]	Q_c [kW]
Toluene	9.64	10.33	166.10	10.58	48.19	228.83	302.41	241.42
Benzene	**9.74**	**10.43**	**165.90**	**10.68**	**49.15**	**228.49**	**301.98**	**241.06**
n-heptane	8.51	9.12	167.80	9.37	42.98	232.97	307.74	245.83
n-octane	8.56	9.16	167.70	9.42	42.77	232.80	307.53	245.65
n-nonane	8.53	9.14	167.70	9.39	42.42	232.89	307.64	245.74
n-decane	8.53	9.13	167.70	9.39	42.27	232.91	307.67	245.76
MDM	7.32	7.84	169.50	8.10	36.51	237.32	313.35	250.46

Table 5. Results for Case 2: CCHP with single-pressure ORC superheated cycle (1P SH).

Fluid	$\eta_{en,ORC}$ [%]	$\eta_{ex,ORC}$ [%]	$\eta_{en,tri}$ [%]	$\eta_{ex,tri}$ [%]	W_{turb} [kW]	Q_e [kW]	Q_a [kW]	Q_c [kW]
Toluene	9.58	10.26	166.20	10.51	47.85	229.06	302.71	241.67
Benzene	**9.71**	**10.40**	**166.00**	**10.65**	**48.99**	**228.59**	**302.10**	**241.17**
n-heptane	8.36	8.95	168.00	9.21	42.19	233.52	308.45	246.42
n-octane	8.38	8.98	168.00	9.23	41.87	233.44	308.35	246.33
n-nonane	8.36	8.95	168.00	9.21	41.55	233.52	308.45	246.41
n-decane	8.35	8.94	168.00	9.20	41.39	233.56	308.50	246.45
MDM	7.10	7.60	169.80	7.86	35.41	238.12	314.38	251.31

Table 6. Results for Case 3: CCHP with single-pressure ORC recuperated cycle (1P REC).

Fluid	$\eta_{en,ORC}$ [%]	$\eta_{ex,ORC}$ [%]	$\eta_{en,tri}$ [%]	$\eta_{ex,tri}$ [%]	W_{turb} [kW]	Q_e [kW]	Q_a [kW]	Q_c [kW]
Toluene	**10.71**	**11.47**	**164.50**	**11.72**	**53.50**	**224.94**	**297.41**	**237.29**
Benzene	10.53	11.28	164.80	11.53	53.18	225.58	298.23	237.97
n-heptane	10.45	11.19	164.90	11.44	52.78	225.88	298.61	238.28
n-octane	10.57	11.32	164.70	11.57	52.83	225.44	298.05	237.82
n-nonane	10.63	11.39	164.60	11.63	52.84	225.22	297.77	237.59
n-decane	10.66	11.41	164.60	11.66	52.81	225.14	297.66	237.50
MDM	10.20	10.92	165.30	11.17	50.87	226.80	299.80	239.27

The performance indexes indicated in Tables 4–9 show that for the six CCHP configurations and the seven organic working fluids analysed, the best performing variant is the CCHP with single-pressure ORC regenerative recuperated superheated cycle (Case 5) with toluene as a working fluid. The achieved energy and exergy efficiency are: 11.24% and 12.04%, respectively, for the ORC, and 163.7% and 12.3%, respectively, for the CCHP. The electricity, cooling and heating productions are 56.2 kW, 222.3 kW and 530.1 kW, respectively. On average for the seven working fluids considered, in terms of ORC energy efficiency Case 5 is 25% more efficient than Case 1 (1P SC). In terms of which organic working fluid is best suited depending on the configuration, benzene performs best for Cases 1 and 2, and toluene for Cases 3–6.

A CCHP with a single-pressure ORC superheated cycle (Case 2) only results in an increase in efficiency if a recovery stage is available downstream of the turbine. The performance indexes show that on average for the seven working fluids considered, in terms of ORC energy efficiency, Case 2 is 1.6% less efficient than Case 1 (1P SC).

Table 7. Results for Case 4: CCHP with single-pressure ORC recuperated superheated cycle (1P REC + SH).

Fluid	$\eta_{en,ORC}$ [%]	$\eta_{ex,ORC}$ [%]	$\eta_{en,tri}$ [%]	$\eta_{ex,tri}$ [%]	W_{turb} [kW]	Q_e [kW]	Q_a [kW]	Q_c [kW]
Toluene	**10.95**	**11.73**	**164.20**	**11.97**	**54.69**	**224.06**	**296.27**	**236.35**
Benzene	10.83	11.60	164.30	11.84	54.63	224.50	296.84	236.82
n-heptane	10.67	11.42	164.60	11.67	53.82	225.10	297.61	237.45
n-octane	10.76	11.52	164.40	11.76	53.73	224.77	297.18	237.10
n-nonane	10.78	11.55	164.40	11.79	53.57	224.68	297.07	237.01
n-decane	10.79	11.55	164.40	11.80	53.47	224.65	297.03	236.97
MDM	10.29	11.02	165.10	11.27	51.32	226.46	299.36	238.91

Table 8. Results for Case 5: CCHP with single-pressure ORC regenerative recuperated superheated cycle (1P REG + REC + SH).

Fluid	$\eta_{en,ORC}$ [%]	$\eta_{ex,ORC}$ [%]	$\eta_{en,tri}$ [%]	$\eta_{ex,tri}$ [%]	W_{turb} [kW]	Q_e [kW]	Q_a [kW]	Q_c [kW]
Toluene	**11.24**	**12.04**	**163.70**	**12.29**	**56.19**	**222.98**	**294.89**	**235.21**
Benzene	11.17	11.96	163.80	12.20	56.41	223.26	295.24	235.50
n-heptane	10.77	11.53	164.40	11.78	54.42	224.71	297.11	237.04
n-octane	10.90	11.68	164.20	11.92	54.51	224.22	296.48	236.53
n-nonane	10.96	10.55	164.10	11.98	54.49	224.02	296.22	236.31
n-decane	10.99	11.77	164.10	12.02	54.49	223.91	296.08	236.19
MDM	10.36	11.09	165.00	11.34	51.67	226.23	299.07	238.66

Table 9. Results for Case 6: CCHP with dual-pressure ORC recuperated superheated cycle (2P REC + SH).

Fluid	$\eta_{en,ORC}$ [%]	$\eta_{ex,ORC}$ [%]	$\eta_{en,tri}$ [%]	$\eta_{ex,tri}$ [%]	W_{turb} [kW]	Q_e [kW]	Q_a [kW]	Q_c [kW]
Toluene	**10.31**	**11.04**	**165.10**	**11.29**	**51.46**	**226.40**	**299.29**	**238.84**
Benzene	10.18	10.90	165.30	11.15	51.30	226.89	299.91	239.36
n-heptane	10.10	10.82	165.40	11.07	50.96	227.15	300.25	239.64
n-octane	10.09	10.81	165.40	11.06	50.39	227.20	300.31	239.69
n-nonane	10.10	10.82	165.40	11.07	50.19	227.16	300.26	239.65
n-decane	10.11	10.83	165.40	11.08	50.11	227.12	300.21	239.61
MDM	9.67	10.36	166.00	10.61	48.21	228.72	302.28	241.31

The main objective in evaporator design is to minimise losses and maximise heat transfer from the solar heat source. This is achieved by introducing multiple pressure levels; as the temperature curves of the heat source and the organic fluid are better adapted to each other (see Figure 8b) the efficiency of the evaporator increases, but also its complexity and cost, as more heat exchangers are introduced. The results obtained for Case 6 (2P REC + SH) show that the fact to include two pressure levels in the evaporator does not imply a performance improvement of the CCHP system in comparison with Case 3 (1P REC), Case 4 (1P REC + SH) and Case 5 (1P REG + REC + SH); in fact, on average, for the seven working fluids considered, in terms of ORC energy efficiency Case 6 is about 8% less efficient than Case 5. This is explained because the temperature of the heat source at the evaporator outlet is constrained by the close loop of SPTCs, which impacts on the capacity of the dual-pressure evaporator to maximise the heat recovery from the solar heat source.

3.2. Parametric Analysis

In this subsection, a parametric approach is conducted for the best pair analysed previously (configuration variant and working fluid) to evaluate the effect of different system parameters on the energy and exergy efficiency of the ORC and on the overall CCHP system performance.

3.2.1. Effect of the Solar Field Outlet Temperature

The selection of an optimal evaporation temperature for the ORC is determined by the heat delivered by the solar field; a weakness of solar parabolic trough technology is the limited outlet temperature of the solar field [27]. This study aims to illustrate the influence of the solar field outlet temperature, varying in the range of 180–260 °C, on the efficiency of the ORC and on the overall trigeneration system. Table 10 and Figure 9 represent the system performance and electrical and thermal generation for each analysed pair.

Table 10. Results of the parametric simulation with the solar field outlet temperature ($T_{1'}$).

Pair	$T_{1'}$ [°C]	$\eta_{en,ORC}$ [%]	$\eta_{ex,ORC}$ [%]	$\eta_{en,tri}$ [%]	$\eta_{ex,tri}$ [%]	W_{turb} [kW]	Q_e [kW]	Q_a [kW]	Q_c [kW]	COP_{cool}	COP_{heat}
Case 1. w/Benzene	180	8.14	8.72	168.70	8.98	40.92	234.87	310.19	247.85	0.7268	1.727
	190	8.98	9.62	167.30	9.87	45.23	231.54	305.91	244.31	0.7266	1.727
	200	9.74	10.43	165.90	10.68	49.15	228.49	301.98	241.06	0.7264	1.726
	220	11.04	11.82	163.60	12.06	56.00	223.08	295.01	235.31	0.7261	1.726
	240	12.08	12.94	161.50	13.18	61.70	218.41	289.06	230.34	0.7256	1.726
	260	12.92	13.83	159.60	14.07	66.43	214.41	283.92	226.09	0.7253	1.725
Case 2. w/Benzene	180	8.12	8.69	168.70	8.95	40.77	234.96	310.30	247.94	0.7268	1.727
	190	8.96	9.59	167.30	9.84	45.07	231.64	306.03	244.41	0.7266	1.727
	200	9.71	10.40	166.00	10.65	48.99	228.59	302.10	241.17	0.7264	1.726
	220	11.01	11.79	163.60	12.04	55.83	223.17	295.12	235.40	0.7261	1.726
	240	12.07	12.92	161.50	13.16	61.54	218.47	289.14	230.41	0.7256	1.726
	260	12.91	13.83	159.60	14.06	66.32	214.42	283.93	226.10	0.7253	1.725
Case 3. w/ Toluene	180	8.75	9.37	167.80	9.63	43.6	232.65	307.33	245.49	0.7266	1.727
	190	9.76	10.45	166.10	10.71	48.71	228.69	302.23	241.27	0.7264	1.726
	200	10.71	11.47	164.50	11.72	53.50	224.94	297.41	237.29	0.7262	1.726
	220	12.42	13.30	161.50	13.53	62.21	217.99	288.52	229.90	0.7256	1.726
	240	13.90	14.88	158.80	15.12	69.89	211.76	280.52	223.28	0.7251	1.725
	260	15.16	16.23	156.30	16.46	76.56	206.19	273.36	217.36	0.7247	1.725
Case 4. w/ Toluene	180	8.96	9.59	167.50	9.85	44.61	231.90	306.36	244.69	0.7266	1.727
	190	9.99	10.70	165.80	10.95	49.81	227.86	301.17	240.40	0.7264	1.726
	200	10.95	11.73	164.20	11.97	54.69	224.06	296.27	236.35	0.7261	1.726
	220	12.69	13.59	161.10	13.82	63.55	217.00	287.24	228.84	0.7255	1.726
	240	14.20	15.20	158.30	15.43	71.36	210.67	279.11	222.11	0.725	1.725
	260	15.44	16.57	155.90	16.80	77.90	205.21	272.11	216.32	0.7246	1.725
Case 5. w/ Toluene	180	9.02	9.66	167.40	9.92	44.96	231.65	306.05	244.43	0.7266	1.727
	190	10.18	10.90	165.50	11.15	50.78	227.18	300.28	239.67	0.7263	1.726
	200	11.24	12.04	163.70	12.29	56.19	222.98	294.89	235.21	0.7261	1.726
	220	13.16	14.09	160.40	14.33	66.01	215.26	285.01	226.99	0.7254	1.725
	240	14.83	15.87	157.40	16.10	74.71	208.36	276.15	219.66	0.7249	1.725
	260	16.27	17.42	154.60	17.64	82.45	202.12	268.15	213.04	0.7243	1.724
Case 6. w/ Toluene	180	8.15	8.73	168.70	8.98	40.57	234.84	310.15	234.84	0.7268	1.727
	190	9.27	9.93	166.80	10.18	46.22	230.48	304.54	230.48	0.7265	1.727
	200	10.31	11.04	165.10	11.29	51.46	226.40	299.29	226.40	0.7263	1.726
	220	12.23	13.10	161.80	13.33	61.24	218.67	289.34	218.67	0.7256	1.726
	240	13.93	14.91	158.70	15.14	70.02	211.65	280.38	211.65	0.7251	1.725
	260	15.45	16.54	155.80	16.77	78.07	205.11	271.98	205.11	0.7246	1.725

As can be observed in Table 9 and Figure 10, higher values of the solar field outlet temperature mean an increase in ORC energy and exergy efficiency, and in CCHP exergy efficiency. This is due to a higher temperature of the heat source causing a higher organic fluid evaporation pressure in the ORC, leading to higher heat recovery efficiency in the evaporator. For Case 5, which is the best performing variant, with the increase in the heat source inlet temperature, the efficiency of the ORC increases from 9.0% to 16.3%. In terms of relative increase for the electricity produced by the turbine, the increase of the heat source inlet temperature of 180–260 °C represents an increase of 83% (from 45.0 kW to 82.5 kW).

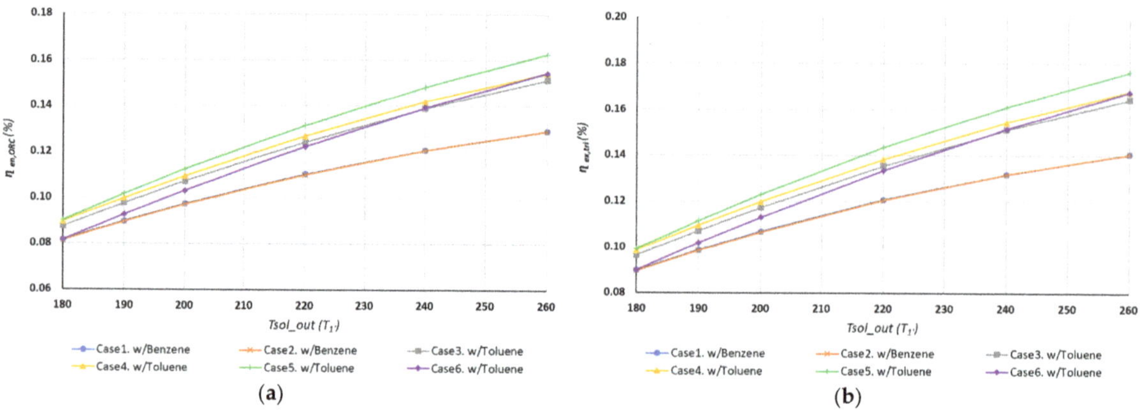

Figure 9. Effect of the solar field outlet temperature on: (**a**) ORC energy efficiency; (**b**) CCHP exergy efficiency.

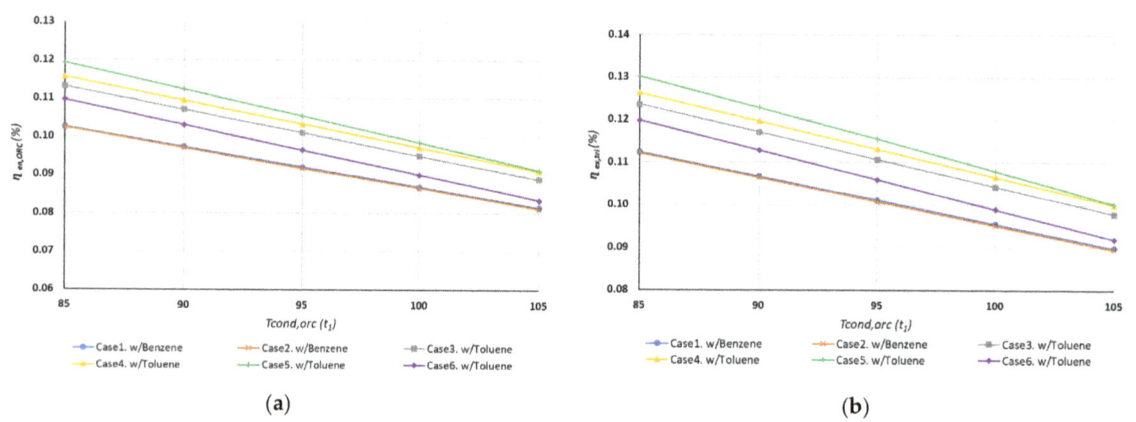

Figure 10. Effect of ORC condensation temperature on: (**a**) ORC energy efficiency; (**b**) CCHP exergy efficiency.

For the CCHP with dual-pressure ORC (Case 6), the relative increase either for the ORC energy efficiency and electricity produced by the turbine with respect to the increase in the heat source inlet temperature of 180–260 °C is significantly greater: 90% for the ORC efficiency (from 8.2% to 15.5%) and 92% for electricity produced by the turbine (from 40.6 kW to 78.1 kW).

3.2.2. Effect of ORC Condensation Temperature

The single-effect absorption heat pump requires a certain heat input in the desorber within a specific temperature range for its operation. This inlet temperature is determined by the condensation temperature of the ORC, so it is important to identify which is the optimal operating temperature based on the production that needs to be prioritised.

In this study, the effect of the ORC condensation temperature is examined from 85 to 105 °C, and system performance and electrical and thermal generation for each analysed pair are presented in Table 11 and Figure 10.

Table 11. Results of the parametric simulation with ORC condensation temperature (T_1).

Pair	T_1 [°C]	$\eta_{en,ORC}$ [%]	$\eta_{ex,ORC}$ [%]	$\eta_{en,tri}$ [%]	$\eta_{ex,tri}$ [%]	W_{turb} [kW]	Q_e [kW]	Q_a [kW]	Q_c [kW]	COP_{cool}	COP_{heat}
Case 1. w/Benzene	85	10.26	10.99	165.20	11.24	51.75	226.57	299.50	239.02	0.7263	1.726
	90	9.74	10.43	165.90	10.68	49.15	228.49	301.98	241.06	0.7264	1.726
	95	9.21	9.87	166.70	10.12	46.55	230.42	304.45	243.11	0.7265	1.727
	100	8.68	9.30	167.50	9.55	43.94	232.34	306.94	245.16	0.7266	1.727
	105	8.15	8.73	168.30	8.99	41.32	234.28	309.43	247.23	0.7267	1.727
Case 2. w/Benzene	85	10.24	10.97	165.20	11.22	51.60	226.65	299.61	239.11	0.7263	1.726
	90	9.71	10.40	166.00	10.65	48.99	228.59	302.10	241.17	0.7264	1.726
	95	9.18	9.83	166.80	10.08	46.37	230.52	304.59	243.23	0.7265	1.727
	100	8.65	9.26	167.60	9.52	43.74	232.46	307.09	245.29	0.7266	1.727
	105	8.12	8.69	168.30	8.95	41.11	234.41	309.60	247.36	0.7267	1.727
Case 3. w/ Toluene	85	11.31	12.11	163.60	12.35	56.45	222.71	294.59	234.92	0.7259	1.726
	90	10.71	11.47	164.50	11.72	53.50	224.94	297.41	237.29	0.7262	1.726
	95	10.11	10.82	165.40	11.07	50.53	227.14	300.23	239.62	0.7263	1.726
	100	9.51	10.18	166.30	10.43	47.55	229.34	303.07	241.97	0.7265	1.726
	105	8.90	9.53	167.20	9.79	44.56	231.55	305.92	244.32	0.7266	1.727
Case 4. w/ Toluene	85	11.56	12.38	163.20	12.62	57.70	221.78	293.40	233.93	0.7259	1.726
	90	10.95	11.73	164.20	11.97	54.69	224.06	296.27	236.35	0.7261	1.726
	95	10.34	11.07	165.10	11.32	51.66	226.30	299.15	238.73	0.7263	1.726
	100	9.72	10.41	166.00	10.66	48.62	228.54	302.04	241.12	0.7264	1.726
	105	9.11	9.75	166.90	10.00	45.57	230.80	304.95	243.52	0.7265	1.727
Case 5. w/ Toluene	85	11.92	12.77	162.70	13.01	59.55	220.46	291.69	232.52	0.7258	1.726
	90	11.24	12.04	163.70	12.29	56.19	222.98	294.89	235.21	0.7261	1.726
	95	10.55	11.30	164.70	11.55	52.77	225.51	298.14	237.89	0.7262	1.726
	100	9.85	10.55	165.80	10.80	49.28	228.09	301.45	240.63	0.7264	1.726
	105	9.13	9.77	166.90	10.03	45.70	230.72	304.85	243.44	0.7265	1.727
Case 6. w/ Toluene	85	10.96	11.73	164.10	11.98	54.65	224.04	296.24	236.33	0.7261	1.726
	90	10.31	11.04	165.10	11.29	51.46	226.40	299.29	238.84	0.7263	1.726
	95	9.66	10.34	166.10	10.59	48.24	228.78	302.35	241.37	0.7264	1.726
	100	9.00	9.64	167.00	9.90	45.00	231.17	305.43	243.92	0.7266	1.727
	105	8.35	8.94	168.00	9.20	41.74	233.57	308.52	246.47	0.7267	1.727

ORC condensation temperature can be a good parameter for controlling the cooling and heating power to be produced by the absorption heat pump. It is observed that as the ORC condensation temperature increases, both the ORC energy efficiency and CCHP exergy efficiency decrease; the lower the condensing pressure, the higher the capacity to extract work from the turbine. For Case 5, with the increase of the ORC condensation temperature, the efficiency of the ORC decreases from 9.1% to 11.9%; in relative terms for the electricity produced by the turbine, the increase in the ORC condensation temperature of 85–105 °C represents a decrease of 23% (from 59.6 kW to 45.7 kW).

Regarding the energy efficiency of the trigeneration system, the effect is the opposite, as the condensation temperature increases the overall efficiency of the system also increases because the heat input to the absorption heat pump desorber is greater, and therefore the heat of the evaporator, absorber and condenser are also greater.

3.3. Optimisation Analysis

The optimisation procedure proposed is based on the optimisation of the analysed operating parameters (see Table 12), and not of the system devices, following strict energy and exergy efficiency criteria. Therefore, a multi-objective optimisation approach is considered for each of the identified best pairs requiring the simultaneous satisfaction of certain objectives, that is the ORC energy efficiency (Equation (1)) and CCHP exergy efficiency (Equation (7)).

Table 12. Optimisation variables.

Parameter	Default Value	Examined Range
Solar field outlet temperature—$T_{1'}$	200 °C	[180–260] °C
ORC Condensation temperature–T_1	90 °C	[85–105] °C

The Pareto front is probably one of the most common approaches used for multi-objective optimisation problems in thermodynamics [28,29]. However, the most straightforward approach to solve these problems is the weighted sum method [30,31], that combines all the multi-objective functions into one scalar by summing the corresponding objectives with some appropriate weights. For the trigeneration system analysis considered in this paper, the bi-objective optimisation is constructed by summing the two beforementioned objectives with some appropriate weights, as follows:

$$\begin{aligned} &MAX\ (MOF = w_1 \cdot \eta_{en,ORC} + w_2 \cdot \eta_{ex,tri}) \\ &0\ \leq w_1,\ w_2\ \leq 1 \\ &w_1 + w_2 = 1 \end{aligned} \tag{17}$$

where w_1 and w_2 are the weighting coefficients for the ORC energy efficiency and CCHP exergy efficiency, respectively. Though any set of optimal solutions can be chosen by selecting the desired values of weighting coefficients, the two objectives are assumed to be of the same importance. The "Conjugate Directions Method" which is supported by EES is used in the bi-objective optimal design (Equation (17)). The results obtained for each of the identified best pair are shown in Table 13.

Table 13. Results of the multi-objective optimisation.

Pair	Opt. Variables		Objectives		Performance Indexes							
	$T_{1'}$ [°C]	T_1 [°C]	$\eta_{en,ORC}$ [%]	$\eta_{ex,tri}$ [%]	$\eta_{ex,ORC}$ [%]	$\eta_{en,tri}$ [%]	W_{turb} [kW]	Q_e [kW]	Q_a [kW]	Q_c [kW]	COP_{cool}	COP_{heat}
Case 1. w/Benzene	260	85	13.33	14.53	14.30	158.90	68.56	218.80	281.85	224.38	0.7252	1.725
Case 2. w/Benzene	260	85	13.36	14.54	14.30	158.90	68.47	212.79	281.84	224.74	0.7252	1.725
Case 3. w/Toluene	260	85	15.67	17.01	16.78	155.50	79.08	204.30	270.94	215.35	0.7245	1.725
Case 4. w/Toluene	260	85	16.02	17.38	17.16	155.00	80.78	203.02	269.30	213.99	0.7244	1.724
Case 5. w/Toluene	260	85	16.82	18.23	18.02	153.80	85.20	200.07	265.51	210.86	0.7241	1.724
Case 6. w/Toluene	260	85	16.00	17.36	17.14	155.00	80.78	203.10	269.39	214.07	0.7244	1.724

The obtained results remark that the optimum design for all the analysed cases is produced for the maximum solar field outlet temperature (260 °C) and the minimum ORC condensation temperature (85 °C). The best performing pair is Case 5 with toluene, presenting values of ORC energy efficiency and CCHP exergy efficiency of 16.82% and 18.23%, respectively. In comparison with nominal design conditions, the optimum design for Case 5 is, in terms of ORC energy efficiency, 50% more efficient.

4. Conclusions

A comprehensive and systematic comparative thermodynamic analysis of six different solar-heated CCHP systems based on ORC and absorption heat pump is conducted. Any configuration can produce electricity, heating and cooling in temperature levels ideal for building or small-medium industry applications. The most suitable CCHP configuration has been identified in terms of system energy and exergy efficiency, as well as the best working fluid for each configuration variant. Through parametric and muti-objective optimisation analysis, it has been possible to determine how the solar field outlet temperature and the ORC condensation temperature affect the performance of the CCHP system for each best pair (configuration variant and working fluid). The main findings of the study are summarised below:

- For the six CCHP configurations and the seven organic working fluids analysed, the best performing variant is the CCHP with single-pressure ORC regenerative recuperated superheated cycle (Case 5) with toluene as a working fluid. The achieved energy and exergy efficiency are: 11.24% and 12.04%, respectively, for the ORC, and 163.7% and 12.3%, respectively, for the CCHP. The electricity, cooling and heating productions are 56.2 kW, 222.3 kW and 530.1 kW, respectively.
- For the seven organic working fluids analysed, benzene performs best for Cases 1 and 2, and toluene for Cases 3–6.
- At nominal conditions and on average for the seven working fluids considered, Case 5 is about 25% more efficient than Case 1, and about 8% more efficient than Case 6 in terms of ORC energy efficiency.
- A CCHP with single-pressure ORC superheated cycle (Case 2) only results in an increase in efficiency if a recovery stage is available downstream of the turbine.
- The use of a dual-pressure evaporator (Case 6) does not imply a performance improvement of the CCHP system if the temperature of the heat source at the evaporator outlet is constrained.
- A higher temperature of the solar heat source causes a higher organic fluid evaporation pressure in the ORC, leading to higher heat recovery efficiency in the evaporator and in CCHP efficiency. For Case 5 with toluene, the electricity produced by the turbine presents an increase of 83% as with the increase of the heat source inlet temperature from 180 to 260 °C.
- As the ORC condensation temperature increases, both the ORC energy efficiency and CCHP exergy efficiency decrease. For Case 5 with toluene, the increase in the ORC condensation temperature from 85 to 105 °C represents a decrease of 23% of the electricity produced by the turbine.
- The optimum design conditions for all the analysed cases are produced for the maximum solar field outlet temperature (260 °C) and the minimum ORC condensation temperature (85 °C). For Case 5 with toluene, in comparison with nominal design conditions, the optimum design is 50% more efficient in terms of ORC energy efficiency.

Author Contributions: Conceptualisation, J.G.-D. and J.D.M.; methodology, J.G.-D. and J.D.M.; software, J.G.-D.; validation, J.D.M.; formal analysis, J.G.-D. and J.D.M.; investigation, J.G.-D.; resources, J.G.-D. and J.D.M.; data curation, J.G.-D. and J.D.M.; writing—original draft preparation, J.G.-D.; writing—review and editing, J.G.-D. and J.D.M.; visualisation, J.G.-D.; supervision, J.D.M.; project administration, J.D.M.; funding acquisition, J.D.M. All authors have read and agreed to the published version of the manuscript.

Funding: The authors would like to acknowledge the financial support of the Regional Research and Development in Technology Programme 2018 (ref. P2018/EMT-4319) in the frame of the ACES2030-CM project.

Data Availability Statement: Data are contained within the article.

Conflicts of Interest: The authors declare no conflict of interest.

Nomenclature

Symbols	
h	heat transfer coefficient, $W/(m^2 K)$
$\dot{m}$	mass flow rate, kg/s
T	temperature, °C
ΔT	temperature difference, °C
η	efficiency
W	electric power, kW
Q	thermal power, kW
$\dot{q}'$	heat rate per SPTC unit length, kW/m
w_{ap}	aperture width of SPTC, m
L	length of SPTC, m
N	number of SPTCs
θ	solar incidence angle on the SPTC, °
Acronyms	
CCHP	combined cooling heating and power
ORC	organic Rankine cycle
SPTC	solar parabolic trough collector
REC	recuperator heat exchanger or recuperated cycle
REG	regenerative cycle
SC	simple cycle
SH	superheater heat exchanger or superheated cycle
DNI	direct normal irradiance
LS	live steam
PP	pinch point
AP	approach point
COP	coefficient of performance
LP	low pressure
H	high pressure
UA	overall heat transfer coefficient
Subscripts	
en	energy
ex	exergy
sol	solar field
0	atmospheric conditions
in	inlet
out	inlet
turb	turbine
cond	heat conduction
conv	heat convection
rad	heat radiation
SolAbs	solar absorption
tri	trigeneration
d	heat pump desorber
a	heat pump absorber
s.he	heat pump solution heat exchanger
c	heat pump condenser
e	heat pump evaporator
Evap.pump	evaporator pump
s.pump	solution pump
cool	heat pump cooling mode
heat	heat pump heating mode

References

1. Al-Sulaiman, F.A.; Hamdullahpur, F.; Dincer, I. Trigeneration: A comprehensive review based on prime movers. *Int. J. Energy Res.* **2011**, *35*, 233–258. [CrossRef]
2. Vélez, F.; Segovia, J.J.; Martín, M.C.; Antolín, G.; Chejnec, F.; Quijano, A. A technical, economical and market review of organic Rankine cycles for the conversion of low-grade heat for power generation. *Energy Convers. Manag.* **2013**, *69*, 209–216. [CrossRef]
3. Quoilin, S.; Van Den Broek, M.; Declaye, S.; Dewallef, P.; Lemort, V. Techno-economic survey of Organic Rankine Cycle (ORC) systems. *Renew. Sustain. Energy Rev.* **2013**, *22*, 168–186. [CrossRef]
4. Villarini, M.; Bocci, E.; Moneti, M.; Di Carlo, A.; Micangeli, A. State of art of small scale solar powered ORC systems: A review of the different typologies and technology perspectives. *Energy Procedia* **2014**, *45*, 257–267. [CrossRef]
5. Li, J.; Ge, Z.; Duan, Y.; Yang, Z.; Liu, Q. Parametric optimization and thermodynamic performance comparison of single-pressure and dual-pressure evaporation organic Rankine cycles. *Appl. Energy* **2018**, *217*, 409–421. [CrossRef]
6. Fernández-Guillamón, A.; Molina-García, Á.; Vera-García, F.; Almendros-Ibáñez, J.A. Organic Rankine Cycle Optimization Performance Analysis Based on Super-Heater Pressure: Comparison of Working Fluids. *Energies* **2021**, *14*, 2548. [CrossRef]
7. Manente, G.; Lazzaretto, A.; Bonamico, E. Design guidelines for the choice between single and dual pressure layouts in organic Rankine cycle (ORC) systems. *Energy* **2017**, *123*, 413–431. [CrossRef]
8. Branchini, L.; De Pascale, A.; Peretto, A. Systematic comparison of ORC configurations by means of comprehensive performance indexes. *Appl. Therm. Eng.* **2013**, *61*, 129–140. [CrossRef]
9. Delgado-Torres, A.M.; García-Rodríguez, L. Analysis and optimization of the low-temperature solar organic Rankine cycle (ORC). *Energy Convers. Manag.* **2010**, *51*, 2846–2856. [CrossRef]
10. Villarini, M.; Bocci, E.; Di Carlo, A.; Sbordone, D. Technical-Economic Analysis of an Innovative Small Scale Solar Thermal—ORC Cogenerative System. In *International Conference on Computational Science and Its Applications Part II LNCS*; Springer: Berlin/Heidelberg, Germany, 2013; Volume 7972, pp. 271–287.
11. Hassoun, A.; Dincer, I. Analysis and performance assessment of a multigenerational system powered by Organic Rankine Cycle for a net zero energy house. *Appl. Therm. Eng.* **2015**, *76*, 25–36. [CrossRef]
12. Khalid, F.; Dincer, I.; Rosen, M.A. Thermoeconomic analysis of a solar-biomass integrated multigeneration system for a community. *Appl. Therm. Eng.* **2017**, *120*, 645–653. [CrossRef]
13. Bellos, E.; Tzivanidis, C. Multi-objective optimization of a solar driven trigeneration system. *Energy* **2018**, *149*, 47–62. [CrossRef]
14. Nasir, M.T.; Ekwonu, M.C.; Park, Y.; Esfahani, J.A.; Kim, K.C. Assessment of a District Trigeneration Biomass Powered Double Organic Rankine Cycle as Primed Mover and Supported Cooling. *Energies* **2021**, *14*, 1030. [CrossRef]
15. García-Domínguez, J.; Blanco-Marigorta, A.M.; Marcos, J.D. Thermodynamic analysis and optimization of a combined cooling, heating, and power system using Organic Rankine Cycles (ORC) and solar parabolic trough collectors. In Proceedings of the 33rd International Conference on Efficiency, Cost, Optimization, Simulation and Environmental Impact of Energy Systems, Osaka, Japan, 29 June–3 July 2020.
16. Al-Sulaiman, F.A.; Hamdullahpur, F.; Dincer, I. Performance comparison of three trigeneration systems using organic rankine cycles. *Energy* **2011**, *36*, 5741–5754. [CrossRef]
17. Al-Sulaiman, F.A.; Hamdullahpur, F.; Dincer, I. Performance assessment of a novel system using parabolic trough solar collectors for combined cooling, heating, and power production. *Renew. Energy* **2012**, *48*, 161–172. [CrossRef]
18. Suleman, F.; Dincer, I.; Agelin-Chaab, M. Development of an integrated renewable energy system for multigeneration. *Energy* **2014**, *78*, 196–204. [CrossRef]
19. Bellos, E.; Tzivanidis, C. Parametric analysis and optimization of a solar driven trigeneration system based on ORC and absorption heat pump. *J. Clean. Prod.* **2017**, *161*, 493–509. [CrossRef]
20. Petela, R. Exergy of undiluted thermal radiation. *Sol. Energy* **2013**, *74*, 469–488. [CrossRef]
21. Forristall, R. *Heat Transfer Analysis and Modeling of a Parabolic Trough Solar Receiver Implemented in Engineering Equation Solver NREL/TP-550-34169*; National Renewable Energy Laboratory: Golden, CO, USA, 2003.
22. Rashid, K. Design, Economics, and Real-Time Optimization of a Solar/Natural Gas Hybrid Power Plant. Ph.D. Thesis, The University of Utah, Salt Lake City, UT, USA, 2019.
23. Liao, X.; Garland, P.; Radermacher, R. The modeling of air-cooled absorption chiller integration in CHP system. *ASME Int. Mech. Eng. Congr. Expo.* **2004**, *47012*, 371–377. [CrossRef]
24. Bakhtiari, B.; Fradette, L.; Legros, R.; Paris, J. A model for analysis and design of H_2O–LiBr absorption heat pumps. *Energy Convers. Manag.* **2011**, *52*, 1439–1448. [CrossRef]
25. Marcos, J.D.; Izquierdo, M.; Palacios, E. New method for COP optimization in water-and air-cooled single and double effect LiBr–water absorption machines. *Int. J. Refrig.* **2011**, *34*, 1348–1359. [CrossRef]
26. Evola, G.; Le Pierre's, N.; Boudehenn, F.; Papillon, P. Proposal and validation of a model for the dynamic simulation of a solar-assisted single-stage LiBr/water absorption chiller. *Int. J. Refrig.* **2013**, *36*, 1015–1028. [CrossRef]
27. Rashid, K.; Safdarnejad, S.M.; Powell, K.M. Process intensification of solar thermal power using hybridization, flexible heat integration, and real-time optimization. *Chem. Eng. Process. Process Intensif.* **2019**, *139*, 155–171. [CrossRef]
28. Spelling, J.; Favrat, D.; Martin, A.; Augsburger, G. Thermoeconomic optimization of a combined-cycle solar tower power plant. *Energy* **2012**, *41*, 113–120. [CrossRef]

29. Baghernejad, A.; Yaghoubi, M.; Jafarpur, K. Exergoeconomic optimization and environmental analysis of a novel solar-trigeneration system for heating, cooling and power production purpose. *Sol. Energy* **2016**, *134*, 165–179. [CrossRef]
30. Kim, I.Y.; de Weck, O.L. Adaptive weighted-sum method for bi-objective optimization: Pareto front generation. *Struct. Multidiscip. Optim.* **2005**, *29*, 149–158. [CrossRef]
31. Zare, V.; Mahmoudi, S.M.S.; Yari, M.; Amidpour, M. Thermoeconomic analysis and optimization of an ammonia-water power/cooling cogeneration cycle. *Energy* **2012**, *47*, 271–283. [CrossRef]

mathematics

MDPI

Article

An Innovative Hybrid Heap-Based and Jellyfish Search Algorithm for Combined Heat and Power Economic Dispatch in Electrical Grids

Ahmed Ginidi [1], Abdallah Elsayed [2], Abdullah Shaheen [1], Ehab Elattar [3] and Ragab El-Sehiemy [4,*]

1 Department of Electrical Engineering, Faculty of Engineering, Suez University, Suez 43533, Egypt; ahmed.ginidi@eng.suezuni.edu.eg (A.G.); abdullahshaheen2015@gmail.com (A.S.)
2 Department of Electrical Engineering, Faculty of Engineering, Damietta University, Damietta 34517, Egypt; am.elsherif@yahoo.com
3 Department of Electrical Engineering, College of Engineering, Taif University, Taif 21944, Saudi Arabia; e.elattar@tu.edu.sa
4 Department of Electrical Engineering, Faculty of Engineering, Kafrelsheikh University, Kafrelsheikh 33516, Egypt
* Correspondence: elsehiemy@eng.kfs.edu.eg

Abstract: This paper proposes a hybrid algorithm that combines two prominent nature-inspired meta-heuristic strategies to solve the combined heat and power (CHP) economic dispatch. In this line, an innovative hybrid heap-based and jellyfish search algorithm (HBJSA) is developed to enhance the performance of two recent algorithms: heap-based algorithm (HBA) and jellyfish search algorithm (JSA). The proposed hybrid HBJSA seeks to make use of the explorative features of HBA and the exploitative features of the JSA to overcome some of the problems found in their standard forms. The proposed hybrid HBJSA, HBA, and JSA are validated and statistically compared by attempting to solve a real-world optimization issue of the CHP economic dispatch. It aims to satisfy the power and heat demands and minimize the whole fuel cost (WFC) of the power and heat generation units. Additionally, a series of operational and electrical constraints such as non-convex feasible operating regions of CHP and valve-point effects of power-only plants, respectively, are considered in solving such a problem. The proposed hybrid HBJSA, HBA, and JSA are employed on two medium systems, which are 24-unit and 48-unit systems, and two large systems, which are 84- and 96-unit systems. The experimental results demonstrate that the proposed hybrid HBJSA outperforms the standard HBA and JSA and other reported techniques when handling the CHP economic dispatch. Otherwise, comparative analyses are carried out to demonstrate the suggested HBJSA's strong stability and robustness in determining the lowest minimum, average, and maximum WFC values compared to the HBA and JSA.

Keywords: heap-based algorithm; jellyfish search algorithm; economic dispatch; combined heat and power plants

Citation: Ginidi, A.; Elsayed, A.; Shaheen, A.; Elattar, E.; El-Sehiemy, R. An Innovative Hybrid Heap-Based and Jellyfish Search Algorithm for Combined Heat and Power Economic Dispatch in Electrical Grids. *Mathematics* **2021**, *9*, 2053. https://doi.org/10.3390/math9172053

Academic Editor: Zbigniew Leonowicz

Received: 3 August 2021
Accepted: 20 August 2021
Published: 26 August 2021

1. Introduction

The energy supply in the globe is shifting toward high efficiency, sustainability, and low carbon content [1]. In conventional power units, a large amount of energy is wasted during the conversion of fossil fuels into electricity because of the low efficiency of these conventional plants. However, by utilizing the CHP economic dispatch, the whole fuel cost (WFC) may be reduced by 10–40%, energy efficiency can be increased to 90%, and greenhouse gases (GHG) can be reduced by roughly 13–18% [2]. The heat and electrical energy in the CHP system can be generated from a single source at the same time. The vital optimization challenge for the CHP economic dispatch is to find the minimum WFC of heat and power supply. There are several constraints that should be considered in the CHP economic dispatch, involving a load balance of the system, capacity limitations of generation plants, the valve-point effect of thermal plants, and the heat and power mutual dependency provided by CHP. Two main categories of optimization approaches are explained to solve the CHP economic dispatch problem in recent research, comprising mathematical and heuristic optimization techniques [3,4].

One such task is the economic dispatch of the power system, which entails coordination, planning, and scheduling generators in an efficient manner. Due to the imposed equality and inequality restrictions, the economic dispatch issue exhibits nonlinear behavior. The economic dispatch problem has been highlighted as a multimodal optimization problem that will be difficult to tackle. Because actual issues are multimodal in nature, gradient methods are inapplicable [5]. In [6], an enhanced multi-objective particle swarm optimizer (MOPSO) model was used to manage a bi-objective dispatch framework in order to enhance the power quality and economic costs. In this study, a deep learning approach has been used to improve wind forecast accuracy where uncertainty analysis is a critical component of any assessment of a wind farm's long-term electricity output [7]. In [8], an improved antlion optimizer was presented to search for potential solutions for the economic dispatch issue in power systems with thermal units in order to minimize the generating fuel costs and guarantee that all restrictions are within functioning ranges. In [9], a modified crow search optimization was applied for solving the economic dispatch considering the environmental impacts and high-voltage direct current systems.

Added to that, the CHP economic dispatch has been solved throughout lots of conventional and mathematical approaches. In [10], a decentralized solution based on bender decomposition (BD) was performed for the optimal schedule of the CHP economic dispatch. The Lagrange relaxation (LR) and LR with surrogate sub-gradient (LRSS) have been employed in [11,12] with two levels to find out the optimal solution for studying the CHP economic dispatch. In [13], sequential quadratic programming (SQP) was combined with the LR method, where the LR technique was applied to the optimal CHP scheduling, and SQP was applied on a portion of the CHP problem to check the validity of the acquired operating point inside the trust region. In [14], the envelope-based branch and bound (EBB) approach was utilized for optimal planning of the CHP.

However, to deal with the non-convex objective function of the CHP economic dispatch and to overcome computational time efforts, heuristic approaches have been applied on the mentioned problem, such as the genetic algorithm (GA) [15], opposition teaching learning-based optimization (OTLBO) [16], differential evolution (DE) [17], multi-player harmony search (MPHS) algorithm [18], cuckoo search (CS) [19], and whale optimization algorithm (WOA) [20]. In [21], a greedy randomized adaptive search procedure (GRASP) method was hybridized with DE optimization and applied for the CHP economic dispatch to increase global search capacity while avoiding converging to local minima. In [22], an advanced mutation mechanism was involved in real coded GA and applied to the CHP economic dispatch for minimizing the operation cost, in order to enhance the convergence characteristics. In [23], an improved GA based on a new crossover and mutation was utilized to solve the CHP economic dispatch problem for handling constraints and applied to four cases for assessing the performance of the approach. In [24], a biogeography-based learning PSO (BLPSO) was carried out to improve the solution accuracy and overcome premature convergence where each particle utilized a migration operator to update itself depending on the best position of the whole particles. In addition, a multi-objective PSO has emerged with non-dominated sorting GA [25], and a modified version of shuffle frog leaping (MVSFL) algorithm [26] has been successfully employed on the CHP economic dispatch with limited small-scale applications, which are 5-unit and 7-unit systems.

The authors of [27] presented a combined optimization approach for power systems, which managed energy with power market and active microgrids in electric vehicle parking lots, diverse CHP economic dispatches, power and heat storage units, and distributed production. In [28], a Manta ray foraging optimizer (MRFO) was incorporated with adaptive constraint handling for solving the CHP economic dispatch, whereas the impact of the inclusion of wind power based on the MRFO was investigated in [29]. Moreover, a two-stage mathematical programming has been proposed in [30] to deal with the nondifferentiable portion of valve-point loading influence and attain a convex operating zone in the CHP economic dispatch problem. In [31], the authors investigated the heat in power equipment and the availability of power flexibility in CHP technology from district heating networks.

Recently, two novel algorithms, heap-based algorithm (HBA) and jellyfish search algorithm (JSA), have been introduced to solve global optimization problems. Firstly, the HBA is a powerful metaheuristic optimization that is inspired from organization hierarchy created by Qamar Askari et al. [32]. Its simplicity and effectiveness enforce the research direction into its promising implementations in solving engineering problems. In [33], the HBA was efficiently utilized for parameter estimation of fuel cells, while it was applied for the CHP economic dispatch in [34] and optimal reactive power dispatch in [35]. Secondly, the standard JSA, inspired from jellyfish movements, was created by J.-S. Chou and D.-N. Truong in January 2021 [36]. In [37], the JSA was employed for a spectrum defragmentation algorithm in an elastic optical network. In [38], the JSA was utilized for efficient power system operation based on optimal power flow, whereas it was effectively applied in distribution networks to integrate distributed generators and the static volt-ampere reactive compensator [39]. In this paper, a novel hybrid heap-based and jellyfish search algorithm (HBJSA) is proposed, which combines the benefits of the standard HBA and standard JSA. Compared with the standard HBA and standard JSA, the proposed HBJSA uses an adjustment mechanism to support explorative and exploitative characteristics. The adjustment mechanism is constructed to boost the explorative features at the start of iterations by enhancing the generated solutions via HBA. Furthermore, towards the conclusion of iterations, it augments and enhances the exploitative features by growing the generated solutions via JSA. The efficiency of the HBA, JSA, and the proposed HBJSA is evaluated for solving the CHP economic dispatch by considering various constraints of heat production and power output balance.

The rest of this paper is structured as follows: the CHP economic dispatch problem is characterized in Section 2, whereas Section 3 includes a description of the standard HBA, the standard JSA, and the proposed hybrid HBJSA. Furthermore, Section 4 presents the outcomes of these algorithms and discussion for simulation, while a conclusion is presented in Section 5 of this work.

2. Problem Formulation

The general form of the CHP economic dispatch problem is described in Figure 1. This figure shows the single line diagram of the 24-unit test system for the CHP economic dispatch problem. As shown, different sources of the CHP, heat only, and power-only units supply power and heat are combined together to satisfy the power and heat demands. Heat production and power output balance means that the total power generation equals the total power load and the total heat generation equals the total heat load.

The objective function of the CHP economic dispatch problem can be illustrated as depicted in the following equation [2]:

$$Min\{WFC\} = Min\left\{\sum_{i=1}^{Npp} C_i\left(P_i^{pp}\right) + \sum_{h=1}^{Nhp} C_h\left(H_h^{hp}\right) + \sum_{k=1}^{NCp} C_k\left(P_k^{Cp}, H_k^{Cp}\right)\right\}(\text{USD/h}) \quad (1)$$

The three terms of costs manifested in Equation (1) are explained in Equations (2)–(4) as in [20]. The cost function of a power-only plant involves quadratic and sinusoidal terms, where the sinusoidal term displays the valve-point impacts as signified in Equation (2). Furthermore, the heat-only cost is formulated in Equation (3), while the CHP cost function is represented in Equation (4).

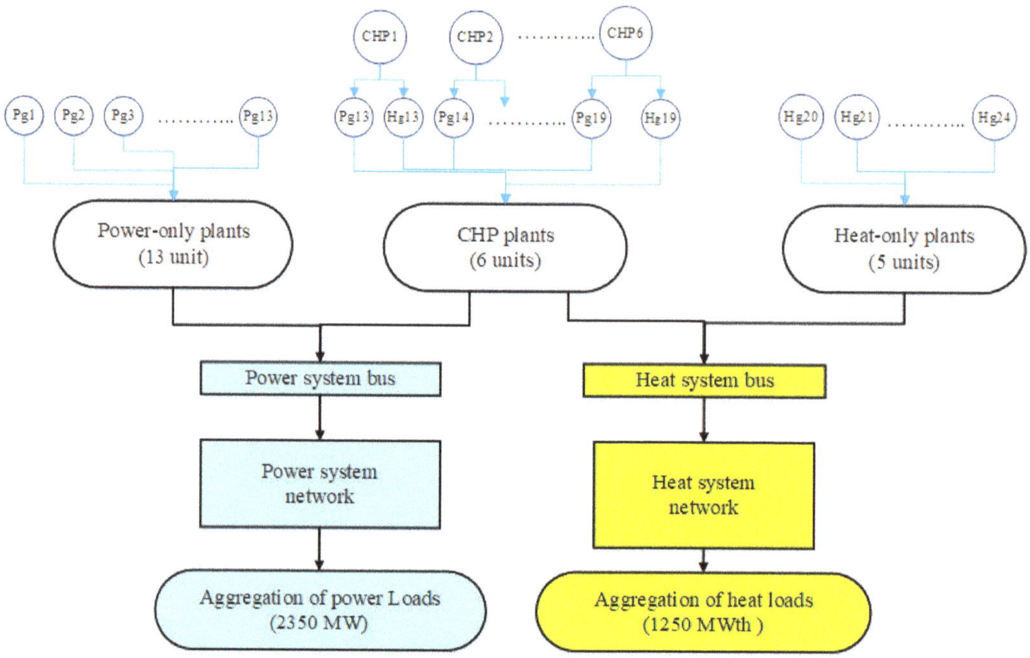

Figure 1. A single line diagram of the CHP economic dispatch problem considering the 24-unit test system.

$$C_i(P_i^{pp}) = a_i(P_i^{pp})^2 + b_i P_i^{pp} + c_i + \left|\lambda_i \sin(\rho_i(P_i^{pp\min} - P_i^{pp}))\right| \text{ (USD/h)} \tag{2}$$

$$C_j(H_j^{hp}) = a_j(H_j^{hp})^2 + b_j P_j^{pp} + c_j \text{ (USD/h)} \tag{3}$$

$$C_k(P_k^{cp}, H_k^{cp}) = a_k(P_k^{cp})^2 + b_k P_k^{pp} + c_k + d_k(H_k^{cp})^2 + e_k H_k^{cp} + f_k H_k^{cp} P_k^{cp} \text{(USD/h)} \tag{4}$$

Diverse constraints for feasible solutions are illustrated for the CHP economic dispatch problem as follows:

$$\sum_{i=1}^{N_{pp}} P_i^{pp} + \sum_{j=1}^{N_{cp}} P_j^{cp} = P_d \tag{5}$$

$$\sum_{j=1}^{N_{cp}} H_j^{cp} + \sum_{k=1}^{N_{hp}} H_k^{hp} = H_d, \tag{6}$$

Furthermore, power-only and heat-only capacity limits are exposed in Equation (7) and Equation (8), respectively. In addition to that, capacity limits of CHP are designated in Equations (9) and (10).

$$P_i^{pp\min} \le P_i^{pp} \le P_i^{pp\max} \quad i = 1, \ldots, N_{pp}, \tag{7}$$

$$H_j^{hp\min} \le H_j^{hp} \le H_j^{hp\max} \quad i = 1, \ldots, N_{hp}, \tag{8}$$

$$P_k^{cp\min}(H_k^{cp}) \le P_k^{cp} \le P_k^{cp\max}(H_k^{cp}) \quad k = 1, \ldots, N_{cp}, \tag{9}$$

$$H_k^{cp\min}(P_k^{cp}) \le H_k^{cp} \le H_k^{cp\max}(P_k^{cp}) \quad k = 1, \ldots, N_{cp}, \tag{10}$$

In the above constraints, Equations (5) and (6) demonstrate the power generated and the power demand balance and the heat generated and the demand balance, respectively.

3. Hybrid HBJSA for CHP Economic Dispatch Problem

3.1. Standard HBA

The standard HBA concept is based on the corporate rank hierarchy (CRH), which states that a team can arrange itself in a hierarchy to fulfill organizational goals [32]. The HBA is classified into three levels: interaction among subordinates, self-contribution of employees and their immediate supervisor, and interaction among colleagues.

In the CRH model, the population is manifested by the full CRH, whereas the heap node is manifested by the search agent. The search agent's fitness is the master of the heap node, and the population index of the search agent is the value of the heap node. The agent position of each search can be updated as:

$$x_i^k(t+1) = B^k + \gamma(2r-1)\left|B^k - x_i^k(t)\right| \tag{11}$$

The kth component of λ vector $\vec{\lambda}$ is represented by:

$$\lambda^k = 2r - 1 \tag{12}$$

γ is computed as follow:

$$\gamma = \left|2 - \frac{(t\ mod \frac{t}{C})}{\frac{t}{4C}}\right| \tag{13}$$

The parameter (C) in Equation (14) controls the variation. However, this parameter will complete in T iterations as follows:

$$C - T^{\max}/25 \tag{14}$$

Added to that, the interaction between colleagues is modeled. As manifested in Equation (15), the position of each agent $(\vec{x_i})$ is updated by its arbitrarily selected colleague $\vec{S_r}$:

$$x_i^k(t+1) = \begin{cases} S_r^k + \gamma\lambda^k\left|S_r^k - x_i^k(t)\right|, & f(\vec{S_r}) < f(\vec{x_i}(t)) \\ x_i^k + \gamma\lambda^k\left|S_r^k - x_i^k(t)\right|, & f(\vec{S_r}) \geq f(\vec{x_i}(t)) \end{cases} \tag{15}$$

where the fitness of the search agent can be represented by f.

Additionally modeled is the self-contribution of each employee, where the position of each agent is updated in this level according to the following equation:

$$x_i^k(t+1) = x_i^k(t) \tag{16}$$

Finally, the position updating equations have been emerged. The roulette wheel probabilities, p_1, p_2, and p_3, are selected to balance the exploration and exploitation processes. The search agent updates its position using Equation (16). Selecting the proportion p_1 is carried out by using Equation (17) as:

$$p_1 = 1 - \frac{t}{T^{max}} \tag{17}$$

The search agent updates its position using Equation (11). Selecting the proportion p_2 is carried by using Equation (18) as:

$$p_2 = p_1 + \frac{1 - p_1}{2} \tag{18}$$

The search agent updates its position using Equation (17). Selecting the proportion p_3 is carried out by using Equation (19) as:

$$p_3 = p_2 + \frac{1 - p_1}{2} = 1 \tag{19}$$

Hence, the general positions' updating mechanism of the HBA is formulated as in Equation (20):

$$x_i^k(t+1) = \begin{cases} x_i^k(t), \quad p \le p_1 \\ B^k + \gamma\lambda^k \left| B^k - x_i^k(t) \right|, p_1 < p < p_2 \\ S_r^k + \gamma\lambda^k \left| S_r^k - x_i^k(t) \right|, p_2 < p \le p_3 \text{ and } f(\vec{S_r}) < f(\vec{x_i}(t)) \\ x_r^k + \gamma\lambda^k \left| S_r^k - x_i^k(t) \right|, p_2 < p \le p_3 \text{ and } f(\vec{S_r}) \ge f(\vec{x_i}(t)) \end{cases} \tag{20}$$

The main steps of the proposed HBA are depicted in Figure 2.

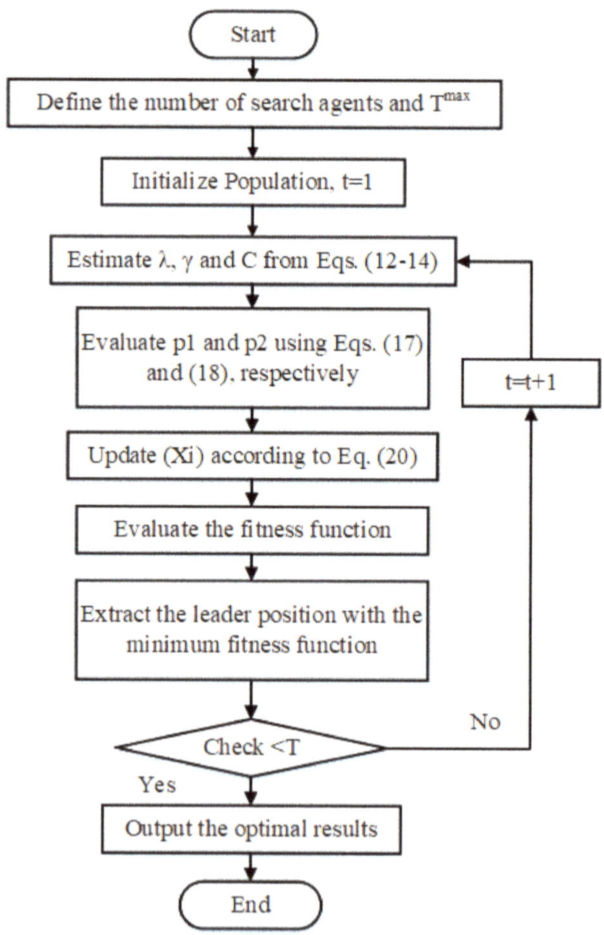

Figure 2. Flowchart of the HBA.

3.2. Standard JSA

The JSA is inspired by the jellyfish movements whether they move in the ocean current or within their swarm [36]. The jellyfish population can be mathematically modeled as:

$$X_i(t+1) = 4P_0(1 - X_i), \quad 0 \leq P_0 \leq 1 \tag{21}$$

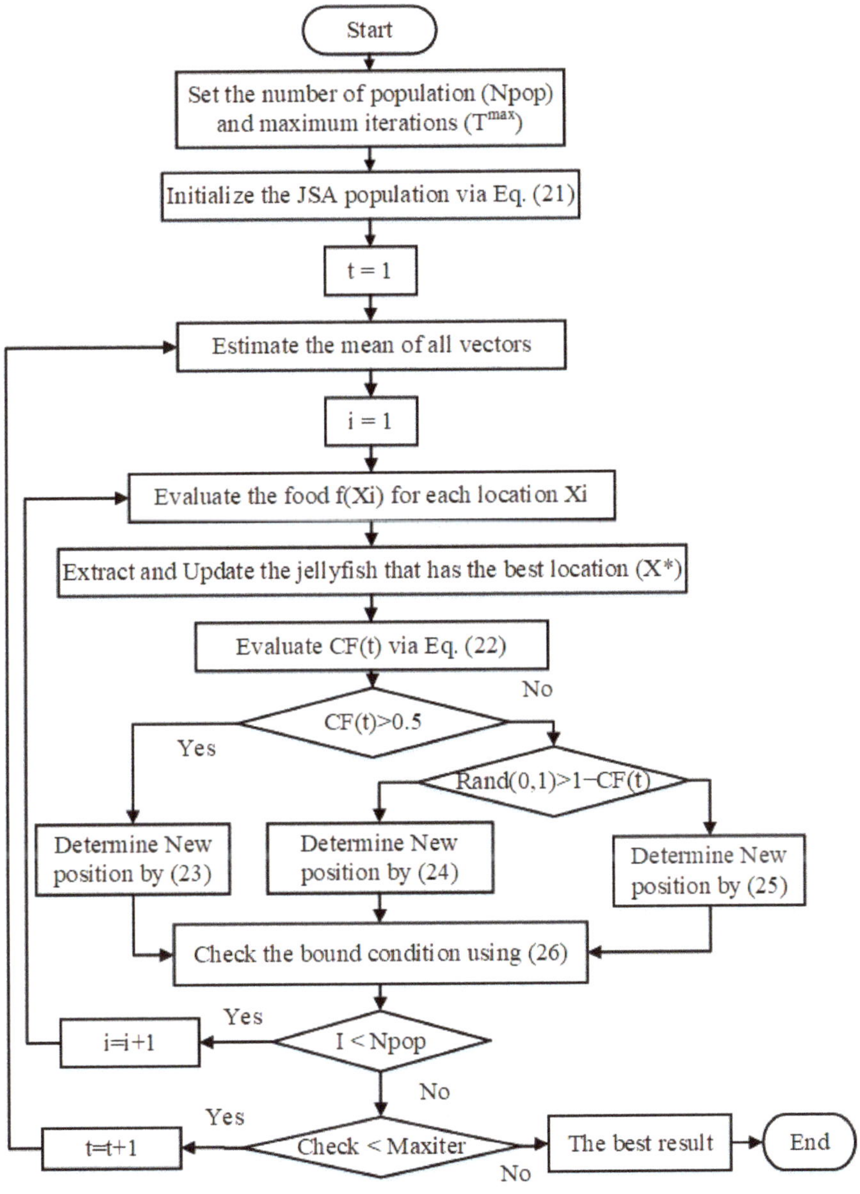

Figure 3. Flowchart of the JSA.

The time control function $CF(t)$ value is assessed as described in Equation (22), and it is varied from 0 to 1 over time:

$$CF(t) = \left| \left(1 - \frac{t}{T^{\max}}\right) \times (2 \times rand(0,1) - 1) \right| \quad (22)$$

If the CF is greater than the constant CO_o (to be 0.5), the new location of each jellyfish can be formulated as demonstrated in Equation (23)

$$X_i(t+1) = R \times (X^* - 3 \times R \times \mu) + X_i(t) \quad (23)$$

If CF value is more than CO_o, each jellyfish location is updated depending on the movement within the swarm, as clarified in Equations (24) and (25).

$$X_i(t+1) = 0.1 \times R \times (U_b - L_b) + X_i(t) \quad (24)$$

$$X_i(t+1) = \begin{cases} X_i(t) + R \times (X_j(t) - X_i(t)) & if\ f(X_i) \geq f(X_j) \\ X_i(t) + R \times (X_i(t) - X_j(t)) & if\ f(X_i) < f(X_j) \end{cases} \quad (25)$$

As soon as a jellyfish moves at the back of the search zone boundaries, it will go back, as is demonstrated in Equation (26), to the reverse boundary.

$$\begin{cases} X'_{i,d} = (X_{i,d} - U_{b,d}) + L_b(d) & if \quad X_{i,d} > U_{b,d} \\ X'_{i,d} = (X_{i,d} - L_{b,d}) + U_b(d) & if \quad X_{i,d} < L_{b,d} \end{cases} \quad (26)$$

where $X_{i,d}$ expresses the ith jellyfish location in dth dimension. The main steps of the JSA are depicted in Figure 3.

3.3. Proposed Hybrid HBJSA

In this sub-section, a hybrid HBJSA is proposed to combine the benefits of the standard HBA and standard JSA. Compared with standard HBA and standard JSA, the proposed HBJSA employs an adjustment mechanism to support the explorative and exploitative characteristics. This mechanism is constructed to boost the explorative feature at the start of iterations by enhancing the generated solutions via HBA. Furthermore, toward the conclusion of iterations, it augments and enhances the exploitative feature by growing the generated solutions via JSA. The adjustment mechanism is executed by employing an adaptive coefficient (φ) designed as follows:

$$\varphi = \frac{t}{2 \times T^{\max}} \quad (27)$$

From this equation, the coefficient (φ) is correlated positively with the number of iterations increases until it reaches 0.5 at the highest quantity of iterations. The more the value of the coefficient (φ) increases, increasing of the generated solutions via JSA will be updated by Equation (28) as follows:

$$x_i^k(t+1) = R \times (Leader^k - 3 \times R \times \mu) + x_i^k(t) \quad (28)$$

where *Leader* is the leader position of the search agents, which achieves the minimum fitness value.

Another point of view for handling the CHP economic dispatch problem, the objective function in Equation (1) is updated to incorporate penalized terms of the power and heat units constraints as follows:

$$OF = WFC + Pen_v \sum_{j=1}^{N_c} B_v \left(P_j^C(H_j^C) - P_j^{CLimit}(H_j^C) \right) \quad (29)$$

where the term (P_j^{CLimit} (H_j^C)) is the power limit to the CHP j heating output; ψ_v is a penalized coefficient for CHP operating violating; Bv equals 1 when there is violation or 0 when there is not. Accordingly, the farthest violated operating point will have a greater penalty.

Figure 4 illustrates the main steps of the proposed hybrid HBJSA for handling the CHP economic dispatch problem. For more information about the proposed HBJSA, the main steps can be summarized as follows:

- Define the parameters of HBJSA.
- Randomly initialize the control parameters that involve the output of power and heat of the committed units and keep it within the accepted boundaries. They are checked versus its acceptable bounds' mechanism. Both units began inside their respective limitations, and if either of them is violated throughout the repetitions, it is reset to the nearest limit.
- Evaluate the fitness function of the CHP problem that minimizes the overall cost using Equation (28).

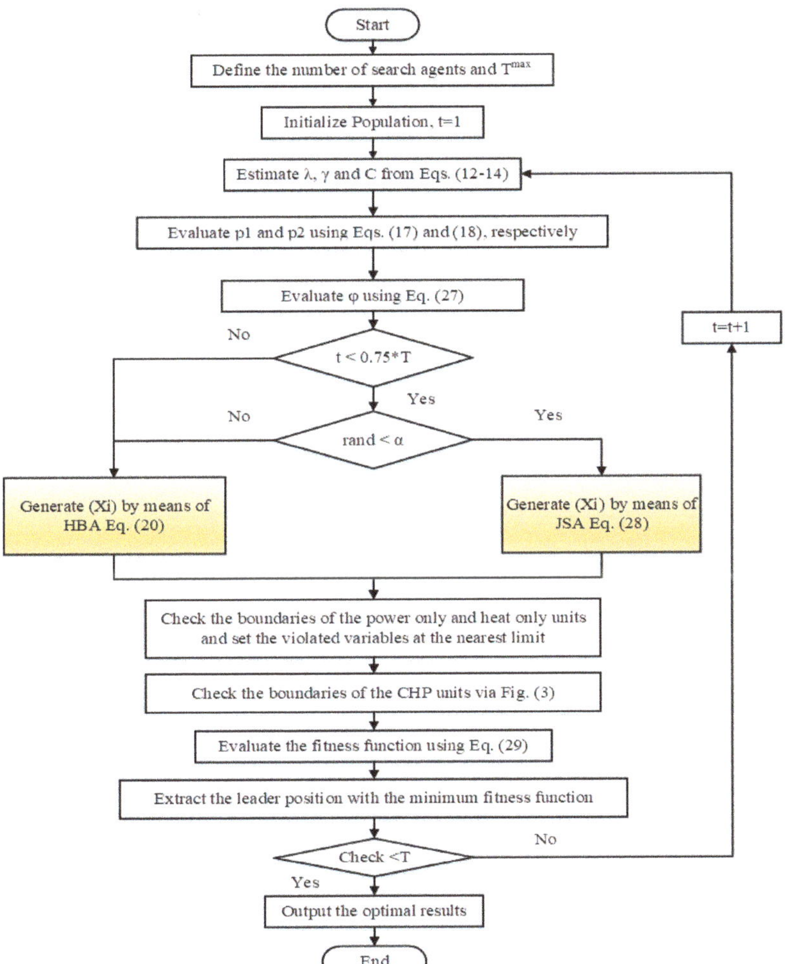

Figure 4. Flowchart of the proposed HBJSA.

Figure 3 shows the second type of mutually dependent CHP unit. They are dealt with as the penalty function that is added in the considered fitness function (OF) in Equation (29). As shown in Figure 3, when an operational point is inside the limits, it has a Bv value of zero, while the infeasible locations have a Bv value of one. On the other side, the greater the penalty amount, the farther the infeasible point is from the nearest border.

As a result, the proposed hybrid HBJSA has a greater capacity for looking for suitable locations. Furthermore, a stopping condition is used in which the ideal result is attained if the maximum number of iterations is reached. HBA penalizes infeasible solutions to varying degrees based on their distance from the next feasible point.

4. Simulation Results

The proposed HBJSA, the standard HBA, and JSA are employed on four test systems. The first two test systems are medium-scale 24-unit and 48-unit systems, whereas the second two test systems are large-scale 84-unit and 96-unit test systems. The number of iterations (T) and population size (n_{pop}), which are the main two parameters of the standard HBA, the standard JSA, and the proposed HBJSA account for 3000 and 100, respectively, for all systems. MatlabR2017b is utilized to carry out the simulations using CPU (2.5 GHz) Intel(R)-Core (TM) i7-7200U and 8 GB of RAM.

4.1. Simulation Results of the 24-Unit Test System

The data for the obtained system are mentioned in [40], which illustrates that 2350 MW and 1250 MWth are the load demand and heat demand, respectively, and it has five heat units, 13 thermal units, and six CHP units. The HBA, JSA, and proposed HBJSA are applied on this test system, and the corresponding MW, MWth for each unit, and WFC are demonstrated in Table 1. It can be manifested that the proposed HBJSA provides the optimal solution for WFC minimization, which accounts for USD 57,968.5399, while the standard HBA and the standard JSA account for USD 57,994.51 and USD 58,739.5241, respectively.

Moreover, convergence characteristics of the proposed HBJSA versus the standard HBA and the standard JSA for the 24-unit test system of the CHP economic dispatch problem are depicted in Figure 5. It is seen from that figure that the proposed hybrid HBJSA is capable of improving the solution quality compared to the standard HBA and the standard JSA. At the last 400 iterations, the proposed hybrid HBJSA provides a higher exploitative feature and, finally, reaches the lowest WFC of USD 57,968.5399. Additionally, the standard HBA, the standard JSA, and the proposed HBJSA effectively achieve all constraints with 100% accuracy, as illustrated in Table 1.

In addition, a comparison between the HBA, JSA, and the proposed HBJSA is conducted in Table 2 for the 24-unit system of CHP economic dispatch with respect to reported techniques such as PSO [40], time-varying acceleration coefficients-PSO (TVAC-PSO) [40], group search optimization (GSO) [41], and improved GSO (IGSO) [42], MRFO [28], and supply demand algorithm (SDA) [34]. In this table, ranking order is evaluated in ascending order based on the minimum WFC. From this table, the proposed hybrid HBJSA achieves the first rank with the lowest WFC. On the other side, the standard HBA occupies the second rank, while the standard JSA occupies the last rank. This table demonstrates that the proposed HBJSA overwhelmed the standard HBA, the standard JSA, and the reported recent techniques for achieving minimum WFC.

Table 1. Comparison between HBA, JSA, and the proposed HBJSA for the 24-unit test system of CHP economic dispatch problem.

Unit	JSA	HBA	HBJSA
P 1	449.27558	538.55874	448.81809
P 2	149.67888	300.2175	299.21886
P 3	202.56201	301.08256	300.72118
P 4	109.86032	159.77793	60.109633
P 5	109.93156	63.21736	159.74512
P 6	159.73642	60.688903	159.77696
P 7	160.00828	160.20653	159.77184
P 8	159.74295	111.5383	60.000007
P 9	109.83449	161.25396	159.75102
P 10	77.389984	40	77.411833
P 11	77.406979	40.000266	40.001098
P 12	92.367201	55.657936	55.008621
P 13	92.395174	55.284533	55.661102
P 14	115.82103	87.944171	85.844198
P 15	40.964462	41.266255	42.751997
P 16	114.87371	84.034893	95.888699
P 17	69.301243	43.143672	44.468374
P 18	10.133787	11.08247	10.046228
P 19	48.715961	35.04403	35.005125
H 14	124.27642	108.69733	107.49154
H 15	75.711188	76.092716	77.376451
H 16	123.80751	106.47628	113.15577
H 17	100.29358	77.714606	78.85075
H 18	40.003606	40.464341	40.020006
H 19	26.210361	20.020468	20.001274
H 20	399.97537	460.53782	453.10933
H 21	59.9258	60	60
H 22	59.902038	60	59.998827
H 23	119.91378	119.99644	119.99649
H 24	119.98035	120	119.99957
Sum (Pg)	2350.0000	2350.0000	2350.0000
Sum (Hg)	1250.0000	1250.0000	1250.0000
WFC	58,739.5241	57,994.5150	57,968.5399

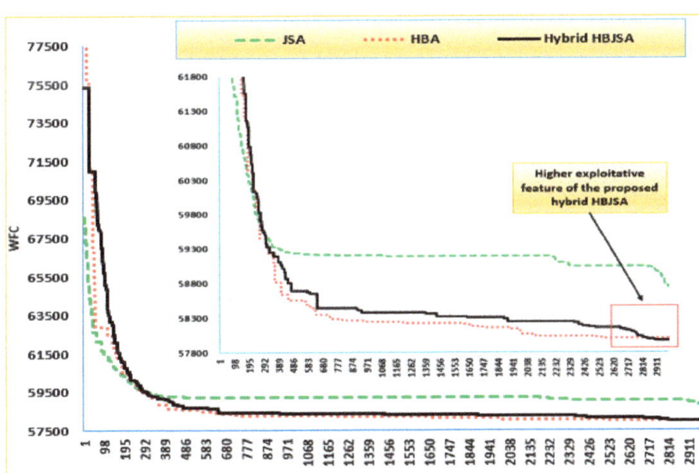

Figure 5. Convergence characteristics of the proposed HBJSA versus the HBA and JSA for the 24-unit system of CHP economic dispatch.

Table 2. Comparison between HBA, JSA, and HBJSA with respect to reported techniques for the 24-unit system of CHP economic dispatch.

Method	Sum (Pg)	Sum (Hg)	WFC	Rank
GSO [41]	2350	1250	58,225.745	7
IGSO [41]	2350	1250	58,049.01	3
PSO [40]	2349.9	1250	59,736.26	8
TVAC-PSO [40]	2350.0002	1250	58,122.74	5
MRFO [28]	2350	1250	58,173.93	6
SDA [34]	2350	1250	58,061.477	4
JSA	2350	1250	58,739.5241	9
HBA	2350	1250	57,994.515	2
Proposed hybrid HBJSA	2350	1250	57968.54	1

4.2. Simulation Results of the 48-Unit Test System

Table 3. Comparison between HBA, JSA, and the proposed HBJSA for the 48-unit test system of CHP economic dispatch problem.

Output	HBA	JSA	HBJSA	Output	HBA	JSA	HBJSA
P 1	628.31969	538.55876	628.31847	P 33	89.741692	93.956738	92.056329
P 2	301.64745	166.31944	298.49068	P 34	45.760605	63.520906	44.496057
P 3	297.56062	299.2078	298.71921	P 35	93.563713	98.369246	95.721078
P 4	60.025209	109.86692	162.17614	P 36	54.117683	64.357184	46.047531
P 5	60.891955	159.73312	60	P 37	10.89958	16.476816	11.179988
P 6	115.51041	109.86683	60.187387	P 38	35	47.149532	36.636756
P 7	60	109.87111	60.000003	H 27	109.52988	130.11379	112.92198
P 8	112.29664	109.86759	61.265074	H 28	77.887623	76.952262	80.359207
P 9	163.79819	109.86802	60.012628	H 29	106.94451	118.12661	112.31166
P 10	84.337266	77.40216	40.077792	H 30	79.211207	102.05398	78.341268
P 11	44.380157	77.405413	77.401957	H 31	42.302628	42.775213	40.213152
P 12	92.442285	92.402577	92.236557	H 32	21.949409	24.616716	20.01776
P 13	92.547707	92.400831	55.000005	H 33	108.80007	112.07165	111.00119
P 14	538.58181	359.03989	628.66001	H 34	79.972231	95.305162	78.407382
P 15	150.5131	149.62114	299.67572	H 35	111.85121	114.54771	112.98988
P 16	302.48049	299.20217	224.39881	H 36	87.186384	96.027118	80.220448
P 17	159.0721	109.86997	160.3726	H 37	40.316118	42.776163	40.506134
P 18	159.48447	110.38702	110.74399	H 38	19.999898	25.522939	20.698728
P 19	60.042597	109.86649	159.34188	H 39	442.02953	399.58888	445.84515
P 20	60.163088	109.86727	60	H 40	59.963929	59.999977	60
P 21	160.27015	109.8666	60.001242	H 41	60	59.999861	59.99996
P 22	158.67694	159.73453	162.0577	H 42	119.87507	119.99992	120
P 23	40	77.402895	78.899748	H 43	119.99995	119.99993	119.99952
P 24	40.430825	77.408107	40	H 44	452.35143	399.52269	446.16766
P 25	55.283719	92.400956	55	H 45	59.829479	59.999808	59.99892
P 26	55.000343	92.649622	55.003382	H 46	59.999794	59.999907	60
P 27	89.427389	126.10612	95.4717	H 47	119.99989	119.99982	120
P 28	43.348156	42.260587	46.55662	H 48	119.99975	119.9999	120
P 29	84.844588	104.74609	94.384639	Sum (Pg)	4700	4700	4700
P 30	44.879351	71.338822	43.869636	Sum (Hg)	2500	2500	2500
P 31	15.371858	16.474563	10.500572	WFC (USD)	116,439.96	117,365.09	116,140.34
P 32	39.288183	45.156162	35.038114				

The data for the obtained system are mentioned in [40], which illustrates that 4700 MW and 2500 MWth are the load demand and heat demand, respectively, and it has 10 heat units, 26 thermal units, and 12 CHP. The HBA, JSA, and proposed HBJSA are applied to this test system, and the corresponding MW, MWth for each unit, and WFC are demonstrated in Table 3. It can be manifested that the proposed HBJSA provides the optimal solution for WFC minimization, which accounts for USD 116,140.34, while the standard HBA and the standard JSA account for USD 116,439.96 and USD 117,365.09, respectively.

Moreover, convergence characteristics of the proposed HBJSA versus the standard HBA and the standard JSA for the 48-unit test system of the CHP economic dispatch problem are depicted in Figure 6. It is seen from that figure that the proposed hybrid HBJSA is capable of improving the solution quality compared to the standard HBA and the standard JSA. After 900 iterations, the suggested hybrid HBJSA delivers more exploitative features and ultimately achieves the lowest WFC of USD 116,140.34. Additionally, the standard HBA, the standard JSA, and the proposed HBJSA effectively achieve the power and heat balance constraints with 100% accuracy, as illustrated in Table 3.

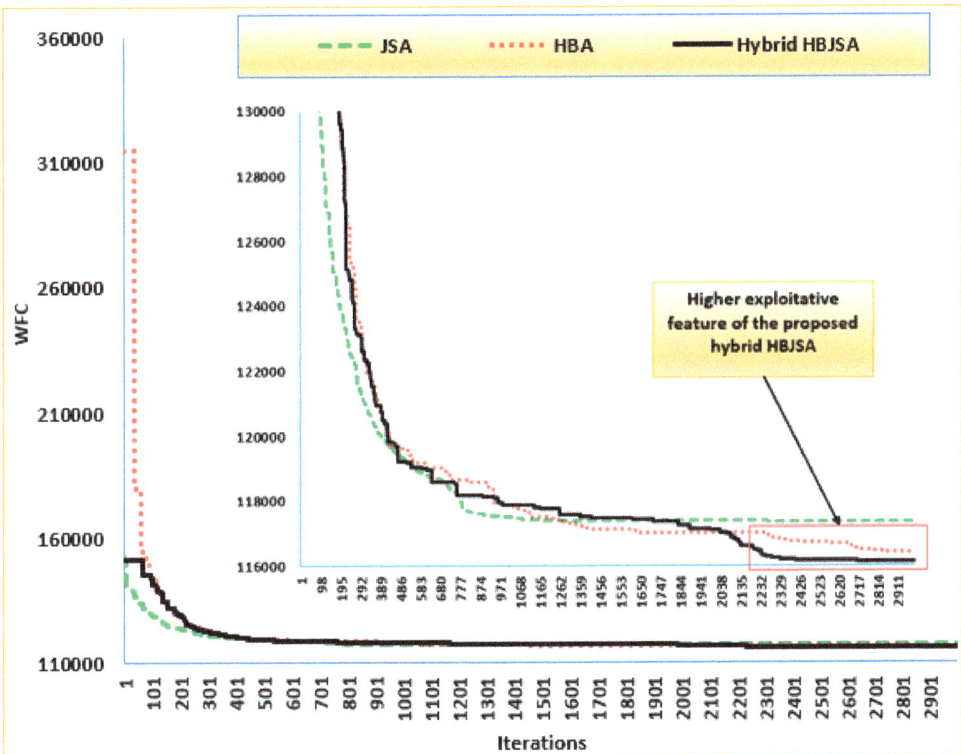

Figure 6. Convergence characteristics of the proposed HBJSA versus the HBA and JSA for the 48-unit system of CHP economic dispatch.

In addition to this, a comparison between the HBA, JSA, and proposed HBJSA is conducted in Table 4 for the 48-unit system of CHP economic dispatch with respect to other reported techniques such as MRFO [28], SDA [34] TVAC-PSO [40], CPSO [40], GSO [43], modified PSO [44], OTLBO [16], and MGSO [43] and gravitational search algorithm (GSA) [45]. Additionally, crow search algorithm (CSA) [46], grey wolf algorithm (GWA) [47], salp swarm algorithm (SSA) [48], multi-verse algorithm (MVA) [49], DE [50], MPA [51–53], civilized swarm optimization CSO [54] and Powell's pattern search PPS [54] are applied to the CHP economic dispatch for this system.

Table 4. Comparison between HBA, JSA, and HBJSA with respect to reported techniques for the 48-unit system of CHP economic dispatch.

Optimizer	WFC (USD)	Rank	Optimizer	WFC (USD)	Rank
SDA [34]	116,620.61	7	Modified PSO [44]	116,465.54	3
MRFO [28]	117,336.9	9	TVAC-PSO [40]	118,962.54	14
CSA [28]	122,953.5	20	GSA [45]	119,775.9	15
GWA [28]	122,583.3	19	MPA [42]	116,860.6	8
SSA [48]	120,174.1	16	CSO and PPS [54]	117,367.09	12
MVA [28]	117,657.9	13	MGSO [43]	117,366.09	11
DE [28]	120,482.7	17	HBA	116,439.96	2
GSO [43]	116,578.475	6	JSA	117,365.09	10
OTLBO [16]	116,579.24	4	Proposed hybrid HBJSA	116,331.21	1
PSO [40]	120,918.92	18			

In this table, ranking order is evaluated in ascending order based on the minimum WFC. From this table, the proposed hybrid HBJSA achieves the first rank with the lowest WFC. On the other side, the standard HBA occupies the second rank, while the standard JSA occupies the tenth rank. This table demonstrates that the proposed HBJSA overwhelmed the standard HBA, the standard JSA, and reported recent techniques for achieving minimum WFC.

4.3. Simulation Results of the 84-Unit Test System

The data for the tested system are mentioned in [20]. The power and heat demands equal 12,700 MW and 5000 MWth, respectively, and it has 20 heat units, 40 thermal units, and 24 CHP. The HBA, JSA, and proposed HBJSA are investigated to this test system, and the corresponding MW, MWth for each unit, and WFC are demonstrated in Table 5. It can be manifested that the proposed HBJSA provides the optimal solution for WFC minimization, which accounts for USD 288,820.7, while the standard HBA and the standard JSA account for USD 289,822.4 and USD 290,323.8, respectively.

Moreover, convergence characteristics of the proposed HBJSA are depicted in Figure 7 versus the standard HBA and the standard JSA for the 84-unit test system of the CHP economic dispatch problem. It is seen from that figure that the proposed hybrid HBJSA is capable of improving the solution quality compared with HBA and JSA. At the last 950 iterations, the proposed hybrid HBJSA provides a higher exploitative feature and, finally, reaches the lowest WFC of USD 288,820.7. Additionally, the standard HBA, the standard JSA, and the proposed HBJSA effectively achieve all constraints with 100% accuracy, as illustrated in Table 5.

In addition, a comparison study between the HBA, JSA, and proposed HBJSA is conducted in Table 6 for the 84-unit system of CHP economic dispatch with respect to reported techniques such as WOA [20], MRFO [28], marine predators algorithm (MPA) [42], improved MPA (IMPA) [42], and SDA [34]. In this table, ranking order is evaluated in ascending order based on the minimum WFC. From this table, the proposed hybrid HBJSA achieves the first rank with the lowest WFC. On the other side, the standard HBA occupies the second rank while the standard JSA occupies the fifth rank. This table demonstrates that the proposed HBJSA overwhelmed the standard HBA, the standard JSA, and reported recent techniques for achieving minimum WFC.

Table 5. Comparison between HBA, JSA, and the proposed HBJSA for the 84-unit test system of CHP economic dispatch problem. (a) Power outputs from power only and CHP units. (b) Heat outputs from CHP and heat-only units.

(a)

Unit	HBA	JSA	HBJSA	Unit	HBA	JSA	HBJSA
Pg1	114	111.1626385	113.9718986	Pg33	181.7205	159.7499057	185.8014921
Pg2	113.11556	112.5520392	112.0171823	Pg34	199.99998	199.9449699	199.988706
Pg3	103.83725	107.7765112	98.99471338	Pg35	182.91598	199.6705009	181.7649853
Pg4	184.80695	179.7375644	179.6906294	Pg36	200	199.991963	199.9921324
Pg5	89.505208	87.78776144	94.40847581	Pg37	109.99936	89.86684403	109.9315908
Pg6	106.64827	139.9948775	138.9511133	Pg38	110	109.9917663	104.3481595
Pg7	256.25545	266.7806527	260.1628409	Pg39	89.839986	94.16960534	109.9999047
Pg8	297.05131	290.8186132	299.6320704	Pg40	550	511.3076161	517.5996621
Pg9	299.99539	284.6025908	288.5380215	Pg41	126.91167	132.3982552	110.3909623
Pg10	130	207.4135385	130	Pg42	126.65154	144.4049507	136.7428407
Pg11	169.30903	243.5827337	242.0952346	Pg43	115.3838	88.99398926	134.0103426
Pg12	306.09411	318.3996454	168.8036849	Pg44	133.27883	105.2551322	117.7366501
Pg13	394.50082	304.5241963	394.2779192	Pg45	42.80197	93.63589835	49.34154951
Pg14	393.7356	304.5184491	394.2937412	Pg46	43.679119	49.07352085	43.51922726
Pg15	305.53666	394.322214	396.1931056	Pg47	77.280238	76.78719074	59.37920101
Pg16	394.45006	304.521412	394.3233195	Pg48	74.818416	55.66222673	50.7813389
Pg17	500	489.2867798	489.4316065	Pg49	99.519272	167.415806	100.5621492
Pg18	490.89205	399.4800955	493.0301498	Pg50	116.0936	170.7899652	105.6288952
Pg19	514.62594	511.432115	511.337944	Pg51	109.31998	167.4168782	114.6635218
Pg20	525.35426	511.3041636	550	Pg52	106.01984	125.329591	133.3378228
Pg21	550	523.3475504	523.296119	Pg53	60.72298	60.42436073	70.08052756
Pg22	548.52995	523.2824448	527.9275831	Pg54	52.580944	79.5458983	67.82445444
Pg23	550	523.2914355	549.9999949	Pg55	43.691272	56.36733616	52.91754938
Pg24	521.61373	523.3169144	536.3000349	Pg56	56.186778	66.27287672	55.44283757
Pg25	522.56351	523.306813	526.3906729	Pg57	12.98492749	10.52629429	24.879583
Pg26	549.31974	523.2792844	523.3095942	Pg58	29.50082157	13.96655832	13.325097
Pg27	14.540184	10.00833641	11.43580483	Pg59	10.22458212	10.24231125	12.523671
Pg28	10.098278	10.00401295	11.56317354	Pg60	13.31866365	18.41685747	12.214388
Pg29	10.909877	10.02394824	10.00000666	Pg61	37.73118881	46.39822013	58.063135
Pg30	96.999939	96.95242919	89.71504585	Pg62	55.41904721	77.24184524	47.952259
Pg31	180.39147	181.241749	188.3825144	Pg63	38.52470481	53.47372752	35.709818
Pg32	189.8298	189.9953502	189.9818368	Pg64	58.3696072	57.21827797	45.08951

(b)

Unit	HBA	JSA	HBJSA	Unit	HBA	JSA	HBJSA
Hg41	130.3604684	133.6418007	121.29	Hg65	397.9644411	347.8697453	409.80933
Hg42	130.2202223	140.381546	136.07204	Hg66	394.0861183	349.5293932	397.46604
Hg43	123.5432156	109.2841022	134.528	Hg67	400.2039596	349.7318595	413.03126
Hg44	134.1352577	118.4055739	125.13589	Hg68	401.4451605	344.0636931	366.65675
Hg45	77.30525496	121.3009135	83.04628	Hg69	59.9999999	59.99488855	59.953414
Hg46	78.02176698	82.83206972	78.035694	Hg70	59.3632427	59.97703456	59.988637
Hg47	107.1826963	106.7556784	88.175722	Hg71	59.86117224	59.99220729	59.466048
Hg48	105.056327	88.52055969	84.30425	Hg72	60	59.81462736	59.836488
Hg49	115.192982	153.2842745	113.88524	Hg73	58.88396654	59.98307532	60
Hg50	124.3738069	155.1483245	118.16811	Hg74	59.5417115	59.99915877	60
Hg51	120.69336	153.2958557	123.68397	Hg75	59.81273203	59.90067694	60
Hg52	118.7925953	129.3907541	133.9953	Hg76	60	59.98083928	59.999977
Hg53	92.88987192	92.63077153	100.95165	Hg77	120	119.9950741	119.99872
Hg54	85.7914299	109.1376617	99.009779	Hg78	120	119.9939719	117.44199
Hg55	78.18666618	89.12490012	86.055856	Hg79	119.9528822	119.9953091	119.99999
Hg56	88.97404025	97.67717242	88.298905	Hg80	120	119.9308343	120
Hg57	41.27968283	40.22573511	46.369847	Hg81	119.9999728	119.9583146	119.99954
Hg58	48.35748487	41.69903249	41.415021	Hg82	119.4423968	119.9958398	120
Hg59	40.09624275	40.10312819	41.063184	Hg83	119.999888	119.9968772	116.18205
Hg60	41.2298315	43.60456849	40.943192	Hg84	111.9564926	119.9957398	119.08122
Hg61	21.18412326	25.17541074	29.921693	Sum (Pg)	12702	12701	12704
Hg62	27.1688682	39.19892893	25.868861	Sum (Hg)	5000	5000	5000
Hg63	21.60258212	28.38461052	19.393777	WFC (USD)	289,822.392	290,323.818	288,820.7
Hg64	25.84708597	30.09746698	21.476288				

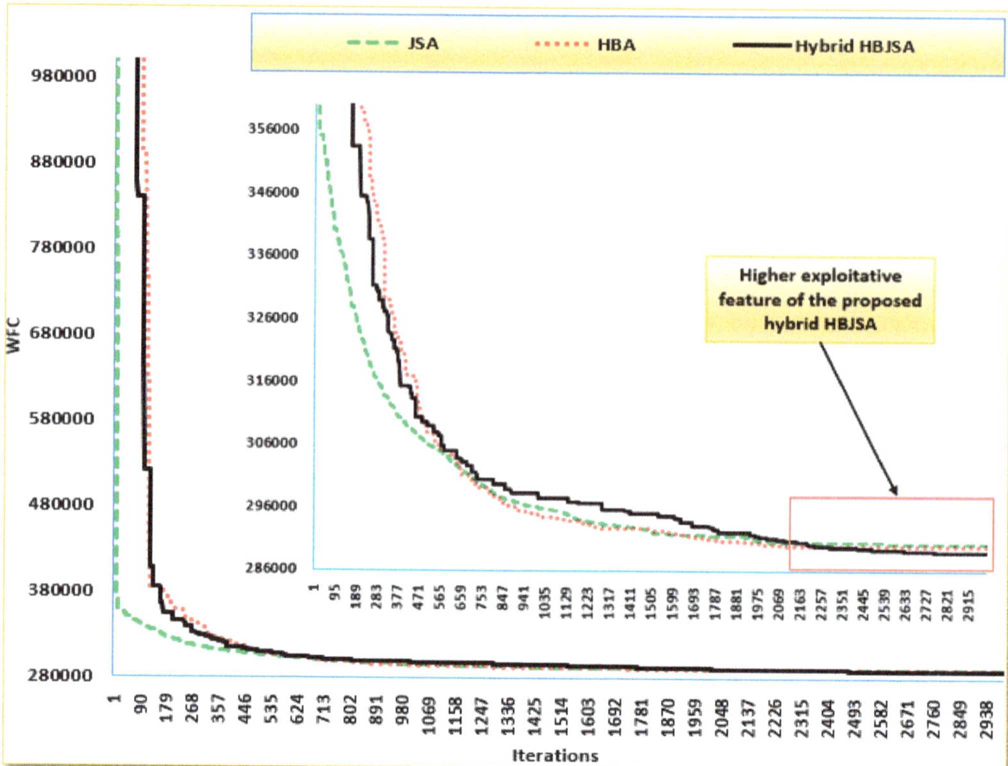

Figure 7. Convergence rates of the proposed HBJSA versus the HBA and JSA for the 84-unit system of CHP economic dispatch.

Table 6. Comparison between HBA, JSA, and HBJSA with respect to reported techniques for the 84-unit system of CHP economic dispatch.

Optimizer	WFC (USD)	Rank
WOA [20]	290,123.97	4
SDA [34]	292,788.5	7
MPA [42]	294,717.7	8
IMPA [42]	289,903.8	3
MRFO [28]	291,225.6	6
HBA	289,822.4	2
JSA	290,323.8	5
Proposed hybrid HBJSA	288,820.7	1

4.4. Simulation Results of the 96-Unit Test System

The data for the obtained system are mentioned in [20], which illustrates that 12700 MW and 5000 MWth are the load demand and heat demand, respectively, and it has 20 heat units, 52 thermal units, and 24 CHP units. The standard HBA, standard JSA, and proposed HBJSA are applied to this test system, and the corresponding MW, MWth for each unit, and WFC are demonstrated in Table 7. It can be manifested that the proposed HBJSA provides the optimal solution for WFC minimization, which accounts for USD 234,836.04, while the standard HBA and the standard JSA account for USD 235,102.65 and USD 235,277.05, respectively.

Table 7. Comparison between HBA, JSA, and the proposed HBJSA for the 96-unit test system of CHP economic dispatch problem. (a) Power outputs from power only and CHP units. (b) Heat outputs from CHP and heat-only units.

(a)

Unit	HBA	JSA	HBJSA	Unit	HBA	JSA	HBJSA
Pg1	537.254715	628.137415	448.717297	Pg39	92.65458803	92.39186824	92.49373328
Pg2	341.5738074	224.3852587	299.2517474	Pg40	448.3942915	448.8085979	442.7202304
Pg3	151.1408777	224.523037	224.1274655	Pg41	297.8196029	150.3797127	299.2207761
Pg4	109.0977628	109.8640106	111.2441868	Pg42	146.2329552	299.2034647	299.2279809
Pg5	64.10866612	109.8714738	62.37611714	Pg43	110.6346294	111.2974761	160.3281579
Pg6	110.0480942	159.7649454	179.2961575	Pg44	161.7767837	110.3086384	60.04191249
Pg7	94.33638081	159.7332315	99.67889937	Pg45	61.66518066	159.5162729	60.0000011
Pg8	60.00000879	110.0519045	117.6713057	Pg46	108.9737886	109.8987507	160.4436803
Pg9	108.2875293	110.3981768	156.6263679	Pg47	110.4947542	109.9034171	60.69590696
Pg10	115.4310336	110.6273109	78.02377234	Pg48	109.9050818	110.2078249	115.9607595
Pg11	49.01342421	77.48642225	78.94705246	Pg49	42.02833221	77.39752348	40.52730593
Pg12	92.02655803	92.39805239	92.1778521	Pg50	72.25178996	77.33236811	76.26902551
Pg13	55.18840658	92.42912952	88.05166676	Pg51	93.0608812	92.76312061	86.46078517
Pg14	360.9754061	358.9389673	538.5168446	Pg52	92.57317097	93.12310103	82.99448914
Pg15	299.3860657	224.471123	224.6561836	Pg53	104.4403419	104.9831519	93.34146328
Pg16	359.9221971	227.8984155	149.8696746	Pg54	47.60359863	45.13159323	52.66559089
Pg17	159.7120077	109.9684494	60	Pg55	88.44497134	100.2247131	106.2115994
Pg18	109.3602599	109.9616018	60	Pg56	50.14389995	51.71950188	65.28739078
Pg19	110.4137284	109.8707063	154.0137736	Pg57	12.03402877	19.00938769	10.16852035
Pg20	101.8946306	110.4372778	109.7933677	Pg58	45.6270289	56.30321085	37.30026155
Pg21	109.8648707	109.96827	60	Pg59	91.41295489	125.4849027	110.4459566
Pg22	179.4592083	109.861271	159.8078912	Pg60	51.86266145	51.90216163	56.35148968
Pg23	40.13034648	114.7915897	40.00785124	Pg61	110.2162351	91.18311611	93.67867133
Pg24	77.22962911	77.44476922	40.00084287	Pg62	47.62594622	54.14716313	42.99511003
Pg25	66.61283247	92.41354325	92.2948894	Pg63	18.272772	34.655934	20.975389
Pg26	91.02277492	92.44381223	94.51903986	Pg64	45.155884	75.627888	35.029188
Pg27	359.4498313	359.4658937	629.8475952	Pg65	88.829361	148.19343	107.83001
Pg28	299.4259674	149.5651638	151.507932	Pg66	43.74986	75.326734	51.732368
Pg29	289.8301184	224.7535524	359.9646486	Pg67	95.340679	100.58341	87.64695
Pg30	161.92495	109.9432458	109.8286727	Pg68	67.297804	42.252459	52.983829
Pg31	107.7669533	109.7440166	60.38342883	Pg69	11.638244	17.202374	10.380566
Pg32	159.4640725	109.899657	109.8614158	Pg70	35.028678	47.915618	37.11649
Pg33	162.5783378	112.3437083	160.9958816	Pg71	87.384128	99.938915	110.78998
Pg34	159.9001838	118.7834711	159.7427634	Pg72	53.523651	67.954604	44.094744
Pg35	60.03756814	109.9579597	109.7936225	Pg73	106.28473	88.257959	99.595259
Pg36	113.7051781	114.8045459	120	Pg74	65.916093	53.319524	62.695199
Pg37	114.3819608	77.49926496	40.00494973	Pg75	12.449246	11.427801	10.796383
Pg38	94.29271341	92.63059511	86.65823591	Pg76	35.002312	47.191073	44.243458

(b)

Unit	HBA	JSA	HBJSA	Unit	HBA	JSA	HBJSA
Hg53	117.8187	118.25717	111.65705	Hg77	385.57459	400.14747	390.26983
Hg54	81.187872	79.425882	85.919458	Hg78	59.99772	59.999812	56.453332
Hg55	108.9226	115.58344	118.85523	Hg79	60	59.922567	59.999913
Hg56	83.024668	85.11551	96.801938	Hg80	118.82035	119.99872	119.42771
Hg57	40.490709	43.852427	40.05269	Hg81	119.99892	119.97171	119.96065
Hg58	24.658925	29.679309	15.297858	Hg82	438.22049	402.77116	436.82909
Hg59	109.94761	129.76254	116.53145	Hg83	60	59.993989	59.982729
Hg60	83.140535	85.274518	89.077417	Hg84	59.858041	59.997954	59.888563
Hg61	120.77066	110.51193	111.89757	Hg85	119.95149	119.99343	119.76019
Hg62	81.226938	87.148267	77.525341	Hg86	118.46015	119.99908	119.9878
Hg63	39.67663	50.561873	44.68415	Hg87	450.50341	399.42417	429.43143
Hg64	23.984424	38.46521	19.947464	Hg88	59.784083	59.977283	60
Hg65	108.20515	142.50587	119.75308	Hg89	59.830588	59.991776	59.830142
Hg66	77.96514	105.49604	85.12105	Hg90	119.99947	119.9983	120
Hg67	112.78318	115.78643	108.31842	Hg91	118.32769	119.99419	119.89929
Hg68	98.469051	76.943631	85.900861	Hg92	451.81766	398.46975	449.27757

Table 7. *Cont.*

(b)							
Unit	**HBA**	**JSA**	**HBJSA**	**Unit**	**HBA**	**JSA**	**HBJSA**
Hg69	40.664318	43.086924	40.158344	Hg93	60	59.998636	59.693249
Hg70	18.721909	25.870776	20.947021	Hg94	59.916361	59.985181	59.957398
Hg71	106.59216	115.42154	121.3902	Hg95	119.99988	119.98799	119.99901
Hg72	86.491633	99.131229	78.521522	Hg96	119.83914	119.99335	119.9259
Hg73	118.09871	108.86068	115.18485	Sum (Pg)	9400	9400	9400
Hg74	97.014286	86.491059	94.56892	Sum (Hg)	5000	5000	5000
Hg75	40.89038	40.610943	37.166946	WFC	235,102.65	235,277.05	234,836.04
Hg76	18.353784	25.540248	24.147382				

Moreover, convergence characteristics of the proposed HBJSA versus the standard HBA and the standard JSA for the 96-unit test system of the CHP economic dispatch problem are depicted in Figure 8. From this figure, the proposed hybrid HBJSA is capable of improving the solution quality compared to the standard HBA and the standard JSA. At the last 1000 iterations, the proposed hybrid HBJSA provides a higher exploitative feature and, finally, reaches the lowest WFC of USD 234,836.04. Additionally, the standard HBA, the standard JSA, and the proposed HBJSA effectively achieve all constraints with 100% accuracy, as illustrated in Table 7.

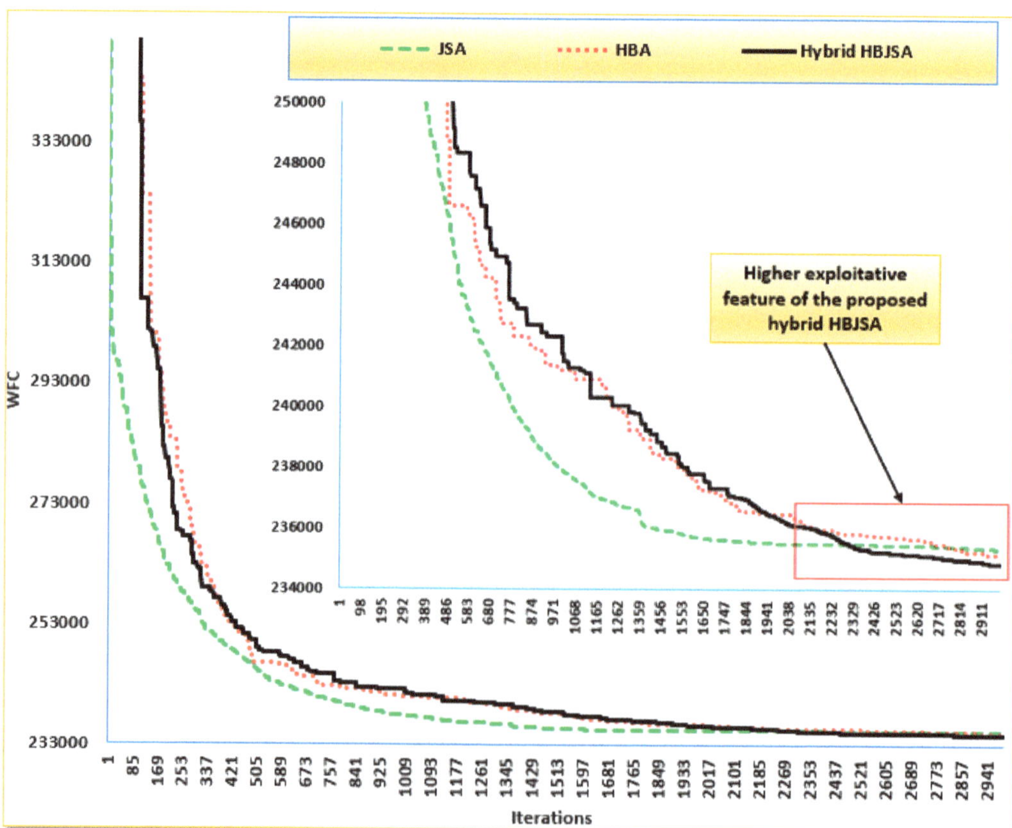

Figure 8. Convergence characteristics of the proposed HBJSA versus the HBA and JSA for the 96-unit system of CHP economic dispatch.

In addition, a comparison study between the standard HBA, JSA, and the proposed HBJSA is conducted in Table 8 for the 96-unit system of CHP economic dispatch with respect to reported techniques such as WVO-PSO [55], WOA [20], MPA [42], IMPA [42], MRFO [29], and SDA [34]. In this table, the ranking order is evaluated in ascending order based on the minimum WFC. From this table, the proposed hybrid HBJSA achieves the first rank with the lowest WFC. On the other side, the standard HBA occupies the second rank, while the standard JSA occupies the fourth rank. Additionally, this table demonstrates that the proposed HBJSA overwhelmed the standard HBA and the standard JSA and reported recent techniques for achieving minimum WFC.

Table 8. Comparison between HBA, JSA, and HBJSA with respect to reported techniques for the 96-unit test system of CHP economic dispatch problem.

Optimizer	Sum (Pg)	Sum (Hg)	WFC (USD)	Rank
WOA [20]	9400.033	5000	236,699.15	8
WVO-PSO [55]	9399.99	4999.99	238,005.79	9
SDA [34]	9400	5000	236,185.18	6
MRFO [29]	9400	5000	235,541.4	5
MPA [42]	9400	5000	236,283.1	7
IMPA [42]	9400	5000	235,260.3	3
HBA	9400	5000	235,102.65	2
JSA	9400	5000	235,277.05	4
Proposed hybrid HBJSA	9400	5000	234,836.04	1

4.5. Statistical Assessment of HBA, JSA, and Proposed Hybrid HBJSA for CHP Economic Dispatch

For all test systems, the proposed hybrid HBJSA, HBA, and JSA are run several times, and the corresponding whiskers box plots are drawn in Figure 9. For the 24-unit system, as shown in Figure 9a, the proposed hybrid HBJSA outperforms HBA and JSA in finding the lower minimum, average, and maximum WFC values. The proposed hybrid HBJSA achieves minimum, average, and maximum WFC values of USD 57,968.539, USD 58,103.95, and USD 58,293.6, respectively. On the other side, the HBA achieves minimum, average, and maximum WFC values of USD 57,994.51, USD 58,111.3, and USD 58,309.416, respectively, whereas the JSA obtains counterparts of USD 58,739.524, USD 58,968.565, and USD 59,125.33, respectively.

For the 48-unit system, as shown in Figure 9b, the proposed hybrid HBJSA outperforms HBA and JSA in finding the lowest minimum WFC value of USD 116,140.335. Compared to the HBA, the proposed hybrid HBJSA obtains lower maximum WFC values of USD 117,848.43 where the HBA obtains USD 117,980.55, while both acquire comparable WFC values of USD 116,952.6 and USD 116,946.22 for the proposed hybrid HBJSA and HBA, respectively. Compared to the JSA, the proposed hybrid HBJSA presents great superiority, since the JSA obtains minimum, average, and maximum WFC values of USD 117,365.09, USD 117,911.105, and USD 118,456.98, respectively.

For the 84-unit system, as shown in Figure 9c, the proposed hybrid HBJSA outperforms HBA and JSA in finding the lower minimum, average, and maximum WFC values. The proposed hybrid HBJSA achieves minimum, average, and maximum WFC values of USD 288,820.68, USD 289,813.827, and USD 291,251.73, respectively. On the other side, the HBA achieves minimum, average, and maximum WFC values of USD 289,822.392, USD 290,891.01, and USD 292,342.51, respectively, whereas the JSA obtains counterparts of USD 290,323.82, USD 292,366.86, and USD 293,747.44, respectively.

For the 96-unit system, as shown in Figure 9d, the proposed hybrid HBJSA outperforms HBA and JSA in finding the lower minimum, average, and maximum WFC values. The proposed hybrid HBJSA achieves minimum, average, and maximum WFC values of USD 234,836.0389, USD 235,646.129, and USD 235,967.06, respectively. On the other side, the HBA achieves minimum, average, and maximum WFC values of USD 235,102.65, USD 2,356,921.613, and USD 239,119.46, respectively, whereas the JSA obtains counterparts of USD 235,277.05, USD 236,688.76, and USD 237,940.189, respectively.

All these comparative assessments illustrate the high stability and robustness of the proposed HBJSA in finding the lowest minimum, average, and maximum WFC value compared with the HBA and JSA.

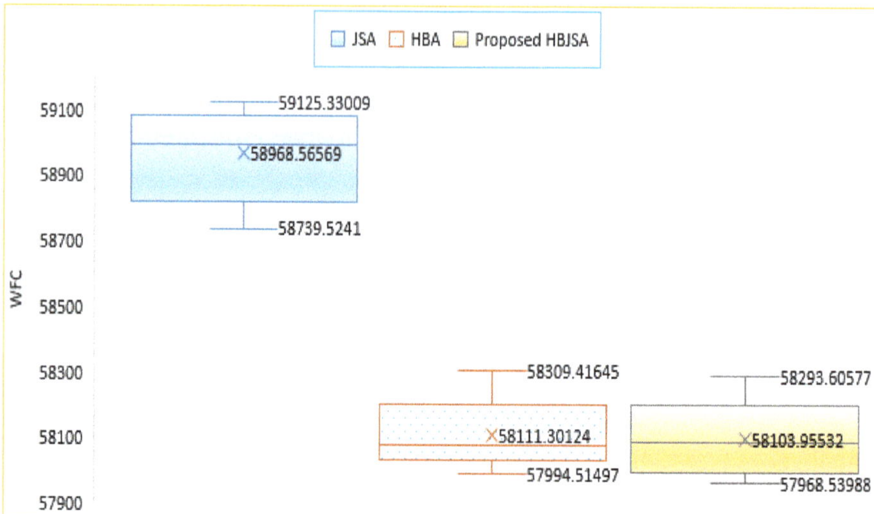

(**a**) 24-unit test system.

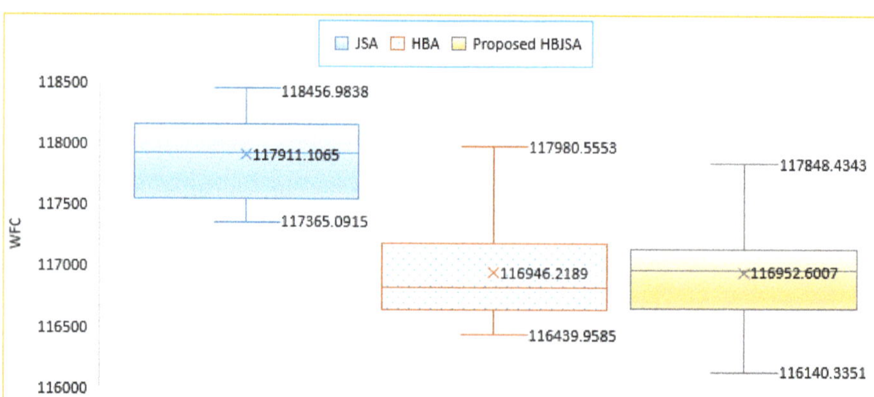

(**b**) 48-unit test system.

Figure 9. *Cont.*

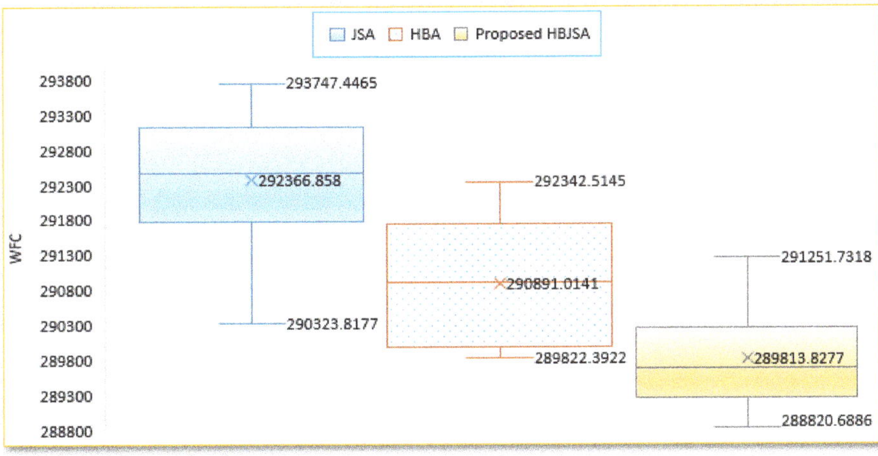

(**c**) 84-unit test system.

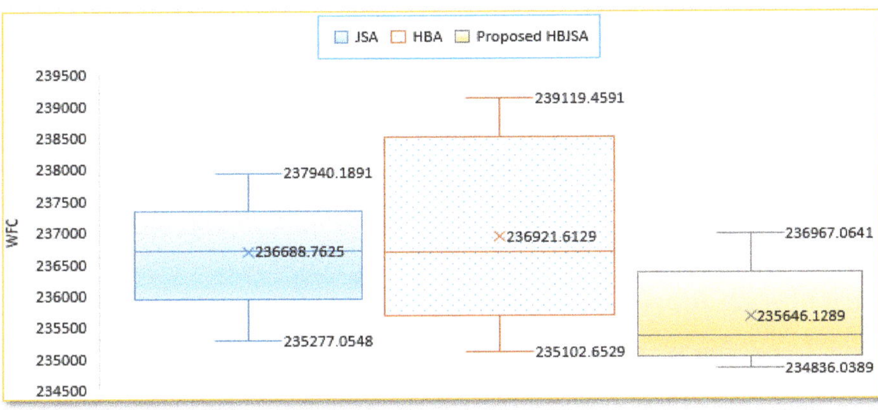

(**d**) 96-unit test system.

Figure 9. Whiskers box plot for the proposed HBJSA versus HBA and JSA for solving the CHP economic dispatch problem.

From these implementations, the practical use of the HJBSA for a larger scale as 84-unit and 96-unit test systems do not require cloud solutions. It requires the input data of the system as follows:

- The data of the power and heat loads.
- The data of the power limits of power-only units.
- The data of the heat production limits of heat-only units.
- The data of the power and heat characteristics curves of the CHP units.

5. Conclusions

In this paper, an innovative hybrid heap-based and jellyfish search algorithm (HBJSA) is presented for solving the CHP economic dispatch problem. The proposed hybrid heap-based and jellyfish search algorithm (HBJSA) combines the benefits of the standard HBA and standard JSA. Compared with standard HBA and standard JSA, the proposed HBJSA uses an adjustment mechanism in order to support the explorative and exploitative characteristics. In the proposed HBJSA, an adjustment mechanism has been constructed to boost the explorative feature at the start of iterations by enhancing the generated solutions via HBA. Furthermore, towards the conclusion of iterations, it augments and enhances the exploitative feature by growing the generated solutions via JSA. Besides, the HBA, JSA, and the proposed HBJSA have been utilized to solve the complex CHP economic dispatch problems with hard constraints, which are the feasible operating area of CHP units and valve-point effects. They are applied on two medium systems, which are 24-unit and 48-unit systems, and two large systems, which are 84-unit and 96-unit systems.

The major contributions of this paper are:

- A novel hybrid HBJSA is proposed, for the first time, in order to enhance the performance of the standard HBA and JSA for solving the CHP economic dispatch problem.
- Significant improvements via the proposed HBJSA are achieved in terms of the solution quality with high exploitative convergence characteristics for all systems studied.
- High superiority of the proposed hybrid HBJSA has been satisfied compared with several competitive algorithms in the literature.
- High robustness and stability of the proposed hybrid HBJSA with respect to standard HBA and JSA in finding the lowest minimum, average, and maximum WFC objectives.

Author Contributions: A.G.: conceptualization, methodology, writing—original draft preparation; A.E.: validation, writing—original draft; A.S.: software, data curation, writing—original draft preparation, visualization, investigation. R.E.-S.: supervision, validation, revision; corresponding author E.E.: writing—reviewing and editing, funding. All authors have read and agreed to the published version of the manuscript.

Funding: This research was funded by Taif University, grant number TURSP-2020/86.

Institutional Review Board Statement: Not applicable.

Informed Consent Statement: Not applicable.

Data Availability Statement: Not applicable.

Acknowledgments: This work was supported by Taif University Researchers Supporting Project number (TURSP-2020/86), Taif University, Taif, Saudi Arabia.

Conflicts of Interest: The authors declare no conflict of interest.

Nomenclature

a_k, b_k, c_k, d_k, e_k and f_k	Cost coefficients of the *k*th unit
a_j; b_j and c_j	Cost coefficients of *j*th heat plant
a_i, b_i and c_i	Cost coefficients of *i*th power plant
WFC	Whole fuel cost
$C_i\left(P_i^{pp}\right)$	Fuel cost of power unit *i*
$C_j\left(H_j^{hp}\right)$	Fuel cost of *j*th heat plant
$C_k\left(P_k^{cp}, H_k^{cp}\right)$	The operational cost of *k*th cogeneration unit
$P_k^{cp_{Limit}}\left(H_k^{cp}\right)$	Power bound for the set heat-output of cogenerator (*k*)
BI	Binary coefficient
N_{pp}	Number of power-only plants
H_d	System heat load
N_{cp}	Number of cogenerators
N_{hp}	Number of heat-only units
P_d	System power load
H^c	Heat output of CHP
P^c	The power output of CHP
ψ_v	Penalty coefficient
λ_i and ρ_i	Valve-point cost coefficients
CRH	Corporate rank hierarchy
t	Current iteration

k	kth vector component
$\mid\ \mid$	Absolute value
$(2r-1)$	kth component of vector $\overrightarrow{\lambda}$
r	Random number from the range [0,1]
f	Fitness of the search agent
p	Produced randomly number [0,1]
C	User-defined parameter which its unit is (iteration)
CO_o	Constant equals 0.5
X_i	ith jellyfish logistic chaotic value
TC	Time control
$CF(t)$	Time control function
t	Iteration number
$T^{\max}$	Maximum iterations' numbers
μ	Mean for all jellyfish locations in the swarm
P_0	The initial jellyfish population, $P_0 \in (0,1), P_0 \notin \{0.0, 0.25, 0.75, 0.5, 1.0\}$.
R	A random number from [0–1]
X^*	Best location of currant jellyfish
f	Objective function
U_b	Search spaces upper limit
L_b	Search spaces lower limit
$X_{i,d}$	ith jellyfish location in dth dimension

References

1. Nazari-Heris, M.; Abapour, S.; Mohammadi-Ivatloo, B. Optimal economic dispatch of FC-CHP based heat and power micro-grids. *Appl. Therm. Eng.* **2017**, *114*, 756–769. [CrossRef]
2. Nazari-Heris, M.; Mohammadi-Ivatloo, B.; Gharehpetian, G.B. A comprehensive review of heuristic optimization algorithms for optimal combined heat and power dispatch from economic and environmental perspectives. *Renew. Sustain. Energy Rev.* **2018**, *81*, 2128–2143. [CrossRef]
3. Nazari-Heris, M.; Mohammadi-Ivatloo, B.; Gharehpetian, G.B.; Shahidehpour, M. Robust Short-Term Scheduling of Integrated Heat and Power Microgrids. *IEEE Syst. J.* **2018**, *13*, 3295–3303. [CrossRef]
4. Perea, E.; Ruiz, N.; Cobelo, I.; Lizuain, Z.; Carrascal, A. A novel optimization algorithm for efficient economic dispatch of Combined Heat and Power devices. *Energy Build.* **2016**, *111*, 507–514. [CrossRef]
5. Singh, N.; Singh, S.; Chopra, V.; Aftab, M.; Hussain, S.; Ustun, T. Chaotic Evolutionary Programming for an Engineering Optimization Problem. *Appl. Sci.* **2021**, *11*, 2717. [CrossRef]
6. Zhang, J.; Xu, Z.; Xu, W.; Zhu, F.; Lyu, X.; Fu, M. Bi-Objective Dispatch of Multi-Energy Virtual Power Plant: Deep-Learning-Based Prediction and Particle Swarm Optimization. *Appl. Sci.* **2019**, *9*, 292. [CrossRef]
7. Shaheen, A.M.; Elattar, E.E.; El-Sehiemy, R.A.; Elsayed, A.M. An Improved Sunflower Optimization Algorithm-Based Monte Carlo Simulation for Efficiency Improvement of Radial Distribution Systems Considering Wind Power Uncertainty. *IEEE Access* **2020**, *9*, 2332–2344. [CrossRef]
8. Dinh, B.H.; Van Pham, T.; Nguyen, T.T.; Sava, G.N.; Duong, M.Q. An Effective Method for Minimizing Electric Generation Costs of Thermal Systems with Complex Constraints and Large Scale. *Appl. Sci.* **2020**, *10*, 3507. [CrossRef]
9. Shaheen, A.M.; Elsayed, A.M.; El-Sehiemy, R.A. Optimal Economic–Environmental Operation for AC-MTDC Grids by Improved Crow Search Algorithm. *IEEE Syst. J.* **2021**, 1–8. [CrossRef]
10. Lin, C.; Wu, W.; Zhang, B.; Sun, Y. Decentralized Solution for Combined Heat and Power Dispatch through Benders Decomposition. *IEEE Trans. Sustain. Energy* **2017**, *8*, 1361–1372. [CrossRef]
11. Guo, T.; Henwood, M.I.; Van Ooijen, M. An algorithm for combined heat and power economic dispatch. *IEEE Trans. Power Syst.* **1996**, *11*, 1778–1784. [CrossRef]
12. Sashirekha, A.; Pasupuleti, J.; Moin, N.; Tan, C. Combined heat and power (CHP) economic dispatch solved using Lagrangian relaxation with surrogate subgradient multiplier updates. *Int. J. Electr. Power Energy Syst.* **2012**, *44*, 421–430. [CrossRef]
13. Chapa, M.A.G.; Galaz, J.R.V. An economic dispatch algorithm for cogeneration systems. In Proceedings of the IEEE Power Engineering Society General Meeting, Denver, CO, USA, 6–10 June 2004; pp. 989–993. [CrossRef]
14. Rong, A.; Lahdelma, R. An efficient envelope-based Branch and Bound algorithm for non-convex combined heat and power production planning. *Eur. J. Oper. Res.* **2007**, *183*, 412–431. [CrossRef]
15. Basu, M. Combined heat and power economic emission dispatch using nondominated sorting genetic algorithm-II. *Int. J. Electr. Power Energy Syst.* **2013**, *53*, 135–141. [CrossRef]
16. Roy, P.; Paul, C.; Sultana, S. Oppositional teaching learning based optimization approach for combined heat and power dispatch. *Int. J. Electr. Power Energy Syst.* **2014**, *57*, 392–403. [CrossRef]

17. Jena, C.; Basu, M.; Panigrahi, C.K. Differential evolution with Gaussian mutation for combined heat and power economic dispatch. *Soft Comput.* **2014**, *20*, 681–688. [CrossRef]
18. Nazari-Heris, M.; Mohammadi-Ivatloo, B.; Asadi, S.; Geem, Z.W. Large-scale combined heat and power economic dispatch using a novel multi-player harmony search method. *Appl. Therm. Eng.* **2019**, *154*, 493–504. [CrossRef]
19. Mehdinejad, M.; Mohammadi-Ivatloo, B.; Dadashzadeh-Bonab, R. Energy production cost minimization in a combined heat and power generation systems using cuckoo optimization algorithm. *Energy Effic.* **2016**, *10*, 81–96. [CrossRef]
20. Nazari-Heris, M.; Mehdinejad, M.; Mohammadi-Ivatloo, B.; Gharehpetian, G.B. Combined heat and power economic dispatch problem solution by implementation of whale optimization method. *Neural Comput. Appl.* **2017**, *31*, 421–436. [CrossRef]
21. Neto, J.X.V.; Reynoso-Meza, G.; Ruppel, T.H.; Mariani, V.C.; Coelho, L.D.S. Solving non-smooth economic dispatch by a new combination of continuous GRASP algorithm and differential evolution. *Int. J. Electr. Power Energy Syst.* **2017**, *84*, 13–24. [CrossRef]
22. Haghrah, A.; Nazari-Heris, M.; Mohammadi-Ivatloo, B. Solving combined heat and power economic dispatch problem using real coded genetic algorithm with improved Mühlenbein mutation. *Appl. Therm. Eng.* **2016**, *99*, 465–475. [CrossRef]
23. Zou, D.; Li, S.; Kong, X.; Ouyang, H.; Li, Z. Solving the combined heat and power economic dispatch problems by an improved genetic algorithm and a new constraint handling strategy. *Appl. Energy* **2019**, *237*, 646–670. [CrossRef]
24. Chen, X.; Li, K.; Xu, B.; Yang, Z. Biogeography-based learning particle swarm optimization for combined heat and power economic dispatch problem. *Knowl.-Based Syst.* **2020**, *208*, 106463. [CrossRef]
25. Sundaram, A. Combined Heat and Power Economic Emission Dispatch Using Hybrid NSGA II-MOPSO Algorithm Incorporating an Effective Constraint Handling Mechanism. *IEEE Access* **2020**, *8*, 13748–13768. [CrossRef]
26. Elattar, E.E. Environmental economic dispatch with heat optimization in the presence of renewable energy based on modified shuffle frog leaping algorithm. *Energy* **2019**, *171*, 256–269. [CrossRef]
27. Bostan, A.; Nazar, M.S.; Shafie-Khah, M.; Catalão, J.P. An integrated optimization framework for combined heat and power units, distributed generation and plug-in electric vehicles. *Energy* **2020**, *202*, 117789. [CrossRef]
28. Shaheen, A.M.; Ginidi, A.R.; El-Sehiemy, R.A.; Ghoneim, S.S.M. Economic Power and Heat Dispatch in Cogeneration Energy Systems Using Manta Ray Foraging Optimizer. *IEEE Access* **2020**, *8*, 208281–208295. [CrossRef]
29. Shaheen, A.M.; Ginidi, A.R.; El-Sehiemy, R.A.; Elattar, E.E. Optimal economic power and heat dispatch in Cogeneration Systems including wind power. *Energy* **2021**, *225*, 120263. [CrossRef]
30. Moradi-Dalvand, M.; Nazari-Heris, M.; Mohammadi-Ivatloo, B.; Galavani, S.; Rabiee, A. A Two-Stage Mathematical Programming Approach for the Solution of Combined Heat and Power Economic Dispatch. *IEEE Syst. J.* **2019**, *14*, 2873–2881. [CrossRef]
31. Yifan, Z.; Wei, H.; Le, Z.; Yong, M.; Lei, C.; Zongxiang, L.; Ling, D. Power and energy flexibility of district heating system and its application in wide-area power and heat dispatch. *Energy* **2019**, *190*, 116426. [CrossRef]
32. Askari, Q.; Saeed, M.; Younas, I. Heap-based optimizer inspired by corporate rank hierarchy for global optimization. *Expert Syst. Appl.* **2020**, *161*, 113702. [CrossRef]
33. Abdel-Basset, M.; Mohamed, R.; Elhoseny, M.; Chakrabortty, R.K.; Ryan, M.J. An efficient heap-based optimization algorithm for parameters identification of proton exchange membrane fuel cells model: Analysis and case studies. *Int. J. Hydrogen Energy* **2021**, *46*, 11908–11925. [CrossRef]
34. Ginidi, A.R.; Elsayed, A.M.; Shaheen, A.M.; Elattar, E.E.; El-Sehiemy, R.A. A Novel Heap based Optimizer for Scheduling of Large-scale Combined Heat and Power Economic Dispatch. *IEEE Access* **2021**, *9*, 83695–83708. [CrossRef]
35. Elsayed, S.K.; Kamel, S.; Selim, A.; Ahmed, M. An Improved Heap-based Optimizer for Optimal Reactive Power Dispatch. *IEEE Access* **2021**, *9*, 58319–58336. [CrossRef]
36. Chou, J.-S.; Truong, D.-N. A novel metaheuristic optimizer inspired by behavior of jellyfish in ocean. *Appl. Math. Comput.* **2020**, *389*, 125535. [CrossRef]
37. Selvakumar, S.; Manivannan, S.S. A Spectrum Defragmentation Algorithm Using Jellyfish Optimization Technique in Elastic Optical Network (EON). *Wirel. Pers. Commun.* **2021**, in press. [CrossRef]
38. Shaheen, A.M.; El-Sehiemy, R.A.; Alharthi, M.M.; Ghoneim, S.S.; Ginidi, A.R. Multi-objective jellyfish search optimizer for efficient power system operation based on multi-dimensional OPF framework. *Energy* **2021**, *237*, 121478. [CrossRef]
39. Shaheen, A.M.; ElSAYED, A.M.; Ginidi, A.R.; Elattar, E.E.; El-Sehiemy, R.A. Effective Automation of Distribution Systems with Joint Integration of DGs/ SVCs considering reconfiguration capability by Jellyfish Search Algorithm. *IEEE Access* **2021**, *9*, 92053–92069. [CrossRef]
40. Mohammadi-Ivatloo, B.; Moradi-Dalvand, M.; Rabiee, A. Combined heat and power economic dispatch problem solution using particle swarm optimization with time varying acceleration coefficients. *Electr. Power Syst. Res.* **2013**, *95*, 9–18. [CrossRef]
41. Hagh, M.T.; Teimourzadeh, S.; Alipour, M.; Aliasghary, P. Improved group search optimization method for solving CHPED in large scale power systems. *Energy Convers. Manag.* **2014**, *80*, 446–456. [CrossRef]
42. Shaheen, A.M.; Elsayed, A.M.; Ginidi, A.R.; El-Sehiemy, R.A.; Alharthi, M.M.; Ghoneim, S.S. A novel improved marine predators algorithm for combined heat and power economic dispatch problem. *Alex. Eng. J.* **2021**, in press. [CrossRef]
43. Davoodi, E.; Zare, K.; Babaei, E. A GSO-based algorithm for combined heat and power dispatch problem with modified scrounger and ranger operators. *Appl. Therm. Eng.* **2017**, *120*, 36–48. [CrossRef]
44. Basu, M. Modified Particle Swarm Optimization for Non-smooth Non-convex Combined Heat and Power Economic Dispatch. *Electr. Power Compon. Syst.* **2015**, *43*, 2146–2155. [CrossRef]

45. Beigvand, S.D.; Abdi, H.; La Scala, M. Combined heat and power economic dispatch problem using gravitational search algorithm. *Electr. Power Syst. Res.* **2016**, *133*, 160–172. [CrossRef]
46. Shaheen, A.M.; El-Sehiemy, R.A.; Elattar, E.E.; Abd-Elrazek, A.S. A Modified Crow Search Optimizer for Solving Non-Linear OPF Problem with Emissions. *IEEE Access* **2021**, *9*, 43107–43120. [CrossRef]
47. Shaheen, A.M.; El-Sehiemy, R.A.; Kharrich, M.; Kamel, S. Transmission Network Planning for Realistic Egyptian Systems via Encircling Prey based Algorithms. *Turk. J. Electr. Eng. Comput. Sci.* **2021**, in press. [CrossRef]
48. Shaheen, A.M.; El-Sehiemy, R.A. A Multiobjective Salp Optimization Algorithm for Techno-Economic-Based Performance Enhancement of Distribution Networks. *IEEE Syst. J.* **2021**, *15*, 1458–1466. [CrossRef]
49. Shaheen, A.M.; El-Sehiemy, R.A. Application of multi-verse optimizer for transmission network expansion planning in power systems. In Proceedings of the 2019 International Conference on Innovative Trends in Computer Engineering (ITCE), Aswan, Egypt, 2–4 February 2019; pp. 371–376. [CrossRef]
50. Shaheen, A.M.; El-Sehiemy, R.A.; Farrag, S.M. A reactive power planning procedure considering iterative identification of VAR candidate buses. *Neural Comput. Appl.* **2017**, *31*, 653–674. [CrossRef]
51. Alharthi, M.; Ghoneim, S.; Elsayed, A.; El-Sehiemy, R.; Shaheen, A.; Ginidi, A. A Multi-Objective Marine Predator Optimizer for Optimal Techno-Economic Operation of AC/DC Grids. *Stud. Inform. Control* **2021**, *30*, 89–99. [CrossRef]
52. Elsayed, A.M.; Shaheen, A.M.; Alharthi, M.M.; Ghoneim, S.S.M.; El-Sehiemy, R.A. Adequate Operation of Hybrid AC/MT-HVDC Power Systems using an Improved Multi-Objective Marine Predators Optimizer. *IEEE Access* **2021**, *9*, 51065–51087. [CrossRef]
53. Shaheen, A.M.; Elsayed, A.M.; El-Sehiemy, R.A.; Kamel, S.; Ghoneim, S.S.M. A modified marine predators optimization algorithm for simultaneous network reconfiguration and distributed generator allocation in distribution systems under different loading conditions. *Eng. Optim.* **2021**, *53*, 1–22. [CrossRef]
54. Narang, N.; Sharma, E.; Dhillon, J. Combined heat and power economic dispatch using integrated civilized swarm optimization and Powell's pattern search method. *Appl. Soft Comput.* **2017**, *52*, 190–202. [CrossRef]
55. Dolatabadi, S.; El-Sehiemy, R.; Ghassemzadeh, S. Scheduling of combined heat and generation outputs in power systems using a new hybrid multi-objective optimization algorithm. *Neural Comput. Appl.* **2019**, *32*, 10741–10757. [CrossRef]

energies

MDPI

Article

Hydro–Connected Floating PV Renewable Energy System and Onshore Wind Potential in Zambia

Kumbuso Joshua Nyoni [1,*], Anesu Maronga [1], Paul Gerard Tuohy [2] and Agabu Shane [3,*]

1 Department of Mechanical and Aerospace Engineering, University of Strathclyde, Glasgow G1 1XJ, UK; anesu@centragrid.com
2 Energy Systems Research Unit (ESRU), University of Strathclyde, Glasgow G1 1XJ, UK; paul.tuohy@strath.ac.uk
3 School of Mines and Mineral Sciences, Copperbelt University, P.O. Box 21692, Kitwe 10101, Zambia
* Correspondence: joshua.kumbuso.nyoni@ieee.org or jnyoni@zesco.co.zm (K.J.N.); agabu.shane@cbu.ac.zm (A.S.); Tel.: +260-977883212 (K.J.N.); +260-977457561 (A.S.)

Citation: Nyoni, K.J.; Maronga, A.; Tuohy, P.G.; Shane, A. Hydro–Connected Floating PV Renewable Energy System and Onshore Wind Potential in Zambia. *Energies* **2021**, *14*, 5330. https://doi.org/10.3390/en14175330

Academic Editor: Zbigniew Leonowicz

Received: 4 July 2021
Accepted: 17 August 2021
Published: 27 August 2021

Abstract: The adoption of a diversification strategy of the energy mix to include low-water consumption technologies, such as floating photovoltaics (FPV) and onshore wind turbines, would improve the resilience of the Zambian hydro-dependent power system, thereby addressing the consequences of climate change and variability. Four major droughts that were experienced in the past fifteen years in the country exacerbated the problems in load management strategies in the recent past. Against this background, a site appraisal methodology was devised for the potential of linking future and existing hydropower sites with wind and FPV. This appraisal was then applied in Zambia to all the thirteen existing hydropower sites, of which three were screened off, and the remaining ten were scored and ranked according to attribute suitability. A design-scoping methodology was then created that aimed to assess the technical parameters of the national electricity grid, hourly generation profiles of existing scenarios, and the potential of variable renewable energy generation. The results at the case study site revealed that the wind and FPV integration reduced the network's real power losses by 5% and improved the magnitude profile of the voltage at nearby network buses. The onshore wind, along with FPV, also added 341 GWh/year to the national energy generation capacity to meet the 4.93 TWh annual energy demand, in the presence of 4.59 TWh of hydro with a virtual battery storage potential of approximately 7.4% of annual hydropower generation. This was achieved at a competitive levelized cost of electricity of GBP 0.055/kWh. Moreover, floating PV is not being presented as a competitor to ground-mounted systems, but rather as a complementary technology in specific applications (i.e., retrofitting on hydro reservoirs). This study should be extended to all viable water bodies, and grid technical studies should be conducted to provide guidelines for large-scale variable renewable energy source (VRES) integration, ultimately contributing to shaping a resilient and sustainable energy transition.

Keywords: energy transition; site appraisal and ranking; time complementarity; onshore wind; levelized cost of electricity; hydro generation; grid integration; floating photovoltaics; energy mix; electrical load; dispatch

1. Introduction

1.1. Overview

Man-made reservoirs currently have a global footprint of not less than 400,000 km^2, theoretically translating into a floating photovoltaic (FPV) potential in the terawatt scale, excluding anchoring and mooring considerations. Mooring involves securing a system of devices on water that are connected with fasteners or wires and anchored to the floor of the water body. The 2017 installed global cumulative PV capacity of 400 GWp is presently exceeded by the FPV global conservative estimate on man-made reservoirs [1]. Floating photovoltaics, otherwise known as "floatovoltaics", originally gained acceptance in Japan owing to limitations in land acquisition and utilization for new power generation projects and thus took advantage of unused water surfaces [2]. Moreover, the new market of FPV swiftly came into fruition as the price of solar photovoltaic modules dropped by 75 percent, between 2010 and 2017, and PV panel efficiency increased from 14 to 21 percent. [3–5]. From the global viewpoint, between 2015 and 2018, more than 100 FPV plants had been installed and commissioned, with a total cumulative equivalent capacity of 1.3 GWp [1,6,7]. With approximately 73 percent of the total global installed capacity in 2018 translating to an equivalent of 950 MW, China had become the FPV systems market leader. The remainder of the installed capacity was distributed among South Korea (6%), the United Kingdom (1%), Japan (16%), and Taiwan (2%), while the rest of the world was represented by 2% at the beginning of 2019. However, no fewer than thirty countries had FPV projects under development [1]. Albeit on a large scale, FPV technology deployment having been initially pioneered by Asian countries (i.e., Thailand, China, Japan, and South Korea), interest had also spread to South America, North America and Europe [8,9]. Consequently, this technology could be embraced by Sub-Saharan African (SSA) countries to complement ground-mounted-based photovoltaics.

According to a recent World Bank and Joint Research Centre (JRC) under the European Commission study, installing floating photovoltaics on 1% of the area of the African hydropower reservoirs corresponds to 101 GWp of FPV potential. This could double the current installed hydropower capacity and increase the electricity output by 58%. Moreover, a 5% and 10% retrofitting of FPV on the reservoirs could translate into 506 GWp and 1011 GWp, respectively, in the African context [1]. Combining solar PV with hydropower installations and hybridizing their output is of keen interest in many countries, in particular for smaller and weaker grids in Sub-Saharan Africa and in places with significant differences in water availability between the dry and wet seasons. The hybrid "hydro + solar PV" plant could behave as a PV + battery plant but can be more affordable and safer while retaining the benefits of hydropower [1,2]. Additionally, FPV presents the added benefit of saving water by decreasing the evaporation in reservoirs. Adding solar capacity (land-based or floating) to existing hydropower plants utilizes the existing transmission infrastructure. Hydropower can smooth the variable output by serving as a storage asset. The FPV brings out resilience by helping manage periods of low water availability [3–5].

1.2. Objectives and Research Contributions

The specific aims of this study are: (1) to document and categorize the potential of FPV and wind near hydropower sites; (2) to develop a selection process based on the documented capabilities of the sites; (3) to develop a systematic scoping design process that can be applied anywhere in the country, region or globe. This will be achieved through:

- i An initial filtering and screening process identifying potential sites for floating PV and onshore wind installations near existing and future hydropower plants.
- ii Development of a ranking and scoring methodology of filtered sites using multi-criteria decision-making and the application of the same.
- iii Development of a methodology for scoping the case study design.
- iv Application of the design methodology on an actual site.
- v Discussion of the findings.

The advent of FPV has been driven mostly by land scarcity for projects, energy security and decarbonization targets, and a loss in PV system efficiency at high operating temperatures. FPV has demonstrated great global market potential in the recent past, with enhanced technological development in photovoltaic modules and a reduction in the levelized cost of energy (LCOE) of PV energy systems [10,11]. Appraising FPV systems and projects has been a challenge, owing to the scarcity of suitable energy simulation tools for approximating the percentage increase in yield due to the cooling effect of the water surface and the different technologies employed for floaters housing the PV modules. However, research [12] correlated different heat loss factors in W/m^2K to the configuration of the floating photovoltaic structure (i.e., free-standing and small/large footprint).

This study aimed to harness the quantitative benefits of coupling FPV and onshore wind facilities with hydropower plants by relating the proximity to the existing infrastructure and grid connection, technical characteristics of the electrical network, the water-saving potential of the hydro reservoir through optimal dispatch strategies, and consequently, reducing the seasonal variations of VRES. Moreover, this study utilized the time complementarity among hydropower, floating photovoltaics and onshore wind power to alleviate the current national estimated power deficit of 810 MW. This has been attributed to reduced hydropower generation, owing to low water levels in hydro reservoirs emanating from climate change-induced droughts in the past six years [13,14]. Furthermore, this study related the integration of VRES to the hydro reservoir water-saving potential by throttling down on hydropower generation in the presence of FPV and wind power. Even though there is growing interest in floating photovoltaics, there has been no systematic appraisal of the techno-economical potential in the Zambian context and Sub-Saharan Africa (SSA). This study provides the first national-level techno-economical site assessment of onshore wind and FPV potential using a combination of validated datasets, geospatial analytical tools, site-specific wind/PV energy production models and VRES grid assessment models near existing and future hydropower plants. Furthermore, this research will help in the implementation of renewable energy technologies, such as floating photovoltaics and onshore wind power, to help increase electricity generation and supply. The study will contribute to closing the data gaps that have existed in this field of study in Zambia. To put this into perspective, the existing national grid code does not address the technical requirements (i.e., rate of frequency change, low/high voltage fault ride-through, the extent of reactive power support, etc.) of integrating VRES into the network. Therefore, this paper also addresses the nature and depth of technical studies that will have to be completed in the future to bridge this gap and thus enhance participation from independent power producers. Moreover, the research will help decision-makers to make timely and informed decisions in this area. The paper will also form a basis for further studies in the academic realm.

Therefore, the authors are highly motivated to contribute to improving the lives of all Zambians and that of neighboring citizens, by enhancing electricity access and increasing the total power generation through the adoption of renewable energy technologies such as onshore wind farms and floating photovoltaics, thus alleviating the energy poverty being faced in the region. Additionally, Zambia has the potential to enhance FOREX (foreign exchange) earnings through power exports with the interconnected SAPP countries, mitigating the chronic trade deficit with which the country has been grappling.

Against this background, the remainder of the paper is structured as follows. The subsequent section looks at the literature review; thereafter, the development of a site assessment, and the screening and ranking methodology employed in this study, are described. This ranged from site identification to the filtering and ranking of sites based on the assigned relative weight and attribute suitability scores as adopted from the literature, industry practice and stakeholder engagement. The developed methodology was then applied to a case study in Zambia. Additionally, the limitations in the site appraisal methods and tools used were highlighted. Having appraised and ranked the sites accordingly, a scoping design methodology was developed to be applied to the site with the most promising potential (i.e., highly ranked site). Furthermore, the results of the detailed case study design and formulated models were examined and discussed. Lastly, the conclusion and recommendations from the research were drawn by referencing the research outcomes, key results, study limitations and further work to be done.

2. Literature Review

2.1. Overview on FPV and Onshore Wind Potential

The evolution of FPV has, in the recent past, included the hydropower industry, owing to the opportunity for retrofitting or installing FPV panels on the abundant water surface area of hydro dams [6,15]. To put this into perspective, hydropower represents a vital aspect of the renewable energy system and covered approximately 16.4 percent of the global electricity generation at the end of 2017, which is equivalent to 1.27 TW and 4.185 TWh of total installed capacity and generated energy, respectively, owing to the increased technological investment in the equatorial regions and China. However, the negative impact of climate change (i.e., noticeable droughts) over the past decade in some regions of the world has necessitated the rapid penetration of solar photovoltaics and wind technologies [11]. The global FPV potential for waterbodies was mapped by research with hydropower capabilities, which included electricity generation and installation capacity in terawatt-hours and gigawatts, respectively. Figure 1 below outlines the total world energy distribution in "GW" at the top and "TWh" at the bottom [16].

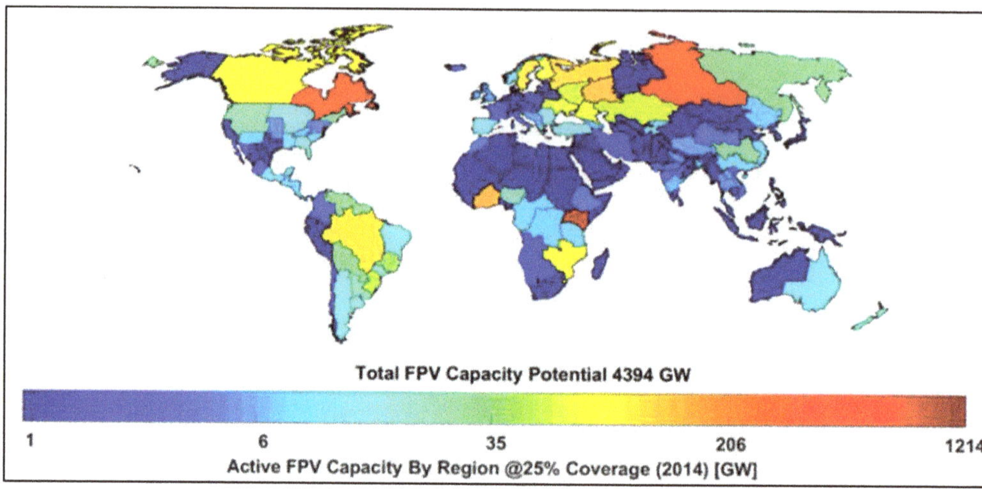

(**a**)

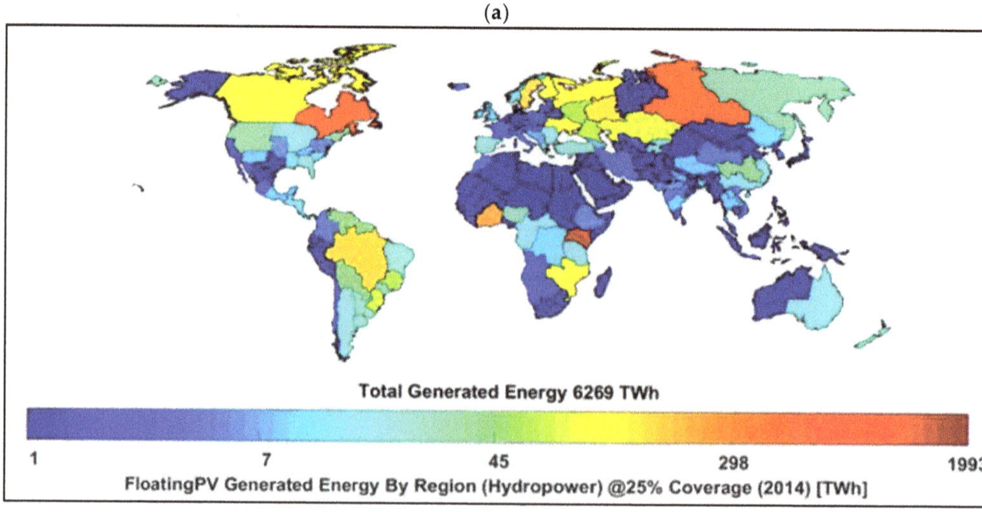

(**b**)

Figure 1. (**a**) Illustrating FPV capacity distribution potential in GW. (**b**) Showing electricity generation in TWh.

The percentage requirement of water-body surface area that matches the capacity of some hydropower plants in Ghana, Brazil, Malaysia, India, Turkey, Egypt, Venezuela, and Zambia is given in Table 1, which compares the various powerplants under consideration.

Table 1. Showing estimated reservoir size and power generation to match hydropower capacity ([1] and authors' compilation). Reproduced from [1], the name of the publisher: ESMAP 2019.

Reservoir/Dam	Country	Dam Size (km^2)	Hydropower (GW)	% Dam Area for FPV to Match Hydropower
Aswan Dam	Egypt	5000	2.0	<1 */<1(0.68)
Attaturk Dam	Turkey	820	2.4	3 */5
Bakun Dam	Malaysia	690	2.4	3 */6
Guri Dam	Venezuela	4250	10.2	2 */4
Itezhi-tezhi Dam	Zambia	113	0.12	1 */2
Kafue Gorge Upper	Zambia	70	0.99	14 */24
Kariba North Bank	Zambia	4354	1.08	<1 */<1(0.42)
Lake Volta	Ghana	8500	1.0	<1 */<1(0.2)
Narmada Dam	India	375	1.5	4 */7
Sobradinho Lake	Brazil	4220	1.0	<1 */<1(0.4)

Note: * means percentage excluding mooring (1 MW covers ~0.01 km^2), including mooring (1MW covers ~0.017 km^2).

To put things into perspective, and by taking Kafue Gorge Upper as an example, Table 1 shows that approximately 14% of the dam area is required to match the existing hydropower capacity of 990 MW, excluding mooring considerations, while the value increases to about 24% by including mooring.

According to research published by Rosa-Clot and Tina, Farfan and Breyer, Cazzaniga, and Nordmann et al., [11,16,17], the potential of large-scale hydro-connected photovoltaics is vastly promising, owing to photovoltaics' technological advancement, including enhanced mooring and anchoring techniques. The Longyangxia power plant in China is an example of a large-scale hydro-PV hybrid energy generation system, with a distribution of 850 MW and 1250 MW of solar PV ground-mounted and hydropower plants, respectively. This energy mix offers a time complementarity in the output by utilizing dispatchable hydropower to reduce the power variations and voltage sags of the system, due to intermittent solar power. The network energy dispatch curve is thus met by the downward or upward throttling of hydropower, depending on whether the photovoltaics output is high or low, respectively, thereby improving the reliability and enhancing the total energy generation of the system [11,18]. Scholarly analyses had brought to light the mutual benefits of FPV systems by not only reducing algae growth and evaporation but also making a reduction in the generation cost of solar PV energy, owing to the lower operating temperatures of PV panels [16,19].

With regard to wind energy reviews, the research by Local-Arantegui and Serrano-Gonzalez [20] has shown a technological evolution toward larger machines (i.e., taller towers, longer blades and high-capacity power generators). To put this into a global perspective, the size of the wind turbines in terms of hub height, rated power and rotor diameter had increased from 30 m hub height, 30 m rotor diameters and 300 kW rated power in the late 1980s, to 87.7 m hub height, 92.7 m rotor diameter and 2.1 MW rated power at the beginning of 2015. This technological evolution has been driven mainly by the process of attaining carbon neutrality, grid code integration adherence, scaling up the process to minimize reliability issues, and further cost reductions owing to the increase in capacity factor of most projects. Moreover, higher wind speeds, and consequently high energy yields, prevail at high altitudes; as such, wind turbine technology has advanced to accommodate longer heights of wind turbines (i.e., an increase in hub height and rotor diameter) [21,22]. According to the global wind energy council and Jin et al., [23], at the beginning of 2015, wind resources had become the largest and most successful renewable technology deployment, with 370 GW of global cumulative capacity. This feat was achieved in approximately 20 years. Many wind turbine configuration types have been addressed in the literature; nonetheless, the doubly fed induction generator (DFIG) configuration stood out in terms of mainstream technological development, owing to its high energy efficiency, low power consumption and low mechanical stress [24]. The evaluation and analysis of the impact of the DFIG on system stability and reliability have become pertinent with the increase in penetration of variable renewable energy source–VRES (i.e., wind) [25–28]. Swarna et al. [29] revealed the reactive power support capability of DFIGs at the wind turbine machine terminals during instances of active power generation curtailment.

2.2. Local Context Perspective

Zambia has great solar thermal and photovoltaic application potential (i.e., 5.5 kWh/m^2/day of average solar insolation, with approximately 3000 sunshine hours per annum) [30], coupled with 13 hydropower plants, accounting for 85 percent of the total installed generation capacity (2800 MW), making the nation better suited to a mix in generation sources. A recent wind resource study conducted by the World Bank revealed great wind-speed potential (i.e., from 6 to 12 m/s) in some parts of Zambia (i.e., Luangwa, Serenje, Muchinga, etc.) for utility-scale wind power generation at heights above sea level between 80 and 200 m. These heights confirm that reasonable wind speeds with the potential for wind energy occur at great altitudes. This resulted in the validation and commissioning of a wind atlas with a mesoscale resolution, based on a 2-year period of accurate wind speed measurement data taken from the 8 meteorological masts [31,32].

Moreover, Zambia aims to become a middle-income nation by the year 2030 (Vision 2030), even though the country is faced with significant challenges in the quest to achieve this feat. Some of the issues faced include a limited infrastructure for electricity evacuation, low electrification rates, and low access to clean energy technologies. With urban and rural access to electricity at 67% and 4%, respectively, translating into a national average of access to electricity of 31%, this leaves approximately 12 million people without access [33,34]. Thus, these unelectrified households depend on other fuel types for energy consumption and utilization (conventional biomass for their cooking and heating, lighting using kerosene and candles). The high dependence on biomass has resulted in huge deforestation of about 250,000–300,000 hectares per annum [35,36]. With a power consumption of approximately 706 kWh per capita, this is below expectations relative to Zambia's economic and social potential, when compared to other resource-rich countries like Namibia, Peru, South Africa and Chile, whose per capita consumption is about 2 to 3 times higher than that of Zambia. At the end of 2016, Zambia had a gross domestic product (GDP) of USD 20.5 billion for a population of about 16 million people and scooped the eighteenth rank in terms of growth economic prospects in Africa. Currently, approximately 69% of the Zambian population has a lack of access to electricity, although the 31% with access usually experience power outages, especially during the drought seasons [34]. The country's estimated 2800 MW of total installed capacity limits economic growth, mostly in drought-ridden years when the expected generation output is lower than normal. This is because eighty-five percent of the total installed capacity is hydropower, which depends on good water resource availability. The country's 3 major hydropower plants (Kafue Gorge, Kariba North Extension and Kariba North) account for 81% of electricity production. This dependency on hydropower can be ascribed to the vast water resource availability, resulting in an estimated hydropower potential of 6000 MW. However, climate change has in the recent past shrunk the dynamics of this potential by making the electrical power system susceptible to droughts. To put this into perspective, four major droughts have been experienced in Zambia in the last fifteen years, with the most recent occurring in the 2015/2016, 2016/2017 and 2019/2020 rainfall seasons. Consequently, the difficulties in load management strategies by the country's power utility companies were exacerbated in the quest to conserve water resources. This led in turn to turmoil in the national GDP, owing to reduced economic activity from the commercial, manufacturing and mining sectors [33,34,37].

This study encourages all stakeholders involved in electricity generation by promoting the use of alternative renewable energy technologies, such as onshore wind and floating photovoltaics, to enhance the capacity of electricity in the country, which is in tune with the perceived outcomes of the Zambia energy policy of 2019. Although there is a need to develop a firm and clear policy framework for effective regulation of these renewable energy technologies, this could help abate risks in project financing and enhance investor confidence. This could be key in transforming Zambia into a prosperous middle-income country by 2030, owing to the technologies' contribution in promoting sustainable and safe electricity generation for economic development and growth. The capacity and provisions to build resilient and better climate models (i.e., global circulation models) and the increase in understanding natural variability would help in enacting sound and well-informed environmental policies that tackle the existing energy challenges faced in the country and prepare for the future at the same time [34].

2.3. Role of Renewable-Energy Hybrid Systems in Energy Transition (Climate Mitigation and Dispatch)

The fight against climate change, through the attainment of carbon neutrality, has been the major motivator toward the adoption of renewable energy systems globally [18,38]. Nevertheless, concerns about system security and stability are amplified by the huge penetration of variable renewable energy sources (VRES), such as wind and solar photovoltaics, into the electrical network grids [39]. The inherent fluctuations in VRES technologies add to the uncertainty and variability in the electric power network and could negatively impact system operations if they are not addressed [40]. Li et al. [41] define uncertainty as an unanticipated change in demand and generation balance from what was forecasted, while variability is an anticipated change in the demand-generation balance. The increase in the penetration of VRES has necessitated the need to understand grid code constraints and electrical network parameters to maintain the integrity, efficiency, and reliability of the power system [42,43]. Large-scale penetration of VRES is one of the main challenges faced in modern electric power systems, owing to the complexities in the interactions between active and reactive power flows in the network, based on system design and connection characteristics, thus impacting dispatch operating costs, network losses and the voltage profile [43].

Certain scholarly analyses [18,44] found an economical operational balance between non-dispatchable (i.e., solar) and dispatchable (i.e., hydro) power sources, hence promoting the penetration of more renewable sources. Due to the benefit of increased system efficiency and enhanced energy supply balance, many countries have adopted hybrid energy systems providing a dynamic mix of two or more energy sources [45,46]. Typical hybrid energy systems include hydro–PV [47,48], hydro–wind–thermal [49], hydro–wind [50,51] and hydro–wind–PV systems [52,53]. A recent study by Maronga et al. [54] evaluated the optimal mix of PV, concentrated solar power (CSP) and storage, for a mining context in Zimbabwe. Previous research [55] found a mix of hydro and photovoltaics to be broadly used in many countries, owing to the vast spread of solar PV as a principal renewable energy source globally, and the swift regulation response of hydropower. Consequently, regions such as SSA (i.e., Zambia) that are rich in both hydropower and solar PV renewable resources are better suited in the development and deployment of hydro–PV energy systems. Regarding the dynamic and optimal mix of renewable energy sources involving solar PV, hydro and wind, research mostly focuses on resource temporal complementarity [56–58], plant operations management [18,53,58], and the optimization of system configuration [48,58–60]. A study by Beluco et al. [56] revealed a reduction in customer power outages because of the time complementarity benefits attributed to the solar PV and hydro hybrid system. Research conducted in Italy by Francois et al. [57] revealed a decrease in energy balance fluctuations, owing to the mix of solar PV and hydro (run-of-river type) power. Kougias et al. [58] were able to relate an improvement in the output of the PV–small hydro energy system by the optimization of the tilt angle and system azimuth.

Studies on hybrid energy systems involving wind, PV, and hydro aim at enhancing reliability and system flexibility by optimally dispatching the available resources. Such scholarly analyses, however, introduce errors in the modeling process by omitting to include the stochastic tendencies of solar PV and wind power [61–63]. Furthermore, by using deterministic and stochastic programming, Wei and Liu [64] tackled the uncertainties of solar PV and wind systems. The enhanced system security, coupled with limitations on the system economy and flexibility resulting from the deterministic inclusion of spinning reserve to the dispatch model, were revealed by Wei and Liu [64] and Liu et al. [53]. Dong et al. [65] and Zou [66] revealed that by adopting a structured multi-scenario perspective, a stochastic optimization problem in nature was able to be converted to a deterministic one, with the inaccuracy in the optimization output being the main trade-off.

In the recent past, the economic coordination of energy systems had employed robust and resilient optimization techniques, owing to its efficiency in excluding large-scale sampling variable datasets and probability models that have a precise distribution [67]. The random nature of VRES necessitates the adaptions between the forecasted and the actual generation of a hybrid system, so as to meet the load curve at any instant in time [52,68]. Researchers [69] had developed a method to track real-time deviations between two consecutive energy-scheduling intervals, while attaching the variability and uncertainty cost of energy. Another research study [64] incorporated the energy curtailment of solar PV and wind as a penalty cost in the scheduling.

3. Methodology

3.1. Site Appraisal and Ranking Methodology

3.1.1. Overview

The decision-making process regarding the suitability and location of sites for variable renewable energy sources (solar PV and wind) utilizes geospatial parameters, mainly involving GIS models in dynamic analysis (i.e., to capture, analyze, store, manage, and manipulate spatial or geographical data) [70–73]. To aid in formulating a ranking and geospatial data interpretation methodology, such GIS modeling is usually paired with multi-criteria decision-making (MCDM) [73]. Moreover, in the late 1990s, literature in the development of VRES models started gaining traction [74,75]. Global interest in the optimal siting of solar PV and wind in the recent past has arisen, due to the quest of attaining carbon neutrality, leading to the development of generic models based on the process shown in Figure 2. Firstly, the input parameter selection, ranging from social-economical, technical, and environmental factors, is completed [76]. For example, ideal wind site considerations typically include the proximity to the existing electrical network (i.e., for easy grid connection), proximity to a good road network, positioning further away from protected zones (i.e., national parks or heritage land), or further away from settlements to prevent noise and flicker, good resource potential (i.e., average wind speeds and capacity factor above sea level), and being further away from the flight path to prevent interference with radar equipment near airports. Unsuitable sites are then excluded from further analysis by scoring against the model input parameters (i.e., sites with low resource potential).

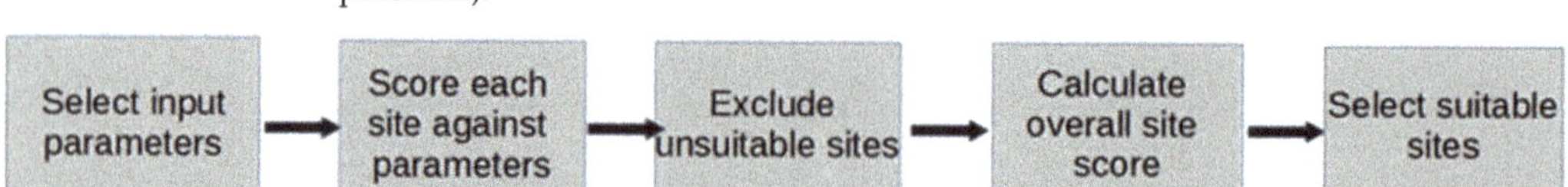

Figure 2. Diagram showing the typical structure of the multicriteria decision method (source: [71]). Reproduced from [71], the name of the publisher: ePrints Soton 2017.

The sites that have the potential for further development and pass the filtering stage are then scored and ranking using the weighted sum method (WSM) to assess their suitability (WSM is given in the equation below) [71,73]:

$$A_i^{WSM} = \sum_{j=1}^{n} w_j\, a_{ij} \text{ for } i = 1, 2, 3, \ldots N \tag{1}$$

where w is the relative parameter weighting, a is the parameter score value, and i is the attribute layer.

3.1.2. Proposed Study Methodology

The proposed study methodology for assessing site suitability was confirmed after stakeholder engagements (i.e., local experts, power utility) and extensive reviews from the literature [77–80]. The sites of interest included 5 reservoir-type, 2 pondage-type and 7 run-of-river (RoR)-type hydro plants, as shown in Table 2.

Table 2. Showing the identification of the hydro sites under study.

No	Hydro Power Station	Coordinates	Rating (MW)	Owner	River	Country	Type	Note
1	Kafue Gorge Upper	15°48′25.0″ S 28°25′16.0″ E	990	Zesco	Kafue	Zambia	Reservoir	Grid
2	Kariba North Bank	16°31′20.0″ S 28°45′42.0″ E	1080	Zesco	Zambezi	Zambia	Reservoir	Grid
3	Kafue Gorge Lower	15°53′46.0″ S 28°33′33.0″ E	750	Zesco	Kafue	Zambia	RoR	Grid
4	Itezhi-Tezhi	15°45′55.0″ S 26°01′05.0″ E	120	Zesco/ITPC	Kafue	Zambia	Reservoir	Grid
5	Lusiwasi Upper	12°59′18.2″ S 30°51′53.6″ E	15	Zesco	Lusiwasi	Zambia	Pondage	Grid
6	Lusiwasi Lower	12°59′18.2″ S 30°51′53.6″ E	12	Zesco	Lusiwasi	Zambia	Pondage	Grid
7	Lunzua	8°48′06.4″ S 31°20′18.3″ E	14.8	Zesco	Lunzua	Zambia	RoR	Grid
8	Musonda	10°42′39.5″ S 28°48′23.1″ E	10	Zesco	Luongo	Zambia	RoR	Grid
9	Chishimba	10°06′29.8″ S 30°55′02.7″ E	6	Zesco	Luombe	Zambia	RoR	Grid
10	Shiwangandu	11°13′10.25″ S 31°45′0.61″ E	1	Zesco	Munshya	Zambia	RoR	Grid
11	Lunsemfwa Hydro	14°29′33.7″ S 29°06′54.6″ E	24	LHPC	Lunsemfwa	Zambia	Reservoir	Grid
12	Mulungushi Hydro	14°43′47.48″ S 28°50′39.22″ E	32	LHPC	Lunsemfwa	Zambia	Reservoir	Grid
13	Victoria Falls	17°55′52.5″ S 25°51′37.9″ E	108	Zesco	Zambezi	Zambia	RoR	Grid
14	Zengamina	11°07′26.0″ S 24°11′32.0″ E	0.7	NWZDT		Zambia	RoR	Off-Grid

The developed methodology for the placement of wind turbines and FPV near hydropower sites is illustrated in the flowchart given in Figure 3, below.

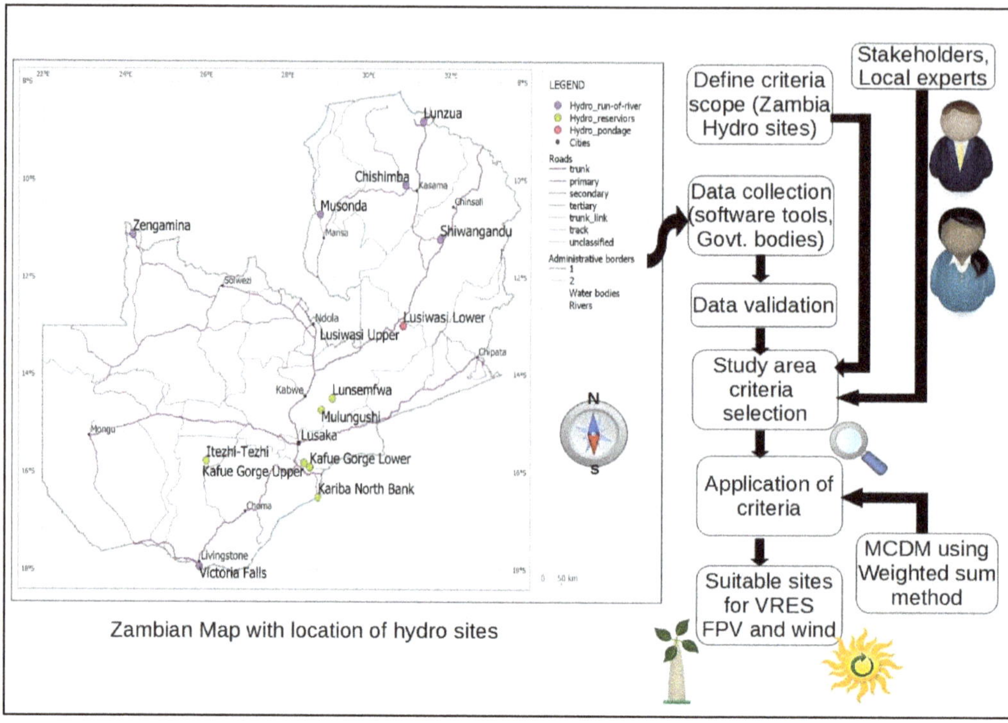

Figure 3. Diagram showing a proposed methodology flowchart of the study area shown.

3.1.3. Criteria Hierarchy Structure

Optimal FPV Site

A two-stage approach was utilized in the selection of FPV sites, namely, screening and filtering (stage 1), and ranking and scoring (stage 2), as shown in Figure 4. The filtering stage looked at the capacity factor, distance to the grid, water surface area and distance to protected zones as the model input parameters [81–83]. The scoring and ranking stage included the relative weight (r.w.) distribution of the energy export (20% r.w.), ease of access (15% total r.w.), demand (5% total r.w.) and floating PV potential (60% total r.w.) [84–87].

Optimal Wind Site

The selection of onshore wind sites utilized a two-stage approach, namely, filtering and screening (stage 1) and scoring and ranking (stage 2), which is similar to the process employed under FPV (please refer to Figure A1 in Appendix A). The filtering stage looked at the distance to the grid, distance to the protected zone, wind speed, capacity factor, the security risk of installation, and noise and flicker considerations due to proximity to buildings and settlements, as the model input parameters [88–95]. The second stage (ranking and scoring) included the distribution of the energy export (20% of r.w.), ease of access (15% of total r.w.), demand (5% of total r.w.) and wind potential (60% of total r.w.).

3.1.4. Site Attribute Suitability Score

Adopted from previous research, three site-attribute suitability tables were developed (shown in Appendix A); these included onshore wind, floating photovoltaics and hybrid suitability, looking at the balanced parameters of FPV and wind [72,77–80,96–105]. Since the wind potential is less pronounced than PV in Zambia, the relative weight for the wind was set lower than that of FPV in the balanced suitability ranking.

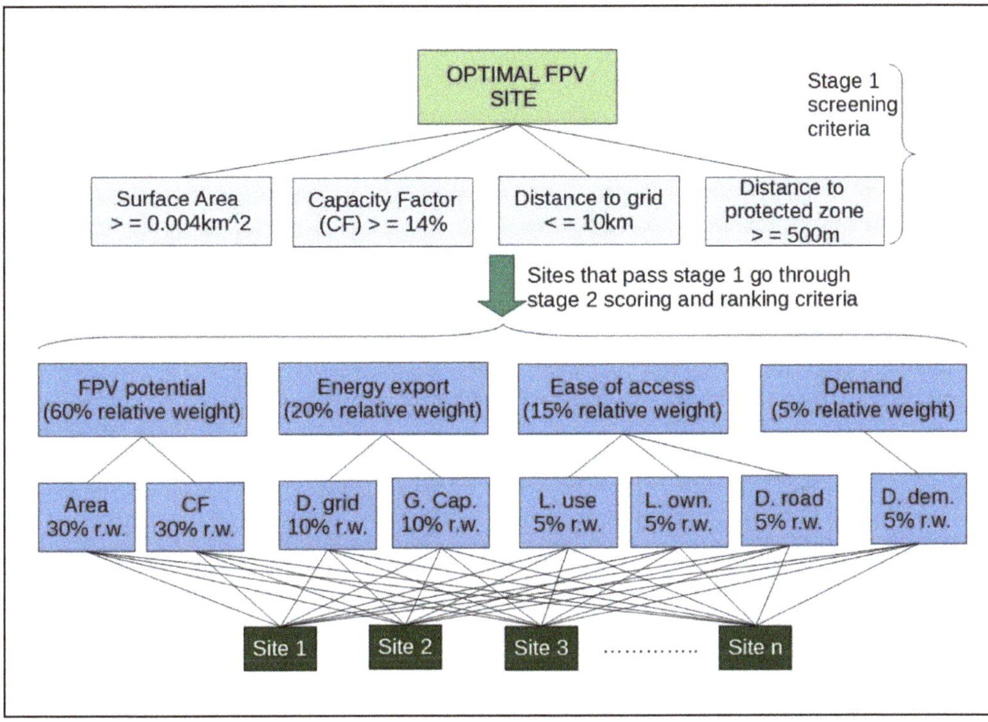

Figure 4. Diagram showing the proposed methodology flowchart of a 2-stage hierarchy structure for optimal floating photovoltaics site selection. Abbreviations: CF—capacity factor, D. grid—distance to grid, G. Cap.—capacity of grid, L. use—land use, L. own—land ownership, D. road—distance to road, D. dem.—distance to demand center, r.w.—relative weight.

3.1.5. Methodology Limitations

The weighted sum method (WSM) was applied without having insight about the assigned relative weight to the attribute layers and the layer combination procedures [73]. On the other hand, the analytic hierarchy procedure (AHP) was able to mitigate raised concerns regarding the WSM [77], although the models remain sensitive to the adopted relative weighting, as evidenced by planning permission refusal for some high-level projects within the United Kingdom. Van Rensburg et al. [106] were able to address the weighting concerns of input parameters by establishing the relationship between significant parameters influencing the quantitative assessment-based decision and the project receiving planning permission. This was then coupled with GIS modeling to assess the geospatial parameters of influence in the UK [71].

To mitigate concerns raised about the weighted sum method, the proposed study looked at a wide range of input parameters that include environmental, social, climate, economic and topographical factors to attain a more pragmatic and acceptable site appraisal (screening and ranking) process. Additionally, this was done with the help of stakeholder engagement, the solicitation of local expert opinions and an extensive literature review in the decision-making process; consequently, this contributed to the reduction in uncertainties when categorizing the attribute suitability scoring scale owing to certain assumptions that were made.

Since there is no commercial floating PV and wind project currently in Zambia, there is an element of bias in the contribution to the study from stakeholders and experts on the renewable energy generation forecast plan and agenda in line with existing policies (i.e.,

Vision 2030, National Energy Policy 2019, and the Seventh National Development Plan). Moreover, the authors acknowledge that the proposed appraisal method is an ongoing process, and hence is prone to some fine-tuning, as stakeholders (i.e., project developers, investors) with specific interests and viewpoints come on board.

3.2. Design Scoping Methodology

3.2.1. Design Methodology Formulation

The proposed energy system at Kafue Gorge Upper will comprise hydro, onshore wind, floating photovoltaics and grid load, as given in the schematic in Figure 5. The schematic shows the existing automatic generation control (AGC), excluding VRES, and the proposed Hydro-FPV-Wind daily dispatch (HFWDD) strategy. The model assumes that all the three sources of generation under consideration are coupled to the same generation bus. Moreover, the model receives inputs from the reservoir height variation "$H_r(t)$", reservoir inflow "$Q_{in}(t)$", hydro generation schedule "$P_{HYg}(t)$", water usage/consumption "$Q_T(t)$", the hydro virtual battery from saved water "$Q_s(t)$", grid load "$P_{LD}(t)$", penstock flowrate "$Q_P(t)$", onshore wind output "$P_{WDg}(t)$" and floating photovoltaic output "$P_{PVg}(t)$".

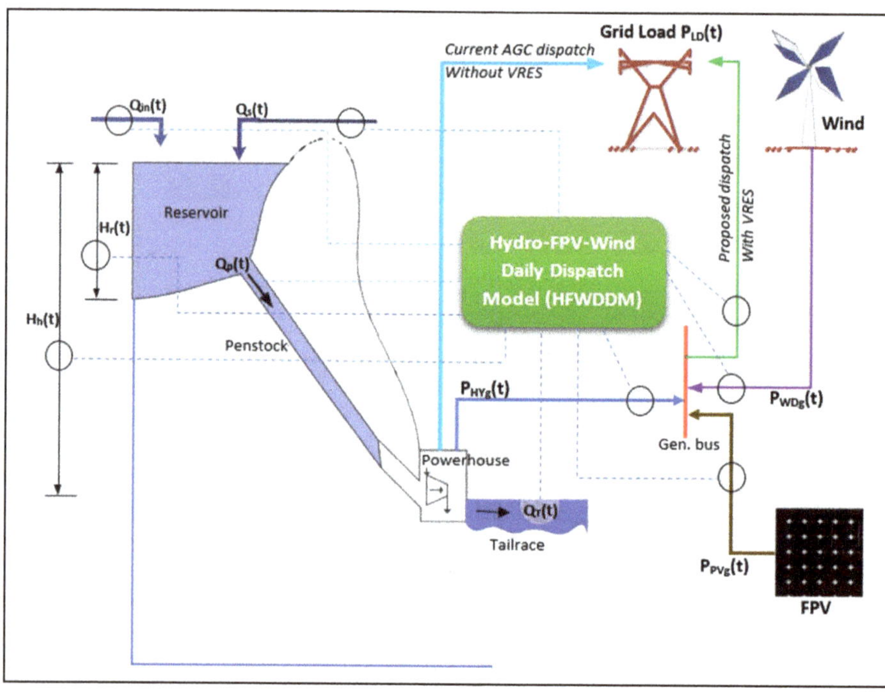

Figure 5. Diagram showing the schematic for the hydro-FPV-wind grid-tied system (adopted from [107]). Reproduced from [107], the name of the publisher: Elsevier 2019.

3.2.2. Hydro-FPV-Wind Daily Dispatch (HFWDD) Model

The objective of the HFWDD model is to balance the seasonal load characteristic curve on the grid by optimally dispatching the three generation sources (i.e., hydro, FPV and wind). This entails developing a two-stage model that addresses the technical parameters of the electrical network for any additional generation and, thereafter, optimizes the energy system using a customized dispatch algorithm (Figure 6). Firstly, the extent of wind and FPV integration on the grid that would negatively impact the network parameters (i.e., power losses, voltage magnitude and stability) is assessed in the two-stage model.

This is in line with previous research [33,42,108–114]. Secondly, seasonal hourly reservoir inflows, water consumption targets, grid demand characteristics and the total generation scenarios (wind, FPV and hydro) are incorporated into the model. Subsequently, the grid load is served by prioritizing the integration of VRES [115] as readily available, followed by a downward regulation of hydro-generation at any moment. This throttling down of hydropower is equivalent to the water-saving potential (virtual hydro battery). Nevertheless, limited reservoir capacity, coupled with a reduction in the grid demand, could present storage challenges in a wet year (which is "rarely experienced"), hence necessitating the opening of the floodgates to get rid of excess water. Similar optimization and dispatch studies of RES were conducted by [111,116–126].

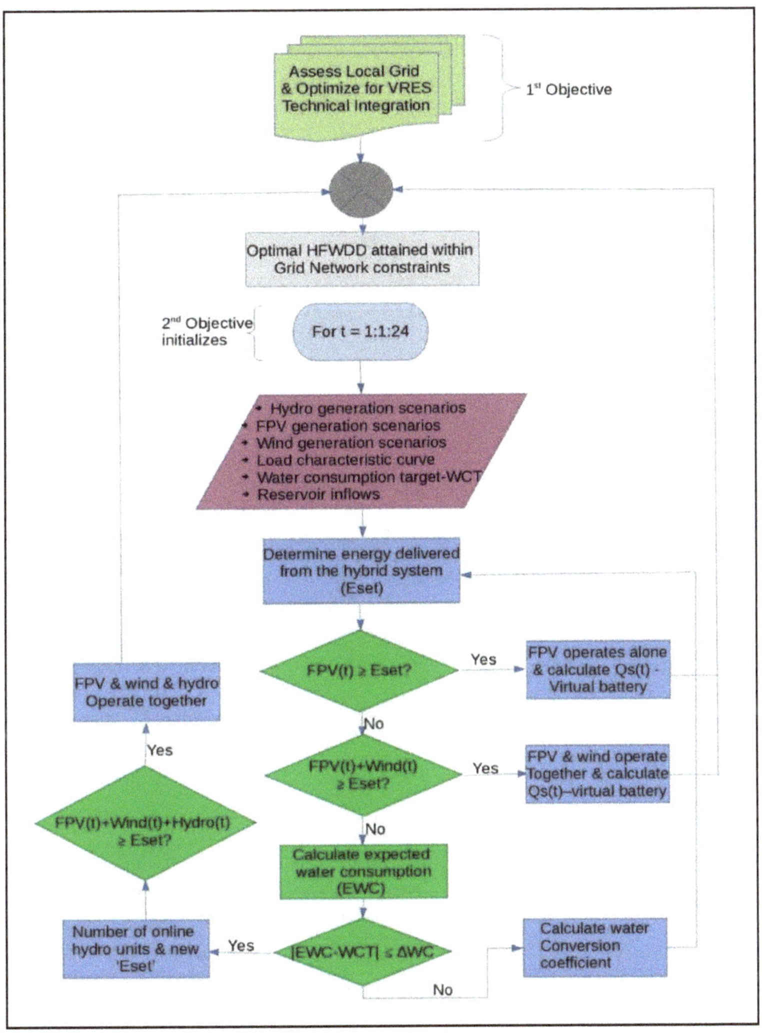

Figure 6. Diagram showing the systematic flow of the decision level to attain optimal hydro-FPV-wind daily dispatch (HFWDD).

Without putting the stochastic nature of wind and FPV power under consideration, the optimization problem is the seasonal daily dispatch on a typical day, based on the minimization of the operating cost of the existing automatic generation controller at the hydropower plant. Additionally, owing to the perceived low operational cost of wind and FPV, the optimization problem also prioritizes the dispatch of VRES over other conventional generation sources:

$$\mathrm{Min}_{k} \rightarrow \mathrm{Conv}_{\mathrm{operate}}(k) \tag{2}$$

where "k" is the dispatch scenario for the day, including hydropower plant status, and Convoperate(.) is the daily operating cost of the power plant.

Virtual storage, as indicated in Figures 5 and 6, was modeled in HomerPro in Section 4.2.4. This looks at the availability of variable renewable energy sources (FPV and wind). Based on this factor, the model calculates how much hydro would have to be ramped down. The ramping down of hydro means that less water is utilized that is then available later (i.e., at night when the sun is not shining, and also at times when the wind is calm)—the dark/calm periods will require the saved water to be utilized.

Based on the relationship between hydrogeneration and the level of the reservoir, the Qs(t) is determined. This also looks at the reservoir rule curves that must not be violated to ensure optimal operation.

However, for a very wet year, which is "rarely experienced", the storage can have limitations in terms of reservoir capacity, and thus excess water is just wasted by opening the flood gates.

Parameter Uncertainty of VRES

Adopted from [59,127], the wind and FPV outputs can be represented as shown below:

$$\text{For wind} \rightarrow P_{WDg,t} \in \left[P_{WDg,t(pre)} - P_{WDg,t(flu)},\ P_{WDg,t(pre)} + P_{WDg,t(flu)}\right] \tag{3}$$

$$\text{For FPV} \rightarrow P_{PVg,t} \in \left[P_{PVg,t(pre)} - P_{PVg,t(flu)},\ P_{PVg,t(pre)} + P_{PVg,t(flu)}\right] \tag{4}$$

where $P_{WDg/PVg,t(pre)}$ is the predicted VRES output, $P_{WDg/PVg,t(flu)}$ is the maximum output fluctuation, and $P_{WDg/PVg,t}$ is the time-dependent power output of the VRES for any given day.

Model Objective Function

Cost parameters are considered for the different generation stages to attain the economical and optimum dispatch scenario "k". The hydro unit's generation cost $\mathrm{Conv}_{\mathrm{operate}}$ ("C_{ope} = in short form" = C_{HYg}) is the first stage. The second stage ($C^{+}{}_{ope}$) mostly includes the hydro unit's adjustment costs $C_{HYg\Delta}$, curtailment costs of FPV and wind, given as $C_{PVg(curt)}$ and $C_{WDg(curt)}$, respectively [128]. Thus, the cost minimization objective function is given as:

$$C_{ope} = C_{HDg} = \sum_{t=1}^{T}\left(a \times P_{HYg,t}^{2} + b \times P_{HYg,t} + c\right) \tag{5}$$

$$\begin{aligned} C^{+}ope &= C_{PVg(curt)} + C_{WDg(curt)} + C_{HYg\Delta} \\ &= \sum_{t=1}^{T}\left[y_{PVg(curt)} \times \left(P_{PVg,t} - P_{PVg,t(inject)}\right) + y_{HYg} \times \Delta P_{HYg,t} + y_{WDg(curt)} \times \left(P_{WDg,t} - P_{WDg,t(inject)}\right)\right] \end{aligned} \tag{6}$$

where the hydro units' power output at time "t" is $P_{HYg,t}$, $P_{WDg,t(inject)}$ and $P_{PVg,t(inject)}$ are the wind and FPV injected power into the grid at time "t", respectively, $ʎ_{HYg}$ is the hydro units adjustment penalty price, $\Delta P_{HYg,t}$ the power output adjustment of hydro units, $ʎ_{WDg(curt)}$ and $ʎ_{PVg(curt)}$ are curtailment penalty prices for wind and FPV, respectively, and "a", "b" and "c" are hydro unit cost coefficients.

HFWDD Model Constraints

Hydro Constraints:

$$P_{HYg(max)} \geq P_{HYg,t} \geq P_{HYg(min)} \tag{7}$$

$$Q_{HYg,t} = y^{b}{}_{HYg} + y^{a}{}_{HYg} \times P_{HYg,t} \tag{8}$$

$$V_{flow(max)} \geq Q_{HYg,t} \geq V_{flow(min)} \quad (9)$$

$$Q_{t+1} = Q_{in,t} - Q_{HYg,t(curt)} - Q_{HYg,t} + Q_{s,t} \quad (10)$$

$$Q^{max} \geq Q_{s,t} \geq Q^{min} \quad (11)$$

$$Q_{s,1} = Q_{s,ini} \quad (12)$$

$$Q_{s,T} = Q_{s,term} \quad (13)$$

where $y^{b}{}_{HYg}$ and $y^{a}{}_{HYg}$ are hydro water conversion coefficients, $Q_{s,term}$ and $Q_{s,ini}$ are final and initial storage values of the reservoir, Q^{max} and Q^{min} are the upper and lower reservoir storage limit at time "t", $Q_{HYg,t}$ is the water consumption at any time "t" of the hydro unit, $P_{HYg,t}$ is the power output of the hydro unit at time "t", $Q_{in,t}$ is the inflow of reservoir at time "t"; $Q_{s,t}$ is the hydro reservoir storage at time "t", $Q_{HYg,t(curt)}$ is the curtailment of the reservoir water, and $V_{flow(max)}$ and $V_{flow(min)}$ are the water consumption upper and lower limits in a given period.

Power Flow Branch Constraints:

$$\sum_{i=1}^{Ni} (f_{bi} \times P_{it}) \leq S_{b(max)} \quad (14)$$

where $S_{b(max)}$ is the branch maximum capacity, "i" is the power system node identifier, "b" is the branch identifier, N_i is the number of system network nodes in total, $P_{i,t}$ is the net active power injected into the ith node. f_{bi} is the sensitivity factor of the bth node.

Power Balance Constraints:

$$P_{WDg,t(pre)} + P_{HYg,t} + P_{PVg,t(pre)} = P_{LD,t} \quad (15)$$

where $P_{LD,t}$ is the grid load of the system at any given time "t".

Onshore Wind Power Constraints:

$$P_{WDg,t} \geq P_{WDg,t(inject)} \geq 0 \quad (16)$$

where $P_{WDg,t}$ is the variable wind generator power output at time "t".

Floating PV Power Constraints:

$$P_{PVg,t} \geq P_{PVg,t(inject)} \geq 0 \quad (17)$$

where $P_{PVg,t}$ the variable FPV power output at time "t".

4. Results and Discussions

4.1. Application of Appraisal and Ranking Methodology

4.1.1. Stage 1—Site Screening

The floating photovoltaics site screening process involved the definition of five criteria that include a distance to protected zones greater than or equal to 500 m, a distance to existing electrical infrastructure less than or equal to 10 km, a capacity factor (CF) greater than or equal to 14%, and a water body surface area greater than or equal to 4000 m^2. Against this benchmark, Zengamina, Victoria, and Lunzua run-of-river sites were excluded on account of having a surface area <4000 m^2 to accommodate a commercially and economically viable FPV project. Further, wind site filtering involved the definition of six criteria that include a distance to protected zones (i.e., national parks) greater than or equal to 500 m, the security risk (i.e., war-prone area) of installation, an average wind speed value at 150 m above ground level greater than or equal to 6 m/s, noise and flicker allowance at five times the rotor diameter (5D), a distance to electrical infrastructure less than or equal to 60 km, and a capacity factor greater than or equal to 26%. Due to the security risk zone bordering the Democratic Republic of Congo, the Zengamina wind site was excluded from the list of potential sites. This is in line with the World Bank findings on mapping security risk-prone areas for the installation of wind validation masts. Table 3 summarizes the stage 1 screening and filtering process for all the FPV and wind sites.

Table 3. Table showing the combined stage 1 screening outcome for both FPV and onshore wind sites.

ID #	Name of Site (FPV and Wind)	Screening Criteria (Yes/No)						
		Capacity Factor	Area	Distance to Protected Zone	Distance to Grid/ Substation	Flicker and Noise Distance	Wind Speed (m/s)	Security Risk
Site 1	KGU	Yes	Yes	Yes	Yes	Yes	Yes	Yes
Site 2	Kariba	Yes	Yes	Yes	Yes	Yes	Yes	Yes
Site 3	KGL FPV	Yes	Yes	Yes	Yes	Yes	Yes	Yes
Site 4	Itezhi-Tezhi	Yes	Yes	Yes	Yes	Yes	Yes	Yes
Site 5	Lusiwasi	Yes	Yes	Yes	Yes	Yes	Yes	Yes
Site 6	Lunzua	Yes	No	Yes	Yes	Yes	Yes	Yes
Site 7	Musonda	Yes	Yes	Yes	Yes	Yes	Yes	Yes
Site 8	Chishimba	Yes	Yes	Yes	Yes	Yes	Yes	Yes
Site 9	Shiwangandu	Yes	Yes	Yes	Yes	Yes	Yes	Yes
Site 10	Lunsemfwa	Yes	Yes	Yes	Yes	Yes	Yes	Yes
Site 11	Mulungushi	Yes	Yes	Yes	Yes	Yes	Yes	Yes
Site 12	Victoria FPV	Yes	No	Yes	Yes	Yes	Yes	Yes
Site 13	Zengamina	Yes	No	Yes	Yes	Yes	Yes	No

4.1.2. Stage 2—Ranking and Scoring

Three ranking and scoring tables were developed; however, only the analysis for the balanced ranking of the hybrid system is presented for simplicity. Table 4 illustrates the site scoring results for a balanced ranking using the weighted sum method (WSM). The distribution of the relative weight for the various attributes is as follows: demand at 5%, ease of access at 15%, energy export at 20%, wind potential at 25% and floating photovoltaics at 35%. Taking "FPV distance to grid" as an example under the "Energy export" attribute layer, the application of the weight sum equation 1 is presented in Figure 7. The results analysis places Kafue Gorge Upper (KGU) and Itezhi-Tezhi at second and first rank, with total attribute values of 86.9% and 90%, respectively, while the least-ranked site is Chishimba, with a total attribute combined value of 70.6%.

Figure 7 presents an example of how to apply the weighted sum method. This looks at the "Energy Export attribute layer", with a focus on the "Distance of the floating photovoltaic plant from the grid". With reference to Table 4, part 1 of Tables A1–A3 in Appendix A, there are 5 attribute layers, namely, (i = 1) "Wind potential", (i = 2) "Floating PV potential", (i = 3) "Energy export", (i = 4) "Ease of access" and (i = 5) "Demand". These attribute layers have the following maximum weight distribution: attribute layer (i = 1) → 25%, (i = 2) → 35%, (i = 3) → 20%, (i = 4) → 15%, (i = 5) → 5%. Therefore, under "select input parameters" in Figure 7, (i = 3) represents the energy export attribute layer, with 20% as total weight. The energy export layer is further broken down in "FPV distance to grid" given a maximum weight of 5%, "Wind distance to grid", also given 5%, and "Grid capacity availability", given 10%. Under the "weigh input parameters" FPV distance to the grid is appearing as the first layer (j = 1) under the energy export attribute layer. This is assigned as "w1" with reference to Equation (1). Under "score each site against parameter", if the site's FPV distance from the grid is less than or equal to 2 km, then according to Table A1 in Appendix A, the suitability score for the site will be 100%. This is assigned as "*a*31" from Equation (1). Therefore, to get the overall site score, the product between the assignment "*w*1" under "weigh input parameters" and the assignment "*a*31" under "score each site against parameter" is calculated. Multiplying "*w*1" by "*a*31" translates into 5% × 100%, yielding a value of 5%. The process is repeated for all other attribute layers and the layers contained underneath. The total site score is the summation of the wind total, FPV total, energy export total, ease of access total and demand total.

Table 4. Table showing balanced scoring and ranking matrix. Analysis based on sources [70,75–78,94–103] (detailed table shown in Appendix A—Tables A2 and A3).

		Wind Potential	FPV Potential	Energy Export	Ease of Access	Demand	Site Total
		Score * (25% Weight)	Score * (35% Weight)	Score * (20% Weight)	Score * (15% Weight)	Score * (5% Weight)	(100% Weight)
Rank #	**Name of Site**	**Wind ∑**	**FPV ∑**	**Export ∑**	**Access ∑**	**Demand ∑**	**Score * Weight**
=1	Itezhi-tezhi FPV/wind site	22.5%	35.0%	12.5%	15.0%	5.0%	90.0%
=2	KGU FPV/wind site	21.3%	30.6%	18.8%	11.3%	5.0%	86.9%
=3	KGL FPV/wind site	25.0%	26.3%	15.0%	13.8%	5.0%	85.0%
=3	Kariba FPV/wind site	18.8%	26.3%	20.0%	15.0%	5.0%	85.0%
=3	Lusiwasi FPV/wind site	20.0%	35.0%	12.5%	15.0%	2.5%	85.0%
=6	Musonda FPV/wind site	22.5%	30.6%	8.8%	15.0%	2.5%	79.4%
=6	Mulungushi FPV/wind site	22.5%	30.6%	10.0%	13.8%	2.5%	79.4%
=8	Shiwangangu FPV/wind site	21.3%	26.3%	12.5%	13.8%	2.5%	76.3%
=9	Lunsemfwa FPV/wind site	25.0%	21.9%	8.8%	15.0%	2.5%	73.1%
=10	Chishimba FPV/wind site	20.0%	21.9%	11.3%	15.0%	2.5%	70.6%

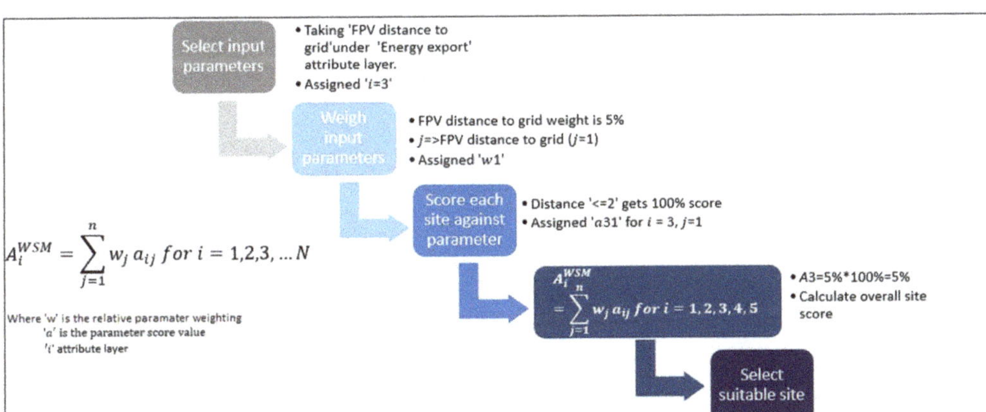

Figure 7. Diagram showing the application of the weighted sum equation under balanced scoring and ranking.

4.2. Application of Design Scoping Methodology—Kafue Gorge Upper Case Study

After appraising the potential wind and FPV sites, the stakeholder (ZESCO Ltd.) was presented with the three ranking matrices (FPV, onshore wind and balanced) of the ten potential sites to choose from. The power utility opted to adopt the balanced ranking for the hybrid energy system, with Kafue Gorge Upper (KGU) as the chosen candidate site for detailed design. Even though Itezhi-Tezhi (ITT) was ranked first over KGU, which was second under the balanced scoring, the latter was chosen over the former owing to the following factors: the presence of a data validation wind mast at KGU, the presence of debris and dead trees in the ITT reservoir, the distance to the demand center (300 km from ITT, compared to 100 km for KGU), less reliability and stability of the grid at ITT, with one 220 kV line emanating from ITT compared to three 330 kV lines from KGU to the grid.

4.2.1. VRES Grid Impact Study

Using the power system analysis toolbox (PSAT), the Zambian electrical power grid was modeled at a 330 kV voltage level comprising a 27–bus system. The model for the existing network had a real and reactive power load distribution of 2383 MW and 1061.8 MWVAr, respectively. The total modeled existing generation was 2530.6 MW real power and 857.4 MVAr reactive power, comprising the following power stations: Itezhi-Tezhi via Nambala, Lunzua via Kasama, Victoria Falls via Mukuni, Maamba Collieries Limited (MCL), PV plant at Lusaka South Multi-Facility Economic Zone (LSMFEZ), Kariba North Bank and Kafue Gorge Upper. For additional generation, 200 MW of VRES was later integrated and modeled, comprising 100 MW FPV and 100 MW wind at the KGU generation bus. The actual PSAT single-line diagram model for the network is shown in Appendix B. The key summary results for the grid impact of VRES are presented below.

Analysis of Existing Network

Figure 8 shows the voltage violations at 11 out of the 27,330 kV buses for the existing network before the addition of compensating equipment and additional generation (VRES, in this case). According to the Zambian grid code limits, the permissible and acceptable voltage should fall in the range between 313.5 and 346.5 kV, which is a tolerance of $+/-5\%$. The total power losses of the existing network were 1475 MW real and -204.3 MVAr reactive power, as shown in the global power summary in Figure 9.

Analysis after VRES Integration

The addition of network compensating equipment corrected all 11 voltage violations with the integration of VRES at the KGU bus, comprising 100 MW wind and 100 MW floating photovoltaics, improving the voltage magnitude profile even further at buses near the KGU generation bus (Figure 10). The integration of 200 MW of VRES contributed to the reduction in the voltage support requirement by the network, owing to the drop in the generated reactive power by the Kafue Gorge hydro plant from 239 to 201 MVAr. The results also showed all the line flows to be within range (less than the 700 MVA maximum line capacity). Moreover, VRES integration also reduced the network's real power losses by 5 percent (from 147 MW to 140 MW), as can be seen in Figure 11.

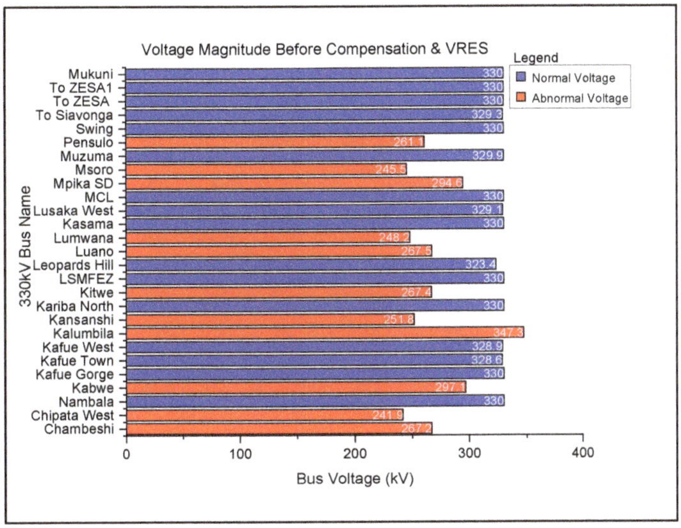

Figure 8. Figure showing the existing 330 kV bus voltage profile without VRES integration and reactive compensation at Luano, Kansanshi, Lumwana, Kitwe, Kalumbila, and Chipata West network buses.

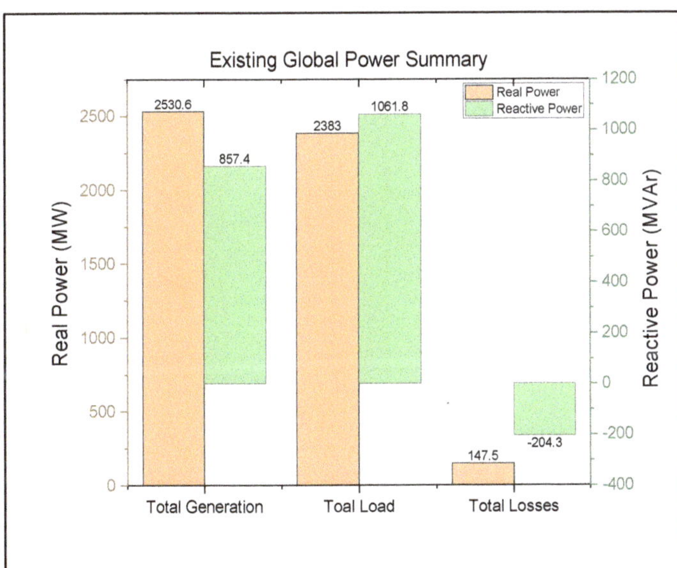

Figure 9. Figure showing the global power summary for the existing network.

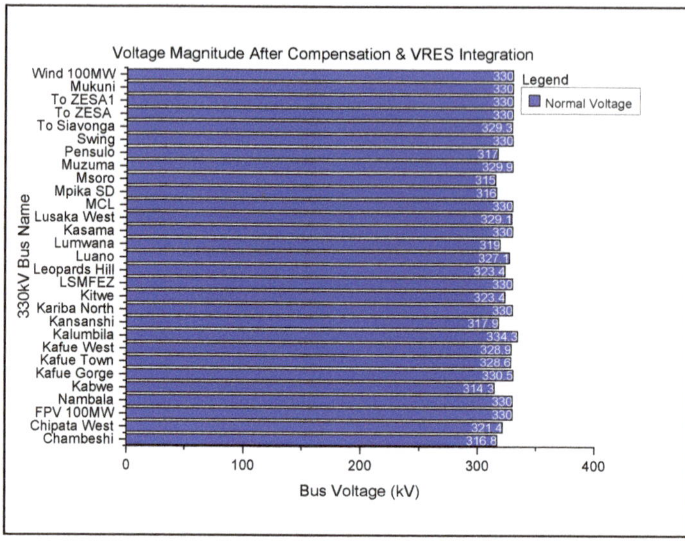

Figure 10. Showing bus voltage after adding network compensation and VRES.

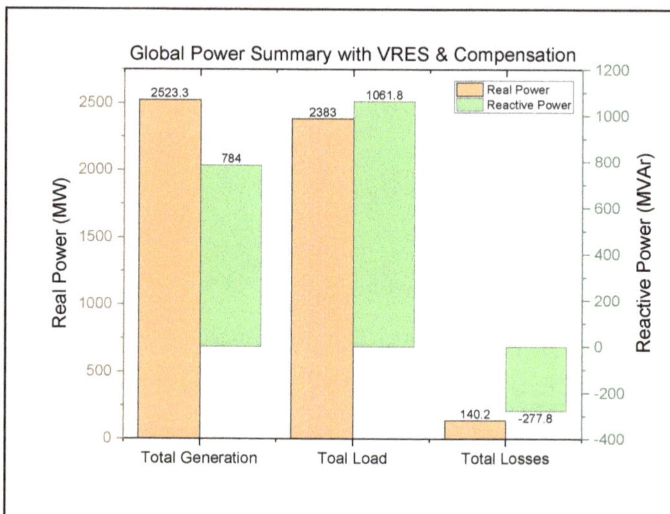

Figure 11. Showing power summary after VRES integration.

4.2.2. Hydro Modeling Results Analysis at KGU

The Kafue Gorge Upper hydro generation was modeled in iHoga using the power plant ratings provided by the national power utility (ZESCO Ltd. located in Zambia). The iHoga model for one turbine has 4% losses in penstock, 0% daily/hourly variability and 85% total turbine efficiency. The key summary results of the model taking a typical winter (i.e., June) and summer (i.e., November) month are presented in Figure 12a–d. In June, the maximum power output of the hydropower plant was 805 MW, corresponding to a discharge rate of 227.8 m^3/s and reservoir level of 974.7 m above sea level, while the minimum generation output was 697 MW, corresponding to a discharge rate and level of 197.9 m^3/s and 974.9 m, respectively. In the month of November, hydro generation output ranged between 648 and 712 MW, corresponding to a discharge rate and level range of

185–201 m^3/s and 974.5–974.7 m, respectively. Figure 12e shows the hydro-generation hourly time-series graphs, serving a fraction of the total national grid load for the first day of each month of January, March, June, September and November. On 1 January, KGU generated 10.69 GWh of hydro, with an evening peak of 1.99 GWh, between 5 p.m. and 9 p.m., to serve about 27% of the total national grid demand for the day. KGU generated 15.6 GWh of hydropower on 1 March, with an evening peak of 3 GWh between 5 p.m. and 9 p.m., to serve about 46% of the total national electrical grid demand for the day. On 1 June, approximately 13.8 GWh of energy was generated, with an evening peak of 2.7 GWh between 5 p.m. and 9 p.m., to serve about 34.4% of the total demand for the day. On 1 September, 14 GWh of hydro-generation was produced, with an evening peak of 2.7GWh to serve 36% of the total grid demand for the day. Further, 1 November yielded 12.6 GWh of hydropower, with an evening peak of 2.4 GWh between 5 p.m. and 9 p.m., to serve 29.6% of the grid demand for the day.

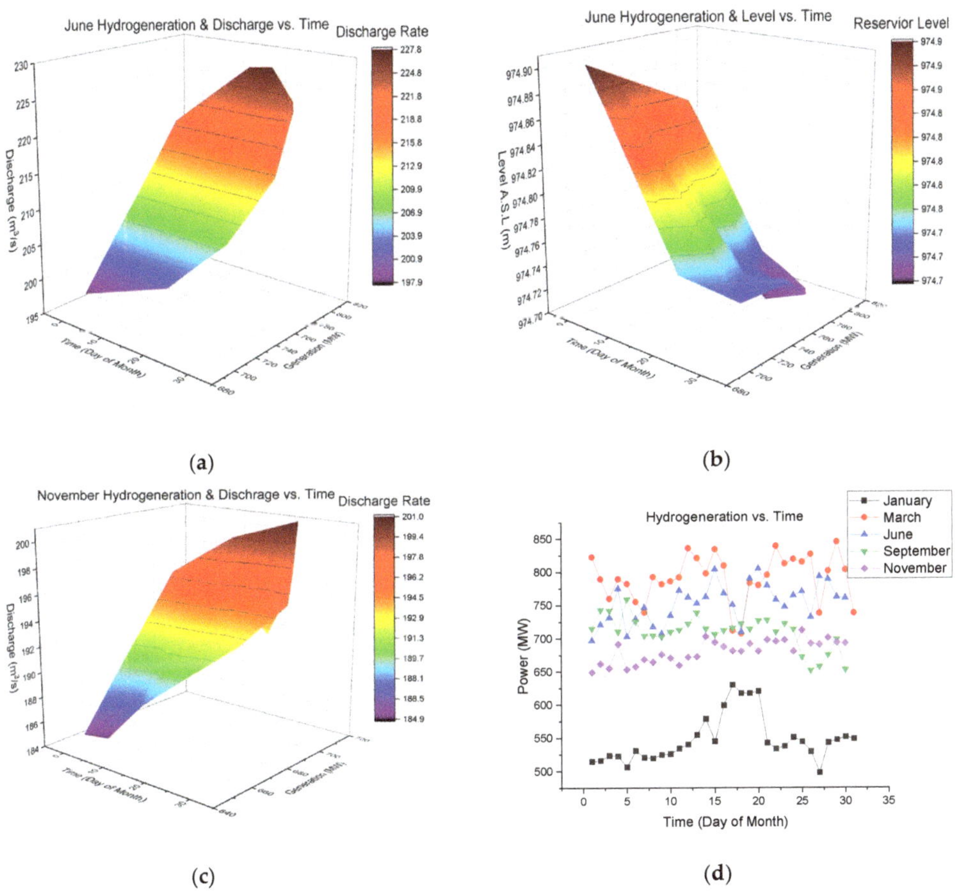

Figure 12. *Cont.*

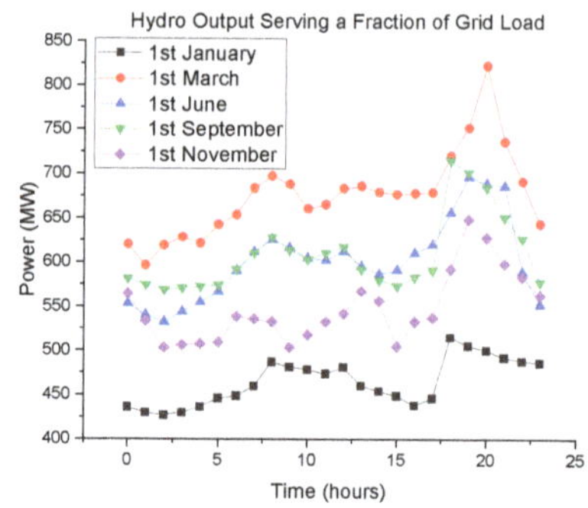

(e)

Figure 12. (**a**) Showing 3D June hydro-generation and discharge rate vs. time. (**b**) 3D June hydro-generation and reservoir level vs. time. (**c**) 3D November hydro-generation and discharge vs. time. (**d**) Day of month time series for hydro-generation vs. time. (**e**) Daily hydro output serving a fraction of grid load.

4.2.3. VRES Modeling and Results Analysis

Results analysis for the modeling and design of the 100 MWac onshore wind and 116MWdc floating photovoltaics at Kafue Gorge Upper hydropower plant is presented.

Floating PV

Detailed design and modeling of the FPV system were performed using the Photovoltaic System (PVSYST) software. A form factor (DC/AC ratio) of 1.16 was adopted for the project, based on industry practice for Southern Africa. The system design comprises a parallel connection of eight sub-arrays. Each sub-array comprises one hundred and twenty series-strings of seventeen PV solar modules (unit photovoltaic module rating of 285 watt-peak with 72 polycrystalline cells) connected to an inverter rated at 500 kWac, with the A/C combiner box linking twenty-five inverters in parallel for 1 sub-array. Firstly, the photovoltaic module and array characteristics were analyzed, based on the results in Figures 13 and 14. The average PV module running temperature of between 10 and 65 °C yielded a minimum of 60 h of operation throughout the year, with a design standard irradiation of 1 kW/m^2, an operating temperature range between 10 and 70 °C, coupled with corresponding module efficiency between 15.8 and 11%, respectively, at a given operating temperature range. However, at all irradiation levels, an increase in efficiency was observed with a decrease in temperature. Additionally, at 1 kW/m^2, the PV module power output at maximum power point (MPP) was found to be 229.1 W (20 percent decrease) and 305.3 W (7 percent increase) at the highest and lowest operating temperatures, respectively.

According to Figure 15, the annual energy yield injected into the grid from the FPV system was 214.4 GWh/year. August had the most FPV energy injection into the grid, with 21.29 GW, while January had the least, with 13.59 GWh. Moreover, August and October had the most average global horizontal irradiation of 187.6 and 193.4 kWh/m^2, respectively, while January and February had the least, with 155.3 and 155.2 kWh/m^2, respectively. October had the highest average ambient temperature of 25.61 °C, while July had the least, at 16.78 °C. This corresponds to a monthly average system efficiency of 11.92 and 12.53 in the hottest and coldest months, respectively.

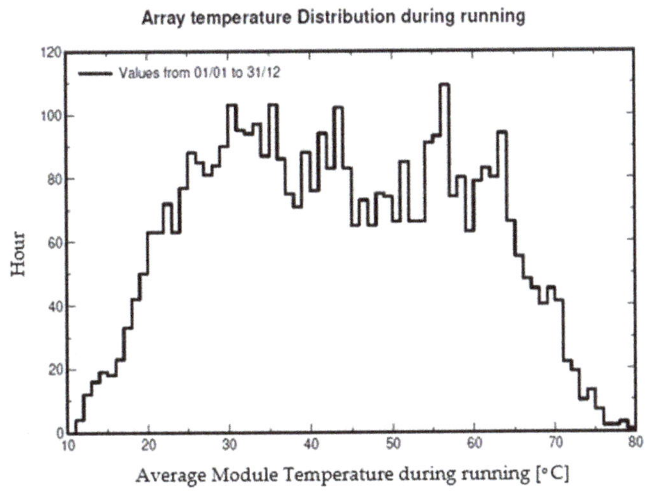

Figure 13. Showing FPV Array operational temperature.

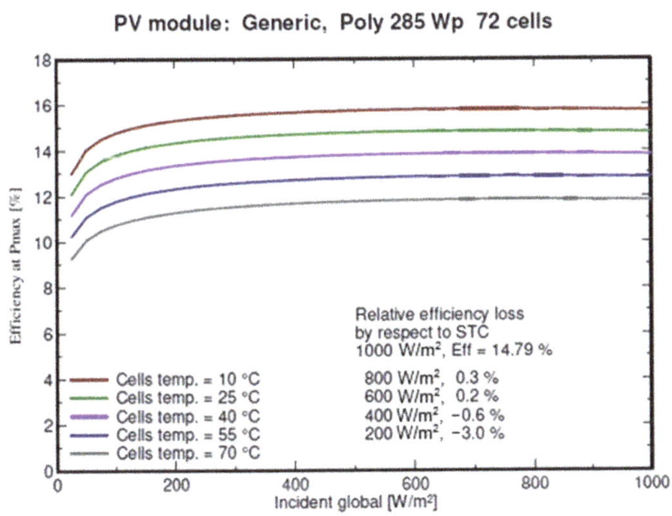

Figure 14. Figure showing FPV efficiency vs. irradiation curves.

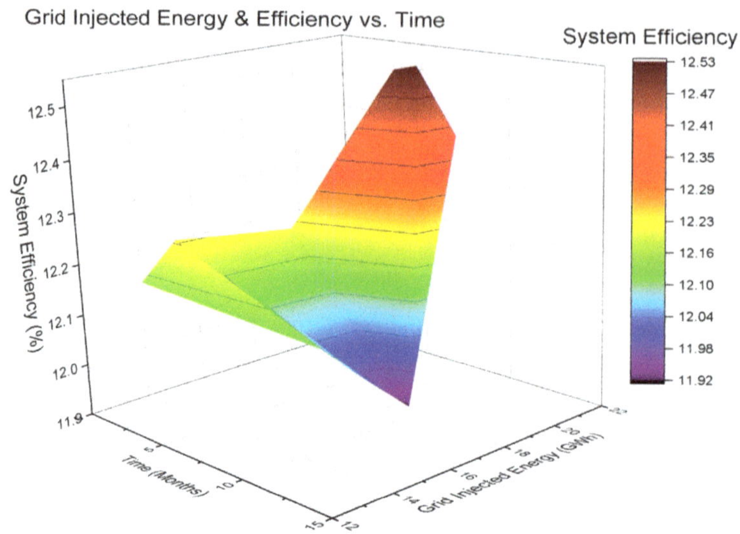

Figure 15. Figure showing FPV energy and efficiency vs. time (months).

It must be mentioned that in the case of generation power being greater than that injected, as indicated in Equation (17), the excess can still be injected as the local and regional demands are far higher. The case study only serves a fraction of the entire national demand; therefore, any excess can still be injected for local (Zambian) consumption or to meet part of the Southern African Power Pool's demand.

System optimization of the FPV system was performed by reviewing the impact of azimuth, pitch, tilt angle and ground-cover ratio (GCR) on grid-injected energy per year. The PVSYST embedded algorithm was used in the system optimization [129]. In the PVSYST software, after the case study location coordinates are entered, the meteorological data is selected from the list of databases. Thereafter, the design and system specifications are selected. The tilt angle values from 0 to 90 degrees are selected, with a sensitivity of 1° intervals, at an azimuth of 0 degrees. The azimuth is then changed to 180° while maintaining the same tilt angle inputs [129]. The objective function is maximizing the energy injected into the grid, which gives the simulation output in GWh. The embedded algorithm carries out a parametric analysis to search for the optimal point and plots the curves accordingly for all the input steps [129]. The ground cover ratio is optimized by looking at the ratio of the active area to the ground area. In PVsyst, the "active area" is the area of one module (length x width), multiplied by the number of modules, while the "ground area" is the area occupied by the PV array. PVSYST maximizes the injected output by tracking this ratio. The closer the ratio is to unity, the lower the injected energy into the grid. Regarding the "pitch", PVSYST will maximize the energy output by increasing the pitch. However, this requires sound engineering judgment in design by factoring in the land constraints for a particular project. Figure 16a shows a 2.6% increase in grid injected energy yield (from a base value of 214.4 to 220 GWh) at a GCR of 5%, while there was a steep decrease in yield of between 80 and 100% of the GCR ratio. Figure 16b shows that the maximum annual yield, between 214 and 215 GWh, is injected into the grid for tilt angles between 10 and 20° for the location in question. Figure 16c reveals that more energy is injected into the grid with every step increase in pitch (i.e., a pitch of 15 m yielded more energy compared to the baseline design value of 3 m). However, for practical considerations on space constraints, the scenario regarding pitch calls for careful analysis because, for a 400% increase (from 3 m to 15 m) in pitch, only a corresponding 2.5% increase in grid injected energy was obtained. Furthermore, a negative and positive sensitivity analysis of the azimuth angle from a baseline value of 0° yielded a reduction in grid-injected energy. This is because the baseline value was already at the optimized azimuth angle, as illustrated in Figure 16d. Additional analysis was able to compare the energy yield and performance ratio (PR) using the PVSYST software adjustments of the albedo and heat loss factor or U–value for floating photovoltaics (0.1 albedo and U-value 31 W/m^2K) and ground-mounted (0.2 and 20 W/m^2K) installations, in line with other research [72]. The results show that a floating PV has better performance (PR of 83.5% and energy yield of 214.4 GWh/y) compared to a ground-mounted system (PR of 79.3% and energy yield of 204.4 GWh/y) at the same location with similar design parameters (i.e., the tilt angle, azimuth, pitch, GCR). PVSYST was also used to evaluate the economics of the floating photovoltaic system. The analysis revealed that the cost of producing 214.4 GWh/year of energy at an investment cost of GBP 0.68/Wp was GBP 0.04/kWh, excluding operation and maintenance (O&M) costs. This FPV LCOE is competitive, with a value of GBP 0.0342/kWh and GBP 0.0335/kWh, obtained by Maronga et al. [54] and RES4Africa [130], respectively, for ground-mounted PV. Homerpro gave a more conservative annual yield of 175 GWh at an LCOE of GBP 0.067/kWh (including O&M); however, this was without factoring in the water albedo and heat loss factor of the PV module's floating island. This goes to show that a reduction in annual energy yield by approximately 18.4% increases the LCOE by almost 40% for the PVSyst and Homerpro cases that are highlighted. The cost summary and energy yield distribution for FPV and wind are summarized in Table 5.

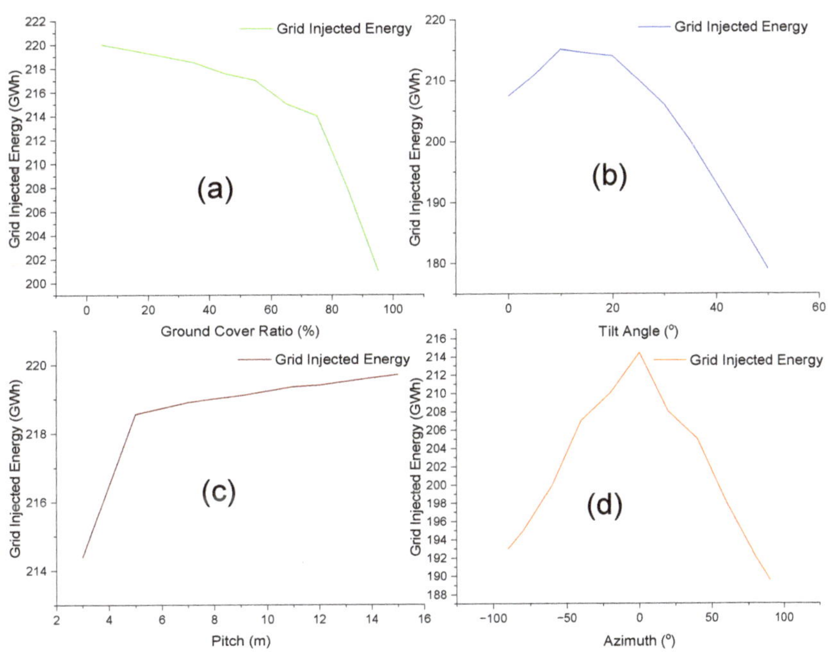

Figure 16. (**a**) Optimization of injected energy vs. ground cover ratio. (**b**) Energy vs. panel tilt angle. (**c**) Energy vs. pitch in meters. (**d**) Injected energy vs. azimuth angle.

Table 5. Table showing the cost and energy production distribution (cost breakdown sources [130,131]).

	FPV (116 MWpdc/100 MWac)	Wind (100 MWac)	FPV + Wind
CAPEX	GBP 80,233,200.00	GBP 109,393,980.00	GBP 189,627,180.00
	GBP 0.69/kWp	GBP 1.09/kWp	GBP 0.88/kWp
OPEX	GBP 1,679,083.20	GBP 2,00,000.00	GBP 3,264,503.20
	GBP 0.017/kWp	GBP 0.02/kWp	GBP 0.017/kWp
GWh/year	214.4 (in PVSyst) @ 100 MWac output. 175 (Homerpro) @ 96 MWpac output.	295 (in Renewables Ninja) @ 100 MW. 166 (in Homperpro) @ 75 MW output	513 (PVSyst + Renewables Ninja). 341 (in Homerpro)

Note: The hydro model in Homerpro had a capital cost of GBP 2.8/Wp with an operation and maintenance cost of GBP 0.017/Wp.

Onshore Wind

Homerpro and Renewables Ninja were used in a complementary fashion in the analysis of the KGU wind-farm output. Owing to the wide coverage of its dataset, Renewables Ninja was used to simulate the output potential of each wind turbine, whose design characteristics included: 129 m hub height, 142 m rotor diameter and 4 MW power rating per turbine. Thereafter, the Renewables Ninja wind speed output was exported to Homerpro to facilitate detailed analysis, to include the practical losses imposed on a typical wind farm with 25 by 4 MW turbines (i.e., wake effects, curtailment losses etc.). With the wind farm capacity density of approximately 6.2 MW/km^2, an optimistic annual energy yield of 294 GWh was registered at the wind farm excluding system losses. However, Homerpro yielded a more conservative annual energy value, with a total of 8174 h of operation. In this scenario, about 167 GWh/year of energy was produced at a competitive LCOE of about GBP 0.07/kWh, as compared to the optimistic forecasted LCOE value of GBP 0.042/kWh obtained in the recent RES4Africa study about Zambia for the 2021/2022 benchmark [130]. The higher LCOE of wind compared to FPV is due to the fact that the resource potential for solar photovoltaics is pronounced, compared to wind in the Zambian context [30–32]. As illustrated from Figure 17a–c below, the total wind energy production on 1st January was 553.87 MWh with a peak energy of 217.54 MWh between 4–8 a.m. On 1 March, the total wind energy generated was 983.18 MWh with a peak energy of 186.1MWh between 6 a.m. and 10 a.m. and 207.2 MWh between 7 p.m. and 11 p.m. The 1st of June yielded 375.17 MWh of wind energy with a peak of 132.27 MWh between 7 a.m. and 12 p.m. and 98.2 MWh between 4 p.m. and 9 p.m. On 1 September, approximately 1654.2 MWh was generated from wind with a peak of 422.2 MWh between 6 a.m. and 11 a.m. Furthermore, 1809.66 MWh of wind was generated on 1 November, with a peak of 457.88 MWh between 6 a.m. and 11 a.m. and 385.8 MWh between 2 p.m. and 7 p.m. Any excess power indicated in Equation (16) is treated in a similar manner to FPV and is injected to meet additional local or regional demand.

4.2.4. Optimal Dispatch of Hydro and VRES

Hybrid System Details

With a Homerpro model (Figure 18) comprising customized virtual storage, a customized hydro initially modeled in iHoga, a PVSYST-based FPV system, and onshore wind values based on Renewables Ninja wind speed data, the following analysis can be made: the Homer Matlab Dispatch was implemented using a customized dispatch algorithm and utilized 5 hydro units, which is equivalent to 700 MW, 100 MWp of FPV, 100 MWp of wind. The Matlab code was used to ascertain a customized dispatch with high VRES penetration. Homerpro calls the Matlab Dispatch at the beginning of each time step in the simulation. The Matlab Dispatch has three input variables, namely, simulation_state, simulation_parameters and custom_variables. The model virtual storage is dependent on available water in the dam and the floating PV and wind potential.

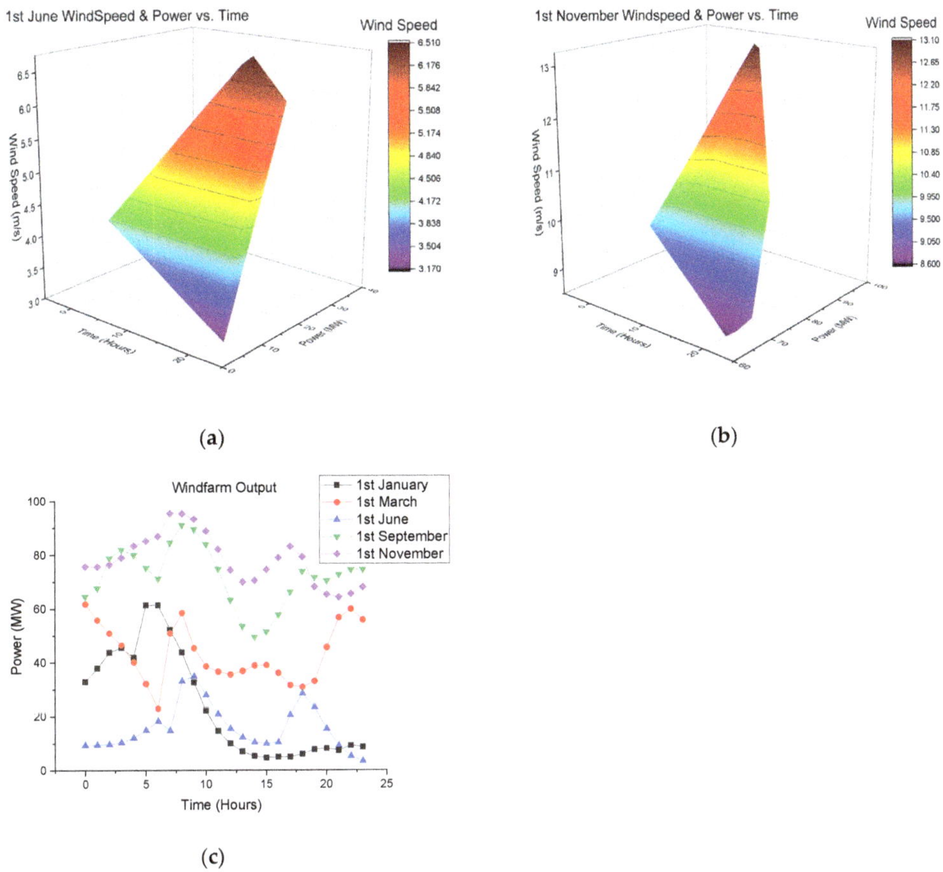

Figure 17. (**a**) June 3D wind speed and power vs. time. (**b**) November 3D wind speed and power vs. time. (**c**) Daily seasonal wind farm output.

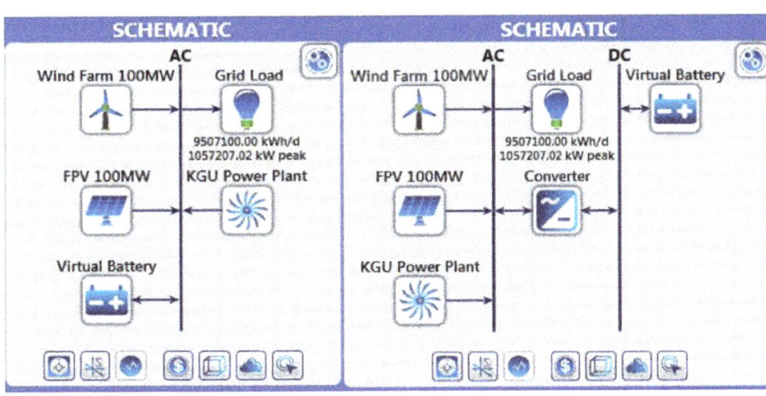

Figure 18. Graphic showing a hybrid energy system schematic in Homerpro.

The Custom Virtual Hydro Battery at Kafue Gorge has a reservoir that can store an assumed maximum capacity of 20 million cubic meters of water (0.5 m rise, assuming it is operating at a minimum elevation of 974 m above sea level), which can discharge over a 173 h (20,000,000 $m^3/(32\ m^3/s \times 60 \times 60)$ period at a rate of 32 m^3/s. The effective head is ~382 m, and the generator efficiency is ~(85–90)%; the power and energy of the Virtual Hydro Battery system during discharging can be calculated as follows:

Discharging:

Power generated = (p) × (g) × (v) × (h) × (eff)

where (p) is the density of water with a value of 1000 kg/m^3, (g) is the gravitational constant of 9.81 m/s^2, (v) is the flow rate in m^3/s, (h) is the head of 382 m, and (eff) is the generator efficiency value of 90%.

Power Generated = 1000 × 9.81 × 32 × 382 × 0.9 ~= 108 MW.

For a 20-million-meter cube of water at a flow rate of 32 m^3/s, the water utilization duration is approximately 173 h for one turbine, based on the plant rating table shown above. However, if more turbines operate to consume the stored water, the duration would be proportional to the number of units in operation. The electrical energy generated over the 173 h is given below.

Energy generated = Power generated × hours of usage

Energy generated = 108,000 kW ×173 h = 18,684,000 kWh (~18.7 GWh)

Charging:

The initial charging assumes having a wet season and thus an abundant water supply, while other charging periods of the virtual battery system involve throttling down on the hydro when there is an availability of floating photovoltaics and onshore wind. The round-trip efficiency of the virtual battery is the efficiency of the turbo-generator unit, including friction losses in the penstock (assumed to be 90% total efficiency). The maximum capacity is the maximum electrical output, divided by the nominal voltage = 18,684,000 × 1000/17,500 = ~1,067,657 amp hours, this assumes the utilization of a generation voltage of 17.5 kV for storage calculations at KGU.

Optimal Daily Dispatch and Reservoir Water Saving

The optimal dispatch of the hybrid energy system at KGU involved the prioritizing of FPV and wind to serve the load and the excess met by hydropower. From Figure 19a below, 1.15 GWh of VRES generation dispatch translated into a reservoir water-saving potential of 9.5% (equivalent to 1.02 GWh of generation) on 1 January. According to Figure 19b, a water-saving potential of 9.7% (equivalent to 1.52 GWh of generation) was realized with a dispatch of 1.67 GWh of VRES on 1 March. Figure 19c shows a water-saving potential of 7.2% with a dispatch of 1.14 GWh of VRES on 1 June. Both 1 September and 1 November yielded better water-saving potentials of 16.8% and 18.7% with VRES dispatch of 2.52 GWh and 2.35 GWh, respectively (Figure 19d,e). Therefore, using the customized Homer-Matlab dispatch code, 4.93 TWh of annual energy consumption was served, translating into 28 percent more demand served when compared to other default dispatch strategies embedded in Homerpro. This load was met by 166 GWh/year of wind, 175 GWh/year of floating photovoltaics, in the presence of 4.59 TWh of hydrogeneration (five out of six 140 MW hydrogenator units with a 10% reserve operating margin per unit) and at a competitive levelized cost of energy of GBP 0.055/kWh. The undispatched hydro unit presents a virtual storage potential of approximately 108 MW by a 7.4% reduction in annual hydropower generation. Moreover, the water saving potential in this study excludes the added benefit of reduced evaporation owing to the presence of retrofitted solar PV panels on the hydro reservoir.

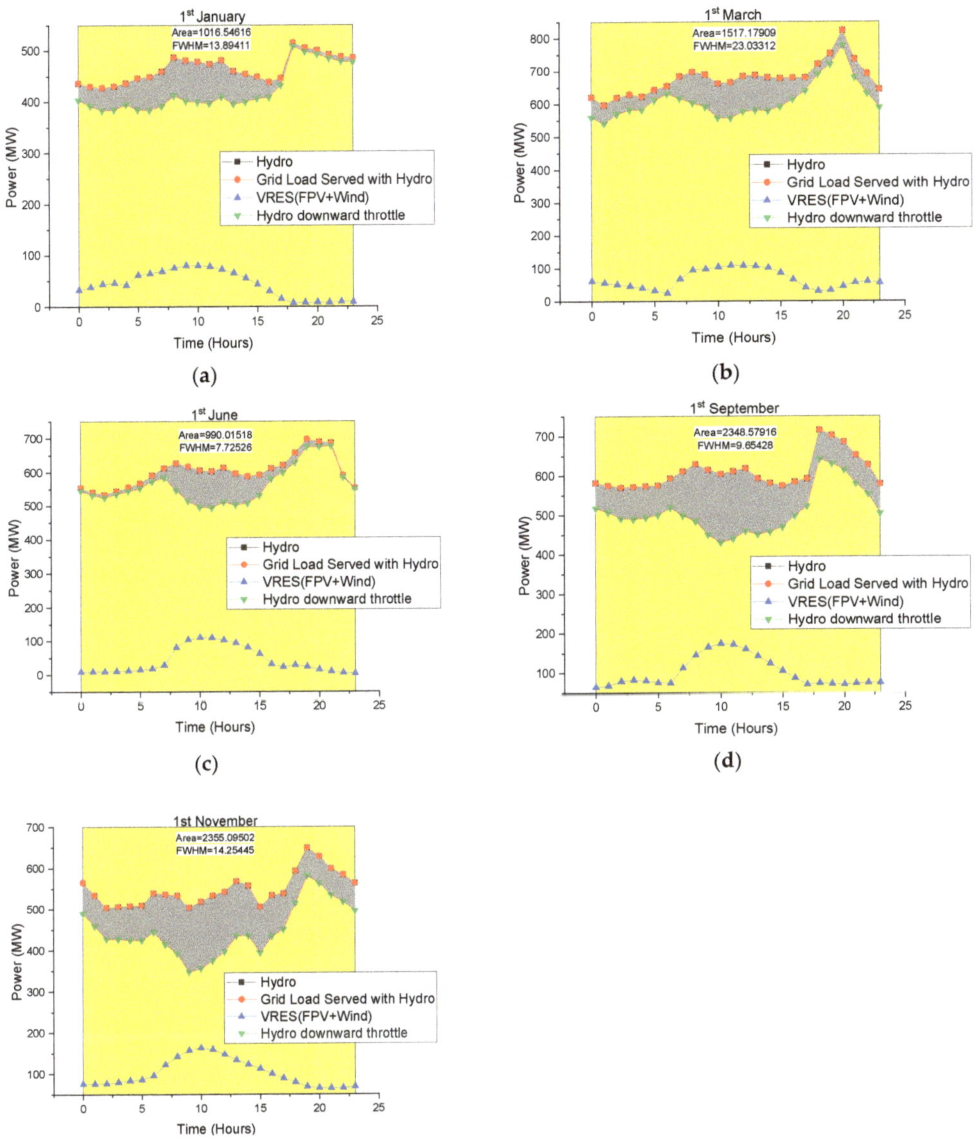

Figure 19. (**a**) Optimal daily dispatch of hydro, FPV and wind on 1 January to serve a fraction of the grid demand. (**b**) Optimal daily dispatch on 1 March. (**c**) Optimal daily dispatch on 1 June. (**d**) Dispatch on 1 September. (**e**) Dispatch on 1 November.

5. Conclusions

This study presented a comprehensive assessment of integrating onshore wind and floating photovoltaics that are adjacent to future and existing hydropower sites in Zambia. All the project objectives were successfully achieved, and these included site appraisal methodology formulation to score and rank possible hydropower sites for the potential addition of onshore wind and retrofitting of floating PV, development and scoping of a case study design methodology and its application. The authors presented an application of the devised screening and ranking multicriteria-based methodology for floating PV and onshore wind, near hydro sites. The extensive data collection for stage 1, filtered off 3 sites (Lunzua, Victoria and Zengamina), thereby presented the remaining 10 sites to the stage 2 scoring and ranking process. This ranking process was developed for three scenarios, which are the balanced hybrid, floating PV and onshore wind models. The three-level scoring and ranking procedure yielded the following results: the balanced ranking placed Itezhi-tezhi and Kafue Gorge Upper (KGU) at first- and second-rank, with total attribute values of 90% and 86.9%, respectively; FPV ranking placed Itezhi-tezhi and Kafue Gorge Upper (KGU) at first- and second-rank, with total attribute values of 95% and 92.5%, respectively; the wind ranking placed Kafue Gorge Lower (KGL) and Kafue Gorge Upper (KGU) at first- and second-rank, with total attribute values of 93.8% and 83.8%, respectively. In all the three scoring and ranking levels, the Chishimba site was ranked the least. This study presents great insight for planners and prospective investors in floating photovoltaics and onshore wind as the factors influencing the suitability of the respective sites can easily be understood.

Moreover, the authors developed a scoping design methodology to be applied at any one of the 10 potential sites. The summarized methodology for the case study application includes assessing the technical parameters of the local electrical grid for integration of variable renewable energy sources (VRES), assessing current seasonal hydro generation and grid electrical demand in a year on an hour-by-hour basis, detailed assessment and design of the VRES (floating photovoltaics and onshore wind) for the chosen case study, assessing the storage potential (implied by throttling down hydro in the presence of VRES for the reservoir type), optimizing daily energy production of the system within grid constraints and ascertaining the levelized cost of the energy of the system.

The results of the case study at Kafue Gorge Upper were promising, with VRES integration potential within grid limits of 341 GWh and 508 GWh per annum, for the conservative and optimistic case, respectively. Furthermore, it is worth noting that the floating PV is not being presented as a competitor to ground-mounted systems, but rather as a complementary technology in specific applications (i.e., retrofitting on hydro reservoirs). Along with providing such benefits as reduced evaporation and algae growth, FPV systems have lower operating temperatures and potentially reduce the costs of solar energy generation. To put this into perspective, the current study using PVSYST showed that floating photovoltaics have a better energy yield compared to a ground-mounted system, as evidenced by a 7.4%, 5.8% and 4.9% increase in energy production for the freestanding, small-footprint and large-footprint FPV configurations, respectively, at a reduced generation cost of GBP 0.04/kWh.

Therefore, floating PV and onshore wind integration could present added techno-economic benefits by fast-tracking new capacity development with opportunities for private investments (IPPs), new opportunities for the Zambian service and manufacturing sectors, power structure decentralization, owing to the wide spread of the renewable resources in the country (i.e., solar PV and wind are more diffused) compared to localized hydropower projects (usually located near large lakes and rivers).

6. Recommendations

The following future work is recommended to add more value and traction to the project research:

- An opportunity to conduct detailed financial and uncertainty analysis to cement project bankability.
- The conducting of pre-feasibility studies (i.e., bathymetry, environmental impact assessment, geotechnical and soil analysis of the reservoir/dam).
- The potential to widen the study scope to include all viable water bodies in the country (both natural and man-made).
- An opportunity to assess the grid impact at the other 9 ranked sites to ascertain overall impact on voltage stability and magnitude profile, network power losses, operating cost differential for different unit commitment scenarios. Moreover, there

is the potential to conduct detailed network analysis to cover: N-1static security assessment, network fault level analysis and protection coordination, short-term and long-term frequency response, the effects of a spinning reserve, transient and dynamic stability performance, and voltage regulation during transience. Consequently, these studies will further provide the necessary technical requirements in the national grid code for large-scale VRES integration.

- Additionally, this study will pave the way for future research in optimizing the dispatch of VRES through an enhanced forecasting model for wind and PV (with the utilization of artificial intelligence and machine learning such as deep neural networks). This will help reduce power balancing costs for large-scale VRES integration.

Author Contributions: Conceptualization, K.J.N. and P.G.T.; software, K.J.N.; methodology, K.J.N. and P.G.T.; formal analysis, K.J.N.; writing—original draft preparation (based on M.Sc. dissertation), K.J.N.; writing—review and editing, A.M., P.G.T. and A.S. All authors have read and agreed to the published version of the manuscript.

Funding: K.J.N. received a scholarship to study for an M.Sc. in Sustainable Engineering: Renewable Energy Systems and the Environment at the University of Strathclyde from the Commonwealth Scholarship Commission (Scholar ID: ZMCS-2019-760).

Institutional Review Board Statement: Not applicable.

Informed Consent Statement: Not applicable.

Data Availability Statement: The data presented in this study is available on request from the corresponding authors.

Acknowledgments: We are grateful for the support and data obtained from industry and would like to acknowledge the contribution of the following: ZESCO Ltd. (Chief Engineer System Studies and Project Design under Distribution Development—George Muyunda and Hydrology Engineer—Kelvin Kabwe) and the Rural Electrification Authority (Chief Executive Officer—Clement Silavwe). Part of this work was conducted under the Energy Systems Research Unit at the University of Strathclyde, Glasgow.

Conflicts of Interest: The author declares no conflict of interest but also acknowledges a Master of Science scholarship from the Commonwealth Scholarship Commission that made it possible to conduct this study.

Appendix A

Appraisal and Ranking Methodology.

Table A1. Showing the site attribute score for FPV, Wind and Hybrid.

Part 1: Balanced Score Attribute Distribution [70,75–78,94–103].

Suitability Score	Wind Potential (25% Relative Weight)			FPV Potential (35% Relative Weight)		Energy Export (20% Relative Weight)			Ease of Access (15% Relative Weight)			Demand (5% Rel. Weight)
	Number of Turbines	Capacity Factor (%)	Slope (%)	Area (km²)	Capacity Factor (%)	FPV Distance to Grid (km)	Wind Distance to Grid (km)	Grid Capacity (MVA)	Land Use	Land Ownership	Distance to Road (km)	Distance to Demand (km)
High (100%)	*>20*	*>=40*	*0–<7*	*>= 10*	*>=14*	*<=2*	*0–<=15*	*>=700*	*No current use*	*Customary /State*	*0–4*	*<=50*
Medium (75%)	*>10–<=20*	*>=35, < 40*	*>=7–<14*	*>=1, <10*	-	*>=5, <2*	*>15–<=30*	*<700–>=400*	-	-	*4–8.*	*50–100*
Low (50%)	*>5–<=10*	*>=30, <35*	*>=14–<20*	*>=0.1, <1*	-	*>=7.5, < 5*	*>30–<=45*	*<400–>=100*	-	-	*8–12.*	*100–150*
Lowest (25%)	*>=1–<5*	*>=26, <30*	*>=20–<=30*	*>= 0.004, <0.1*	-	*<=10, < 7.5*	*>45–<=60*	*<100–>=1*	*Agriculture*	*Private*	*12–16.*	*150–200*
Unsuitable	*0*	*<26*	*>30*	*<0.004*		*>10*	*>60*	*<1*	*Protected/sensitive land*	-	-	*>=200*

Part 2: Floating Photovoltaic (FPV) Attribute Score Distribution [70,75–78,94–103].

Suitability Score	FPV Potential (60% Relative Weight)		Energy Export (20% Relative Weight)		Ease of Access (15% Relative Weight)			Demand (5% Relative Weight)
	Area (km²)	Capacity Factor (%)	FPV Distance to Grid (km)	Grid Capacity (MVA)	Land Use	Land Ownership	Distance to Road (km)	Distance to Demand (km)
High (100%)	*>= 10*	*>=14*	*<=2*	*>=700*	*No current use*	*Customary/State*	*0–<=4*	*<=50*
Medium (75%)	*>=1, <10*		*>=5, <2*	*<700–>=400*	-	-	*>4–<=8.*	*>50–<=100*
Low (50%)	*>=0.1, <1*		*>=7.5, < 5*	*<400–>=100*	-	-	*>8–<=12.*	*>100–<=150*
Lowest (25%)	*>= 0.004, <0.1*		*<=10, < 7.5*	*<100–>=1*	*Agriculture*	*Private*	*>12–<=16.*	*>150–<=200*
Unsuitable	*<0.004*		*>10*	*<1*	*Protected/sensitive land*	-	-	*>200*

Table A1. *Cont.*

Part 3: Onshore Wind Attribute Score Distribution [70,75–78,94–103].									
	Wind Potential (60% Relative Weight)			**Energy Export (20% Relative Weight)**		**Ease of Access (15% Relative Weight)**			**Demand (5% Relative Weight)**
	Number of Turbines	**Capacity Factor (%)**	**Slope (%)**	**Wind Distance to Grid (km)**	**Grid Capacity (MVA)**	**Land Use**	**Land Ownership**	**Distance to Road (km)**	**Distance to Demand (km)**
High (100%)	*>20*	*>=40*	*0–<7*	*0–<=15*	*>=700*	*No current use*	*Customary/State*	*0–<=4*	*<=50*
Medium (75%)	*>10–<=20*	*>=35, < 40*	*>=7–<14*	*>15–<=30*	*<700–>=400*	-	-	*>4–<=8.*	*50–100*
Low (50%)	*>5–<=10*	*>=30, <35*	*>=14–<20*	*>30–<=45*	*<400–>=100*	-	-	*>8–<=12.*	*100–150*
Lowest (25%)	*>=1–<5*	*>=26, <30*	*>=20–<=30*	*>45–<=60*	*<100–>=1*	*Agriculture*	*Private*	*>12–<=16.*	*150–200*
Unsuitable	*0*	*<26*	*>30*	*>60*	*<1*	*Protected/sensitive land*	-	-	*>=200*

Table A2. Showing balanced scoring and ranking matrix results.

		Wind Potential (25% Weight)				**FPV Potential (35% weight)**		
		Score * Weight (10%)	**Score * Weight (10%)**	**Score * Weight (5%)**	**Σ**	**Score * Weight (17.5%)**	**Score * Weight (17.5%)**	**Σ**
Rank #	**Name of Site**	**Number of Turbines**	**Capacity Factor**	**Slope**	**Wind Total**	**Area**	**Capacity Factor**	**FPV Total**
=1′	Itezhi-tezhi FPV/wind site	10.0%	7.5%	5.0%	22.5%	17.5%	17.5%	35.0%
=2′	KGU FPV/wind site	10.0%	7.5%	3.3%	21.3%	13.1%	17.5%	30.6%
=3′	KGL FPV/wind site	10.0%	10.0%	5.0%	25.0%	8.8%	17.5%	26.3%
=3′	Kariba FPV/wind site	10.0%	5.0%	3.3%	18.8%	8.8%	17.5%	26.3%
=3′	Lusiwasi FPV/ wind site	10.0%	5.0%	5.0%	20.0%	17.5%	17.5%	35.0%
=6′	Musonda FPV/wind site	10.0%	7.5%	5.0%	22.5%	13.1%	17.5%	30.6%
=6′	Mulungushi FPV/wind site	10.0%	7.5%	5.0%	22.5%	13.1%	17.5%	30.6%
=8′	Shiwangangu FPV/ wind site	10.0%	7.5%	3.8%	21.3%	8.8%	17.5%	26.3%
=9′	Lunsemfwa FPV/wind site	10.0%	10.0%	5.0%	25.0%	4.4%	17.5%	21.9%
=10′	Chishimba FPV/wind site	10.0%	5.0%	5.0%	20.0%	4.4%	17.5%	21.9%

Table A3. Showing continuation of balanced scoring and ranking matrix results.

		Energy Export (20% Weight)				Ease of Access (15% Weight)				Demand (5% Weight)		Site Total (100% Weight)
		Score * Weight (5%)	Score * Weight (5%)	Score * Weight (10%)	Σ	Score * Weight (5%)	Score * Weight (5%)	Score * Weight (5%)	Σ	Score * Weight (5%)	Σ	Σ
Rank#	Name of Site	FPV Distance to Grid	Wind Distance to Grid	Grid Capacity (10%)	Total	Land Use (5%)	Land Ownership	Distance to Road	Total	Distance to Demand	Demand Total	Score * Weight
=1′	Itezhi-tezhi FPV/wind site	5.0%	2.5%	5.0%	12.5%	5.0%	5.0%	5.0%	15.0%	5.0%	5.0%	90.0%
=2′	KGU FPV/wind site	5.0%	3.8%	10.0%	18.8%	5.0%	5.0%	1.3%	11.3%	5.0%	5.0%	86.9%
=3′	KGL FPV/wind site	2.5%	2.5%	10.0%	15.0%	5.0%	5.0%	3.8%	13.8%	5.0%	5.0%	85.0%
=3′	Kariba FPV/wind site	5.0%	5.0%	10.0%	20.0%	5.0%	5.0%	5.0%	15.0%	5.0%	5.0%	85.0%
=3′	Lusiwasi FPV/ wind site	5.0%	5.0%	2.5%	12.5%	5.0%	5.0%	5.0%	15.0%	2.5%	2.5%	85.0%
=6′	Musonda FPV/wind site	1.3%	5.0%	2.5%	8.8%	5.0%	5.0%	5.0%	15.0%	2.5%	2.5%	79.4%
=6′	Mulungushi FPV/wind site	2.5%	5.0%	2.5%	10.0%	5.0%	5.0%	3.8%	13.8%	2.5%	2.5%	79.4%
=8′	Shiwangangu FPV/ wind site	5.0%	5.0%	2.5%	12.5%	5.0%	5.0%	3.8%	13.8%	2.5%	2.5%	76.3%
=9′	Lunsemfwa FPV/wind site	5.0%	1.3%	2.5%	8.8%	5.0%	5.0%	5.0%	15.0%	2.5%	2.5%	73.1%
=10′	Chishimba FPV/wind site	5.0%	3.8%	2.5%	11.3%	5.0%	5.0%	5.0%	15.0%	2.5%	2.5%	70.6%

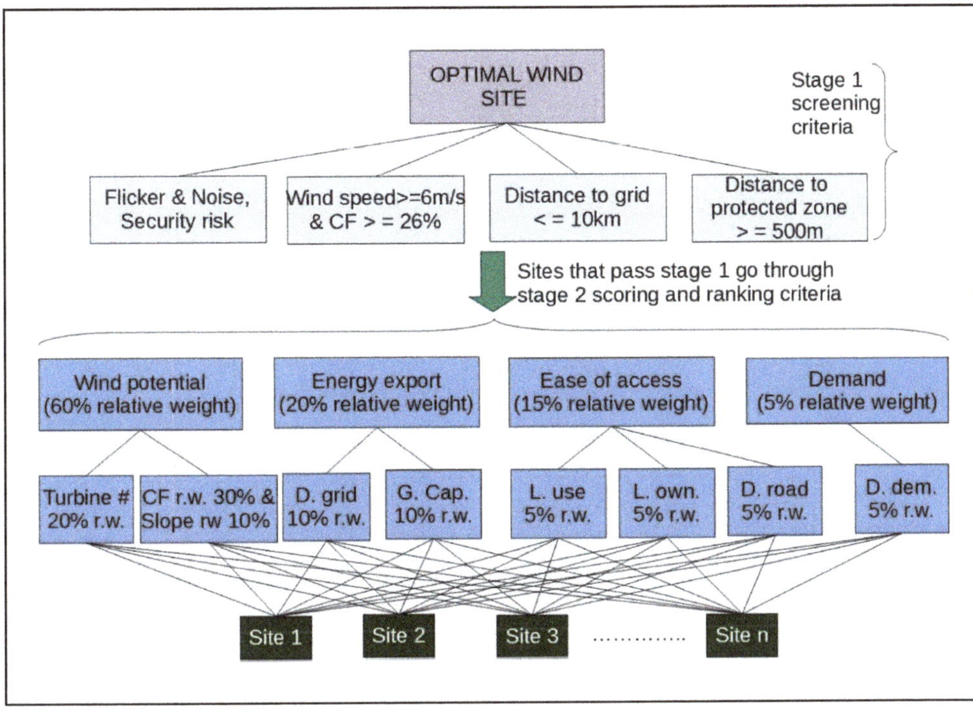

Figure A1. Graphic showing the 2-stage hierarchy structure for optimal onshore wind site selection. Abbreviations: Turbines#—number of turbines, CF—capacity factor, r.w.—relative weight, D. grid—distance to grid, G. cap.—grid capacity, L. use—land use, L. own—land ownership, D. road—distance to road, D. dem.—distance to demand center, r.w.—relative weight.

Appendix B

ZESCO Power System Analysis Toolbox Models.

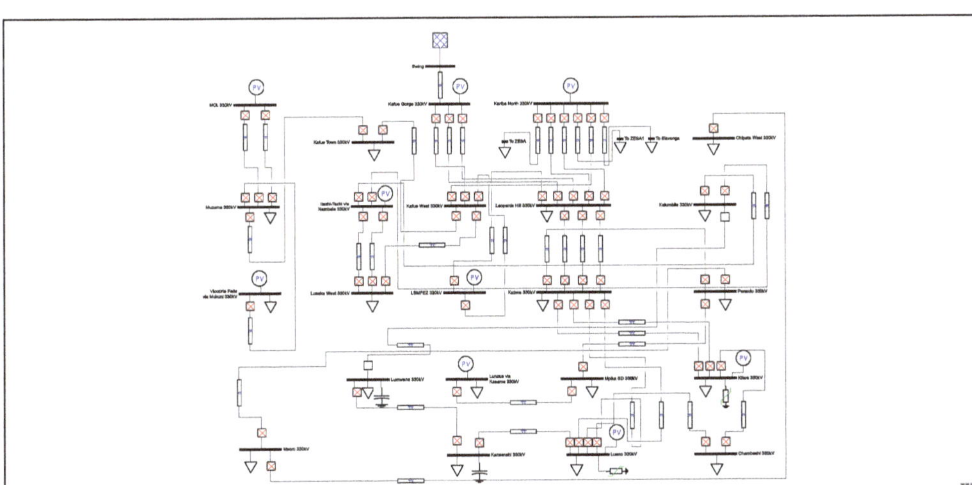

Figure A2. Layout showing the PSAT model for the existing 330 kV Zambian network.

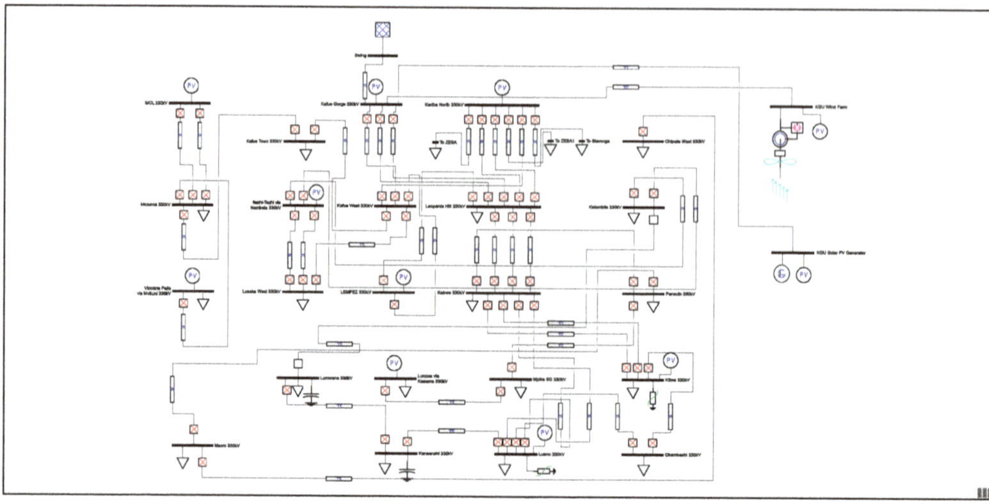

Figure A3. Layout showing the PSAT FPV and wind integration model for the 330 kV Zambian network.

Table A4. Showing the power system modeling data.

Base MVA	100				Shunt Compensation		
					Station	Rating (MVAr)	
					Kitwe	20	
330kV Line Parameters					Lumwana	20	
R (p.u/km)	X (p.u/km)	B (p.u/km)			Kansanshi	20	
0.00004	0.000315	0.003708					
					SVCs		
220kV Line Parameters					Station	Rating (MVA)	
R (p.u/km)	X (p.u/km)	B (p.u/km)			Kitwe	35	
0.000115	0.000682	0.001701			Luano	80	
132kV Line Parameters							
R (p.u/km)	X (p.u/km)	B (p.u/km)					
0.001224	0.002365	0.000486					
FROM	TO	km	R(p.u)	X(p.u)	B(p.u)	Voltage Level (kV)	Rating (MVA)
Kafue Gorge	Leopards Hill	47	0.00188	0.014805	0.174276	330	700
Kafue Gorge	Kafue West	43	0.00172	0.013545	0.159444	330	700
Kariba North	Leopards Hill	123	0.00492	0.038745	0.456084	330	700
Leopards Hill	Kabwe	97	0.00388	0.030555	0.359676	330	700

Table A4. *Cont.*

Kabwe	Kitwe	211	0.00844	0.066465	0.782388	330	700
Kabwe	Luano	247	0.00988	0.077805	0.915876	330	700
Kabwe	Pensulo	298	0.01192	0.09387	1.104984	330	700
Kitwe	Luano	40	0.0016	0.0126	0.14832	330	700
Kitwe	Chambishi	21.5	0.00086	0.0067725	0.079722	330	700
Chambishi	Luano	21.5	0.00086	0.0067725	0.079722	330	700
Luano	Kansanshi	196	0.00784	0.06174	0.726768	330	700
Kansanshi	Lumwana	72	0.00288	0.02268	0.266976	330	700
Kafue West	Lusaka West	51	0.00204	0.016065	0.189108	330	700
Kafue West	Kafue Town	3	0.00012	0.000945	0.011124	330	700
Kafue West	Leopards Hill	53	0.00212	0.016695	0.196524	330	700
VicFalls	Muzuma	159	0.018285	0.108438	0.270459	220	230
Muzuma	Kafue Town	189	0.021735	0.128898	0.321489	220	230
Luano	Michelo	44	0.00506	0.030008	0.074844	220	375
Michelo	Karavia	8	0.00092	0.005456	0.013608	220	375
Leopards Hill	Roma	28	0.034272	0.06622	0.013608	132	85
Leopards Hill	Coventry	28	0.034272	0.06622	0.013608	132	85
Roma	Lusaka West	21	0.025704	0.049665	0.010206	132	85
Lusaka West	Roma	21	0.025704	0.049665	0.010206	132	85
Transformers							
Station	Qty	x(p.u)	Rating	Ratio			
Kafue Town	1	0.185	60	220/88			
Kafue Town	1	0.1707	60	330/88			
Lusaka West	1	0.056	125	330/132			
Leopard Hill	2	0.056	125	330/132			
Kitwe	6	0.042	125	330/220			
Luano	4	0.042	125	330/220			

References

1. World Bank Group, ESMAP, and SERIS. *Where Sun Meets Water: Floating Solar Handbook for Practitioners*; World Bank: Washington, DC, USA, 2019. Available online: https://www.esmap.org/where_sun_meets_water_handbook (accessed on 10 September 2020).
2. Obane, H.; Nagai, Y.; Asano, K. Assessing land use and potential conflict in solar and onshore wind energy in Japan. *Renew. Energy* **2020**, *160*, 842–851. [CrossRef]
3. Thi, N.D.A. *The Evolution of Floating Solar Photovoltaics*; Research Gate: Berlin, Germany, 2017.
4. Trapani, K.; Santafé, M.R. A review of floating photovoltaic installations: 2007–2013. *Prog. Photovolt. Res. Appl.* **2014**, *23*, 524–532. [CrossRef]
5. Ferrer-Gisbert, C.; Ferrán-Gozálvez, J.J.; Redón-Santafé, M.; Gisbert, P.F.; Sanchez-Romero, F.-J.; Torregrosa-Soler, J.B. A new photovoltaic floating cover system for water reservoirs. *Renew. Energy* **2013**, *60*, 63–70. [CrossRef]
6. Mesbahi, M.; Minamino, S. Top 70 Floating Solar PV Plants. Available online: https://www.solarplaza.com/channels/top-10s/11761/top-70-floating-solar-pv-plants/ (accessed on 20 June 2020).
7. Tsanova, T. Floating solar installs to top 1.5 GW in 2019. Available online: https://renewablesnow.com/news/to-the-point-floatingsolar-installs-to-top-15-gw-in-2019-613278/ (accessed on 27 May 2021).
8. Ranjbaran, P.; Yousefi, H.; Gharehpetian, G.B.; Astaraei, F.R. A review on floating photovoltaic (FPV) power generation units. *Renew. Sustain. Energy Rev.* **2019**, *110*, 332–347. [CrossRef]

9. Sulaeman, S.; Brown, E.; Quispe-Abad, R.; Müller, N. Floating PV system as an alternative pathway to the amazon dam underproduction. *Renew. Sustain. Energy Rev.* **2020**, *135*, 110082. [CrossRef]
10. Oliveira-Pinto, S.; Stokkermans, J. Assessment of the potential of different floating solar technologies—Overview and analysis of different case studies. *Energy Convers. Manag.* **2020**, *211*, 112747. [CrossRef]
11. Rosa-Clot, M. *Floating PV Plants*; Academic Press: Cambridge, MA, USA, 2020.
12. Liu, H.; Krishna, V.; Leung, J.L.; Reindl, T.; Zhao, L. Field experience and performance analysis of floating PV technologies in the tropics. *Prog. Photovolt. Res. Appl.* **2018**, *26*, 957–967. [CrossRef]
13. ZESCO Limited. Statement on Electricity Deficit. 2020. Available online: https://www.reuters.com/article/zambia-electricity/update-1-zambian-power-firm-zesco-says-electricity-deficit-growing-idUSL8N2B626D (accessed on 11 September 2020).
14. Ministry of Energy (MOE). Official Opening of the Fifth Session of the Twelfth National Assembly by His Excellency, Mr. Edgar Chagwa Lungu, President of the Republic of Zambia on Friday 11 September 2020. Available online: https://www.moe.gov.zm/?wpfb_dl=82 (accessed on 11 September 2020).
15. Stiubiener, U.; da Silva, T.C.; Trigoso, F.B.M.; Benedito, R.D.S.; Teixeira, J.C. PV power generation on hydro dam's reservoirs in Brazil: A way to improve operational flexibility. *Renew. Energy* **2020**, *150*, 765–776. [CrossRef]
16. Farfan, J.; Breyer, C. Combining Floating Solar Photovoltaic Power Plants and Hydropower Reservoirs: A Virtual Battery of Great Global Potential. *Energy Procedia* **2018**, *155*, 403–411. [CrossRef]
17. Nordmann, T.; Vontobel, T.; Clavadetscher, L.; Boström, T.; Remlo, H. Large Scale Hybrid PV Hydro Electricity Production in Floating Devices on Water. *Forest* **2009**, *1*, 4189–4192. [CrossRef]
18. Ming, B.; Liu, P.; Cheng, L.; Zhou, Y.; Wang, X. Optimal daily generation scheduling of large hydro–photovoltaic hybrid power plants. *Energy Convers. Manag.* **2018**, *171*, 528–540. [CrossRef]
19. Spencer, R.S.; Macknick, J.; Aznar, A.; Warren, A.; Reese, M.O. Floating Photovoltaic Systems: Assessing the Technical Potential of Photovoltaic Systems on Man-Made Water Bodies in the Continental United States. *Environ. Sci. Technol.* **2018**, *53*, 1680–1689. [CrossRef]
20. Serrano-González, J.; Lacal-Arántegui, R. Technological evolution of onshore wind turbines—A market-based analysis. *Wind. Energy* **2016**, *19*, 2171–2187. [CrossRef]
21. Sliz-Szkliniarz, B.; Eberbach, J.; Hoffmann, B.; Fortin, M. Assessing the cost of onshore wind development scenarios: Modelling of spatial and temporal distribution of wind power for the case of Poland. *Renew. Sustain. Energy Rev.* **2019**, *109*, 514–531. [CrossRef]
22. Ladenburg, J.; Hevia-Koch, P.; Petrović, S.; Knapp, L. The offshore-onshore conundrum: Preferences for wind energy considering spatial data in Denmark. *Renew. Sustain. Energy Rev.* **2020**, *121*, 109711. [CrossRef]
23. Jin, X.; Rong, Y.; Zhong, X. Wind turbine manufacturing industry in China: Current situation and problems. *Renew. Sustain. Energy Rev.* **2014**, *33*, 729–735. [CrossRef]
24. Hughes, F.M.; Anaya-Lara, O.; Ramtharan, G.; Jenkins, N.; Strbac, G. Influence of tower shadow and wind turbulence on the performance of power system stabilizers for DFIG-based wind farms. *IEEE Trans. Energy Convers.* **2008**, *23*, 519–528. [CrossRef]
25. Liu, C.; Xu, D.; Zhu, N.; Blaabjerg, F.; Chen, M. DC-Voltage Fluctuation Elimination Through a DC-Capacitor Current Control for DFIG Converters Under Unbalanced Grid Voltage Conditions. *IEEE Trans. Power Electron.* **2012**, *28*, 3206–3218. [CrossRef]
26. Jiang, L.; Chi, Y.; Qin, H.; Pei, Z.; Li, Q.; Liu, M.; Bai, J.; Wang, W.; Feng, S.; Kong, W.; et al. Wind energy in China. *IEEE Power Energy Mag.* **2011**, *9*, 36–46. [CrossRef]
27. Feng, Y.; Lin, H.; Ho, S.L.; Yan, J.; Dong, J.; Fang, S.; Huang, Y. Overview of wind power generation in China: Status and development. *Renew. Sustain. Energy Rev.* **2015**, *50*, 847–858. [CrossRef]
28. Vittal, E.; O'Malley, M.; Keane, A. Rotor Angle Stability with High Penetrations of Wind Generation. *IEEE Trans. Power Syst.* **2011**, *27*, 353–362. [CrossRef]
29. Swarna, K.S.V.; Vinayagam, A.; Yang Khoo, S.; Stojcevski, A. Impacts of Integration of Wind and Solar PV in a Typical Power Network. In Proceedings of the 2015 International Conference on Sustainable Energy and Environmental Engineering, Shenzhen, China, 20–21 December 2015.
30. International Renewable Energy Agency (IRENA). *Renewables Readiness Assessment (RRA)*; IRENA: New York, NY, USA, 2013.
31. ESMAP (World Bank). *Wind Resource Mapping in Zambia—Mesoscale Wind Modelling Report, July 2015*; World Bank: Washington, DC, USA, 2015.
32. ESMAP (World Bank). *Wind Resource Mapping in Zambia—12 Month Site Resource Report, May 2018*; World Bank: Washington, DC, USA, 2018.
33. CSO—Central Statistical Office, Republic of Zambia. 2019. Available online: https://knoema.com/atlas/sources/CSO-ZM (accessed on 27 May 2021).
34. USAID Southern Africa Energy Program. *Zambia Power Sector Assessment*; USAID: Pretoria, South Africa, 2018.
35. Chidumayo, E.N. *Handbook on Miombo Ecology and Management*; Stockholm Environmental Institute: Stockholm, Sweden, 1996.
36. Shane, A.; Gheewala, S.H.; Fungtammasan, B.; Silalertruksa, T.; Bonnet, S.; Phiri, S. Bioenergy resource assessment for Zambia. *Renew. Sustain. Energy Rev.* **2016**, *53*, 93–104. [CrossRef]
37. ZESCO. Zambia Electricity Supply Corporation. 2020. Available online: https://www.zesco.co.zm/ (accessed on 20 May 2021).
38. Shah, R.; Mithulananthan, N.; Bansal, R.; Ramachandaramurthy, V. A review of key power system stability challenges for large-scale PV integration. *Renew. Sustain. Energy Rev.* **2015**, *41*, 1423–1436. [CrossRef]

39. Ding, M.; Xu, Z.; Wang, W.; Wang, X.; Song, Y.; Chen, D. A review on China's large-scale PV integration: Progress, challenges and recommendations. *Renew. Sustain. Energy Rev.* **2016**, *53*, 639–652. [CrossRef]
40. Mararakanye, N.; Bekker, B. Renewable energy integration impacts within the context of generator type, penetration level and grid characteristics. *Renew. Sustain. Energy Rev.* **2019**, *108*, 441–451. [CrossRef]
41. Li, J.; Liu, F.; Li, Z.; Shao, C.; Liu, X. Grid-side flexibility of power systems in integrating large-scale renewable generations: A critical review on concepts, formulations and solution approaches. *Renew. Sustain. Energy Rev.* **2018**, *93*, 272–284. [CrossRef]
42. Matschoss, P.; Bayer, B.; Thomas, H.; Marian, A. The German incentive regulation and its practical impact on the grid integration of renewable energy systems. *Renew. Energy* **2018**, *134*, 727–738. [CrossRef]
43. Rauf, H.; Gull, M.S.; Arshad, N. Complementing hydroelectric power with floating solar PV for daytime peak electricity demand. *Renew. Energy* **2020**, *162*, 1227–1242. [CrossRef]
44. Chen, Y.; Wei, W.; Liu, F.; Mei, S. Distributionally robust hydro-thermal-wind economic dispatch. *Appl. Energy* **2016**, *173*, 511–519. [CrossRef]
45. Deshmukh, M.; Deshmukh, S. Modeling of hybrid renewable energy systems. *Renew. Sustain. Energy Rev.* **2008**, *12*, 235–249. [CrossRef]
46. Paska, J.; Biczel, P.; Kłos, M. Hybrid power systems—An effective way of utilising primary energy sources. *Renew. Energy* **2009**, *34*, 2414–2421. [CrossRef]
47. Campana, P.E.; Li, H.; Yan, J. Dynamic modelling of a PV pumping system with special consideration on water demand. *Appl. Energy* **2013**, *112*, 635–645. [CrossRef]
48. Glasnovic, Z.; Margeta, J. Optimal sizing of photovoltaic-hydro power plant. *Prog. Photovolt. Res. Appl.* **2009**, *17*, 542–553. [CrossRef]
49. Chen, J.; Zhuang, Y.; Li, Y.; Wang, P.; Zhao, Y.; Zhang, C. Risk-aware short term hydro-wind-thermal scheduling using a probability interval optimization model. *Appl. Energy* **2017**, *189*, 534–554. [CrossRef]
50. Bayón, L.; Grau, J.M.; Ruiz, M.; Suárez, P. A comparative economic study of two configurations of hydro-wind power plants. *Energy* **2016**, *112*, 8–16. [CrossRef]
51. Portero, U.; Velázquez, S.; Carta, J.A. Sizing of a wind-hydro system using a reversible hydraulic facility with seawater. A case study in the Canary Islands. *Energy Convers. Manag.* **2015**, *106*, 1251–1263. [CrossRef]
52. Wang, X.; Mei, Y.; Kong, Y.; Lin, Y.; Wang, H. Improved multi-objective model and analysis of the coordinated operation of a hydro-wind-photovoltaic system. *Energy* **2017**, *134*, 813–839. [CrossRef]
53. Liu, Y.; Tan, S.; Jiang, C. Interval optimal scheduling of hydro-PV-wind hybrid system considering firm generation coordination. *IET Renew. Power Gener.* **2016**, *11*, 63–72. [CrossRef]
54. Maronga, A.; Nyoni, K.J.; Tuohy, P.; Shane, A. Modelling and assessing CSP and PV systems technical and economic performances to supply power to a mining context in Zimbabwe. *Energies* **2021**, *14*, 3740. [CrossRef]
55. Shabani, M.; Mahmoudimehr, J. Techno-economic role of PV tracking technology in a hybrid PV-hydroelectric standalone power system. *Appl. Energy* **2018**, *212*, 84–108. [CrossRef]
56. Beluco, A.; de Souza, P.K.; Krenzinger, A. A method to evaluate the effect of complementarity in time between hydro and solar energy on the performance of hybrid hydro PV generating plants. *Renew. Energy* **2012**, *45*, 24–30. [CrossRef]
57. François, B.; Borga, M.; Creutin, J.; Hingray, B.; Raynaud, D.; Sauterleute, J. Complementarity between solar and hydro power: Sensitivity study to climate characteristics in Northern-Italy. *Renew. Energy* **2016**, *86*, 543–553. [CrossRef]
58. Kougias, I.; Szabo, S.; Monforti-Ferrario, F.; Huld, T.; Bodis, K. A methodology for optimization of the complementarity between small-hydropower plants and solar PV systems. *Renew. Energy* **2016**, *87*, 1023–1030. [CrossRef]
59. Jurasz, J.; Ciapała, B. Integrating photovoltaics into energy systems by using a run-off-river power plant with pondage to smooth energy exchange with the power gird. *Appl. Energy* **2017**, *198*, 21–35. [CrossRef]
60. Fang, W.; Huang, Q.; Huang, S.; Yang, J.; Meng, E.; Li, Y. Optimal sizing of utility-scale photovoltaic power generation complementarily operating with hydropower: A case study of the world's largest hydro-photovoltaic plant. *Energy Convers. Manag.* **2017**, *136*, 161–172. [CrossRef]
61. Chen, L.; Liu, Y. Scheduling strategy of hybrid wind-photovoltaic-hydro power generation system. In Proceedings of the International Conference on Sustainable Power Generation and Supply (SUPERGEN 2012), Hangzhou, China, 8–9 September 2012.
62. Yang, X.; Wang, W.; Xue, B.; Huang, Q. On short-term united optimal operation of wind power, thermal power and waterpower. *J. Hydroelectr. Eng.* **2013**, *4*, 199–203.
63. Yang, X.; Chen, Q.; Wang, M.; Zhang, L. Cooperating control for wind farm and hydro power plant based on the fruit fly optimization. *Proc. Chin. Soc. Electr. Eng.* **2017**, *37*, 5286–5293.
64. Wei, P.; Liu, Y. The Integration of Wind-Solar-Hydropower Generation in Enabling Economic Robust Dispatch. *Math. Probl. Eng.* **2019**, *2019*, 4634131. [CrossRef]
65. Dong, W.; Wang, Q.; Yang, L. A coordinated dispatching model for a distribution utility and virtual power plants with wind/photovoltaic/hydro generators. *Autom. Electr. Power Syst.* **2015**, *39*, 75–81.
66. Zou, Y.; Yang, L. Synergetic dispatch models of a wind/PV/hydro virtual power plant based on representative scenario set. *Power Syst. Technol.* **2015**, *39*, 1855–1859.
67. Liu, Y.; Guo, L.; Wang, C. Economic dispatch of microgrid based on two stage robust optimization. *Proc. CSEE* **2018**, *38*, 4013–4022.

68. Reddy, S.S. Optimal scheduling of wind-thermal power system using clustered adaptive teaching learning based optimization. *Electr. Eng.* **2017**, *99*, 535–550. [CrossRef]
69. Reddy, S.S.; Bijwe, P. Real time economic dispatch considering renewable energy resources. *Renew. Energy* **2015**, *83*, 1215–1226. [CrossRef]
70. Ali, S.; Taweekun, J.; Techato, K.; Waewsak, J.; Gyawali, S. GIS based site suitability assessment for wind and solar farms in Songkhla, Thailand. *Renew. Energy* **2018**, *132*, 1360–1372. [CrossRef]
71. Harper, M.; Anderson, B.; James, P.; Bahaj, A. Identifying suitable locations for onshore wind turbines using a GIS-MCDA approach. In Proceedings of the 16th International Conference on Sustainable Energy Technologies, Bologna, Italy, 17–20 July 2017.
72. Sunak, Y.; Höfer, T.; Siddique, H.; Madlener, R.; De Doncker, R.W. *A GIS-Based Decision Support System for the Optimal Siting of Wind Farm Projects*; E.ON Energy Research Center: Aachen, Germany, 2015.
73. Malczewski, J. GIS-based land-use suitability analysis: A critical overview. *Prog. Plan.* **2004**, *62*, 3–65. [CrossRef]
74. Voivontas, D.; Assimacopoulos, D.; Mourelatos, A.; Corominas, J. Evaluation of Renewable Energy potential using a GIS decision support system. *Renew. Energy* **1998**, *13*, 333–344. [CrossRef]
75. Baban, S.M.; Parry, T. Developing and applying a GIS-assisted approach to locating wind farms in the UK. *Renew. Energy* **2001**, *24*, 59–71. [CrossRef]
76. Gigović, L.; Pamučar, D.; Božanić, D.; Ljubojević, S. Application of the GIS-DANP-MABAC multi-criteria model for selecting the location of wind farms: A case study of Vojvodina, Serbia. *Renew. Energy* **2017**, *103*, 501–521. [CrossRef]
77. Watson, J.J.; Hudson, M.D. Regional Scale wind farm and solar farm suitability assessment using GIS-assisted multi-criteria evaluation. *Landsc. Urban Plan.* **2015**, *138*, 20–31. [CrossRef]
78. Höfer, T.; Sunak, Y.; Siddique, H.; Madlener, R. Wind farm siting using a spatial Analytic Hierarchy Process approach: A case study of the Städteregion Aachen. *Appl. Energy* **2016**, *163*, 222–243. [CrossRef]
79. Latinopoulos, D.; Kechagia, K. A GIS-based multi-criteria evaluation for wind farm site selection. A regional scale application in Greece. *Renew. Energy* **2015**, *78*, 550–560. [CrossRef]
80. Uyan, M. GIS-based solar farms site selection using analytic hierarchy process (AHP) in Karapinar region, Konya/Turkey. *Renew. Sustain. Energy Rev.* **2013**, *28*, 11–17. [CrossRef]
81. Aragonés-Beltrán, P.; Chaparro-González, F.; Pastor-Ferrando, J.; Rodríguez-Pozo, F. An ANP-based approach for the selection of photovoltaic solar power plant investment projects. *Renew. Sustain. Energy Rev.* **2010**, *14*, 249–264. [CrossRef]
82. Rojas-Zerpa, J.C.; Yusta, J.M. Application of multicriteria decision methods for electric supply planning in rural and remote areas. *Renew. Sustain. Energy Rev.* **2015**, *52*, 557–571. [CrossRef]
83. Campos-Guzmán, V.; García-Cáscales, M.S.; Espinosa, N.; Urbina, A. Life Cycle Analysis with Multi-Criteria Decision Making: A review of approaches for the sustainability evaluation of renewable energy technologies. *Renew. Sustain. Energy Rev.* **2019**, *104*, 343–366. [CrossRef]
84. Rediske, G.; Siluk, J.C.M.; Michels, L.; Rigo, P.; Rosa, C.B.; Cugler, G. Multi-criteria decision-making model for assessment of large photovoltaic farms in Brazil. *Energy* **2020**, *197*, 117167. [CrossRef]
85. Ridha, H.M.; Gomes, C.; Hizam, H.; Ahmadipour, M.; Heidari, A.A.; Chen, H. Multi-objective optimization and multi-criteria decision-making methods for optimal design of standalone photovoltaic system: A comprehensive review. *Renew. Sustain. Energy Rev.* **2011**, *135*, 110202. [CrossRef]
86. Charabi, Y.; Gastli, A. PV site suitability analysis using GIS-based spatial fuzzy multi-criteria evaluation. *Renew. Energy* **2011**, *36*, 2554–2561. [CrossRef]
87. Kim, S.; Lee, Y.; Moon, H.-R. Siting criteria and feasibility analysis for PV power generation projects using road facilities. *Renew. Sustain. Energy Rev.* **2018**, *81*, 3061–3069. [CrossRef]
88. Dhiman, H.S.; Deb, D. Fuzzy TOPSIS and fuzzy COPRAS based multi-criteria decision making for hybrid wind farms. *Energy* **2020**, *202*, 117755. [CrossRef]
89. Deveci, M.; Cali, U.; Kucuksari, S.; Erdogan, N. Interval type-2 fuzzy sets based multi-criteria decision-making model for offshore wind farm development in Ireland. *Energy* **2020**, *198*, 117317. [CrossRef]
90. Mahdy, M.; Bahaj, A. Multi criteria decision analysis for offshore wind energy potential in Egypt. *Renew. Energy* **2018**, *118*, 278–289. [CrossRef]
91. Noorollahi, Y.; Yousefi, H.; Mohammadi, M. Multi-criteria decision support system for wind farm site selection using GIS. *Sustain. Energy Technol. Assess.* **2016**, *13*, 38–50. [CrossRef]
92. Moradi, S.; Yousefi, H.; Noorollahi, Y.; Rosso, D. Multi-criteria decision support system for wind farm site selection and sensitivity analysis: Case study of Alborz Province, Iran. *Energy Strat. Rev.* **2020**, *29*, 100478. [CrossRef]
93. Cunden, T.S.; Doorga, J.; Lollchund, M.R.; Rughooputh, S.D. Multi-level constraints wind farms siting for a complex terrain in a tropical region using MCDM approach coupled with GIS. *Energy* **2020**, *211*, 118533. [CrossRef]
94. Supciller, A.A.; Toprak, F. Selection of wind turbines with multi-criteria decision making techniques involving neutrosophic numbers: A case from Turkey. *Energy* **2020**, *207*, 118237. [CrossRef]
95. Xu, Y.; Li, Y.; Zheng, L.; Cui, L.; Li, S.; Li, W.; Cai, Y. Site selection of wind farms using GIS and multi-criteria decision making method in Wafangdian, China. *Energy* **2020**, *207*, 118222. [CrossRef]

96. Jangid, J.; Bera, A.K.; Joseph, M.; Singh, V.; Singh, T.; Pradhan, B.; Das, S. Potential zones identification for harvesting wind energy resources in desert region of India—A multi criteria evaluation approach using remote sensing and GIS. *Renew. Sustain. Energy Rev.* **2016**, *65*, 1–10. [CrossRef]
97. Janke, J.R. Multicriteria GIS modeling of wind and solar farms in Colorado. *Renew. Energy* **2010**, *35*, 2228–2234. [CrossRef]
98. Chimres, N.; Wongwises, S. Critical review of the current status of solar energy in Thailand. *Renew. Sustain. Energy Rev.* **2016**, *58*, 198–207. [CrossRef]
99. Al Garni, H.Z.; Awasthi, A. Solar PV power plant site selection using a GIS-AHP based approach with application in Saudi Arabia. *Appl. Energy* **2017**, *206*, 1225–1240. [CrossRef]
100. Krewitt, W.; Nitsch, J. The potential for electricity generation from on-shore wind energy under the constraints of nature conservation: A case study for two regions in Germany. *Renew. Energy* **2003**, *28*, 1645–1655. [CrossRef]
101. Ho, C.; Sims, C.; Christian, J. Evaluation of Glare at the Ivanpah Solar Electric Generating System. *Energy Procedia* **2015**, *69*, 1296–1305. [CrossRef]
102. Yue, C.-D.; Wang, S.-S. GIS-based evaluation of multifarious local renewable energy sources: A case study of the Chigu area of southwestern Taiwan. *Energy Policy* **2006**, *34*, 730–742. [CrossRef]
103. Tanavud, C.; Yongchalermchai, C.; Bennui, A.; Densreeserekul, O. Assessment of flood risk in Hat Yai Municipality, Southern Thailand, using GIS. *J. Nat. Disaster Sci.* **2004**, *26*, 1–14. [CrossRef]
104. Ghenai, C.; Albawab, M.; Bettayeb, M. Sustainability indicators for renewable energy systems using multi-criteria decision-making model and extended SWARA/ARAS hybrid method. *Renew. Energy* **2019**, *146*, 580–597. [CrossRef]
105. Shao, M.; Han, Z.; Sun, J.; Xiao, C.; Zhang, S.; Zhao, Y. A review of multi-criteria decision making applications for renewable energy site selection. *Renew. Energy* **2020**, *157*, 377–403. [CrossRef]
106. Van Rensburg, T.M.; Kelley, H.; Jeserich, N. What influences the probability of wind farm planning approval: Evidence from Ireland. *Ecol. Econ.* **2015**, *111*, 12–22. [CrossRef]
107. Sichilalu, S.; Wamalwa, F.; Akinlabi, E.T. Optimal control of wind-hydrokinetic pumpback hydropower plant con-strained with ecological water flows. *Renew. Energy* **2019**, *138*, 54–69. [CrossRef]
108. Asaduz-Zaman, M.; Rahaman, M.H.; Reza, M.S.; Islam, M.M. A Method for Distributed Generator Dispatch Strategy in Distribution Network. *J. Electr. Eng.* **2018**, *6*, 261–270. [CrossRef]
109. Emmanuel, M.; Rayudu, R. Evolution of dispatchable photovoltaic system integration with the electric power network for smart grid applications: A review. *Renew. Sustain. Energy Rev.* **2017**, *67*, 207–224. [CrossRef]
110. Loudiyi, K.; Berrada, A.; Svendsen, H.G.; Mentesidi, K. Grid code status for wind farms interconnection in Northern Africa and Spain: Descriptions and recommendations for Northern Africa. *Renew. Sustain. Energy Rev.* **2018**, *81*, 2584–2598. [CrossRef]
111. Shafiullah, G.; Oo, A.M.; Ali, A.S.; Wolfs, P. Potential challenges of integrating large-scale wind energy into the power grid–A review. *Renew. Sustain. Energy Rev.* **2013**, *20*, 306–321. [CrossRef]
112. Saqib, M.A.; Saleem, A.Z. Power-quality issues and the need for reactive-power compensation in the grid integration of wind power. *Renew. Sustain. Energy Rev.* **2015**, *43*, 51–64. [CrossRef]
113. Rona, B.; Güler, O. Power system integration of wind farms and analysis of grid code requirements. *Renew. Sustain. Energy Rev.* **2015**, *49*, 100–107. [CrossRef]
114. Mohseni, M.; Islam, S. Review of international grid codes for wind power integration: Diversity, technology and a case for global standard. *Renew. Sustain. Energy Rev.* **2012**, *16*, 3876–3890. [CrossRef]
115. Shaker, H.; Zareipour, H.; Wood, D. Impacts of large-scale wind and solar power integration on California's net electrical load. *Renew. Sustain. Energy Rev.* **2016**, *58*, 761–774. [CrossRef]
116. Zappa, W.; Broek, M.V.D. Analysing the potential of integrating wind and solar power in Europe using spatial optimisation under various scenarios. *Renew. Sustain. Energy Rev.* **2018**, *94*, 1192–1216. [CrossRef]
117. Viviescas, C.; Lima, L.; Diuana, F.A.; Vasquez, E.; Ludovique, C.; Silva, G.N.; Huback, V.; Magalar, L.; Szklo, A.; Lucena, A.F.; et al. Contribution of Variable Renewable Energy to increase energy security in Latin America: Complementarity and climate change impacts on wind and solar resources. *Renew. Sustain. Energy Rev.* **2019**, *113*, 109232. [CrossRef]
118. Liu, L.; Wang, Z.; Wang, Y.; Wang, J.; Chang, R.; He, G.; Tang, W.; Gao, Z.; Li, J.; Liu, C.; et al. Optimizing wind/solar combinations at finer scales to mitigate renewable energy variability in China. *Renew. Sustain. Energy Rev.* **2020**, *132*, 110151. [CrossRef]
119. Sterl, S.; Donk, P.; Willems, P.; Thiery, W. Turbines of the Caribbean: Decarbonising Suriname's electricity mix through hydro-supported integration of wind power. *Renew. Sustain. Energy Rev.* **2020**, *134*, 110352. [CrossRef]
120. Nghitevelekwa, K.; Bansal, R. A review of generation dispatch with large-scale photovoltaic systems. *Renew. Sustain. Energy Rev.* **2018**, *81*, 615–624. [CrossRef]
121. Boqiang, R.; Chuanwen, J. A review on the economic dispatch and risk management considering wind power in the power market. *Renew. Sustain. Energy Rev.* **2009**, *13*, 2169–2174. [CrossRef]
122. Cheng, C.; Liu, B.; Chau, K.-W.; Li, G.; Liao, S. China's small hydropower and its dispatching management. *Renew. Sustain. Energy Rev.* **2015**, *42*, 43–55. [CrossRef]
123. Das, H.S.; Yatim, A.; Tan, C.W.; Lau, K.Y. Proposition of a PV/tidal powered micro-hydro and diesel hybrid system: A southern Bangladesh focus. *Renew. Sustain. Energy Rev.* **2016**, *53*, 1137–1148. [CrossRef]

124. Dallinger, B.; Schwabeneder, D.; Lettner, G.; Auer, H. Socio-economic benefit and profitability analyses of Austrian hydro storage power plants supporting increasing renewable electricity generation in Central Europe. *Renew. Sustain. Energy Rev.* **2019**, *107*, 482–496. [CrossRef]
125. Schmidt, J.; Cancella, R.; Junior, A.O.P. The effect of windpower on long-term variability of combined hydro-wind resources: The case of Brazil. *Renew. Sustain. Energy Rev.* **2016**, *55*, 131–141. [CrossRef]
126. Upadhyay, S.; Sharma, M. A review on configurations, control and sizing methodologies of hybrid energy systems. *Renew. Sustain. Energy Rev.* **2014**, *38*, 47–63. [CrossRef]
127. Mukhopadhyay, B.; Das, D. Multi-objective dynamic and static reconfiguration with optimized allocation of PV-DG and battery energy storage system. *Renew. Sustain. Energy Rev.* **2020**, *124*, 109777. [CrossRef]
128. Bird, L.; Lew, D.; Milligan, M.; Carlini, E.M.; Estanqueiro, A.; Flynn, D.; Gomez-Lazaro, E.; Holttinen, H.; Menemenlis, N.; Orths, A.; et al. Wind and solar energy curtailment: A review of international experience. *Renew. Sustain. Energy Rev.* **2016**, *65*, 577–586. [CrossRef]
129. PVSyst Help Manual. Available online: https://www.pvsyst.com/help/ (accessed on 1 July 2020).
130. RES4Africa. Integration of Variable Renewable Energy Sources in the National Electric System of Zambia. Available online: https://www.res4africa.org/wp-content/uploads/2020/10/R4AF-Integration-of-Variable-Renewable-Energy-Sources-in-the-National-Electric-System-of-Zambia.pdf (accessed on 30 June 2021).
131. International Renewable Energy Agency—IRENA. *Renewable Power Generation Costs in 2019*; IRENA: Abu Dhabi, United Arab Emirates, 2020.

mathematics

MDPI

Article

Hour-Ahead Photovoltaic Output Forecasting Using Wavelet-ANFIS

Chao-Rong Chen [1], Faouzi Brice Ouedraogo [2,*], Yu-Ming Chang [1], Devita Ayu Larasati [1] and Shih-Wei Tan [3]

1 Department of Electrical Engineering, National Taipei University of Technology, Taipei 10608, Taiwan; crchen@ntut.edu.tw (C.-R.C.); peter8606@gmail.com (Y.-M.C.); devitaayularasati@gmail.com (D.A.L.)
2 International Program of Electrical Engineering and Computer Science, National Taipei University of Technology, Taipei 10608, Taiwan
3 Department of Electrical Engineering, National Taiwan Ocean University, Keelung City 202301, Taiwan; tanshwei@mail.ntou.edu.tw
* Correspondence: ouedfao@gmail.com; Tel.: +886-2-27712171 (ext. 2110)

Abstract: The operational challenge of a photovoltaic (PV) integrated system is the uncertainty (irregularity) of the future power output. The integration and correct operation can be carried out with accurate forecasting of the PV output power. A distinct artificial intelligence method was employed in the present study to forecast the PV output power and investigate the accuracy using endogenous data. Discrete wavelet transforms were used to decompose PV output power into approximate and detailed components. The decomposed PV output was fed into an adaptive neuro-fuzzy inference system (ANFIS) input model to forecast the short-term PV power output. Various wavelet mother functions were also investigated, including Haar, Daubechies, Coiflets, and Symlets. The proposed model performance was highly correlated to the input set and wavelet mother function. The statistical performance of the wavelet-ANFIS was found to have better efficiency compared with the ANFIS and ANN models. In addition, wavelet-ANFIS coif2 and sym4 offer the best precision among all the studied models. The result highlights that the combination of wavelet decomposition and the ANFIS model can be a helpful tool for accurate short-term PV output forecasting and yield better efficiency and performance than the conventional model.

Keywords: PV forecasting; ANFIS; wavelet-ANFIS; wavelet decomposition; mother wavelet function

Citation: Chen, C.-R.; Ouedraogo, F.B.; Chang, Y.-M.; Larasati, D.A.; Tan, S.-W. Hour-Ahead Photovoltaic Output Forecasting Using Wavelet-ANFIS. *Mathematics* **2021**, *9*, 2438. https://doi.org/10.3390/math9192438

Academic Editors: Zbigniew Leonowicz, Arsalan Najafi and Michał Jasinski

Received: 23 August 2021
Accepted: 24 September 2021
Published: 1 October 2021

1. Introduction

Estimating and forecasting solar power output has played an influential and critical role in integrated system management and the optimal operation of solar power in high-demand periods. Therefore, this subject is of interest both to academia and to power companies. The forecasting of upcoming events is the backbone of crisis management; as soon as this target can be accomplished, integrated system management becomes accessible [1]. Several published studies have proposed methods to forecast the future power output. Exploiting each of these proposed prediction methods generally leads to some error. Accurate prediction of PV output can provide helpful information for power management in an integrated grid [1–3].

The booming power forecasting methods proposed for solar power systems in the last decade can be divided into statistical, artificial intelligence, fuzzy inference, and hybrid methods [4]. Statistical methods, for example, auto-regressive moving average and auto-regressive integrated moving average, have been used in power system prediction [5,6]. Artificial intelligence (AI) methods can be used to minimize research costs and reduce computing time as reliable alternative methods for forecasting the performance of complex systems [7]. Among others, artificial intelligence methods, multilayer perceptron neural networks [8–10], radial basis function neural networks (RBF NN) [11], physical hybrid artificial neural networks [12], recurrent neural networks [13], deep neural networks [14], fuzzy logic (FL) [15] and ANFIS [16–18] have been developed to forecast the output power of PV sources. Artificial neural network (ANN) is among the most effective techniques to forecast solar power system output. ANN modeling of PV power forecasting has recently been studied by researchers [19,20]. Numerous studies have demonstrated the efficiency and robustness of the ANN approach for cases characterized with high non-linearity and mapping function due to its effectiveness. ANN has a powerful learning ability even with different noise levels, the ability to simplify complex functions, wide fault tolerance, and good accomplishment for data uncertainty [19–22]. The backpropagation learning using different variants algorithms, such as the Bayesian regularization, the scaled conjugate gradient, and the Levenberg–Marquardt algorithms, with tremendous advantage as mentioned above, were used in the ANN by Basurto et al. [22].

Meanwhile, fuzzy inference systems (FIS) offer more advantages than mathematical ones. The inference operation is close to human thinking logic and can efficiently handle complex non-linear systems. [23–25]. Moreover, FIS with a backpropagation algorithm that is geared towards collecting input and output betters the effectiveness. The FIS modeling and recognition toolbox creates Takagi–Sugeno fuzzy methods from data through product space fuzzy clustering [23]. FIS has become popular in the field of engineering in the last decade, finding a considerable variety of applications—for instance, forecasting, systems pattern recognition, control theory, and power systems. Yona et al. [24] suggested a fuzzy technique to forecast hourly PV-generated power. Besides the above asset, FIS can be merged with ANN to create ANFIS [25–29]. Since Jang [25] first introduced ANFIS in the early 1990s [25], its application in power system forecasting has become rather recent. Different power system forecasting techniques have been developed to improve integrated system management [26–31]. Yaïci et al. [26] highlight that ANFIS can yield high reliability forecasting of the performance of this type of energy system. However, time series decomposition would effectively improve the ANFIS model by obtaining usable knowledge at various resolution levels to enhance the forecasting performance.

Research literature that uses wavelet and ANFIS to forecast PV-generated power, setting the model input data as PV power output, has not been found during the investigation of other proposed methods. Nevertheless, various published studies in the PV system field and other fields developed forecasting models based on the conjunction of wavelet decomposition and ANFIS. One model uses the variability reduction index, gene expression programming, wavelet transform, and ANFIS to assess the produced power for a batch of PV systems spread over one square kilometer, utilizing solar irradiance data and weather conditions [28]. Because of the irregularity and complexity of PV power, the research model [28] yielded good forecast results, but it is not easy to improve the prediction accuracy. Osorio et al. [29] proposed a short-term electricity market prices forecasting model integrating wavelet transforms and ANFIS. In addition, another research paper [30] presents a model using wavelet decomposition and ANFIS to forecast water levels. Notwithstanding, this method [30] should not include the reconstructed wavelet. Their accuracy should be calculated taking data after wavelet decomposition into consideration. Stefenon et al. [31] showed in their study that the time-series decomposition with wavelet transform improves the ANFIS robustness and accuracy and reduces training time. However, they did not propose a study of the other wavelet mother functions.

As illustrated above, in many studies, numerical weather predictions (NWP), such as temperature, relative humidity, irradiance, cloud cover, wind speed, and direction, are utilized as input to forecast the output PV power. Besides, the numerical weather data correlation with PV power is very low in the case study due to their low variation in the study area. The present study proposed endogenous data (the recorded current and past time series of the generated PV plant) as input; that is, the power generation data are directly taken, implicitly incorporating the results of NWP. Accordingly, the PV-generated power data cover the real situation on-site. In this study, two- and three-hours-previous PV output data are used to forecast solar peak time output data at different time horizons from 10 min to 60 min. The proposed idea comprises wavelet transform on the input data, training, and a testing stage using ANFIS. Compared to the existing literature, the main contributions of this paper are:

- A new method of combined PV power forecasting based on the decomposition at different resolution levels to optimize the weight determination.
- A study of different combinations of wavelet mother functions is proposed to find the most suitable for PV power time series.
- The combination of wavelet decomposition and ANFIS would have a strong learning ability and handle non-linear sequences regarding the chaotic and high non-linearity PV power output.
- The use of 2 and 3 h of data each day to forecast 10 min to 60 min ahead to increase the forecasting accuracy and reduce the computation time.
- Up to 30 shuffles are conducted to have initial random weighting and capture their diversity for a reliable and robust proposed combination. Moreover, the reconstructed wavelet features are used to calculate the accuracy of the proposed method.

This manuscript is organized as follows: The wavelet transforms (decomposition and reconstruction) model is presented in Section 2; the overview of basic ANFIS methodology is provided in Section 3; Section 4 gives an overview of the case study. In Section 5, the forecasting results are presented with different mother wavelets, such as haar, Daubechies, Symlets, and Coiffets wavelets, on forecasting accuracy. The discussion and conclusion are presented in Sections 6 and 7, respectively.

2. Wavelet Transform

Generally, PV power output data can usually contain non-linear and dynamic features in pick and fluctuations [32], which are among the principal attributes that influence endogenous PV output power forecasting accuracy. In actual conditions, low and high-frequency signals are included in PV power data [33], which can affect the learning process of the artificial intelligence method. However, the outliers and behaviors at each frequency are more easily forecasted. Accordingly, signal decomposition techniques such as wavelet transforms can be used to forecast each frequency behavior. Consequently, the latter can improve the PV-generated forecasting accuracy. Wavelet transform is a well-documented effective "scalable" technique for time series data analysis with a higher time-frequency localization. Conventionally, discrete wavelet transforms (DWT) are generally used to increase the calculation efficiency of the prediction model. DWT of time series data can be computed as [34]:

$$DWT_{(m,n)} = 2^{-\left(\frac{m}{2}\right)} \sum_{t=0}^{T-1} f(t)\psi\left(\frac{t - n2^n}{2^m}\right) \tag{1}$$

where $\psi(t)$, a continuous function, designates the wavelet mother function, $f(t)$ is the time-series function signal, and m and n are integers used to manage the wavelet dilatation and translation, respectively. Various wavelet mother functions are available, such as Harr, Daubechies, Symlets, Coiflets, Biorthogonal, Morlet, or Mexican Hat. The most popular wavelet mother functions, including Haar, Daubechies (db3), Coiflets (coif2), and Symlets (sym4), are illustrated in Figure 1.

The Mallat technique [34] is utilized as a fast-DWT method to equilibrate the wave-length and smoothness in the current literature. This computation requires less time as it is the least complex process of calculating the DWT. The original time-series function $f(t)$ signal can be reconstructed as the following [34]:

$$f(t) = \overline{T}(t) + \sum_{m=1}^{M} \sum_{n=0}^{2^{M-n}-1} DWT_{(m,n)} 2^{-\left(\frac{m}{2}\right)} \left(\frac{t - n2^n}{2^m}\right) \tag{2}$$

the integers m and n are, respectively, in the range: $1 < m < M$ and $0 < n < 2^{M-n} - 1$. $\overline{T}(t)$ designates the approximation subsignal at the M level. A simple format of the signal reconstruction can be computed as [34]:

$$f(t) = \overline{T}(t) \sum_{m=1}^{M} W_m(t) \tag{3}$$

in which $W_m(t)$ are the details subsignals that can apprehend the interpretation of the value of small features present in the data.

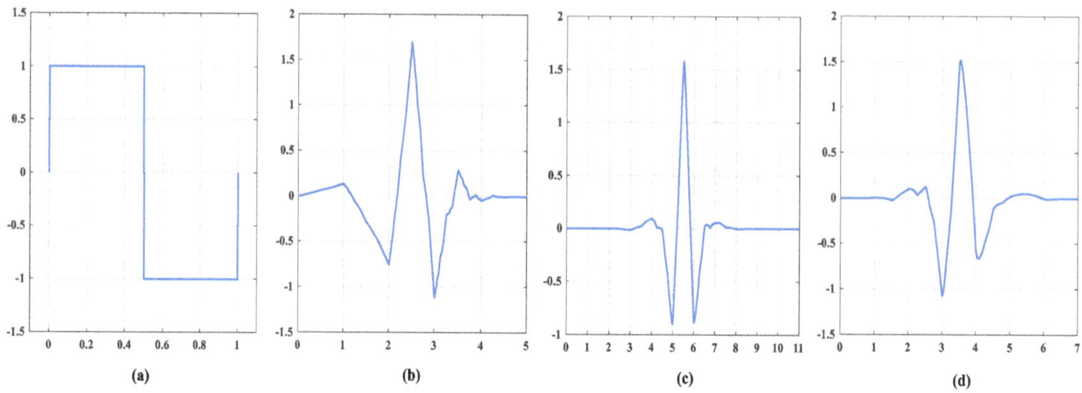

Figure 1. Sample illustration of various wavelet mother functions: (**a**) haar, (**b**) db3, (**c**) coif2, and (**d**) sym4.

3. Adaptive Neuro-Fuzzy Inference System

ANFIS is an intelligent system that combines the strong points of the principles of fuzzy logic and neural networks into a hybrid technique. ANFIS is considered as a mere data learning algorithm that uses fuzzy logic to process data inputs into the desired output through using strongly interconnected neural network processing elements and weighted information connections to map inputs into output. This approach can simulate complex non-linear mappings, and it is suitable for an accurate short-term predictions [35–37].

As illustrated in Figure 2, the ANFIS architecture uses five layers relying on Takagi–Sugeno–Kang (TSK) rules inference system. Each layer carries out distinct functions. There are five layers: the fuzzification layer, rule reasoning layer, normalization layer, defuzzification layer, and output layer. The premise and consequent parameters are the essential parameters of ANFIS. The premise parameters $\{\alpha_k, \beta_k, \gamma_k\}$ are included in the designed membership functions of the fuzzification layer, which is the input layer of the ANFIS network used to generate input spaces by retrieving patterns in the input data. The consequent parameters $\{\rho_k, \sigma_k, \tau_k\}$ correspond to the parameter sets of the defuzzification layer.

The premise and the consequent parameters are optimized through training. A hybrid algorithm optimizes the parameter sets of the ANFIS forecasting system in the proposed approach.

ANFIS is explained by assuming two input variables (z_1 and z_2) and a unique output variable ($\hat{y}$). The rule illustrating the relationship associating the input, the membership function, and the output can be expressed using the if–then TSK fuzzy rules illustrated in the following conditional statements [25]:

$$\text{if } z_1 \text{ is } D_1 \text{ and } z_2 \text{ is } E_1, \text{ then } \hat{y}_1 = \rho_1 z_1 + \sigma_1 z_2 + \tau_1 \tag{4}$$

$$\text{if } z_1 \text{ is } D_2 \text{ and } z_2 \text{ is } E_2, \text{ then } \hat{y}_2 = \rho_2 z_1 + \sigma_2 z_2 + \tau_2 \tag{5}$$

where D_1, D_2, E_1 and E_2 correspond to the fuzzy sets, also called linguistic labels. The function of every layer is presented as below:

Fuzzification layer (layer 1): Each individual node is adaptive in this layer. The input variables' membership functions are mapped into fuzzy sets. The function is assigned to every node as described:

$$L_k^1 = \mu_{A_k}(z_1),\ \text{for } k = 1,\ 2 \tag{6}$$

$$L_k^1 = \mu_{B_{k-2}}(z_2),\ \text{for } k = 3,\ 4 \tag{7}$$

where L_k^1 represents the output of the node kth, μ_{A_k} and μ_{B_k} are the membership function, there exist various membership functions. The generalized bell function can be written as:

$$\mu_{A_k}(z_1) = \frac{1}{1 + \left|\frac{z_1 - \gamma_k}{\alpha_k}\right|^{2\beta_k}} \quad 0 \le \mu_{A_k}(z_1) \le 1 \tag{8}$$

where α_k, β_k, and γ_k represent premise parameters used to change the shape of the membership function.

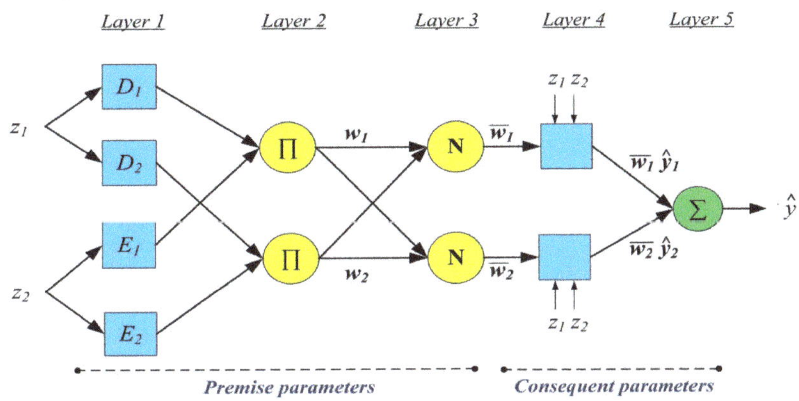

Figure 2. Basic ANFIS diagram.

Rule layer (layer 2): Each node in the layer is shown with a non-adaptive node, labeled as Π in Figure 2, and all the incoming signals are multiplied to compute the output:

$$L_k^2 = w_k = \mu_k(z_1).\mu_{B_k}(z_2),\ \text{for } k = 1,\ 2 \tag{9}$$

where the layer 2 output is given by L_k^2 and w_k is weight strength for the *kth* TSK rule. Every node illustrates the weight strength of a rule.

Normalization layer (layer 3): the activity level of each rule is calculated by each node. The kth node is normalized as w_k. The *kth* rule determines the weight strength, then is divided by the sum of all weight strength rules and can be computed by:

$$L_k^3 = \overline{w}_k = \frac{w_k}{w_1 + w_2},\ \text{for } k = 1,\ 2 \tag{10}$$

where L_k^3 stands for the layer 3 output and $\overline{w}_k$ the normalized weight strength.

Defuzzification layer (layer 4): The nodes of this layer are adaptable. This layer function attributes to a single node the contribution of the *kth* rule inference to the fifth layer. L_k^4 is computed as the defuzzification layer output using the consequent parameters and is defined as:

$$L_k^4 = \overline{w}_k\hat{y}_k = \overline{w}_k(\rho_k z_1 + \sigma_k z_2 + \tau_k) \tag{11}$$

Output layer (layer 5): Only one fixed node composes this layer. The whole output is computed by the sum of the arriving signals from the previous layer as:

$$\hat{y} = L_k^5 = \sum_k \overline{w}_k \hat{y}_k = \left(\sum_k w_k \hat{y}_k\right) / \left(\sum_k w_k\right) \tag{12}$$

4. Case Study

In power systems, for enhancing the grid performance, its stability should be guaranteed at all costs. To achieve grid stability, the appropriate scheduling of the spinning reserves and demand response is vital. With the increase in the adoption of PV as a source of energy, its intermittent nature may impact grid stability. There is a need to design forecasting models with good accuracy in forecasting the PV output power to mitigate issues with the grid stability and make PV a reliable energy source. This study used all-year solar power output data from 2017 to 2020, averaging the values for 10 min interval data. The data used for the current study were recorded at a solar power plant close to Taichung in the middle of Taiwan. The first 900 days' measurement data were employed as input training data, and the last 185 last days' data were employed to assess the trained model. The model has been designed to forecast 10 min, 30 min, and 60 min ahead each day during PV generation peak time.

The proposed forecasting algorithm consists of two parts. In the first part, the input dataset of ANFIS training data are decomposed using DWT. At that point, the function operated to decompose the training dataset is applied to the test dataset for test data decomposition. The study has also investigated the accuracy of well-known wavelet mother functions. A statistical study of different wavelet mother functions is conducted to evaluate each function's performance, including haar, db2, db3, db5, db8, coif1, coif2, coif3, coif5, sym4, sym6, and sym8.

The second part corresponds to the training and testing steps using ANFIS. Before developing the model, the optimal selection of input model numbers is essential because it can significantly reduce the computational time and cost. Two different input pattern numbers are presented in this work. The main settings of the ANFIS network include the types of input and the output membership function, the number of input and output membership functions, the number of iterations (epochs), and optimization methods such as hybrid learning.

The MATLAB toolbox is used to generate the ANFIS model for the studied data. The resulting equation of each rule is obtained by applying the linear least square assessment. Fuzzy c-means (fcm) was employed as a data clustering technique. Every data point is incumbent to a cluster and determined by membership level. The number of clusters has been set to 12, with 4 partition matrix exponents in 0.01 steps. First, the optimal dataset is carried out so that the generation of the initial FIS becomes possible. In this study, a maximum of 200 epochs has been used to achieve accurate prediction.

The training data were trained to determine the parameters of the TSK-type FIS build on the hybrid learning algorithm, which comprises the integration of the least square estimator and backpropagation gradient descent, as described by Table 1. After the training stage, the developed forecasting models were performed, and the efficiency was calculated with different evaluation criteria. The architecture of the ANFIS model developed is presented in Figure 3, with n inputs ($k = 12$ or $k = 18$), 1 output ($\hat{y}$), and 12 fuzzy rules ($r = 12$). When $k = 12$, the inputs represent every 10 min PV generated data of the last 2 h; meanwhile, $k = 18$ employs the last 3 h of data. The 12 conditional statements are used as a rule-base. The output ($\hat{y}_i$) gives the forecasted value of PV power. Data for the same training period are used to forecast the coming 10 min, 30 min, and 60 min of each day. That is, using 2- or 3 h data to train the model (from 9 am to 11 am or 8 am to 11 am in other words $t - 12$, and $t - 18$, separately) and forecast 11:10 ($t + 1$), 11:30 ($t + 3$), and 12:00 ($t + 6$) PV power. The training and testing flowchart are, respectively summarized in Figures 4 and 5.

The asterisk (*) in Figure 5 indicates that the test dataset wavelet transforms are acquired by applying the wavelet transform function used to decompose the training dataset.

Table 1. The hybrid learning methodology for ANFIS.

	Forward	**Backward**
$\alpha_k, \beta_k, \gamma_k$	Fixed	Gradient Descent
ρ_k, σ_k, τ_k	Least-squares estimator	Fixed
Signals	Node outputs	Error signals

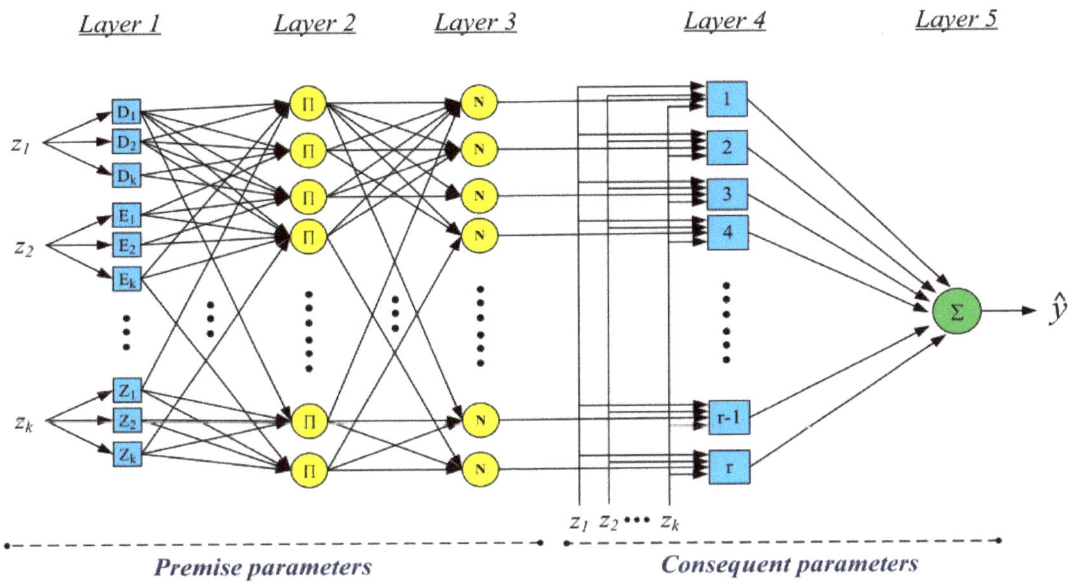

Figure 3. The proposed idea for ANFIS architecture.

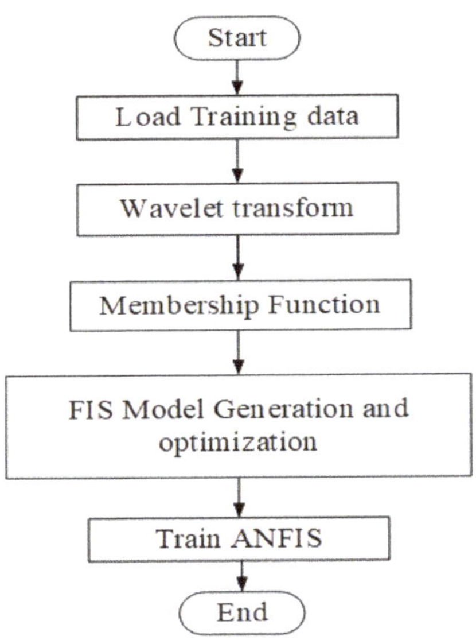

Figure 4. Wavelet-ANFIS training flowchart.

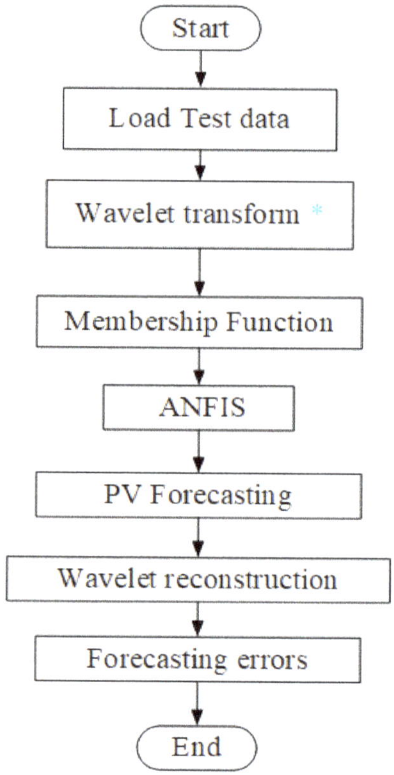

Figure 5. Wavelet-ANFIS testing flowchart. The asterisk (*) indicates that the test dataset wavelet transforms are acquired by applying the wavelet transform function used to decompose the training dataset.

The same datasets were used for an ANN model as a comparison to establish the effectiveness of the current idea. The ANN model's input pattern contained 12 or 16 inputs. The output layer was designed to forecast the power output value as described above for the ANFIS model.

5. Results

5.1. Forecasting Accuracy Evaluation

Various standard error metrics were exercised to evaluate the proposed PV output power prediction strategy. The actual and forecast sequence are represented, respectively, by $\dot{y}_i$ and $\hat{y}_i$ with N time steps. $\overline{y}_i$ represents the maximum recorded PV power in the test dataset. Normalize root means square error ($nRMSE$ (%)), mean absolute percentage error ($MAPE$ (%)), means absolute error (MAE (kWh)), root mean square error ($RMSE$ (kWh)), and the standard deviation (STD (kWh) criteria are used to assess the accuracy of the different forecasting model. The metrics mentioned above are computed as below:

$$nRMSE(\%) = \sqrt{\frac{1}{N}\sum_{i=1}^{N}\left|\frac{\dot{y}_i - \hat{y}_i}{\overline{y}_i}\right|^2} \times 100\% \tag{13}$$

$$MAPE(\%) = \frac{1}{N}\sum_{i=1}^{N}\left|\frac{\dot{y}_i - \hat{y}_i}{\overline{y}_i}\right| \times 100\% \tag{14}$$

$$MAE(\text{kWh}) = \frac{1}{N}\sum_{i=1}^{N}|\dot{y}_i - \hat{y}_i| \tag{15}$$

$$RMSE(\text{kWh}) = \sqrt{\frac{1}{N}\sum_{i=1}^{N}(\dot{y}_i - \hat{y}_i)^2} \tag{16}$$

$$STD(\text{kWh}) = \sqrt{\frac{1}{N-1}\sum_{i=1}^{N}(\dot{y}_i - \hat{y}_i)^2} \tag{17}$$

5.2. Wavelet Study Results

The performance of the wavelet-ANFIS model for various wavelets mother functions for the input datasets is computed. Figure 6 shows the PV output of the original in blue compared with the wavelet output with different mother functions (Haar, Daubechies(db3), coiflets (coif2), and symlets (sym4)). Table 2 shows the forecasting result with different wavelet mother functions. It is discernible from Table 2 that wavelet-ANFIS models using coif2, sym4, and sym6 were more highly accurate than the other mother wavelets. The wavelet-ANFIS models with sym4 gave the highest efficiency in an overall view, indicating that wavelet decomposition utilizing the mother wavelet, sym4, can noticeably enhance ANFIS forecasting models' accuracy compared with other mother wavelets functions. It should be mentioned that the value of Table 2 is the average value computed from 30-time differently shuffled data points to have a strong comparison.

5.3. Forecasting Results

The forecasting evaluation analysis was performed by analyzing the results using the forecasting accuracy evaluation metrics mentioned in Section 5.1.

All the mother wavelets described previously have been computed, and the subsequent outcomes are presented in Table 2. It is essential to mention that after the forecasting output is obtained, the reconstitution function (2) is used to reconstruct the output. The forecasting error is computed with $\dot{y}_i$, which represents the original value of the PV power output before the wavelet decomposition.

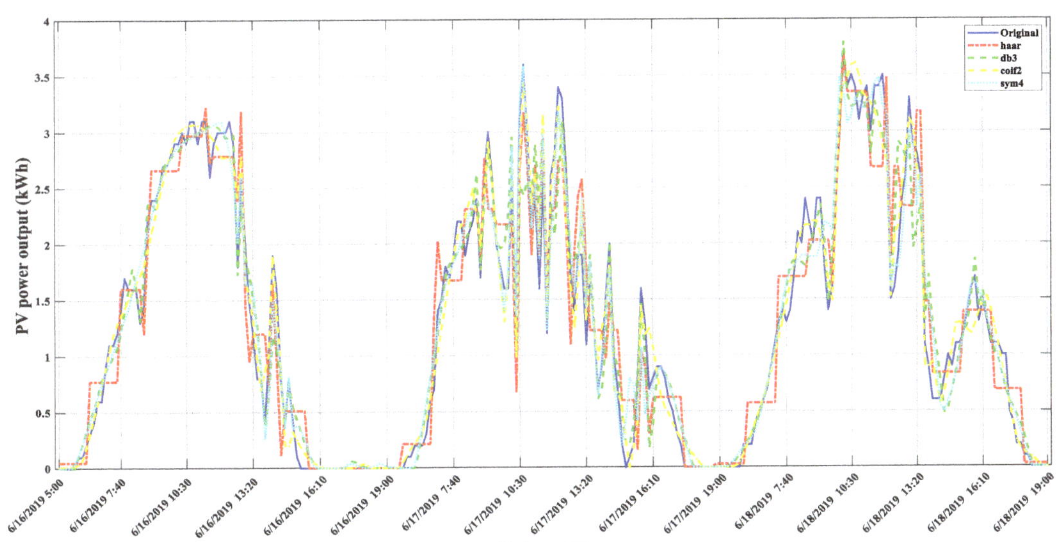

Figure 6. The original PV power output and different wavelet mother functions.

Table 2. Different wavelet-ANFIS with mother function performance.

		12 Input Patterns			18 Input Patterns		
		Forecasting Time					
	RMSE (kWh)	10 Min	30 Min	60 Min	10 Min	30 Min	60 Min
Wavelet mother function	Haar	3.5965×10^{-4}	3.8168×10^{-4}	3.9033×10^{-3}	3.7218×10^{-3}	4.3430×10^{-3}	5.9257×10^{-3}
	db2	2.3432×10^{-4}	2.5778×10^{-4}	2.7985×10^{-4}	1.6028×10^{-4}	2.1587×10^{-4}	3.1221×10^{-4}
	db3	2.2701×10^{-4}	2.2992×10^{-4}	2.5655×10^{-4}	1.5331×10^{-4}	2.0677×10^{-4}	2.1517×10^{-4}
	db5	1.9905×10^{-4}	2.1077×10^{-4}	2.3834×10^{-4}	1.7060×10^{-4}	1.8702×10^{-4}	2.5660×10^{-4}
	db8	2.2789×10^{-4}	2.8417×10^{-4}	2.9929×10^{-4}	1.5424×10^{-4}	1.9433×10^{-4}	2.9495×10^{-4}
	coif1	2.1413×10^{-4}	2.3070×10^{-4}	2.3842×10^{-4}	1.7586×10^{-4}	2.0476×10^{-4}	2.3773×10^{-4}
	coif2	1.8734×10^{-4}	2.1022×10^{-4}	2.2619×10^{-4}	2.1707×10^{-4}	2.4265×10^{-4}	2.7823×10^{-4}
	coif3	2.1759×10^{-4}	2.3643×10^{-4}	2.4524×10^{-4}	1.7536×10^{-4}	2.3334×10^{-4}	2.8807×10^{-4}
	coif5	2.3467×10^{-4}	2.8025×10^{-4}	3.2647×10^{-4}	2.1136×10^{-4}	2.7625×10^{-4}	3.3846×10^{-4}
	sym4	2.2402×10^{-4}	2.3770×10^{-4}	2.7530×10^{-4}	1.5365×10^{-4}	1.8049×10^{-4}	2.1273×10^{-4}
	sym6	2.6657×10^{-4}	2.8111×10^{-4}	2.9734×10^{-4}	1.4681×10^{-4}	2.2755×10^{-4}	2.5019×10^{-4}
	sym8	2.0610×10^{-4}	2.4349×10^{-4}	3.1602×10^{-4}	1.9454×10^{-4}	2.1813×10^{-4}	2.9410×10^{-4}

Different forecasting methods have been completed. Firstly, wavelet-ANFIS was devised to show the effectiveness of wavelet decomposition. The decomposition significantly improves the efficiency of the forecasting method. The same forecasting cases were than conducted with ANFIS without wavelet decomposition. The statistical performance has likewise been computed with the ANN forecasting model. The ANN and ANFIS model have similar input patterns. They have inputs in the range of 12 and 18, an additional 2 hidden layers with 20 nodes at each hidden layer, and one output. The Bayesian regularization algorithm of the backpropagation method is used to update the weighting value. The maximum iteration epochs are equal to 800.

Table 3 gives the comparison of wavelet-ANFIS, ANFIS, and ANN models of the different prediction scenarios. Up to 30 scenarios are conducted for each forecasting method to randomize the initial weight. Each scenario is obtained by shuffling the available data to capture their diversity for a reliable and robust comparison. The average value of all cases RMSE is illustrated in Table 2. Table 3 summarized all evaluation criteria (nRMSE, MAPE, MAE, RMSE, and STD) of the wavelet-ANFIS, ANFIS, and ANN model.

Table 3. The performance of the Wavelet-ANFIS compared with ANFIS and ANN model.

		12 Input Patterns					18 Input Patterns				
Inputs	Forecasting Time	nRMSE (%)	MAPE (%)	MAE (kWh)	RMSE (kWh)	STD (kWh)	nRMSE (%)	MAPE (%)	MAE (kWh)	RMSE (kWh)	STD (kWh)
Wavelet-ANFIS	10 min	4.4604×10^{-3}	2.4436×10^{-3}	1.0263×10^{-4}	1.8734×10^{-4}	1.8718×10^{-4}	3.4954×10^{-3}	1.8751×10^{-3}	7.8754×10^{-5}	1.4681×10^{-4}	1.4681×10^{-4}
	30 min	5.0183×10^{-3}	2.8624×10^{-3}	1.2022×10^{-4}	2.1077×10^{-4}	2.1048×10^{-4}	4.2973×10^{-3}	2.2672×10^{-3}	9.5222×10^{-5}	1.8049×10^{-4}	1.8049×10^{-4}
	60 min	5.3727×10^{-3}	3.0646×10^{-3}	1.2871×10^{-4}	2.2565×10^{-4}	2.2550×10^{-4}	5.0651×10^{-3}	2.6137×10^{-3}	1.0977×10^{-4}	2.1273×10^{-4}	2.1273×10^{-4}
ANFIS	10 min	1.5059×10^{-2}	8.7031×10^{-3}	3.6553×10^{-4}	6.3248×10^{-4}	6.3196×10^{-4}	1.3616×10^{-2}	6.5940×10^{-3}	2.7695×10^{-4}	5.7185×10^{-4}	5.7216×10^{-4}
	30 min	2.0272×10^{-2}	1.2034×10^{-2}	5.0543×10^{-4}	8.5144×10^{-4}	8.5028×10^{-4}	1.5735×10^{-2}	7.3180×10^{-3}	3.0736×10^{-4}	6.6085×10^{-4}	6.6011×10^{-4}
	60 min	2.5262×10^{-2}	1.4211×10^{-2}	5.9684×10^{-4}	1.0610×10^{-3}	1.0603×10^{-3}	2.3642×10^{-2}	1.1736×10^{-2}	4.9290×10^{-4}	9.9294×10^{-4}	9.9184×10^{-4}
ANN	10 min	3.8131×10^{-2}	2.9975×10^{-2}	1.2589×10^{-3}	1.6015×10^{-3}	1.6065×10^{-3}	0.49695	0.35041	1.4717×10^{-2}	2.0872×10^{-2}	1.8214×10^{-2}
	30 min	4.6381×10^{-2}	3.7486×10^{-2}	1.5744×10^{-3}	1.9480×10^{-3}	1.9539×10^{-3}	0.52487	0.36033	1.5134×10^{-2}	2.2045×10^{-2}	1.8964×10^{-2}
	60 min	5.4582×10^{-2}	3.9186×10^{-2}	1.6458×10^{-3}	2.2924×10^{-3}	2.2951×10^{-3}	1.0199	0.71746	3.0133×10^{-2}	4.2837×10^{-2}	3.8449×10^{-2}

6. Discussion

The model simulation results for different wavelet decomposition mother functions and two different input patterns are listed in Table 2. All the examinations are estimated by the index RMSE (kWh) defined in (13). One can observe that the mother function coif2 yielded the highest efficiency among mother wavelets in the forecasting model with 12 input patterns. Regarding the 18 input patterns, the sym4 yielded the lowest amount of forecasting errors in the overall view. Nevertheless, for the 10 min forecasting, sym6 yielded the lowest number of errors, followed by the sym4 wavelet model.

Furthermore, Table 2 also indicates that the best accuracy of 10, 30, and 60 min PV power forecasting are obtained with 18 input patterns. Meanwhile, the comparison in Table 2 demonstrates that DWT decomposition computing mother wavelets, coif2, sym4, and sym6 can considerably increase ANFIS models' effectiveness compared with the ANN one.

The comparisons between the forecasts obtained using wavelet-ANFIS, the ANFIS, and ANN are listed in Table 3. The maximums forecasting RMSE are obtained for 60 min, 2.2565×10^{-4}, 1.0610×10^{-3}, and 2.2924×10^{-3} kWh for the wavelet-ANFIS, ANFIS, ANN model, respectively. In other words, the maximum forecasting error of the ANN model was about 10.16 and 4.70 times higher than the wavelet-ANFIS and ANFIS model, separately. Such results show that the wavelet-ANFIS and ANFIS models have better performance in PV output forecasting. In addition, the MAPE, MAE, and nRMSE wavelet-ANFIS and ANFIS led to better efficiency compared with the ANN prediction model in all of the forecasting scenarios presented in this paper. The results in Table 3 are the average of 30 different shuffles, and it can be concluded that in the short-term, forecasting the PV power using the ANFIS model is more efficient than doing so using the ANN.

For more investigation, the comparison between wavelet-ANFIS and ANFIS that is presented in Table 3 indicates that the 10, 30, and 60 min for the ANFIS model with 12 input patterns forecasting RMSE are, respectively, equal to 6.3196×10^{-4}, 8.5144×10^{-4}, and 1.0610×10^{-3} kWh, which are about 3.38, 4.04, and 4.70 times higher than the wavelet-model, separately. Meanwhile, the RMSE for the previous scenarios with 18 input patterns using the ANFIS model yielded, respectively, results about 3.90, 3.66, and 4.67 times the RMSE higher than the wavelet-ANFIS model RMSE for the exact same scenarios. From Figure 7, which illustrates the error distribution of the ANFIS and wavelet-ANFIS model, it can be observed that the 10, 30, and 60 min forecasts using wavelet-ANFIS model have more error-index computations in (18) close to zero.

$$error(\text{kWh}) = \dot{y}_i - \hat{y}_i \quad (18)$$

In summary, from Tables 2 and 3 and Figure 7, the models using wavelet components used by DWT as inputs can result in greater efficiency than the ANFIS models without wavelet transforms. It can be concluded that the forecasts using the wavelet-ANFIS models are closer to the actual PV power output values than those with the ANFIS and the ANN models. Such a result indicates that the wavelet decomposition has the capacity to enhance the effectiveness of the ANFIS forecasting PV power models.

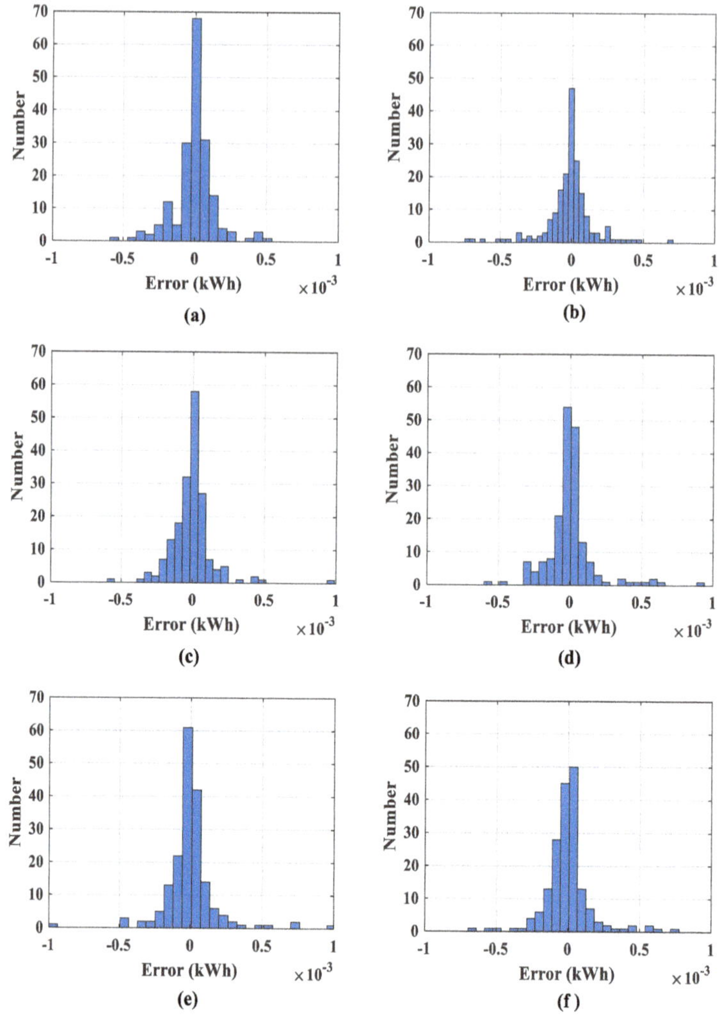

Figure 7. The error distribution of wavelet-ANFIS and ANFIS with 18 input patterns: (**a**) 10 min ahead forecasting using wavelet-ANFIS model; (**b**) 10 min ahead using ANFIS model; (**c**) 30 min ahead forecasting using wavelet-ANFIS model; (**d**) 30 min ahead using ANFIS model; (**e**) 60 min ahead forecasting using wavelet-ANFIS model; (**f**) 60 min ahead using the ANFIS model.

7. Conclusions

An extensive study on applying the wavelet-ANFIS method for PV output forecasting is presented in this paper. The specific aims are to develop and evaluate a properly endogenous method for PV output forecasting in Taiwan. In this work, we compared wavelet-ANFIS, ANFIS, and ANN based on miscellaneous performance indexes, including RMSE, nRMSE, MAE, MAPE and standard deviation. The results highlight that ANFIS obtains better results than the ANN model. Furthermore, the wavelet-ANFIS model yields a better accuracy in the PV output power than the ANFIS model and the ANN model. Comparison of the wavelet-based model shows that wavelet-ANFIS-coif2 and sym4 yield better performance than all other mother functions when the last 2 and 3 h of generated PV power are used as the input of the proposed model, separately. The wavelet-ANFIS model used the last 3 h of data sym4 decomposed to yield the best overall accuracy for 10,

30, and 60 min ahead of PV power. All the same, forecasting ahead by 60 min has yielded high accuracy. The outcomes of this research demonstrate that the conjunction of DWT decomposition and the ANFIS model could meaningfully better the reliability of models used in the short-term forecasting of PV output power. The results achieved in this paper demonstrate that the connective forecast with discrete wavelet decomposition and ANFIS could be an outstanding tool for the short-term forecasting of PV output power.

Author Contributions: Conceptualization, C.-R.C. and F.B.O.; methodology, C.-R.C., F.B.O., Y.-M.C., D.A.L. and S.-W.T.; software, C.-R.C., F.B.O., Y.-M.C. and D.A.L.; validation, C.-R.C., F.B.O., Y.-M.C., D.A.L. and S.-W.T.; formal analysis, C.-R.C., F.B.O., Y.-M.C., D.A.L. and S.-W.T.; investigation, C.-R.C., F.B.O., Y.-M.C. and D.A.L.; resources, C.-R.C., and S.-W.T.; data curation, F.B.O. and Y.-M.C.; writing—original draft preparation, C.-R.C., F.B.O. and Y.-M.C.; writing—review and editing, C.-R.C., F.B.O. and S.-W.T.; visualization C.-R.C., F.B.O., Y.-M.C. and D.A.L.; supervision, C.-R.C. and S.-W.T.; project administration, C.-R.C. and S.-W.T.; funding acquisition, C.-R.C. and S.-W.T. All authors have read and agreed to the published version of the manuscript.

Funding: This research was partly supported by the University System of Taipei Joint Research Program, Project no. USTP-NTUT-NTOU-107-04.

Institutional Review Board Statement: The study did not involve humans or animals.

Informed Consent Statement: The study did not involve humans.

Data Availability Statement: Not applicable.

Acknowledgments: This study is partly supported by the University System of Taipei Joint Research Program, Project no. USTP-NTUT-NTOU-107-04.

Conflicts of Interest: The authors declare no conflict of interest.

References

1. Nestle, D.; Ringelstein, J. Integration of DER into distribution grid operation and decentralized energy management. *Smart Grids Eur.* **2009**, *19*, 3–25.
2. Allen, J.; Halberstadt, A.; Powers, J.; El-Farra, N.H. An Optimization-Based Supervisory Control and Coordination Approach for Solar-Load Balancing in Building Energy Management. *Mathematics* **2020**, *8*, 1215. [CrossRef]
3. Elavarasan, R.M.; Selvamanohar, L.; Raju, K.; Vijayaraghavan, R.R.; Subburaj, R.; Nurunnabi, M.; Khan, I.A.; Afridhis, S.; Hariharan, A.; Pugazhendhi, R.; et al. A Holistic Review of the Present and Future Drivers of the Renewable Energy Mix in Maharashtra, State of India. *Sustainability* **2020**, *12*, 6596. [CrossRef]
4. Ahmed, R.; Sreeram, V.; Mishra, Y.; Arif, M.D. A review and evaluation of the state-of-the-art in PV solar power forecasting: Techniques and optimization. *Renew. Sustain. Energy Rev.* **2020**, *124*, 109792. [CrossRef]
5. Xie, T.; Zhang, G.; Liu, H.; Liu, F.; Du, P. A Hybrid Forecasting Method for Solar Output Power Based on Variational Mode Decomposition, Deep Belief Networks and Auto-Regressive Moving Average. *Appl. Sci.* **2018**, *8*, 1901. [CrossRef]
6. Mo, J.Y.; Jeon, W. How does energy storage increase the efficiency of an electricity market with integrated wind and solar power generation?—A case study of korea. *Sustainability* **2017**, *9*, 1797. [CrossRef]
7. Kosko, B. *Neural Networks and Fuzzy Systems, a Dynamical Systems Approach*; Prentice-Hall: Hoboken, NJ, USA, 1991.
8. Nespoli, A.; Ogliari, E.; Leva, S.; Massi Pavan, A.; Mellit, A.; Lughi, V.; Dolara, A. Day-Ahead Photovoltaic Forecasting: A Comparison of the Most Effective Techniques. *Energies* **2019**, *12*, 1621. [CrossRef]
9. Mellit, A.; Sağlam, S.; Kalogirou, S.A. Artificial neural network-based model for estimating the produced power of a photovoltaic module. *Renew. Energy* **2013**. [CrossRef]
10. Ouedraogo, F.B.; Zhang, X.-Y.; Chen, C.-R.; Lee, C.-Y. Hour-Ahead Power Generating Forecasting of Photovoltaic Plants Using Artificial Neural Networks Days Tuning. In Proceedings of the 2020 IEEE International Conference on Systems, Man, and Cybernetics (SMC), Toronto, ON, Canada, 11–14 October 2020; pp. 3874–3879. [CrossRef]
11. Huang, C.-M.; Chen, S.-J.; Yang, S.-P.; Kuo, C.-J. One-day-ahead hourly forecasting for photovoltaic power generation using an intelligent method with weather-based forecasting models. *IET Gener. Transm. Distrib.* **2015**, *9*, 1874–1882. [CrossRef]
12. Ogliari, E.; Niccolai, A.; Leva, S.; Zich, R.E. Computational Intelligence Techniques Applied to the Day Ahead PV Output Power Forecast: PHANN, SNO and Mixed. *Energies* **2018**, *11*, 1487. [CrossRef]
13. Li, G.; Wang, H.; Zhang, S.; Xin, J.; Liu, H. Recurrent Neural Networks Based Photovoltaic Power Forecasting Approach. *Energies* **2019**, *12*, 2538. [CrossRef]
14. Mellit, A.; Pavan, A.M.; Lughi, V. Deep learning neural networks for short-term photovoltaic power forecasting. *Renew. Energy* **2021**, *172*, 276–288. [CrossRef]

15. Yang, H.; Huang, C.; Huang, Y.; Pai, Y. A Weather-Based Hybrid Method for 1-Day Ahead Hourly Forecasting of PV Power Output. *IEEE Trans. Sustain. Energy* **2014**, *5*, 917–926. [CrossRef]
16. Osório, G.J.; Gonçalves, J.N.D.L.; Lujano-Rojas, J.M.; Catalão, J.P.S. Enhanced Forecasting Approach for Electricity Market Prices and Wind Power Data Series in the Short-Term. *Energies* **2016**, *9*, 693. [CrossRef]
17. Dong, W.; Yang, Q.; Fang, X. Multi-Step Ahead Wind Power Generation Prediction Based on Hybrid Machine Learning Techniques. *Energies* **2018**, *11*, 1975. [CrossRef]
18. Fernandez-Jimenez, L.; Muñoz-Jimenez, A.; Falces, A.; Mendoza-Villena, M.; Garcia-Garrido, E.; Lara-Santillan, P.; Zorzano-Alba, E.; Zorzano-Santamaria, P. Short-term power forecasting system for photovoltaic plants. *Renew. Energy* **2012**, *44*, 311–317. [CrossRef]
19. Singla, P.; Duhan, M.; Saroha, S. A comprehensive review and analysis of solar forecasting techniques. *Front. Energy* **2021**. [CrossRef]
20. Antonanzas, J.; Osorio, N.; Escobar, R.; Urraca, R.; de Pison, F.M.; Antonanzas-Torres, F. Review of photovoltaic power forecasting. *Sol. Energy* **2016**, *136*, 78–111. [CrossRef]
21. Chung, Y.T.; Tse, C.H. The forecast of the electrical energy generated by photovoltaic systems using neural network method. *Int. Conf. Electron. Control. Eng.* **2011**, 2758–2761. [CrossRef]
22. Basurto, N.; Arroyo, Á.; Vega, R.; Quintián, H.; Calvo-Rolle, J.L.; Herrero, Á. A hybrid intelligent system to forecast solar energy production. *Comput. Electr. Eng.* **2019**, *78*, 373–387. [CrossRef]
23. Takagi, T.; Sugeno, T. Fuzzy identification of system and its applications to modelling and control. *IEEE Trans. Syst. Man Cybern.* **1985**, *15*, 116–132. [CrossRef]
24. Yona, A.; Senjyu, T.; Funabashi, T.; Kim, C.H. Determination method of insolation prediction with fuzzy and applying neural network for long-term ahead PV power output correction. *IEEE Trans. Sustain. Energy* **2013**, *4*, 527–533. [CrossRef]
25. Jang, J. ANFIS: Adaptive-network-based fuzzy inference systems. *IEEE Trans. Syst. Man Cybern.* **1993**, *23*, 665–685. [CrossRef]
26. Yaïci, W.; Entchev, E. Adaptive Neuro-Fuzzy Inference System modelling for performance prediction of solar thermal energy system. *Renew. Energy* **2016**, *86*, 302–315. [CrossRef]
27. Solgi, A.; Nourani, V.; Pourhaghi, A. Forecasting daily precipitation using hybrid model of wavelet-artificial neural network and comparison with adaptive neurofuzzy inference system (case study: Verayneh station, Nahavand). *Adv. Civ. Eng.* **2014**, *2014*, 279368. [CrossRef]
28. Al-Hilfi, H.A.H.; Abu-Siada, A.; Shahnia, F. Combined ANFIS–Wavelet Technique to Improve the Estimation Accuracy of the Power Output of Neighboring PV Systems during Cloud Events. *Energies* **2020**, *13*, 1613. [CrossRef]
29. Osório, G.J.; Lotfi, M.; Shafie-khah, M.; Campos, V.M.A.; Catalão, J.P.S. Hybrid Forecasting Model for Short-Term Electricity Market Prices with Renewable Integration. *Sustainability* **2018**, *11*, 57. [CrossRef]
30. Seo, Y.; Kim, S.; Kisi, O.; Singh, V.P. Daily water level forecasting using wavelet decomposition and artificial intelligence techniques. *J. Hydrol.* **2015**, *520*, 224–243. [CrossRef]
31. Stefenon, S.F.; Ribeiro, M.H.D.M.; Nied, A.; Mariani, V.C.; Coelho, L.D.S.; da Rocha, D.F.M.; Grebogi, R.B.; Ruano, A.E.D.B. Wavelet group method of data handling for fault prediction in electrical power insulators. *Int. J. Electr. Power Energy Syst.* **2020**, *123*, 106269. [CrossRef]
32. Wu, L.; Cheng, J.Z.; Li, S.; Lei, B.; Wang, T.; Ni, D. FUIQA: Fetal ultrasound image quality assessment with deep convolutional networks. *IEEE Trans. Cybernet.* **2017**, *47*, 1336–1349. [CrossRef]
33. Lin, F.; Lu, K.; Ke, T. Probabilistic wavelet fuzzy neural network based reactive power control for grid-connected three-phase PV system during grid faults. *Renew. Energy* **2016**, *92*, 437–449. [CrossRef]
34. Mallat, S.G. A Theory for Multiresolution Signal Decomposition: The Wavelet Representation. *IEEE Trans. Pattern Anal. Mach. Intell.* **1989**, *11*, 674–693. [CrossRef]
35. Semero, Y.K.; Zhang, D.; Zhang, J. A PSO-ANFIS based Hybrid Approach for Short Term PV Power Prediction in Microgrids. *Electr. Power Compon. Syst.* **2018**, *46*, 95–103. [CrossRef]
36. Gao, W.; Moayedi, H.; Shahsavar, A. The feasibility of genetic programming and ANFIS in prediction energetic performance of a building integrated photovoltaic thermal (BIPVT) system. *Sol. Energy* **2019**, *183*, 293–305. [CrossRef]
37. De Jesús, D.A.R.; Mandal, P.; Velez-Reyes, M.; Chakraborty, S.; Senjyu, T. Data Fusion Based Hybrid Deep Neural Network Method for Solar PV Power Forecasting. In Proceedings of the 51st North American Power Symposium (NAPS), Wichita, KS, USA, 13–15 October 2019. [CrossRef]

energies

Article

Comparative Analysis of Combined Heating Systems Involving the Use of Renewable Energy for Greenhouse Heating

Chung-Geon Lee [1], La-Hoon Cho [2], Seok-Jun Kim [2], Sun-Yong Park [2] and Dae-Hyun Kim [3,*]

1 Agriculture and Life Sciences Research Institute, Kangwon National University, Hyoja 2 Dong 192-1, Chuncheon-si 24341, Korea; cndrjs9605@kangwon.ac.kr
2 Department of Interdisciplinary Program in Smart Agriculture, Kangwon National University, Hyoja 2 Dong 192-1, Chuncheon-si 24341, Korea; jjola1991@kangwon.ac.kr (L.-H.C.); ksj91@kangwon.ac.kr (S.-J.K.); psy0712@kangwon.ac.kr (S.-Y.P.)
3 Department of Biosystems Engineering, Kangwon National University, Hyoja 2 Dong 192-1, Chuncheon-si 24341, Korea
* Correspondence: daekim@kangwon.ac.kr; Tel.: +82-33-250-6496; Fax: +82-33-259-5561

Abstract: The continued use of fossil fuels is contributing to severe environmental pollution and the establishment of an abnormal climate; consequently, alternative renewable energy sources are being actively investigated worldwide. Further, following global trends, numerous countermeasures aimed at improving carbon neutrality, promoting sustainable agriculture, and reducing fossil fuel dependence are being implemented in the Republic of Korea. Therefore, this study was conducted to investigate the application of renewable energies for greenhouse heating in the Republic of Korea. Three hybrid systems, numbered 1–3, were constructed using a pellet boiler, hydrothermal heat pump, and solar heat collection system, respectively. Thereafter, the heating performance, combined heat efficiency, energy consumption per temperature lifting, and energy cost per temperature lifting of the systems were compared. The combined thermal efficiency results showed no significant differences. However, in terms of energy consumption and cost, hybrid system 1 demonstrated 25.7 and 24.1% savings, respectively, compared with the other systems. Moreover, based on economic analysis via the net present value and life cycle cost analysis methods, the system reduced costs by 29.2 and 27.7%, respectively, compared with conventional fossil fuel boilers. Thus, hybrid system 1 was identified as the most economical system.

Keywords: hydrothermal heat pump; pellet boiler; solar heat collection; hybrid heating system; greenhouse

Citation: Lee, C.-G.; Cho, L.-H.; Kim, S.-J.; Park, S.-Y.; Kim, D.-H. Comparative Analysis of Combined Heating Systems Involving the Use of Renewable Energy for Greenhouse Heating. *Energies* **2021**, *14*, 6603. https://doi.org/10.3390/en14206603

Academic Editors: Zbigniew Leonowicz, Michał Jasinski and Arsalan Najafi

Received: 16 September 2021
Accepted: 7 October 2021
Published: 13 October 2021

1. Introduction

Carbon emissions are a global concern, and countermeasures to combat this issue are being implemented globally. Numerous international treaties, such as the United Nations Framework Convention on Climate Change, Kyoto Protocol, and Paris Agreement, aim to reduce greenhouse gas emissions worldwide. Accordingly, most advanced countries have set targets to achieve carbon neutrality by 2050 [1–6]. The increased focus on carbon neutrality has prompted extensive research on new and renewable energy technologies. For example, to actively participate in pollution reduction and voluntarily reduce estimated carbon emissions by 37% by 2030, the Republic of Korea (ROK) announced the "2030 Greenhouse Gas Reduction Roadmap" in 2018 [7]. The roadmap included fixed targets for greenhouse gas emission reduction for each industrial sector. For the agricultural sector in particular, the target was to reduce total carbon emissions by 1%. The Energy Consumption Survey conducted in the ROK in 2017 revealed that the fossil fuel dependence of the agriculture, forestry, and fishery sectors was 97.5% [8]. Moreover, the use of fossil fuels in heating systems and in powering agricultural equipment accounted for 53.4% of all the energy consumption associated with the agricultural sector [9]. Accordingly, the

ROK is promoting the "Agricultural New and Renewable Energy Utilization Efficiency Project" to accelerate the utilization of new and renewable energy sources as well as energy-saving technologies.

In accordance with this project, heating systems using pellet boilers and geothermal heat pumps are being supplied to farms in the ROK. However, several post-installation problems are associated with these thermal energy supply systems. On the one hand, pellet boilers are difficult to maintain and have fluctuating operational costs because of variations in the cost of the wood pellets that are used as fuel [10–14]. Consequently, some farmers revert to using fossil fuel boilers. On the other hand, hydrothermal heat pumps are known to have high thermal efficiency and high heat source stability in the ROK (Republic of Korea) compared with those of air heat source pumps [15,16]. Specifically, the hydrothermal heat pumps that are supplied to farmers under the project primarily use groundwater as a heat source. However, the supply of heat, in this case, is unstable, as it depends on the amount of groundwater available [17,18]. Moreover, hydrothermal heat pumps require exorbitant initial investment costs.

Owing to these issues, farmers are hesitant to rely on alternative energy sources, and the utilization of new and renewable energy in greenhouses has remained unsuccessful.

First of all, domestic and foreign papers that applied pellet boilers to greenhouse heating include the following. Carlo Bibbiani et al. (2016) [12] judged the applicability of boilers using fossil fuels and wood fuels for greenhouse heating. In the Italian peninsula, the electric power energy load was estimated at 30 W/m^2 (south) to 175 W/m^2 (northern), but it was estimated to be 75,362.4–1,967,796 kW/m^2·yr depending on the outside temperature. The flue gas produced by the boiler contains a large amount of CO_2, so recycling it has the advantage of leading to increased plant production, but biomass boilers emit more NO_x, SO_x, VOC, PM, and ash than fossil fuels. There are disadvantages. When using a scrubber and flue gas control device, it is possible to use a more stable wood pellet boiler, and considering this, it is highly valuable to replace with a wood biomass boiler in the case of 5–100 €/m^2 depending on the greenhouse area. Kang et al. (2013) [19] designed and manufactured a wood pellet boiler to obtain basic data for practical application of a wood pellet boiler system for greenhouse heating. In order to estimate the heating efficiency according to the change in heat capacity of 75,000, 100,000, and 120,000 kcal/h, the heating efficiency test was performed by controlling the amount of air flowing into the wood pellets and burner. The thermal efficiencies of 75,000, 100,000, and 120,000 kcal/h were 80.2%, 84.2%, and 81.6%, respectively, and the highest thermal efficiency was reported at 100,000 kcal/h.

Second, there are countless papers on greenhouse heating using a heat pump, but the following studies using solar heat as an auxiliary heat source are typical for heating systems that form a heat pump and a hybrid system. Hassanien et al. (2018) [20] studied a heat pump system using a vacuum tube type solar collector as an auxiliary heat source for greenhouse heating. The internal temperature for heating was set at 14 °C, and it was possible to cover 62%, 40%, and 78% of the required heating load in October, March, and April, respectively. In addition, the thermal efficiency of the vacuum tube type solar collector was 0.49 and the COP of the heat pump was 4.24. However, in January, the required heating load could not be fully met, and it was judged that it was due to heat loss in the thermal storage tank. Kwon et al. (2013) [21] developed a system that improves the performance of a heat pump by selectively using solar surplus heat and external air heat in the greenhouse as heat sources, and reduces carbon dioxide fertilization costs by delaying greenhouse ventilation. Using this system, the heating performance coefficient in the internal circulation mode was about 3.35, which was improved compared to 2.46 in the night external circulation mode and 2.67 in the daytime external circulation mode. However, as the greenhouse was operated without ventilation, the light transmittance was only 62% due to excessive moisture and moisture condensation. In a horticultural environment, light transmittance is the most important factor. It was mentioned that the part showing the effect of decreasing light transmittance is considered to have room for improvement. As such, research on a heat pump system using a pellet boiler or solar heat as an auxiliary heat source has been conducted in Korea and around the world. However, it was limited to judging the suitability or application of greenhouse heating using renewable energy. Therefore, this study was intended to study the continuous use of greenhouse heating using renewable energy or a new hybrid system.

The aforementioned solar thermal system acts as an excellent heat source under moderate climatic conditions [22]; however, its application in agriculture is challenging because of the limited land area that can be used for agriculture in the ROK. In addition, the economic feasibility of applying the solar heat collection system to agriculture without government subsidies in Korea is limited [23]. Additionally, studies on the application of solar heat in agriculture in the ROK are scarce; information regarding its performance and economic properties as a renewable energy source in agriculture is lacking.

Therefore, to ensure the use of renewable energy for heating greenhouses in ROK, this study was conducted to construct hybrid systems using new and renewable energies and to test the applicability of these systems in agriculture in the ROK. Further, based on multilateral comparative analyses and economic analyses of greenhouse heating, the most suitable system was identified.

2. Materials and Methods

2.1. Greenhouse Design

A glass greenhouse with an area of approximately 90 m^2 at Kangwon University in Chuncheon-si, Gangwon-do, Korea was selected in this study. This greenhouse, which is equipped with an insulating curtain as well as ventilation facilities, features a double structure with an additional internal greenhouse. The experiments were conducted with the internal greenhouse closed. The floor area of the internal greenhouse and the covering area was 68.37 and 121.44 m^2, respectively. The heating load of the experimental system was calculated based on these areas, and the capacity of each piece of equipment used in the heating system was also selected in consideration of these areas (Figure 1).

(a)

Figure 1. *Cont.*

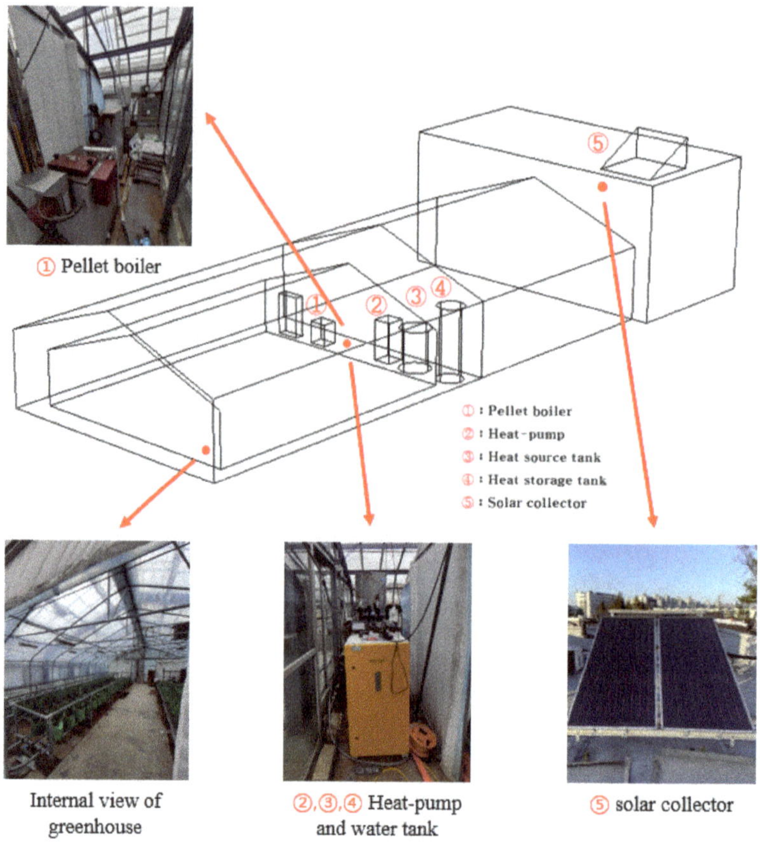

(b)

Figure 1. Greenhouse overview. (**a**) Photo of experimental greenhouse, and (**b**) Heating device installation overview.

2.2. Thermal Energy Supply Systems

The thermal energy supply systems tested in the greenhouse included a 20,000 kcal/h-class pellet boiler (KN-23D, Kyuwon Tech, Gyeongsan, Korea), a 3RT-class hydrothermal heat pump (3RT, Inergy Technologies, Gwangsan-gu, Korea), and a 4.04 m^2 solar heat collecting plate with a heat collection capacity of 2230 kcal/m^2·day. Three hybrid systems were designed using these systems. The thermal energy supply device comprised a fan coil unit and a tube rail. Hybrid system 1 consisted of a hydrothermal heat pump system with a pellet boiler as the heat source. In hybrid system 2, a pellet boiler was used as the main heat source, and a solar collector was used as an auxiliary heat source for the heat storage tank. Hybrid system 3 consisted of a heat pump system that included all the heating systems, with a pellet boiler and a solar panel as the main heat source and the auxiliary heat source, respectively. The combined heating system is as shown in Figure 2.

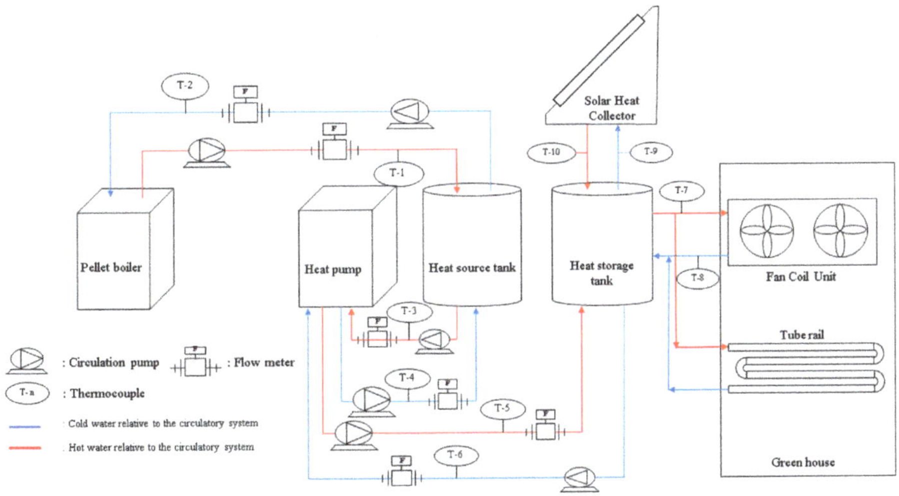

Figure 2. Schematic diagram of the combined heating system.

2.3. Data Measurement

In this study, a thermocouple (GTPK-02-17, GILTRON, Seoul, Korea), sensor-type flowmeter (VVX25, SIKA Dr. Siebert & Kore GmbH & Co., Kaufungen, Germany), and turbine-type flowmeter (VTH40, SIKA Dr. Siebert & KGmb GmbH & Co., Baden-Wüttemberg, Germany) were used for data collection. The data were recorded using a data logger (GL840, GRAPHTEC Co., Tokyo, Japan), and the heat energy transfer amount was calculated by measuring the temperature and flow rate in each closed-loop system. A load-cell scale (HPS-300A, CAS Co., Seoul, Korea), an integrated watt-hour meter (LD3410DRM-080, LS ELECTRIC Co., Seoul, Korea), and a solar radiation meter (Li-200R LI-COR, Inc., Lincoln, NE, USA) were used to determine the amount of input energy. For the pellet boiler, hydrothermal heat pump, and solar heating system, the input energy was the consumption of the pellets as fuel, power consumption by the compressor, and the amount of collected solar heat, respectively. The amount of input energy was measured before heating started at 17:00, which was the standard time when the experiments commenced each day. In the experimental groups involving solar heat, the input energy was measured every day from 15:00 to noon the next day, excluding three hours (12:00–15:00) for solar heat storage. To calculate the coefficient of performance (COP) of the heat pump, the amount of instantaneous power was measured using a power analyzer (DW-6092, LUTRON ELECTRONIC ENTERPRISE C, Taipei, Taiwan).

2.4. Auto Control System

The automatic control system used in this study was the Farmos program (JINONG Co., Ltd., An-yang, Korea). This automatic control system allows communication between mobile devices and PCs, making it possible to check the operation status of various actuators, including the main heating pump, and to set manual and automatic operations. The control logic was configured such that the main heating pump and heat source supply pump could detect the temperature of the heat storage tank and the heat source tank, and accordingly set the lower and upper temperature ranges required to attain the required temperature range.

2.5. Overview of Experiments

Experiments using hybrid system 1 were conducted from 2 to 4 March 2020, and experiments using hybrid systems 2 and 3 were conducted from 12 to 17 November 2020. The heating water temperature was 55 °C, which is the maximum discharge temperature of the heat pump that was used in the experiment. In hybrid systems 1 and 3, the heat source temperature was set to 20–25 °C. The reference heating temperature of the greenhouse was based on melon, which requires a nighttime growth temperature of 18–22 °C and is considered a high-temperature crop compared with other crops that are grown in Korea. Further, all the experiments were conducted via automatic control based on the control logic of the designed test method and were based on the on/off control of the thermal energy supply device.

2.6. Experimental Methods

2.6.1. Hybrid System 1

The overview and control logic corresponding to hybrid system 1 are shown in Figures 3 and 4, respectively. Specifically, the operation of hybrid system 1, which enabled the detection of the temperature of the heat source tank, heat storage tank, and greenhouse, began after the set temperatures of the heat source tank and the heat storage tank were achieved and maintained before 17:00. The detected room temperature was used as input for the control logic, which determines the presence or absence of heating. To heat the greenhouse, energy was released from the heat storage tank; hence, its temperature decreased. The operation of the hydrothermal heat pump supplemented the heat from the heat storage tank based on the measured temperature of the heat storage tank. Once the temperature of the heat source tank reached the lower limit of the temperature range of the heat source, the pellet boiler was then utilized as a heat source and was operated by controlling it between the circulation pump and the heat source tank. Thus, the control logic was implemented during the realization of the experiment. Further, the daily experiments were concluded at the same time as that of the energy input check, which was performed at 17:00 the next day.

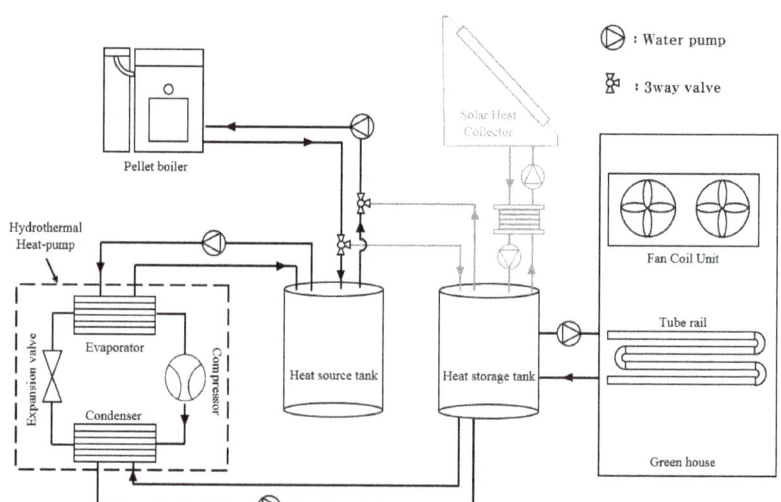

Figure 3. Schematic of hybrid system 1.

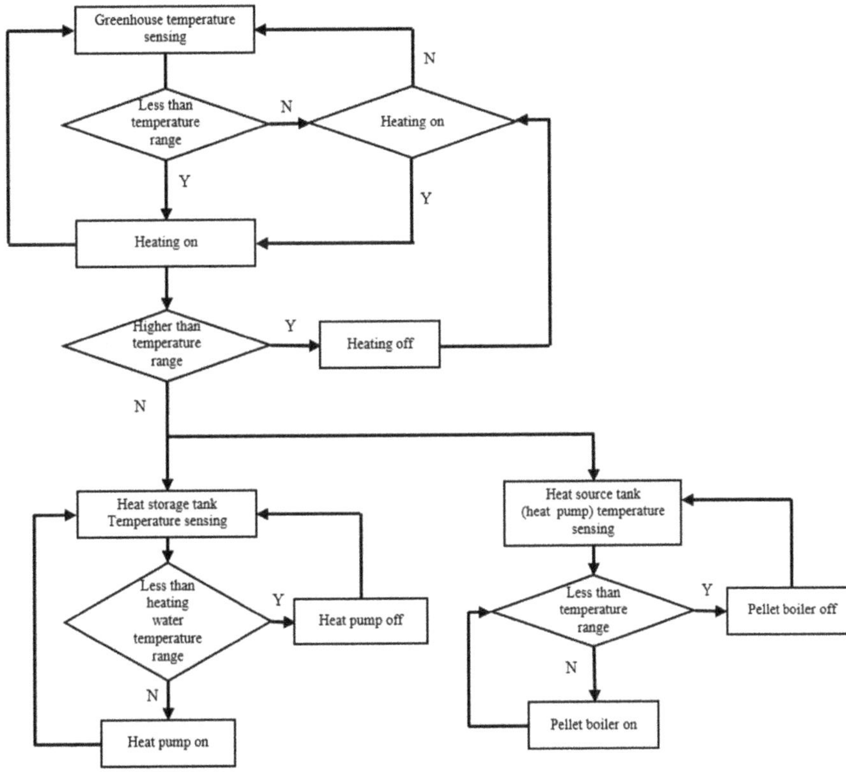

Figure 4. Control logic of hybrid system 1.

2.6.2. Hybrid System 2

The overview and control logic of hybrid system 2 are shown in Figures 5 and 6, respectively. The experiment involving hybrid system 2 commenced at sunrise with the determination of the solar radiation. When the temperature of the circulating water inside the collector reached approximately 60–70 °C, the circulation pump between the heat exchanger and heat storage tank was operated to collect solar heat. This process continued until approximately 15:00. Further, the system continuously checked whether the heating temperature varied within the fixed range by measuring the temperature of the internal greenhouse. Heating was activated when the ambient temperature fell below the lower limit of the range. Continuous heating was performed up to the upper limit temperature of the set temperature range, and once it was determined that heating was not required, the system ceased heating activity. The collected solar heat served as an auxiliary heat source and supplemented the heat lost from the heat storage tank during the day. Additionally, the collected solar heat was used to increase the temperature of the heat storage tank to more than 55 °C, which represents the standard temperature for heating water, when the solar radiation intensity was high.

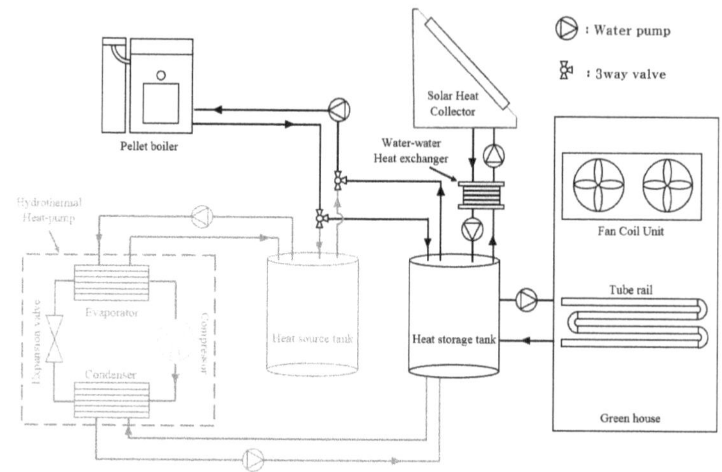

Figure 5. Schematic of hybrid system 2.

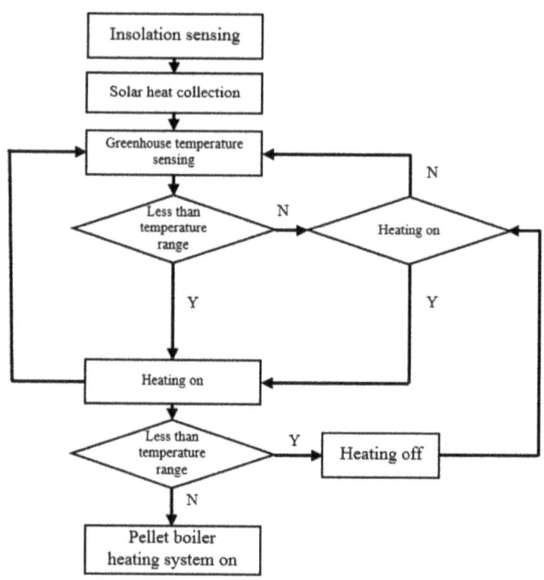

Figure 6. Control logic of hybrid system 2.

2.6.3. Hybrid System 3

The overview and control logic of hybrid system 3 are shown in Figures 7 and 8, respectively. The operation of hybrid system 3 was based on the experimental method corresponding to hybrid system 1, with an additional solar thermal collection system. In this system, heat storage was initiated when more than a certain amount of solar radiation was detected after sunrise. Further, the heat storage process lasted from approximately 12:00 to 15:00, and once the temperature fell below the required value at 17:00, heating was initiated, with the stored solar heat utilized first. The subsequent methodology was similar to that corresponding to hybrid system 1.

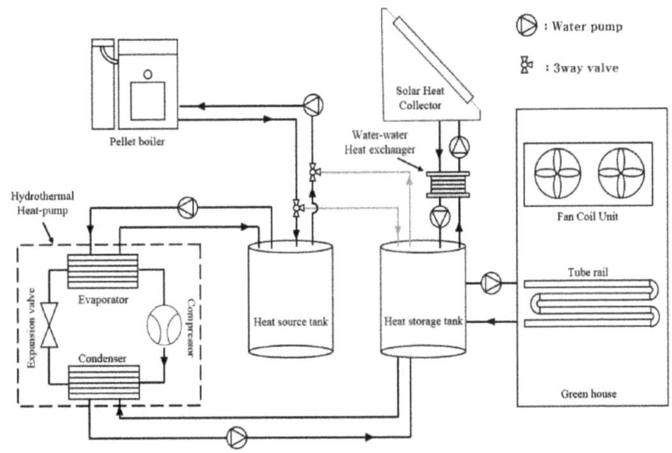

Figure 7. Schematic of hybrid system 3.

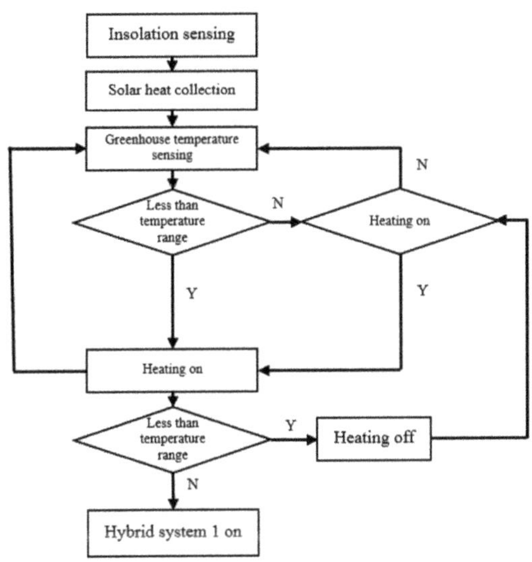

Figure 8. Control logic of hybrid system 3.

2.7. Analysis

2.7.1. Thermal Energy Calculation

Heat transfer was calculated to obtain the thermal efficiency of the different hybrid systems. The amount of heat transferred was calculated using the temperature difference (ΔT), flow rate (m), and specific heat capacity of water (C_p) obtained from each closed-loop system. The calculation was performed according to Equation (1).

$$Q = \Delta T \times m \times C_p \times 3600, \quad (1)$$

where Q represents the total caloric energy (kcal/h), ΔT represents the temperature difference in in/out water (°C), m represents mass flow (kg/s), and C_p represents the specific heat capacity of water (kcal/kg·°C).

2.7.2. Determination of the Coefficient of Performance of the Heat Pump

The unit coefficient of performance (COP) was calculated by dividing the amount of heat transferred (kcal/h) from the heat pump to the heat storage tank (Q_{hst}, $Q_{\text{heat storage tank}}$) because of compressor power consumption (P_{cpc}, $P_{\text{compressor power consumption}}$). The calculation was conducted by converting the amount of heat transferred into kilowatts-per-hour or converting the power consumption of the compressor into kilocalories-per-hour.

$$\text{COP} = \frac{Q_{hst}}{P_{cpc}} \tag{2}$$

2.7.3. Energy Consumption and Cost Analysis

Energy consumption and energy cost are essential factors for the comparative analysis of the developed hybrid systems. Energy consumption was also used to calculate the efficiency of the combined thermal systems. The energy consumption of the individual systems was determined as pellet consumption, power consumption, and solar heat collection. Specifically, pellet consumption was calculated by multiplying the daily consumption amount (kg), which was obtained using a scale (HPS-300A, CAS Co., Ltd., Yang Ju, Korea), by the low-level heating amount (kcal/kg) of the pellet. Power consumption (kcal/h) was calculated by multiplying the amount of electricity measured using a watt-hour meter with a unit conversion factor, and solar heat collection was calculated by multiplying the solar heat collector efficiency by the solar radiation measured using an insolation meter (Equation (3)).

$$E_{total} = (\text{Pellet consumption} \times \text{LHV}) + (\text{Power consumption} \times \text{UCF}) + (\text{insolation} \times \eta), \tag{3}$$

where E_{total} represents total energy consumption (kcal), LHV represents the lower heating value (kcal/kg), UCF is the unit conversion factor (power to calories), and η represents solar collector efficiency.

The energy consumption cost was calculated based on pellet and power consumption. The standard price per unit energy was based on the wood pellet unit price (0.31 USD/kg) as announced by the Korea Forest Service in June 2019 and the Korea Electric Power Corporation (KEPCO) electricity bill calculation table (0.042 USD/kWh). The energy consumption cost, based on energy consumption and energy cost, was calculated according to Equation (4).

$$EP_{total} = (\text{Pellet consumption} \times \text{PP}) + (\text{Power consumption} \times \text{EC}), \tag{4}$$

where EP_{total} represents energy consumption cost, (USD, \$), PP represents pellet cost, (USD/kg), and EC represents electricity charge (USD/kWh).

2.7.4. Combined Thermal Efficiency Analysis

The combined thermal efficiencies of the different hybrid systems were calculated to compare and identify the system with optimal thermal efficiency. The calculations were performed according to Equation (5), which considers the total energy consumption corresponding to each system (E_{input}) and the amount of heat transferred to the heat storage tank (E_{ouput}).

$$\eta_{combined} = \frac{E_{output}}{E_{input}}, \tag{5}$$

where E_{output} represents energy transferred to the heat storage tank, and E_{input} represents the input energy (pellet consumption, power consumption, and solar heat collection).

2.8. Economic Analysis

2.8.1. Heating Load Calculation

The total cost incurred by each system over 10 years can be predicted by dividing the total energy required for 10 years, which was calculated based on the heating load, by the energy cost per unit energy. Specifically, the heating load was calculated according to Equation (6), while the cover area heat flux, ventilation area heat flux, and floor area heat flux were calculated using Equations (7)–(9), respectively [24].

$$Qg = A_g \times [qt + qv] + A_s \times qs \times fw \tag{6}$$

$$qt = ht \times (T_s - T_a) \times (1 - fr) \tag{7}$$

$$qv = hv \times (T_s - T_a) \tag{8}$$

$$qs = hs \times (T_s - T_a) \tag{9}$$

The different parameters in these equations are defined in Table 1.

Table 1. Factors required for the calculation of the heating load.

Parameter	Definition		Unit
Qg	Maximum heating load		kcal/h
Ag	Covered area of a greenhouse		m^2
As	Floor area of a greenhouse		m^2
qt	Transmission heat load for unit covered area of the greenhouse		kcal/m^2
qv	Ventilation heat load for unit covered area of the greenhouse		kcal/m^2
qs	Ground heat transfer load for unit covered area of the greenhouse		kcal/m^2
fw	Correction factor according to wind velocity		-
ht	Thermal perfusion ratio	4.50	kcal/m^2·h·°C
Ts	Greenhouse inside set temperature		°C
Ta	Set ambient air temperature		°C
Tg	Ground temperature		°C
Sf	Safety factor	1.2	-
fr	Reduction rate of heating cover	0.71 (Triple screen)	-
hv	Ventilation heat coefficient	0.2	kcal/m^2·h·°C
hs	Ground heat transfer coefficients	0.244	kcal/m^2·h·°C

The total energy consumption could be calculated by multiplying the maximum daily heating load by the 10-year durability period of the device as shown in Equation (10); a 12 h non-sunlight period was assumed.

$$\text{TEC (kcal)} = \text{DAHL} \times \text{daily data excluding July and August} \times 12\text{ h} \times 10\text{ year}, \tag{10}$$

where TEC implies total energy consumption and DAHL implies daily average heating load.

2.8.2. Economic Analysis (Net Present Value)

The net present value method was used for the comparative economic analysis of the developed hybrid systems. In this regard, the cultivation area was assumed to be 1000 m^2, and the factors considered in the net present value method of economic analysis included the durability life of the device, initial investment cost, interest rate, operating cost, and depreciation amount. Further, the initial investment cost was analyzed in two parts based on the total project cost (IIC) and the actual required project cost (IIC_{self}) borne by the farmers according to the Renewable Energy Use Efficiency Project conducted in Korea. The initial investment cost calculation method using the present value method was expressed as shown in Equation (11).

$$\mathrm{TPW} = \mathrm{IIC} \times \mathrm{CRF} \times \mathrm{DP}, \tag{11}$$

where TPW implies total present worth, IIC implies initial investment costs, CRF implies capital recovery factor, and DP implies durability period.

The capital recovery factor (CRF) used in the net present value method was based on the straight-line depreciation method, and the cash flow was assumed to follow the same trend. The resulting CRF was calculated according to Equation (12), and the interest rate was calculated according to Equation (13) based on a nominal interest rate (2%).

$$\mathrm{CRF} = \frac{i \times (1+i)^n}{(1+i)^n - 1} \tag{12}$$

where i represents the nominal interest rate and n represents the applicable year.

$$i = [(1+\mathrm{r}) \times (1+\mathrm{p})] - 1 \tag{13}$$

where i represents the nominal interest rate, r represents the real interest rate, and p represents the inflation rate.

Interest expenses, income tax, and annual operating expenses were calculated using the nominal interest rate, and the total expenses incurred during the durability period, including initial investment expenses, were calculated using Equation (14) [25].

$$TC_{10yr} = 0\mathrm{TPW} + \left[\left\{\sum_{n=1}^{10} DR \times (1+i)^n + \sum_{n=1}^{10} IT \times (1+i)^n + \sum_{n=1}^{10} AE \times (1+i)^n\right\} \times 10\right] \tag{14}$$

where TC_{10yr} represents total cost for 10 years, DR represents debt return, IT represents income tax, AE represents annual expenses, i represents nominal interest rate, and n represents the applicable year.

The cost per unit energy for each system was determined using the total cost for 10 years calculated above and the total energy required for 10 years calculated using Equation (10). Subsequently, Equation (15), which was used to calculate energy cost (EC), was derived as follows:

$$\mathrm{EC} = \frac{TC_{10yr}}{TEC}. \tag{15}$$

2.8.3. Economic Analysis (Life Cycle Cost)

The life cycle cost analysis method offers the possibility to calculate the total cost incurred during the life cycle of a device. The components considered for the life cycle cost analysis included initial investment cost, considering self-pay, maintenance and repair cost, and total fuel cost, which were then multiplied by the hourly fuel cost for 10 years of heating time required for the complex heating system. The formula that was used to calculate the life cycle cost based on the abovementioned factors was expressed as shown in Equation (16).

$$\mathrm{LCC} = \mathrm{IIC}_{self} + (\mathrm{IIC} \times MCR) + [\frac{TEC}{Q_g} \times \left(Fuel\ cost \times \frac{Q_g}{LHV\ with\ used\ fuelratio}\right)], \tag{16}$$

where MCR implies maintenance cost ratio.

The values of the different factors that were used in the economic analysis are listed in Table 2. The Table 2 is shown the initial investment cost & installation cost for pellet boiler (KN-23D, KYUWON Co., Gyeong-san, Korea) and a heat pump (COMPORT-A-03, Innergie Technologies Inc., Gwang-ju, Korea) and solar collector (KNSC-003, KANGNAM Co., Kwang-ju, Korea) recommended by manufacturers.

Table 2. Factors used in economic analysis.

Division	Hybrid 1 (Pellet Boiler/ Heat Pump)	Hybrid 2 (Pellet Boiler/ Solar Collector)	Hybrid 3 (Pellet Boiler/ Heat Pump/ Solar Collector)
Generating capacity	100,000 kcal/h (1 EA)/ 15 RT (2 EA)	100,000 kcal/h (1 EA)/ 2231 kcal/m^2 day (40 EA)	50,000 kcal/h (1 EA)/ 30 RT (1 EA)/ 2231 kcal/m^2 day (40 EA)
Initial investment cost total cost (USD, $)	47,817	64,332	45,795
Initial investment cost self-pay (USD, $)	12,386	28,227	28,901
Debt ratio (%)	13	9.6	13
Interest rate on agricultural loans when using renewable energy		Fixed interest rate of 2%	
Annual average inflation (%)		2%	
Effective income tax rate (%)		15%	
Durability period (yr)		10	
Depreciation method		Straight-line	

2.9. *Statistical Analysis*

Statistical analysis was conducted to confirm the significance of the comparative analysis between the experimental groups. The statistical program, SAS v9.4 (SAS Institute, Inc., Cary, NC, USA), was used, and the analysis was performed using Duncan's multiple range test. Given that the heating experiments were conducted in connection with external weather, repeating them was challenging. Therefore, the effective data obtained during the experiments were assumed to be a single data sample per day.

3. Results

3.1. *Experiment Schedule and Heating Performance Comparison*

Each experiment was conducted for three days. The weather data during this experimental period are listed in Table 3.

Table 3. Weather data during experimental period.

Experiment	Period	Temperature (°C)		Wind Speed (m/s)		Humidity (g/m^3)		Insolation (MJ/m^2)	
		Max	Min	Max	Min	Max	Min	Max	Min
Hybrid 1 55 °C	2020-03-02~2020-03-04	14.5	−1.3	4.9	0.0	5.8	2.5	2.7	1.85
Hybrid 2 55 °C	2020-11-15~2020-11-17	22.3	5.7	2.3	0.0	9.6	5.4	1.7	1.3
Hybrid 3 55 °C	2020-11-12~2020-11-14	25.6	5.5	2.7	0.0	7.6	3.9	1.9	1.8

The results of the comparison of the heating performances of the hybrid systems are listed in Table 4. Overall, the results satisfied the heating temperature setting range. However, hybrid system 1 showed the highest temperature increase, even though the outside temperatures of hybrid systems 2 and 3 were higher.

Table 4. Summary of experimental results.

System at 55 °C	Average Ambient Temperature for Heating Period (°C)	Lifting Temperature Average (°C)	Average Greenhouse Temperature (°C)
Hybrid 1	5.8	16.06	20.76
Hybrid 2	11.2	10.83	21.41
Hybrid 3	12.7	13.43	21.97

The variation of the temperature of the greenhouse under the different heating systems is shown in Figure 9, from which it is evident that when the outdoor temperature decreased, the indoor temperature also decreased. Further, an abnormal state with severe fluctuations in room temperature, which was considered to occur because the system was heated above the set heating temperature via simple on/off control, was observed during the experiment. In hybrid systems 2 and 3, in which solar heat was used, the indoor temperature decreased with the outside temperature, despite the relatively high outside temperature. Additionally, solar heat increased the temperature of the heat storage tank to 57 °C, but could not go above the reference temperature of 55 °C. This phenomenon was attributed to the relatively small amount of insolation due to seasonal characteristics. Consequently, the necessity for additional solar heat collection facilities was established. Although the hybrid systems showed excellent heating capacities, they suffered from some unwanted phenomena, including seasonal effects or failure to maintain the temperature of the heat storage tank evenly when solar heat was being used. However, as the heating capacities in the experiments involving the use of solar heat were generally lower than that corresponding to hybrid system 1, the application of the solar heat collection system changed the temperature of the heat storage tank, thereby reducing heating performance and decreasing the indoor temperature.

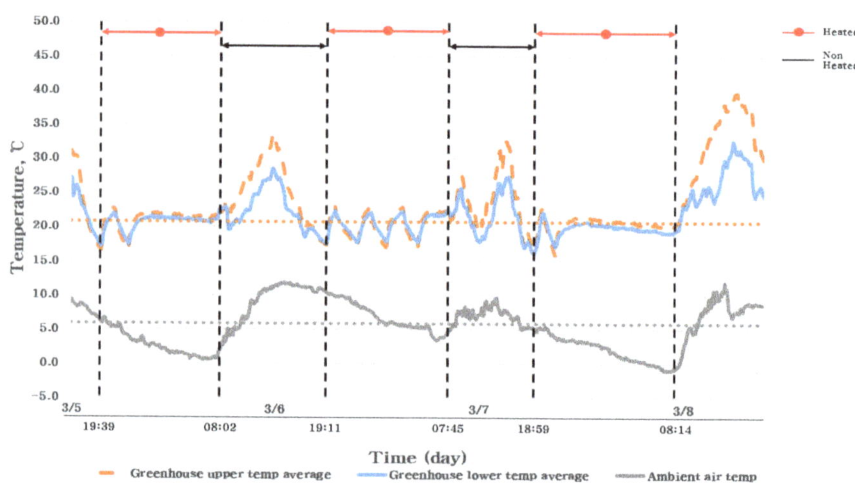

(a)

Figure 9. *Cont.*

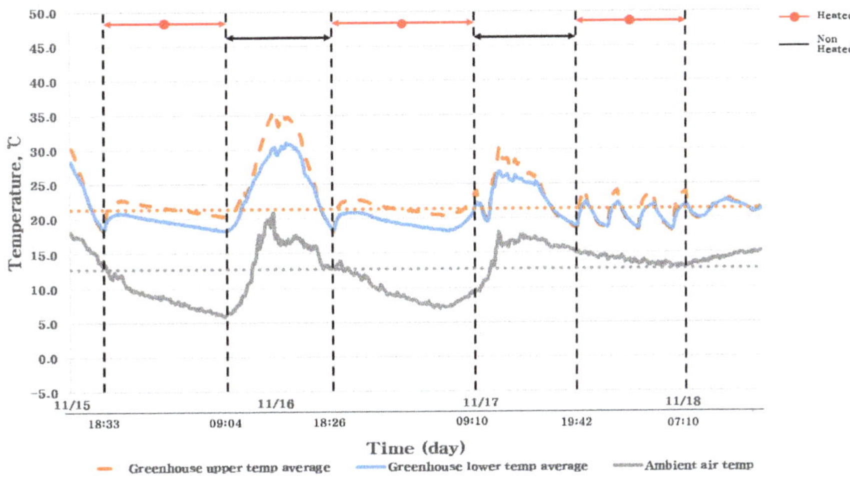

(b)

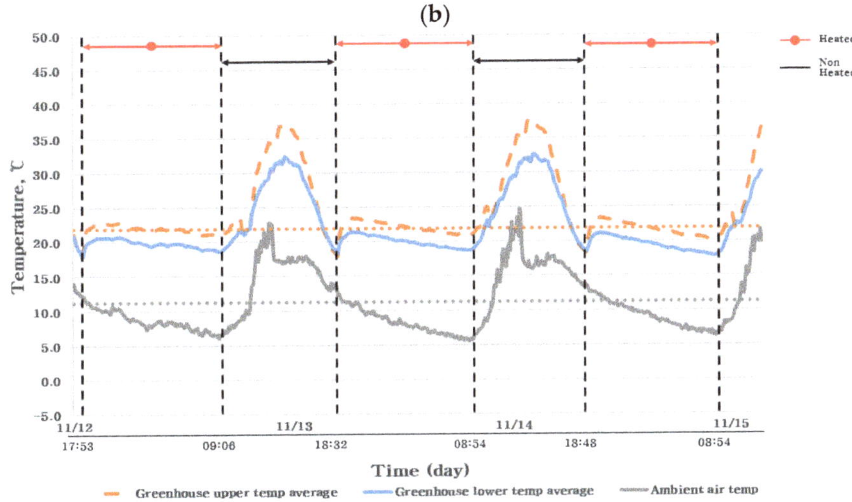

(c)

Figure 9. Experimental results showing variation of greenhouse temperature. (**a**) Hybrid system 1, (**b**) hybrid system 2, and (**c**) hybrid system 3, all at a standard temperature of 55 °C.

3.2. Results of Combined Thermal Efficiency

A comparison of the combined thermal efficiencies is shown in Figure 10 and Tables 5 and 6. The thermal efficiency was highest at 68.1% in hybrid system 3. Considering hybrid systems 1 and 3, which involved the use of the hydrothermal heat pump, the average and maximum COP were 2.73 and 3.47 and 2.29 and 3.16, respectively. For hybrid system 2, which involved the use of the heat pump, the average COP was 2 because of the partial load operation of the inverter using the PID control that was built into the heat pump used in this study. Further, statistical analysis showed that all three experimental groups, which did not show any significant differences, could be classified under group A.

Table 5. Comparative analysis combined thermal efficiency.

Device at 55 °C	Input Calorie (kcal/h)	Heating Calorie (kcal/h)	Combined Thermal Efficiency (%)	Combined Thermal Efficiency (%), Average max/min		Average COP/ Maximum COP
Hybrid system 1	16,075	10,679	66.4	63.8		2.29/3.16
	18,640	10,812	58.0			
	15,016	10,076	67.1	67.1	58.0	
Hybrid system 2	12,749	8485	69.5	59.1		-
	13,445	8036	62.8			
	21,960	9893	45.1	69.5	45.1	
Hybrid system 3	12,286	8703	70.8	68.1		2.73/3.47
	12,728	8811	69.2			
	13,535	8717	64.4	70.8	64.4	

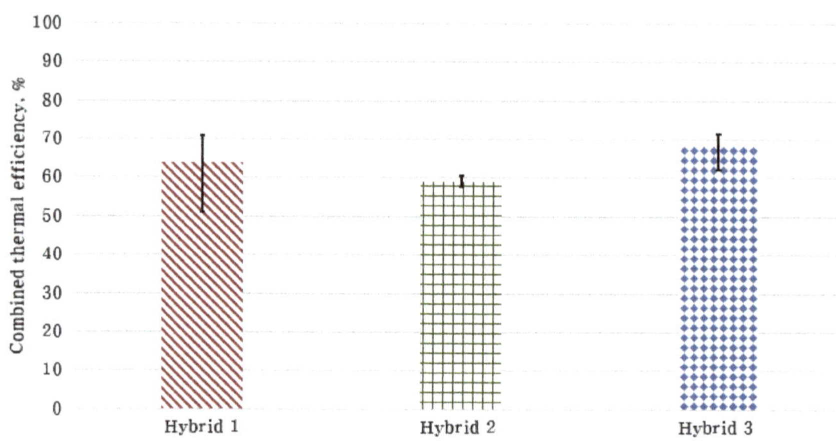

Figure 10. Comparison results of combined thermal efficiency.

Table 6. Statistical analysis results of combined thermal efficiency.

Duncan Grouping	Mean	N	System at 55 °C
A	68.1	3	Hybrid system 3
A	63.8	3	Hybrid system 1
A	59.1	3	Hybrid system 2

3.3. *Variation of Energy Consumption with Increase in Temperature*

A comparison of the energy consumption of the different hybrid systems as a function of increasing temperature is shown in Figure 11 and Table 7. The aim of studying these effects was to convert the energy consumption into a unit for comparison so as to recur any uncertainties arising from the fact that the different groups have different experimental dates. The comparison results revealed that hybrid system 1 consumed the lowest amount of energy, followed by hybrid systems 3 and lastly hybrid system 2. These results could be attributed to the proportion of power usage associated with the use of heat pumps in hybrid systems 1 and 3. For hybrid system 2, which involved the use of pellets and solar heat, the energy consumption with increasing temperature was expected to be the highest because of the high dependence on the pellets. Additionally, the energy-saving effect of hybrid system 1 was the greatest, as statistical analysis showed that this system had the lowest energy consumption. Thus, hybrid system 1 and the other experimental groups were classified under different groups and showed significant differences in this regard.

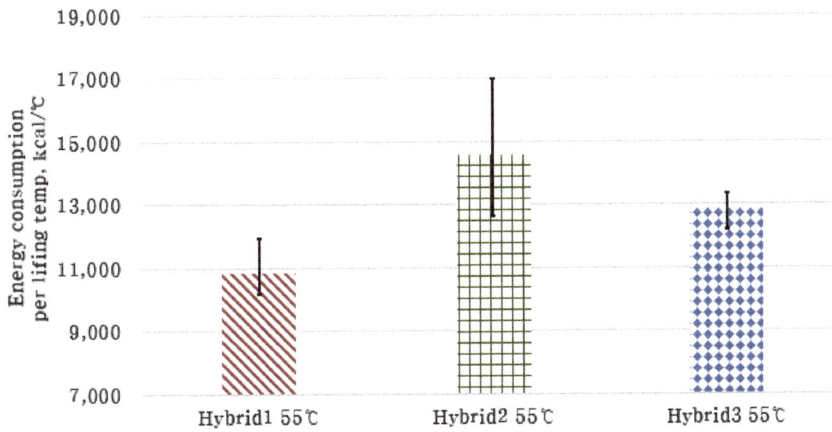

Figure 11. Comparison of energy consumption as a function of increasing temperature.

Table 7. Results of the statistical analysis of energy consumption as a function of increasing temperature.

Duncan Grouping	Mean	N	System at 55 °C
A	14,584.0	3	Hybrid 2
B	12,849.7	3	Hybrid 3
C	10,830.8	3	Hybrid 1

3.4. Results Corresponding to the Variation of Energy Cost with Increasing Temperature

A comparison of the energy cost with increasing temperature is shown in Figure 12 and Table 8. Hybrid system 1 exhibited the lowest application energy cost at 0.86 USD/h, whereas hybrid systems 3 and 2 showed higher energy costs at 0.98 and 1.13 USD/h, respectively. Unlike energy consumption, the energy cost was obtained by converting energy consumption. However, the cost served as a correction value; thus, it differed from energy consumption. Statistical analysis showed that hybrid system 2 could be classified under group A, while hybrid systems 1 and 3 could be classified under group B. The energy consumption cost corresponding to hybrid system 1 was significantly lower than that corresponding to hybrid system 2 but showed no significant difference compared with that corresponding to hybrid system 3. Additionally, the energy consumption and used energy cost tended to be similar. For the sake of comparison, the ratio of the energy consumption in each system is shown in Table 9.

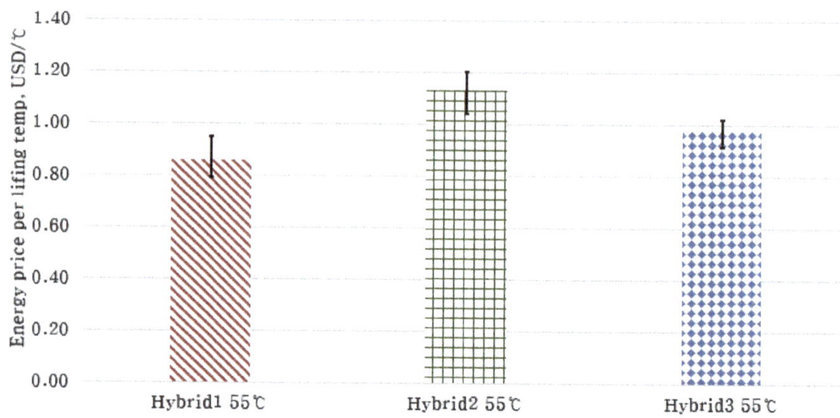

Figure 12. Comparison of average cost of fuel consumption with increasing temperature.

Table 8. Statistical analysis results corresponding to the variation of energy consumption with increasing temperature.

Duncan Grouping	Mean	N	System at 55 °C
A	1.13	3	Hybrid 2
B	0.98	3	Hybrid 3
B	0.86	3	Hybrid 1

Table 9. Used fuel ratio of hybrid systems 1, 2, and 3.

	Hybrid System 1 (Pellet:Power) (%)	Hybrid System 2 (Pellet:Solar) (%)	Hybrid System 3 (Pellet:Power:Solar) (%)
Day 1 experiment	83.3:16.7	95.7:4.3	82.8:12.3:4.9
Day 2 experiment	79.9:20.1	95.1:4.9	82.7:12.2:5.3
Day 3 experiment	84.4:15.6	96.0:4.0	78.5:18.3:3.4
Average ratio	82.5:17.5	95.6:4.4	81.1:14.2:4.6

The energy consumption ratio revealed that no significant difference existed between hybrid systems 1 and 3 with respect to pellet consumption; the difference in power consumption due to solar heat was approximately 3.3% on average. Further, when comparing hybrid systems 2 and 3, the amount of solar heat used was 4.5–4.6%, showing no significant difference. Thus, higher power consumption led to lower energy consumption and cost. The absolute energy consumption figures corresponding to hybrid systems 2 and 3 increased because of the use of solar heat; however, considering the three systems, hybrid system 1 showed superior performance in terms of energy consumption and cost.

3.5. Results of Comparative Economic Analysis (Net Present Value)

For a comparative analysis of economic feasibility, the developed hybrid systems were compared with conventional fossil fuel boilers (kerosene and diesel boilers). Additionally, the results obtained when only self-pay was considered when applying the government subsidy program that is currently in place in the ROK was compared with that based on the total project cost. Further, the results of the comparative analysis of economic feasibility using the net present value method are presented in Table 10.

Table 10. Results of economic analysis based on the net present value method for each system.

Division	Kerosene Boiler	Diesel Boiler	Hybrid 1 (P.B + H.P)	Hybrid 2 (P.B + S.C)	Hybrid 3 (P.B + H.P + S.C)
Initial investment cost (USD, $)	5055	5055	47,817	64,332	45,795
Debt ratio (%)	0	0	13	13	9.6
Initial investment cost—Self-pay (USD, $)	5055	5055	12,386	28,227	28,901
Debt ratio (%)	0	0	50	20.7	21.4
Capital recovery factor			0.12		
TC_{10yr} of total investment cost (USD, $)/10 years)	196,139.79	209,114.17	183,879.34	184,834.68	201,293.06
TC_{10yr} of self-pay (USD, $)/10 years)	196,139.79	209,114.17	139,416.16	163,957.79	156,836.28
TEC (Mcal/10 years)			1,511,842		
EC total investment cost (KRW/Mcal)	0.129	0.138	0.121	0.122	0.133
EC self-pay (KRW/Mcal)	0.129	0.138	0.092	0.108	0.103

Regarding the total project cost, hybrid systems 1 and 2 showed an energy cost reduction effect of approximately 5.8–6.5% compared to that of the kerosene boiler, whereas hybrid 3 was less economical than the standard kerosene boiler. Furthermore, hybrid systems 1 and 2 exhibited no significant differences. However, when implementing self-payment by applying the government subsidy project in Korea, hybrid system 1 led to 29.2% cost savings compared with that of kerosene boilers, showing the highest economic feasibility. In this regard, all systems built in this study were more economically feasible than kerosene boilers.

3.6. Results of Comparative Economic Analysis (Life Cycle Cost)

Based on the results of the economic analysis performed using the net present value method, it was difficult to confirm whether the observed dependence on the initial investment cost was large or whether the difference in operating expenses affected the economic feasibility of the systems. Therefore, we examined the economic feasibility of the hybrid systems and compared them, taking operating costs into consideration, by performing life cycle cost analysis. The results thus obtained are presented in Table 11.

Comparison performed using life cycle cost analysis showed that hybrid system 1 exhibited the best cost reduction effect (27.7%) as compared with that of the kerosene boiler. Further, kerosene and diesel boilers were found to have low economic feasibility due to their excessive fuel costs. Additionally, hybrid systems 1, 2, and 3 all showed higher economic feasibility than that of the fossil fuel boilers; however, despite exhibiting the lowest operating cost, hybrid system 3 was less feasible than hybrid system 1 because of its high initial investment cost. Consequently, hybrid system 1 was judged to be the best system overall.

Table 11. Comparative results of economic analysis based on life cycle cost analysis for each system.

Division	Kerosene Boiler	Diesel Boiler	Hybrid 1 (P.B [1] + H.P [2])	Hybrid 2 (P.B + S.C [3])	Hybrid 3 (P.B + H.P + S.C)
Fuel using ratio	100%	100%	P.B 82% H.P 18%	96%	P.B 82% H.P 14%
TEC			151,184,200 kcal		
Heating load per hour			97,262 kcal/h		
Total heating time required per 10 years			1554.4		
Fuel cost	Kerosene, 0.92 USD/L; diesel, 1.09 USD/L; Pellet, 373 KRW/kg; electricity, 50 KRW/kWh; solar: 0 KRW				
Fuel cost per hour (USD/h)	10.94	11.79	7.06	7.26	6.87
Operating expenses required for 10 years (total investment cost) (USD, $)	170,517.19	183,725.14	114,555.86	117,488.20	113,244.86
Initial investment cost self-pay (USD, $)	5055.61	5055.61	12,386	28,227	28,901
LCC (USD, $)	175,572.80	188,780.75	126,942.11	145,715.37	142,146.11

[1] P.B: Pellet Boiler; [2] H.P: Heat Pump; [3] S.C: Solar Collector.

4. Conclusions

In this study, different hybrid systems for the heating of greenhouses in the ROK were built using available renewable energy sources. Their heating performances, combined thermal efficiencies, energy consumption characteristics, and energy costs as a function of increasing temperature were analyzed and compared. Additionally, the practical applicability of the developed hybrid systems was evaluated by performing a comparative analysis of their economic feasibility with respect to fossil fuel boilers. All the systems showed similar heating performance. Specifically, hybrid system 3 showed the best performance in terms of the combined thermal efficiency. However, the differences between the systems in this regard were not significant. Hence, comparing them with respect to combined thermal efficiency was challenging. Additionally, given that the results of the combined thermal efficiency analysis showed a tendency to change with changes in the external temperature, an appropriate balance between the thermal insulation of the systems and the thermal energy supply system was necessary. Hybrid system 1 showed a 25.7% reduction in energy consumption and a 24.1% reduction in energy cost with increasing temperature compared with those of the other systems. Thus, its performance was the best when considering the three hybrid systems. Further, the practical applicability of the developed hybrid systems was evaluated by performing economic analysis. Subsequently, the net present value approach and the life cycle cost analysis method were implemented. In the net present value approach, when considering only self-pay, hybrid system 1 showed a cost reduction effect of 29.2% compared with that of a kerosene boiler, and in the life cycle cost analysis, which included operating costs and initial investment costs; it showed a cost reduction effect of 27.7% compared with that of the kerosene boiler. Thus, it was judged to be the best system. Hybrid systems 2 and 3 showed higher economic efficiency than that of fossil fuel boilers; however, they were less efficient than hybrid system 1. In addition, hybrid systems 2 and 3 have seasonal restrictions on the use of solar heat collection systems. It can be used in spring and autumn when the outside temperature does not drop below freezing. However, it is difficult to use in winter due to the freezing problem of the collector. Therefore, hybrid system 1 is suitable for heating applications using renewable energy, and it is applicable to greenhouses in the ROK. In addition, Hybrid System 1 is judged to be applicable not only to ROK, but also to countries with a climate environment similar to that of ROK, or to greenhouses of small and medium size requiring heating anywhere in the world.

Author Contributions: Conception and design of study, C.-G.L., L.-H.C., S.-Y.P., S.-J.K. and D.-H.K.; experiments and experimental equipment configuration, C.-G.L., L.-H.C., S.-Y.P. and S.-J.K.; acquisition of data, C.-G.L. and L.-H.C.; analysis and/or interpretation of data, C.-G.L., L.-H.C., S.-Y.P. and S.-J.K.; writing of original draft, C.-G.L., L.-H.C., S.-Y.P., S.-J.K. and D.-H.K.; revising the manuscript critically for important intellectual content, C.-G.L. and D.-H.K.; approval of the version of the manuscript to be published, C.-G.L., L.-H.C., S.-Y.P., S.-J.K. and D.-H.K.; review & editing, supervision, funding acquisition, D.-H.K. All authors have read and agreed to the published version of the manuscript.

Funding: This work was supported by the Korea Institute of Planning and Evaluation for Technology in Food, Agriculture and Forestry (IPET) through the Smart Farm Innovation Technology Development Program funded by Ministry of Agriculture, Food and Rural Affairs (MAFRA) (grant number 421040-04).

Institutional Review Board Statement: Not applicable.

Informed Consent Statement: Not applicable.

Conflicts of Interest: The authors declare no conflict of interest.

References

1. The Government of the Republic of Korea. 2050 Carbon Neutral Strategy of the Republic of Korea: Towards a Sustainable and Green Society. 2020. Available online: https://unfccc.int/sites/default/files/resource/LTS1_RKorea.pdf (accessed on 12 December 2020).
2. Zhang, R.; Hanaoka, T. Deployment of electric vehicles in China to meet the carbon neutral target by 2060: Provincial disparities in energy systems, CO_2 emissions, and cost effectiveness. *Resour. Conserv. Recycl.* **2021**, *170*, 105622. [CrossRef]
3. Potrč, S.; Čuček, L.; Martin, M.; Kravanja, Z. Sustainable renewable energy supply networks optimization—The gradual transition to a renewable energy system within the European Union by 2050. *Renew. Sustain. Energy Rev.* **2021**, *146*, 111186. [CrossRef]
4. Hirvonen, J.; Heljo, J.; Jokisalo, J.; Kurvinen, A.; Saari, A.; Niemelä, T.; Sankelo, P.; Kosonen, R. Emissions and power demand in optimal energy retrofit scenarios of the Finnish building stock by 2050. *Sustain. Cities Soc.* **2021**, *70*, 102896. [CrossRef]
5. Zhang, C.; Hu, M.; Sprecher, B.; Yang, X.; Zhong, X.; Li, C.; Tukker, A. Recycling potential in building energy renovation: A prospective study of the Dutch residential building stock up to 2050. *J. Clean. Prod.* **2021**, *301*, 126835. [CrossRef]
6. Sithole, H.; Cockerill, T.; Hughes, K.; Ingham, D.; Ma, L.; Porter, R.; Pourkashanian, M. Developing an optimal electricity generation mix for the UK 2050 future. *Energy* **2016**, *100*, 363–373. [CrossRef]
7. Ministry of Environment, Republic of Korea. 2030 Greenhouse Gas Reduction Roadmap. 2018. Available online: https://www.gihoo.or.kr/portal/kr/images/download/2030gas_loadmap.pdf (accessed on 20 July 2020).
8. Kim, S.; Jeon, Y. *2017 Energy Consumption Survey*; Ministry of Trade Industry & Energy Republic of Korea & Korea Energy Economics Institute: Sejog City, Korea, 2018.
9. Lee, Y. *Agricultural and Livestock Income Data Book for Improving Agricultural Management*; Rural Development Administration: Wansan-gu, Korea, 2014; ISBN 9788948002331.
10. Chau, J.; Sowlati, T.; Sokhansanj, S.; Preto, F.; Melin, S.; Bi, X. Economic sensitivity of wood biomass utilization for greenhouse heating application. *Appl. Energy* **2009**, *86*, 616–621. [CrossRef]
11. Karner, K.; Dißauer, C.; Enigl, M.; Strasser, C.; Schmid, E. Environmental trade-offs between residential oil-fired and wood pellet heating systems: Forecast scenarios for Austria until 2030. *Renew. Sustain. Energy Rev.* **2017**, *80*, 868–879. [CrossRef]
12. Bibbiani, C.; Fantozzi, F.; Gargari, C.; Campiotti, C.A.; Schettini, E.; Vox, G. Wood Biomass as Sustainable Energy for Greenhouses Heating in Italy. *Agric. Agric. Sci. Procedia* **2016**, *8*, 637–645. [CrossRef]
13. McKenney, D.W.; Yemshanov, D.; Fraleigh, S.; Allen, D.; Preto, F. An economic assessment of the use of short-rotation coppice woody biomass to heat greenhouses in southern Canada. *Biomass Bioenergy* **2011**, *35*, 374–384. [CrossRef]
14. Kim, Y.; Park, H.; Kang, C.; Park, G.; Jeong, E.; Lee, W.; Park, M. *Prospect of Demand and Supply of energy in the Agricultural Sector and Strategies for Introducing Clean Energy Farming System*; Rural Development Administration: Jeollabuk-do, Korea, 2010; ISBN 97889601326109352.
15. Maddah, S.; Goodarzi, M.; Safaei, M.R. Comparative study of the performance of air and geothermal sources of heat pumps cycle operating with various refrigerants and vapor injection. *Alex. Eng. J.* **2020**, *59*, 4037–4047. [CrossRef]
16. De Swardt, C.A.; Meyer, J. A performance comparison between an air-source and a ground-source reversible heat pump. *Int. J. Energy Res.* **2001**, *25*, 899–910. [CrossRef]

17. Ministry of Environment, Republic of Korea & K-Water (Korea Water Resources Corporation). 2019 Groundwater Annual Report. 2019. Available online: http://www.kogga.or.kr/site3/data/da_law_view.asp?intSeq=14772&strboardId=pdskwa&page=1 (accessed on 20 August 2021).
18. Nam, H.; Kim, Y.; Seo, J.; Shin, Y. A Study on the Variation of ground water temperature for development of ground water source heat pump. *Trans. Korea Soc. Geotherm. Energy Eng.* **2004**, *11*, 80.
19. Kang, Y.K.; Ryou, Y.S.; Kim, J.G.; Kim, Y.H.; Jang, J.K. Heating effects of the Greenhouse for Mango heated by the Wood Pellet Boiler. *Korean Soc. New Renew. Energy* **2013**, *5*, 86.
20. Hassanien, R.H.E.; Li, M.; Tang, Y. The evacuated tube solar collector assisted heat pump for heating greenhouses. *Energy Build.* **2018**, *169*, 305–318. [CrossRef]
21. Kwon, J.K.; Kang, G.C.; Moon, J.P.; Kang, Y.K.; Kil Kim, C.; Lee, S.J. Performance Improvement of an Air Source Heat Pump by Storage of Surplus Solar Energy in Greenhouse. *Prot. Hortic. Plant Fact.* **2013**, *22*, 328–334. [CrossRef]
22. Tian, Z.; Zhang, S.; Deng, J.; Fan, J.; Huang, J.; Kong, W.; Perers, B.; Furbo, S. Large-scale solar district heating plants in Danish smart thermal grid: Developments and recent trends. *Energy Convers. Manag.* **2019**, *189*, 67–80. [CrossRef]
23. Ko, M.-J.; Kim, Y.-S. Applicability Estimation of Solar Water Heater according to Solar Radiation and Water Supply Temperature in Korea. *J. Korean Soc. Living Environ. Syst.* **2017**, *24*, 579–587. [CrossRef]
24. Ahn, H. *Facility Horticulture Automation*; Munundang: Seoul, Korea, 2008; pp. 128–138. ISBN 9788973934997.
25. Kim, D.H. A review of desalting process techniques and economic analysis of the recovery of salts from retentates. *Desalination* **2011**, *270*, 1–8. [CrossRef]

designs

MDPI

Article

The Effect of Oxygenated Turpentine Oil Additive in Diesel Fuel on the Performance and Emission Characteristics in One-Cylinder DI Engines

Asep Kadarohman [1], Fitri Khoerunnisa [1], Syazwana Sapee [2], Ratnaningsih Eko Sardjono [1], Izuan Izzudin [2], Hendrawan [1], Rizalman Mamat [2], Ahmad Fitri Yusop [2], Erdiwansyah [2,3] and Talal Yusaf [4,*]

1 Department of Chemistry, Indonesia University of Education, Bandung 40154, Indonesia; asep@ieu.edu.id (A.K.); fitri.khoerunnisa@gmail.com (F.K.); Ratnaningsih@iue.edu.id (R.E.S.); Hendrawan@iue.edu.in (H.)
2 Faculty of Mechanical Engineering, Universiti Malaysia Pahang, Pekan 26600, Malaysia; Syazwana.sapee@gmail.com (S.S.); Izuan@ump.edu.my (I.I.); rizalman@ump.edu.my (R.M.); Fitriy@ump.edu.my (A.F.Y.); Erdi.wansyah@yahoo.co.id (E.)
3 Fakultas Teknik, Universitas Serambi Mekkah, Banda Aceh 23249, Indonesia
4 School of Engineering and Technology, Central Queensland University, Brisbane, QLD 4009, Australia
* Correspondence: t.yusaf@cqu.edu.au

Citation: Kadarohman, A.; Khoerunnisa, F.; Sapee, S.; Eko Sardjono, R.; Izzudin, I.; Hendrawan; Mamat, R.; Yusop, A.F.; Erdiwansyah; Yusaf, T. The Effect of Oxygenated Turpentine Oil Additive in Diesel Fuel on the Performance and Emission Characteristics in One-Cylinder DI Engines. *Designs* **2021**, *5*, 73. https://doi.org/10.3390/designs5040073

Academic Editors: Zbigniew Leonowicz, Arsalan Najafi and Michał Jasinski

Received: 18 October 2021
Accepted: 10 November 2021
Published: 17 November 2021

Publisher's Note: MDPI stays neutral with regard to jurisdictional claims in published maps and institutional affiliations.

Abstract: A study on the application of oxygenated turpentine oil as a bio-additive in diesel fuel was conducted. The purpose of this research was to investigate the effect of oxygenated turpentine oil additive in diesel fuel on the performance and emission characteristics in diesel engines. Oxygenated turpentine oil is obtained from the oxidation process of turpentine oil. In this experimental study, the influences of oxygenated turpentine oil-diesel blended fuel OT0.2 (0.2% vol oxygenated turpentine oil and 99.8% vol diesel) were compared with pure diesel on engine performance, and emission characteristics were examined in a one-cylinder four-stroke CI engine. The test was performed at two engine loads (25% and 50%) and seven engine speeds (from 1200–2400 rpm with intervals of 200 rpm). The physiochemical characteristics of test fuels were acquired. The engine indicated power, indicated torque, fuel flow rate, and emissions (carbon dioxide, CO_2; carbon monoxide, CO; and nitrogen oxide, NO_X) were examined. The results revealed that the engine power shows slight increments of 0.7–1.1%, whereas the engine torque slightly decreased with oxygenated turpentine usage compared to pure diesel in most conditions. Furthermore, a reduction in NO_X emission decreased by about 0.3–66% with the addition of oxygenated turpentine in diesel compared to diesel. However, usage of OT0.2 decreased fuel flow rate in most speeds at low load but gave a similar value to diesel at 50% load. CO emissions slightly increased with an average of 1.2% compared to diesel while CO_2 emissions increased up to 37.5% than diesel. The high-water content, low cetane number, and low heating value of oxygenated turpentine oil were the reasons for the inverse effect found in the engine performances.

Keywords: bio-additive; oxygenated turpentine oil; diesel fuel; diesel engine performance; emission

1. Introduction

Diesel fuel is produced from the distillation process of petroleum and is used to fuel diesel engines. Diesel engine usage has become more popular today ever since it was founded in 1983 by Rudolf Diesel. The popularity of diesel engines comes from the advantage of having a low-cost of fuel compared to gasoline engines. There is a wide use of diesel engines from transportation to industrial applications [1–4]. As a result, the amount of harmful gas emissions emitted from diesel fuel combustion such as CO, NO_X, and hydrocarbon (HC) will be increased as well [5–8]. Consequently, this has led to adverse impacts on human health and on the environment. For this reason, a lot of studies have been conducted to minimize harmful gases emitted from diesel engines [9–11].

Mixing diesel fuel with additives is one of the many attempts to reduce emissions from diesel combustion, as well as a way to optimize fuel consumption of the engine. There are many compounds used as diesel fuel additives such as organometals, nitrates, oxygenates (compounds rich with oxygen), and natural matters (bioactive) [12–15]. Organometals and nitrates have been known to increase the burning efficiency of diesel fuel. However, it is also discovered that those additives may result in additional emissions of NO_X that is harmful to humans [16–18]. On the other hand, oxygenates and bio-additives are known to be more environmentally friendly. Nayyar et al. [19] in their recent work stated that the addition of compounds rich in oxygen (oxygenates) into diesel fuel could reduce smoke and NO_X production by 61.85% and 8.07%, respectively. This finding was also supported by other research [20–23], which explained how soot reduction is linearly related to the increasing oxygen mass fraction in the fuel. Other researchers who used oxygenated additives reported enhancement in its application [24–27].

Turpentine oil is often referred to as spirits of turpentine in the form of volatile liquid, derived from the distillation of tree sap species belonging to the pine genus. It is colorless (liquid), has a distinctive smell, and is flammable [28–30]. In general, the physical and chemical properties of boiling turpentine oil is 149–180 °C, insoluble in water, density 0.9, flash point 30–46, auto ignition temperature 220–225 °C (International Program on Chemical Safety and the European Commission, 2002) [31–34]. It contains monoterpenes with C 10 carbon atoms. Turpentine oil is generally composed of a mixture of unsaturated isomers, bicyclic hydrocarbons namely α-pinene, β-pinene, and δ-carene as presented in Figure 1 [28].

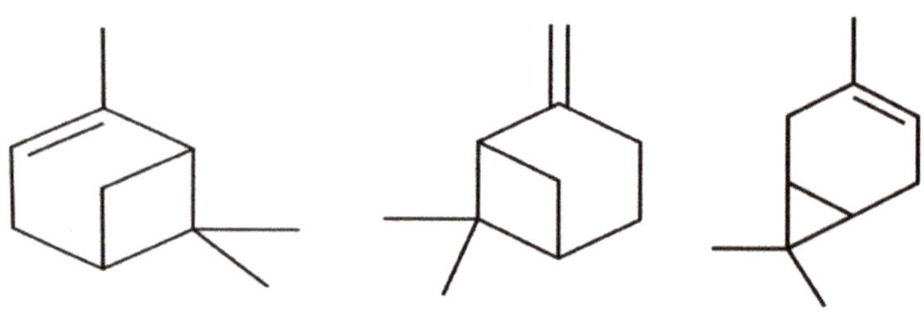

α-Pinene β-Pinene δ-carene

Figure 1. Chemical structure of the main component of turpentine oil.

From the work of Polonowski et al. [35], it is reported that diesel fuel with 5% of pure turpentine oil could reduce smoke production and reduce fuel consumption. This was in line with Butkus' finding that 5% of oxidized turpentine oil was the best diesel fuel additive [36–38]. Furthermore, Kadarohman et al. [39] found that terpene compounds contained in clove oil (0.2%) were largely contributed to make a better mixture between bio-additive and diesel fuel, which lead to rapid combustion and shorter ignition delay in combustion of diesel engines. This discovery is interesting for further investigation on the influence of terpene compounds addition in diesel fuel [40–42].

The four atom C rings on α-pinene and β-pinene have high spatial strain that are reactive. The presence of a double bond causes α-pinene to easily undergo an oxidation reaction when there is air contact, then forms a hydroperoxyl compound which has intermediate molecules that are reactive [43]. The cyclic structure in turpentine oil will effectively disrupt the Van der Waals interaction between the carbon chains of diesel fuel, consequently leading to the diesel oil molecules becoming easier to evaporate, hence accelerating the combustion process [39,44]. The reactive nature of turpentine oil constituents is also expected to accelerate the combustion of diesel fuel. Song et al. [45] suggested that the addition of oxygen enriched additives into diesel fuel has a significant role to increase the cetane number of the fuel. Choi and Reitz [46] mentioned that oxygen atoms in fuel play a major role in oxidizing soot and CO gas.

For this reason, efforts to speed up and to refine the combustion process of diesel fuel can be carried out by enriching the levels of oxygen atoms contained in turpentine oil through the oxidation of the double bonds formed in the compound. In this paper, the effects of additive oxygenated turpentine oil-diesel (0.2% vol and 99.8% vol) on the performance and emission in one-cylinder diesel engines were tested. The experiment was performed at different engine speeds and two engine loads (25% and 50%). The physiochemical characteristics of test fuels were determined. Moreover, the effects of tested fuels upon indicated power, indicated torque, flue flow rate, and emissions characteristics were systematically observed.

2. Materials and Methods

2.1. Materials

In this study, the diesel used was pure diesel Euro2M from Malaysia. Turpentine oil and oxygen gas were from Brataco and Sangkuriang companies, Indonesia. Turpentine oil was oxygenated via the oxidation process. It was carried out by reflux method using a cylindrical column reactor with length and diameter dimensions of 30 cm and 2 cm, respectively. The 15 mL turpentine oil was aerated by oxygen gas with flow rate of 3 L/min and heated by an electrical wire heater at 90–100 °C for 3 h. The oxidation procedure was conducted at Life Science Laboratory, Department of Chemistry, Indonesia University of Education, Bandung Indonesia. Oxygenated turpentine oil as a bio-additive was dissolved in diesel fuel at a volume percent level of 0.2% (note as OT0.2) by a manual direct blending method using a mechanical stirrer IKA RW20 with blending speed 700 rpm for 15 min at room temperature. The characterizations of diesel, turpentine oil, and oxygenated turpentine oil were done by gas chromatography—mass spectrometry GC-MS QP5050A. Diesel fuel and OT0.2 were examined on one-cylinder DI engines in order to obtain their performance and emission.

2.2. Experiment Setup

The test engine was a YANMAR TF120M one-cylinder DI diesel engine with a 17.7 compression ratio. The specifications of the engine and schematic diagram of the set up for this test are shown in Table 1 and Figure 2, respectively. The data were recorded by data acquisition system TFX Engineering, which consisted of in-cylinder pressure and crank angle sensors. Furthermore, exhaust gas temperature and ambient temperature were measured using K-type thermocouples, that were recorded using a TC-08 thermocouple data logger by Pico Technology. The thermocouple was installed at the exhaust manifold. The emissions were measured using KANE Auto 4-1 series exhaust gas analysers. The experiment was conducted with seven speeds from 1200 to 2400 rpm with intervals of 200 rpm and two engine loads at 25% and 50%. The test fuels used were diesel as base line and oxygenated turpentine oil-diesel (0.2% vol and 99.8% vol). The data were recorded under steady state conditions. The engine power, engine torque, the fuel flow rate, and the emissions (CO, CO_2, and NO_x) were measured. The experiment was conducted at Universiti Malaysia Pahang (UMP), Kuantan, Malaysia.

Table 1. Engine Specifications.

Description	Specification
Engine model	YANMAR TF120M
Engine year	2016
Engine type	Horizontal, 4-cycle, 4 stroke, diesel engine
Number of cylinders	1
Continuous power output (kW)	7.82 kW at 2400 rpm
Rated power output (kW)	8.94 kW at 2400 rpm
Bore x Stroke (mm)	92 × 96
Displacement (L)	0.638
Injection timing	17° BTDC

Table 1. *Cont.*

Description	Specification
Compression ratio	17.7
Combustion system	Direct injection
Aspiration	Natural aspiration
Cooling system	Water-cooled
Starting system	Manual (Hand) Starting

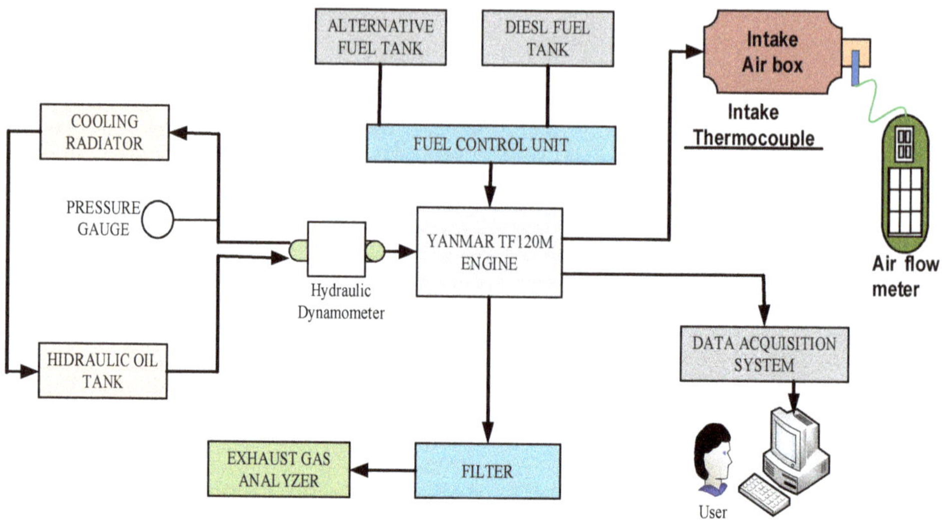

Figure 2. Schematic diagram of diesel engine test set up.

3. Results and Discussion

3.1. Physiochemical Properties

The performed research on physical properties showed that the bio-additive fuel blends were in full compliance with the standard of American Society for Testing and Materials, ASTM D975 specifications for diesel fuel. The physical properties of diesel and OT0.2 were presented in Table 2.

Table 2. Physical properties of test fuels.

Parameters	Diesel Fuel	OT0.2	ASTM D975 Limit	
			Min	Max
Specific Gravity at 25 °C (g/mL)	0.8452	0.8549	-	-
Specific Gravity at 15.55 °C (g/mL)	0.8522	-	0.848	0.87
API Gravity	34.5408	-	-	-
Anilin point (°F)	156.2	159.2	129.6	-
Index Diesel	53.9527	51.3883	-	-
Viscosities (cSt)	3.7215	4.4625	1.3	4.5
Flash Point (°C)	61.89	-	60	80

The diesel, turpentine, and oxygenated enriched turpentine used in this experiment were characterized by GC-MS. Figure 3 shows the chromatogram of diesel fuel, turpentine, and oxygenated turpentine that provides the information of its chemical components and composition. In particular, diesel fuel consisted of saturated hydrocarbons such as normal paraffins, is paraffins, and cycloparaffins. The main components of diesel fuel are hexadecane (n-cetane), pristane (2,6,10,14-tetramethylpentadecane), and is paraffins

(Figure 3a), in line with previous studies [47]. The chemical constituents of diesel fuel are listed in Table 3.

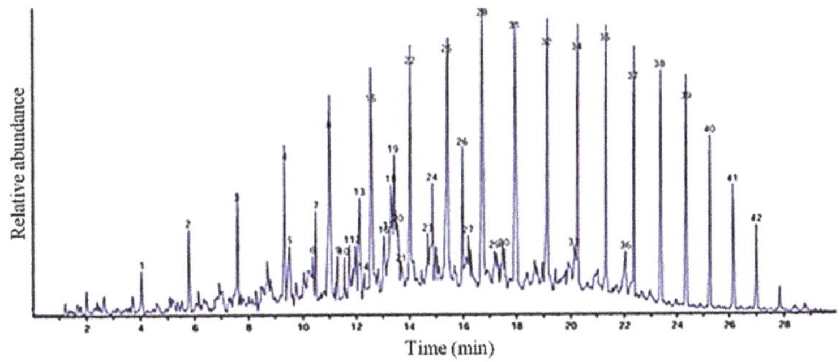

(a)

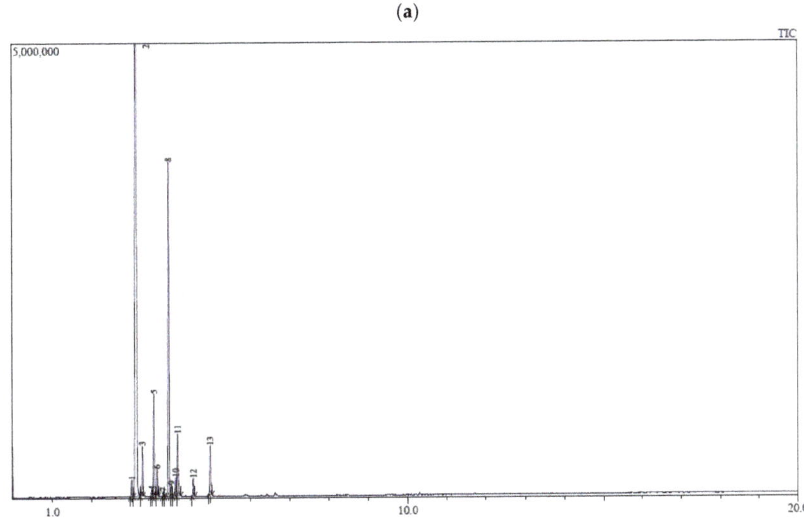

(b)

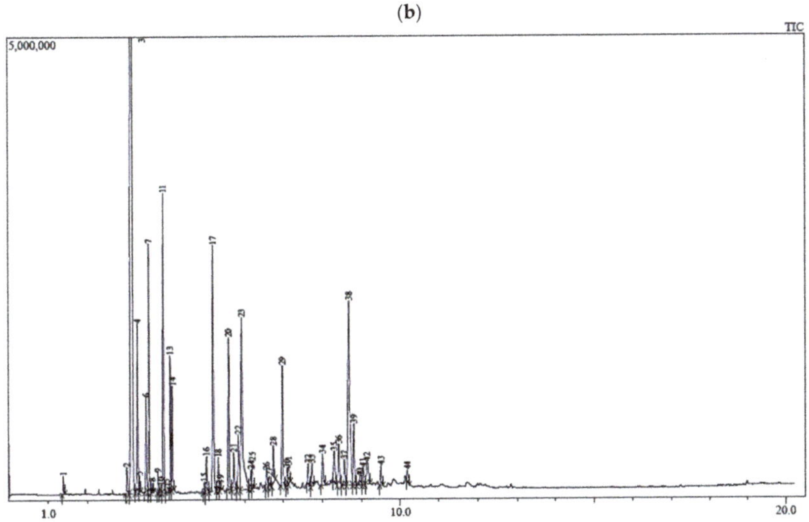

(c)

Figure 3. Chromatogram of diesel fuel (**a**), turpentine (**b**), and oxygenated turpentine (**c**).

Table 3. Chemical constituents of diesel fuel.

Peak No	Molecular Formula	Name	Retention Time (Min)	Conc. (%)	Structure
22	$C_{15}H_{32}$	Pentadecane	14.050	5.27	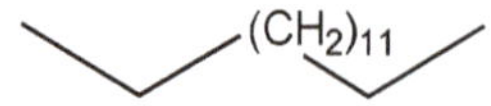
26	$C_{19}H_{40}$	2,6,10,14-tetramethylpentadecane (pristane)	15.993	2.44	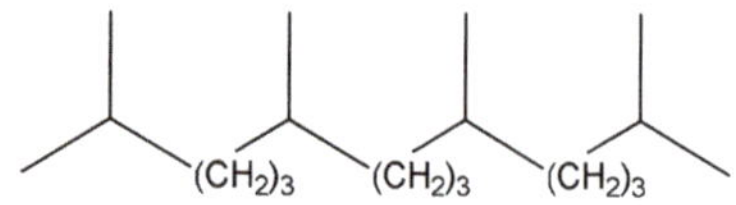
28	$C_{16}H_{34}$	Hexadecane (n-cetane)	16.769	10.67	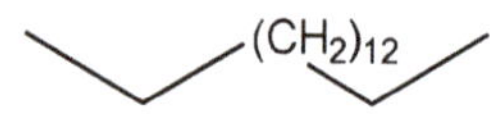
31	$C_{18}H_{38}$	n-octadecane	17.980	7.47	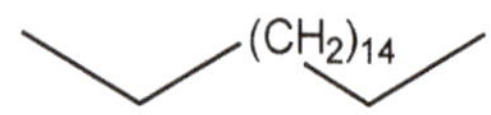
32	$C_{19}H_{40}$	n-nonadecane	19.151	5.37	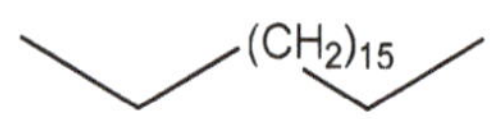
34	$C_{21}H_{44}$	n-heneicosane	20.284	4.84	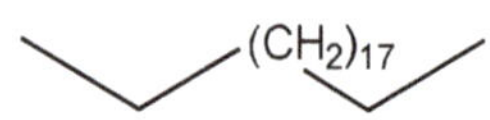
35	$C_{22}H_{46}$	n-docosane	21.365	4.81	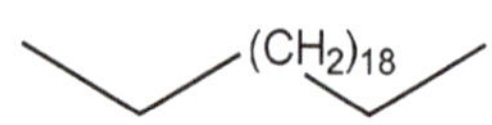
37	$C_{23}H_{48}$	n-tricosane	22.404	4.07	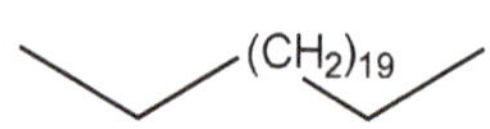

On the other hand, turpentine contains at least 12 compounds as shown in Figure 3b, that are predominantly composed of α-pinene (61.61%), δ-carene (19.70%), β-pinene (4.8%), limonene (3.58%), and camphene (2.25%), with based mass fragment at retention times of 3.127, 3.950, 3.568, 4.712 and 3.267 min, respectively. These results align with previous studies [48]. All chemical compounds of turpentine including its structure and composition is listed in Table 4.

Interestingly, oxidation treatment led to remarkable modifications of turpentine in term of chemical constituents and composition. Figure 3c demonstrates the chemical constituents of oxygenated turpentine where at least 44 compounds were detected by GCMS. In particular, the oxidation process of turpentine yields new compounds with various composition. After oxidation, the composition of major constituents of turpentine experienced a significant reduction, i.e., α-pinene (32.68%), δ-carene (5.77%), β-pinene (4.44%), and limonene (1.93%). This presents new oxygenated compounds with significant composition such as α-pinene-oxide, patchcoulane, trans-verbenol, verbenone, and α-champholene aldehyde at retention times of 5.213, 8.684, 5.932, 6.974, and 5.604, respectively. Details of chemical constituents of oxygenated turpentine are summarized in Table 5. Additionally, the mass fragment of major chemical components of oxygenated turpentine is shown in Figure 4. The oxygenated products contain more oxygen related functional groups, i.e., hydroxyl (-OH), aldehyde (-HC=O), and ketone (-C=O). These results indicated the effectiveness of selected oxidation procedures of turpentine where the predominant oxygenated compounds came from the oxidation of α-pinene and δ-carene as the most major constituents of turpentine.

Table 4. Major chemical constituents of turpentine.

Peak No	Molecular Formula	Name	Retention Time (Min)	Conc. (%)	Structure
2	$C_{10}H_{16}$	Alpha-pinene	3.127	61.81	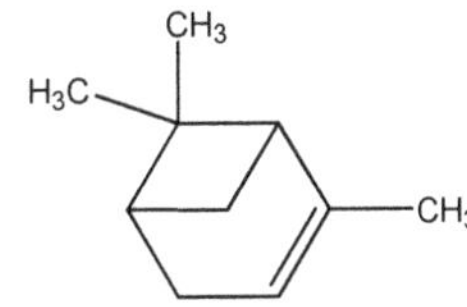
3	$C_{10}H_{16}$	Camphene	3.267	2.25	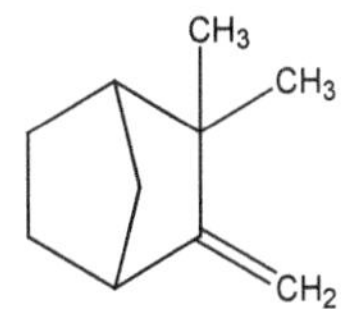
5	$C_{10}H_{16}$	Beta-pinene	3.568	4.80	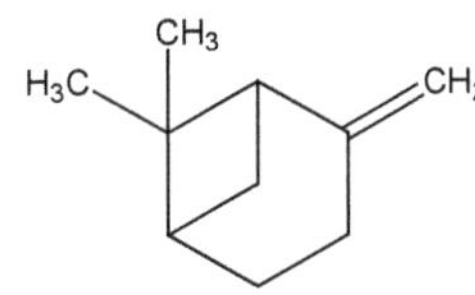
8	$C_{10}H_{16}$	Delta 3 Carene	3.950	19.70	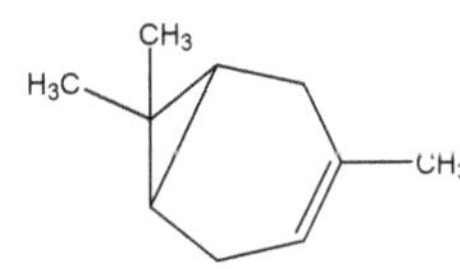
11	$C_{10}H_{16}$	dl-Limonene	4.172	3.58	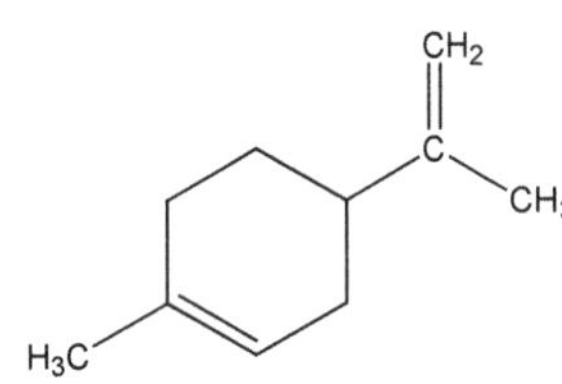
13	$C_{10}H_{16}$	Alpha-terpinolene	4.993	2.49	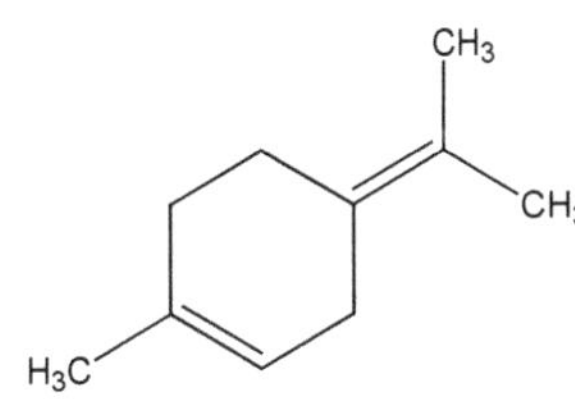

Table 5. Major chemical constituents of oxygenated turpentine.

Peak No	Molecular Formula	Name	Retention Time (Min)	Conc. (%)	Structure
3	$C_{10}H_{16}$	Alpha-pinene	3.143	32.68	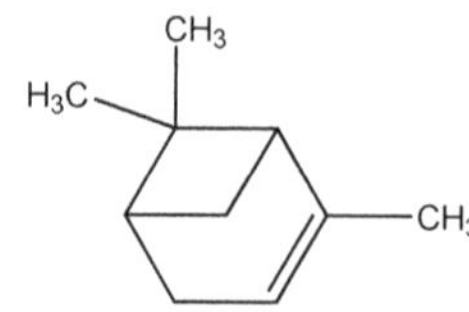
4	$C_{10}H_{16}$	Camphene	3.272	2.94	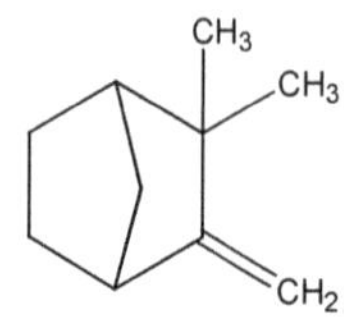
7	$C_{10}H_{16}$	Beta-pinene	3.571	4.44	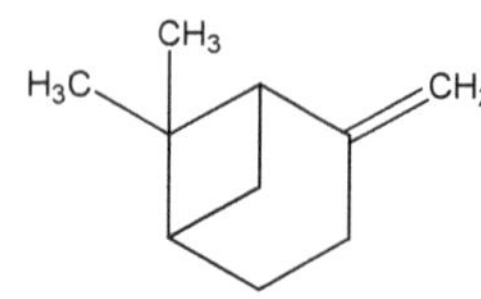
11	$C_{10}H_{16}$	Delta 3 Carene	3.944	5.77	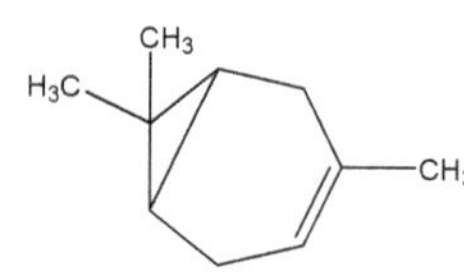
14	$C_{10}H_{16}$	dl-Limonene	4.173	1.93	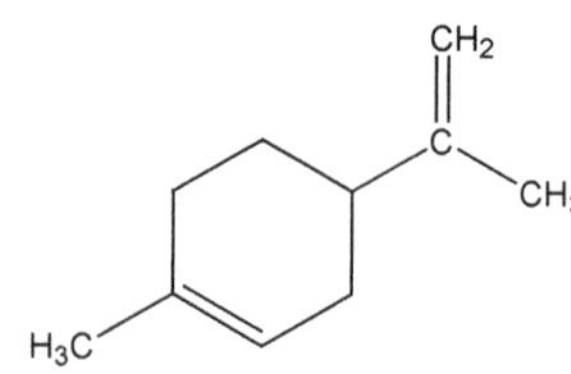
17	$C_{10}H_{16}O$	Alpha-pinene oxide	5.213	6.15	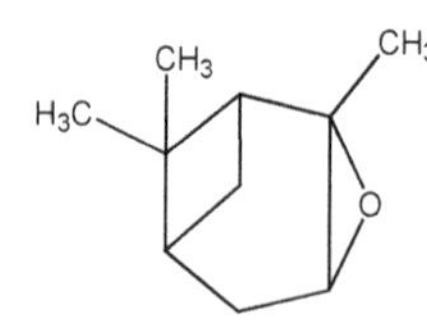
20	$C_{10}H_{16}O$	Alpha-campholene aldehyde	5.604	3.59	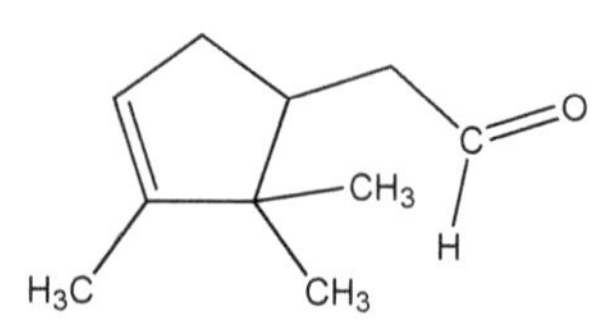
23	$C_{10}H_{16}O$	Trans-verbenol	5.932	6.66	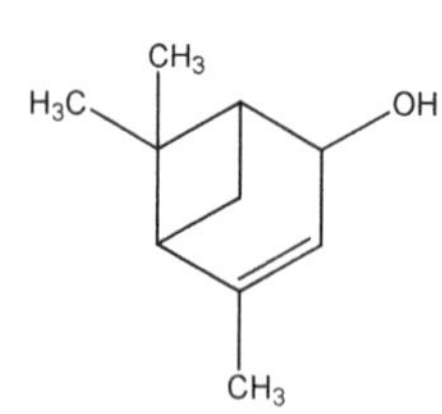

Table 5. *Cont.*

Peak No	Molecular Formula	Name	Retention Time (Min)	Conc. (%)	Structure
29	$C_{10}H_{14}O$	Verbenone	6.974	3.11	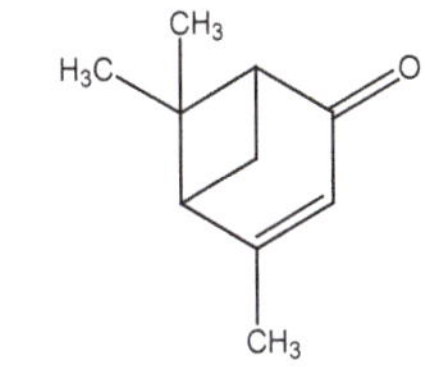
38	$C_{15}H_{26}$	Patchoulane	8.684	8.29	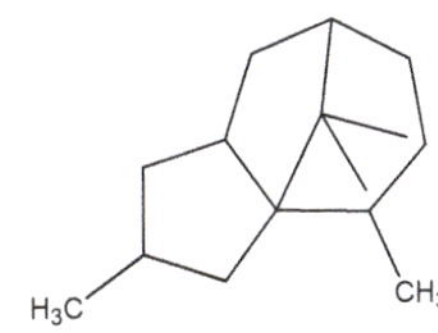

3.2. Engine Performance

Figures 5 and 6 show comparison results of indicated power and indicated torque at various engine speeds and loads, respectively. The power and torque depended on the fuel supplied and engine operating conditions. In this study, at the maximum engine speed of 2400 rpm, the indicated power of the engine slightly increased with the addition of an oxygenated additive compared to diesel fuel. The average increment when an additive was introduced into diesel was 0.7% to 1.1%. The higher oxygen content in oxygenated turpentine improved the in-cylinder combustion reaction process, hence producing higher power than diesel [33,49,50]. Another reason is due to higher fuel mass flow used for additive fuel. The increments were supported by a few studies that used oxygenated additives in the fuel [51–53]. On the other hand, the torque profile for low and high loads of oxygenated turpentine was found to be lower than diesel. The decrement is due to the increase in mass and flow resistance and the decrease in volumetric efficiency [33,54].

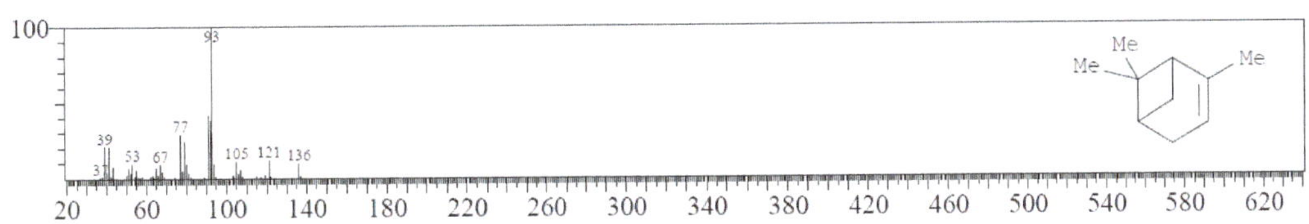

(**a**)

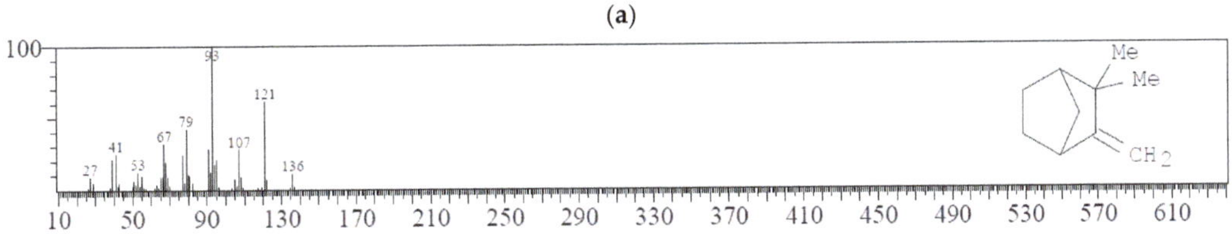

(**b**)

Figure 4. *Cont.*

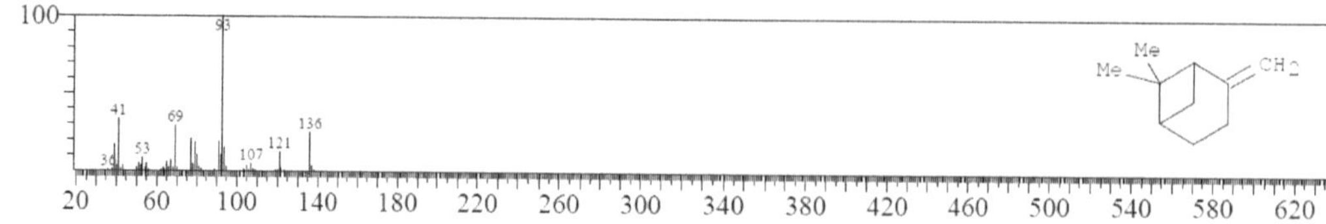

(c)

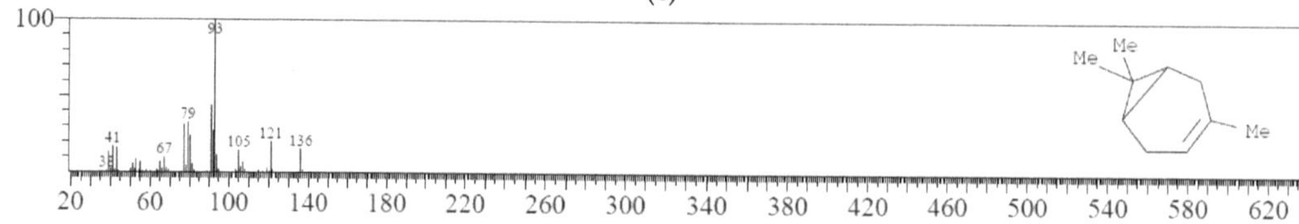

(d)

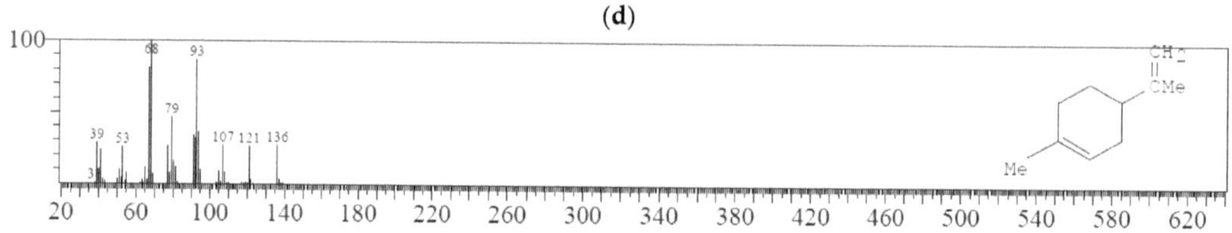

(e)

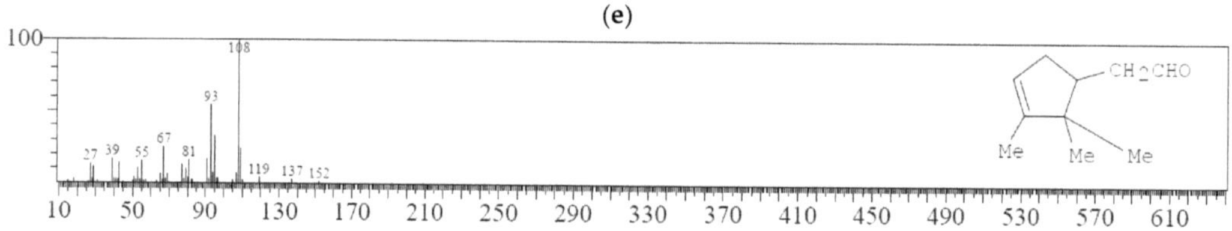

(f)

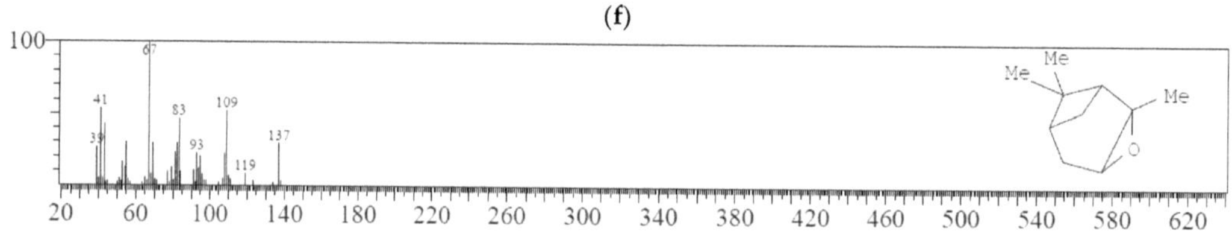

(g)

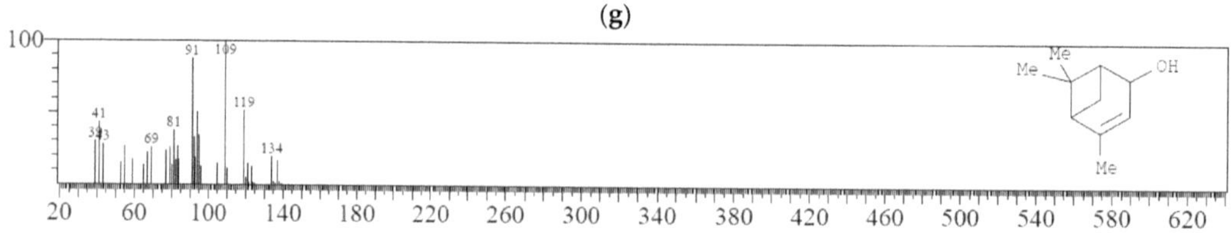

(h)

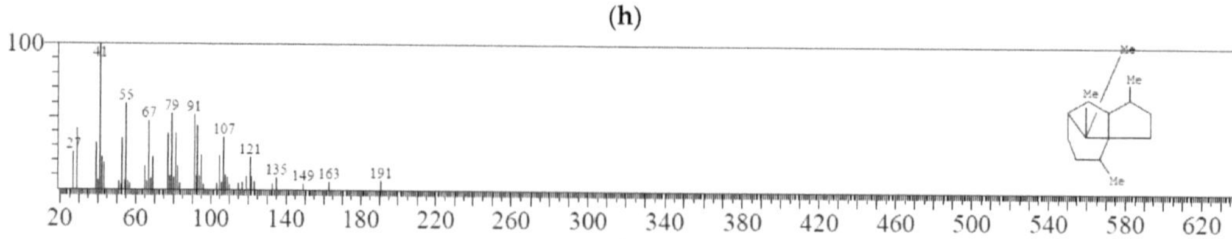

(i)

Figure 4. Mass fragments of α-pinene (**a**), camphene (**b**), β-pinene (**c**), δ-carene (**d**), limonene (**e**), α-champholene aldehyde (**f**), α-pinene oxide (**g**), trans-verbenol (**h**), and Patchoulane (**i**) for oxygenated turpentine.

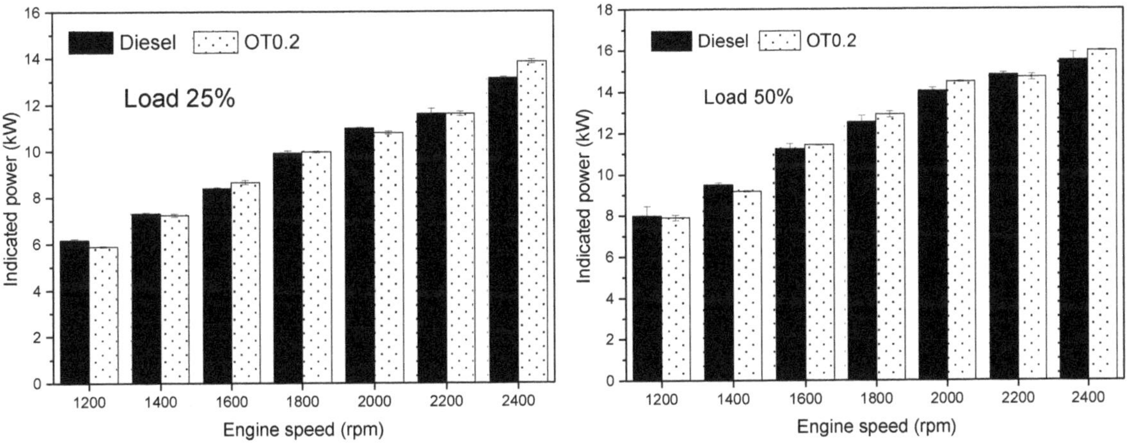

Figure 5. Indicated power at various engine speeds.

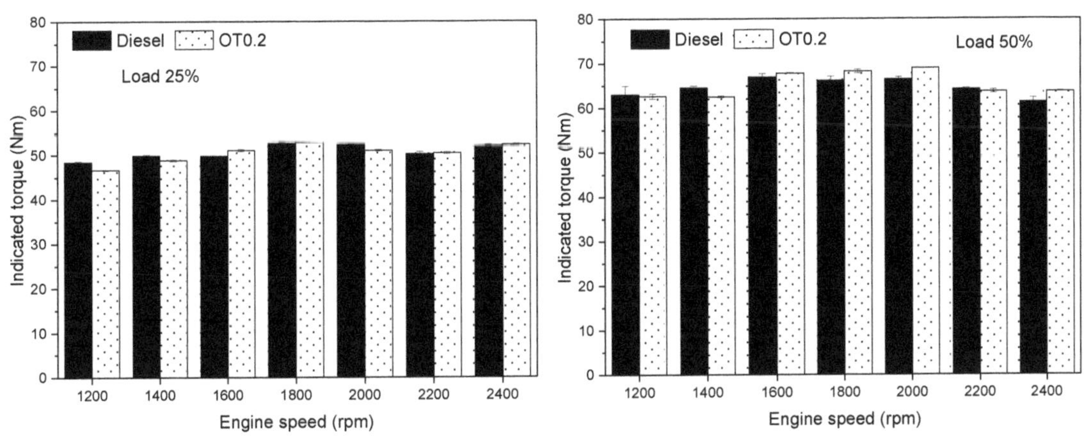

Figure 6. Indicated torque at various engine speeds.

Figure 7 presents the variations of fuel flow rates at different speeds and loads for diesel and oxygenated turpentine oil-diesel fuel. Mostly, the flow rate increases with the increment of engine speed and load. At low load cases, at most engine speeds, there are decrement of fuel flow rate of additive fuel compared to diesel. The enhancement rate of fuel flow while using an additive are between 5 to 9.09% compared to diesel. However, at 50% load, oxygenated turpentine oil-diesel fuel shows a slight increment of fuel flow rates with diesel at most engine speeds. The percentage of increment of fuel flow rate while using an additive at 50% load are in the range of 0.42 to 10.67%, compared to diesel. It is due to the lower heating value of oxygenated turpentine oil—diesel that requires higher fuel consumption. In contrast, at medium load with high speeds, 2200 and 2400 rpm, fuel flow rate of oxygenated turpentine oil-diesel fuel shows reduction up to 4.6% compared to diesel.

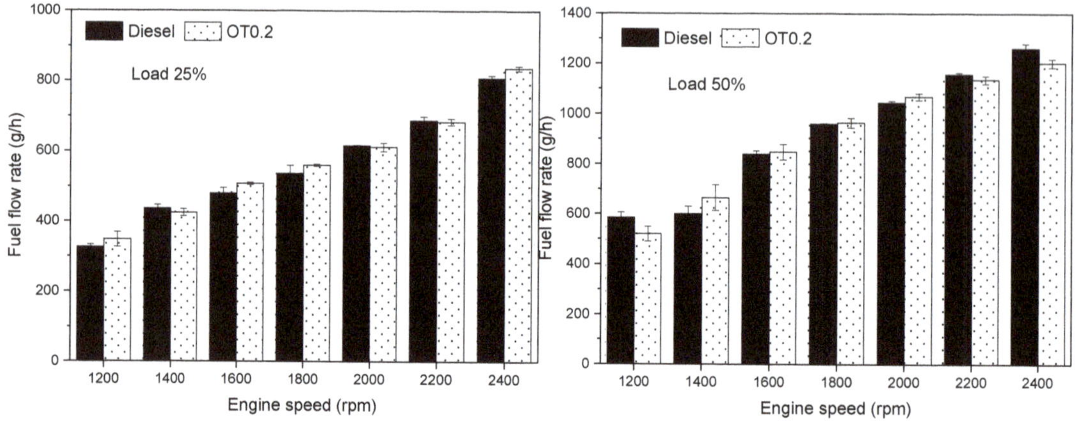

Figure 7. Fuel flow rate at various engine speeds.

3.3. Gas Emissions

In general, carbon monoxide emission shows a decline pattern when oxygenated additives are introduced into diesel fuel [55–57]. Reduction in CO occurs due to oxygenated characteristics of fuel and well-flammability properties of the oxygenated additive. Furthermore, the higher latent heat of evaporation in oxygenated-based fuels compared to diesel allows lower intake of manifold temperature and enhances the volumetric efficiency [58–60]. Figure 8 presents variations in CO emission emitted from the combustion diesel engine using diesel and oxygenated turpentine oil-diesel fuel. In this study, the lowest value of CO emission was found at low engine speeds for both engine loads. Mostly, an increase in engine speed leads to an increase in CO emission. In most operating conditions, CO emission shows a slight increment of 1.2% on average compared to diesel. This is parallel with previous statements. Several studies also reported the same decrements relative to diesel fuel when oxygenated fuel was added into diesel [61–64]. However, at 1600 rpm engine speed, the percentage of CO was increased for both load cases.

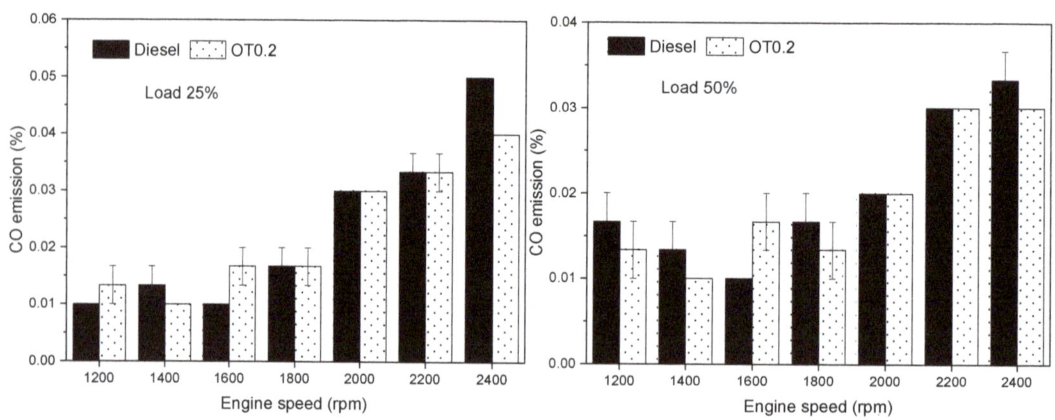

Figure 8. CO emission at various engine speeds.

CO_2 emission is a product of complete combustion. Theoretically, the combustion of hydrocarbon-based fuel should form only two elements, namely CO_2 and H_2O. Figure 9 shows the CO_2 emissions for diesel and oxygenated turpentine oil-diesel fuel at low and medium loads. For both load cases, there are slight increments of CO_2 emissions compared to diesel at most engine speeds. The average increment was 0–37.5% and 0–18% for 25% load and 50% load, respectively. The increase in CO_2 emissions compared to diesel fuel is due to higher average carbon content per energy in oxygenated turpentine. The high oxygen content of additives also leads to an increment of CO_2. The increment aligns with the reported studies [22,56,65].

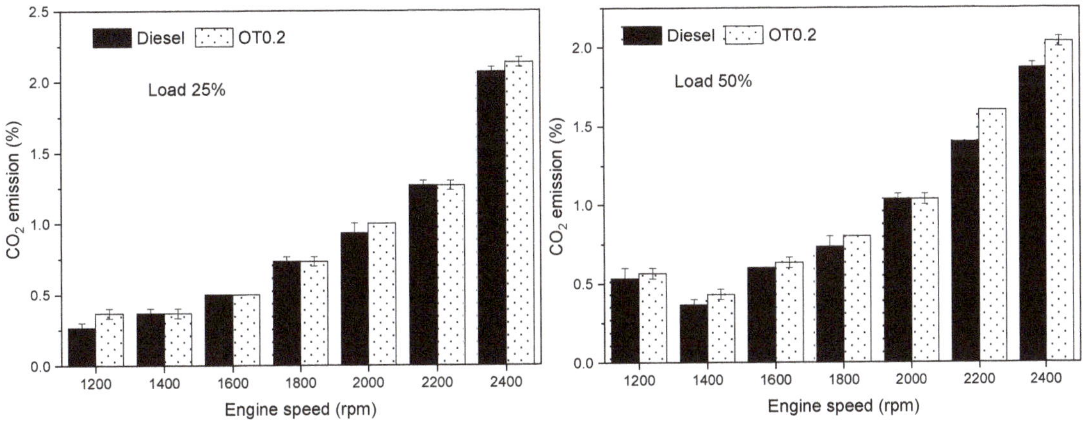

Figure 9. CO_2 emission at various engine speeds.

The major concern of emissions from compression ignition engines is NO_X. Formation of NO_X is strongly related to combustion temperature. It is also connected to engine operation conditions including engine speeds and engine loads, as well as fuel-to-air ratio. Nitrogen reacts with oxygen inside the combustion chamber at high temperatures. At temperatures above 1600 °C, NO_X formation occurs and increases rapidly with increments of temperature [66]. Moreover, NO_X formation happens in the presence of CH radicals at the flame front [67–69]. In this study, generally, there are slight increments of NO_X emission using additives compared to diesel as shown in Figure 10. At 25% load, the range of increment for NO_X emission was 0.5–66% compared to diesel. At 50% load, increments of NO_X emission was 0.3–7.9% relative to diesel. The increment of NO_X formation is due to higher oxygen content in oxygenated turpentine oil-diesel fuel compared to diesel fuel. A similar finding was reported on oxygenated fuel addition in diesel [56,70].

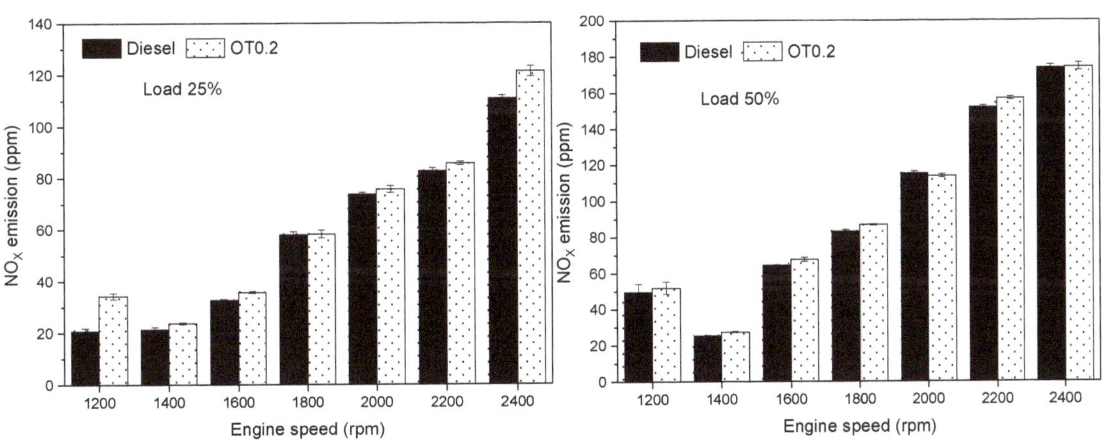

Figure 10. NO_X emission at various engine speeds.

4. Conclusions

The performance and emission of a one-cylinder DI engine using pure diesel and oxygenated turpentine oil-diesel (0.2% vol and 99.8% vol) blend was studied. The addition of oxygenated turpentine into diesel affected the physicochemical properties of the blends including specific gravity, density, aniline point, viscosities, flash point, and stability. The acquired results lead to major conclusions as drawn below.

- The engine power shows slight increments, 0.7–1.1%, whereas the engine torque was slightly decreased using oxygenated turpentine oil-diesel fuel compared to diesel fuel in most conditions.
- The fuel flow rate was lower for OT0.2 compared to diesel in most conditions for low load. The enhancement rate of fuel flow while using an additive is between 5 and 9.09 percent.
- CO emission shows a slight increment when OT0.2 was used, 1.2% on average compared to diesel.
- CO_2 emission increases with OT0.2 usage in diesel fuel up to 37.5%.
- NO_X emission decreased by about 0.3–66% in addition to oxygenated turpentine in diesel compared to diesel fuel.

Therefore, there are a few recommendations for future work and research that could improve and broaden the scope of this experiment. The following suggestions could be used to improve and gain a better understanding of the additive's performance and emission.

- The load applied to the engine could be increased to a high-level load;
- For wider understanding of the effect of oxygenated turpentine to the performance and emission, a larger volume of additive could be tested;
- The application of the additive could be tested in higher power and different types of engines.

Author Contributions: Conceptualization, F.K.; methodology, S.S. and E.; validation, A.K., R.E.S. and F.K.; formal analysis, I.I.; resources, H.; writing—original draft preparation, S.S.; writing—review and editing, E.; supervision, R.M. and A.F.Y.; project administration, T.Y.; funding acquisition, A.K. All authors have read and agreed to the published version of the manuscript.

Funding: This research was funded by the Ministry of Research, Technology, and Higher Education of Indonesia for the competitive grant of World Class Professor (WCP) program scheme–A (No.123.2/D2.3/KP/2018) and University Malaysia Pahang (UMP) for financial support through the short-term research grant scheme (RDU172204, RDU130131, RDU1703314 and RDU1603126).

Institutional Review Board Statement: Not applicable.

Informed Consent Statement: Not applicable.

Data Availability Statement: Not applicable.

Conflicts of Interest: The authors declare no conflict of interest.

Nomenclature

OT0.2	Diesel fuel + 0.2% of oxygenated turpentine oil
CO	Carbon monoxide
CO_2	Carbon dioxide
NO_X	Nitrogen oxides
HC	Hydro carbons
DI	Direct injection

References

1. Kaimal, V.K.; Vijayabalan, P. A study on synthesis of energy fuel from waste plastic and assessment of its potential as an alternative fuel for diesel engines. *Waste Manag.* **2016**, *51*, 91–96. [CrossRef]
2. Jayed, M.; Masjuki, H.; Kalam, M.; Mahlia, T.; Husnawan, M.; Liaquat, A. Prospects of dedicated biodiesel engine vehicles in Malaysia and Indonesia. *Renew. Sustain. Energy Rev.* **2011**, *15*, 220–235. [CrossRef]
3. Pilusa, T.J. The use of modified tyre derived fuel for compression ignition engines. *Waste Manag.* **2017**, *60*, 451–459. [CrossRef]
4. Faussone, G.C. Transportation fuel from plastic: Two cases of study. *Waste Manag.* **2018**, *73*, 416–423. [CrossRef] [PubMed]
5. Shim, E.; Park, H.; Bae, C. Intake air strategy for low HC and CO emissions in dual-fuel (CNG-diesel) premixed charge compression ignition engine. *Appl. Energy* **2018**, *225*, 1068–1077. [CrossRef]
6. Ospina, G.; Selim, M.Y.; Al Omari, S.A.; Ali, M.I.H.; Hussien, A.M. Engine roughness and exhaust emissions of a diesel engine fueled with three biofuels. *Renew. Energy* **2018**, *134*, 1465–1472. [CrossRef]
7. Amin, N.A.S.; Talebian-Kiakalaieh, A. Reduction of CO_2 emission by INCAM model in Malaysia biomass power plants during the year 2016. *Waste Manag.* **2018**, *73*, 256–264. [CrossRef] [PubMed]
8. Jensen, M.B.; Møller, J.; Mønster, J.; Scheutz, C. Quantification of greenhouse gas emissions from a biological waste treatment facility. *Waste Manag.* **2017**, *67*, 375–384. [CrossRef] [PubMed]
9. How, H.; Masjuki, H.; Kalam, M.; Teoh, Y.; Chuah, H. Effect of Calophyllum Inophyllum biodiesel-diesel blends on combustion, performance, exhaust particulate matter and gaseous emissions in a multi-cylinder diesel engine. *Fuel* **2018**, *227*, 154–164. [CrossRef]
10. Hupponen, M.; Grönman, K.; Horttanainen, M. Areas on which to focus when seeking to reduce the greenhouse gas emissions of commercial waste management. A case study of a hypermarket, Finland. *Waste Manag.* **2018**, *76*, 1–18. [CrossRef]
11. Röder, M.; Thornley, P. Waste wood as bioenergy feedstock. Climate change impacts and related emission uncertainties from waste wood based energy systems in the UK. *Waste Manag.* **2018**, *74*, 241–252. [CrossRef]
12. Pan, M.; Huang, R.; Liao, J.; Ouyang, T.; Zheng, Z.; Lv, D.; Huang, H. Effect of EGR dilution on combustion, performance and emission characteristics of a diesel engine fueled with n-pentanol and 2-ethylhexyl nitrate additive. *Energy Convers. Manag.* **2018**, *176*, 246–255. [CrossRef]
13. Allen, E.; Browne, J.; Hynes, S.; Murphy, J.D. The potential of algae blooms to produce renewable gaseous fuel. *Waste Manag.* **2013**, *33*, 2425–2433. [CrossRef]
14. Karmee, S.K. A spent coffee grounds based biorefinery for the production of biofuels, biopolymers, antioxidants and biocomposites. *Waste Manag.* **2018**, *72*, 240–254. [CrossRef] [PubMed]
15. Wang, Y.; Dai, L.; Fan, L.; Cao, L.; Zhou, Y.; Zhao, Y.; Liu, Y.; Ruan, R. Catalytic co-pyrolysis of waste vegetable oil and high density polyethylene for hydrocarbon fuel production. *Waste Manag.* **2017**, *61*, 276–282. [CrossRef] [PubMed]
16. da Silva, M.A.V.; Ferreira, B.L.G.; Marques, L.G.d.; Murta, A.L.S.; de Freitas, M.A.V. Comparative study of NOx emissions of biodiesel-diesel blends from soybean, palm and waste frying oils using methyl and ethyl transesterification routes. *Fuel* **2017**, *194*, 144–156. [CrossRef]
17. El Sheikh, K.; Khan, M.J.H.; Hamid, M.D.; Shrestha, S.; Ali, B.S.; Ryabov, G.A.; Dolgushin, L.A.; Hussain, M.A.; Bukharkina, T.V.; Golerova, E.A. Advances in reduction of NOx and N2O emission formation in an oxy-fired fluidized bed boiler. *Chin. J. Chem. Eng.* **2018**, *27*, 426–443. [CrossRef]
18. Hu, N.; Tan, J.; Wang, X.; Zhang, X.; Yu, P. Volatile organic compound emissions from an engine fueled with an ethanol-biodiesel-diesel blend. *J. Energy Inst.* **2017**, *90*, 101–109. [CrossRef]
19. Nayyar, A.; Sharma, D.; Soni, S.L.; Bhardwaj, B.; Augustine, M. Modeling and experimental investigation for performance and emissions on a diesel engine using bio-oxygenated ternary fuel blends. *Energy* **2019**, *168*, 136–150. [CrossRef]
20. Manigandan, S.; Gunasekar, P.; Devipriya, J.; Nithya, S. Emission and injection characteristics of corn biodiesel blends in diesel engine. *Fuel* **2019**, *235*, 723–735. [CrossRef]
21. Mueller, C.J.; Pitz, W.J.; Pickett, L.M.; Martin, G.C.; Siebers, D.L.; Westbrook, C.K. *Effects of Oxygenates on Soot Processes in DI Diesel Engines: Experiments and Numerical Simulations*; SAE Technical Paper 0148-7191; SAE International: Warrendale, PA, USA, 2003.
22. Awad, O.I.; Mamat, R.; Ali, O.M.; Yusri, I.M.; Abdullah, A.A.; Yusop, A.F.; Noor, M.M. The effect of adding fusel oil to diesel on the performance and the emissions characteristics in a single cylinder CI engine. *J. Energy Inst.* **2017**, *90*, 382–396. [CrossRef]
23. Couth, R.; Trois, C. Carbon emissions reduction strategies in Africa from improved waste management: A review. *Waste Manag.* **2010**, *30*, 2336–2346. [CrossRef] [PubMed]
24. Damodharan, D.; Sathiyagnanam, A.P.; Rana, D.; Saravanan, S.; Kumar, B.R.; Sethuramasamyraja, B. Effective utilization of waste plastic oil in a direct injection diesel engine using high carbon alcohols as oxygenated additives for cleaner emissions. *Energy Convers. Manag.* **2018**, *166*, 81–97. [CrossRef]
25. Chong, C.T.; Ng, J.-H.; Ahmad, S.; Rajoo, S. Oxygenated palm biodiesel: Ignition, combustion and emissions quantification in a light-duty diesel engine. *Energy Convers. Manag.* **2015**, *101*, 317–325. [CrossRef]
26. Ahmed, S.; Krumpelt, M. Hydrogen from hydrocarbon fuels for fuel cells. *Int. J. Hydrog. Energy* **2001**, *26*, 291–301. [CrossRef]
27. Szybist, J.P.; Song, J.; Alam, M.; Boehman, A.L. Biodiesel combustion, emissions and emission control. *Fuel Process. Technol.* **2007**, *88*, 679–691. [CrossRef]
28. Sastrohamidjojo, H. *Kimia Minyak Atsiri*; Universitas Gadjah Mada: Yogyakarta, Indonesia, 2004.

29. Dubey, P.; Gupta, R. Influences of dual bio-fuel (Jatropha biodiesel and turpentine oil) on single cylinder variable compression ratio diesel engine. *Renew. Energy* **2018**, *115*, 1294–1302. [CrossRef]
30. Dubey, P.; Gupta, R. Effects of dual bio-fuel (Jatropha biodiesel and turpentine oil) on a single cylinder naturally aspirated diesel engine without EGR. *Appl. Therm. Eng.* **2017**, *115*, 1137–1147. [CrossRef]
31. Karthikeyan, R.; Mahalakshmi, N.V. Performance and emission characteristics of a turpentine–diesel dual fuel engine. *Energy* **2007**, *32*, 1202–1209. [CrossRef]
32. Esplugas, S.; Bila, D.M.; Krause, L.G.T.; Dezotti, M. Ozonation and advanced oxidation technologies to remove endocrine disrupting chemicals (EDCs) and pharmaceuticals and personal care products (PPCPs) in water effluents. *J. Hazard. Mater.* **2007**, *149*, 631–642. [CrossRef]
33. Anand, B.P.; Saravanan, C.; Srinivasan, C.A. Performance and exhaust emission of turpentine oil powered direct injection diesel engine. *Renew. Energy* **2010**, *35*, 1179–1184. [CrossRef]
34. Anandavelu, K.; Alagumurthi, N.; Saravanan, C. Performance and emission evaluation of low heat rejection direct injection diesel engine fueled by diesel–turpentine oil blends. In Proceedings of the ASME 2010 International Mechanical Engineering Congress and Exposition, Vancouver, BC, Canada, 12–18 November 2010; pp. 1581–1587.
35. Polonowski, C.J.; Mathur, V.K.; Naber, J.; Blough, J.R. *Accelerometer Based Sensing of Combustion in a High Speed HPCR Diesel Engine*; SAE Technical Paper 0148-7191; SAE International: Warrendale, PA, USA, 2007.
36. Akbarian, E.; Najafi, B. A novel fuel containing glycerol triacetate additive, biodiesel and diesel blends to improve dual-fuelled diesel engines performance and exhaust emissions. *Fuel* **2019**, *236*, 666–676. [CrossRef]
37. Jiaqiang, E.; Zhang, Z.; Chen, J.; Pham, M.; Zhao, X.; Peng, Q.; Zhang, B.; Yin, Z. Performance and emission evaluation of a marine diesel engine fueled by water biodiesel-diesel emulsion blends with a fuel additive of a cerium oxide nanoparticle. *Energy Convers. Manag.* **2018**, *169*, 194–205.
38. Butkus, A.; Pukalskas, S.; Bogdanovičius, Z. The influence of turpentine additive on the ecological parameters of diesel engines. *Transport* **2007**, *22*, 80–82. [CrossRef]
39. Kadarohman, A.; Khoerunisa, F.H.; Astuti, R.M. Potency of clove oil and turpentine oil as a diesel fuel bioadditives and their performance on one cylinder engine. In Proceedings of the International Seminar on Chemistry, Jatinangor, Indonesia, 30–31 October 2008; pp. 30–31.
40. Godiño, J.A.V.; Aguilar, F.J.J.-E.; García, M.T. Simulation of HCCI combustion in air-cooled off-road engines fuelled with diesel and biodiesel. *J. Energy Inst.* **2018**, *91*, 549–562. [CrossRef]
41. Bednarski, M.; Orliński, P.; Wojs, M.K.; Sikora, M. Evaluation of methods for determining the combustion ignition delay in a diesel engine powered by liquid biofuel. *J. Energy Inst.* **2018**, *92*, 1107–1114. [CrossRef]
42. Kannan, M.; Karthikeyan, R.; Deepanraj, B.; Baskaran, R. Feasibility and performance study of turpentine fueled DI diesel engine operated under HCCI combustion mode. *J. Mech. Sci. Technol.* **2014**, *28*, 729–737. [CrossRef]
43. Berndt, T.; Böge, O.; Stratmann, F. Gas-phase ozonolysis of α-pinene: Gaseous products and particle formation. *Atmos. Environ.* **2003**, *37*, 3933–3945. [CrossRef]
44. Kadarohman, A.; Rohman, I.; Kusrini, R.; Astuti, R.M. Combustion characteristics of diesel fuel on one cylinder diesel engine using clove oil, eugenol, and eugenyl acetate as fuel bio-additives. *Fuel* **2012**, *98*, 73–79. [CrossRef]
45. Song, J.; Cheenkachorn, K.; Wang, J.; Perez, J.; Boehman, A.L.; Young, P.J.; Waller, F.J. Effect of oxygenated fuel on combustion and emissions in a light-duty turbo diesel engine. *Energy Fuels* **2002**, *16*, 294–301. [CrossRef]
46. Choi, C.-Y.; Reitz, R.D. An experimental study on the effects of oxygenated fuel blends and multiple injection strategies on DI diesel engine emissions. *Fuel* **1999**, *78*, 1303–1317. [CrossRef]
47. Fuel, D. *Diesel Fuel And Exhaust Emissions*; WHO Library Cataloguing: Geneva, Switzerland, 1996.
48. Jonsson, Å.M.; Hallquist, M.; Ljungström, E. Impact of humidity on the ozone initiated oxidation of limonene, Δ3-carene, and α-pinene. *Environ. Sci. Technol.* **2006**, *40*, 188–194. [CrossRef] [PubMed]
49. Singh, P.; Chauhan, S.; Goel, V. Assessment of diesel engine combustion, performance and emission characteristics fuelled with dual fuel blends. *Renew. Energy* **2018**, *125*, 501–510. [CrossRef]
50. Tamilselvan, P.; Nallusamy, N.; Rajkumar, S. A comprehensive review on performance, combustion and emission characteristics of biodiesel fuelled diesel engines. *Renew. Sustain. Energy Rev.* **2017**, *79*, 1134–1159. [CrossRef]
51. Soudagar, M.E.M.; Nik-Ghazali, N.-N.; Kalam, M.A.; Badruddin, I.; Banapurmath, N.; Akram, N. The effect of nano-additives in diesel-biodiesel fuel blends: A comprehensive review on stability, engine performance and emission characteristics. *Energy Convers. Manag.* **2018**, *178*, 146–177. [CrossRef]
52. Vigneswaran, R.; Annamalai, K.; Dhinesh, B.; Krishnamoorthy, R. Experimental investigation of unmodified diesel engine performance, combustion and emission with multipurpose additive along with water-in-diesel emulsion fuel. *Energy Convers. Manag.* **2018**, *172*, 370–380. [CrossRef]
53. Sahoo, R.R.; Jain, A. Experimental analysis of nanofuel additives with magnetic fuel conditioning for diesel engine performance and emissions. *Fuel* **2019**, *236*, 365–372. [CrossRef]
54. Och, S.H.; Moura, L.M.; Mariani, V.C.; Coelho, L.d.S.; Velasquez, J.A.; Domingues, E. Volumetric efficiency optimization of a single-cylinder DI diesel engine using differential evolution algorithm. *Appl. Therm. Eng.* **2016**, *108*, 660–669. [CrossRef]
55. Huang, H.; Li, Z.; Teng, W.; Zhou, C.; Huang, R.; Liu, H.; Pan, M. Influence of n-butanol-diesel-PODE3-4 fuels coupled pilot injection strategy on combustion and emission characteristics of diesel engine. *Fuel* **2019**, *236*, 313–324. [CrossRef]

56. Örs, I.; Sarıkoç, S.; Atabani, A.; Ünalan, S.; Akansu, S. The effects on performance, combustion and emission characteristics of DICI engine fuelled with TiO2 nanoparticles addition in diesel/biodiesel/n-butanol blends. *Fuel* **2018**, *234*, 177–188. [CrossRef]
57. Yadav, A.K.; Dewangan, A.; Mallick, A. Effect of n-Butanol and Diethyl Ether on Performance and Emission Characteristics of a Diesel Engine Fueled with Diesel–Pongamia Biodiesel Blend. *J. Energy Eng.* **2018**, *144*, 4018062. [CrossRef]
58. Basha, J.S. Impact of Carbon Nanotubes and Di-Ethyl Ether as additives with biodiesel emulsion fuels in a diesel engine–An experimental investigation. *J. Energy Inst.* **2016**, *91*, 289–303. [CrossRef]
59. Sharma, G.; Lal, H. Effects of ethanol-gasoline blends on engine performance and exhaust emissions in a spark ignition. *Int. J. Emerg. Technol.* **2015**, *6*, 184.
60. Bilgin, A.; Durgun, O.; Sahin, Z. The effects of diesel-ethanol blends on diesel engine performance. *Energy Sources* **2002**, *24*, 431–440. [CrossRef]
61. Jamrozik, A.; Tutak, W.; Pyrc, M.; Gruca, M.; Kočiško, M. Study on co-combustion of diesel fuel with oxygenated alcohols in a compression ignition dual-fuel engine. *Fuel* **2018**, *221*, 329–345. [CrossRef]
62. Hagos, F.Y.; Ali, O.M.; Mamat, R.; Abdullah, A.A. Effect of emulsification and blending on the oxygenation and substitution of diesel fuel for compression ignition engine. *Renew. Sustain. Energy Rev.* **2017**, *75*, 1281–1294. [CrossRef]
63. Redel-Macías, M.; Pinzi, S.; Leiva-Candia, D.; López, I.; Dorado, M. Ternary blends of diesel fuel oxygenated with ethanol and castor oil for diesel engines. *Energy Procedia* **2017**, *142*, 855–860. [CrossRef]
64. Park, W.; Park, S.; Reitz, R.D.; Kurtz, E. The effect of oxygenated fuel properties on diesel spray combustion and soot formation. *Combust. Flame* **2017**, *180*, 276–283. [CrossRef]
65. Algayyim, S.J.M.; Wandel, A.P.; Yusaf, T.; Al-Lwayzy, S.; Hamawand, I. Impact of butanol-acetone mixture as a fuel additive on diesel engine performance and emissions. *Fuel* **2018**, *227*, 118–126. [CrossRef]
66. Cybulski, A.; Moulijn, J.A. *Structured Catalysts and Reactors*; CRC: Boca Raton, FL, USA, 2005.
67. Naha, S.; Aggarwal, S.K. Fuel effects on NOx emissions in partially premixed flames. *Combust. Flame* **2004**, *139*, 90–105. [CrossRef]
68. Winer, W.O.; Bergles, A.E.; Klutke, G.A.; Wang, K.K.; Finnie, I.; Welty, J.R.; Bryant, M.D.; Yang, H.T.; Mow, V.C.; Leckie, F.A. *Nanoindentation*; Mechanical Engineering Series; Springer: New York, NY, USA, 2011.
69. Balaji, G.; Cheralathan, M. Study of antioxidant effect on oxidation stability and emissions in a methyl ester of neem oil fuelled DI diesel engine. *J. Energy Inst.* **2014**, *87*, 188–195. [CrossRef]
70. Huang, H.; Liu, Q.; Teng, W.; Pan, M.; Liu, C.; Wang, Q. Improvement of combustion performance and emissions in diesel engines by fueling n-butanol/diesel/PODE3–4 mixtures. *Appl. Energy* **2018**, *227*, 38–48. [CrossRef]

energies

MDPI

Article

Two-Stage Robust and Economic Scheduling for Electricity-Heat Integrated Energy System under Wind Power Uncertainty

Ruijie Liu [1], Zhejing Bao [1,*], Jun Zheng [2], Lingxia Lu [1] and Miao Yu [1]

1 College of Electrical Engineering, Zhejiang University, Hangzhou 310027, China; liurj@zju.edu.cn (R.L.); lulingxia@zju.edu.cn (L.L.); zjuyumiao@zju.edu.cn (M.Y.)
2 Research Institute of Zhejiang University-Taizhou, Taizhou 318000, China; dbzj@netease.com
* Correspondence: zjbao@zju.edu.cn

Abstract: As renewable energy increasingly penetrates into electricity-heat integrated energy system (IES), the severe challenges arise for system reliability under uncertain generations. A two-stage approach consisting of pre-scheduling and re-dispatching coordination is introduced for IES under wind power uncertainty. In pre-scheduling coordination framework, with the forecasted wind power, the robust and economic generations and reserves are optimized. In re-dispatching, the coordination of electric generators and combined heat and power (CHP) unit, constrained by the pre-scheduled results, are implemented to absorb the uncertain wind power prediction error. The dynamics of building and heat network is modeled to characterize their inherent thermal storage capability, being utilized in enhancing the flexibility and improving the economics of IES operation; accordingly, the multi-timescale of heating and electric networks is considered in pre-scheduling and re-dispatching coordination. In simulations, it is shown that the approach could improve the economics and robustness of IES under wind power uncertainty by taking advantage of thermal storage properties of building and heat network, and the reserves of electricity and heat are discussed when generators have different inertia constants and ramping rates.

Keywords: multi-timescale; integrated energy system (IES); robust; scheduling; uncertainty

Citation: Liu, R.; Bao, Z.; Zheng, J.; Lu, L.; Yu, M. Two-Stage Robust and Economic Scheduling for Electricity-Heat Integrated Energy System under Wind Power Uncertainty. *Energies* **2021**, *14*, 8434. https://doi.org/10.3390/en14248434

Academic Editor: Zbigniew Leonowicz

Received: 12 November 2021
Accepted: 11 December 2021
Published: 14 December 2021

1. Introduction

With the enhancement of coupling between multi-type energy sources, integrated energy system (IES) has drawn the increasing attention. In IES, combined heat and power (CHP) unit, as a significant component, generates electricity and heat simultaneously, leading to the higher energy utilization efficiency. With the growth in utilization of CHP unit, its heat-led mode has caused serious wind abandonment especially in winter heating periods. This becomes a key issue limiting wind power penetration. Moreover, the strong intermittency and uncertainty of wind power make its precise forecasting difficult to achieve; as a result, the current wind power prediction error is usually up to 25% to 40% [1], imposing serious challenges to the secure and stable operation of IES.

Many studies have been conducted to improve the flexibility of electricity and heat coupled IES under wind power uncertainty. The maximum flexibility of a combined heat and power system with thermal energy storage is discussed [2], where the robustness of the system under renewable energy resources uncertainty is not considered. A chance-constrained programming-based scheduling is proposed [3], with the joint operation optimization of battery energy storage and heat storage tank integrated. However, the distribution of uncertainty is assumed to be known, which is inconsistent with engineering practice. Two-stage scheduling is a commonly used approach to deal with wind power uncertainty. The two stages are implemented day-ahead and in real time, based on day-ahead wind power prediction and wind power realization, respectively. In the first pre-scheduling stage, the factors such as units' startup and shutdown, heat storage tank capacity, etc. need to be determined in advance. Furthermore, in the second stage, namely re-dispatching, the decision should be amended to compensate for wind prediction error, such as units' generations and so on. In [4], a minimax regret model based two-stage robust scheduling for IES is introduced, where the electrical and thermal load tracking strategies are applied to attenuate the uncertainty. However, its heat balance cannot be guaranteed under wind power uncertainty. A two-stage robust operation strategy is explored [5]. The decisions of day-ahead thermal storage charging/discharging are in the first stage, and the decisions of CCHP and auxiliary boiler output are in the second stage to compensate the first stage operation and follow the uncertainty realization. However, its timescale for the second-stage is too long. The real-time robustness of the system cannot be guaranteed due to the randomness of wind power changes. A scenario-based stochastic multi-energy scheduling is developed in [6], where the scenario-independent and scenario-relative two-stage decisions are made in an optimization model with various energy storage considered. However, of all the mentioned research above, the energy storage units are installed to alleviate uncertain wind power, which is not recommended since it might cause extra costs and failures. Taking full advantage of thermal energy storage of heating system can solve this problem.

Distinct from electric system, heat energy supply and demand balance are kept not instantaneously but during a period, since a few minutes to several hours are taken for hot water to carry heat energy from source to load through pipeline [7]. Moreover, the real-time heat power imbalance is reflected in the variation of water temperature, the operational limits of which are within a wider range. Thus, contrary to the electric transmission network, the heat network could serve as a natural thermal storage, bringing great flexibility to wind power absorption in IES. Several scholars have noticed the potential role of heat network and endeavored to implement coordinated scheduling by considering the dynamics of thermal energy transmission. The unit commitment in IES is studied [8], where temperature quasi-dynamics of heat network is modeled to characterize its heat storage capacity in CF-VT (constant mass flow and variable temperature) strategy. The intra-day power dispatching of IES is explored [7], integrating the heat network dynamics under the variable mass flow and variable temperature (VF-VT) strategy. A dispatching model of IES is proposed, considering thermal energy storage of pipelines and the detailed heat transfer constraints [9].

Furthermore, the heat load at an instant is usually pre-given as a constant in previous research on heat and electricity coupled scheduling. However, since the building has the potential of thermal energy storage and could offer a source of flexibility to absorb wind power, it is necessary to model the heat load and then integrate it into the coordinated scheduling. The storage capacity of building could be illustrated as follows: similar to heat network, the thermal inertia of building is reflected in thermal transmission dynamics, usually lasting for a period that could not be ignored. In building, the instantaneous imbalance between heat power supply and demand is allowed, resulting in indoor temperature changes; and the indoor temperature meeting human comfort requirements is usually given as an interval. In addition, modeling the building's heat load by considering the heat loss and comfortable requirement for indoor temperature could improve the practicality of the approach. In a few studies on heat and electricity coupled scheduling, the dynamic heat load model of buildings is established. Feasible region method is proposed to formulate the flexibility of IES [10] and the IES scheduling with demand response is explored [11], where the first-order equivalent thermal parameter method is employed to model the heat dissipation of building. The thermal modeling of dwelling is established through equivalent thermal resistance and capacitance, and the expected thermal discomfort metric is defined to quantify user's discomfort level [12].

In order to accommodate uncertain wind power in two stages of pre-scheduling with its forecasting value and re-dispatching with its uncertain realization, a two-stage robust and economical scheduling methodology of IES is developed. In this method, the natural storage capacity of heat network and buildings, and the reserves and generations of electric generators and CHP unit are utilized. The main contributions of the paper are summarized as follows:

- The dynamic models of heat network and building are established to describe their thermal energy storage characteristics. Their storage capacities are utilized and integrated in the two-stage robust and economic scheduling to accommodate wind power. The thermal transmission dynamics in the seat network represents the water temperature variations with respect to time variable and pipeline position, considering heat loss, heat propagation and time-delay of flowing. In order to make a practical description of the thermal demand from building, the heat dissipation process from indoor to outdoor is modeled, and the standard effective temperature (SET) is applied to measure the degree of user's thermal comfort.
- Inspired by the robust scheduling of power system [13], the approach for accommodating uncertain wind power in heat and electricity coupled IES is developed, specifically in two stages, i.e., in pre-scheduling with its forecasted value and in re-dispatching with its realization, by utilizing the coupled reserves of electricity generators and co-generator CHP unit, along with the natural thermal storage of heat network and building. With the predicted wind power, the robust generation and reserves of electric generators and CHP unit are optimized in pre-scheduling, where the dynamics of building and heat network are integrated to absorb wind power. With the uncertain realization of wind power, prediction error is alleviated in re-dispatching constrained by the reserves optimized in pre-scheduling. In pre-scheduling, the feasibility constraints of re-dispatching solution under uncertain realization are considered, with zero-sum game between the re-dispatching and wind power uncertainty formulated as a max–min problem.
- In the electricity-heat IES, the generators with different inertia constants and reserve costs are considered, leading to two situations: the reserve supply by CHP unit or not. Due to the strong coupling between electricity supply and heat provision of CHP unit, when the CHP generator takes on the electricity reserve, there exists the security margins of temperature in heat network and building to provide heat reserve. From this perspective, the natural energy storage capacity of heat pipeline and building could be utilized to serve as heat reserve in electricity-heat coupled IES. When no reserve is required from CHP unit due to its high cost, the temperatures in heat network and building could reach their bounds, induced by the objective of economic operation.

The remainder of this paper is organized as follows. Section 2 describes the framework of two-stage robust economic scheduling approach. In Section 3, the detailed formulations of robust economic scheduling problem in electricity-heat coupled IES are illustrated, where the heat transmission dynamics in heat network and building are modeled. Simulation results are presented in Section 4 to demonstrate the effectiveness of the proposed approach. Conclusions are finally given in Section 5.

2. Framework Description

2.1. Uncertainty Set

In robust and economic scheduling, uncertainty set is used to define the possible range of uncertain variables. Excessive description of uncertainty may lead to conservatism, i.e., higher operational cost; while insufficient consideration of uncertainty cannot guarantee the operational reliability under uncertain realizations. Since it is almost impossible for all the predicted values simultaneously reaching the boundaries, budget uncertainty set is adopted to describe and restrict wind power uncertainty [13], formulated as:

$$\boldsymbol{P}_{\mathbf{wind}} = \{P_{\text{wind},t} = P^{\text{m}}_{\text{wind},t} + v^{+}_{t} P^{\text{u}}_{\text{wind},t} - v^{-}_{t} P^{\text{u}}_{\text{wind},t}, v^{+}_{t}, v^{-}_{t} \in \{0,1\}, v^{+}_{t} + v^{-}_{t} \leq 1, \forall t; \sum_{t}(v^{+}_{t} + v^{-}_{t}) \leq \Gamma\} \quad (1)$$

$$\begin{aligned} P^{\mathrm{m}}_{\mathrm{wind},t} &= 0.5(P^{\min}_{\mathrm{wind},t} + P^{\max}_{\mathrm{wind},t}) \\ P^{\mathrm{u}}_{\mathrm{wind},t} &= 0.5(P^{\max}_{\mathrm{wind},t} - P^{\min}_{\mathrm{wind},t}) \end{aligned} \quad (2)$$

where $\boldsymbol{P}_{\mathbf{wind}}$ is the uncertainty set of wind power; $P_{\mathrm{wind},t}$ is the wind power at time t; $P^{\min}_{\mathrm{wind},t}$ and $P^{\max}_{\mathrm{wind},t}$ are the upper and lower bounds of prediction interval; the uncertainty budge Γ is the number of uncertainty variables reaching the boundaries, influencing the conservativeness. According to the central limit theorem, the value of Γ is given as [13]

$$\Gamma = N\mu + \Phi^{-1}(\alpha)\sqrt{N}\sigma \quad (3)$$

where μ and σ are the expected value and standard deviation of $\frac{|P_{\mathrm{wind},t} - P^{\mathrm{m}}_{\mathrm{wind},t}|}{P^{\mathrm{u}}_{\mathrm{wind},t}}$; N is the number of uncertain variables; $\Phi(\cdot)$ is the cumulative probability distribution function of standard normal distribution; α is the confidence level. All the parameters in (1)–(3) can be obtained by the historical and predicted values of wind power.

2.2. Coordinated Framework and Multi-Timescale

To alleviate wind power uncertainty, before and after the real value of wind power is observed, the scheduling is divided into two stages, i.e., pre-scheduling and re-dispatching, the framework of which is displayed in Figure 1. In pre-scheduling stage, the robust and economic generations and reserves of electric generators and CHP unit are optimized based on the predicted wind power, where the robust feasibility constraint is considered. In re-dispatching stage, with the wind power realization, the coordinated generations of electric generators and CHP unit are optimized within the pre-scheduled reserves to compensate the wind power forecasting error. In the pre-scheduling and re-dispatching, the optimization is implemented over a time window rather than just at an instant in order to describe the thermal dynamics. In the proposed approach, two kinds of generators are included, the CHP unit and electricity generator that only generates electric power.

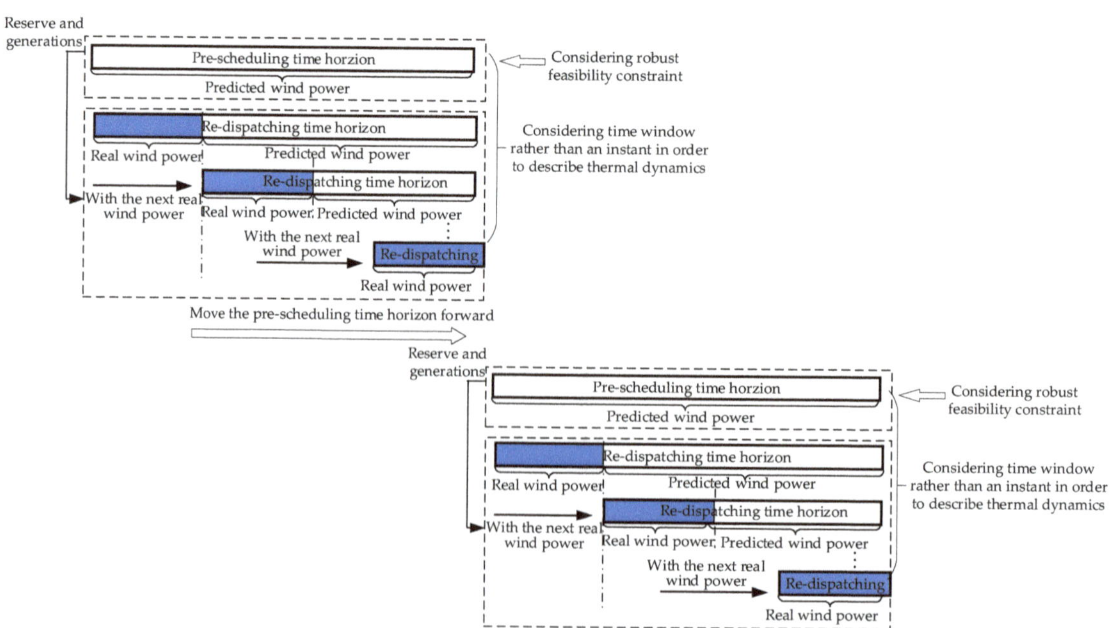

Figure 1. Coordinated framework of pre-scheduling and re-dispatching.

The pre-scheduling problem is formulated as (4) [13]:

$$\begin{cases} \min\ \boldsymbol{a}^{\mathrm{T}}\boldsymbol{x} \\ \text{s.t.}\ \ \boldsymbol{A}\boldsymbol{x}+\boldsymbol{C}\boldsymbol{w}^{\mathrm{p}}\leq \boldsymbol{b} \\ \qquad \boldsymbol{Y}(\boldsymbol{x},\boldsymbol{w})\neq\varnothing\ \ \forall\boldsymbol{w}\in\boldsymbol{W} \end{cases} \tag{4}$$

where $\boldsymbol{a}$ is the cost coefficient vector; $\boldsymbol{A}\boldsymbol{x}+\boldsymbol{C}\boldsymbol{w}^{\mathrm{p}}\leq \boldsymbol{b}$ represents the constraints for economic operation; $\boldsymbol{w}^{\mathrm{p}}$ is the predictive wind power; $\boldsymbol{Y}$ denotes the feasible region of re-dispatching strategy, which is a function of the uncertain wind power realization $\boldsymbol{w}$ and pre-scheduling decision $\boldsymbol{x}$; the uncertain set of wind power is defined as $\boldsymbol{W}$, determined by (1)–(3). In order to ensure the secure operation of IES under the uncertain wind power prediction error, the pre-scheduling strategy $\boldsymbol{x}$, composed by the optimal coordinated generations and reserves of electric generators and CHP unit, needs to ensure that there exists a feasible re-dispatching strategy $\boldsymbol{y}$ for any $\boldsymbol{w}$ under the pre-scheduled reserves. Insufficient reserves may lead to the infeasibility of re-dispatching and drive the system to an unsecure operating state. On the contrary, excessive reserves may result in higher operational cost. Therefore, as shown in pre-scheduling problem (4), $\boldsymbol{Y}(\boldsymbol{x},\boldsymbol{w})\neq\varnothing$ gives the constraint that the appropriate reserves should satisfy. The detailed formulation of $\boldsymbol{Y}$ will be illustrated in Section 2.3, where zero-sum game describes the relation between uncertain wind power realization and re-dispatching.

The re-dispatching optimization can be expressed as:

$$\begin{cases} \min\ \boldsymbol{d}^{\mathrm{T}}\boldsymbol{y} \\ \text{s.t.}\ \ \boldsymbol{A}\boldsymbol{x}+\boldsymbol{B}\boldsymbol{y}+\boldsymbol{C}\boldsymbol{w}^{\mathrm{r(p)}}\leq \boldsymbol{b} \end{cases} \tag{5}$$

As shown in Figure 1, in the re-dispatching time horizon, considering that the current dispatching strategy may have some impact on the subsequent operation states of IES because of the long transient process of heat network, not only the real wind power at the current moment but the predicted values at following instants are used, denoted as $\boldsymbol{w}^{\mathrm{r(p)}}$ in (5). Among the re-dispatched strategies over the time horizon, only the current one is implemented on the IES.

In the pre-scheduling and re-dispatching of electricity-heat coupled IES, two timescales are considered for electricity and heat networks coordinate in the unified framework, shown in Figures 2 and 3. It is assumed that the time resolution of wind power prediction is Δt. The smaller time resolution of real wind power fluctuation is defined as $\Delta\tau$. The inertia of CHP unit is assumed to be smaller and the time resolution of ramping up/ down is set as $\Delta\tau$. The electricity generators having different inertia are considered, respectively with the different ramping time resolution $\Delta\tau$ and Δt.

In pre-scheduling, the timescale of the variables in power grid, such as the electricity power of CHP unit and electric generators, is same as that of the predicted wind power; the timescale of the variables in heat network is $\Delta\tau$, such as the heating power by CHP unit, the temperatures of flowing water and insulation layer, heat load, indoor temperature and so on; the timescale of the reserves from CHP unit and the electric generator with smaller inertia is $\Delta\tau$ and that of the reserves from the electric generator with larger inertia is Δt. In re-dispatching, the time resolution of real wind power is $\Delta\tau$; the timescale of the generations from CHP unit and the electric generator with smaller inertia is $\Delta\tau$ since they are dispatched to follow wind power fluctuation, while that of the power of the electric generator with larger inertia is Δt because of its slower ramping rate.

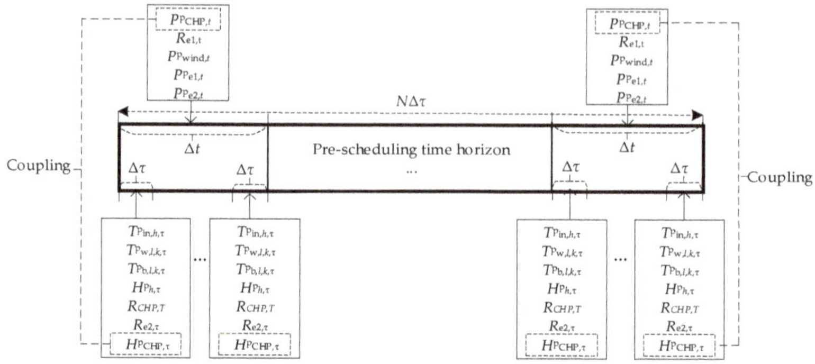

Figure 2. Multi-timescale coordination in pre-scheduling.

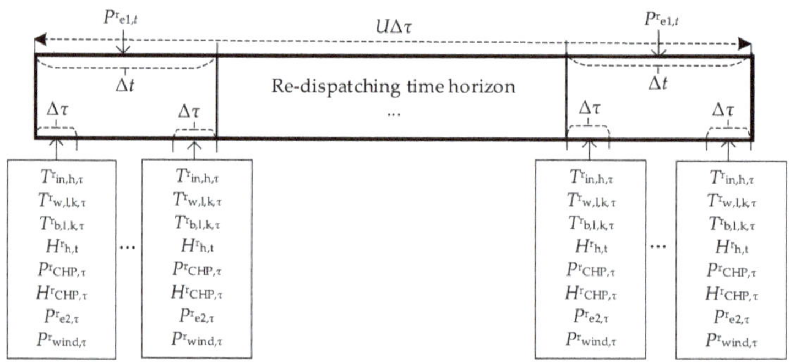

Figure 3. Multi-timescale coordination in re-dispatching.

2.3. Robust Feasibility Constraint of Re-Dispatching in Pre-Scheduling

In pre-scheduling, besides the electricity and heat power generations, the optimal reserves of electric generators and CHP unit are derived to ensure the feasibility of re-dispatching under uncertain wind power, while both economy and robustness are guaranteed. The dispatcher and the uncertain wind power act as the two sides of zero-sum game [13], i.e., uncertainty intends to violate the secure constrains of the system as much as possible, resulting in more reserve required; on the contrary, facing the uncertainty, the dispatcher tries to maintain secure operation through dispatching strategy constrained by the reserve, which is expected as low as possible due to the objective of economic operation.

With a given pre-scheduling strategy $\boldsymbol{x}^*$, the indicator $S(\boldsymbol{x}^*, \boldsymbol{w}^\star)$ is formulated in (6) to reflect the feasibility of re-dispatching strategy $\boldsymbol{y}$ under wind power realization $\boldsymbol{w}^*$ [13].

$$\begin{cases} S(\boldsymbol{x}^*, \boldsymbol{w}^\star) = \min\limits_{\boldsymbol{y},\boldsymbol{r}^+,\boldsymbol{r}^-} (\mathbf{1}^{\mathrm{T}}\boldsymbol{r}^+ + \mathbf{1}^{\mathrm{T}}\boldsymbol{r}^-) \\ \text{s.t. } \boldsymbol{Ax}^* + \boldsymbol{By} + \boldsymbol{Cw}^* + \boldsymbol{Ir}^+ - \boldsymbol{Ir}^- \leq \boldsymbol{b} \\ \qquad \boldsymbol{r}^+ \geq 0, \boldsymbol{r}^- \geq 0 \end{cases} \tag{6}$$

where $\boldsymbol{r}^+ \geq 0$ and $\boldsymbol{r}^- \geq 0$ are the introduced slack variables. If re-dispatching is feasible, there must exist a solution where $S(\boldsymbol{x}^*, \boldsymbol{w}^\star) = 0$; on the contrary, if re-dispatching is unfeasible, $S(\boldsymbol{x}^*, \boldsymbol{w}^\star) > 0$.

The most unfavorable wind power realization $\boldsymbol{w}$ intends to maximize S, and the zero-sum game between the dispatcher and wind power uncertainty is expressed as

$$\begin{cases} S(\boldsymbol{x}^*) = \max\limits_{\boldsymbol{w}} \min\limits_{\boldsymbol{y},\boldsymbol{r}^+,\boldsymbol{r}^-} (\mathbf{1}^{\mathrm{T}}\boldsymbol{r}^+ + \mathbf{1}^{\mathrm{T}}\boldsymbol{r}^-) \\ \text{s.t. } \boldsymbol{Ax}^* + \boldsymbol{By} + \boldsymbol{Cw} + \boldsymbol{Ir}^+ - \boldsymbol{Ir}^- \leq \boldsymbol{b} \\ \qquad \boldsymbol{w} \in \boldsymbol{W} \end{cases} \tag{7}$$

With the pre-scheduling strategy $\boldsymbol{x}^*$, $S(\boldsymbol{x}^*) = 0$ suggests that there exists the feasible re-dispatching strategy $\boldsymbol{y}$ under any wind power realization $\boldsymbol{w}$; $S(\boldsymbol{x}^*) > 0$ indicates that the most unfavorable wind power realization $\boldsymbol{w}$ could make re-dispatching problem infeasible. By transforming the inner layer into its dual problem, the problem in (7) can be converted into a single-layer optimization:

$$\begin{cases} S(\boldsymbol{x}^*) = \max(\boldsymbol{o}^{\mathrm{T}}(\boldsymbol{b} - \boldsymbol{A}\boldsymbol{x}^*) - \boldsymbol{o}^{\mathrm{T}}\boldsymbol{C}\boldsymbol{w}) \\ \text{s.t. } \boldsymbol{o}^{\mathrm{T}}\boldsymbol{B} \leq \boldsymbol{0}^{\mathrm{T}} \\ \qquad -\boldsymbol{1}^{\mathrm{T}} \leq \boldsymbol{o}^{\mathrm{T}} \leq \boldsymbol{0}^{\mathrm{T}} \\ \qquad \boldsymbol{w} \in \boldsymbol{W} \end{cases} \tag{8}$$

where $\boldsymbol{o}$ is dual variable. The optimization (8) is a mixed integer programming problem containing quadratic form $\boldsymbol{o}^{\mathrm{T}}\boldsymbol{C}\boldsymbol{w}$ in its objective function. It can be solved by many solvers such as Cplex.

Up to now, it is almost impossible to give the explicit expression of constraint $\boldsymbol{Y}(\boldsymbol{x}, \boldsymbol{w}) \neq \varnothing$. However, with the solution of (8), the iterative solving approach of pre-scheduling strategy $\boldsymbol{x}^*$ is proposed to guarantee its robust feasibility [13], illustrated as follows:

$S(\boldsymbol{x})$ is linearly approximated at $\boldsymbol{x}^*$

$$S(\boldsymbol{x}) \approx S(\boldsymbol{x}^*) - (\boldsymbol{o}^*)^{\mathrm{T}}\boldsymbol{A}(\boldsymbol{x} - \boldsymbol{x}^*) \tag{9}$$

where $\boldsymbol{o}^*$ is the optimal solution of (8). Then, the robust feasible $\boldsymbol{x}$ should satisfy the constraint.

$$(\boldsymbol{o}^*)^{\mathrm{T}}\boldsymbol{A}\boldsymbol{x} \geq S(\boldsymbol{x}^*) + (\boldsymbol{o}^*)^{\mathrm{T}}\boldsymbol{A}\boldsymbol{x}^* \tag{10}$$

In order to replace $\boldsymbol{Y}(\boldsymbol{x}, \boldsymbol{w}) \neq \varnothing$, constraint (10) is gradually added to the optimization problem (8) until $S(\boldsymbol{x}^*) = 0$. Then a robust feasible solution $\boldsymbol{x}^*$ can be derived.

2.4. Procedure of Two-Stage Robust Economic Scheduling

The procedure of two-stage robust economic scheduling is illustrated as follows.

Step 1. Set initial parameter $k = 0$, $\boldsymbol{x}_0 = \boldsymbol{0}$, $\boldsymbol{o}_0 = \boldsymbol{0}$, $\boldsymbol{x}^*$, where k denotes the iterative step;

Step 2. The pre-scheduling problem (4) is solved, where the robust feasible constraint $\boldsymbol{Y}(\boldsymbol{x}, \boldsymbol{w}) \neq \varnothing$ is replaced by the constraints $(\boldsymbol{o}_l)^{\mathrm{T}}\boldsymbol{A}\boldsymbol{x} \geq S_l + (\boldsymbol{o}_l)^{\mathrm{T}}\boldsymbol{A}\boldsymbol{x}_l$, $0 \leq l \leq k$; and then $k = k + 1$.

Step 3. With the obtained pre-scheduling strategy $\boldsymbol{x}^*$, calculate $S(\boldsymbol{x}^*)$ according to (8). If $S(\boldsymbol{x}^*) = 0$, the pre-scheduling ends and the re-dispatching in (5) is optimized with the real wind power; otherwise, derive the constraint $(\boldsymbol{o}_k)^{\mathrm{T}}\boldsymbol{A}\boldsymbol{x} \geq S_k + (\boldsymbol{o}_k)^{\mathrm{T}}\boldsymbol{A}\boldsymbol{x}_k$ where $S_k = S(\boldsymbol{x}^*)$, $\boldsymbol{o}_k = \boldsymbol{o}^*$, $\boldsymbol{x}_k = \boldsymbol{x}^*$ and go to Step 2.

3. Model Formulation

3.1. Pre-Scheduling Model

3.1.1. Optimization Objective

The optimization objective in pre-scheduling is to minimize the total costs during the time horizon, including the costs for operations and reserves of CHP unit and electric generators. It is formulated as:

$$\min\left(\sum_{t=1}^{N\Delta\tau/\Delta t}\left(a_{\mathrm{e1}}\cdot P^{\mathrm{P}}_{\mathrm{e1},t} + a_{\mathrm{e2}}\cdot P^{\mathrm{P}}_{\mathrm{e2},t} + a_{\mathrm{CHP}}\cdot P^{\mathrm{P}}_{\mathrm{CHP},t} + q_{\mathrm{e1}}\cdot R_{\mathrm{e1},t}\right)\cdot\Delta t + \sum_{\tau=1}^{N}\left(q_{\mathrm{CHP}}\cdot R_{\mathrm{CHP},\tau} + q_{\mathrm{e2}}\cdot R_{\mathrm{e2},\tau}\right)\cdot\Delta\tau\right) \tag{11}$$

3.1.2. Optimization Constraints

(1) CHP unit

There are usually two types of CHP units, back-pressure turbine and extraction condensing turbine [14]. For the former, the heat-to-electricity ratio is constant and the relation between electricity and heat power is linear; for the latter, the heat-to-electricity ratio varies in a wide range by pumping rate, with the lower energy efficiency and more flexibility compared with the former. In the paper, in order to show the flexibility brought by thermal storage of heat network and building, the CHP unit with the fixed heat-to-electricity ratio is chosen to study. Its operational characteristics is described as:

$$P^{\mathrm{P}}_{\mathrm{CHP},t} = K \cdot H^{\mathrm{P}}_{\mathrm{CHP},\tau}, \tau \in t \tag{12}$$

as shown in (12), the CHP unit combines the different timescale τ and t.

CHP unit generation is constrained by the ramping rate in MW/$\Delta\tau$. Since the timescale of CHP unit generated electric power is Δt in pre-scheduling, its ramping rate constraint is described as:

$$P^{\mathrm{down}}_{\mathrm{CHP}} \cdot \frac{\Delta t}{\Delta\tau} \leq P^{\mathrm{P}}_{\mathrm{CHP},t+1} - P^{\mathrm{P}}_{\mathrm{CHP},t} \leq P^{\mathrm{up}}_{\mathrm{CHP}} \cdot \frac{\Delta t}{\Delta\tau}, t = 1, \cdots, \frac{N\Delta t}{\Delta\tau} - 1 \tag{13}$$

The scheduled electric power output of CHP is bounded by its upper and lower limits considering reserve:

$$P^{\min}_{\mathrm{CHP}} + R_{\mathrm{CHP},\tau} \leq P^{\mathrm{P}}_{\mathrm{CHP},t} \leq P^{\max}_{\mathrm{CHP}} - R_{\mathrm{CHP},\tau}, t = 1, \cdots, \frac{N\Delta t}{\Delta\tau}, \tau \in t \tag{14}$$

The scheduled reserve cannot exceed its upper limit:

$$0 \leq R_{\mathrm{CHP},\tau} \leq R^{\max}_{\mathrm{CHP}} \tag{15}$$

(2) Electricity network

For electric network modeling, DC power flow model is employed for simplicity. The active power flow $P_{mn,t}$ through bus m to bus n is described as

$$P^{\mathrm{P}}_{mn,t} = -b_{mn}\left(\theta^{\mathrm{P}}_{m,t} - \theta^{\mathrm{P}}_{n,t}\right), t = 1, \cdots, \frac{N\Delta t}{\Delta\tau} \tag{16}$$

where b_{mn} is the reactance of line from m to n; $\theta_{m,t}$ and $\theta_{n,t}$ are the voltage phase angle at bus m and n at the time t respectively.

For each bus m, power balance constraint should be satisfied

$$P^{\mathrm{P}}_{\mathrm{inject},m,t} - P^{\mathrm{P}}_{\mathrm{load},m,t} + \sum_{n \in O_{m,\mathrm{branch}}} P^{\mathrm{P}}_{mn,t} = 0, t = 1, \cdots, \frac{N\Delta t}{\Delta\tau} \tag{17}$$

where $P^{\mathrm{P}}_{\mathrm{inject},m,t}$ and $P^{p}_{\mathrm{load},m,t}$ are power injection and load at bus m at time t respectively; $O_{m,\mathrm{branch}}$ denotes the set of buses directly connected with bus m; active flow $P_{mn,t}$ is the active power though the line between bus m and n, which is positive when inflow to bus m and negative when outflow from bus n.

Moreover, a slack node d is defined, whose voltage phase angle remains zero:

$$\theta^{\mathrm{P}}_{d,t} = 0, t = 1, \cdots, \frac{N\Delta t}{\Delta\tau} \tag{18}$$

The voltage phase angle at a bus should be kept within its upper and lower limits:

$$\theta^{\min}_{m} \leq \theta^{\mathrm{P}}_{m,t} \leq \theta^{\max}_{m}, t = 1, \cdots, \frac{N\Delta t}{\Delta\tau} \tag{19}$$

The power flow is limited by its transmission line capacity:

$$P^{\min}_{mn} \leq P^{\mathrm{P}}_{mn,t} \leq P^{\max}_{mn}, t = 1, \cdots, \frac{N\Delta t}{\Delta\tau} \tag{20}$$

Similar to CHP unit, the output of electric generators is also constrained by the ramping rate and the upper and lower bounds:

$$P_{\mathrm{e1}}^{\mathrm{down}} \leq P_{\mathrm{e1},t+1}^{\mathrm{p}} - P_{\mathrm{e1},t}^{\mathrm{p}} \leq P_{\mathrm{e1}}^{\mathrm{up}}, t = 1, \cdots, \frac{N\Delta t}{\Delta \tau} - 1 \tag{21}$$

$$P_{\mathrm{e1}}^{\mathrm{min}} + R_{\mathrm{e1},t} \leq P_{\mathrm{e1},t}^{\mathrm{p}} \leq P_{\mathrm{e1}}^{\mathrm{max}} - R_{\mathrm{e1},t}, t = 1, \cdots, \frac{N\Delta t}{\Delta \tau} \tag{22}$$

$$P_{\mathrm{e2}}^{\mathrm{down}} \cdot \frac{\Delta t}{\Delta \tau} \leq P_{\mathrm{e2},t+1}^{\mathrm{p}} - P_{\mathrm{e2},t}^{\mathrm{p}} \leq P_{\mathrm{e2}}^{\mathrm{up}} \cdot \frac{\Delta t}{\Delta \tau}, t = 1, \cdots, \frac{N\Delta t}{\Delta \tau} - 1 \tag{23}$$

$$P_{\mathrm{e2}}^{\mathrm{min}} + R_{\mathrm{e2},\tau} \leq P_{\mathrm{e2},t}^{\mathrm{p}} \leq P_{\mathrm{e2}}^{\mathrm{max}} - R_{\mathrm{e2},\tau}, t = 1, \cdots, \frac{N\Delta t}{\Delta \tau}, \tau \in t \tag{24}$$

The reserve should also be kept in its physical limit:

$$0 \leq R_{\mathrm{e1},t} \leq R_{\mathrm{e1}}^{\mathrm{max}} \tag{25}$$

$$0 \leq R_{\mathrm{e2},\tau} \leq R_{\mathrm{e2}}^{\mathrm{max}} \tag{26}$$

(3) Heat network

Heat network is mainly composed of heat exchanger stations, pipelines and loads. Heat energy is extracted from the heat station, carried by hot water and distributed to heat consumers through heat pipelines. Heat transfer time, usually varying from hours to days, cannot be ignored [7]. In the paper, the thermal transmission dynamics of heat network is described since steady thermal model cannot reflect its energy storage property. The pressure dynamics are faster than thermal dynamics, with little impact on temperature distribution, so it is not considered here.

(a) Heat pipeline

For a pipeline, in radial direction, hot water dissipates heat energy to insulation and the surrounding soil; while in axial direction, hot water transfers heat energy downwards through water flowing. Consequently, the water temperature in the pipeline varies with the time τ and the position variable x along the pipeline, representing its temporal and spatial characteristics. The most commonly used strategy CF-VT in north China is considered.

The pipeline is divided equally into small segments of length Δx. For the segment k, the thermal transmission dynamical model is established by including the heat dissipation to the surrounding soil and heat transferring to the adjacent segment, and then the heat energy delivering dynamics through the pipeline is modelled considering the pipeline topology.

The thermal resistance between hot water and insulation layer can be calculated

$$R_{\mathrm{wb}} = \frac{1}{h_{\mathrm{wp}} D_{\mathrm{in}}} + \frac{1}{2\lambda_{\mathrm{b}}} \ln(\frac{D_{\mathrm{out}}}{D_{\mathrm{in}}}) \tag{27}$$

The thermal resistance between insulation layer and soil layer can be described as

$$R_{\mathrm{bs}} = \frac{1}{2\lambda_{\mathrm{s}}} \ln\left[\frac{2Z}{D_{\mathrm{out}}} + \sqrt{\left(\frac{2Z}{D_{\mathrm{out}}}\right)^2 - 1}\right] \tag{28}$$

The insulation layer absorbs heat energy from the hot water and then dissipates it to the soil layer. The heat dissipation of the insulation layer can be expressed as

$$C_b \frac{T^f_{b,l,k,\tau+1} - T^f_{b,l,k,\tau}}{\Delta\tau} = \frac{\pi\Delta x}{R_{wb}}(T^f_{w,l,k,\tau} - T^f_{b,l,k,\tau}) - \frac{\pi\Delta x}{R_{bs}}(T^f_{b,l,k,\tau} - T_s), \tau = 1,\cdots,N, k = 1,\cdots,L/\Delta x \quad (29)$$

where $C_b = c_b \frac{\pi}{4}(D_{out}{}^2 - D_{in}{}^2)\Delta x \rho_b$.

In additional to the heat energy dissipation to insulation layer, heat energy is transferred to the adjacent next segment at the same time, which is modeled as

$$m_w \frac{\partial T_w}{\partial \tau} + M_l \mathbf{C}_w \frac{\partial T_w}{\partial x} = \frac{\pi}{R_{wb}}(T_b - T_w) \quad (30)$$

Using the finite difference approximation method, (30) can be reformulated as

$$T^p_{w,l,k+1,\tau+1} = \frac{\frac{\pi}{4} D_{in}{}^2 \rho_w c_w \Delta x T^p_{w,l,k+1,\tau} + M_l c_w \Delta\tau T^p_{w,l,k,\tau+1} + \frac{\pi}{R_{wb}} \Delta x \Delta\tau T^p_{b,l,k+1,\tau+1}}{\frac{\pi}{4} D_{in}{}^2 \rho_w c_w \Delta x + M_l c_w \Delta\tau + \frac{\pi}{R_{wb}} \Delta x \Delta\tau} \quad (31)$$

$$\tau = 1,\cdots,N-1, k = 1,\cdots,L/\Delta x - 1$$

The water temperature is limited by the upper and lower bounds

$$T^{min}_{hn} \le T^p_{w,l,k,\tau} \le T^{max}_{hn}, \tau = 1,\cdots,N, k = 1,\cdots,L/\Delta x \quad (32)$$

(b) Hydraulics

In order to model the hydraulics of the pipeline, the following assumptions are made: (1) Water is continuous and uncompressible, and according to mass conservation law, the mass flow entering into a node is equal to the mass flow leaving the node; (2) there is no heat energy loss at the mixed node; (3) when the flowing water meets at the crossing node, the water temperature uniformly mixes instantly.

The mass flow balance at the mixed node i is expressed as

$$\sum_{l \in O_{i,pipe+}} M_l = \sum_{j \in O_{i,pipe-}} M_j \quad (33)$$

where $O_{i,pipe+}$ and $O_{i,pipe-}$ are the sets of pipelines that flows into and out of node i.

At the crossing node i, the water temperature after mixing is given as

$$T^p_{w,j,1,\tau} = \frac{\sum\limits_{l \in O_{k,pipe+}} M_l T^p_{w,l,L/\Delta x,\tau}}{\sum\limits_{l \in O_{k,pipe+}} M_l}, j \in O_{i,pipe-} \quad (34)$$

(c) Heat exchanger

Absorbing the heat energy produced by CHP unit, the heat exchanger heats the water at the terminal end of return pipeline, which then flows out of the exchanger to the beginning end of supply pipeline. The heat exchange station is simplified to a node r, and the heat energy exchange is formulated as

$$\left(\sum_{j \in O_{r,pipe-}} T^p_{w,j,1,\tau+1} M_j - \sum_{l \in O_{r,pipe+}} T^p_{w,l,L/\Delta x,\tau} M_l \right) c_w = \eta_{ex} H^p_{CHP,\tau}, \tau = 1,\ldots,N \quad (35)$$

where η_{ex} is the heat energy utilization coefficient of heat exchanger.

(d) Heat load

The heat load refers to the heat power which is absorbed from the heat network to maintain the temperature of the building within the human comfort range, while the heat dissipation from the building interior to exterior is considered.

Similar to the heat exchange station, the heat load is simplified to be a node g. It absorbs heat energy from the hot water flowing in supply network. The heat exchange at the heat load is expressed as

$$\left(\sum_{i\in O_{\mathrm{g,pipe+}}} T^{\mathrm{p}}_{\mathrm{w},i,l/\Delta x,\tau} M_m - \sum_{j\in O_{\mathrm{g,pipe-}}} T^{\mathrm{p}}_{\mathrm{w},j,1,\tau+1} M_n\right) c_{\mathrm{w}} = \frac{H^{\mathrm{p}}_{h,\tau}}{\eta_{\mathrm{load}}}, \tau = 1,\ldots,N, h = 1,\ldots,E \tag{36}$$

Due to the difference between indoor and outdoor temperature, the heat power dissipates from indoors to outdoors. It is assumed that the heat dissipation power is linearly proportional to the temperature difference, which is formulated as

$$H^{\mathrm{p}}_{\mathrm{dis},h,\tau} = A_h S_h (T^{\mathrm{p}}_{\mathrm{in},h,\tau} - T^{\mathrm{p}}_{\mathrm{out},h,\tau}), \tau = 1,\ldots,N, h = 1,\ldots,E \tag{37}$$

where the heat transfer coefficient A_h of building h is related to the structure of the fences such as windows and walls; the outdoor temperature $T^{p}_{out,h,\tau}$ of building h at time τ in pre-scheduling is a known parameter.
Considering the heat energy absorption from the heat network and the heat energy dissipation from indoors to outdoors, the indoor temperature of the building can be expressed as

$$T^{\mathrm{p}}_{\mathrm{in},h,\tau+1} = T^{\mathrm{p}}_{\mathrm{in},h,\tau} + \frac{(H^{\mathrm{p}}_{h,\tau} - H^{\mathrm{p}}_{\mathrm{dis},h,\tau})\Delta\tau}{F_h G_h}, \tau = 1,\ldots,N, h = 1,\ldots,E \tag{38}$$

The indoor temperature should be restricted within a certain range in order to guarantee thermal comfort of users. SET established by the ASHRAE (American Society of Heating, Refrigerating and Air-Conditioning Engineers) is introduced in the paper for its universality and concision. The range of comfortable indoor temperature has the following constraints [15]:

$$22.2 \le T^{\mathrm{p}}_{\mathrm{in},h,\tau} \le 25.6, \tau = 1,\ldots,N, h = 1,\ldots,E \tag{39}$$

3.2. *Re-Dispatching*

3.2.1. Optimization Objective

The optimization objective of re-dispatching is to minimize the total operational costs of CHP unit and electric generators during the time horizon, formulated as:

$$\min\left(\sum_{\tau=1}^{U} (a_{\mathrm{CHP}} \cdot P^{\mathrm{r}}_{\mathrm{CHP},\tau} + a_{\mathrm{e2}} \cdot P^{\mathrm{r}}_{\mathrm{e2},\tau}) \cdot \Delta\tau + \sum_{t=1}^{U\Delta\tau/\Delta t} a_{\mathrm{e1}} \cdot P^{\mathrm{r}}_{\mathrm{e1},t} \cdot \Delta t\right) \tag{40}$$

3.2.2. Optimization Constraints

With the pre-scheduled reserve, the outputs of generators are appropriately adjusted to adapt to the wind power realization in re-dispatching. Similar to Section 3.2, the operational constraints of re-dispatching are illustrated below.

CHP unit constraints are given in (41)–(43):

$$P^{\mathrm{r}}_{\mathrm{CHP},\tau} = K H^{\mathrm{r}}_{\mathrm{CHP},\tau}, \tau = 1,\ldots,U \tag{41}$$

$$P^{\mathrm{down}}_{\mathrm{CHP}} \le P^{\mathrm{r}}_{\mathrm{CHP},\tau+1} - P^{\mathrm{r}}_{\mathrm{CHP},\tau} \le P^{\mathrm{up}}_{\mathrm{CHP}}, \tau = 1,\ldots,U-1 \tag{42}$$

$$P^{\mathrm{p}}_{\mathrm{CHP},\tau} - R_{\mathrm{CHP},\tau} \le P^{\mathrm{r}}_{\mathrm{CHP},\tau} \le P^{\mathrm{p}}_{\mathrm{CHP},\tau} + R_{\mathrm{CHP},\tau}, \tau = 1,\ldots,U \tag{43}$$

Electricity network constraints are represented in (44)–(52):

$$P^{\mathrm{r}}_{mn,\tau} = -b_{mn}(\theta^{\mathrm{r}}_{m,\tau} - \theta^{\mathrm{r}}_{n,\tau}), \tau = 1,\ldots,U \tag{44}$$

$$P^{\mathrm{r}}_{\mathrm{inject},m,\tau} - P^{\mathrm{r}}_{\mathrm{load},m,\tau} + \sum_{n \in O_m} P^{\mathrm{r}}_{mn,\tau} = 0, \tau = 1, \ldots, U \tag{45}$$

$$\theta^{\mathrm{r}}_{d,\tau} = 0, \tau = 1, \ldots, U \tag{46}$$

$$\theta^{\min}_{m} \leq \theta^{\mathrm{r}}_{m,\tau} \leq \theta^{\max}_{m}, \tau = 1, \ldots, U \tag{47}$$

$$P^{\min}_{mn} \leq P^{\mathrm{r}}_{mn,\tau} \leq P^{\max}_{mn}, \tau = 1, \ldots, U \tag{48}$$

$$P^{\mathrm{down}}_{\mathrm{e1}} \leq P^{\mathrm{r}}_{\mathrm{e1},t+1} - P^{\mathrm{r}}_{\mathrm{e1},t} \leq P^{\mathrm{up}}_{\mathrm{e1}}, t = 1, \ldots, \frac{U\Delta t}{\Delta \tau} - 1 \tag{49}$$

$$P^{\mathrm{p}}_{\mathrm{e1},t} - R_{\mathrm{e1},t} \leq P^{\mathrm{r}}_{\mathrm{e1},t} \leq P^{\mathrm{p}}_{\mathrm{e1},t} + R_{\mathrm{e1},t}, t = 1, \ldots, \frac{U\Delta t}{\Delta \tau} \tag{50}$$

$$P^{\mathrm{down}}_{\mathrm{e2}} \leq P^{\mathrm{r}}_{\mathrm{e2},\tau+1} - P^{\mathrm{r}}_{\mathrm{e2},\tau} \leq P^{\mathrm{up}}_{\mathrm{e2}}, \tau = 1, \ldots, U - 1 \tag{51}$$

$$P^{\mathrm{p}}_{\mathrm{e2},t} - R_{\mathrm{e2},\tau} \leq P^{\mathrm{r}}_{\mathrm{e2},\tau} \leq P^{\mathrm{p}}_{\mathrm{e2},\tau} + R_{\mathrm{e2},\tau}, \tau = 1, \ldots, U \tag{52}$$

The heat network constraints in (27)–(29), (31)–(39) are included in re-dispatching optimization, with the superscript 'p' replaced by 'r'.

4. Simulation Results

To illustrate the effectiveness of the two-stage robust and economic scheduling methodology for electricity and heat coupled IES in accommodating uncertain wind power, an electricity-heat IES, with an IEEE 9-bus, 9-branch electricity network and a 3-building, 12-pipeline heat network, is established in Figure 4. The two electricity and heat networks are coupled by CHP unit and heat exchanger. Electricity generator G1 with larger inertial time constant is attached to Bus 1, and the electricity generator G2 with smaller inertial time constant is installed at Bus 7. The parameters involved in the simulation are given in Table A1.

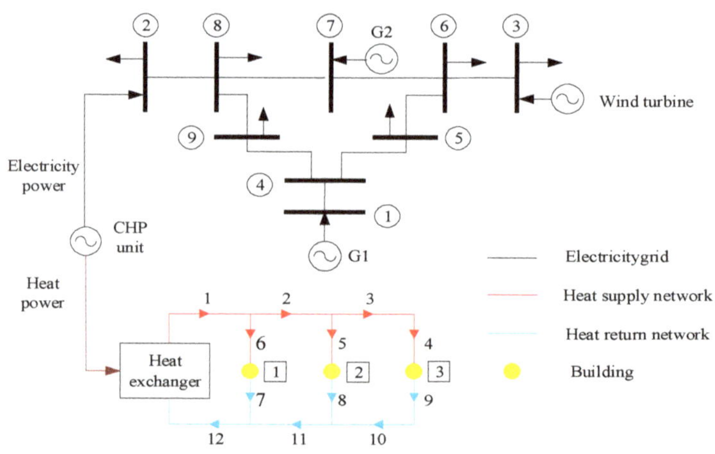

Figure 4. Diagram of electricity-heat IES.

Several cases are implemented to explore the effectiveness of thermal energy storage properties of buildings and heat network in improving system flexibility and enhancing wind power absorption. The cases with different considered factors, such as storage capacity of heat network and buildings, time-of-use price of CHP unit generation, ramping speed limit of generators, are listed in Table 1, where $\surd$ denotes considering the factor while $\times$ denotes not. In Cases I, II and IV, the cost coefficients remain unchanged during the scheduling; while the cost coefficient of CHP generation varies in Case III, as shown in Table 2. The real and forecasted wind power values are depicted in Figure A1. The electricity load and outside temperature are drawn in Figures A2 and A3.

Table 1. Simulation cases and the considered factors.

Case	Thermal Storage Properties of Heat Network and Buildings	Time-of-Use Price of CHP Generation	Ramping Speed Limit		
			CHP (MW/$\triangle\tau$)	G1 (MW/$\triangle t$)	G2 (MW/$\triangle\tau$)
Case I	√	×	0.3	0.3	0.03
Case II	√	×	0.3	0.3	0.3
Case III	√	√	0.3	0.3	0.03
Case IV	×	×	0.3	0.3	0.03

Table 2. Cost coefficients of generators.

Cost Coefficient	Case	CHP (\$/(MW·min))	G1 (\$/(MW·min))	G2 (\$/(MW·min))
Generation cost coefficient	Cases I, II and IV	1	0.67	0.8
	Case III	0.67 (12:00–14:00) 1.33 (14:00–16:00)	0.87	1.2
Reserve cost coefficient	All Cases	2	1.33	1.6

4.1. Case I

From Figures A2 and A3, it can be observed that during the scheduling horizon, the thermal demand gradually increases as the outdoor temperature drops; on the contrary, the electrical load gradually decreases. The pre-scheduled electricity generations and the balance between electricity supply and demand are depicted in Figure 5. The water temperature of heat network is drawn in Figure 6. It can be seen, in order to achieve economic operation, the generator G1 with the cheapest cost coefficient is scheduled to generate the most electric power; the G2 generation is very small in order to satisfy the down reserve requirement; the CHP unit with the highest cost coefficient is scheduled to satisfy the heat demand by the lowest generation, which can be indicated by the phenomenon shown in Figure 6 that the water temperature at the return Pipeline 7 at the end of the pre-scheduling period is close to its lower limit.

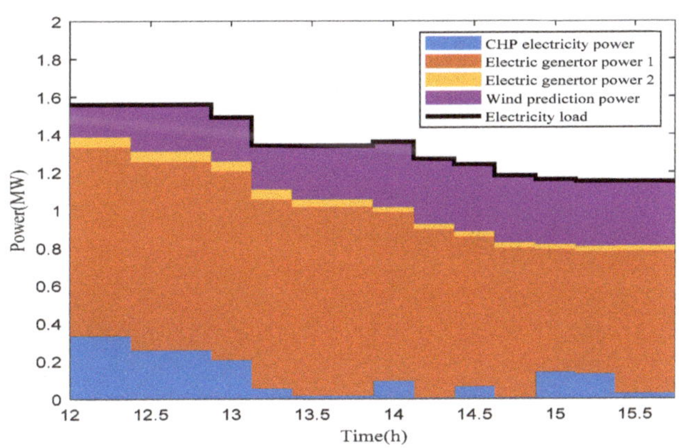

Figure 5. Pre-scheduled electricity generations and the balance between supply and demand in Case I.

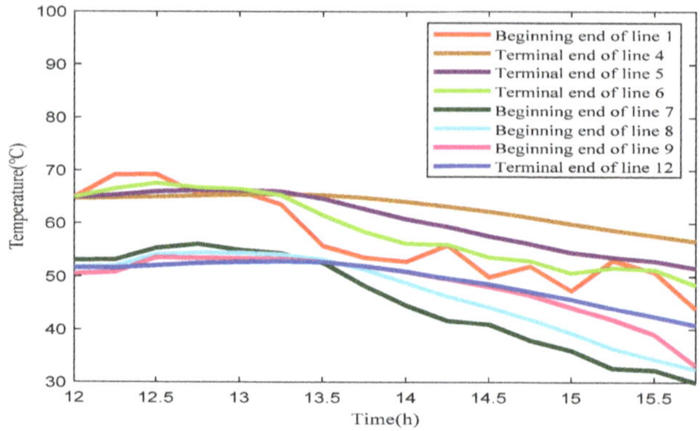

Figure 6. Pre-scheduled water temperatures in heat network in Case I.

The temperature at the beginning of Line 1, i.e., the outlet of heat station, fluctuates greatly, for it is closely related with CHP unit generation. In order to maintain the real-time supply and demand balance of electricity load, the heat power output of CHP unit with the fixed heat-to-electricity ratio has a higher volatility. However, after the heat transferring through the pipelines, the temperature curves at the inlet of the load and the heat station, such as the beginning of Lines 4–6 and 12, becomes smooth. Consequently, from this perspective, the large volume of water in the pipeline serves as energy storage and the fluctuation of instantaneous load or wind power can be smoothed by heat network.

The imbalance between heat supply and demand is depicted in Figure 7. From 12:00 to 13:00, CHP unit generates more heat power than the total building dissipation. Further, the surplus heat energy, depicted in green, is stored into the pipeline and buildings. From 13:00 to 16:00, since the electricity demand decreases and at the same time the heat load increases, the stored heat energy drawn in yellow, is discharged to satisfy the heat demand, ensuring the indoor temperature in comfortable range. The indoor temperatures of buildings are shown in Figure 8.

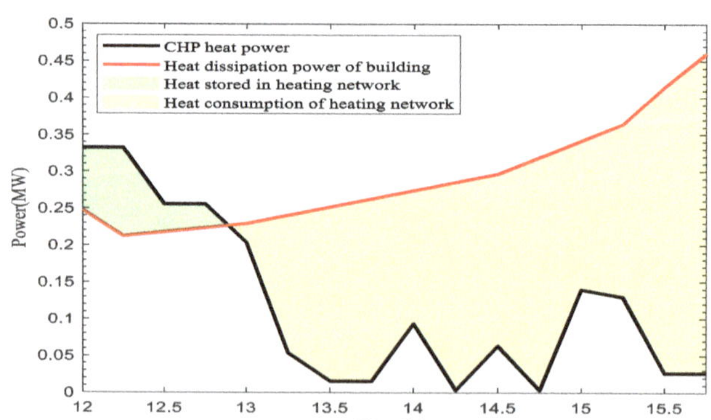

Figure 7. Pre-scheduled heat energy charging and discharging in Case I.

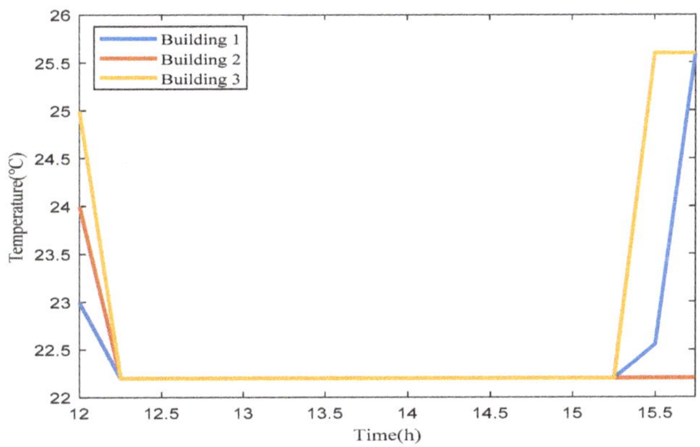

Figure 8. Pre-scheduled building indoor temperature in Case I.

Due to predicted wind power errors, CHP unit and electric generators need to keep some reserve in advance. If the robust and economical scheduling methodology is not considered, when there is wind power prediction error, the system cannot meet the electric and heat balance for lack of reserve, which leads to infeasibility. On the contrary, when robust and economical scheduling methodology is taken into consideration, the system can operate safely and stably and also absorb all wind power. The costs are manifested in Table 3.

Table 3. Comparison of reserve and operation cost between Cases I and II.

Cost	Case I	Case II
Reserve cost ($)	12.9600	11.5200
Operation cost ($)	178.4432	176.7158
Total cost ($)	191.4023	188.2358

The reserves of generators are shown in Figure 9. At each instant, the sum of the electricity power reserves of all generators is 0.03 MW, equal to the wind power uncertainty. It is worth noting that the reserve of G1 is always 0. It is because that the timescale $\triangle t$ of G1 ramping speed is larger than the time resolution $\triangle \tau$ of wind power fluctuation. In order to achieve economic operation, G2 with the cheaper reserve cost should be given more priority to provide reserve, rather than CHP unit. CHP unit will take on the remaining reserve when G2 cannot accommodate the uncertain wind power, constrained by its ramping speed 0.03 MW/$\triangle \tau$. Considering the strong coupling of electricity generation and heat supply of CHP unit, the heat energy reserve is also required, which is represented in security margin of the water temperature and indoor temperature. As shown in Figure 6, at the beginning end of Line 7, i.e., the outlet of Load 1, the temperature reaches its lower bound at 15:45 with no secure water temperature margin left; but the indoor temperature of Building 1 is also 25.6 °C, much higher than the lower bound, which means it can supply heat reserve. The indoor temperature of Building 2 is equal to its lower limit, but the water temperature at the beginning end of Line 8 has some distance from the limit. The excess heat is stored in the heating pipelines and buildings, providing the heat reserves when CHP unit is scheduled to take on the reserves.

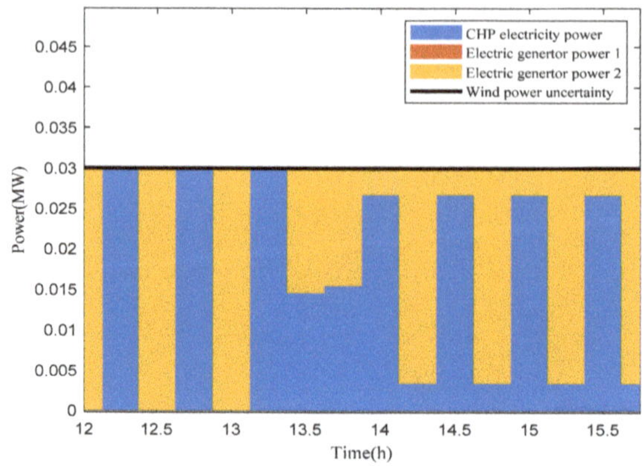

Figure 9. Reserves of generators and wind power uncertainty in Case I.

4.2. Case II

Compared with Case I, the ramping speed of G2 changes from 0.03 MW/$\triangle\tau$ to 0.3 MW/$\triangle\tau$ in Case II. The pre-scheduled electricity generation and the balance of supply and demand is depicted in Figure 10. The reserves of generators are shown in Figure 11. It can be seen that only G2 offers 0.03 MW reserve to alleviate the wind power uncertainty. Considering that G2 has sufficient reserve capacity and its reserve cost is cheaper, reserve is completely supplied by G2. From the time 12:00 to 14:00, G1 generation reaches its upper limit and then G2 with the medium operational cost is scheduled to satisfy the remaining electricity load; since no reserve is required from CHP unit and simultaneously the CHP unit generation cost is the most expensive, there is no scheduled CHP unit generation, and the pipelines and buildings are in the heat energy release state, causing the water temperature and indoor temperature to drop. From the time 13:00 to 16:45, G2 generation is kept as 0.03 MW in order to provide sufficient downward reserve; and most of the electricity demand is satisfied by G1 to achieve the economic operation.

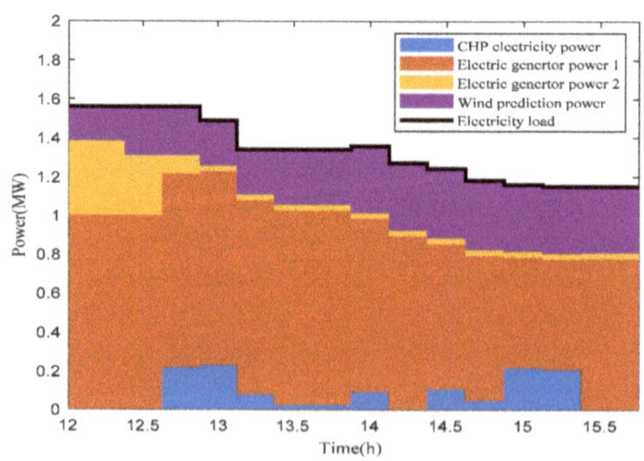

Figure 10. Pre- scheduled electricity generations and the balance between supply and demand in Case II.

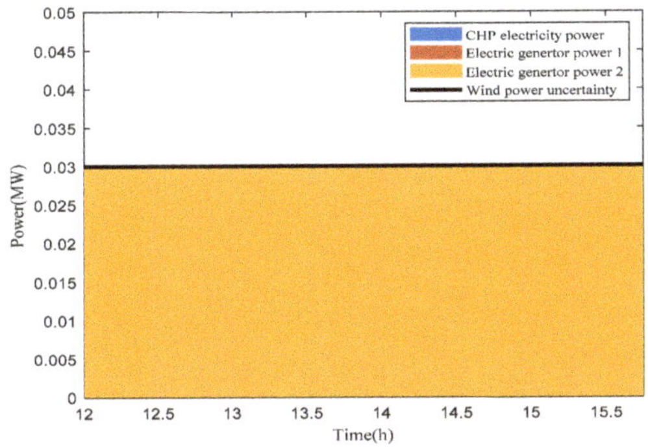

Figure 11. Reserves of generators and wind power uncertainty in Case II.

Because there is no need for CHP unit to supply reserve to accommodate the fluctuation of wind power, as shown in Figure 12, the indoor temperatures of buildings and the water temperatures at the outlets of heat loads all reach at their lower bounds, induced by the optimization objective of economic operation. The comparison of reserve and operational cost between Cases I and II are listed in Table 3. Owing to the enhanced ramping speed of G2, more G2 electricity generations and reserves in Case II decrease the corresponding cost, compared with Case I. Moreover, Due to the less CHP unit generation, the heat network always stays in the state of releasing energy. As a result, the water temperatures in the pipelines and indoor temperature in buildings show the downward trend, which can be seen in Figures 13 and 14.

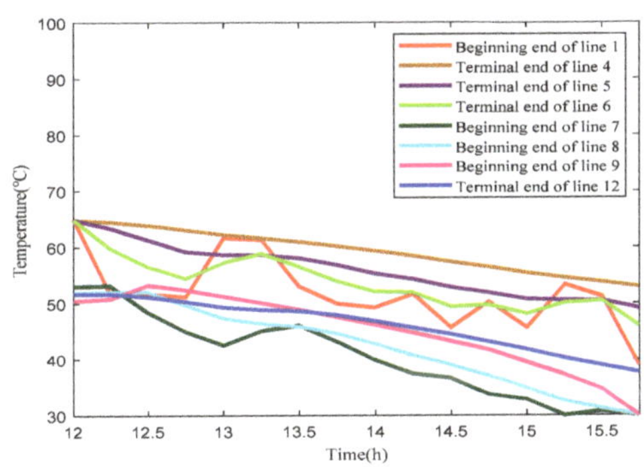

Figure 12. Pre-scheduled water temperature of heat network in Case II.

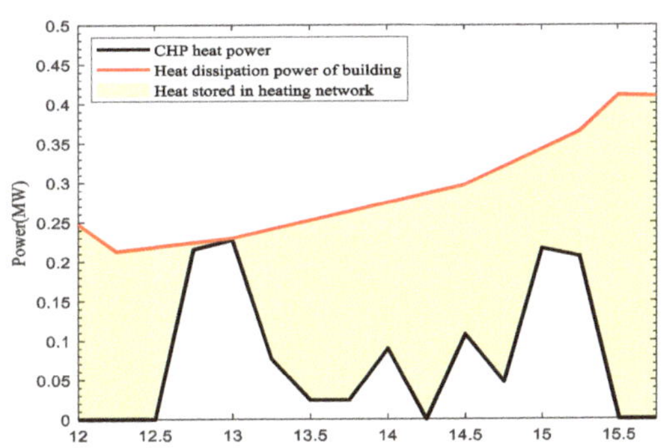

Figure 13. Pre-scheduled heat energy charging and discharging in Case II.

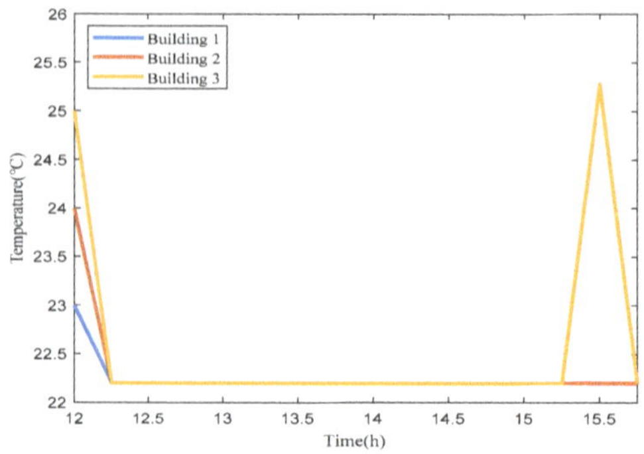

Figure 14. Pre-scheduled building indoor temperature in Case II.

4.3. Case III

The pre-scheduled electricity generations are drawn in Figure 15. Comparison of electricity generations during different time periods are given in Table 4. From 12:00–14:00, more CHP unit generation is scheduled owing to the lower price; while from 14:15–16:00, G1 generates more electricity. The scheduling results represent the operational economics.

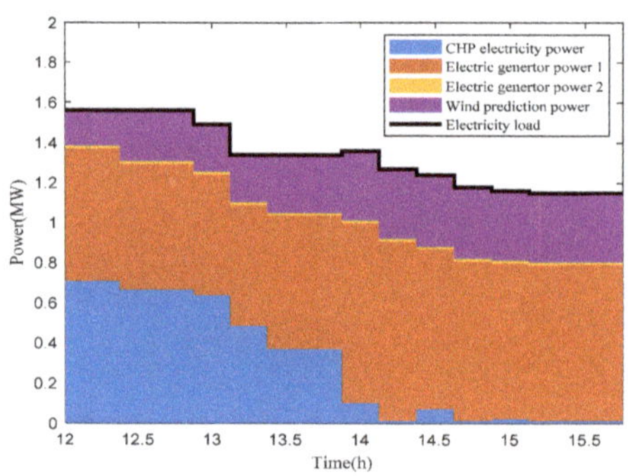

Figure 15. Pre-scheduled electricity generations and the balance between supply and demand in Case III.

Table 4. Comparison of electricity generations under time-of-use price.

Case	Average Output of CHP Unit (MW)		Average Output of G1 and G2 (MW)	
	12:00–14:00	**14:00–16:00**	**12:00–14:00**	**14:00–16:00**
Case I	0.1831	0.0607	1.0497	0.7998
Case III	0.5790	0.0350	0.6538	0.8255

Pre-scheduled water temperature of heat network is displayed in Figure 16. It can be seen that, the outlet temperature of the heat station drops first and then rises, consistent with the load inlet temperature. However, the turning point for outlet of the heat station appears at 12:15–13:15 while the load outlet at 13:15–14:30. This shows that it takes some time for hot water flowing from heat station to load. Heat production and consumption are not balanced in real time. It is very necessary to conduct transient analysis on the heat network. The water temperatures in Figure 16 are higher than those in Figures 6 and 12, caused by the cheaper cost coefficient of CHP unit generation during 12:00–14:00.

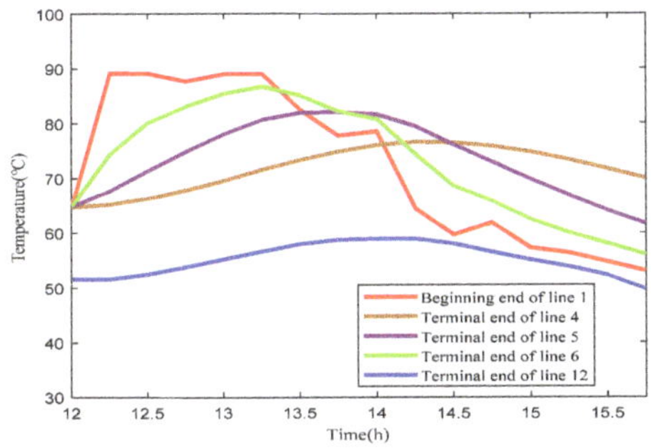

Figure 16. Pre-scheduled water temperature of heat network in Case III.

4.4. Case IV

In Case IV, both time-of-use price and the thermal storage capacity of heat network and buildings are not concerned. Assuming that the temperature of the building is kept at 23 °C constantly, and the heat dissipation from indoor to outdoor is regarded as heat load. Heat network is simplified to be three heat load nodes. The heat supply and demand are balanced instantaneously for no heat reserve is supplied. The wind power is absorbed as much as possible under the condition of satisfying the operational constraints. The absorbed and abandoned wind power is depicted in Figure 17.

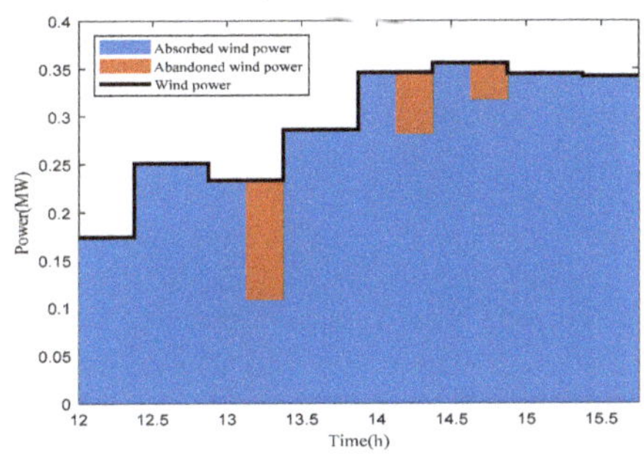

Figure 17. Absorbed and abandoned wind power in Case IV.

The electricity generators G1 and G2 are limited to their ramping speed, and the CHP unit working in constant heat-to-electricity ratio mode is subjected to real-time heat balance. It is difficult for all of them to respond to the wind power fluctuation, inevitably leading to wind abandonment. Wind power abandonment occurs at three instants and the maximum abandonment appears at 13:15, with about 53.18% wind power abandoned; on the contrary, the wind absorption is 100% if the thermal storage capacity is considered, as shown in Cases I, II and III. What is more, it will lead to infeasibility under wind power uncertainty for there is no heat reserve in Case IV.

4.5. Discussions on Robustness and Economics

To verify the robustness of the proposed method, 10,000 scenarios are generated by Monte Carlo sampling within the wind prediction boundaries to simulate the real wind power uncertainty. The price coefficient is chosen the same as Cases I, II and IV, listed in Table 2. According to (3), the appropriate value $\Gamma = 6$ is chosen, whose corresponding pre-scheduled electric power generation and reserve results have been given in Figures 5 and 9, and heat energy results have been depicted in Figure 7. With the pre-scheduled results of generations and reserves, the re-dispatching problem could derive the feasible solutions of re-dispatched generations under all the 10,000 uncertain wind power realizations. The robustness of the proposed pre-scheduling and re-dispatching coordination approach is validated. As shown in Table 5, if the smaller value $\Gamma = 2$ is chosen, the feasibility of re-dispatching problem under any uncertain wind power realization might not be guaranteed, with the unfeasibility proportion about 5.55% among the 10,000 wind power realizations, although the total cost could be decreased. On the contrary, with the larger value $\Gamma = 16$, the derived total cost in pre-scheduling optimization increases and even worse the computing burden is significantly exacerbated, consuming the computational time several hundred times as much as $\Gamma = 6$ due to the less feasibility domain.

Table 5. Comparisons in robustness, economics and computational costs.

Solutions		Operation Cost ($)	Reserve Cost ($)	Total Cost ($)	Infeasibility Proportion of Re-Dispatching	Iterations
Robust economic scheduling	$\Gamma = 2$	178.2558	12.96	191.2158	5.55%	787
	$\Gamma = 6$	178.4432	12.96	191.4023	0%	528
	$\Gamma = 16$	178.5876	12.96	191.5476	0%	256
Interval optimization	The worst optimal	182.0079	24.75	206.7579	0%	65,536
	The best optimal	172.4928	23.0630	195.5558	0%	65,536
Scenario-based optimization	N1 = 1000, N2 = 100	177.1415	5.7776	182.9191	100%	100
	N1 = 10,000, N2 = 1000	177.2181	12.1895	189.4076	65.57%	1000

In order to evaluate the superiority of the proposed approach, the interval optimization [16–18] and scenario-based optimization methods [19–21] are employed for comparison. In the interval optimization method, the robust feasibility constraints are added iteratively as described in Section 2.3 to solve the pre-scheduling problem. The best and worst optimal solutions of interval optimization considering robust feasibility constraints are given in Table 5. It is shown that the robustness can be guaranteed, but their total costs are respectively 2.17% and 8.02% higher compared with the robust economic scheduling. In the scenario-based optimization method, the wind power scenarios are sampled by the Monte Carlo simulations, and reduced by the backward reduction method. In Table 5, N1 and N2 mean the scenario number before and after the scenario reduction. For the reduced scenarios, the feasible iterations are carried out. The total costs of scenario-based optimization methods are lower, but the safe operation of the system cannot be guaranteed when facing the uncertain wind power realization. Under the pre-scheduling results derived by 1000 scenarios-based optimization, the infeasibility proportion of re-dispatching is still about 65.57%. Both of the methods, i.e., the interval optimization and the scenario-based optimization, will face the combinatorial explosion problem as time horizon increases, indicated by the iterations in Table 5. Although the scene reduction can decrease the number of scenarios, its calculation time also cannot be ignored.

Consequently, the robust economic scheduling approach can ensure the feasibility of re-dispatching problem under any uncertain wind power realization while ensuring the economics of scheduling solution. Furthermore, the combination explosion problem can be solved.

5. Conclusions

A two-stage robust economic scheduling is proposed for electricity-heat IES to cope with the wind power uncertainty. In pre-scheduling, while ensuring the economics of scheduling results, the sufficient regulation margin subjected to the operational bounds is reserved for the possible uncertain wind power realization by considering the robust feasibility constraint of re-dispatching. With the pre-scheduling solution, for electric system the appropriate generation reserve is kept to achieve the flexibility, and for heat system, the inherent thermal energy storage of buildings and heat network is utilized to compensate for the fluctuations of heat power generation from CHP unit caused by wind power uncertainty. The thermal storage capability is characterized by modeling the dynamics of building and heat network. The simulation indicates that the proposed approach could enhance the flexibility of heat and electricity coupled system in wind power accommodation. Furthermore, the appropriate choosing of uncertain budget could achieve the robustness and economics of scheduling results at the lower computational cost. The optimization objective of the total cost derived by the proposed approach is lower than the traditional interval optimization. The proposed approach is more robust compared with the traditional scenario-based optimization method. Gradient explosion problem existing in the two traditional methods could be avoided by the proposed approach. The superiority of the proposed approach can be demonstrated.

In cold seasons, with large heat demand and abundant wind power, the fixed heat-power ratio of CHP unit could cause problems in wind accommodation and more importantly, the safe operation of the system, especially in Northeast China. The proposed method can effectively enhance wind power accommodation and achieve robustness and economics of scheduling. In the future research, the uncertainties of electricity demand and heat network parameters will be further considered.

Author Contributions: Conceptualization, R.L. and Z.B.; methodology, R.L., Z.B.; software, R.L.; validation, R.L., Z.B.; formal analysis, R.L., Z.B., J.Z., L.L., M.Y.; investigation, R.L., Z.B., J.Z., L.L., M.Y.; resources, R.L., Z.B., J.Z., L.L., M.Y.; data curation, R.L.; writing—original draft preparation, R.L.; writing—review and editing, R.L., Z.B.; visualization, R.L., Z.B.; supervision, Z.B.; project administration, Z.B., J.Z., L.L., M.Y.; funding acquisition, Z.B., J.Z., L.L., M.Y. All authors have read and agreed to the published version of the manuscript.

Funding: This work was funded by the Key Research and Development Program of Zhejiang Province, Grant Number 2021C01113, Zhejiang Provincial Natural Science Foundation of China, Grant Number LGG22F030008, and National Natural Science Foundation of China, Grant Number 51777182.

Institutional Review Board Statement: Not applicable.

Informed Consent Statement: Not applicable.

Conflicts of Interest: The authors declare no conflict of interest.

Nomenclature

Indices and Sets

O	the set of branches or pipelines directly connected with a bus or node
τ	time index for heat network
d	index of electric network slack node
g	node index of heat load
i	index of heat network node
j, l	index of heat pipeline
k	index of pipeline segment
m, n	indices of electric network buses
r	node index of heat exchange station
t	time index for electricity network

Superscript and Subscript

CHP	CHP (combined heat and power) unit
b	insulation layer
dis	heat dissipation of building
down	downward ramping
e1	electric generator with large system inertia
e2	electric generator with small system inertia
ex	heat exchange station
in	inner layer of pipeline or indoors
inject	power injection
load	electric or heat load

out	outer layer of pipeline or outdoors
p	pre-scheduled value or wind power predicted value
r	re-dispatched value or real wind power value
s	soil layer
up	upward ramping
w	water flow layer
wind	wind power

Input Parameters

A	heat transfer coefficient of building fence [W/(m^2·°C)]
D	diameter of pipeline [m]
E	total number of buildings
F	equivalent specific heat capacity of building [J/(kg·°C)]
G	equivalent mass of building [kg]
L	length of the pipeline [m]
M	mass flow rate in pipeline [kg/s]
N/U	time horizon length of pre-scheduling/re-dispatching
R	thermal resistance [(m·K)/W]
S	area of the fence [m^2]
Z	buried depth of pipeline [m]
λ	thermal conductivity [W/(m·K)]
ρ	density [kg/m^3]
η	energy utilization coefficient
a	price coefficient of operation [$/(MW·min)]
b	reactance [pu]
c	specific heat capacity [J/(kg·K)]
h	convection heat transfer rate [W/(m^2·K)]
k	heat-to-electricity ratio of CHP unit
q	price coefficient of reserve [$/(MW·min)]
$\triangle t$	timescale of heat network [min]
$\triangle \tau$	timescale of electricity network [min]
$\triangle x$	segment length of heat pipeline [m]

Decision Variables

$H^{\mathrm{p}}_{\mathrm{CHP},\tau}$, $H^{\mathrm{r}}_{\mathrm{CHP},\tau}$	heat power generation of CHP unit at time τ in pre-scheduling/ re-dispatching [W]
$H^{\mathrm{p}}_{\mathrm{dis},h,\tau}$, $H^{\mathrm{r}}_{\mathrm{dis},h,\tau}$	heat power dissipation of building h at time τ in pre-scheduling/ re-dispatching [W]
$H^{\mathrm{p}}_{h,\tau}$, $H^{\mathrm{r}}_{h,\tau}$	heat power absorbed by building h at time τ in pre-scheduling/ re-dispatching [W]
$P^{\mathrm{p}}_{\mathrm{CHP},t}$, $P^{\mathrm{r}}_{\mathrm{CHP},\tau}$	electric power generation of CHP unit at time t in pre-scheduling/ τ in re-dispatching [W]
$P^{\mathrm{p}}_{\mathrm{e1},t}$, $P^{\mathrm{r}}_{\mathrm{e1},t}$	electric power generation of electric Generator 1 at time t in pre-scheduling/ re-dispatching [W]
$P^{\mathrm{p}}_{\mathrm{e2},t}$, $P^{\mathrm{r}}_{\mathrm{e2},\tau}$	electric power generation of electric Generator 2 at time t in pre-scheduling/ τ in re-dispatching [W]
$R_{\mathrm{e1},t}$, $R_{\mathrm{e2},\tau}$, $R_{\mathrm{CHP},\tau}$	reserve of electric Generator 1/electric Generator 1/CHP unit at time t/τ [W]
$T^{\mathrm{p}}_{\mathrm{b},l,k,\tau}$, $T^{\mathrm{p}}_{\mathrm{w},l,k,\tau}$	temperature of insulation layer/water flow at k-th segment of pipeline l at time τ in pre-scheduling [K]
$T^{\mathrm{r}}_{\mathrm{b},l,k,\tau}$, $T^{\mathrm{r}}_{\mathrm{w},l,k,\tau}$	temperature of insulation layer/water flow at k-th segment of pipeline l at time τ in re-dispatching [K]
$T^{\mathrm{p}}_{\mathrm{in},h,\tau}$, $T^{\mathrm{r}}_{\mathrm{in},h,\tau}$	indoor temperature of building h at time τ in pre-scheduling/ re-dispatching [°C]
$\theta^{\mathrm{p}}_{m,t}$, $\theta^{\mathrm{r}}_{m,\tau}$	voltage phase angle at bus m at time t in pre-scheduling/ τ in re-dispatching [rad]

Appendix A

Table A1. Simulation parameters.

Parameters	Value	Parameters	Value	Parameters	Value
Δt/min	30	η_{ex}	1	$A_h S_h/(\text{MW}\cdot{}^\circ\text{C})^{-1}$	7.5×10^{-3}
$\Delta \tau$/min	15	η_{load}	1	$F_h G_h/(\text{MJ}\cdot{}^\circ\text{C})^{-1}$	7.5×10^{-2}
Δx/m	50	P_{mn}^{min}/MW	3	P_{CHP}^{min}/MW	0
N	16	P_{mn}^{max}/MW	0	P_{CHP}^{max}/MW	1
E	3	$\lambda_b/\text{W}\cdot(\text{m}\cdot\text{K})^{-1}$	0.033	P_{e1}^{min}/MW	0
K	1	Z/m	1	P_{e1}^{max}/MW	1
$T_{max}^{hn}/{}^\circ$C	90	$\lambda_s/\text{W}\cdot(\text{m}\cdot\text{K})^{-1}$	0.31	P_{e2}^{min}/MW	0
$T_{min}^{hn}/{}^\circ$C	30	$c_b/\text{J}\cdot(\text{kg}\cdot{}^\circ\text{C})^{-1}$	1380	P_{e2}^{max}/MW	1
$c_w/\text{J}\cdot(\text{kg}\cdot{}^\circ\text{C})^{-1}$	4200	$\rho_b/(\text{kg}\cdot\text{m}^{-3})$	1000	$\theta_{max}/{}^\circ$	5
$\rho_w/(\text{kg}\cdot\text{m}^{-3})$	1000	$\alpha_s/\text{W}\ (\text{m}^2\cdot\text{K})^{-1}$	14	$\theta_{min}/{}^\circ$	−5
D_{in}/m	0.4	$M_{4,\tau}/(\text{kg}\cdot\text{s})^{-1}$	1.5	R_{CHP}^{max}/MW	0.3
D_{out}/m	0.6	$M_{5,\tau}/(\text{kg}\cdot\text{s})^{-1}$	1.5	R_{e1}^{max}/MW	0.3
$h_{wp}/\text{W}\cdot(\text{m}^2\cdot\text{K})^{-1}$	7000	$M_{6,\tau}/(\text{kg}\cdot\text{s})^{-1}$	1.5	R_{e2}^{max}/MW	0.3

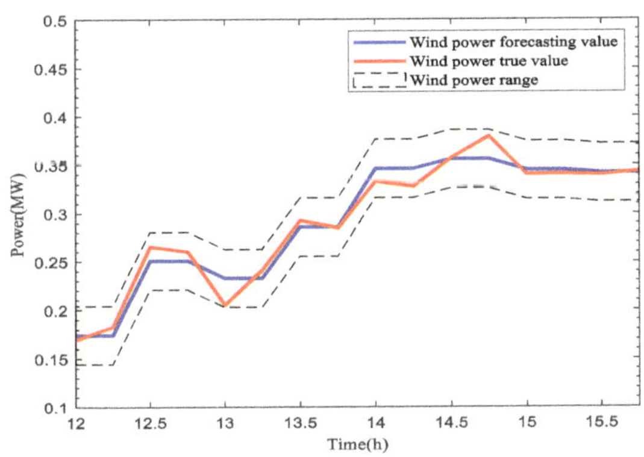

Figure A1. Wind power forecasting and true value under ±0.03 uncertainty.

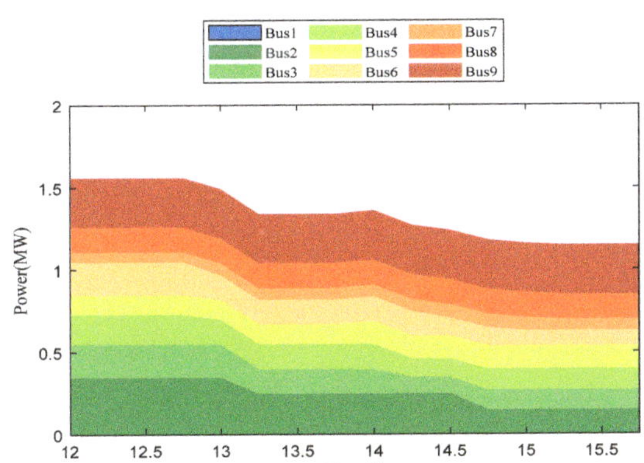

Figure A2. Electric load of each bus in power system.

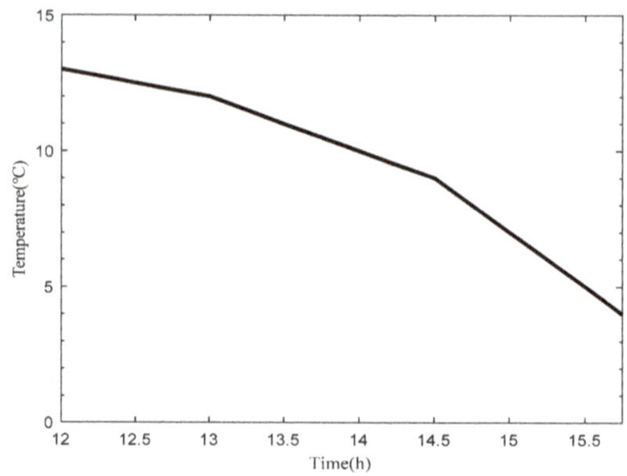

Figure A3. Outside temperature of buildings.

References

1. Tan, L.; Han, J.; Zhang, H. Ultra-Short-Term Wind Power Prediction by Salp Swarm Algorithm-Based Optimizing Extreme Learning Machine. *IEEE Access* **2020**, *8*, 44470–44484. [CrossRef]
2. Nuytten, T.; Claessens, B.; Paredis, K.; Van Bael, J.; Six, D. Flexibility of a Combined Heat and Power System with Thermal Energy Storage for District Heating. *Appl. Energy* **2013**, *104*, 583–591. [CrossRef]
3. Li, Y.; Wang, C.; Li, G.; Wang, J.; Zhao, D.; Chen, C. Improving Operational Flexibility of Integrated Energy System with Uncertain Renewable Generations Considering Thermal Inertia of Buildings. *Energy Convers. Manag.* **2020**, *207*, 112526. [CrossRef]
4. Wang, L.; Li, Q.; Sun, M.; Wang, G. Robust Optimisation Scheduling of CCHP Systems with Multi-Energy Based on Minimax Regret Criterion. *IET Gener. Transm. Distrib.* **2016**, *10*, 2194–2201. [CrossRef]
5. Zhang, C.; Dong, Z.; Xu, Y.; Chen, Y. A Two-Stage Robust Operation Approach for Combined Cooling, Heat and Power Systems. In Proceedings of the 2017 IEEE Conference on Energy Internet and Energy System Integration (EI2), Beijing, China, 26–28 November 2017; pp. 1–6.
6. Li, Y.; Zou, Y.; Tan, Y.; Cao, Y.; Liu, X.; Shahidehpour, M.; Tian, S.; Bu, F. Optimal Stochastic Operation of Integrated Low-Carbon Electric Power, Natural Gas, and Heat Delivery System. *IEEE Trans. Sustain. Energy* **2018**, *9*, 273–283. [CrossRef]
7. Li, Z.; Wu, W.; Shahidehpour, M.; Wang, J.; Zhang, B. Combined Heat and Power Dispatch Considering Pipeline Energy Storage of District Heating Network. *IEEE Trans. Sustain. Energy* **2016**, *7*, 12–22. [CrossRef]
8. Li, Z.; Wu, W.; Wang, J.; Zhang, B.; Zheng, T. Transmission-Constrained Unit Commitment Considering Combined Electricity and District Heating Networks. *IEEE Trans. Sustain. Energy* **2016**, *7*, 480–492. [CrossRef]
9. Dai, Y.; Chen, L.; Min, Y.; Chen, Q.; Hao, J.; Hu, K.; Xu, F. Dispatch Model for CHP With Pipeline and Building Thermal Energy Storage Considering Heat Transfer Process. *IEEE Trans. Sustain. Energy* **2019**, *10*, 192–203. [CrossRef]
10. Pan, Z.; Guo, Q.; Sun, H. Feasible Region Method Based Integrated Heat and Electricity Dispatch Considering Building Thermal Inertia. *Appl. Energy* **2017**, *192*, 395–407. [CrossRef]
11. Chen, Y.; Xu, Y.; Li, Z.; Feng, X. Optimally Coordinated Dispatch of Combined-Heat-and-Electrical Network with Demand Response. *IET Gener. Transm. Distrib.* **2019**, *13*, 2216–2225. [CrossRef]
12. Good, N.; Karangelos, E.; Navarro-Espinosa, A.; Mancarella, P. Optimization Under Uncertainty of Thermal Storage-Based Flexible Demand Response With Quantification of Residential Users' Discomfort. *IEEE Trans. Smart Grid* **2015**, *6*, 2333–2342. [CrossRef]
13. Wei, W.; Liu, F.; Mei, S. Robust and Economical Scheduling Methodology for Power Systems Part One Theoretical Foundations. *Autom. Electr. Power Syst.* **2013**, *37*, 37–43. (In Chinese) [CrossRef]
14. Gao, S.; Jurasz, J.; Li, H.; Corsetti, E.; Yan, J. Potential Benefits from Participating in Day-Ahead and Regulation Markets for CHPs. *Appl. Energy* **2022**, *306*, 117974. [CrossRef]
15. Gagge, A.P.; Fobelets, A.P.; Berglund, L.G. A Standard Predictive Index of Human Response to the Thermal Environment. *ASHRAE Trans.* **1986**, *92*, 2B.
16. Liu, Y.; Jiang, C.; Shen, J.; Hu, J. Coordination of Hydro Units with Wind Power Generation Using Interval Optimization. *IEEE Trans. Sustain. Energy* **2015**, *6*, 443–453. [CrossRef]
17. Huang, H.; Li, F.; Mishra, Y. Modeling Dynamic Demand Response Using Monte Carlo Simulation and Interval Mathematics for Boundary Estimation. *IEEE Trans. Smart Grid* **2015**, *6*, 2704–2713. [CrossRef]
18. Bai, L.; Li, F.; Cui, H.; Jiang, T.; Sun, H.; Zhu, J. Interval Optimization Based Operating Strategy for Gas-Electricity Integrated Energy Systems Considering Demand Response and Wind Uncertainty. *Appl. Energy* **2016**, *167*, 270–279. [CrossRef]
19. Heitsch, H.; Römisch, W. Scenario Reduction Algorithms in Stochastic Programming. *Comput. Optim. Appl.* **2003**, *24*, 187–206. [CrossRef]

20. Razali, N.M.M.; Hashim, A.H. Backward Reduction Application for Minimizing Wind Power Scenarios in Stochastic Programming. In Proceedings of the 2010 4th International Power Engineering and Optimization Conference (PEOCO), Shah Alam, Malaysia, 23–24 June 2010; pp. 430–434.
21. Shu, Z.; Jirutitijaroen, P. Latin Hypercube Sampling Techniques for Power Systems Reliability Analysis with Renewable Energy Sources. *IEEE Trans. Power Syst.* **2011**, *26*, 2066–2073.

mathematics

MDPI

Article

Optimal Sizing of Stand-Alone Microgrids Based on Recent Metaheuristic Algorithms

Ahmed A. Zaki Diab [1,*], Ali M. El-Rifaie [2,*], Magdy M. Zaky [3] and Mohamed A. Tolba [4,5,*]

1 Electrical Engineering Department, Faculty of Engineering, Minia University, Minia 61111, Egypt
2 College of Engineering and Technology, American University of the Middle East, Egaila 54200, Kuwait
3 Nuclear Research Center, Engineering Department, ETRR-2, Atomic Energy Authority (EAEA), Cairo 11787, Egypt; zaky_magdy@yahoo.com
4 Nuclear Research Center, Reactors Department, Egyptian Atomic Energy Authority, Cairo 11787, Egypt
5 Electrical Power Systems Department, Moscow Power Engineering Institute, 111250 Moscow, Russia
* Correspondence: a.diab@mu.edu.eg (A.A.Z.D.); ali.el-rifaie@aum.edu.kW (A.M.E.-R.); matolba@ieee.org (M.A.T.)

Abstract: Scientists have been paying more attention to the shortage of water and energy sources all over the world, especially in the Middle East and North Africa (MENA). In this article, a microgrid configuration of a photovoltaic (PV) plant with fuel cell (FC) and battery storage systems has been optimally designed. A real case study in Egypt in Dobaa region of supplying safety loads at a nuclear power plant during emergency cases is considered, where the load characteristics and the location data have been taken into consideration. Recently, many optimization algorithms have been developed by researchers, and these algorithms differ from one another in their performance and effectiveness. On the other hand, there are recent optimization algorithms that were not used to solve the problem of microgrids design in order to evaluate their performance and effectiveness. Optimization algorithms of equilibrium optimizer (EQ), bat optimization (BAT), and black-hole-based optimization (BHB) algorithms have been applied and compared in this paper. The optimization algorithms are individually used to optimize and size the energy systems to minimize the cost. The energy systems have been modeled and evaluated using MATLAB.

Keywords: isolated microgrids; cost of energy (COE); loss of power supply probability (LPSP); optimization techniques

Citation: Diab, A.A.Z.; El-Rifaie, A.M.; Zaky, M.M.; Tolba, M.A. Optimal Sizing of Stand-Alone Microgrids Based on Recent Metaheuristic Algorithms. *Mathematics* **2022**, *10*, 140. https://doi.org/10.3390/math10010140

Academic Editor: Zbigniew Leonowicz

Received: 9 December 2021
Accepted: 27 December 2021
Published: 4 January 2022

Publisher's Note: MDPI stays neutral with regard to jurisdictional claims in published maps and institutional affiliations.

1. Introduction

Recently, Egypt has shown interest and determination to be one of the worldwide energy producers. In 2030, Egypt plans to increase its renewable energy production to 30% of its demand to support the rising population and growth [1–3]. Egypt's location provides it with an excellent average irradiance all over the year. In addition, the wind energy atlas shows a great ability to depend on wind energy. In the last decade, the total installed capacity of new and renewable energy sources of wind and solar power plants has been raised from 1157 MW in 2017/2018 to 2247 MW in 2018/2019 with an increase of 94.2%, as reported in the 2018/2019 annual report of the Egyptian electricity holding company [4,5]. Likewise, the total energy generated from wind and solar sources, which are connected to the unified national grid, has been increased from 2871 GWh in 2017/2018 to 4543 GWh in 2018/2019 with a growth rate of 58.2% [1–6]; on the other hand, solar energy generation has increased by 184% reaching a value of 1525 GWh in 2019 [1–6].

Several renewable energy configurations have been designed and evaluated for such cases. Different configurations that are based on solar, wind, and fuel cells have been introduced [7–9]. Solar PV energy is a great source of clean energy in Egypt source; the high average irradiance all over the year [4], as well as the low costs of both operation and maintenance, led to a remarkable increase in the investments in PV plants to be the safest source in distant zones [7–9]. PV plants became the main energy source in most of the presented system configurations in Egypt as well as other countries [1–9]. Wind energy, storage batteries, geothermal, wave, tidal energies, and fuel cells are other sources that can be used with PV forming a hybrid system. The necessity of hybridization of other energy sources with PV sources is due to the variation in the generated solar energy with many factors such as meteorological conditions. Another suggested case study of a microgrid is to feed the nuclear power plants (NPPs) during emergencies, enhancing the electrical system safety and NPP reliability.

Design of emergency power systems provides electrically independent and physically separated power distribution divisions. One is designated odd and the other even. The distribution design provides power to redundant station loads and prevents failures or damage of one division cascaded to the others. Integration of all power sources in a microgrid arrangement enhances the safety of operation, normal shutdown, and unplanned shutdown, as well as overall plant safety. In addition, it mitigates the negative impact of emergency power absence on the environment. Solar and/or wind energy may supply the services and emergency load, while fuel cells can be used as storage devices.

Backup diesel generators have been used for compensating the lack of solar energy in the shortage periods [10]. However, the dependency on fossil fuels is still the main problem besides environmental conditions [3]. On the other side, batteries storage units are the first decision as a traditional storage component. The battery storage units assist the system to be both stable and reliable. More attention has been paid to reducing the cost of battery units and raising its efficiency and lifetime. Researchers have proposed the usage of battery units to improve the power quality of power systems that are being interconnected with renewable energy sources [7–10]. Fuel cells have been utilized as a reliable storage device with acceptable efficiency [11]. Fuel cells have distinguished advantages compared to the battery units, which include lower cost and less negative effect on the environment. Water-based FC with a combined electrolyzer unit is the standard type that is used with renewable energy sources. From the reported papers [7–12], it could be noted that the battery storage units increase the cost of the COE of all configured systems. Moreover, the grid-connected hybrid systems, in most cases, have the best COE. The reasons may be summarized in the lower cost of kWh obtained from the grid compared to the initial costs obtained from renewable energy sources. However, in recent years, acceptable rate reduction in the initial costs of renewable energy sources increased the chances of using such sources.

Several research studies have been carried to develop a reliable procedure to optimize the configuration of hybrid energy systems. Few reported attempts considered real case studies, while others focused on the techniques and methodologies [12–21]. Great efforts have been made to better manage the uncertainties of renewable energy systems (RES), cost of energy (COE), and load demands (LD) by various recent research studies. In [22], the authors proposed the management strategy of RES uncertainties, the electricity price, and LD based on a hybrid stochastic/robust (HSR) optimizer in different scenarios, which have the advantage of improving the convergence characteristics. The authors in [23] developed a distributed robust optimization approach to overcome the restrictions of the dispatchable flexible resources taking into account the uncertainties from RES and LD based on different constraints that can be appropriate in piratical schemes with tacking the transmission loss in the consideration.

In 2020 [12], a hybrid configuration composed of PV plant, WT plant, battery units, and diesel generators was designed, involving a comprehensive comparison between the different possible configurations. The simulation and optimization process has been done using HOMER® (Hybrid Optimization of Multiple Energy Resources, Boulder County, CO, USA) and NEPLAN® (NEPLAN AG, Zurich, Switzerland) platforms. A case study in Egypt has been considered for evaluating the designed configurations to determine the best configuration involving renewable and conventional energy sources. The results showed that the most effective design is the interconnected grid system with PV and diesel generators without any storage devices with a COE of 0.124 USD/kWh. A procedure for designing an isolated microgrid in Con Dao Island in Vietnam has been presented in [13]. The results through HOMMER show that a reliable operation of the designed microgrid.

Away from the fixed configurational platforms, many optimization algorithms have been proposed and applied for determining the optimal configuration of the microgrids. In [14], the optimal sizing of the energy storage system using the state of energy model was reported for an active distribution network. The results show an effective reduction in the possible error in the sizing optimization considering the case of insufficient data at the planning stage. In [15], a valuable effort has been made to present a method for optimizing the size of battery and ultracapacitor hybrid storage systems. The technique can be used for plug-in electric vehicles (PEVs) and smart grids. The presented energy management method had been applied in real-time applications with Markov chain and stochastic dynamic programming (SDP) algorithm. Moreover, a village in Egypt has been recognized as a case study. A complete system from PV, wind, and diesel generators with battery storage units has been developed to feed people with electricity [10]. A fuel cell and renewable energy sources have been combined in an energy hybrid system [11]. Various optimization algorithms of water cycle optimizer, hybrid particle swarm, whale optimizer, and moth-flame optimizer have been applied for designing different configurations of microgrids involving photovoltaic, diesel generators with battery storage units or hydro-electric pumped storage considering real data have been presented in [6,11]. The results show that the whale optimization algorithm and whale optimization algorithm gives the best results regarding the best COE and convergence characteristics for the specified case of study. A grid-connected photovoltaic and wind turbine hybrid system has been designed with the application of the methods of GA and PSO as presented in [16]. The results showed that the COE waw minimized by continuing feeding to the load demand. A techno-economic of a stand-alone hybrid system involving hybrid pumped and battery storage with photovoltaic has been presented with the application of the algorithms of GA, firefly algorithm, and grey wolf optimizer in [17]. In [17], a case of study for feeding a low load has been considered. The results prove the ability of the grey wolf optimizer to minimize the COE of the system. To improve the energy-use efficiency in a case study related to agricultural fields, GA has been applied to optimize the configuration of a hybrid energy system for reducing environmental impacts [18]. In Spain, a PV/WT/Biomass/H2/fuel cell hybrid system based on model predictive control and genetic algorithm results in a COE of 0.123 USD/kWh [19].

The application of the optimization techniques is essential for finding solutions and optimal configuration of the energy systems in many fields, such as agricultural, milling industry, nuclear power plant systems, and flood control operations, to find the optimal configuration with satisfying the constraints [20,21]. The optimization leads to finding the optimum and finest solutions between reasonable alternates, which ensure satisfying the considered problem constraints. Additionally, the complex task, the multidisciplinary problem of designing the microgrid systems considering many variables and constraints, leads to implementing it as an optimization problem to achieve one or more objectives such as minimizing cost, minimizing loss of power supply probability, and/or maximizing the energy reliability.

This paper presents a comprehensive comparison between the performances of three metaheuristic methods of equilibrium optimizer (EQ), bat optimization (BAT), and black-hole-based optimization (BHB) to optimize and get the techno-economic optimal configuration of microgrids to evaluate their effectiveness. Acceptable convergence characteristics of the three optimization techniques have been proved considering other optimization problems. However, no attempt has been made to present a comprehensive comparison between the performance of the three algorithms to optimize the sizing of such a hybrid energy system of this paper. Therefore, for a closer look at their performance, the application of these methods was considered to optimize the hybrid energy system (PV plant, FC systems, and battery storage systems) considering a real case study of Egypt in the Dobaa region. Moreover, statistical tests were performed to evaluate the robustness of the three applied algorithms.

The article is developed as follows: Section 2 illustrates methods involving the system configurations, complete microgrid mathematical model, energy management methodology, the sizing of microgrid problem formalization, the applied optimization techniques, and the case of study. The numeric results and discussions are executed in Section 3. The last section contributes to the conclusion of the proposed work.

2. Methods

2.1. System Configurations

The configuration of a stand-alone microgrid is illustrated in Figure 1. This is a general configuration that contains a PV power plant with FC. Moreover, a battery was included as a storage device. This system is designed to introduce an essential solution in remote areas.

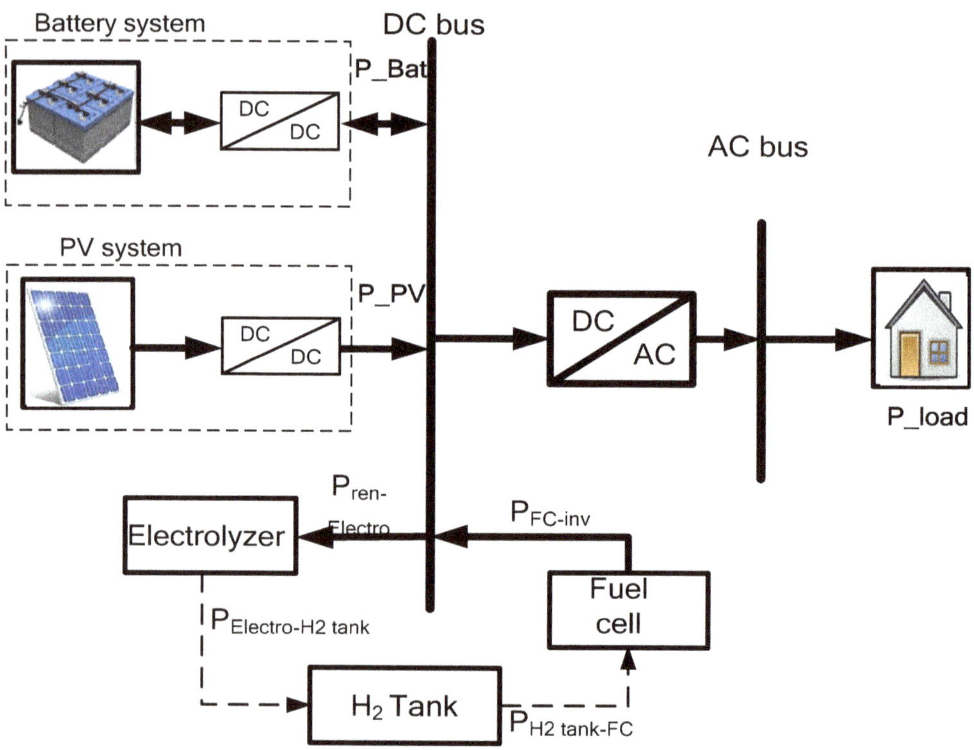

Figure 1. Arrangement of the studied microgrid.

2.2. Complete Microgrid Mathematical Model

2.2.1. Solar System

The solar system is modeled considering the variations of the produced power from the PV solar system with both irradiance and temperature. The model of the produced power is illustrated using Equations (1) and (2) [10,11].

$$P_{PV}(t) = N_{PV} P_{PV\_rated} \eta_{PV} \eta_{Wire} \times \frac{G(t)}{G_{nom}} (1 - \beta_T (T_C(t) - T_{C\_nom})) \quad (1)$$

where N_{PV} and P_{PV_rated} indicate the PV modules number and the nominal power of each, while η_{PV} and η_{wire} represent the efficiencies of the PV and connected wire, respectively. Moreover, *G(t)* and G_{nom} represent solar irradiance at the operating temperature and the standard one of 1000 W/m^2, respectively. B_T, T_C, and T_{C_nom} indicate temperature coefficient module, cell temperature, and standard temperature of 25 °C, respectively. Furthermore, the cell temperature is estimated as follows:

$$T_C(t) - T_{ambient} = G(t)\frac{T_{Test}}{800} \quad (2)$$

where $T_{ambient}$ and T_{Test} denote the module's ambient and tested temperatures, respectively.

2.2.2. Battery Storage Unit

Battery storage units have been used to mitigate the variation of the PV and wind-generated power, which vary with many uncontrolled operation conditions, for instance, irradiance, temperature, and wind speed. The usage of storage units will be useful to control power flow to loads. Several studies reported that the lead-acid battery has a wide usage for such applications because of its availability and competitive price with respect to other battery types. Factors affecting the battery bank sizing include lifetime, temperature as well as depth of discharge (DOD) [10,24]. The capacity boundary of the battery banks can be weighed considering the state of charge (SOC).

Moreover, SOC respecting the storage banks is determined using the charging and discharging energy. The state of charge is estimated using time as per Equations (3)–(6). There are two modes of operation: the first one is the charging mode, while the other is the discharging mode. In the period of charging, the charged energy is calculated as follows:

$$E_{CH}(t) = (\frac{P_{WT}(t) - P_{load}(t)}{\eta_{conv}} + P_{PV}(t)) \times \eta_{CH} \times \Delta t \quad (3)$$

where $E_{CH}(t)$ represents the charged energy during Δt, which is one hour at instant t. $P_{load}(t)$ represents the load power at the same instant t, while η_{conv} and η_{CH} are the efficiencies of converter and charging, respectively. Moreover, the state of charge is estimated as

$$SOC(t) = SOC(t-1)(1-\sigma) + E_{CH}(t) \quad (4)$$

where $SOC\ (t)$ and $SOC(t-1)$ denote battery SOCs at two instances of t and $t-1$, one-to-one. Additionally, σ denotes the rate of self-discharging. While the discharging mode energy and SOC can be calculated as follows:

$$E_{DIS}(t) = (\frac{P_{load}(t) - P_{WT}(t)}{\eta_{conv}} - P_{PV}(t)) \times \eta_{DIS} \times \Delta t \quad (5)$$

$$SOC(t) = SOC(t-1)(1-\sigma) - E_{DIS}(t) \quad (6)$$

where $E_{DIS}(t)$ represents the discharging energy at a time t. At the same time, η_{DIS} denotes the discharging efficiency of the battery.

2.2.3. Electrolyzer

The electrolyzer is implemented based on the water-electrolysis concept, producing hydrogen and oxygen by flowing a DC-current among two electrodes. Subsequently, the hydrogen is gathered on all sides of the anode surface. According to the water electrolyzer mentioned by [25–27], the generated hydrogen is gathered at a pressure of 30 bar. This value is very high as compared with the produced one from the reactant-pressure to supply the proton exchange membrane fuel cell (PEMFC). In many studies, the generated hydrogen from the electrolyzer can be supplied to the hydrogen tank, or its pressure is increased to 200 bar by a compressor to boost the energy stored density [25]. The other studies stated that the hydrogen evaluated from the electrolyzer is applied to a low-pressure tank until it is charged; therefore, a compressor is utilized to force the stored hydrogen to a high-pressure tank. Hence, the energy depleted by the compressor is decreased because it is not in the process of launching the whole time [25,26]. In this work, the generated hydrogen is applied to the hydrogen tank. The electrolyzer can be simulated via the transmitted power from the DC-bus to the hydrogen tank, and it can be expressed by the following formula [25–27],

$$P_{Electro-\text{tank}} = P_{ren-Electro} \times \eta_{Electro} \tag{7}$$

where $P_{Electro-\text{tank}}$ is the output electrolyzer power that is injected into the hydrogen tank. $P_{ren-Electro}$ is the output electrolyzer power. The $\eta_{Electro}$ is the electrolyzer efficiency.

2.2.4. Hydrogen Tank

In the proposed work, the hydrogen tank is simulated via the energy-stored quantity (E_{tank}) in the hydrogen tank at a time t. The E_{tank} can be formulated as follows [25]:

$$E_{\text{tank}}(t) = E_{\text{tank}}(t-1) + \left(P_{Electro-\text{tank}}(t) - \frac{P_{\text{tank}-FC}(t)}{\eta_{storage}}\right) \times \Delta t \tag{8}$$

where $P_{\text{tank}-FC}(t)$ is the equivalent power that pulled out from the hydrogen tank and applied to the fuel cell (FC), and $\eta_{storage}$ is the efficiency of the storage tank, and it indicates the losses, and it can be provided as 95% for all operating scenarios [25]. Δt is the interval of the simulation process, and it is deemed to be one hour in the proposed work.

The mass of stored hydrogen M_{tank} in the tank can be expressed by the following equation [25–27]:

$$M_{\text{tank}}(t) = \frac{E_{\text{tank}}(t)}{HHV_{H_2}} \tag{9}$$

where HHV_{H2} is the higher heating value of hydrogen (HHV). In line with [26], the amount of HHV is deemed as 39.7 kWh/m^2. The energy stored in the tank comes among a predefined upper and lower limit of the tank capacity. According to some issues related to hydrogen nature, there is a recommendation that the low quantity (lower limit) of stored hydrogen is not discharged and can be considered here by 5%. Consequently,

$$M_{\text{tank,min}} \leq M_{\text{tank}}(t) \leq M_{\text{tank,max}} \tag{10}$$

where M_{tank} is ranged between the upper $M_{\text{tank,max}}$ and lower $M_{\text{tank,min}}$ limits of hydrogen tank at time t.

2.2.5. Fuel Cell (FC)

The electrolysis of the hydrogen FC is working in reverse to generate electric current when the hydrogen is recombined with the oxygen. However, the PEMFC is manufactured in large generating sizes to be reliable with a short-power release time of around 1–3 s [25].

In this work, the efficiency of the FC (η_{FC}) is considered to be 50% constant value. Therefore, the produced power can be estimated in a simple way based on the input power $P_{\text{tank}-FC}$ and η_{FC} of the FC by the following formula:

$$P_{FC-inv} = P_{\text{tank}-FC} \times \eta_{FC} \tag{11}$$

2.2.6. DC/AC Converter

The main roles of the inverter are to convert the produced DC power from renewable sources and the FC source into AC power, besides exceeding the supplied power to the grid. In this study and based on [28], the inverter efficiency (η_{inv}) is supposed to be 90% as a constant value. Hence, the output power of the inverter can be estimated using the following equation:

$$P_{inv-AC} = (P_{FC-inv} + P_{ren-inv}) \times \eta_{inv} \tag{12}$$

2.3. *Energy Flow Scenarios*

The methodology of the energy management system in a microgrid is planned to ensure continued energy covering the load demand. The energy management can be concluded in the following six scenarios. The first three scenarios are related to the generated energy from PV being less than the demand load; in this case, the battery and FC should be operated to cover the load demand energy. The other three scenarios are related to the generated energy from PV being larger than the demand load; in this case, the battery and FC should be operated to store the extra energy.

2.3.1. Case 1

Scenario I:

The battery should operate in the discharging mode of the battery to feed the load demand when renewable energy does not cover it. So, in this case, the load power shearing is as follows:

$$P_{load}(t) \times \Delta t = (P_{PV}(t) \times \eta_{conv}) \times \Delta t + E_{DIS}(t) \times \eta_{conv} \tag{13}$$

Scenario II:

Continuously; if the renewable energy and battery do not cover the load demand, this is the time of FC operation. The generated power from the FC is estimated as

$$P_{FC}(t) \times \Delta t = (P_{load}(t) - P_{PV}(t) \times \eta_{conv}) \times \Delta t - E_{DIS}(t) \times \eta_{conv} \tag{14}$$

Scenario III:

Continuously; if the renewable energy, battery, and FC, does not suffice the load needs, there is a shortage in the energy to supply the load needs. The LPS should be calculated to be minimized. Moreover, a solution of DG may be used.

2.3.2. Case 2

Scenario IV:

On the other hand, when the generated energy from renewables exceeds the load demand, the battery will be operated in charging mode. The power fellow in this scenario will be demonstrated as follows:

$$P_{load}(t) \times \Delta t = (P_{PV}(t) \times \Delta t - E_{CH}(t)) \times \eta_{conv} \tag{15}$$

Scenario V:

Sometimes the battery is full in the time of the renewable energy is exceeds the load demand. In this scenario, the extra energy will be stored in the FC tank. The power fellow in this interval is expressed as:

$$P_{load}(t) \times \Delta t = (P_{PV}(t) \times \Delta t - P_{Electro-\text{tank}}(t)) \times \eta_{conv} \tag{16}$$

Scenario VI:

Sometimes, the battery is full in the time of the renewable energy is exceeds the load demand. The extra energy is supplied to the dummy load in this scenario. The dummy load in this interval is expressed as

$$P_{dummy}(t) \times \Delta t = (P_{PV}(t) \times \eta_{conv} - P_{load}(t)) \times \Delta t - E_{CH}(t)) \times \eta_{conv} - P_{FC-inv}(t) \times \eta_{conv} \tag{17}$$

All scenarios can be visualized as the flowchart of Figure 2.

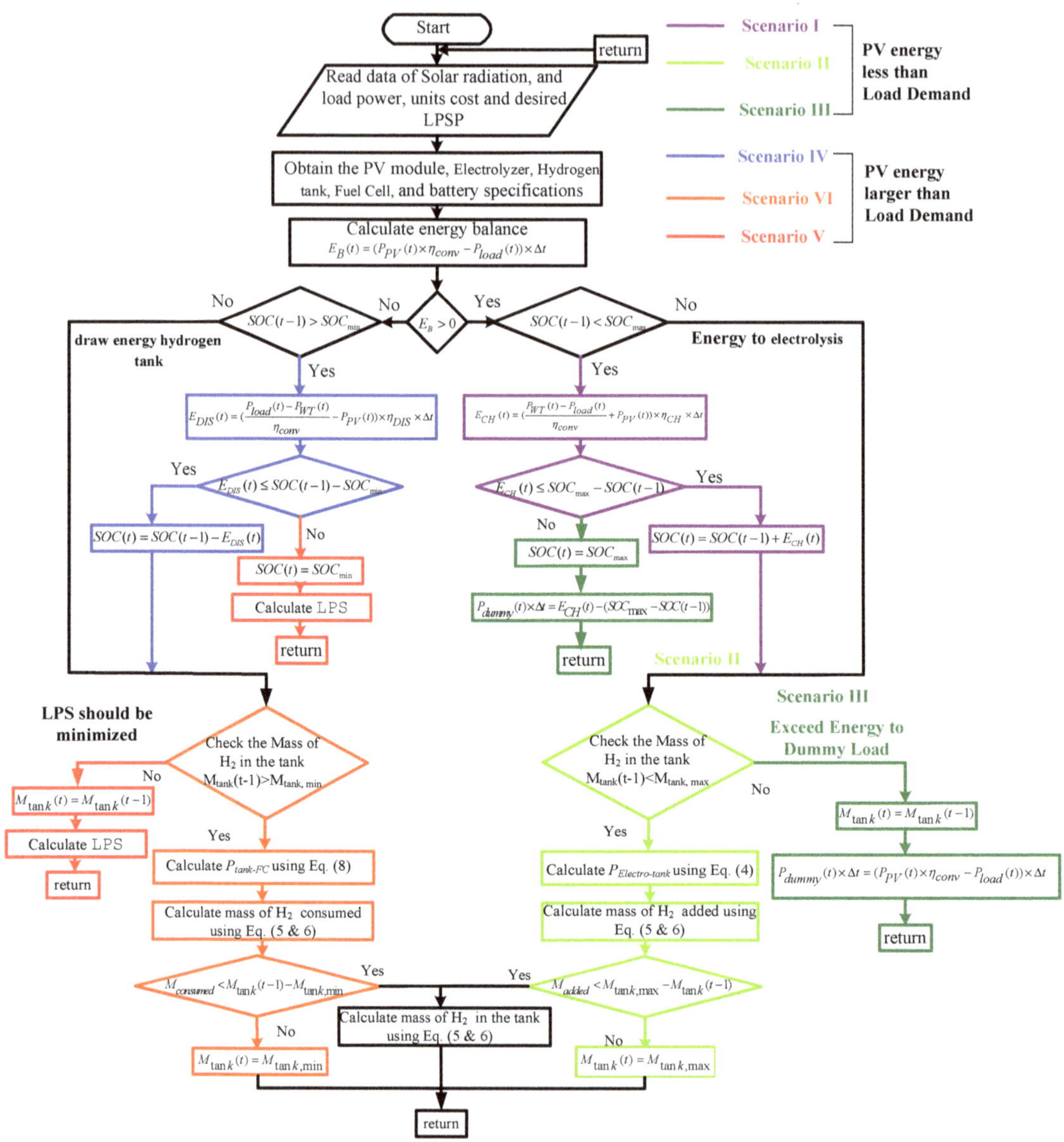

Figure 2. Scenarios of energy management flowchart.

2.4. Sizing of Microgrid Problem Formalization

2.4.1. Optimized Objective Function Indices

There are many indices that should be minimized to ensure the excellent design of the microgrid. Three indices have been considered, which are (1) the cost of rnergy (COC), (2) loss of power supply probability (LPSP), and (3) the dummy load (P_{dummy}) [10,29,30]. Therefore, the objective function is composed of the four indices with a weighted ratio. The system will be designed in order to minimize the weighted objective function.

(1) Cost of Energy (COC)

The net present cost (NPC) is utilized to estimate the whole cost of the hybrid micro-grid. The system annual cost of investment C_{ann_tot} can be expressed by the following formula:

$$C_{ann\_tot} = C_{ann\_cap} + C_{ann\_rep} + C_{ann\_oper\&maint} \quad (18)$$

where, C_{ann_cap}, C_{ann_rep}, and $C_{ann\_oper\&maint}$ are the annual costs of the system components, system components' replacements, operation and maintenance, respectively.

(a) The Annual Capital Cost of the Microgrid System

The capital recovery factor (CRF) is utilized changing the initial investment to annual capital costs based on the following equation:

$$CRF(r, M_{sys}) = \frac{r \times (1+r)^{M_{sys}}}{(1+r)^{M_{sys}} - 1} \quad (19)$$

where r and M_{sys} are the rate of interest (%) and the life span of the whole hybrid system under study.

The annual capital cost of the individual subsystems is evaluated using the next expressions,

$$\begin{cases} C_{ann\_cap\_PV} = C_{cap\_PV} * CRF(r, M_{PV}) \\ C_{ann\_cap\_FC} = C_{cap\_FC} * CRF(r, M_{FC}) \\ C_{ann\_cap\_batt} = C_{cap\_batt} * CRF(r, M_{batt}) \\ C_{ann\_cap\_conv} = C_{cap\_conv} * CRF(r, M_{conv}) \end{cases} \quad (20)$$

where C_{cap_PV}, C_{cap_batt}, C_{cap_FC}, and C_{cap_conv} are the initial capital cost of PV system integration, the initial cost of the battery bank, and the initial cost of all components of FC, respectively. M_{PV}, M_{batt}, and M_{FC} are the lifetime of PV modules, battery banks, and FC, consequently.

Therefore, the annual capital investment cost of the hybrid system is formulated as follows:

$$\begin{aligned} C_{ann\_cap} &= C_{ann\_cap\_PV} \\ &+ C_{ann\_cap\_batt} \\ &+ C_{ann\_cap\_FC} \\ &+ C_{ann\_cap\_conv} \end{aligned} \quad (21)$$

where $C_{ann\_cap\_PV}$, $C_{ann\_cap\_batt}$, $C_{ann\_cap\_FC}$, and $C_{ann\_cap\_conv}$ are the annual-capital-cost share of the integration of the PV, FC, battery bank, and converter, consequently.

(b) The Operation and Maintenance Cost

The operation and maintenance cost of the proposed scheme is estimated in the following form:

$$\begin{aligned} C_{oper\&main} &= C_{oper\&main\_PV} * t_{PV} \\ &+ C_{oper\&main\_batt} * t_{batt} \\ &+ C_{oper\&main\_conv} * t_{conv} \\ &+ C_{oper\&main\_FC} * t_{FC} \end{aligned} \quad (22)$$

where, $C_{oper\&main\_PV}$, $C_{oper\&main\_batt}$, $C_{oper\&main\_conv}$, and $C_{oper\&main\_FC}$ are the operation and maintenance costs of PV, battery banks, converter, and FC per unit time, respectively. t_{PV}, t_{batt}, t_{conv}, and t_{FC} are the operating time of PV, battery banks, converter, and FC, respectively.

(c) The Annual Replacement Cost

The replacement cost of the hybrid system components during its lifetime is determined by the next formula [10–27],

$$C_{rep} = \sum_{j=1}^{n_{rep}} K_{C_{rep}} C_u \left(\frac{1+i}{1+r}\right)^{jM_{sys}/(n_{rep}+1)} \quad (23)$$

where, i, $K_{C_{rep}}$, C_u, and n_{rep} are the inflation-rate of replacement, the unit's size utilized in the system, the cost of the replaced units, besides the number of the replacements during the project time M_{sys}.

Hence, the net present cost (NPC) is expressed as follows,

$$NPC = \frac{C_{ann\_tot}}{CRF} \tag{24}$$

The cost of energy (COE) is defined as the generated electrical energy cost from the hybrid system in (USD/kWh). The COE is expressed as

$$COE = \frac{C_{ann\_tot}}{\sum_{h=1}^{h=8760} P_{load}} = \frac{NPC}{\sum_{h=1}^{h=8760} P_{load}} * CRF \tag{25}$$

(2) Loss of Power Supply Probability (LPSP)

The LPSP is defined as a design factor. It takes the measurements of the insufficient operation probability of the power supply in the case of the hybrid microgrid is unsuccessful in satisfying the energy demand. The loss of power supply $LPS(t)$ is given by the following formula:

$$\begin{aligned} LPS(t) &= P_{Load}(t) \times \Delta t \\ &-(P_{PV}(t) \times \eta_{conv}) \times \Delta t \\ &-E_{DIS}(t) \times \eta_{conv} \\ &-E_{FC}(t) \times \eta_{conv} \end{aligned} \tag{26}$$

LPSP is known as a practical index to estimate the reliability in the issues of the optimum capacity of hybrid renewable energy systems. Thus, the LPSP is evaluated based on the summation of the $LPS(t)$ overall load demand within all study periods, and it can be mathematically written as follows:

$$LPSP = \frac{\sum_{t=1}^{8760} LPS(t)}{\sum_{t=1}^{8760} P_{load}(t) \times \Delta t} \tag{27}$$

2.4.2. The Proposed Objective Function

The objective function (OF) of this work is proposed to minimize the COE, LPSP, in addition to the dummy load (P_{dummy}) based on the optimization technique. Subsequently, the OF is created in the following expression:

$$OF = \psi_1 * COE + \psi_2 * LPSP + \psi_3 * P_{dummy} \tag{28}$$

In this work, the weighting factors Ψ_1, Ψ_2, and Ψ_3 are selected based on trial and error to achieve the best results. This selection is considered according to the following conditions; weighting factors summation are equal to unity, their values ϵ (0, 1), and the Ψ_2 value must be higher than Ψ_1, and Ψ_3 values for ensuring the whole system reliability. Therefore, the produced Ψ_1, Ψ_2, and Ψ_3 values are 0.2, 0.6, and 0.2, respectively.

2.4.3. Design of Constrains for Optimization

In off-grid hybrid systems, the operation of the system components must be considered based on constraints. To ensure the optimal operation of the system to confirm the condition in Equation (29), the generated energy at a time (t) is balanced with the energy consumed by the load. This constraint can be expressed as follows:

$$\begin{aligned} P_{Load}(t) \times \Delta t &= (P_{PV}(t) \times \eta_{conv} \\ &+P_{WT}(t) + P_{DG}(t)) \times \Delta t \\ &+E_{batt}(t) \times \eta_{conv} \\ &+E_{FC}(t) \times \eta_{conv} \end{aligned} \tag{29}$$

Equation (30) is considered as one of the optimization policies to assure that the power storage per hour by hydrogen tank at a time should be performed within limits, and it can be represented as

$$M_{\text{tank,min}} \leq M_{\text{tank}}(t) \leq M_{\text{tank,max}} \tag{30}$$

Avoiding the over/under charging issues, the SOS of the battery bank is constrained between its maximum $SOC_{\max}$ and minimum $SOC_{\min}$ values. Where the $SOC_{\max}$ is considered the full size of the battery, on the other hand, the $SOC_{\min}$ is subjected to the depth of discharge. This condition can be illustrated by Equations (31) and (32):

$$SOC(t+1) = SOC(t)(1-\sigma) \tag{31}$$

$$SOC_{\min} \leq SOC(t) \leq SOC_{\max} \tag{32}$$

The LPSP can be lower than the system indicator of predefined reliability (β_L). In this work, the β_L is equal to 0.04 [10]. This can be formulated as follows:

$$LPSP \leq \beta_L \tag{33}$$

2.5. Optimization Algorithms

2.5.1. Bat Optimization (BAT)

Bat algorithm (BA) is inspired by the echolocation manner of bats detecting their foods [31]. Only mammals can detect their prey based on sonar waves "echolocation" avoiding barriers in the darkness. Based on the echolocation attitude, bats can determine the distance of object "prey" as follows [31]:

- Based on the echolocation characteristics, the bat can detect their prey and recognize the variation between food and surrounding barriers.
- The flying bat is in a random form toward a place (x_i) via velocity (v_i) and frequency (F_{min}), wide wavelength (λ), and loudness (L_o) for detecting the prey. It has the capability to control its pulses λ/F_{min} based on pulse rate emission r via a range of 0–1 to approximate its goal.
- Based on [31], the loudness has been supposed to be ranged from maximum value (L_o) to a minimum value (L_{min}).

The virtual movements of *ith* bat can be evaluated based on x_i and v_i in D-dimension. Additionally, the values of x_i and v_i are upgrading at each iteration, and the process can be defined as follows:

$$F_i = F_{\min} + \beta \cdot (F_{\max} - F_{\min}) \tag{34}$$

$$v_i^t = v_i^{t-1} + (x_i^{t-1} - x^*) F_i \tag{35}$$

$$x_i^t = x_i^{t-1} + v_i^t \tag{36}$$

where $\beta \in [0, 1]$ is a random vector that is extracted from a uniform distribution, and x^* is the current global best solution from all bats. Locally, if a current best solution is so far, then it is generated new solutions based on a local walk randomly as formulated below:

$$x_{new} = x_{old} + \varepsilon \cdot L^t \tag{37}$$

where $\varepsilon \in [-1, 1]$ is a random number, and L^t is the average of loudness at t time. In the case of the bat being very close to prey, it reduces its loudness increasing the emitted rate of pulses. It can be assumed that when a bat detected its prey, then the $L_o = 1$ and $L_{min} = 0$, and it can be formulated as

$$L_i^{t+1} = \alpha \cdot L_i^t; \quad r_i^{t+1} = r_i^o [1 - \exp(-\gamma \cdot t)] \tag{38}$$

where α and γ are constant values that are $\alpha > 0$ and $\gamma < 1$. In addition,

$$L_i^{t+1} \to 0; \quad r_i^t \to r^o \ as \ t \to \infty \tag{39}$$

Note that the initial values of L_o and r_i can range between 0 and 1.

The steps of the BAT technique process can be deduced in the flowchart in Figure 3.

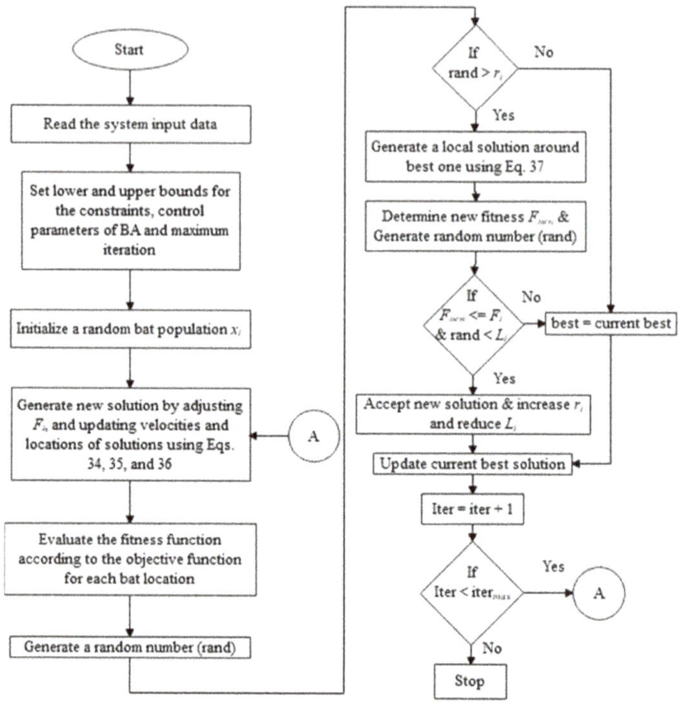

Figure 3. Flowchart of BAT technique.

2.5.2. Black-Hole-Based Optimization Technique (BHB)

The BHB technique was inspired by the black-hole phenomenon [32]. It is a population-based algorithm, where a "black-hole" is known as the best solution/candidate of the population at every iteration, and the other solutions are known as "normal stars". The basic election of the black hole is one of the genuine candidates of the population. The black hole attracts all solutions based on their current placement, including a random number. The normal stars can be pulled around the black hole after the initializing process. In addition, the black hole will allow the too-closed stars to be gone forever.

The proposed BHB can be formulated in the following process [32]:

Process 1: Initializing. For stars, a population size "N" has been generated with random sittings in a research space that is distributed in limited upper and lower boundaries.

Process 2: Run the program performing all constraints for every star of the population. If the satisfied star constraints, then it is workable; otherwise, it is not workable.

Process 3: Evaluate the fitness function for every workable star.

Process 4: Record the best fitness/star as the black-hole X_{BH} star.

Process 5: Start with count $t = 1$.

Process 6: Alter the sitting of every star based on the following Equation (40):

$$\begin{aligned} X_i^{t+1} &= X_i^t + rand_i(0,1) * (X_{BH} - X_i^t); \\ i &= 1,2,3,\ldots,N \end{aligned} \tag{40}$$

where t is the iteration number, X_i is the sitting of the star at iteration t, and X_{BH} is the sitting of the black-hole in the search space.

Process 7: If a star gets a sitting where the parameters of the proposed design are lower than previous, then update their sittings as the following:

$$X_{BH}^{t+1} = \left\{ \begin{array}{l} X_i^{t+1} \rightarrow if\left[F(X_i^{t+1}) < F(X_{BH}^t)\right] \\ X_{BH}^{t+1} \rightarrow if\left[F(X_i^{t+1}) \geq F(X_{BH}^t)\right] \end{array} \right\} \tag{41}$$

Process 8: Calculate the radius of events horizon R based on the following Equation (42):

$$R = \frac{F(X_{BH})}{\sum\limits_{i=1}^{N} F(X_i)} \tag{42}$$

where the $F(X_{BH})$ is the fitness content of the black hole, and the $F(X_i)$ is the fitness content of ith star.

Process 9: If a star passes the event horizon R of the black hole, alter it with a new one in a random sitting in the space. Otherwise, go to **process 6**.

Process 10: Raise generation count $t = t + 1$.

Process 11: If $t \leq t_{max}$, start again from **process 6**. Otherwise, **stop**.

According to the above process, the flowchart of the proposed BHB technique is shown in Figure 4.

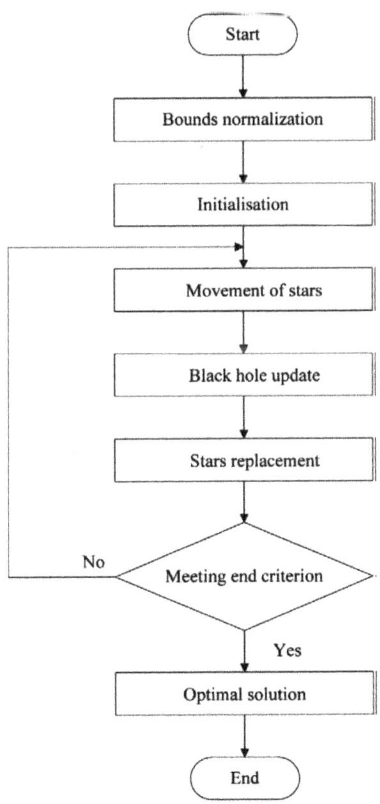

Figure 4. Flowchart of BHB technique.

2.5.3. Equilibrium Optimizer (EQ)

The EQ technique was created by Farmarzi in 2020 [33]. It is a meta-heuristic technique that simulates the dynamics and equilibrium mass balance schemes. This technique can be initiated with concentrations of *ith* particles "C" via special number and dimensions as recorded in the following equation:

$$C_i = B_l + r \cdot (B_u - B_l) \tag{43}$$

where the B_l and B_u are the lower and upper boundaries of the search space, the $r \in (0, 1)$ is uniform random.

Based on evaluating the fitness function, the "C" can be upgraded via computing the equilibrium solutions to deduce the best candidates. The upgrading process of the EQ technique can be deduced in the following form:

$$C_{new} = C_{eq} + \frac{G}{\mu}(1 - E) + (C - C_{eq}) \cdot E \tag{44}$$

where the exponential term "E" and the rate of generation "G" can be known as

$$E = a_1 sign(m1 - 0.5)(e^{-\lambda t} - 1) \tag{45}$$

$$t = (1 - \frac{iter}{\max\_iter})^{a_2(\frac{iter}{\max\_iter})} \tag{46}$$

$$G = \left\{ \begin{array}{c} 0.5r_1(C_{eq} - C)E \rightarrow if \quad r_2 \geq GP \\ 0 \rightarrow if \quad r_2 < GP \end{array} \right\} \tag{47}$$

where a_1 and a_2 are constants and equals 2 and 1, respectively, $m1 \in (0, 1)$ is random vector, *iter* and max_*iter* are the iteration number and the maximum one respectively. r_1 and $r_2 \in [0, 1]$ are random numbers, $GP = 0.5$ and it is the generated probability.

Within every upgrading, the proposed fitness function is calculated for every particles' concentration to evaluate their states and to include the best so far particles. Based on Equation (44), the upgrading process of every concentration particle re-generated depends on the sharing of three sections. The first section is random, and it is extracted from the equilibrium pool. The other sections focus on the variations in concentrations. The last two sections are in charge of the exploitation accuracy and the global searching in the research space to deduce the optimum issues, respectively.

Figure 5 illustrates the flowchart of the proposed EQ technique.

2.6. Case Study

To evaluate the energy management system based on various optimization algorithms, a real study case was introduced with the purpose of designing a hybrid renewable energy system, and it was selected in the Dobaa region in Egypt. The microgrid has been designed for the emergency operation of the projected nuclear station of Dobaa in Egypt at Geographical coordinates of 30.040566, 26.806641 (30°02′26″, 26°48′24″) [34]. For the emergency operation, the microgrid should be disconnected from the electrical grid. The location of the site, as mentioned, is in the Dobaa nuclear station, as shown in Figure 6. The data of horizontal solar radiation are presented in Figure 7. As the temperature represents an essential factor for the PV performance, the monthly average temperature is shown in Figure 8. The estimated emergency average load demand per month and the estimated load curve per day are introduced in Figures 9 and 10. The load curve was calculated and estimated based on the expected load of the plant in an emergency. It should be noted the residential loads in the period from 7 pm to 10 pm. Other domestic facilities were recorded. The average load and the maximum load demand were 265 kW and 420 kW, respectively.

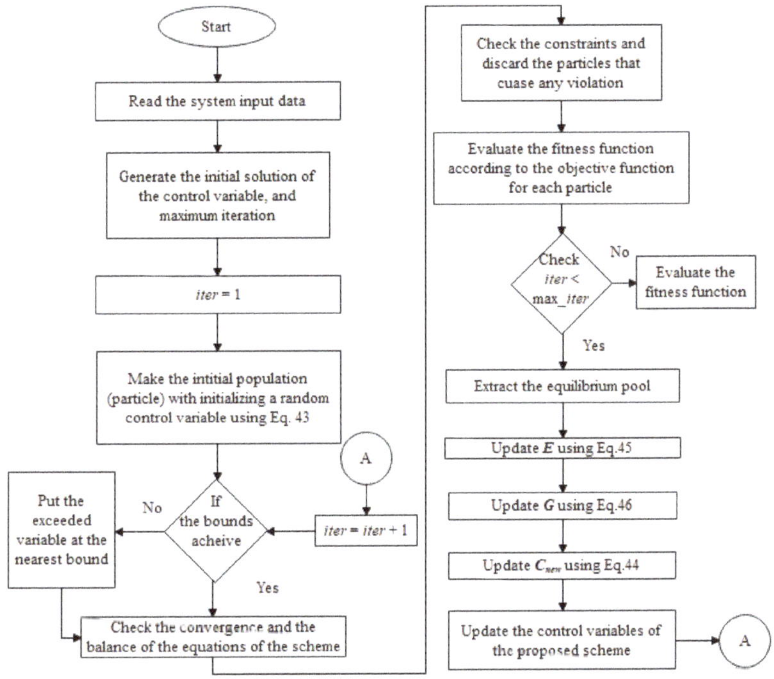

Figure 5. Flowchart of EQ Technique.

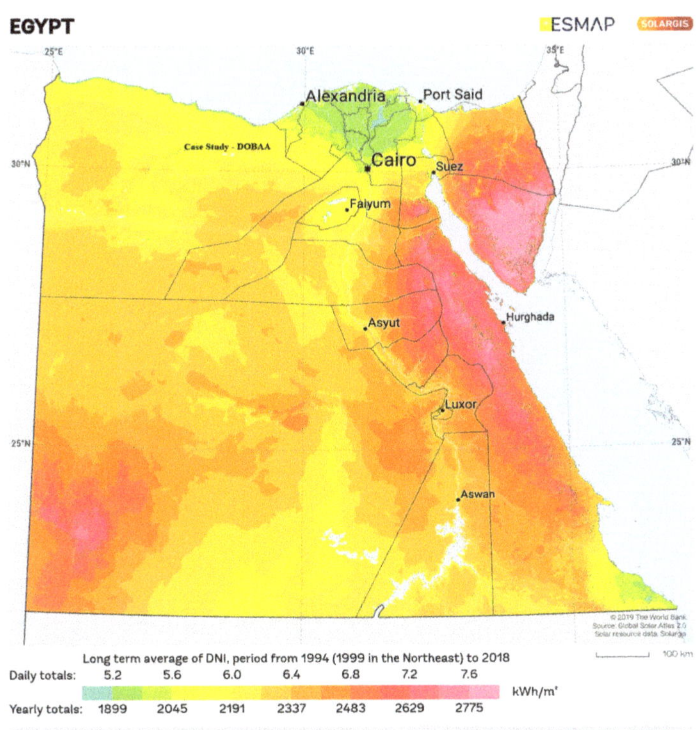

Figure 6. Location of the studied microgrid.

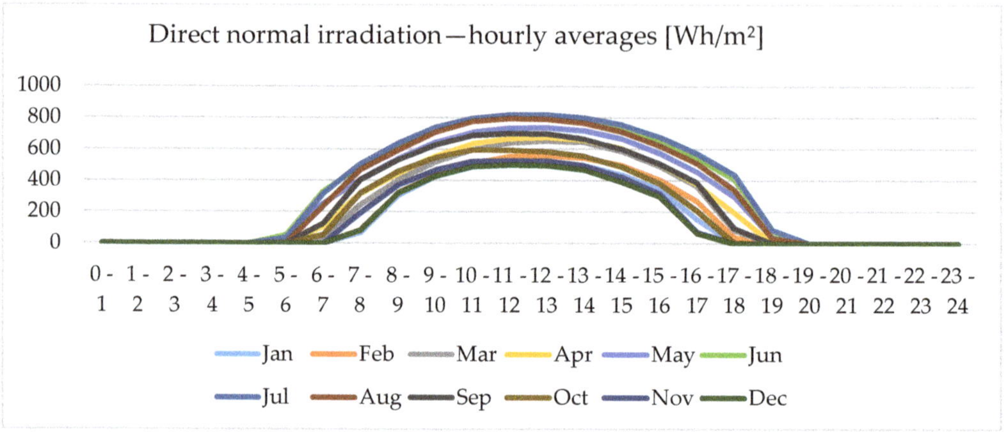

Figure 7. Solar radiation for the study area.

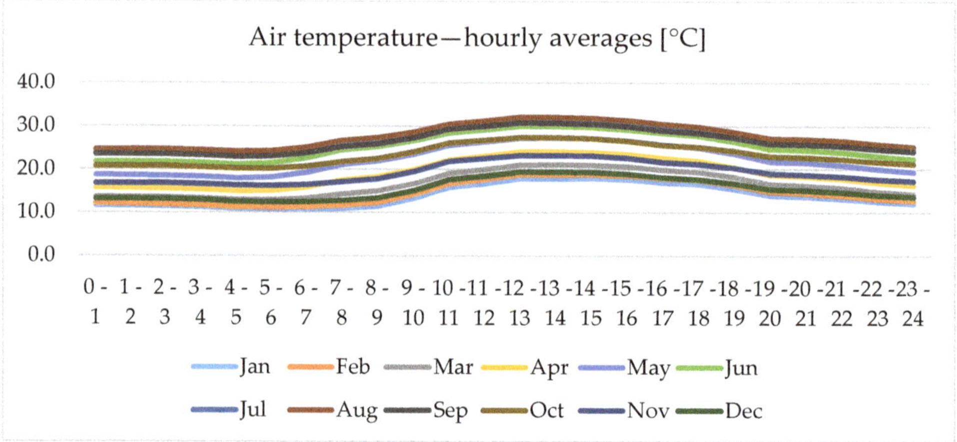

Figure 8. Temperature for the study area.

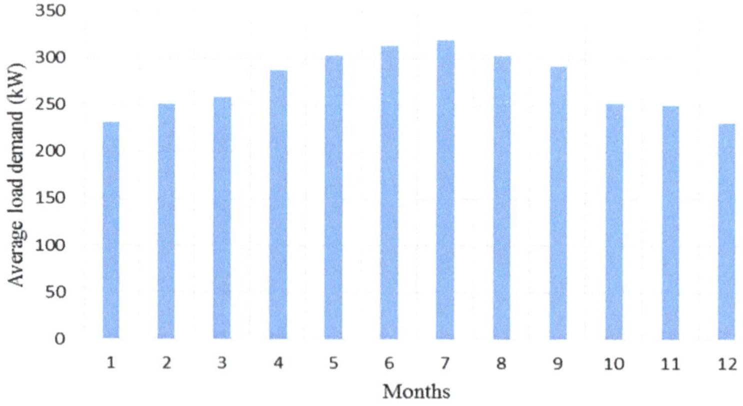

Figure 9. The average load demand per month.

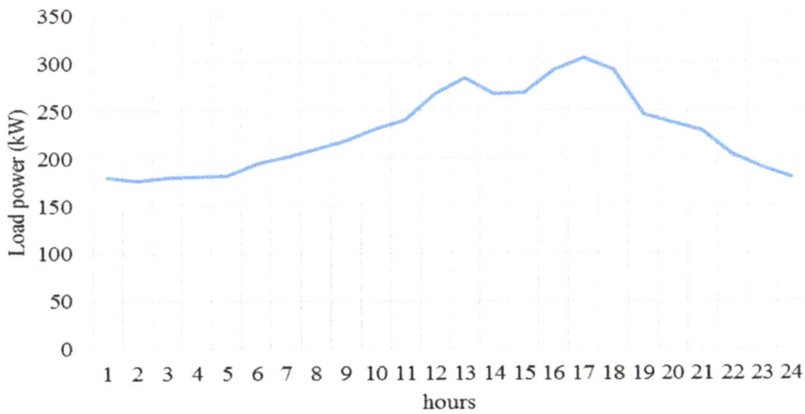

Figure 10. The load curve per day of the study area.

3. Results and Discussions

MATLAB package was used in order to determine the optimal configuration-based individual optimization algorithms. For each algorithm, the maximum iterations and search agents number were considered to be 200 and 30, respectively. According to this work, the capacity of the proposed microgrid (hybrid system) is realized as PV rated power and their modules number, the mass of hydrogen tank, number of battery units, the rated power of the electrolyzer and fuel cell. The optimization algorithm should determine the configuration of the energy system to minimize the objective function. The data specifications of the system components can be found in [11] and are given in Table 1.

Table 1. The system components' descriptions [11].

Component	Capital Cost (USD/Unit)	Replacement Cost (USD/Unit)	O&M (USD/Unit–yr)	Lifetime (yr)	Efficiency (%)	Unit
PV array	7000	6000	20	20	0.15	1 kW
Electrolyzer	2000	1500	25	20	75	1 kW
Hydrogen tank	1300	1200	15	20	95	1 kg
Fuel cell	3000	2500	175	5	50	1 kW
Battery bank	146.5	102.55	-	10	86	12 V (50 Ah)
DC/AC converter	800	750	8	15	90	1 kW

3.1. The Optimal Configuration of Energy System

Table 2 displays the comprehensive outcomes of the optimization procedures of BAT algorithm, equilibrium optimizer, and black hole algorithm. The minimum value of each is highlighted in the table for visualizing the best results. The minimum best objective function was obtained by the EQ algorithm. Moreover, the best COE is 0.289129, which is obtained by BAT algorithm, while LPSP and dummy Load indices with BAT algorithm are 0.045548 and 0.113331, respectively, which indicate more load is not covered in certain periods and other periods; the dummy load index shows that more surprise energy is dissipated. The results of the EQ algorithm for LPSP and dummy load are 0.043986 and 0.113607, respectively. This indicates that although the COE is higher than that of the BAT algorithm, the LPSP of 0.0436418 was obtained, which results from more uninterrupted energy to the load.

Table 2. Optimization parameters of microgrid configurations based on the three algorithms.

Items		BAT	EQ	BHB
Best objective function		0.112231	**0.1074**	0.108078
Best solution	**PV (kw)**	339.6977	**339.6308**	348.8188
	Battery	4695.026	**4909.019**	4893.064
	Electrolyzer (kw)	459.6776	**452.9702**	491.7788
	Tank (kg)	88.80917	**87.25968**	107.2796
	FC (kw)	34.73013	**34.10566**	35.4967
COE		0.289129	**0.291437**	0.302562
NPV		546067.9	**550426**	571437.7
LPSP		0.045548	**0.043986**	0.039623
Dummy load		0.113331	**0.113607**	0.118959

The net present values are 546067.9, 550426, and 571437.7 for BAT, EQ, and BHB optimizers, respectively. Although the net present value of EQ is higher than that of BAT by 0.9664%, the recommended design is that of the EQ algorithm. This recommendation is because of the decrease of the LPSP with the EQ algorithm rather than that of the BAT algorithm. In general, the two designs based on BAT and EQ algorithms can be considered concerning the priority of LPSP or COE. Figures 11 and 12 visualize the obtained results of the various configurations of the microgrid based on the three algorithms.

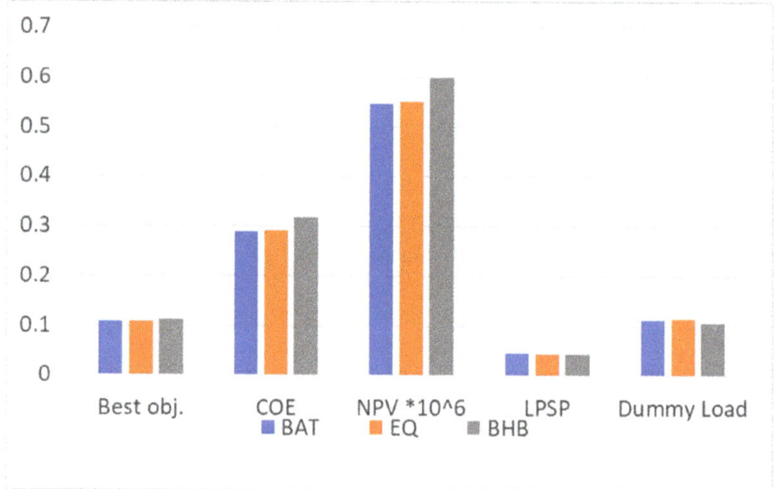

Figure 11. Indices of the energy system based on various algorithms.

3.2. *Performance of Different Algorithms*

Figure 13 illustrates the convergence curves for the proposed BAT, EQ, and BHB techniques. From this figure, the proposed optimizers can achieve the optimal values of the recommended objective function to be 0.112231, 0.1074, and 0.108078 for BAT, EQ, and BHB algorithms, respectively, which indicated that the EQ has the best minimum objective function. Nevertheless, Figure 13 demonstrates that the EQ is the fastest one compared with BAT and BHB optimization techniques. The detailed results of the three algorithms are presented in Section 3.3.

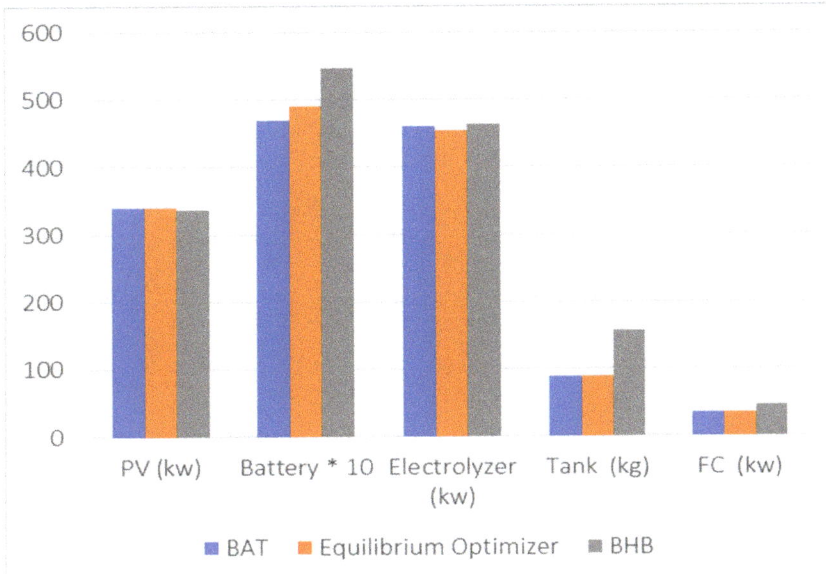

Figure 12. Configuration of the energy system based on various algorithms.

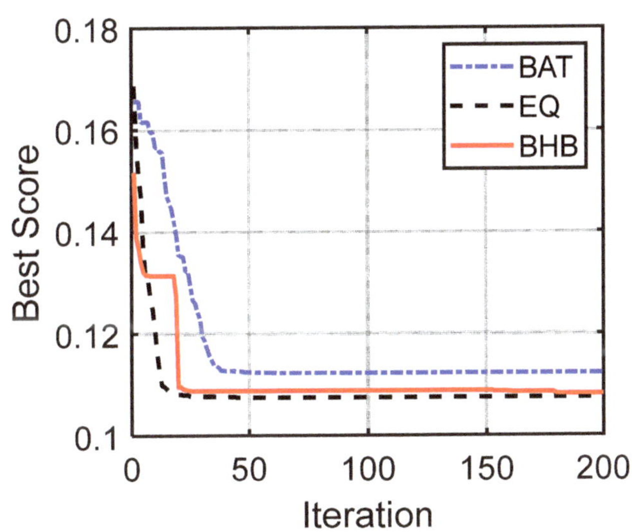

Figure 13. Convergence curves of the three algorithms.

3.3. Statistical Results

The BAT algorithm was implemented for 30 individual runs. The statistical results of the BAT, EQ, and BHB algorithms are listed in Table 3. The results show that the minimum and maximum obtained cost functions of the BAT algorithm are 0.112231 and 0.13043, respectively, while the stranded division and the average obtained results of the BAT algorithm are 0.117865 and 0.469301, respectively. On the other hand, the statistical results show that the obtained minimum and maximum values of the cost functions of the EQ algorithm are 0.1074 and 0.112216, respectively, while the stranded division and the average values EQ algorithm are 0.223387 and 0.110812, respectively. Moreover, the table shows that the minimum and maximum obtained cost functions of BHB are 0.108078 and 0.124731, respectively. In comparison, the stranded division and the average results of BHB are 0.41938 and 0.114019, respectively. The obtained results show that the best-obtained results are acceptable for all algorithms, but EQ and BHB algorithms are better algorithms rather than BAT. Furthermore, the statistical indices show that the divisions between the results of individual runs of the BAT algorithm are bigger than those of the EQ and BHB.

Table 3. Statistical results of the three algorithms of BAT, EQ, and BHB.

Items	BAT	EQ (Best Results)	BHB
cost_min	0.112231	**0.1074**	0.108078
cost_maxworst	0.13043	**0.112216**	0.124731
cost_mean	0.117865	**0.110812**	0.114019
cost_median	0.117716	**0.112215**	0.113443
cost_sD	0.469301	**0.223387**	0.41938
RE	1.506026	**0.762435**	1.649028
MAE	0.005634	**0.003412**	0.005941
RMSE	0.007282	**0.004053**	0.007231
eff	95.36117	**96.95946**	94.91016
p	1.73×10^{-6}	$\mathbf{1.82 \times 10^{-5}}$	1.73×10^{-6}
h	1	**1**	1

The convergence curves of the 30 individual runs are shown in Figure 14 for the three algorithms. Furthermore, the boxplot of the best objective functions is shown in Figure 13. On the other side, the Wilcoxon signed-rank test was performed to validate the application of the BAT, EQ, and BHB algorithms. The results show that the value of the rank h is 1 for the three algorithms, which indicates that the test rejects the null hypothesis of zero median. Moreover, the p is 1.73×10^{-6}, 1.82×10^{-5}, and 1.73×10^{-6}, which proves the robustness of the BAT, EQ, and BHB algorithms, respectively. The box plot has been plotted for the three algorithms for more visualization of the performance of the three algorithms, as shown in Figure 15. The figure demonstrates the superiority of the EQ algorithm for optimizing the size of the microgrid for the considered case study.

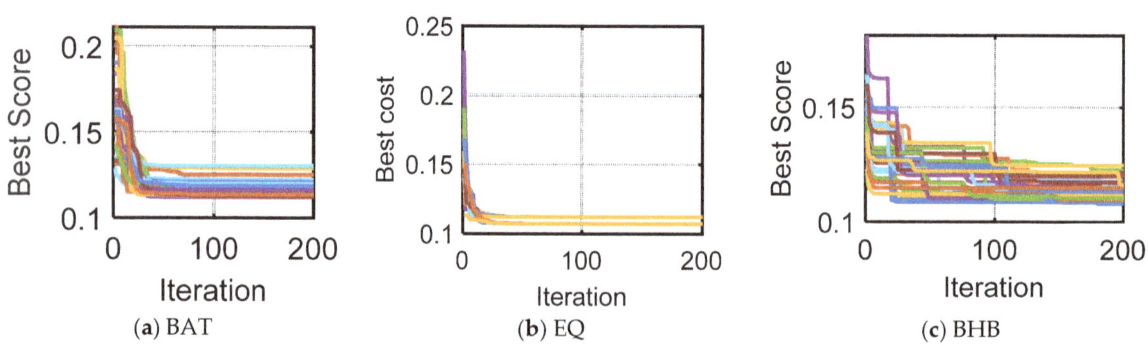

Figure 14. Convergence curves of the three algorithms over 30 runs: (**a**) 30-run convergence curves of BAT, (**b**) 30-run convergence curves of EQ, and (**c**) 30-run convergence curves of BHB.

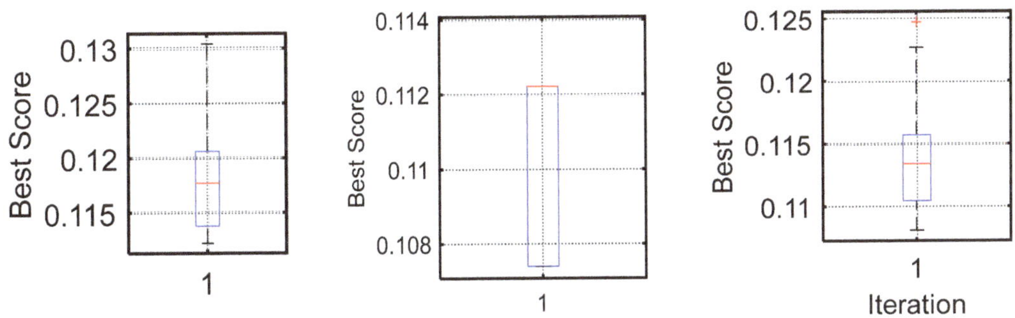

(**a**) BAT (**b**) EQ (**c**) BHB

Figure 15. Box plots of the three algorithms over 30 runs: (**a**) box plot of BAT, (**b**) box plot of EQ, and (**c**) box plot of BHB.

3.4. Operation of the Microgrid

The operation of the microgrid is investigated in this part. The performance of the microgrid is shown in Figure 16. Figure 16 illustrates the generated power change per hour of proposed hybrid system components at the optimal case of the EQ technique. Figure 16a shows the load demand (P_{load}), the hole generated power from the renewable energy (P_{PV}), and the difference between the renewable generation and the load (P_{diff}). Moreover, Figure 16b illustrates the battery charge, discharge, tank, and fuel cell during the operation period.

Furthermore, Figure 16c displays the dummy load and LPSP. Because of the design conditions, it is not easy to realize the optimization requirements even though keeping LPSP or dummy load to be zero. Through the hourly period of the high generated power from renewables, the extra energy is utilized to charge the battery and fill the hydrogen tank. For more clearance of the energy management concept behindhand the optimization techniques, the simulated numeric results are focused on the hybrid system performance for one day at the optimal conditions of their operation.

Figure 17 displays the results for a certain day regarding the optimum capacity from the EQ optimizer. According to Figure 17a, it is noticeable that the per diem curve of load demand includes two topmost points. The first one is nearby at 1 pm, where the temperature is high during this time. Hence, all existing equipment is required to be in service to decrease the air temperature. The second topmost point is nearby at 6 pm after the sunset. Within the nighttime till the early hours, the generated power from the renewable sources is very low; consequently, the battery and full cell are operating to cover the electricity demand. If the same conditions are still, the LPSP, as shown in Figure 17c, has a value. During the sunrise, around 6 am, the generated power from the PV station increases. Consequently, the electric energy over the load needs is utilized to charge the battery and fill up the hydrogen tank. Through the daytime between 2 pm and 6 pm, the generated power from the nontraditional sources is higher than the load needs. Therefore, the excess power is used to fill the tank and charge the battery, as shown in Figure 17b. The hydrogen increased until the tank amounted to its maximum limit and the battery charging. When the battery and the tank also are fully charged, the excess power is pushed into the dummy load, as shown by Figure 17c.

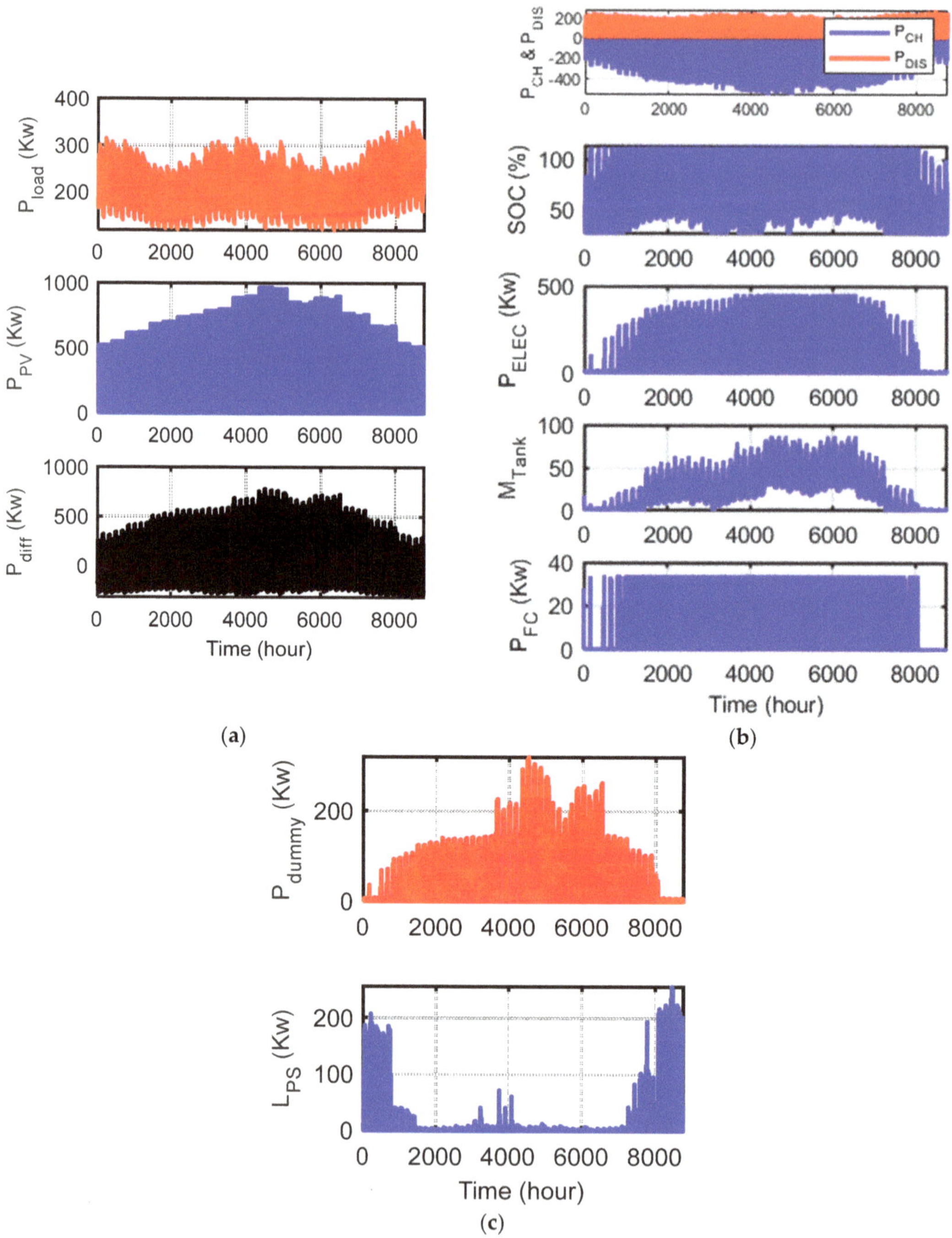

Figure 16. The results for the operation of the microgrid over one year considering the optimal configuration based on EQ technique. (**a**) Load, PV, and the different power. (**b**) Performance of the charging and discharging of the storage units for the battery and FC. (**c**) Dummy load and LPSP.

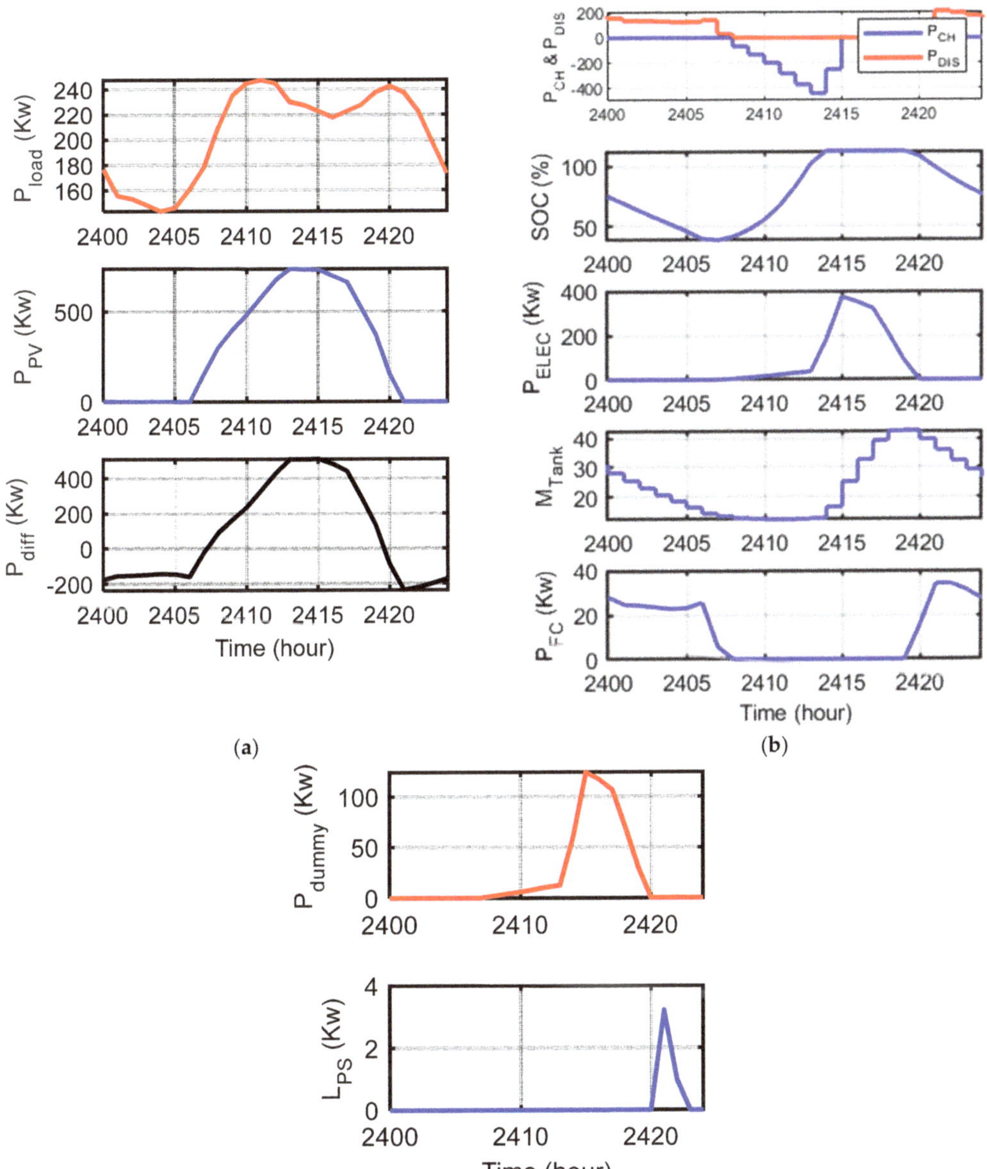

Figure 17. Numeric results of the microgrid operation for one working day via optimal configuration using EQ technique. (**a**) Load, PV, and the different power. (**b**) Performance of the charging and discharging of the storage units for the battery and FC. (**c**) Dummy load and LPSP.

4. Conclusions and Future Directions

In this paper, a stand-alone microgrid has been designed to feed emergency loads of a nuclear power plant using recent optimization algorithms of equilibrium optimizer (EQO), bat optimization (BAT), and black hole (BH). A comprehensive comparison between the ability and performance of the algorithms was conducted to solve the problem of microgrids design. A configuration of a microgrid consisting of a PV plant with FC and battery storage systems was optimally designed, and the possibility of integrating with a nuclear power plant to enhance the emergency power supplies was studied. The optimization algorithms are individually used to optimize and size the energy systems to minimize the cost and ensure the optimized microgrid's reliability. The energy systems were modeled and evaluated in MATLAB.

The results show that the EQ algorithm has a better performance than the other algorithms considering the best objective function value. The objective function was improved to 0.1074 using the EQ algorithm, while its values were 0.112231 and 0.108078 with BAT and BHB. However, the COE of the EQ-based results is higher than the BAT algorithm, while it is lower than the BHB. On the other hand, the reliability index of the EQ algorithm is better than the BAT algorithm, which is the main reason to increase the COE of the EQ algorithm. The results of BHB indicate that the LPSP is a smaller one with respect to the EQ and BAT, while the dummy load of the BHB is higher than those of the BAT and EQ algorithms. Finally, the designed microgrid based on the EQ and BHB is recommended based on the obtained results of the simulation of the microgrid operation and statistical analyses. It should be remarked that the storage energy cost is considered one of the main reasons to increase the COE and affect the system reliability. Using other renewable energy sources such as bioenergy or wind energy may enhance the system performance. Therefore, in future work, different configurations of off-grid and grid-connected microgrids should be designed to include a wind power plant, bioenergy, and/or diesel generator to increase the system's reliability and reduce the COE.

Author Contributions: Conceptualization, A.A.Z.D. and M.M.Z.; Methodology, A.A.Z.D. and M.A.T.; Software, A.A.Z.D.; Data curation, A.A.Z.D., A.M.E.-R. and M.A.T.; Formal analysis, A.A.Z.D., A.M.E.-R. and M.A.T.; Visualization, A.A.Z.D., A.M.E.-R. and M.A.T.; Investigation, A.M.E.-R. and M.M.Z.; Analysis, A.A.Z.D. and A.M.E.-R.; Writing—original draft, A.A.Z.D. and M.A.T.; Writing—review and editing, A.A.Z.D., A.M.E.-R. and M.M.Z. All authors have read and agreed to the published version of the manuscript.

Funding: This research received no external funding.

Institutional Review Board Statement: Not applicable.

Informed Consent Statement: Not applicable.

Data Availability Statement: Not applicable.

Acknowledgments: The researchers (Ahmed A. Zaki Diab and Mohamed A. Tolba) are funded by a full scholarship (Mission 2020/21 and Mission 2019/20) from the Ministry of Higher Education of Egypt. However, the current research work is not funded by the mentioned Ministry in Egypt or any other organization/foundation.

Conflicts of Interest: The authors declare no conflict of interest. Non-financial competing interest.

References

1. The Arab Republic of Egypt, Ministry of Electricity and Energy. Egyptian Electricity Holding Company_Annual Report 2015–2016. Available online: http://www.moee.gov.eg/english_new/report.aspx (accessed on 10 April 2021).
2. Sultan, M.H.; Kuznetsov, O.N.; Diab, A.A.Z. Site selection of large-scale grid-connected solar PV system in Egypt. In *2018 IEEE Conference of Russian Young Researchers in Electrical and Electronic Engineering (EIConRus)*; IEEE: New York, NY, USA, 2018; pp. 813–818.
3. Sultan, H.M.; Diab, A.A.Z.; Oleg, N.K.; Irina, S.Z. Design and evaluation of PV-wind hybrid system with hydroelectric pumped storage on the National Power System of Egypt. *Glob. Energy Interconnect.* **2018**, *1*, 301–311.
4. The Official Website of the Ministry of Electricity and Energy. Available online: http://www.moee.gov.eg (accessed on 10 April 2021).
5. Tolba, M.A.; Diab, A.A.Z.; Tulsky, V.N.; Abdelaziz, A.Y. LVCI approach for optimal allocation of distributed generations and capacitor banks in distribution grids based on moth–flame optimization algorithm. *Electr. Eng.* **2018**, *100*, 2059–2084. [CrossRef]
6. Diab, A.A.Z.; Sultan, H.M.; Kuznetsov, O.N. Optimal sizing of hybrid solar/wind/hydroelectric pumped storage energy system in Egypt based on different meta-heuristic techniques. *Environ. Sci. Pollut. Res.* **2019**, *27*, 32318–32340. [CrossRef] [PubMed]
7. Shouman, R.; El Shenawy, E.T.; Khattab, N.M. Market financial analysis and cost performance for photovoltaic technology through international and national perspective with case study for Egypt. *Renew. Sustain. Energy Rev.* **2016**, *57*, 540–549. [CrossRef]

8. Kamel, O.M.; Abdelaziz, A.Y.; Zaki Diab, A.A. Damping Oscillation Techniques for Wind Farm DFIG Integrated into Inter-Connected Power System. *Electr. Power Compon. Syst.* **2020**, *48*, 1551–1570. [CrossRef]
9. Ngan, M.S.; Tan, C.W. Assessment of economic viability for PV/wind/diesel hybrid energy system in Southern Peninsular Malaysia. *Renew. Sustain. Energy Rev.* **2012**, *16*, 634–647. [CrossRef]
10. Diab, A.A.Z.; Sultan, H.M.; Mohamed, I.S.; Kuznetsov, O.N.; Do, T.D. Application of different optimization algorithms for optimal sizing of PV/wind/diesel/battery storage stand-alone hybrid microgrid. *IEEE Access* **2019**, *7*, 119223–119245. [CrossRef]
11. Diab, A.A.Z.; El-ajmi, S.I.; Sultan, H.M.; Hassan, Y.B. Modified farmland fertility optimization algorithm for optimal design of a grid-connected hybrid renewable energy system with fuel cell storage: Case study of Ataka, Egypt. *Int. J. Adv. Comput. Sci. Appl.* **2019**, *10*, 119–132. [CrossRef]
12. Rezk, H.; Al-Dhaifallah, M.; Hassan, Y.B.; Ziedan, H.A. Optimization and Energy Management of Hybrid Photovoltaic-Diesel-Battery System to Pump and Desalinate Water at Isolated Regions. *IEEE Access* **2020**, *8*, 102512–102529. [CrossRef]
13. Tran, Q.T.; Davies, K.; Sepasi, S. Isolation Microgrid Design for Remote Areas with the Integration of Renewable Energy: A Case Study of Con Dao Island in Vietnam. *Clean Technol.* **2021**, *3*, 804–820. [CrossRef]
14. Xiang, Y.; Han, W.; Zhang, J.; Liu, J.; Liu, Y. Optimal sizing of energy storage system in active distribution networks using Fourier–Legendre series based state of energy function. *IEEE Trans. Power Syst.* **2017**, *33*, 2313–2315. [CrossRef]
15. Lu, X.; Wang, H. Optimal sizing and energy management for cost-effective PEV hybrid energy storage systems. *IEEE Trans. Ind. Inform.* **2019**, *16*, 3407–3416. [CrossRef]
16. Gonzalez, A.; Riba, J.R.; Rius, A.; Puig, R. Optimal sizing of a hybrid grid-connected photovoltaic and wind power system. *Appl. Energy* **2015**, *154*, 752–762. [CrossRef]
17. Biswas, A.; Kumar, A. Techno-Economic Optimization of a Stand-alone PV/PHS/Battery systems for very low load situation. *Int. J. Renew. Energy Res.* **2017**, *7*, 844–856.
18. Kaab, A.; Sharifi, M.; Mobli, H.; Nabavi-Pelesaraei, A.; Chau, K.W. Use of optimization techniques for energy use efficiency and environmental life cycle assessment modification in sugarcane production. *Energy* **2019**, *181*, 1298–1320. [CrossRef]
19. Acevedo-Arenas, C.Y.; Correcher, A.; Sánchez-Díaz, C.; Ariza, E.; Alfonso-Solar, D.; Vargas-Salgado, C.; Petit-Suárez, J.F. MPC for optimal dispatch of an AC-linked hybrid PV/wind/biomass/H2 system incorporating demand response. *Energy Convers. Manag.* **2019**, *186*, 241–257. [CrossRef]
20. Ciardiello, A.; Rosso, F.; Dell'Olmo, J.; Ciancio, V.; Ferrero, M.; Salata, F. Multi-objective approach to the optimization of shape and envelope in building energy design. *Appl. Energy* **2020**, *280*, 115984. [CrossRef]
21. Gassar, A.A.; Koo, C.; Kim, T.W.; Cha, S.H. Performance Optimization Studies on Heating, Cooling and Lighting Energy Systems of Buildings during the Design Stage: A Review. *Sustainability* **2021**, *13*, 9815. [CrossRef]
22. Chang, X.; Xu, Y.; Gu, W.; Sun, H.; Chow, M.-Y.; Yi, Z. Accelerated Distributed Hybrid Stochastic/Robust Energy Management of Smart Grids. *IEEE Trans. Ind. Inform.* **2021**, *17*, 5335–5347. [CrossRef]
23. Chang, X.; Xu, Y.; Sun, H.; Khan, I. A distributed robust optimization approach for the economic dispatch of flexible resources. *Int. J. Electr. Power Energy Syst.* **2021**, *124*, 106360. [CrossRef]
24. Ekren, O.; Ekren, B.Y. Size optimization of a PV/wind hybrid energy conversion system with battery storage using simulated annealing. *Appl. Energy* **2010**, *87*, 592–598. [CrossRef]
25. Garcia, R.S.; Weisser, D. A wind–diesel system with hydrogen storage: Joint optimization of design and dispatch. *Renew. Energy* **2006**, *31*, 2296–2320. [CrossRef]
26. El-Sharkh, M.Y.; Tanrioven, M.; Rahman, A.; Alam, M.S. Cost related sensitivity analysis for optimal operation of a grid-parallel PEM fuel cell power plant. *J. Power Sources* **2006**, *161*, 1198–1207. [CrossRef]
27. Kashefi, K.; Riahy, G.H.; Kouhsari, S.M. Optimal design of a reliable hydrogen-based stand-alone wind/PV generating system, considering component outages. *Renew. Energy* **2009**, *34*, 2380–2390. [CrossRef]
28. Khan, M.J.; Iqbal, M.T. Pre-feasibility study of stand-alone hybrid energy systems for applications in Newfoundland. *Renew. Energy* **2005**, *30*, 835–854. [CrossRef]
29. Mohamed, M.; Eltamaly, A.; Alolah, A.; Hatata, Y.A. A novel framework-based cuckoo search algorithm for sizing and optimization of grid-independent hybrid renewable energy systems. *Int. J. Green Energy* **2018**, *16*, 86–100. [CrossRef]
30. Ramli, M.A.M.; Bouchekara, H.R.E.H.; Alghamdi, A.S. Optimal sizing of PV/wind/diesel hybrid microgrid system using multiobjective self-adaptive differential evolution algorithm. *Renew. Energy* **2018**, *121*, 400–411. [CrossRef]
31. Yang, X.-S. A new meta heuristic bat-Inspired algorithm. In *Nature Inspired Cooperative Strategies for Optimization (NISCO 2010)*; Gonzalez, J.R., Ed.; Studies in Computational Intelligence; Springer: Berlin/Heidelberg, Germany, 2010; pp. 65–74.
32. Bouchekara, H.R.E.H. Optimal design of electromagnetic devices using a black-Hole-Based optimization technique. *IEEE Trans. Magn.* **2013**, *49*, 5709–5714. [CrossRef]
33. Faramarzi, A.; Heidarinejad, M.; Stephens, B.; Mirjalili, S. Equilibrium optimizer: A novel optimization algorithm. *Knowl.-Based Syst.* **2020**, *191*, 2020. [CrossRef]
34. Available online: Solargis.com (accessed on 15 April 2021).

energies

MDPI

Article

Day-Ahead Optimal Scheduling of an Integrated Energy System Based on a Piecewise Self-Adaptive Particle Swarm Optimization Algorithm

Jiming Chen [1], Ke Ning [1,*], Xingzhi Xin [2], Fuhao Shi [3], Qing Zhang [4] and Chaolin Li [4]

1 College of New Energy, China University of Petroleum (East China), Qingdao 266580, China; jimingchen@126.com
2 Electric Power Branch Company of Shengli Oilfield, SINOPEC, Dongying 257001, China; n369949307@gmail.com
3 Public Service Center of Shengli Oilfield, SINOPEC, Dongying 257001, China; blackwoodgaming@163.com
4 Qingdao Ruinengda Electrical Technology Limited Company, Qingdao 266580, China; s19150010@s.upc.edu.cn (Q.Z.); wang3yan2fei1@163.com (C.L.)
* Correspondence: ningke4228@163.com; Tel.: +86-178-6426-6015

Abstract: The interdependency of electric and natural gas systems is becoming stronger. The challenge of how to meet various energy demands in an integrated energy system (IES) with minimal cost has drawn considerable attention. The optimal scheduling of IESs is an ideal method to solve this problem. In this study, a day-ahead optimal scheduling model for IES that included an electrical system, a natural gas system, and an energy hub (EH), was established. The proposed EH contained detailed models of the fuel cell (FC) and power to gas (P2G) system. Considering that the optimal scheduling of an IES is a non-convex complex optimal problem, a piecewise self-adaptive particle swarm optimization (PCAPSO) algorithm based on multistage chaotic mapping was proposed to solve it. The objective was to minimize the operating cost of the IES. Three operation scenarios were designed to analyze the operation characteristics of the system under different coupling conditions. The simulation results showed that the PCAPSO algorithm improved the convergence rate and stability compared to the original PSO. An analysis of the results demonstrated the economics of an IES with the proposed EHs and the advantage of cooperation between the FC and P2G system.

Keywords: integrated energy system; optimal scheduling; piecewise self-adaptive; chaotic mapping; fuel cell

Citation: Chen, J.; Ning, K.; Xin, X.; Shi, F.; Zhang, Q.; Li, C. Day-Ahead Optimal Scheduling of an Integrated Energy System Based on a Piecewise Self-Adaptive Particle Swarm Optimization Algorithm. *Energies* **2022**, *15*, 690. https://doi.org/10.3390/en15030690

Academic Editors: Zbigniew Leonowicz, Michał Jasinski and Arsalan Najafi

Received: 24 December 2021
Accepted: 15 January 2022
Published: 18 January 2022

1. Introduction

An integrated energy system (IES) can couple various forms of energy, such as electricity and natural gas, to meet the demands of users for multiple energy sources. Additionally, IESs are capable of realizing the complementary utilization of energy, reducing the operation cost of the system, promoting the absorption of solar/wind power [1,2], improving energy efficiency, and mitigating pollution emissions, which makes them a viable option to solve the energy and environmental problems [3]. Optimal energy flow (OEF) calculation provides a basis for the economic operation of IESs. Furthermore, the daily fluctuation of wind power and load should be taken into consideration [4], which requires day-ahead planning of the optimal energy flow of the IES.

The modeling of the energy hub (EH) and its internal components is one of the key points in current research on day-ahead optimal energy flow. In [5], taking the industrial production process (IPP) as a control variable of optimal scheduling, a universal extension EH model was proposed. The results demonstrate that such a method reduces the operation cost. Analogously, a digester can be added to the EH to interconnect the EH with a biogas–electric multi-energy system to form a biogas–solar–wind complementary model, which improves the absorption rate of renewable energy and reduces the operation cost of the IES [6]. A concentrating solar power plant model that couples the power grid and the heating network, making the internal components of the EH more diversified, was established in [7]. The results showed that the EH can lower the operation cost and promote the wind power penetration of the IES. Considering nitrogen and ammonia cycles, Xu et al. used a power to ammonia system to replace the P2G system in a general EH, which improved the operating efficiency and economics of the EH [8]. This topology strengthened the interconnection between electrical and thermal systems but weakened the interconnection between electrical and gas systems. An energy router was added into the EH in [9] to form a new EH model structure, and an optimal energy management strategy was proposed. The case studies demonstrated that the interconnection of two EHs could effectively reduce the operation cost of the IES. Ju et al. designed a novel structure of a P2G-based virtual power plant. The P2G model was divided into two parts: electrolysis for hydrogen production and the synthesis of methane [10]. In [11], an energy hub that incorporates emerging distributed energy resources as well as energy storage devices was proposed. The results showed that the operation cost was reduced with multiple energy hubs. A literature review shows that the research on EHs mainly focuses on the elaboration and enrichment of the model, which can improve the energy utilization rate of the IES and reduce its operation cost.

On the other hand, exploring the optimal algorithms suitable for the OEF problem is also important research in the field of IESs. Numerical algorithms, such as mixed-integer linear programming [7,12,13], Benders decomposition [14], and second-order cone programming [15], are fast in convergence. However, the implementation of the numerical algorithm is complicated, and they may not converge when the objective function is discontinuous or contains multiple extreme points. Many intelligent optimization algorithms are used to solve the OEF problem of the IES, such as the genetic algorithm [16–18], teaching–learning-based optimization algorithm [19,20], whale optimization algorithm [21,22], particle swarm optimization (PSO) algorithm, etc. PSO, proposed by Eberhart and Kennedy [23], does not require the continuity and convexity of the objective functions and has a strong adaptability to the uncertainty of computational data. Nevertheless, premature convergence and falling into local optima are disadvantages of PSO [24,25]. In [26], a modified crisscross PSO and improved binary PSO technique was proposed, in which the crisscross search has horizontal and vertical crossover operators to explore the search space in every dimension and mitigate the stagnancy problem. The results showed that the modified crisscross PSO reduced the local optimal problem of PSO and the computational effort of the algorithm. Mellouk et al. proposed a new parallel hybrid genetic algorithm–particle swarm optimization algorithm to solve the optimization problem, which had a convergence time and solution quality that were better than those of ordinary PSO [27]. Combining PSO with other algorithms can also overcome the weaknesses of the original PSO. Bao et al. integrated a heuristic PSO into the decomposition-based sequential multi-energy flow calculation to effectively solve a scheduling model with highly nonlinear multi-energy flow constraints [28]. In [29], the PSO algorithm and niche technology were combined to form a nonlinear decreasing inertia weighting strategy to prevent the algorithm from falling into local optima. The authors in [30] proposed a distributed algorithm that combined PSO and the interior point method. The above work indicates that the modification of the PSO algorithm or its combination with other algorithms can feasibly solve the OEF problem of IESs.

Little of the above literature jointly considered detailed models of the FC and P2G system, and none of it mentioned the effect of their operation on the energy flow distribution, renewable energy consumption, and economics of IESs. In the literature, the deeply modified PSO algorithm and the mathematical optimization algorithm are improved in efficiency; however, their complexity or computation effort is also increased accordingly.

Based on the above discussion, we developed a piecewise self-adaptive particle swarm optimization (PCAPSO) algorithm based on chaotic mapping, which updates the inertia weight factor utilizing the random numbers generated by different types of chaotic mappings during the iteration, to prevent the algorithm from falling into local optima and to enhance its ability to jump out of local optima. In addition, a novel model of energy hubs (EH) was established, which includes detailed models of the fuel cell (FC) and P2G system, and the joint operation of the P2G system, FC, and hydrogen storage tank (HST) was studied. To investigate the influence of multiple EHs on the IES, three EHs were added to the electricity–gas energy system. The operation costs of the systems with different numbers of EHs were compared.

The rest of this paper is organized as follows. In Section 2, the novel EH model is proposed, and the mathematical models of the coupling components are presented. The models of the electrical system and natural gas system are presented in this section as well. In Section 3, the constraints of each system are introduced. The PCAPSO algorithm based on chaotic mapping is constructed and the objective function is presented. Case studies for the day-ahead scheduling of OEF for IESs are provided in Section 4. Finally, the conclusions are drawn in Section 5.

2. Electric–Natural Gas IES Considering New Structural EH

In this section, the structure and formula of EHs are explained first. Then the detailed mathematical models of each coupling component, specifically the two-stage model of the P2G system and co-generation model of the FC, are presented. Thirdly, the steady-state energy flow model of the electric and natural gas system is introduced.

2.1. EH Model

In an IES, the EH is an important structure that couples different energies to provide input and output interfaces for each energy sub-system. The schematic diagram of the proposed EH is shown in Figure 1. The EH mainly includes a battery storage system (BSS), power to hydrogen (P2H) system, hydrogen to gas (H2G) system, hydrogen storage tank (HST), fuel cell (FC), electrical chillers (EC), gas boiler (GB), micro-turbine (MT), and an external renewable energy source (RES). The P2G model consists of an electrolytic cell and a synthetic cell. The electrolytic cell can cooperate with the HST and FC to integrate hydrogen energy flow into the EH, thus enhancing the flexibility of the energy supply. The input energy mainly includes electric energy and natural gas, while the output energy includes electric, thermal, cooling energy, and natural gas.

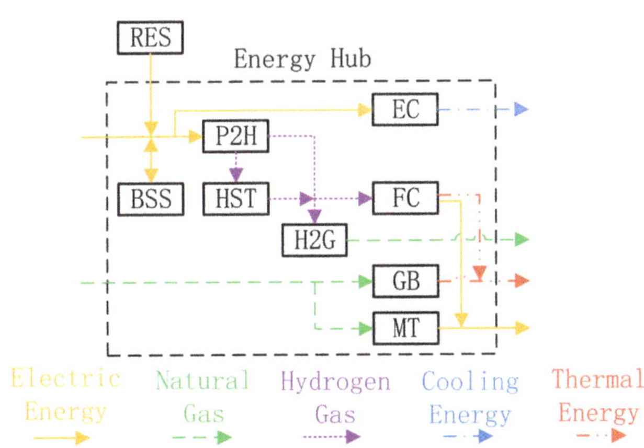

Figure 1. Structure of proposed EH.

The relationship between the input and output ports of the EH can be expressed by a coupling matrix [20]:

$$L = C \cdot P \tag{1}$$

$$\begin{bmatrix} L_\alpha \\ L_\beta \\ \vdots \\ L_\zeta \end{bmatrix} = \begin{bmatrix} c_{\alpha\alpha} & c_{\beta\alpha} & \cdots & c_{\zeta\alpha} \\ c_{\alpha\beta} & c_{\beta\beta} & \cdots & c_{\zeta\beta} \\ \vdots & \vdots & \ddots & \vdots \\ c_{\alpha\zeta} & c_{\beta\zeta} & \cdots & c_{\zeta\zeta} \end{bmatrix} \begin{bmatrix} P_\alpha \\ P_\beta \\ \vdots \\ P_\zeta \end{bmatrix} \tag{2}$$

where $\boldsymbol{L}$ is the output matrix; $\boldsymbol{P}$ is the input matrix; and $\boldsymbol{C}$ is the coupling matrix; $\alpha, \beta, \ldots, \zeta$ represent the forms of energy, such as electricity, natural gas, thermal energy, cold energy, etc. Each element in $\boldsymbol{C}$ can be a constant or an input-output formulation of a coupling component.

2.2. Modeling of Coupling Components

2.2.1. Power to Gas (P2G) System

The P2G system is an energy coupling device capable of converting electrical energy into natural gas using two main processes: electrolysis and synthesis. In the electrolysis process, water is used as a feedstock to produce hydrogen. The hydrogen is then synthesized with carbon dioxide to produce synthetic natural gas. The chemical reaction equations corresponding to the above processes are [31]:

$$\begin{gathered} 2H_2O \rightarrow 2H_2 + O_2 \\ CO_2 + 4H_2 \rightarrow CH_4 + 2H_2O \end{gathered} \tag{3}$$

The mathematical model of the P2G system is refined into two processes, P2H and H2G. In this case, the hydrogen produced by the electrolytic cell can be used not only as the reactant of synthetic natural gas, but also as the fuel material for the FC; in addition, it can be fed into the HST for storage, thus enhancing the coupling between the P2G system, HST, and FC. The conversion equations are:

$$F_{\mathrm{P2H}} = \eta_{\mathrm{P2H}} P_{\mathrm{P2H}} \tag{4}$$

$$F_{\mathrm{H2G}} = \eta_{\mathrm{H2G}} F_{\mathrm{P2H}} \tag{5}$$

where F_{P2H} is the hydrogen flow generated by electrolysis; η_{P2H} is the operation efficiency of the electrolytic cell; P_{P2H} is the electric power input into the electrolytic cell; F_{H2G} is the flow of synthesized natural gas; and η_{H2G} is the synthesis efficiency.

2.2.2. Co-Generation Model of the Fuel Cell (FC)

The FC is a device that can convert the chemical energy in hydrogen into electricity and thermal energy. The FC output voltage is affected by various internal overvoltages [32]:

$$V_{\mathrm{FC}} = E_{\mathrm{Nernst}} - \varsigma_{\mathrm{act}} - \varsigma_{\mathrm{ohm}} - \varsigma_{\mathrm{diff}} \tag{6}$$

where E_{Nernst} is the Nernst voltage; η_{act} is the activation polarization overvoltage; η_{ohm} is the ohmic overvoltage; and η_{diff} is the concentration overvoltage. The expressions for the above overvoltages are [32,33]:

$$\begin{gathered} E_{\mathrm{Nernst}} = 1.299 - 0.85 \times 10^{-3}(T_{\mathrm{FC}} - 298.15) \\ +4.3085 \times 10^{-5} T_{\mathrm{FC}} \cdot [\ln(p_{\mathrm{H}}) + 0.5\ln(p_{\mathrm{O}})] \end{gathered} \tag{7}$$

where T_{FC} is the operation temperature of the FC; p_{H} is the pressure of hydrogen fed into the FC; and p_{O} is the pressure of oxygen fed into the FC.

$$\varsigma_{\mathrm{act}} = \xi_1 + \xi_2 T_{\mathrm{FC}} + \xi_4 T_{\mathrm{FC}} \ln I + \xi_3 T_{\mathrm{FC}} \ln\left[\frac{p_{\mathrm{O}}}{5.08 \times 10^6 \exp(-498/T_{\mathrm{FC}})}\right] \tag{8}$$

In Equation (8), I is the current density of FC; $\xi_1 = -0.9514$; $\xi_2 = 0.00286 + 0.0002 \ln A_{FC} + 4.3 \times 10^{-5} \ln\{p_H/[1.09 \times 10^6 \exp(77/T_{FC})]\}$; A_{FC} is the effective area of the FC; $\xi_3 = 7.4 \times 10^{-5}$; and $\xi_4 = -1.87 \times 10^{-4}$.

$$\varsigma_{ohm} = IR_{int} \tag{9}$$

In Equation (9), R_{int} is the internal resistance of the FC.

$$\varsigma_{diff} = m \exp(nI) \tag{10}$$

In Equation (10), n is the porosity function of the gas diffusion layer, which is a constant in this paper, and m is the conductivity function of the electrolyte, which is a temperature-related function as follows [34]:

$$\begin{cases} 1.1 \times 10^{-4} - 1.2 \times 10^{-6}(T_{FC} - 273.15), T_{FC} \geq 312.15\text{K} \\ 3.3 \times 10^{-3} - 8.2 \times 10^{-5}(T_{FC} - 273.15), T_{FC} < 312.15\text{K} \end{cases} \tag{11}$$

If N_1 FCs are in parallel and N_2 FCs are in series, then the output voltage and current of the FC stack are:

$$V_{out} = N_2 V_{FC} \tag{12}$$

$$I = \frac{2FW_H}{N_1 M_H} \tag{13}$$

where F is the Faraday constant, 96,485 C/mol; W_H is the rate of hydrogen fed into the FC stack; and M_H is the molar mass of hydrogen. The active power and thermal energy output of the FC can be obtained as:

$$P_{FC} = V_{out} I = \frac{2FW_H}{M_H}(E_{Nernst} - \varsigma_{act} - \varsigma_{ohm} - \varsigma_{diff}) \tag{14}$$

$$Q_{FC} = I(N_1 E_{Nernst} - V_{out}) = \frac{2FW_H}{M_H}(\varsigma_{act} + \varsigma_{ohm} + \varsigma_{diff}) \tag{15}$$

2.2.3. Hydrogen Storage Tank (HST)

The HST is mainly used to store hydrogen converted from surplus renewable energy through the P2H process. On the one hand, the stored hydrogen can be used in the P2G process to synthesize natural gas; on the other hand, it can be used as fuel for the FC. The storage state of the HST at time t can be expressed as [10]:

$$S_{HST,t} = S_{HST,T} + \sum_{t=1}^{T} \eta_{HST}\left(F_{HST,t}^{P2H} - F_{HST,t}^{FC} - F_{HST,t}^{H2G}\right) \tag{16}$$

where $S_{HST,T}$ is the hydrogen storage capacity of the HST at initial time; η_{HST} is the operation efficiency of the HST; $F_{HST,t}^{P2H}$ is the hydrogen fed into the HST from the P2H system at time t; $F_{HST,t}^{FC}$ is the hydrogen fed into the FC from the HST at time t; and $F_{HST,t}^{H2G}$ is the hydrogen fed into the H2G system from the HST at time t.

2.3. Modeling of Power System

2.3.1. Electric Network

The steady-state power flow model is assumed [35].

2.3.2. Battery Storage System (BSS)

The energy storage state of the BSS is generally represented by its state of charge (*SOC*). According to the different operating characteristics, the *SOC* can be divided into two states: the charging process and the discharging process [10]. For charging:

$$SOC(t) = (1-\delta)SOC(t-1) + \frac{P_c \Delta t \eta_c}{E_c} \tag{17}$$

where $SOC(t)$ is the SOC of the BSS at time t; δ is the self-discharge rate; $SOC(t-1)$ is the SOC of the BSS at time $t-1$; P_c is the charging power of the BSS; Δt is the time interval; η_c is the charging efficiency of the BSS; and E_c is the rated capacity of the BSS.

For discharging:

$$SOC(t) = (1-\delta)SOC(t-1) - \frac{P_d \Delta t}{E_c \eta_d} \tag{18}$$

where P_d is the discharge power of the BSS and η_d is the discharge efficiency of the BSS.

2.4. Modeling of Natural Gas System

The model of the natural gas system is described in [35].

2.5. Energy Conversion Relationship of the Proposed EH

Combining Equation (2) with the mathematical models of the coupling devices, the energy conversion relationship between input and output can be expressed as:

$$\begin{bmatrix} L_e \\ L_{ngs} \\ L_h \\ L_c \end{bmatrix} = \begin{bmatrix} 0 & v_{g1}\eta_{MT} + \frac{2F\eta_{P2H}}{M_H}(E_{Nernst} - \varsigma_{act} - \varsigma_{ohm} - \varsigma_{diff}) \\ v_{e1}\eta_{P2H}\cdot\eta_{P2G} & 0 \\ 0 & v_{g2}\eta_{GB} + \frac{2F\eta_{P2H}}{M_H}(\varsigma_{act} + \varsigma_{ohm} + \varsigma_{diff}) \\ v_{e2}\eta_{EC} & 0 \end{bmatrix} \begin{bmatrix} P_e \\ P_{ngs} \end{bmatrix} \tag{19}$$

where η_{MT}, η_{GB}, and η_{EC} are the conversion efficiency of MT, GB, and EC, respectively; v_{g1} and v_{g2} are dispatch factors of natural gas; v_{e1} and v_{e2} are dispatch factors of electric power; P_e is electric power input into the EH; P_{ngs} is natural gas input into the EH; L_e is electric power output from the EH; L_{ngs} is natural gas output from the EH; L_h is thermal energy output from the EH; and L_c is cold energy output from the EH.

3. PCAPSO Algorithm for the Optimal Scheduling of IES

In this section, the constraints of the subsystems and components are explained. The objective function of the day-ahead scheduling of optimal energy flow for an IES is presented to evaluate the operation cost of IESs. After that, the piecewise self-adaptive particle swarm optimization based on chaotic mapping is proposed to solve the optimization problem.

3.1. System Constraints

3.1.1. Power System Constraints

Considering that coal-fired generators are the main power sources of a power system [36], the power system constraints mainly include active power constraints, reactive power constraints, node voltage constraints, and other constraints.

3.1.2. Natural Gas System Constraints

The constraints in a natural gas system mainly include node pressure constraint, pipeline flow constraint, and compressor inlet and outlet pressure constraints.

3.1.3. Battery Storage System (BSS) Constraints

The main constraints of the BSS are the electric quantity constraint and power constraint [9,37].

The electric quantity constraint is given by:

$$SOC_{min} < SOC(t) < SOC_{max} \tag{20}$$

where SOC_{min} and SOC_{max} are the lower and upper limit of SOC, respectively.

The power constraints are given by:

$$\begin{cases} P_{c,\max}(t) = \min\left\{P_{\max,C}, \frac{E_c \cdot [SOC_{\max} - (1-\delta)SOC(t-1)]}{\Delta t \eta_c}\right\} \\ P_{d,\max}(t) = \min\left\{P_{\max,D}, \frac{E_c \cdot \eta_d \cdot [(1-\delta)SOC(t-1) - SOC_{\min}]}{\Delta t}\right\} \end{cases} \quad (21)$$

where $P_{c,\max}(t)$ is the maximum charging power of the battery at time t; $P_{\max,C}$ is the maximum permissible continuous charging power of the battery; $P_{d,\max}(t)$ is the maximum discharge power of the battery at time t; and $P_{\max,D}$ is the maximum permissible continuous discharge power of the battery. If the rated power of the battery is P_n, then:

$$\begin{cases} P_{\max,C} = N_{c,\max} P_n \\ P_{\max,D} = N_{d,\max} P_n \end{cases} \quad (22)$$

where $N_{c,\max}$ and $N_{d,\max}$ are the maximum charge and discharge multiples, respectively.

3.1.4. Hydrogen Storage Tank (HST) Constraints

The constraints of the HST are given by:

$$S_{HST,\min} \leq S_{HST,t} \leq S_{HST,\max} \quad (23)$$

$$F_{HST,\min} \leq F_{HST,t} \leq F_{HST,\max} \quad (24)$$

where $S_{HST,\min}$ and $S_{HST,\max}$ are the minimum and maximum values of the hydrogen reserves in the HST, respectively; $F_{HST,t}$ is the amount of hydrogen input/output into the HST at time t; and $F_{HST,\min}$ and $F_{HST,\max}$ are the minimum and maximum values of hydrogen input/output in the HST, respectively.

3.2. *PCASO Optimization Model and Algorithm*

3.2.1. Objective Function

The objective function includes the electric cost, natural gas cost of compressor, operation cost of the FC and BSS, and the profit from selling heat, as:

$$\min \sum_{t=1}^{24} \left(C_{\text{electric}}(t) + C_{\text{natural gas}}(t) + C_{FC}(t) + C_{BSS}(t) - C_{\text{sell}}(t) \right) \quad (25)$$

where $C_{\text{electric}}(t)$ is the operation cost of the electrical system at time t; $C_{\text{natural gas}}(t)$ is the operation cost of the natural gas system at time t; $C_{FC}(t)$ is the operation cost of the FC at time t; $C_{BSS}(t)$ is the operation cost of the BSS at time t; and $C_{\text{sell}}(t)$ is the profit from selling heat and cold energy at time t. Each of these variables can be expressed as follows:

$$C_{\text{electric}}(t) = \sum_{i=1}^{N_G} \left(a_i(t) P_{G,i}^2(t) + b_i(t) P_{G,i}(t) + c_i(t) \right) \quad (26)$$

$$C_{\text{natural gas}}(t) = \sum_{i=1}^{N_{com}} \left(c_{com,i}(t) \tau_{com,i}(t) \right) \quad (27)$$

$$C_{FC}(t) = \sum_{i=1}^{N_{FC}} \left(c_{FC,i}(t) P_{FC,i}(t) - s_{FC,i}(t) Q_{FC,i}(t) \right) \quad (28)$$

$$C_{BSS}(t) = \sum_{i=1}^{N_{BSS}} \left(c_{BSS,i}(t) P_{BSS,i}(t) \right) \quad (29)$$

$$C_{\text{sell}}(t) = s_{EC,i}(t) \sum_{i=1}^{N_{EC}} Q_{EC,i}(t) + s_{AC,i}(t) \sum_{i=1}^{N_{GB}} Q_{GB,i}(t) \quad (30)$$

where N_G, N_{com}, N_{FC}, N_{BSS}, N_{EC}, and N_{GB} are the number of generators, compressors, FCs, BSSs, ECs, and GBs, respectively; $a_i(t)$, $b_i(t)$, and $c_i(t)$ are the cost coefficients of the ith generator at time t; $P_{G,i}(t)$ is the active power generated by the ith generator at time t; $c_{com,i}(t)$ is the cost coefficient of the ith compressor at time t; $\tau_{com,i}(t)$ is the natural gas flow fed into the ith compressor at time t; $c_{FC,i}(t)$ and $s_{FC,i}(t)$ are the cost and profit coefficients, respectively, of the ith FC at time t; $P_{FC,i}(t)$ and $Q_{FC,i}(t)$ are the electricity and thermal energy, respectively, produced by the ith FC at time t; $c_{BSS,i}(t)$ is the cost coefficient of the ith BSS at time t; $P_{BSS,i}(t)$ is the charge or discharge power of the ith BSS at time t; $s_{EC,i}(t)$ is the profit coefficient of the ith EC at time t; $s_{GB,i}(t)$ is the profit coefficient of the ith GB at time t; $Q_{EC,i}(t)$ is the cold energy produced by the ith EC at time t; and $Q_{GB,i}(t)$ is the thermal energy produced by the ith GB at time t.

3.2.2. Optimization Algorithm

In this paper, by simplifying and modifying the algorithm in reference [25], a piecewise self-adaptive particle swarm optimization (PCAPSO) based on chaotic mapping is proposed.

Chaotic mapping is a typical collection of nonlinear mappings. Random sequences generated by chaotic mappings are often used in intelligent optimization algorithms. In PCAPSO, the nonlinear inertial weight factor is divided into two sections: search and escape. This piecewise inertial weight factor based on the use of different mappings can accelerate the iterative convergence and evade local optima. The basic working principle is that the search section of the inertial weight factor is used in a regular iterative process, while the escape section is used when particles are trapped in local optima. Based on this mechanism, Gaussian mapping and logistic mapping are used in the search and escape sections, respectively. The definition of Gaussian mapping is as follows:

$$\begin{aligned} &z(1) = \text{rand} \\ &z(k+1) = \begin{cases} 0, & z(k) = 0 \\ \text{mod}(1/z(k), 1), & z(k) \neq 0 \end{cases} \end{aligned} \tag{31}$$

where $z(1)$ is the first number generated by Gaussian mapping; $z(k)$ and $z(k+1)$ is the kth and (k+1)th number generated by Gaussian mapping; rand represents random number, and mod (a, b) represents obtaining the remainder after a is divided by b. Thus, the search section of the nonlinear inertia weight factor can be expressed as:

$$\omega_{\text{search}}(k) = z(k) \cdot \omega_{\min} + \frac{(\omega_{\max} - \omega_{\min})}{k_{\max}} \cdot k \tag{32}$$

where ω_{max} and ω_{min} are the maximum and minimum of the inertia weight factor, respectively; k_{max} is the maximum of iteration; and $z(k)$ is a random number generated by the Gaussian mapping.

The definition of logistic mapping is as follows:

$$\begin{cases} r(0) = rand, & k = 0 \\ r(k+1) = \mu \cdot r(k) \cdot (1 - r(k)), & 0 < k \end{cases} \tag{33}$$

where $r(0) \neq \{0, 0.25, 0.5, 0.75, 1\}$ and $\mu \in [0, 4]$. Correspondingly, the escape section of the nonlinear inertia weight factor can be expressed as:

$$\omega_{\text{esc}}(k) = r(k) c_{\text{mag}} + c_{\text{offset}} \tag{34}$$

where c_{mag} and c_{offset} are the magnitude and offset coefficient, respectively, and $r(k)$ is a random number generated by the logistic mapping.

The algorithm flowchart of PCAPSO is shown in Figure 2.

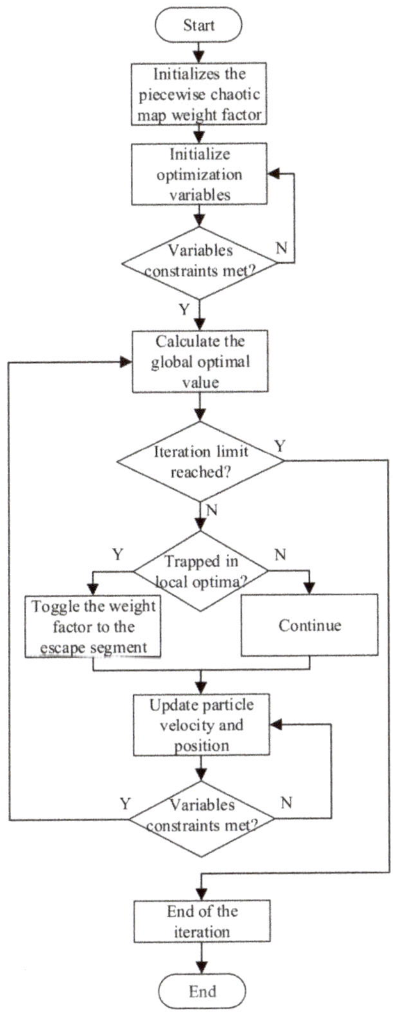

Figure 2. Flowchart of PCAPSO.

4. Case Study

In this section, the testing system and the coupling mode between subsystems under different operation conditions are introduced. Based on the simulation results, the improvement of the proposed algorithm is demonstrated, and the operation characteristics and economics of the optimized IESs with different numbers of EHs are analyzed.

4.1. System Description

The simulation system was composed of a modified IEEE 30-bus system, the 48-bus natural gas system referred from [35], and EHs. The topologies of the two subsystems are depicted in Figures 3 and 4. Three EHs were connected to electric buses 3, 4, 5, 10, 12, 16, 20, 26, and 29 in the electrical system and the natural gas nodes 2, 8, 10, 13, 14, 19, 21, 31, and 40 in the natural gas system. The coupling relationship inside the IES is shown in Figure 5. The detailed internal structure of the EH is shown in Figure 1 in Section 2.

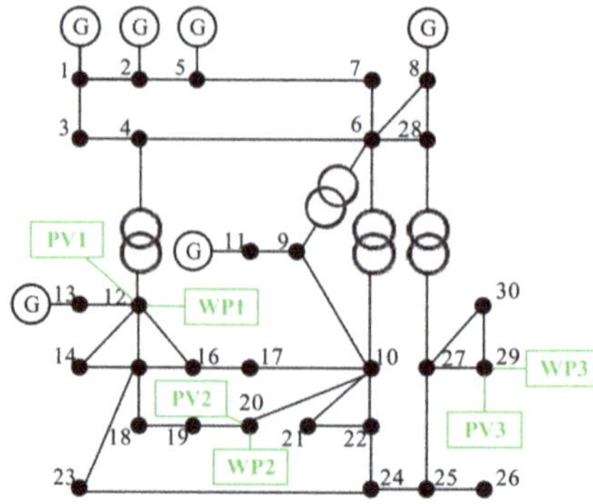

Figure 3. Topology of the electrical system.

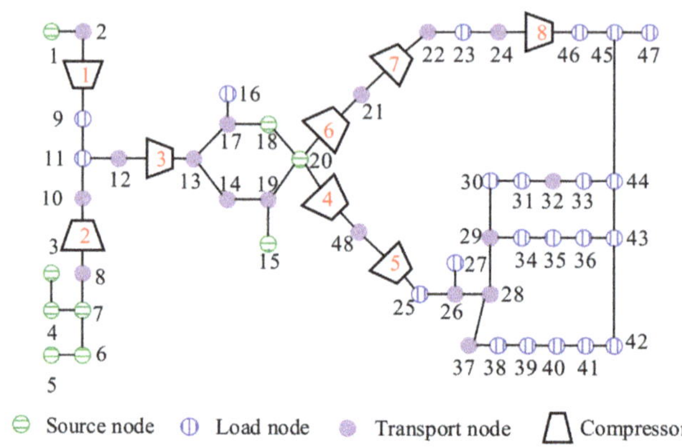

Figure 4. Topology of the natural gas system.

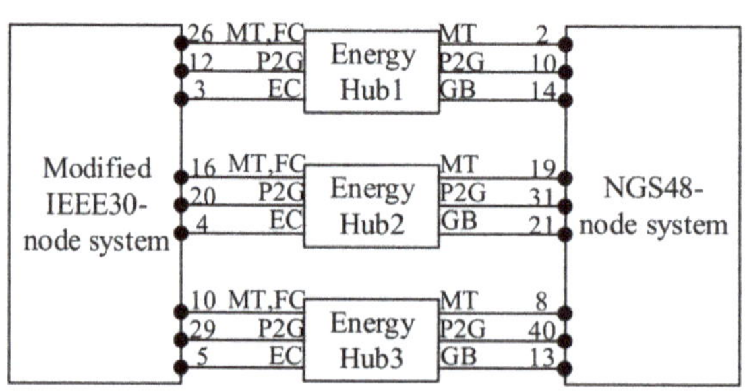

Figure 5. Coupling relationship of IES.

In the modified IEEE 30-bus system, three groups of photovoltaic power (PV) and wind power (WP) were included, and load fluctuation [38–40] was considered. It was assumed that the output of WP and PV in every group was the same. The curves are shown in Figure 6. The load curves of nodes 10, 12, 16, 20, 26, and 29 are shown in Figure 7. In this simulation, the time-of-use (TOU) price was taken into account in both the electrical and natural gas systems. The price was set to the valley price from 00:00 to 7:00 h. At 8:00 h and from 12:00 to 18:00 h, it was set to the normal price. From 9:00 to 11:00 h and from 19:00 to 23:00 h, the price was set to the peak price. The TOU for the IES is displayed in Table 1.

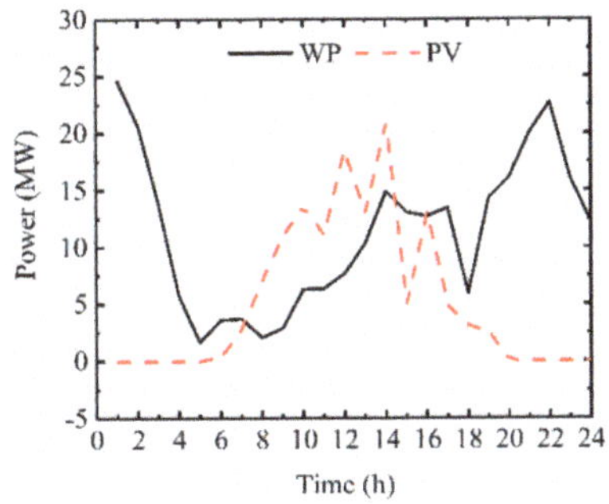

Figure 6. Daily output curves of WP and PV.

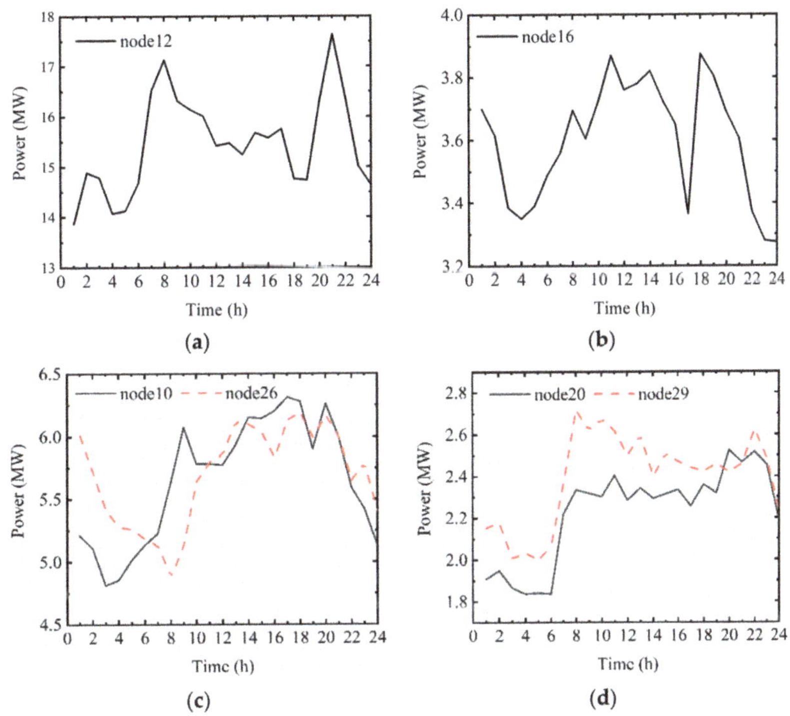

Figure 7. Daily fluctuation curves of loads: (**a**) node 12, (**b**) node 16, (**c**) node 10 and node 26, and (**d**) node 20 and node 29.

Table 1. Time-of-use prices for the IES.

Coefficient \ Price Type		Peak Price/¥	Normal Price/¥	Valley Price/¥
Electrical system	$a(t)$	0.045	0.0375	0.03
	$b(t)$	2.1	1.75	1.4
Natural gas system	$C_{com}(t)$	0.24	0.2	0.16

4.2. Effectiveness of OEF Using PCAPSO Based on Chaotic Mapping

A new optimal algorithm called PCAPSO was proposed and used to solve the OEF problem of IES. To validate the effectiveness of the proposed optimization methodology, the original PSO and PCAPSO were used to optimize the above simulation system three times, the results of which are shown in Table 2. The iteration curves are depicted in Figure 8.

Table 2. Comparison of optimal results with different optimization algorithms.

Algorithm \ Calculation Label	No.1/$\times 10^3$¥	No.2/$\times 10^3$¥	No.3/$\times 10^3$¥	Average/$\times 10^3$¥	Standard Deviation/$\times 10^3$¥
PCAPSO	89.59	89.52	88.96	89.36	0.24
PSO	90.67	92.14	91.56	91.46	0.52

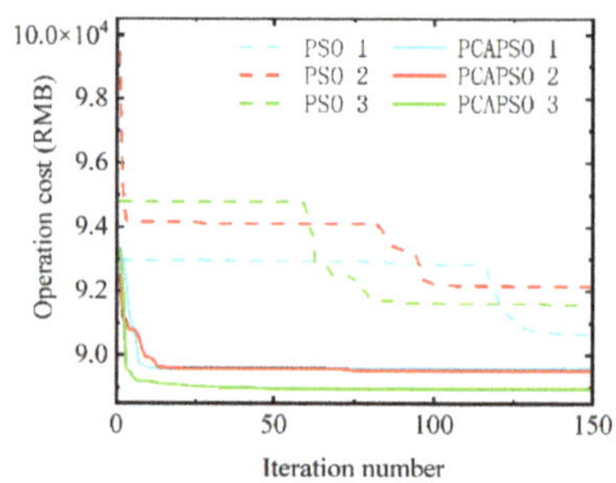

Figure 8. Iteration curves with different optimization algorithms.

As shown in Table 2, the optimal cost of PCAPSO was lower than that of PSO. Therefore, PCAPSO based on piecewise-chaotic mapping was less likely to fall into local optima than PSO. The average value of the three optimization results obtained by using PCAPSO was also lower than that of PSO. Since the standard deviation of PCAPSO was much lower than that of PSO, it can be concluded that the stability of PCAPSO is better than that of PSO. Comparing the three iteration curves of PSO and PCAPSO in Figure 8, it can be observed that each optimization with PSO fell into a local optimum at least once in the first half of the iteration, and the optimization results did not converge to the optimal value when reaching the maximum number of iterations. However, the optimization with PCAPSO barely fell into local optima during entire iteration. This is because in the search section, the randomness of the Gaussian mapping ensures the global searching ability of the inertia weight factor, and its weakening fluctuation characteristic causes the volatility of the inertia weight factor to weaken regularly, which ensures the transition of the inertia weight factor from the global search to the local search. However, once PCAPSO detects that the optimization has fallen into local optima, the escape section is activated; in the escape section, the stationary and stronger volatility of the logistic mapping allows the inertia weight factor to keep its ability of global search, which can help the algorithm to jump out of the local optima. On the other hand, these results also demonstrate that the convergence rate of PCAPSO was faster than that of PSO. The results in Table 2 and Figure 8 show that PCAPSO outperformed PSO in computational efficiency, stability, and convergence accuracy.

4.3. Operation Characteristics and Economic Analysis of IES in Different Scenarios

To illustrate the effectiveness and economics of the proposed EH model and IES framework, three operation scenarios of an IES according to the different operation states of the EHs were designed and analyzed:

Scenario 1: Only EH1 is operational, and the fuel cell in EH1 is operational.

Scenario 2: All three Ehs are operational, but the fuel cells in each EH are deactivated.

Scenario 3: All three Ehs are operational, and the fuel cells in each EH are also operational.

The remaining IES operation parameters were the same in the three scenarios.

Figure 9 shows the daily operation cost of the aforementioned scenarios. The daily operation cost of the IES in Scenario 1 was more than that of Scenario 3, which indicates that with more operational Ehs, the operation cost of the Ehs increased; however, this can be compensated for by selling the cooling and heat power produced by the Ecs, GBs, MTs, and FCs. The profits obtained from selling heat or cooling power are presented in Figure 10, which also shows the components of the sold energy. The daily operation cost of the IES in Scenario 2 was slightly larger than that of Scenario 3. This shows that due to the co-generation function of the FC, the profit of sold heat is further increased when the FC in each EH is operational.

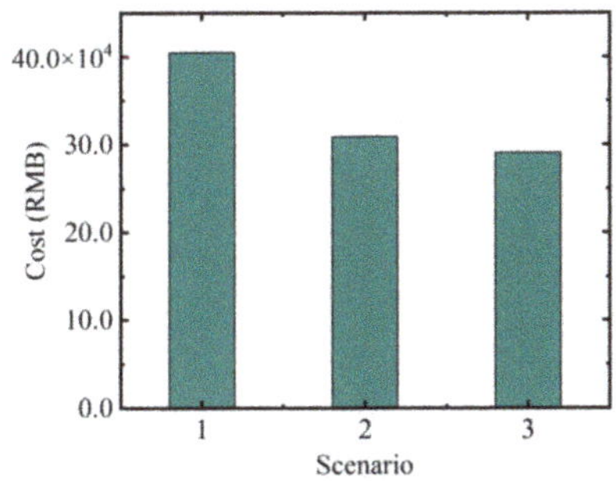

Figure 9. Operation cost of the IES in different scenarios.

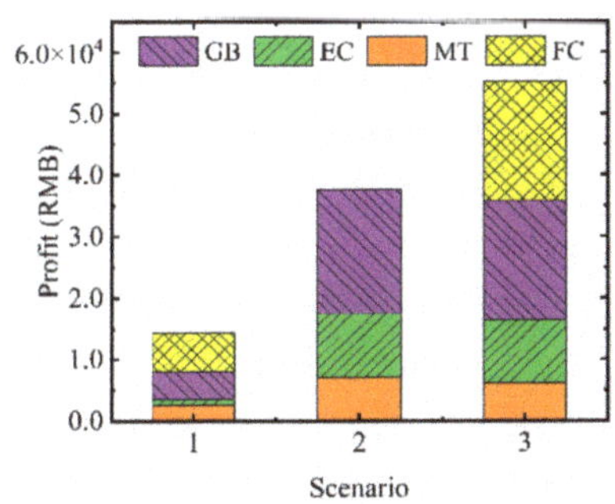

Figure 10. Profit from selling heat/cooling power in the different scenarios.

Table 3 describes the load fluctuations of node 10, 16, and 26 in the three different scenarios. Comparing the data from Scenarios 2 and 3, it can be seen that the peak loads were reduced up to 10%, the load valleys were filled up by P2H, and the standard deviations of the loads were reduced up to 77% when the FC was operational. These changes reflect the effect of FC–HST cooperation in peak load shifting.

Table 3. Load fluctuation of the IES in different scenarios.

	Node 10			Node 16			Node 26		
	Peak Load/MW	Valley Load/MW	Standard Deviation/MW	Peak Load/MW	Valley Load/MW	Standard Deviation/MW	Peak Load/MW	Valley Load/MW	Standard Deviation/MW
Scenario 1	6.31	4.81	0.48	3.88	3.28	0.19	5.68	5.39	0.09
Scenario 2	6.31	4.81	0.48	3.88	3.28	0.19	6.19	4.88	0.38
Scenario 3	5.65	5.24	0.11	3.59	3.23	0.11	5.68	5.39	0.09

Figure 11 displays the electrical (PFC) and heat power (QFC) produced by the FC in each EH and the change of the hydrogen storage capacity of the corresponding HST in Scenario 3. As can be seen from Figure 11a, from 3:00 to 10:00 h, due to the light load and cheap electricity price, the IES purchases electricity from the upstream grid to supplement the hydrogen in the HST1 by electrolysis, which explains why PFC1 is negative during this period. At 11:00, both the load and the price of electricity increase. FC1 then uses the hydrogen supplied by the HST1 to generate heat and power, and the electric energy produced is used to smooth the load fluctuations and reduce the pressure on the power grid. The heat produced is then sold to the heat market to offset the operation cost of the system. At the end of the day, the hydrogen reserves in HST1 are slightly larger than those at the beginning of the day, which ensures the continuous supply of hydrogen. The situation in Figure 11b,c is similar to that in 11a.

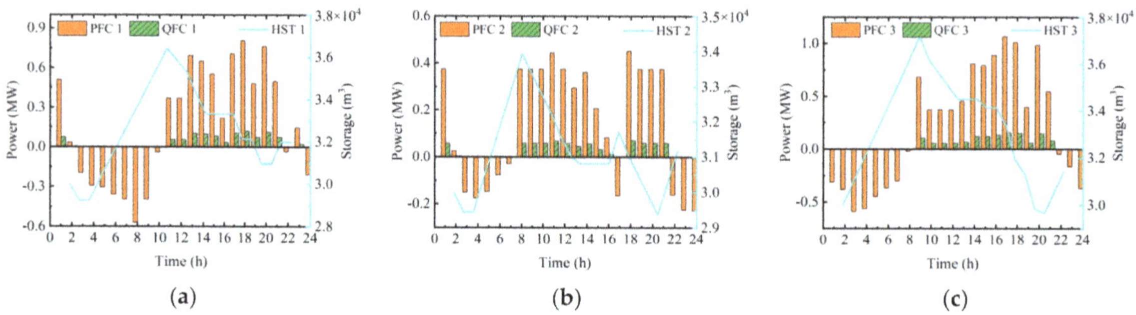

Figure 11. Output power of FCs and states of HSTs in Scenario 3: (**a**) FC1 and state of HST1, (**b**) FC2 and state of HST2, and (**c**) FC3 and state of HST3.

Figures 12–14 present the optimal schedules of the coupling nodes. Comparing the optimal schedule of identical coupling nodes in Scenarios 2 and 3, again the load fluctuations are reduced with operational FCs. Hence, the results show that the FC is capable of peak load shifting if it cooperates with the HST, and its operation cost can be offset by selling heat. Adding more Ehs into an IES not only strengthens the coupling between subsystems, but also improves the economics of IES operation. In addition, PCAPSO has good applicability in solving IES optimization scheduling problems in different scenarios.

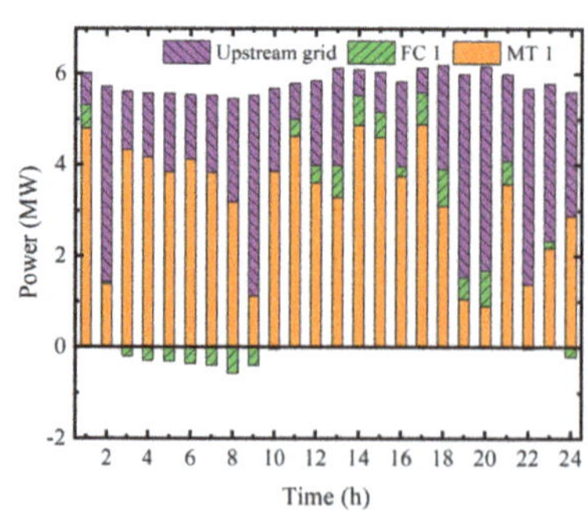

Figure 12. Optimal schedule of the coupling node (node 26) in scenario 1.

The optimal energy structures of the nodes connected with sustainable energy in scenario 3 are displayed in Figure 15. As shown in Figure 15a, when the output of WP1 and PV1 exceeds the load demand, surplus electric energy is used to charge the BSS1 (e.g., 10:00–14:00 h). If the charging power or *SOC* reaches the upper limit during the charging process, the excess electric energy is transferred to the electrolytic cell of the FC for hydrogen production (e.g., 15:00–17:00 h). When the output of WP1 and PV1 is not enough to meet the load demand, the BSS compensates the power deficiency by discharging (e.g., 5:00–7:00 h). If the *SOC* of BSS is less than the minimum limit at this time, the additional load must be borne by the upstream grid (e.g., 8:00–10:00 h). At the end of the day, the *SOC*1 is slightly higher than that at the beginning of the day, which ensures the continuous supply of electric power from the BSS1. Given the above discussion, with the integration of the BSS and FC into the IES, the consumption rate of renewable energy increases and the amount of electricity drawn by the load from the upstream grid decreases, which reduces the operating cost of the IES.

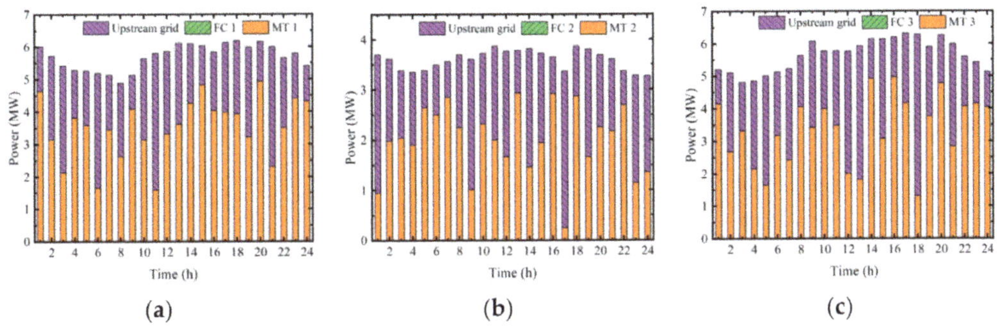

Figure 13. Optimal schedules of the coupling nodes in scenario 2: (**a**) node 26, (**b**) node 16, and (**c**) node 10.

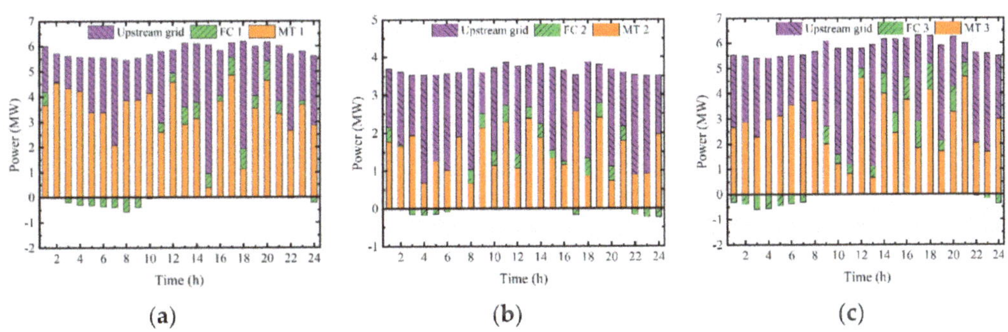

Figure 14. Optimal schedules of the coupling nodes in scenario 3: (**a**) node 26, (**b**) node 16, and (**c**) node 10.

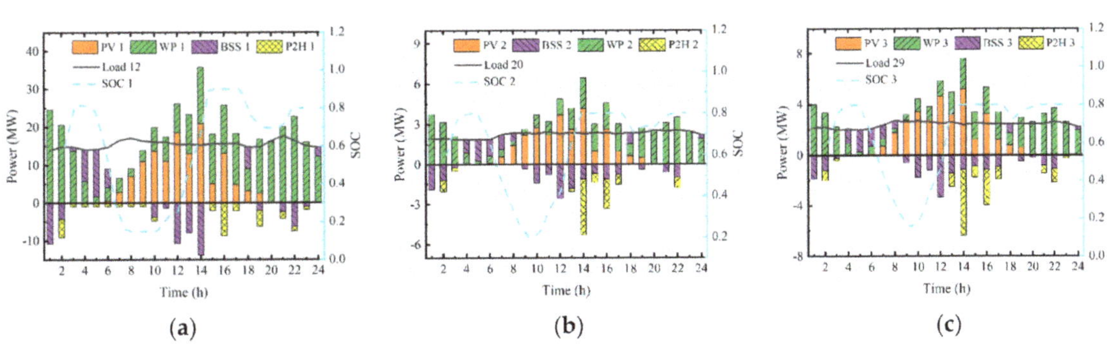

Figure 15. Optimal energy structures of nodes connected with sustainable energy in scenario 3: (**a**) node 12, (**b**) node 20, and (**c**) node 29.

5. Conclusions

Considering the detailed models of the FC and P2G system, this study developed a new framework of EH for the optimal day-ahead operation of IES. The IES model consisted of electrical and natural gas systems, presented in Section 4, and an EH, presented in Section 2. The day-ahead scheduling of the IES was carried out considering the nodal pressures, voltage constraints, and energy flow constraints of each sub-network. The objective function of the IES day-ahead scheduling was to minimize the daily operating cost of the IES. To solve the optimization problem, PCAPSO based on multistage chaotic mapping was proposed in this paper. The major contributions of this work are:

(1) Developing a new efficient PCAPSO algorithm based on multistage chaotic mapping.
(2) Integrating detailed models of FC and P2G into the EH to form a novel framework of the EH, in which the joint operation of the FC, P2G system, and HST is considered and hydrogen is introduced.
(3) Establishing a new scheduling strategy to reduce the operation cost of IES. In this strategy, when the electric load is light, P2H and HST cooperate to absorb excess wind/solar energy and transform it into hydrogen, which can then be used as the starting material for the H2G process to synthesize natural gas. When the electric load is heavy, hydrogen transformed from the P2H process or stored in the HST can be used as the fuel in the FC to generate electricity.

Numerical tests were performed on an IES including the modified IEEE 30-bus and natural gas 48-bus systems. The key findings are as follows:

(1) The PCAPSO based on piecewise-chaotic mapping is less likely to fall into local optima than PSO. In addition, the stability and convergence rate of PCAPSO are better than those of PSO.
(2) The FC cooperating with the HST and P2H system is capable of peak load shifting when operated as a co-generation device, and its operation cost can be compensated by selling heat.
(3) The integration of the BSS and FC into the IES can increase the consumption rate of renewable energy and decrease the amount of electricity drawn from the upstream grid.
(4) The proposed EH, which contains the detailed model of the FC and P2G, is more in line with actual operations than the EH described in [3,12,20]. With the insertion of multiple EHs, the system coupling is enhanced, and the stability and economics of IES operation are improved.

Future work should include detailed modeling of the natural gas system considering the hydrodynamic properties of natural gas. In addition, the thermal network could be coupled with electrical and natural gas systems to improve the flexibility and diversity of IES operation. On the other hand, the introduction of a new optimization algorithm with a high computational efficiency and convergence rate may also be an important area of research in day-ahead optimal scheduling of IESs.

Author Contributions: All authors have cooperated in the preparation of this work. Conceptualization, J.C. and K.N.; methodology, J.C. and K.N.; software, K.N.; validation, K.N. and C.L.; formal analysis, K.N. and X.X.; writing—original draft preparation, K.N.; writing—review and editing, K.N., F.S. and Q.Z.; visualization, Q.Z. and C.L.; project administration, J.C. All authors have read and agreed to the published version of the manuscript.

Funding: This research did not receive any specific grant from funding agencies in the public, commercial, or not-for-profit sectors.

Data Availability Statement: Not applicable.

Conflicts of Interest: The authors declare no conflict of interest.

References

1. Hoseinzadeh, S.; Ghasemi, M.H.; Heyns, S. Application of Hybrid Systems in Solution of Low Power Generation at Hot Seasons for Micro Hydro Systems. *Renew. Energy* **2020**, *160*, 323–332. [CrossRef]
2. Tomin, N.; Shakirov, V.; Kozlov, A.; Sidorov, D.; Kurbatsky, V.; Rehtanz, C.; Lora, E.E.S. Design and Optimal Energy Management of Community Microgrids with Flexible Renewable Energy Sources. *Renew. Energy* **2022**, *183*, 903–921. [CrossRef]
3. Sun, Y.; Zhang, B.; Ge, L.; Sidorov, D.; Wang, J.; Xu, Z. Day-Ahead Optimization Schedule for Gas-Electric Integrated Energy System Based on Second-Order Cone Programming. *CSEE J. Power Energy Syst.* **2020**, *6*, 142–151. [CrossRef]
4. Ghasemi, A.; Banejad, M.; Rahimiyan, M. Integrated Energy Scheduling under Uncertainty in a Micro Energy Grid. *Transm. Distrib. IET Gener.* **2018**, *12*, 2887–2896. [CrossRef]
5. Zhang, Y.; Wang, X.; He, J.; Xu, Y.; Pei, W. Optimization of Distributed Integrated Multi-Energy System Considering Industrial Process Based on Energy Hub. *J. Mod. Power Syst. Clean Energy* **2020**, *8*, 863–873. [CrossRef]
6. Zhou, B.; Xu, D.; Li, C.; Chung, C.Y.; Cao, Y.; Chan, K.W.; Wu, Q. Optimal Scheduling of Biogas–Solar–Wind Renewable Portfolio for Multicarrier Energy Supplies. *IEEE Trans. Power Syst.* **2018**, *33*, 6229–6239. [CrossRef]
7. Chen, C.; Liu, S.; Lin, Z.; Yang, L.; Wu, X.; Qiu, W.; Gao, Q.; Zhu, T.; Xing, F.; Zhang, J. Optimal Coordinative Operation Strategy of the Electric–Thermal–Gas Integrated Energy System Considering CSP Plant. *IET Energy Syst. Integr.* **2020**, *2*, 187–195. [CrossRef]
8. Xu, D.; Zhou, B.; Wu, Q.; Chung, C.Y.; Li, C.; Huang, S.; Chen, S. Integrated Modelling and Enhanced Utilization of Power-to-Ammonia for High Renewable Penetrated Multi-Energy Systems. *IEEE Trans. Power Syst.* **2020**, *35*, 4769–4780. [CrossRef]
9. Li, P.; Sheng, W.; Duan, Q.; Li, Z.; Zhu, C.; Zhang, X. A Lyapunov Optimization-Based Energy Management Strategy for Energy Hub With Energy Router. *IEEE Trans. Smart Grid* **2020**, *11*, 4860–4870. [CrossRef]
10. Ju, L.; Zhao, R.; Tan, Q.; Lu, Y.; Tan, Q.; Wang, W. A Multi-Objective Robust Scheduling Model and Solution Algorithm for a Novel Virtual Power Plant Connected with Power-to-Gas and Gas Storage Tank Considering Uncertainty and Demand Response. *Appl. Energy* **2019**, *250*, 1336–1355. [CrossRef]
11. Geng, S.; Vrakopoulou, M.; Hiskens, I.A. Optimal Capacity Design and Operation of Energy Hub Systems. *Proc. IEEE* **2020**, *108*, 1475–1495. [CrossRef]
12. Shao, C.; Wang, X.; Shahidehpour, M.; Wang, X.; Wang, B. An MILP-Based Optimal Power Flow in Multicarrier Energy Systems. *IEEE Trans. Sustain. Energy* **2017**, *8*, 239–248. [CrossRef]
13. Li, C.; Yang, H.; Shahidehpour, M.; Xu, Z.; Zhou, B.; Cao, Y.; Zeng, L. Optimal Planning of Islanded Integrated Energy System With Solar-Biogas Energy Supply. *IEEE Trans. Sustain. Energy* **2020**, *11*, 2437–2448. [CrossRef]
14. Yang, W.; Liu, W.; Chung, C.Y.; Wen, F. Coordinated Planning Strategy for Integrated Energy Systems in a District Energy Sector. *IEEE Trans. Sustain. Energy* **2020**, *11*, 1807–1819. [CrossRef]
15. Wang, C.; Wei, W.; Wang, J.; Bai, L.; Liang, Y.; Bi, T. Convex Optimization Based Distributed Optimal Gas-Power Flow Calculation. *IEEE Trans. Sustain. Energy* **2018**, *9*, 1145–1156. [CrossRef]
16. Wang, Y.; Wang, Y.; Huang, Y.; Yu, H.; Du, R.; Zhang, F.; Zhang, F.; Zhu, J. Optimal Scheduling of the Regional Integrated Energy System Considering Economy and Environment. *IEEE Trans. Sustain. Energy* **2019**, *10*, 1939–1949. [CrossRef]
17. Eladl, A.A.; El-Afifi, M.I.; Saeed, M.A.; El-Saadawi, M.M. Optimal Operation of Energy Hubs Integrated with Renewable Energy Sources and Storage Devices Considering CO2 Emissions. *Int. J. Electr. Power Energy Syst.* **2020**, *117*, 105719. [CrossRef]
18. Chen, Y.; Wang, Y.; Ma, J. Multi-Objective Optimal Energy Management for the Integrated Electrical and Natural Gas Network with Combined Cooling, Heat and Power Plants. *Energies* **2018**, *11*, 734. [CrossRef]
19. Keihan Asl, D.; Seifi, A.R.; Rastegar, M.; Mohammadi, M. Optimal Energy Flow in Integrated Energy Distribution Systems Considering Unbalanced Operation of Power Distribution Systems. *Int. J. Electr. Power Energy Syst.* **2020**, *121*, 106132. [CrossRef]
20. Shabanpour-Haghighi, A.; Seifi, A.R. Energy Flow Optimization in Multicarrier Systems. *IEEE Trans. Ind. Inf.* **2015**, *11*, 1067–1077. [CrossRef]
21. Massrur, H.R.; Niknam, T.; Fotuhi-Firuzabad, M. Day-Ahead Energy Management Framework for a Networked Gas–Heat–Electricity Microgrid. *Transm. Distrib. IET Gener.* **2019**, *13*, 4617–4629. [CrossRef]
22. Tan, Y.; Wang, X.; Zheng, Y. A New Modeling and Solution Method for Optimal Energy Flow in Electricity-gas Integrated Energy System. *Int. J. Energy Res.* **2019**, *43*, 4322–4343. [CrossRef]
23. Eberhart, R.; Kennedy, J. A New Optimizer Using Particle Swarm Theory. In Proceedings of the Sixth International Symposium on Micro Machine and Human Science, Nagoya, Japan, 4–6 October 1995; pp. 39–43.
24. Wang, D.; Tan, D.; Liu, L. Particle Swarm Optimization Algorithm: An Overview. *Soft Comput.* **2018**, *22*, 387–408. [CrossRef]
25. Liu, H.; Zhang, X.-W.; Tu, L.-P. A Modified Particle Swarm Optimization Using Adaptive Strategy. *Expert Syst. Appl.* **2020**, *152*, 113353. [CrossRef]
26. Patwal, R.S.; Narang, N. Multi-Objective Generation Scheduling of Integrated Energy System Using Fuzzy Based Surrogate Worth Trade-off Approach. *Renew. Energy* **2020**, *156*, 864–882. [CrossRef]
27. Mellouk, L.; Ghazi, M.; Aaroud, A.; Boulmalf, M.; Benhaddou, D.; Zine-Dine, K. Design and Energy Management Optimization for Hybrid Renewable Energy System- Case Study: Laayoune Region. *Renew. Energy* **2019**, *139*, 621–634. [CrossRef]
28. Bao, Z.; Chen, D.; Wu, L.; Guo, X. Optimal Inter- and Intra-Hour Scheduling of Islanded Integrated-Energy System Considering Linepack of Gas Pipelines. *Energy* **2019**, *171*, 326–340. [CrossRef]
29. Zou, Y.; Yang, G.; Zheng, H.; Yi, J.; Hu, R. Dispatching for Integrated Energy System Based on Improved Niche PSO Algorithm. *Proc. CSU-EPSA* **2020**, *32*, 47–52,60.

30. Liu, X.; Wang, Q.; Wu, C. A Stackelberg Game Approach for Heterogeneous Energy Market in Integrated Energy System. *Int. J. Energy Res.* **2021**, *45*, 1038–1054. [CrossRef]
31. Mirzaei, M.A.; Oskouei, M.Z.; Mohammadi-Ivatloo, B.; Loni, A.; Zare, K.; Marzband, M.; Shafiee, M. Integrated Energy Hub System Based on Power-to-Gas and Compressed Air Energy Storage Technologies in the Presence of Multiple Shiftable Loads. *Transm. Distrib. IET Gener.* **2020**, *14*, 2510–2519. [CrossRef]
32. Chang, H.; Xu, X.; Shen, J.; Shu, S.; Tu, Z. Performance Analysis of a Micro-Combined Heating and Power System with PEM Fuel Cell as a Prime Mover for a Typical Household in North China. *Int. J. Hydrogen Energy* **2019**, *44*, 24965–24976. [CrossRef]
33. Sui, Q.; Ma, X.; Wei, F.; Lin, X.; Guo, X.; Sun, J.; Li, Z. Day-ahead Dispatching Optimization Strategy for Energy Network Considering Fuel Cell Thermal-electric Comprehensive Utilization. *Proc. Chin. Soc. Electr. Eng.* **2019**, *39*, 1603–1613.
34. Al-Baghdadi, M. Modelling of Proton Exchange Membrane Fuel Cell Performance Based on Semi-Empirical Equations. *Renew. Energy* **2005**, *30*, 1587–1599. [CrossRef]
35. Chen, J.; Yu, Q.; Li, Q.; Lin, Z.; Li, C. Probabilistic Energy Flow Analysis of MCE System Considering Various Coupling Units and the Uncertainty of Distribution Generators. *IEEE Access* **2019**, *7*, 100394–100405. [CrossRef]
36. Hoseinzadeh, S.; Stephan Heyns, P. Advanced Energy, Exergy, and Environmental (3E) Analyses and Optimization of a Coal-Fired 400 MW Thermal Power Plant. *J. Energy Resour. Technol.* **2021**, *143*, 082106. [CrossRef]
37. Dou, C.; Mi, X.; Ma, K.; Xu, S. Coordinated Operation of Multi-Energy Microgrid with Flexible Load. *J. Renew. Sustain. Energy* **2019**, *11*, 054101. [CrossRef]
38. Hussain, S.; Ahmed, M.A.; Kim, Y.-C. Efficient Power Management Algorithm Based on Fuzzy Logic Inference for Electric Vehicles Parking Lot. *IEEE Access* **2019**, *7*, 65467–65485. [CrossRef]
39. Hussain, S.; Ahmed, M.A.; Lee, K.-B.; Kim, Y.-C. Fuzzy Logic Weight Based Charging Scheme for Optimal Distribution of Charging Power among Electric Vehicles in a Parking Lot. *Energies* **2020**, *13*, 3119. [CrossRef]
40. Hussain, S.; Lee, K.-B.; Ahmed, M.A.; Hayes, B.; Kim, Y.-C. Two-Stage Fuzzy Logic Inference Algorithm for Maximizing the Quality of Performance under the Operational Constraints of Power Grid in Electric Vehicle Parking Lots. *Energies* **2020**, *13*, 4634. [CrossRef]

energies

MDPI

Review

The Evolution of Knowledge and Trends within the Building Energy Efficiency Field of Knowledge

Talita Mariane Cristino [1], Antonio Faria Neto [1,*], Frédéric Wurtz [2] and Benoit Delinchant [2]

[1] School of Engineering, São Paulo State University (UNESP), Guaratinguetá 12516-410, Brazil; talita.cristino@unesp.br

[2] University Grenoble Alpes, CNRS, Grenoble INP, G2Elab, F-38000 Grenoble, France; frederic.wurtz@g2elab.grenoble-inp.fr (F.W.); benoit.delinchant@g2elab.grenoble-inp.fr (B.D.)

* Correspondence: antfarianeto@gmail.com

Abstract: The building sector is responsible for 50% of worldwide energy consumption and 40% of CO_2 emissions. Consequently, a lot of research on Building Energy Efficiency has been carried out over recent years, covering the most varied topics. While many of these themes are no longer of interest to the scientific community, others flourish. Thus, reading trends within a field of knowledge is wise since it allows resources to be directed towards the most promising topics. However, there is a paucity of research on trend analysis in this field. Therefore, this article aims to analyse the evolution of the Building Energy Efficiency field of knowledge, identifying the recurrent themes and pointing out their trends, supported by statistical methods. Such an analysis relied on more than 9000 authors' keywords collected from 2000 articles from the Scopus database and classified into 30 topics/themes. A frequency distribution of these themes enabled us to distinguish those most published as well as those whose academic interest has cooled down. This field of knowledge has evolved over three distinct phases, throughout which, eight themes presented an upward trend. These findings can assist researchers in optimising time and resources, investigating the topics with growing interest, and possibilities for new contributions.

Keywords: energy efficiency; energy saving; building energy efficiency; trend analysis; Mann-Kendall test; clustering

Citation: Cristino, T.M.; Neto, A.F.; Wurtz, F.; Delinchant, B. The Evolution of Knowledge and Trends within the Building Energy Efficiency Field of Knowledge. *Energies* **2022**, *15*, 691. https://doi.org/10.3390/en15030691

Academic Editor: Zbigniew Leonowicz

Received: 5 November 2021
Accepted: 1 January 2022
Published: 18 January 2022

1. Introduction

Energy consumption has increased over the last decades [1]. Such an increase brings several concerns, such as the necessity to develop alternative energy sources and reduce environmental impacts due to greenhouse gas emissions [2,3]. The building sector has overtaken the industrial sector and has played an important role in such a scenario, being responsible for more than 50% of the total energy consumption and 40% of the total CO_2 emissions [4]. In Europe, according to the European Commission [5], only residential buildings are responsible for approximately 40% of energy consumption and 36% of CO_2 emissions. Therefore, it was necessary to find a way to decrease such energy consumption without affecting economic development, as well as the comfort of the building's occupants [6,7]. The way that researchers found to achieve such a goal was to increase the efficiency of processes and products [8,9]. This put the building sector on the target of important public policies.

Hence, a lot of research on Building Energy Efficiency (BEE) has been carried out. From 2000 to 2018, more than 14,000 papers were published dealing with several themes concerned with BEE. From this total, a little more than 100 papers reviewed specific themes: building energy modelling [10–34]; building envelope [21–25]; building energy performance [35–42]; sustainability [43–53]; building information modelling [54–69]; thermal comfort [70–77]; thermal storage [78,79]; zero energy buildings [49,80–87]; building integrated photovoltaics [88–90]; green buildings [91–93]; occupancy behaviour [94–98]; smart buildings [99–101]; lighting [102,103]; and regulations [104].

Unfortunately, only a few of these reviews dealt with trend analysis and yet, in a restricted way, only considered technical aspects within the respective themes, missing the big picture: building energy modelling [15]; building energy performance [38]; building information modelling [65,66,68]; building integrated photovoltaics [89]; and smart buildings [101].

Therefore, this article aims to fill this gap by answering important questions. Has the BEE field of knowledge reached maturity? What are the recurrent themes within this field of knowledge? What is the trend of each of them? How has the relationship between such themes changed over time?

Identifying the currently relevant topics, regardless of the field of knowledge, is wise because resources are becoming increasingly scarce, and this means they can be directed towards the most promising themes. This also allows researchers to be able to optimise resources, investigate the topics with growing interest, and increase the possibility for new contributions.

This article is organised as follows: Section 2 provides an extensive description of the methodological approach used to answer the research questions. Section 3 shows the results of the research. The article concludes with Section 4, discussing the main findings.

2. Materials and Methods

The deductive method was the methodological approach chosen to carry out this research. This was well-explained by Davidavičienė [105]. The deductive method starts with a set of theory-driven research questions, which guide the data collection and their analysis. Such questions were precisely posed in the previous section.

The analyses carried out in this research were based on the authors' keywords of a significant sample of relevant articles addressing Building Energy Efficiency (BEE). Therefore, this section will describe the procedures used to collect and manipulate such keywords. Figure 1 illustrates the methodological flow of this research.

2.1. Research Problem Formulation

This is the stage in which the general orientation of the research was established. In this step, the research theme was formally defined and well delimited. Furthermore, the gaps in the field of knowledge under investigation were stated and the questions that had to be answered in order to fill such a gap were posed [106].

Energy Efficiency is a huge field of knowledge with several intertwined branches. Building Energy Efficiency (BEE) is one of these branches, which constitutes several themes.

Since the building sector is a great energy consumer, overlapping the industrial sector, much has been written about Building Energy Efficiency. However, no article has studied, in a methodological manner, the evolution of this field of knowledge, the themes within it, or how they relate to each other. Thus, in order to fill this gap, it was necessary to answer the research questions as posed in Section 1.

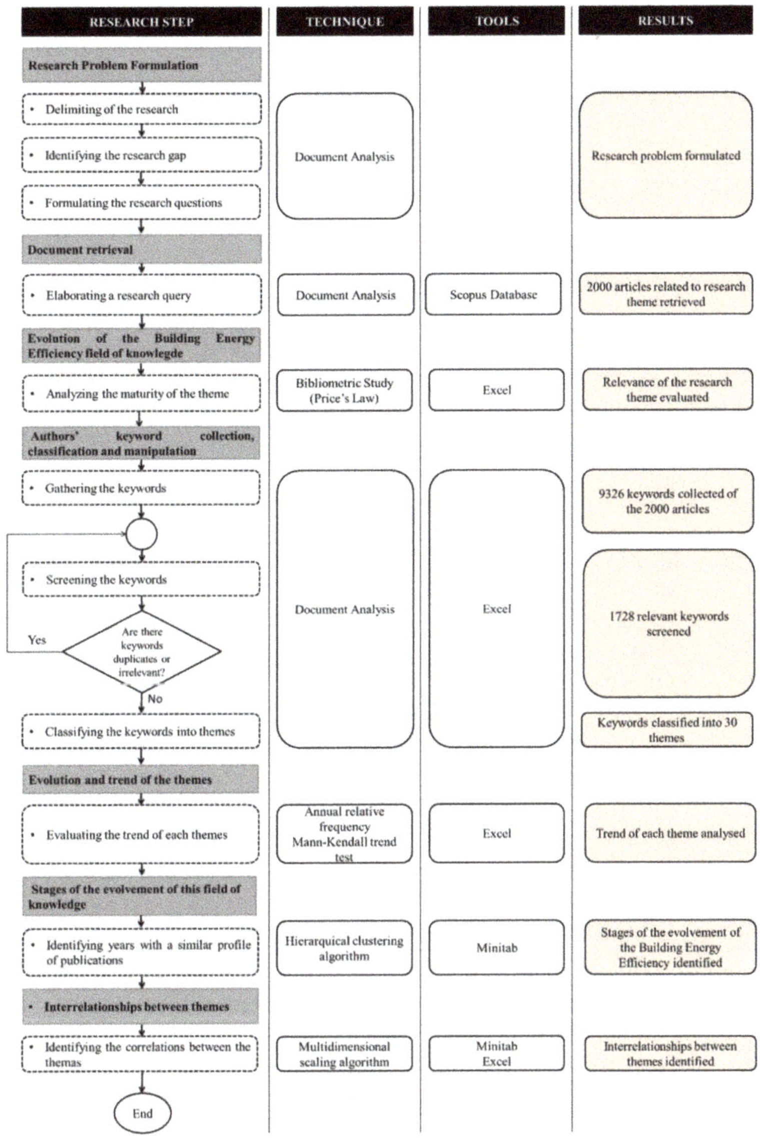

Figure 1. Procedural flow diagram (source: authors).

2.2. *Document Retrieval*

The main goal of this step relied on retrieving a significant sample of publications addressing relevant issues in the Building Energy Efficiency field of knowledge. In order to accomplish such a goal, it was necessary to define the database from which the documents would be gathered and, most importantly, design the proper query to get the job done.

The publications were retrieved from the SCOPUS database since it has a wide coverage of high-impact journals, and "it is the largest database of abstracts and citation literature peer review" [107]. Only journal articles from 2000 to 2018 were considered because, before then, only a few articles presented the authors' keywords, which provided the background of the method employed.

The construction of a query is a process of association of several terms concerned with a core theme. Such terms could be the keywords of the first sample of articles related to the field of knowledge under analysis. A very simple query can be used to capture the articles in which keywords will be used, to formulate the ultimate query.

In this case, the term "Building Energy Efficiency" was used to formulate the very first query, retrieving 893 publications and resulting in more than 3000 keywords, illustrated as a word cloud in Figure 2.

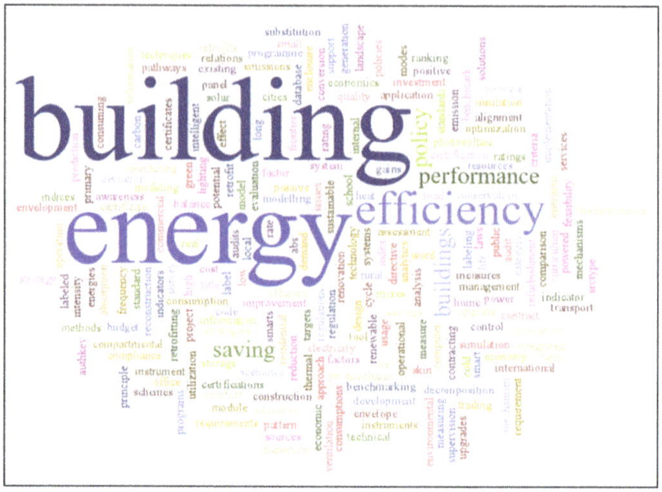

Figure 2. Word Cloud used to build the search terms (source: authors).

Figure 2 is a graphical representation of the word frequency that greater prominence to the key terms that appeared more frequently in the 893 publications previously gathered. The larger the word in the visual, the more common the word was among the keywords. Thus, words like building, energy, efficiency were prevalent, whilst others like pattern, technical, measuring were incidental. Therefore, not all 3000 keywords fit with the current research. Thus, after examining these keywords, only 682 of them were considered suitable to the scope of this research and were used in the construction of the ultimate query, as shown in Figure 3.

TITLE-ABS ("BUILDING ENERGY CONSERVATION" OR "Building energy" OR "Building energy analysis" OR " Building energy consumption analysis" OR " Building energy demand" OR "Building energy efficiency" OR "Building energy efficiency certification" OR "Building energy efficiency design" OR "building energy efficiency labeled project" OR " Building energy efficiency potential" OR "Building energy efficiency ratings" OR "Building energy management" OR "Building energy management system" OR "Building energy model" OR "Building energy modeling" OR "Building energy modelling" OR "Building energy performance" OR "Building energy performance certification" OR "Building energy policy" OR "Building energy regulation" OR "Building energy simulation" OR "Building energy simulation computer technology" OR "building energy Policy mixes" OR "building energy Policy pathways" OR "building energy Policy scenarios" OR "Renewable building energies" OR "Renewable building energy" OR "Renewable building energy resources" OR "Renewable building energy sources" OR

....

OR "Renewable building energy system" OR "Renewable building energy use" OR " Renewable building energy resources" OR "Renewable building energy technology") AND SRCTYPE(j) AND PUBYEAR > 1999 AND PUBYEAR < 2019

Figure 3. Fragment of the query applied in the search on SCOPUS database (source: authors).

Figure 3 illustrates just a fragment of the query used to collect the articles scrutinised by this research. As can be seen at the beginning of the fragment, were selected articles containing one or more of the key terms in its title and/or abstract. At the bottom of the Figure 3 the type of source selected (only articles) and the year of publication can be seen.

The search using this query returned more than 14,000 articles, indexed on the Scopus database. However, most of them have never been cited and, as the number of citations is recognised as a quality standard [106,108], only the documents cited five or more times were considered eligible for further analysis. Thus, the number of selected articles dropped to nearly 2000.

2.3. The Evolution of the Building Energy Efficiency Field of Knowledge

According to Price's Law [109], every field of knowledge is a dynamic structure that evolves over time, with a few publications covering distinct themes within it, until maturity when it is consolidated and there are only a few left to explore. Thus, it is important to investigate which stage of development this field has reached.

Price [109] stated that the number of publications related to a field of knowledge can be used to characterise its evolution. Indeed, scientific publications meet the scientific community's demand for new finds. Thus, it is fair to say that scientific productivity is directly associated with scientific community interest in themes within this field. Based on that, Price [109] established a law, which states that the several stages of evolution of a field of knowledge can be characterised by the growth of the number of publications over the years.

Based on that, the evolution of the Building Energy Efficiency field of knowledge was assessed by means of Price's fundamental law.

2.4. Authors' Keyword Collection, Classification, and Manipulation

This step shows the data structure used to store 9326 keywords, collected from the 2000 articles that sourced this research; the classification procedure of these keywords into themes; and the data manipulations necessary to feed the following step.

Initially, the keywords and the respective articles were stored in a matrix, as illustrated in Table 1.

Table 1. Articles and their keywords.

Metadata Identification	Keyword$_1$	Keyword$_2$	...	Keyword$_3$
a_1 [authors; ... ; year]	kw_1^1	kw_1^2	...	$kw_1^{n_1}$
⋮	⋮	⋮		⋮
a_i [authors; ... ; year]	kw_i^1	kw_i^2	...	$kw_i^{n_i}$
⋮	⋮	⋮		⋮
a_{2000} [authors; ... ; year]	kw_{2000}^1	kw_{2000}^2	...	$kw_{2000}^{n_{2000}}$

a_i = article; kw_i^j = jth keyword extracted from the ith article.

Each row of Table 1 represents one of the 2000 articles ($a_1, a_i, a_{2000})$). Each row stores the metadata for the articles (article title, authorship, publication date, and keywords). Special attention was given to the keywords, which number varies from article to article. For data processing purposes, each keyword is represented as kw_i^j meaning the jth keyword from the ith article.

The keywords from Table 1 were screened, looking for those relevant to Building Energy Efficiency, as well as for insights about their classification into groups. As a result, 7598 keywords were discarded and the remaining 1728 were grouped into 30 distinct themes, as illustrated in Table 2.

Table 2. Keywords classified into 30 themes.

Theme$_1$	...	**Theme$_2$**	...	**Theme$_3$**
kw_1^1	...	kw_1^j	...	kw_1^{30}
$\vdots$		$\vdots$		$\vdots$
$kw_{k_1}^2$	...	$kw_{k_2}^j$	...	$kw_{k_n}^{30}$

Table 2 classifies the keywords, stores in Table 1, into thirty categories, or themes (the number of categories was defined a priori). Each column of Table 2 is assigned to a theme, and each theme has a different number of keywords assigned to it.

The task of accounting for the number of times a given theme appears in the literature was made easy by combining Tables 1 and 2, which resulted in Table 3.

Table 3. Themes addressed by each article.

Article	**Theme$_1$**	...	**Theme$_2$**	...	**Theme$_3$**
a_1 [authors; ... ; year]	X		-		X
$\vdots$	$\vdots$		$\vdots$		$\vdots$
a_i [authors; ... ; year]	-		X		X
$\vdots$	$\vdots$		$\vdots$		$\vdots$
a_{2000} [authors; ... ; year]	X		-		X

Table 3 is very similar to Table 1, except for the fact that in Table 1 each article is associated which its own keywords, whilst in Table 3 each article is associated with one, or more, the themes previously defined. Thus, Table 3 can be read in two ways: by row or by column. When read by row, it is possible to see the themes addressed by each article. When read by column, it is possible to see which theme is present in which article. Thus, the presence of a theme in the literature in a given year can be assessed by counting the number of articles addressing such a theme that year. Therefore, in order to build a distribution of themes over a given period, it was enough to restrict Table 3 to articles published over such a period.

2.5. Evolution and Trend of the Themes

The participation of a theme within a given period can be assessed by the percentage of the articles published within this period that address such a theme, as shown in (1).

$$Theme_i\% = \frac{\#articles\ adressing\ theme_i}{total\ articles\ published\ in\ the\ year.} \quad (1)$$

Thus, the higher *Theme$_i$%*, the more important is the theme. Based on this, it was possible to study the evolution of the themes over time.

The evolution of a theme was defined as the participation of such a theme in the literature over the years covered by the analysis. Thus, if the participation of a theme increases over time, it can be said that the theme shows an upward trend. Conversely, if the participation decreases, the theme shows a downward trend. There are also occasions in which the trend is stable over the years.

In order to avoid subjectivity, the trend analysis was supported by a nonparametric statistical procedure, called the Mann-Kendall test for trend [110].

Table 4 shows a fragment of a table used to show the evolution and trend of the themes.

Table 4. Annual participation of the themes, followed by their trends.

Theme	2000	2001	2002	2003	2004	2005	...	2015	2016	2017	2018	Annual Participation	Trend (α = 5%)
BEV	0%	9%	0%	0%	6%	0%	-	3%	2%	2%	3%		↔
SMB	8%	0%	14%	0%	6%	5%	-	6%	6%	4%	8%		↔
TOB	8%	0%	14%	6%	6%	5%	-	5%	9%	7%	4%		↔
BMS	8%	0%	10%	0%	0%	0%	-	0%	1%	2%	2%		↔
BAC	0%	0%	10%	0%	6%	0%	-	7%	6%	4%	7%		↑
...													...
BIM	8%	0%	10%	6%	0%	0%	-	15%	12%	18%	12%		↑
OCB	8%	0%	0%	0%	0%	0%	-	5%	4%	11%	5%		↑
BRF	0%	0%	5%	0%	0%	5%	-	7%	6%	9%	7%		↑
ZEB	0%	0%	5%	0%	0%	5%	-	9%	6%	9%	13%		↑

Table 4 presents the evolution of each theme over the period under analysis, both in figures and graphically, as well as their trends according to the Mann-Kendall test with 5% of significance. The rows represent each of the thirty themes and, the columns store the participation of each of them in the literature calculated according to Equation (1). A bar graph illustrates the calculations, whilst ↑ means an upward trend, ↓ means a downward trend and ↔ means no-trend.

2.6. Stages of the Evolution of this Field of Knowledge

Table 4 can also be read from the point of view of the columns, i.e., from the point of view of the years. By doing that, it was possible to establish the profile of the years based on the themes addressed by the articles published during these years. This opened up the possibility of comparing the years and identifying certain patterns that allowed them to be grouped into clusters, demarcating evolutionary stages of this field of knowledge.

2.7. Interrelationships between Themes

According to Sun and Latora [111], the development of a field of knowledge is marked by the flow of knowledge between several themes or subareas of this field. That is why it is important to study the relationship between the themes.

Yet according to Sun and Latora [111], the knowledge flows with more intensity between synergetic themes, and the most influential themes are those most interconnected, around which the field develops itself. The relationship between the themes can be represented by means of an abstract two-dimensional plot resulting from a multidimensional scaling [112].

3. Results

The following section presents the outcomes of the complete analysis carried out to answer the research questions.

3.1. The Evolution of the Building Energy Efficiency Field of Knowledge

According to Price's Law [109], the scientific production concerned with a field of knowledge grows exponentially until it reaches a point of inflection and, afterwards, a threshold value around which it stabilises, meaning that this field has reached its maturity. The aspect of the curve that represents the evolution of publications goes from exponential to logistics, signalling that the scientific community's interest in this field has cooled down.

According to Dabi et al. [113]: "The main hypothesis of Price's law is that the development of science follows an exponential growth. The growth of a scientific domain goes through four phases". The first phase is the precursors' phase. According to Dabi et al. [113] "during this phase only a small number of researchers begin publishing". The second phase is the proper exponential growth. "During this phase, the expansion of the field attracts many researchers as many aspects of the subject still have to be explored" [113]. In the third phase, the body of knowledge is consolidated and the growth of scientific production becomes linear [113]. The next phase, according to Dabi et al. [113], "corresponds to the collapse of the domain and is marked by a decrease in the number of the publications". The aspect of the curve transforms from exponential to logistical, reaching a ceiling value after passing through an inflection point. Therefore, in order to perform the Price's Law analysis, the frequency distribution of the publications addressing BEE is presented in Figure 4.

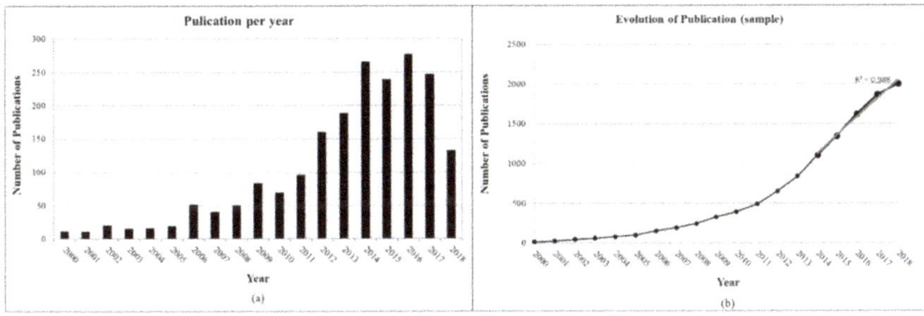

Figure 4. Frequency distribution of publications addressing Building Energy Efficiency. (**a**) Discrete, (**b**) cumulative (source: authors).

Figure 4a shows the number of publications on a yearly basis, whilst Figure 4b shows the cumulative version, on which compliance with Price's Law is investigated.

The first phase roughly extends to 2005. The second phase is from 2005 to 2014. The number of publications fits well with an exponential function since the statistic R^2 is very close to 1.00. The third phase extends from 2014 to 2018. The growth of scientific production becomes linear (R^2 = 0.988). There is no statistical evidence that an inflection point has been reached yet. It is worth mentioning that only articles with 5 or more citations were considered and it is well known that the older the article, the more cited it is. Thus, it is likely that the number of articles during the later years will increase, reinforcing the linear trend of the plot for the final years even more. Therefore, the maturity of this field of knowledge has not yet been reached, leaving several aspects to be explored.

3.2. Authors' Keyword Collection, Classification, and Manipulation

From the 9326 keywords collected, only 1728 were useful for the purposes of this study. They were, naturally, classified into 30 categories (or themes): building automation and control (BAC), building energy modelling (BEM), building envelope (BEV), building information modelling (BIM), building integrated photovoltaics (BIP), building management systems (BMS), building retrofitting (BRF), data analysis techniques (DAT), decision making (DMK), energy management systems (EMS), environmental (ENV), energy performance software (EPS), energy storage (EST), green building (GRB), heat pumping systems (HPS), heating-ventilation-air-conditioning (HVAC), life cycle assessment (LCA), lighting (LIG), occupancy behaviour (OCB), regulations (REG), renewable energy sources (RNE), smart buildings (SMB), smart grids (SMG), sustainability (SUS), thermal comfort (THC), thermal storage (THS), types of building (TOB), windows (WIN), water heating (WTH), and near zero/zero energy building (ZEB). Figure 5 shows a schematic representation of such a classification. Figure 5 is a pictorial representation of the thirty categories along with the number of keywords classified into each of them.

In the majority of cases, the classification of a keyword into a given category was straightforward, like 'green building' (classified under the Green Building theme or group) for instance. However, there were cases in which a keyword could be coded into more than one theme. In such cases, the classification demanded some extra work. It was necessary to read the title and abstract and, in some cases, the introduction of the articles from which the keyword was collected, to decide which theme it fitted best.

A keyword was classified into a unique theme but a theme could cluster several keywords with similar meanings, in such a way that each theme represents a homogenous group. It is worth mentioning that an article can have keywords classified into different themes.

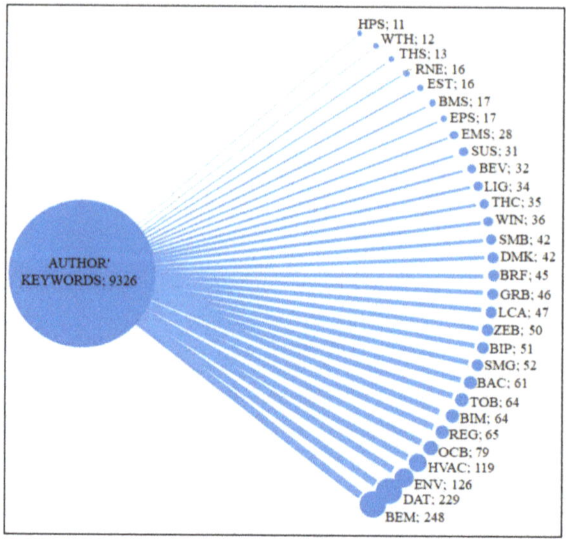

Figure 5. Keywords category (theme) and their number of members (source: authors).

3.3. Associating the Articles with the Themes

Once the keywords were classified into themes, the next step was to associate the 2000 articles captured for this research, with the themes. Some articles addressed only one theme, while others addressed more than one. The presence of a theme in a given period was used as a measure of its relevance and it was estimated by counting the number of articles in which the theme appeared during a period.

3.4. Evolution and Trend of the Themes

Before studying the evolution and trend of the themes it is worth discussing their relevance over the period under investigation.

The relevance of a theme can be derived from the number of articles that address it over the period considered [111]. Thus, Table 5 presents the themes ranked according to their relevance.

Table 5. The total number of articles dealing with each theme.

Themes		N	%	Interdisciplinarity
BEM	Building energy modelling	505	25.3	30
DAT	Data analysis techniques	337	16.9	27
BIM	Building information modelling	244	12.2	22
HVAC	Heating-ventilation-air-conditioning	159	8.0	30
ENV	Environmental	157	7.9	27
GRB	Green building	144	7.2	27
TOB	Types of buildings	130	6.5	27
ZEB	Zero energy building	128	6.4	28
SMB	Smart buildings	121	6.1	26
THC	Thermal comfort	121	6.1	28
REG	Regulations	117	5.9	26
SUS	Sustainability	114	5.7	23
BRF	Building retrofitting	105	5.3	26
OCB	Occupant behaviour	103	5.2	26
BAC	Building automation and control	99	5.0	26

Table 5. *Cont.*

	Themes	N	%	Interdisciplinarity
LCA	Life cycle assessment	89	4.5	24
SEM	Energy management systems	87	4.4	25
BIP	Building integrated photovoltaics	83	4.2	23
SMG	Smart grids	81	4.1	23
EPS	Energy performance software	75	3.8	23
BEV	Building envelope	52	2.6	25
WIN	Windows	52	2.6	17
LIG	Lighting	52	2.6	24
DMK	Decision making	49	2.5	27
RNE	Renewable energy sources	48	2.4	26
BMS	Building management systems	36	1.8	22
THS	Thermal storage	25	1.3	14
EST	Energy storage	24	1.2	17
HPS	Heat pumping systems	17	0.9	12
WTH	Water heating	12	0.6	10

Table 5 shows the absolute number and percentage of articles addressing each of the thirty themes. Therefore, it can be seen that the three largest themes are BEM, DAT, and BIM, which are present in more than 54% of the articles captured for this research. Eleven themes are addressed by less than 4% of the articles, meaning that the interest in them is small, so that they will be neglected for further analysis (grey background). However, it is worth mentioning that some of them are indirectly of interest for the ZEB and BRF themes. The themes BMS, BEV, EPS, EST, LIG, RNE, THS, and WIN could be still focused on the recent research under the umbrella of other themes with increasing.

Table 5 also shows the interdisciplinary character of the research carried out in the BEE field of knowledge. For instance, from the 2000 articles collected for this study, 505 (25.3%) address the theme BEM and the other 29 themes. According to Sun and Latora [111], such interaction can reflect the exchange of knowledge across themes. It is possible to infer that the strength of such an interaction depends on the number of publications sharing the themes.

Table 5 provides a static view of the BEE field of knowledge. It shows the most relevant themes within the field but it does not show the evolution and trend of each theme. Thus, Table 6 presents the trend of each theme, allowing investigation as to whether a given theme has a perennial presence or is just incidental in the literature. A theme can be analysed as to when it emerged, if it is still active or vanished, and when its apogee was. Table 6 presents the annual participation of each theme in the literature, summarising their trend in the last column.

Eight themes are in an upward trend: BAC, EMS, DAT, BEM, BIM, OCB, BRF and ZEB. It can be seen that the themes BAC, EMS and DAT reached a maximum in the early 2000s, while the others peaked in the late 2010s. The development of the internet and image processing software packages explain the remarkable growth of the theme BIM [112,114]. Once the stock of old buildings far surpasses the stock of new buildings everywhere in the world, the only way to achieve the current energy-saving standards is by retrofitting them, which explains the growing interest of the scientific community in the BRF theme. The raising of the theme OCB can be explained because the scientific community has realised that the success of energy-efficient projects are significantly influenced by human factors [115,116].

Table 6. The annual relative frequency of articles that address each of the thirty themes.

Then	2000	2001	2002	2003	2004	2005	2006	2007	2008	2009	2010	2011	2012	2013	2014	2015	2016	2017	2018	Annual Participation	Trend (α = 5%)
BEV	0%	9%	0%	0%	6%	0%	2%	0%	6%	2%	4%	2%	4%	2%	2%	3%	2%	2%	3%		↔
SMB	8%	0%	14%	0%	6%	5%	8%	7%	8%	7%	9%	4%	6%	6%	6%	6%	6%	4%	8%		↔
TOB	8%	0%	14%	6%	6%	5%	0%	7%	10%	4%	9%	7%	6%	6%	7%	5%	9%	7%	4%		↔
BMS	8%	0%	10%	0%	0%	0%	2%	2%	2%	2%	0%	1%	3%	2%	3%	0%	1%	2%	2%		↔
BAC	0%	0%	10%	0%	6%	0%	0%	5%	0%	1%	6%	4%	4%	6%	6%	7%	6%	4%	7%		↑
EMS	0%	0%	10%	0%	0%	0%	0%	2%	0%	1%	1%	3%	3%	5%	4%	5%	7%	6%	5%		↑
GRB	8%	0%	0%	19%	0%	15%	2%	15%	14%	6%	7%	10%	6%	7%	7%	10%	6%	5%	5%		↔
HVAC	0%	0%	14%	19%	6%	5%	12%	15%	18%	12%	4%	11%	9%	5%	6%	7%	8%	7%	7%		↔
WIN	0%	0%	5%	6%	0%	0%	4%	2%	0%	2%	1%	4%	1%	3%	4%	4%	2%	2%	2%		↔
RNE	0%	0%	0%	6%	0%	0%	0%	0%	2%	0%	0%	2%	3%	4%	2%	3%	4%	1%	5%		↔
DMK	0%	0%	0%	0%	6%	0%	4%	0%	5%	1%	4%	1%	3%	2%	4%	2%	3%	1%	3%		↔
DAT	0%	0%	0%	6%	12%	25%	6%	10%	8%	11%	19%	14%	19%	17%	19%	19%	20%	18%	20%		↑
SUS	0%	0%	0%	0%	0%	15%	10%	12%	4%	6%	13%	5%	8%	9%	2%	4%	6%	6%	2%		↓
LCA	0%	0%	5%	6%	0%	10%	8%	7%	4%	1%	4%	7%	3%	5%	3%	3%	5%	7%	5%		↔
EST	0%	0%	0%	0%	0%	5%	2%	2%	0%	0%	0%	2%	1%	3%	0%	1%	2%	1%	2%		↔
LIG	0%	0%	0%	0%	0%	5%	8%	5%	4%	1%	1%	3%	2%	3%	3%	4%	3%	2%	1%		↓
THC	8%	0%	0%	13%	0%	5%	4%	17%	12%	10%	3%	2%	7%	5%	5%	9%	5%	6%	6%		↔
BIP	0%	0%	5%	0%	6%	0%	4%	7%	6%	7%	1%	5%	2%	5%	3%	5%	4%	3%	7%		↔

Table 6. *Cont.*

Then	2000	2001	2002	2003	2004	2005	2006	2007	2008	2009	2010	2011	2012	2013	2014	2015	2016	2017	2018	Annual Participation	Trend (α = 5%)
WTH	0%	0%	0%	0%	0%	0%	2%	0%	2%	0%	0%	0%	0%	0%	1%	1%	1%	0%	1%		↔
REG	8%	0%	0%	0%	12%	10%	2%	10%	6%	13%	9%	8%	9%	7%	5%	4%	5%	3%	4%		↔
ENV	8%	0%	5%	0%	6%	10%	8%	5%	12%	12%	10%	8%	5%	9%	7%	9%	8%	8%	5%		↔
EPS	0%	0%	0%	0%	0%	0%	0%	0%	2%	2%	3%	8%	3%	3%	4%	5%	4%	4%	8%		↔
HPS	0%	0%	0%	0%	0%	0%	0%	0%	0%	1%	1%	0%	1%	2%	0%	1%	1%	1%	2%		↔
THS	0%	0%	0%	0%	0%	0%	0%	0%	0%	0%	0%	0%	1%	3%	1%	2%	2%	2%	2%		↔
BEM	8%	0%	24%	6%	0%	20%	17%	22%	25%	23%	14%	27%	28%	22%	28%	25%	30%	27%	29%		↑
SMG	0%	0%	0%	0%	0%	0%	0%	0%	2%	5%	0%	2%	5%	3%	6%	3%	7%	3%	6%		↔
BIM	8%	0%	10%	6%	0%	0%	2%	5%	8%	8%	13%	8%	10%	13%	15%	15%	12%	18%	12%		↑
OCB	8%	0%	0%	0%	0%	0%	6%	5%	0%	2%	1%	7%	5%	4%	6%	5%	4%	11%	5%		↑
BRF	0%	0%	5%	0%	0%	5%	0%	0%	2%	1%	1%	2%	6%	5%	6%	7%	6%	9%	7%		↑
ZEB	0%	0%	5%	0%	0%	5%	0%	0%	2%	5%	3%	3%	7%	6%	6%	9%	6%	9%	13%		↑

Since there are many consecrated statistical methods, which have been waiting for the development of informatics to become popular, it is expected that DAT will keep growing for a while, even within other fields of knowledge. According to Cristino et al. [108] the data analysis techniques mentioned by the papers within this field of knowledge can be roughly clustered into the following categories: regression analysis, descriptive statistics, multivariate analysis, computational intelligence, inferential statistics, and design of experiments.

There is no statistical evidence of a particular trend for SMB, TOB, GRB, HVAC, LCA, THC, BIP, REG, ENV and SMG.

The theme SUS shows a downward trend. The themes concerned with environmental issues (ENV, SUS and LCA) reached their maximum in the second half of the 2000s and have decreased since then, showing that interest in these subjects cooled down.

The volume of publications addressing each theme, as well as the interaction between them, defines the evolution of a field of knowledge. As these variables change over time, it is possible to infer that such an evolution is marked by distinct phases. Thus, the next step in this study is to identify such stages.

3.5. Stages of the Evolution of this Field of Knowledge

The evolution of a field of knowledge is marked by a sequence of periods with a similar profile of publications. Thus, reading Table 6 from the columns' point of view, it is possible to see the profile of the years according to the themes published and look for a pattern.

One of the ways to identify similarities between multivariate observations is to apply clustering techniques [112,117]. Thus, the space of the columns in Table 6 was submitted to a hierarchical clustering algorithm, leading to the dendrogram presented in Figure 6.

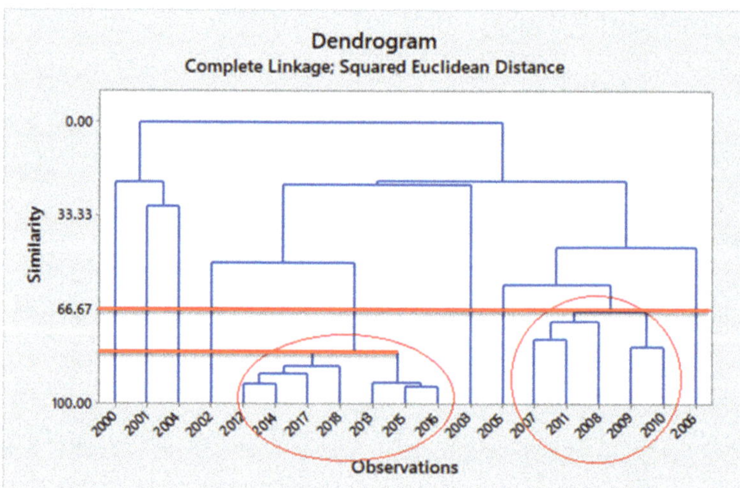

Figure 6. Years grouped according to the profile of themes (source: authors).

It is worth mentioning that a dendrogram is a tree diagram that shows hierarchical relationships between similar objects [118], which, in this case, are the years.

Therefore, the dendrogram shows in Figure 6 two well-defined clusters. One of these clusters groups the years 2007–2011, at a similarity level of 66.7, and the other, the years 2012–2018, at a similarity level larger than 80. The years ranging from 2000 to 2006 are very heterogeneous. This suggests that the period covered by this research could be divided into three phases. Figure 7 shows the profile of each of these phases. Figure 7 presents the annual participation of the themes in the literature for the three evolutionary periods determined by the cluster analysis.

During the first period (2000–2006), the scientific community's gaze was scattered over 26 themes, differently distributed over the whole period. In 2000, ten themes were presented in the literature; in 2001, only one theme (BEV); in 2002, this number increased to 14; in 2003 and 2004 decreased to 10; in 2005, increased to 15; and in 2006, the number of different themes presented in the literature reached its maximum, 20.

The participation of the themes in the literature varied over the years. In 2000, ten themes shared the same participation in the literature (10%); in 2002, the theme BEM stood out (24%); in 2003 two themes were highlighted, GRB and HVAC with 19%; in 2004, other two themes stood out, but this time, with 12% of participation (DAT, REG); in 2005, the theme DAT increased its participation to 25%, and, in 2006 the theme BEM stood out with 17% of participation in the literature.

The low number of themes in 2001 is due to the fact that only the articles that reached five or more citations were considered, which leads to the conclusion that the production of articles addressing the Building Envelope was the most consistent in 2001.

Therefore, it can be seen that the evolution of this field of knowledge over this period did not exhibit any pattern. The second period (2007–2011) is the shortest of the three periods (five years). It presented more themes consolidated than the previous one. Twenty-nine themes had been explored over this period, and 15 of them were present in all five years of this period. In 2007, 20 different themes were present in the literature; in 2008, 23 themes; in 2009, 26; in 2010, 24; and in 2011, 27. Thus, it is fair to conclude that the scientific community's interest in this field of knowledge became more consistent.

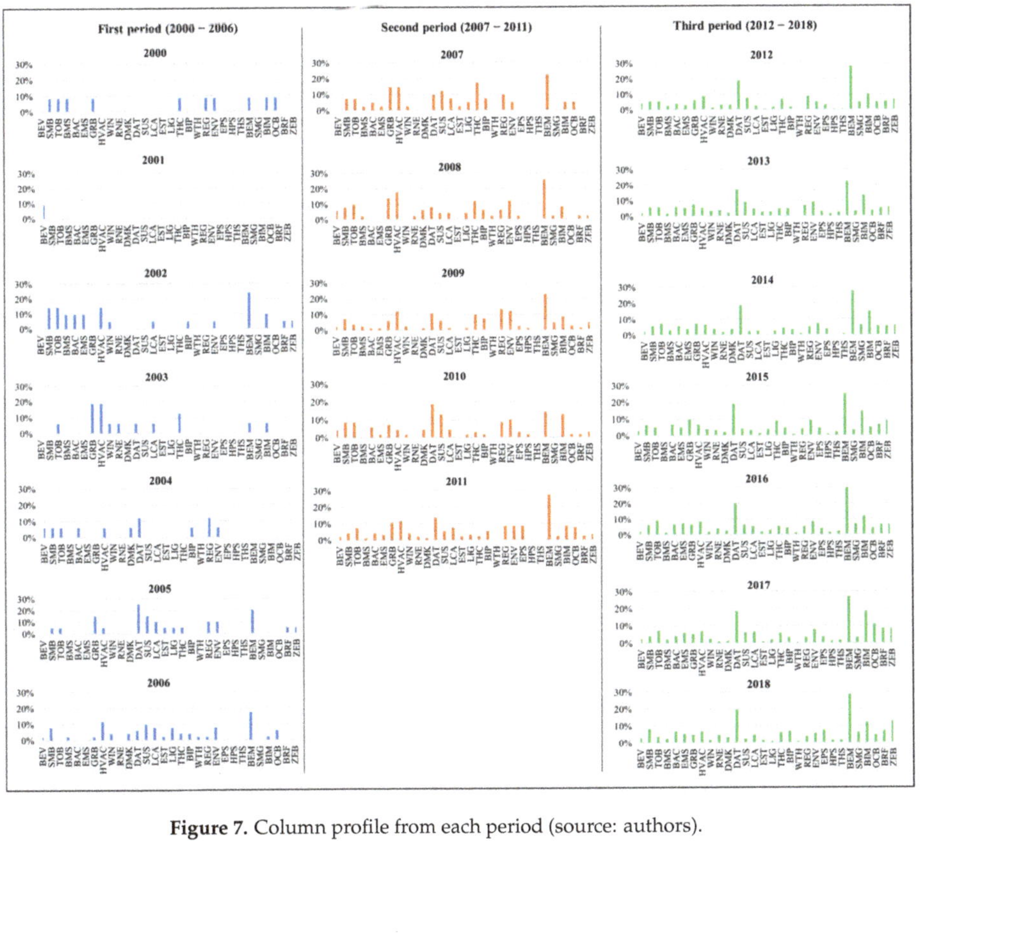

Figure 7. Column profile from each period (source: authors).

It was in this period that the themes BIP, ENV, REG and THC reached their greatest participation in the literature. However, the theme BEM was by far the one most present in the literature, closely followed by DAT. The participation of the themes BAC, BRF, EMS, OCB, SMG and ZEB had a neglectable participation in the literature over this period, while the participation of the themes BEV, BMS, GRB, HVAC, SMB, SUS and TOB shrank.

In the third period (2012–2018), all the thirty themes had been explored, 29 themes in 2012, 2013 and 2015; 28 themes in 2014; 30 themes in 2016, 2017 and 2018. Thus, it can be said that the scientific interest in this field of knowledge increased even more over this period.

The participation of the themes BEM, BIM, BRF, DAT, OCB and ZEB had increased and, according to statistical analysis, they are in an upward trend. participation of other themes like BAC, EMS and SMG had increased as well, but not enough signalise an upward trend. The interest for the themes ENV, GRB, HVAC, REG and SUS had decreased. The other themes remained stable.

3.6. Interrelationships between Themes

According to Sun and Latora [111], the interaction between themes within a field of knowledge reflects the flow of knowledge between the sub-areas of this field. Thus, in order to understand the evolution of this field, it is fundamentally important to define and study the interaction between the themes.

Many articles address multiple themes at once. What indicates an interaction between themes? The interaction between the themes i and j can be assessed by means of Equation (2).

$$\lambda_{ij} = 100 \cdot \frac{N_{ij}}{N_p} \quad (2)$$

where N_{ij} is the number of articles that concurrently address the themes i and j, and N_p is the number of articles for the considered period (N_1 = 149, N_2 = 342 and N_3 = 1509).Thus, λ_{ij} is the percentage of the articles produced during the period under investigation that addresses the themes i and j. Figure 8 presents a graphical representation of the model used to account for the interactions between themes.

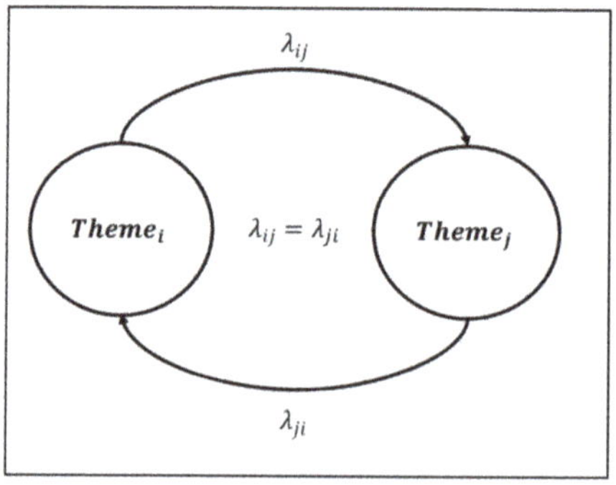

Figure 8. Model of the interrelationship between themes (source: authors).

Based on Figure 8, it can be seen that λ_{ij} can be stored in a symmetric matrix, called interaction or interrelationship matrix. According to Equation (2), such a matrix varies depending on the evolutionary period. Figure 9 shows the interrelationship matrix for each period. The darker the fill colour, the greater the interaction between themes i and j.

Observing the matrix for the first period, it can be said that, during this period, this field of knowledge was driven in large part by themes concerned with sustainable development and thermal comfort. Also, it can be noticed that the greatest interaction occurred between HVAC-THC and LCA-SUS. It is possible to observe the emergence of the relationship between the themes BEM-DAT, which would increase until the end of the third period.

During the second period, the interest of the scientific community revolved more around the interaction between BEM-DAT; BEM-HVAC; GRB-SUS; HVAC-TOB and HVAC-THC.

The interaction between BEM-DAT is remarkable; it is by far the largest one, not only over the third period, but over the whole period covered by this research. Therefore, these two themes have been the great engine for developing the research on Building Energy Efficiency.

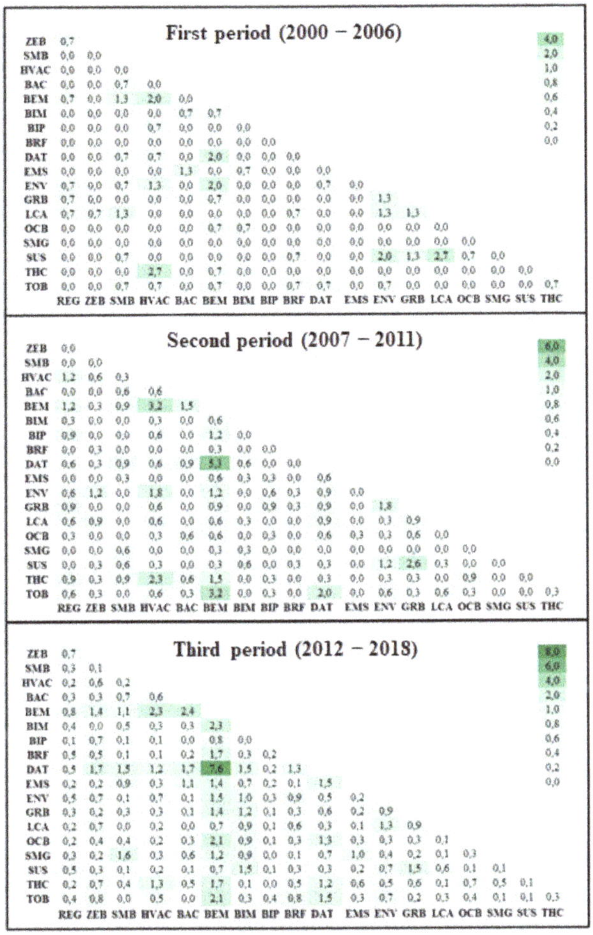

Figure 9. λ_{ij} for each period (source: authors).

Since it is difficult to analyse and understand the interaction between the themes only by examining the interrelationship matrices in Figure 9, a visual representation of such matrices is valuable. Such a representation can be obtained by means of a data analysis technique known as multidimensional scaling [119], which allows the representation of the interrelationship of the themes in an abstract, two-dimensional Cartesian plot, as illustrated in Figure 10. Although such a representation is not absolutely perfect, it gives some insight into the interaction between themes. For instance, the greater the interaction between themes, the closer they are in the plot, forming clusters of synergetic themes. In other words: the closer the themes, the greater the flow of knowledge between them.

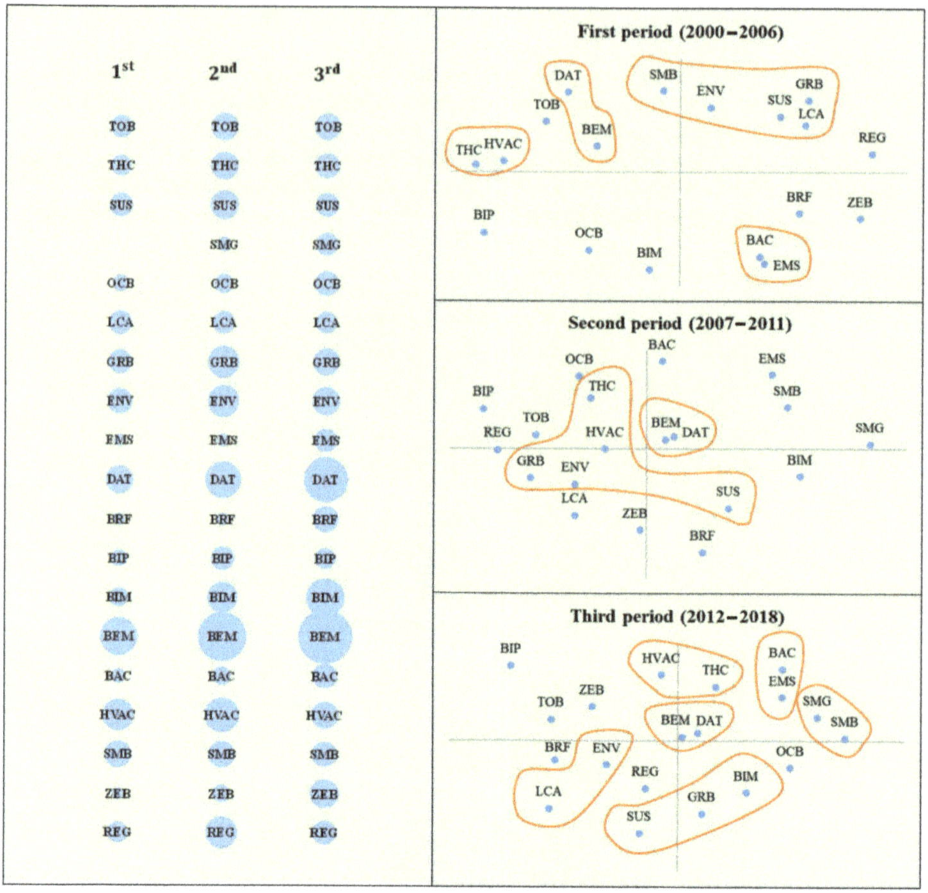

Figure 10. Evolution of the themes and relationship between them in each period (source: authors).

The left side of Figure 10 shows the participation of the themes over the three evolutionary stages. On the right side, three plots represent the interrelationship matrix between the themes for each of the three evolutionary stages.

The clusters shown in Figure 10 only include the themes for which the $\lambda_{ij} > 1.0$. The distance between clusters and elements was assessed according to the nearest neighbour strategy [118].

Observing the plots for the three evolutionary periods, it should be noted that the themes have clustered around the origin of the plot as this field evolved. In general, although it can distort the representation of the themes in the plot, the more central a theme, the greater the interaction with the others.

In the first evolutionary period of this field of knowledge, significant interaction between themes related to thermal comfort (THC-HVAC), themes concerned with environmental/sustainability issues (SUS-ENV-LCA-GRB-SMB), and themes addressing modelling and data analysis techniques (BEM-DAT) can be seen.

A number of clusters dropped from the first to the second evolutionary period. The cluster BEM-DAT remained and came closer to the centre of the plot. They are cross themes. Some articles are devoted to revisiting a given theme and have an interest in comparing the results emerging from different data analysis techniques, in such a way that the modelling and data analysis become the kernel of the paper instead of being tools by means of which better results can be achieved. Such articles give little attention to the aspects concerned with Building Energy Efficiency, which are only the background and data source, while their main purpose is data analysis.

Still, within the second period, the themes LCA and SMB leave the environment/sustainability cluster because of the lack of interaction with the other themes. The interaction of the remaining themes with the thermal comfort cluster increased, resulting in the formation of a new cluster.

The cluster BAC-EMS was extinct by this stage. Since the participation of both themes in the literature increased over this period, it is fair to assume that both themes developed in isolation, without sharing knowledge.

The number of isolated themes in this period was the largest amongst the evolutionary stages. Thus, it can be said that, during this period, the exchange of knowledge was the smallest.

The third stage is the one with the largest number of clusters and the smallest number of isolated themes. It can be considered the period with the greatest flow of knowledge between sub-areas within this field of knowledge.

The clusters THC-HVAC and BAC-EMS, from the first evolutionary stage, have been re-established, meaning that the themes within each cluster restarted, triggering knowledge production in each other.

The cluster BEM-DAT is even closer to the centre of the plot in this stage. According to the interaction matrix for the third period, in Figure 9, this cluster interacts with all the themes ($\lambda_{ij} > 1.0$) except the themes BIP and REG.

The cluster concerned with environment/sustainability in the first period was broken into three small clusters (LCA-ENV, SMB-SMG, and SUS-GRB-BIM), suggesting the exchange of more specialised knowledge. The flow of knowledge between themes related to sustainability and information modelling is noteworthy. As the latter theme shows an upward trend, it is quite possible that its development increases the knowledge production of themes related to green buildings and sustainability.

The themes BIP, TOB, REG, ZEB, BRF, and OCB developed in isolation over all three evolutionary stages. The latter three are in an upward trend, according to the trend analysis previously presented. Thus, a clear relation between trend and evolutionary development of a theme within the Building Energy Efficiency field of knowledge could not be seen.

4. Conclusions

After analysing 2000 articles concerned with Building Energy Efficiency, this paper shows that this field of knowledge has not yet reached maturity. Thus, much remains to be studied, meaning that investment in research is still needed.

This research identified thirty recurrent themes within this field of knowledge. However, only nineteen of these themes are statistically significant. According to the Mann-Kendall trend test, eight out of these themes show a clear upward trend, one a downward trend, and ten do not show any clear evidence for a particular trend.

This study shows that the evolution of this field of knowledge passed through three stages, whose dynamics were clearly explained, as well as the changes in the patterns of cross-fertilisation.

This research shows that energy modelling, along with data analysis techniques, have been influencing this field of knowledge since its beginning and they have been instigating production in other areas within this field. Therefore, themes like Building Energy Modelling and Data Analysis Techniques are in an upward trend and still very far from maturity, constituting good research opportunities.

The scientific community's gaze is on other themes with low connections, like Occupancy Behaviour, Building Information Modelling, Zero Energy Buildings, and Building Retrofitting. All of these themes have increased in importance and seem to be new frontiers of this field of knowledge.

Considering the Occupancy Behaviour, topics like eco-feedback, gamification, behaviour, and advanced building automation systems have not been adequately addressed.

Building Information Modelling is a very recent research front, therefore there is a great interest among the scientific community in this field, which signalise that there is a great potential for research on integrating BIM with technologies like monitoring systems, thermography, geographic information systems.

The concept of Zero Energy Building has drawn the attention of the scientific community. However, there are few studies focusing on the feasibility of Zero Energy Building in a diversity of climates and the integration with information technology.

Building Energy Retrofitting provides substantial opportunities to reduce the energy consumption of the building sector. There are a lot of research opportunities such as identifying and designing optimal cost-effective energy retrofitting strategies, combining retrofitting with Building Energy Modelling, Retrofitting, and Occupancy Behaviour.

Furthermore, it is worth mentioning that some of the themes are indirectly of interest for the Zero Energy Building and Building Energy Retrofitting themes. The themes Building Management Systems, Building Envelope, Energy Performance Software, Energy Storage, Lighting, Renewable Energy Sources, Thermal Storage, and Windows could be still focused on the recent research under the umbrella of themes with increasing trends.

These findings allow the researchers to optimise time and resources by investigating the themes with growing interest and possibilities for new contributions, as the scientific community directs its efforts towards cutting edge themes and topics. Furthermore, they improve the understanding of the laws governing the development of a field of knowledge, impacting the formulation of research strategies.

This research identified the interaction between themes but does not propose a mechanism to objectively identify the direction of the flow of knowledge. Such information can improve the understanding of the development of the field of knowledge. Therefore, it is suggested that future research addresses this unanswered aspect of the current research.

Author Contributions: Conceptualisation, T.M.C. and A.F.N.; methodology, T.M.C. and A.F.N.; software, T.M.C.; validation, T.M.C.; formal analysis, T.M.C.; investigation, T.M.C. and A.F.N.; resources, A.F.N.; data curation, A.F.N.; writing—original draft, T.M.C.; writing—review & editing, A.F.N., F.W. and B.D.; visualisation, T.M.C.; supervision, A.F.N. and B.D.; project administration, T.M.C. and F.W.; funding acquisition, T.M.C., A.F.N., F.W. and B.D. All authors have read and agreed to the published version of the manuscript.

Funding: This research has been partially supported by the CDP Eco-SESA, receiving funds from the French National Research Agency in the framework of the "Investissements d'avenir" programme [ANR-15-IDEX-02]; by the Coordination for the Improvement of Higher Education Personnel—Brazil (CAPES) [Finance Code 001]; and by the São Paulo Research Foundation (FAPESP) [grant numbers 2021/01423-9 and 2019/17937-1].

Institutional Review Board Statement: Not applicable.

Informed Consent Statement: Not applicable.

Data Availability Statement: Not applicable.

Conflicts of Interest: The authors declare no conflict of interest.

Abbreviations

The following abbreviations are used in this research:

BAC	Building Automation and Control
BEE	Building Energy Efficiency
BEM	Building Energy Modelling
BEV	Building Envelope
BIM	Building Information Modelling
BIP	Building Integrated Photovoltaics
BMS	Building Management Systems

BRF	Building Retrofitting
DAT	Data Analysis Techniques
DMK	Decision Making
EMS	Energy Management Systems
ENV	Environmental
EPS	Energy Performance Software
EST	Energy Storage
GRB	Green Building
HPS	Heat Pumping Systems
HVAC	Heating-Ventilation-Air-Conditioning
LCA	Life Cycle Assessment
LIG	Lighting
OCB	Occupancy Behaviour
REG	Regulations
RNE	Renewable Energy Sources
SMB	Smart Buildings
SMG	Smart Grids
SUS	Sustainability
THC	Thermal Comfort
THS	Thermal Storage
TOB	Types Of Building
WIN	Windows
WTH	Water Heating
ZEB	Near Zero/Zero Energy Building

References

1. Zhang, Y.; Wang, Y. Barriers' and policies' analysis of China's building energy efficiency. *Energy Policy* **2013**, *62*, 768–773. [CrossRef]
2. De Oliveira, K.B.; dos Santos, E.F.; Neto, A.F.; Santos, V.H.D.M.; de Oliveira, O.J. Guidelines for efficient and sustainable energy management in hospital buildings. *J. Clean. Prod.* **2021**, *329*, 129644. [CrossRef]
3. Curtius, H.C. The adoption of building-integrated photovoltaics: Barriers and facilitators. *Renew. Energy* **2018**, *126*, 783–790. [CrossRef]
4. International Energy Agency (IEA). Energy Efficiency: Buildings—The Global Exchange for Energy Efficiency Policies, Data and Analysis. 2019. Available online: https://www.iea.org/topics/energyefficiency/buildings (accessed on 27 May 2019).
5. Buildings—European Commission, (n.d.). Available online: https://ec.europa.eu/energy/en/topics/energy-efficiency/buildings (accessed on 20 September 2019).
6. Bond, S. Barriers and drivers to green buildings in Australia and New Zealand. *J. Prop. Invest. Finance* **2011**, *29*, 494–509. [CrossRef]
7. Greenough, R.; Tosoratti, P. Low carbon buildings: A solution to landlord-tenant problems? *J. Prop. Invest. Finance* **2014**, *32*, 415–423. [CrossRef]
8. Zhou, L.; Li, J.; Chiang, Y.H. Promoting energy efficient building in China through clean development mechanism. *Energy Policy* **2013**, *57*, 338–346. [CrossRef]
9. Persson, J.; Grönkvist, S. Drivers for and barriers to low-energy buildings in Sweden. *J. Clean. Prod.* **2015**, *109*, 296–304. [CrossRef]
10. Zhao, H.-X.; Magoulès, F. A review on the prediction of building energy consumption. *Renew. Sustain. Energy Rev.* **2012**, *16*, 3586–3592. [CrossRef]
11. Bourdic, L.; Salat, S. Building energy models and assessment systems at the district and city scales: A review. *Build. Res. Inf.* **2012**, *40*, 518–526. [CrossRef]
12. Tian, W. A review of sensitivity analysis methods in building energy analysis. *Renew. Sustain. Energy Rev.* **2013**, *20*, 411–419. [CrossRef]
13. Coakley, D.; Raftery, P.; Keane, M. A review of methods to match building energy simulation models to measured data. *Renew. Sustain. Energy Rev.* **2014**, *37*, 123–141. [CrossRef]
14. Li, X.; Wen, J. Review of building energy modeling for control and operation. *Renew. Sustain. Energy Rev.* **2014**, *37*, 517–537. [CrossRef]
15. Reinhart, C.F.; Davila, C.C. Urban building energy modeling—A review of a nascent field. *Build. Environ.* **2016**, *97*, 196–202. [CrossRef]
16. Harish, V.S.K.V.; Kumar, A. A review on modeling and simulation of building energy systems. *Renew. Sustain. Energy Rev.* **2016**, *56*, 1272–1292. [CrossRef]

17. Whiffen, T.; Naylor, S.; Hill, J.; Smith, L.; Callan, P.; Gillott, M.; Wood, C.; Riffat, S. A concept review of power line communication in building energy management systems for the small to medium sized non-domestic built environment. *Renew. Sustain. Energy Rev.* **2016**, *64*, 618–633. [CrossRef]
18. Molina-Solana, M.; Ros, M.; Ruiz, M.D.; Gómez-Romero, J.; Martin-Bautista, M. Data science for building energy management: A review. *Renew. Sustain. Energy Rev.* **2017**, *70*, 598–609. [CrossRef]
19. Li, W.; Zhou, Y.; Cetin, K.; Eom, J.; Wang, Y.; Chen, G.; Zhang, X. Modeling urban building energy use: A review of modeling approaches and procedures. *Energy* **2017**, *141*, 2445–2457. [CrossRef]
20. Tian, Z.; Zhang, X.; Jin, X.; Zhou, X.; Si, B.; Shi, X. Towards adoption of building energy simulation and optimization for passive building design: A survey and a review. *Energy Build.* **2018**, *158*, 1306–1316. [CrossRef]
21. Sadineni, S.B.; Madala, S.; Boehm, R.F. Passive building energy savings: A review of building envelope components. *Renew. Sustain. Energy Rev.* **2011**, *15*, 3617–3631. [CrossRef]
22. Friess, W.A.; Rakhshanbabanari, K. A review of passive envelope measures for improved building energy efficiency in the UAE. *Renew. Sustain. Energy Rev.* **2017**, *72*, 485–496. [CrossRef]
23. Rao, V.V.; Parameshwaran, R.; Ram, V.V. PCM-mortar based construction materials for energy efficient buildings: A review on research trends. *Energy Build.* **2018**, *158*, 95–122. [CrossRef]
24. Kheiri, F. A review on optimization methods applied in energy-efficient building geometry and envelope design. *Renew. Sustain. Energy Rev.* **2018**, *92*, 897–920. [CrossRef]
25. Rashidi, S.; Esfahani, J.A.; Karimi, N. Porous materials in building energy technologies—A review of the applications, modelling and experiments. *Renew. Sustain. Energy Rev.* **2018**, *91*, 229–247. [CrossRef]
26. Kazmi, A.H.; O'Grady, M.; Delaney, D.; Ruzzelli, A.G.; O'Hare, G.M.P. A Review of Wireless-Sensor-Network-Enabled Building Energy Management Systems. *ACM Trans. Sens. Netw.* **2014**, *10*, 1–43. [CrossRef]
27. Pérez-Lombard, L.; Ortiz, J.; Pout, C. A review on buildings energy consumption information. *Energy Build.* **2008**, *40*, 394–398. [CrossRef]
28. Fumo, N. A review on the basics of building energy estimation. *Renew. Sustain. Energy Rev.* **2014**, *31*, 53–60. [CrossRef]
29. Deb, C.; Zhang, F.; Yang, J.; Lee, S.E.; Shah, K.W. A review on time series forecasting techniques for building energy consumption. *Renew. Sustain. Energy Rev.* **2017**, *74*, 902–924. [CrossRef]
30. Wang, Z.; Srinivasan, R. A review of artificial intelligence based building energy use prediction: Contrasting the capabilities of single and ensemble prediction models. *Renew. Sustain. Energy Rev.* **2017**, *75*, 796–808. [CrossRef]
31. Amasyali, K.; El-Gohary, N.M. A review of data-driven building energy consumption prediction studies. *Renew. Sustain. Energy Rev.* **2018**, *81*, 1192–1205. [CrossRef]
32. Wei, Y.; Zhang, X.; Shi, Y.; Xia, L.; Pan, S.; Wu, J.; Han, M.; Zhao, X. A review of data-driven approaches for prediction and classification of building energy consumption. *Renew. Sustain. Energy Rev.* **2018**, *82*, 1027–1047. [CrossRef]
33. Ahmad, T.; Chen, H.; Guo, Y.; Wang, J. A comprehensive overview on the data driven and large scale based approaches for forecasting of building energy demand: A review. *Energy Build.* **2018**, *165*, 301–320. [CrossRef]
34. Shi, X.; Tian, Z.; Chen, W.; Si, B.; Jin, X. A review on building energy efficient design optimization from the perspective of architects. *Renew. Sustain. Energy Rev.* **2016**, *65*, 872–884. [CrossRef]
35. Tian, W.; Heo, Y.; de Wilde, P.; Li, Z.; Yan, D.; Park, C.S.; Feng, X.; Augenbroe, G. A review of uncertainty analysis in building energy assessment. *Renew. Sustain. Energy Rev.* **2018**, *93*, 285–301. [CrossRef]
36. Zhu, N.; Ma, Z.; Wang, S. Dynamic characteristics and energy performance of buildings using phase change materials: A review. *Energy Convers. Manag.* **2009**, *50*, 3169–3181. [CrossRef]
37. Chung, W. Review of building energy-use performance benchmarking methodologies. *Appl. Energy* **2011**, *88*, 1470–1479. [CrossRef]
38. Hauge, Å.L.; Thomsen, J.; Berker, T. User evaluations of energy efficient buildings: Literature review and further research. *Adv. Build. Energy Res.* **2011**, *5*, 109–127. [CrossRef]
39. Luck, K. Energy efficient building services for tempering performance-oriented interior spaces—A literature review. *J. Clean. Prod.* **2012**, *22*, 1–10. [CrossRef]
40. Abu Bakar, N.N.; Hassan, M.Y.; Abdullah, H.; Rahman, H.A.; Abdullah, P.; Hussin, F.; Bandi, M. Energy efficiency index as an indicator for measuring building energy performance: A review. *Renew. Sustain. Energy Rev.* **2015**, *44*, 1–11. [CrossRef]
41. Loonen, R.C.G.M.; Favoino, F.; Hensen, J.L.M.; Overend, M. Review of current status, Requirements and opportunities for building performance simulation of adaptive facades. *J. Build. Perform. Simul.* **2017**, *10*, 205–223. [CrossRef]
42. Song, M.; Niu, F.; Mao, N.; Hu, Y.; Deng, S.S. Review on building energy performance improvement using phase change materials. *Energy Build.* **2018**, *158*, 776–793. [CrossRef]
43. Fuentes, E.; Arce, L.; Salom, J. A review of domestic hot water consumption profiles for application in systems and buildings energy performance analysis. *Renew. Sustain. Energy Rev.* **2018**, *81*, 1530–1547. [CrossRef]
44. Shaikh, P.H.; Nor, N.B.M.; Sahito, A.A.; Nallagownden, P.; Elamvazuthi, I.; Shaikh, M. A review on optimized control systems for building energy and comfort management of smart sustainable buildings. *Renew. Sustain. Energy Rev.* **2014**, *34*, 409–429. [CrossRef]
45. Evins, R. A review of computational optimisation methods applied to sustainable building design. *Renew. Sustain. Energy Rev.* **2013**, *22*, 230–245. [CrossRef]

46. John, G.; Clements-Croome, D.; Jeronimidis, G. Sustainable building solutions: A review of lessons from the natural world. *Build. Environ.* **2005**, *40*, 319–328. [CrossRef]
47. Lemougna, P.N.; Melo, U.F.C.; Kamseu, E.; Tchamba, A.B. Laterite Based Stabilized Products for Sustainable Building Applications in Tropical Countries: Review and Prospects for the Case of Cameroon. *Sustainability* **2011**, *3*, 293–305. [CrossRef]
48. Cabeza, L.F.; Barreneche, C.; Miró, L.; Martínez, M.; Fernandez, A.I.; Urge-Vorsatz, D. Affordable construction towards sustainable buildings: Review on embodied energy in building materials. *Curr. Opin. Environ. Sustain.* **2013**, *5*, 229–236. [CrossRef]
49. Huedo, P.; López-Mesa, B.; Mesa, M.B.L. Review of tools to assist in the selection of sustainable building assemblies. *Inf. Constr.* **2013**, *65*, 77–88. [CrossRef]
50. Alotaibi, S.S.; Riffat, S. Vacuum insulated panels for sustainable buildings: A review of research and applications. *Int. J. Energy Res.* **2013**, *38*, 1–19. [CrossRef]
51. Asdrubali, F.; D'Alessandro, F.; Schiavoni, S. A review of unconventional sustainable building insulation materials. *Sustain. Mater. Technol.* **2015**, *4*, 1–17. [CrossRef]
52. Nielsen, A.N.; Jensen, R.L.; Larsen, T.S.; Nissen, S.B. Early stage decision support for sustainable building renovation—A review. *Build. Environ.* **2016**, *103*, 165–181. [CrossRef]
53. Ju, C.; Ning, Y.; Pan, W. A review of interdependence of sustainable building. *Environ. Impact Assess. Rev.* **2016**, *56*, 120–127. [CrossRef]
54. Shaikh, P.H.; Nor, N.B.M.; Sahito, A.A.; Nallagownden, P.; Elamvazuthi, I.; Shaikh, M. Building energy for sustainable development in Malaysia: A review. *Renew. Sustain. Energy Rev.* **2017**, *75*, 1392–1403. [CrossRef]
55. Merschbrock, C.; Munkvold, B. A Research review on building information modeling in construction—An area ripe for IS research. *Commun. Assoc. Inf. Syst.* **2012**, *31*, 207–228. [CrossRef]
56. Volk, R.; Stengel, J.; Schultmann, F. Building Information Modeling (BIM) for existing buildings—Literature review and future needs. *Autom. Constr.* **2014**, *38*, 109–127. [CrossRef]
57. Shou, W.; Wang, J.; Wang, X.; Chong, H.Y. A Comparative Review of Building Information Modelling Implementation in Building and Infrastructure Industries. *Arch. Comput. Methods Eng.* **2015**, *22*, 291–308. [CrossRef]
58. Cho, Y.K.; Ham, Y.; Golpavar-Fard, M. 3D as-is building energy modeling and diagnostics: A review of the state-of-the-art. *Adv. Eng. Inform.* **2015**, *29*, 184–195. [CrossRef]
59. Lopez, R.; Chong, H.-Y.; Wang, X.; Graham, J. Technical Review: Analysis and Appraisal of Four-Dimensional Building Information Modeling Usability in Construction and Engineering Projects. *J. Constr. Eng. Manag.* **2016**, *142*, 06015005. [CrossRef]
60. Parn, E.; Edwards, D.; Sing, C.P. The building information modelling trajectory in facilities management: A review. *Autom. Constr.* **2017**, *75*, 45–55. [CrossRef]
61. Liu, X.; Wang, X.; Wright, G.; Cheng, J.C.P.; Li, X.; Liu, R. A State-of-the-Art Review on the Integration of Building Information Modeling (BIM) and Geographic Information System (GIS). *ISPRS Int. J. Geo-Inf.* **2017**, *6*, 53. [CrossRef]
62. Santos, R.; Costa, A.A.; Grilo, A. Bibliometric analysis and review of Building Information Modelling literature published between 2005 and 2015. *Autom. Constr.* **2017**, *80*, 118–136. [CrossRef]
63. Chong, H.-Y.; Lee, C.-Y.; Wang, X. A mixed review of the adoption of Building Information Modelling (BIM) for sustainability. *J. Clean. Prod.* **2017**, *142*, 4114–4126. [CrossRef]
64. Lu, Y.; Wu, Z.; Chang, R.; Li, Y. Building Information Modeling (BIM) for green buildings: A critical review and future directions. *Autom. Constr.* **2017**, *83*, 134–148. [CrossRef]
65. Eleftheriadis, S.; Mumovic, D.; Greening, P. Life cycle energy efficiency in building structures: A review of current developments and future outlooks based on BIM capabilities. *Renew. Sustain. Energy Rev.* **2017**, *67*, 811–825. [CrossRef]
66. Zhou, W.; Qin, H.; Qiu, J.; Fan, H.; Lai, J.; Wang, K.; Wang, L. Building information modelling review with potential applications in tunnel engineering of China. *R. Soc. Open Sci.* **2017**, *4*, 170174. [CrossRef]
67. Zhou, Y.; Ding, L.; Rao, Y.; Luo, H.; Medjdoub, B.; Zhong, H. Formulating project-level building information modeling evaluation framework from the perspectives of organizations: A review. *Autom. Constr.* **2017**, *81*, 44–55. [CrossRef]
68. Martinez-Aires, M.D.; López-Alonso, M.; Martínez-Rojas, M. Building information modeling and safety management: A systematic review. *Saf. Sci.* **2018**, *101*, 11–18. [CrossRef]
69. Saieg, P.; Sotelino, E.D.; Nascimento, D.; Caiado, R.G.G. Interactions of building information modeling, lean and sustainability on the architectural, engineering and construction industry: A systematic review. *J. Clean. Prod.* **2018**, *174*, 788–806. [CrossRef]
70. Sanhudo, L.; Ramos, N.M.M.; Martins, J.P.; Almeida, R.M.S.F.; Barreira, E.; Simões, M.L.; Cardoso, V. Building information modeling for energy retrofitting—A review. *Renew. Sustain. Energy Rev.* **2018**, *89*, 249–260. [CrossRef]
71. Geetha, N.B.; Velraj, R. Passive cooling methods for energy efficient buildings with and without thermal energy storage—A review. *Energy Educ. Sci. Technol. Part A Energy Sci. Res.* **2012**, *29*, 913–946.
72. Yang, L.; Yan, H.; Lam, J.C. Thermal comfort and building energy consumption implications—A review. *Appl. Energy* **2014**, *115*, 164–173. [CrossRef]
73. Guillén-Lambea, S.; Soria, B.R.; Marín, J.M. Review of European ventilation strategies to meet the cooling and heating demands of nearly zero energy buildings (nZEB)/Passivhaus. Comparison with the USA. *Renew. Sustain. Energy Rev.* **2016**, *62*, 561–574. [CrossRef]
74. Saffari, M.; de Gracia, A.; Ushak, S.; Cabeza, L.F. Passive cooling of buildings with phase change materials using whole-building energy simulation tools: A review. *Renew. Sustain. Energy Rev.* **2017**, *80*, 1239–1255. [CrossRef]

75. Johra, H.; Heiselberg, P. Influence of internal thermal mass on the indoor thermal dynamics and integration of phase change materials in furniture for building energy storage: A review. *Renew. Sustain. Energy Rev.* **2017**, *69*, 19–32. [CrossRef]
76. Tian, W.; Han, X.; Zuo, W.; Sohn, M.D. Building energy simulation coupled with CFD for indoor environment: A critical review and recent applications. *Energy Build.* **2018**, *165*, 184–199. [CrossRef]
77. Pérez-Lombard, L.; Ortiz, J.; Coronel, J.F.; Maestre, I.R. A review of HVAC systems requirements in building energy regulations. *Energy Build.* **2011**, *43*, 255–268. [CrossRef]
78. Mirsadeghi, M.; Cóstola, D.; Blocken, B.; Hensen, J.L.M. Review of external convective heat transfer coefficient models in building energy simulation programs: Implementation and uncertainty. *Appl. Therm. Eng.* **2013**, *56*, 134–151. [CrossRef]
79. Soares, N.; Costa, J.J.; Gaspar, A.R.; Santos, P. Review of passive PCM latent heat thermal energy storage systems towards buildings' energy efficiency. *Energy Build.* **2013**, *59*, 82–103. [CrossRef]
80. Lizana, J.; Chacartegui, R.; Padura, A.B.; Valverde, J.M. Advances in thermal energy storage materials and their applications towards zero energy buildings: A critical review. *Appl. Energy* **2017**, *203*, 219–239. [CrossRef]
81. Marszal, A.J.; Heiselberg, P.; Bourrelle, J.S.; Musall, E.; Voss, K.; Sartori, I.; Napolitano, A. Zero Energy Building—A review of definitions and calculation methodologies. *Energy Build.* **2011**, *43*, 971–979. [CrossRef]
82. Li, D.H.; Yang, L.; Lam, J.C. Zero energy buildings and sustainable development implications—A review. *Energy* **2013**, *54*, 1–10. [CrossRef]
83. Liu, Z.; Zhang, L.; Gong, G.; Li, H.; Tang, G. Review of solar thermoelectric cooling technologies for use in zero energy buildings. *Energy Build.* **2015**, *102*, 207–216. [CrossRef]
84. Chastas, P.; Theodosiou, T.; Bikas, D. Embodied energy in residential buildings-towards the nearly zero energy building: A literature review. *Build. Environ.* **2016**, *105*, 267–282. [CrossRef]
85. Kivimaa, P.; Martiskainen, M. Innovation, low energy buildings and intermediaries in Europe: Systematic case study review. *Energy Effic.* **2017**, *11*, 31–51. [CrossRef]
86. Ferrara, M.; Monetti, V.; Fabrizio, E. Cost-Optimal Analysis for Nearly Zero Energy Buildings Design and Optimization: A Critical Review. *Energies* **2018**, *11*, 1478. [CrossRef]
87. Wang, Y.; Kuckelkorn, J.; Liu, Y. A state of art review on methodologies for control strategies in low energy buildings in the period from 2006 to 2016. *Energy Build.* **2017**, *147*, 27–40. [CrossRef]
88. Sartori, I.; Hestnes, A.G. Energy use in the life cycle of conventional and low-energy buildings: A review article. *Energy Build.* **2007**, *39*, 249–257. [CrossRef]
89. Heinstein, P.; Ballif, C.; Perret-Aebi, L.-E. Building Integrated Photovoltaics (BIPV): Review, Potentials, Barriers and Myths. *Green* **2013**, *3*, 125–156. [CrossRef]
90. Ikkurti, H.P.; Saha, S. A comprehensive techno-economic review of microinverters for Building Integrated Photovoltaics (BIPV). *Renew. Sustain. Energy Rev.* **2015**, *47*, 997–1006. [CrossRef]
91. Agathokleous, R.; Kalogirou, S.A. Double skin facades (DSF) and building integrated photovoltaics (BIPV): A review of configurations and heat transfer characteristics. *Renew. Energy* **2016**, *89*, 743–756. [CrossRef]
92. Raji, B.; Tenpierik, M.J.; van den Dobbelsteen, A. The impact of greening systems on building energy performance: A literature review. *Renew. Sustain. Energy Rev.* **2015**, *45*, 610–623. [CrossRef]
93. Zuo, J.; Pullen, S.; Rameezdeen, R.; Bennetts, H.; Wang, Y.; Mao, G.; Zhou, Z.; Du, H.; Duan, H. Green building evaluation from a life-cycle perspective in Australia: A critical review. *Renew. Sustain. Energy Rev.* **2017**, *70*, 358–368. [CrossRef]
94. Yuan, Y.; Yu, X.; Yang, X.; Xiao, Y.; Xiang, B.; Wang, Y. Bionic building energy efficiency and bionic green architecture: A review. *Renew. Sustain. Energy Rev.* **2017**, *74*, 771–787. [CrossRef]
95. Del Zendeh, E.; Wu, S.; Lee, A.; Zhou, Y. The impact of occupants' behaviours on building energy analysis: A research review. *Renew. Sustain. Energy Rev.* **2017**, *80*, 1061–1071. [CrossRef]
96. D'Oca, S.; Hong, T.; Langevin, J. The human dimensions of energy use in buildings: A review. *Renew. Sustain. Energy Rev.* **2018**, *81*, 731–742. [CrossRef]
97. Hong, T.; Chen, Y.; Belafi, Z.; D'Oca, S. Occupant behavior models: A critical review of implementation and representation approaches in building performance simulation programs. *Build. Simul.* **2018**, *11*, 1–14. [CrossRef]
98. Zhang, Y.; Bai, X.; Mills, F.P.; Pezzey, J. Rethinking the role of occupant behavior in building energy performance: A review. *Energy Build.* **2018**, *172*, 279–294. [CrossRef]
99. Ralegaonkar, R.; Gupta, R. Review of intelligent building construction: A passive solar architecture approach. *Renew. Sustain. Energy Rev.* **2010**, *14*, 2238–2242. [CrossRef]
100. Clements-Croome, D. Sustainable intelligent buildings for people: A review. *Intell. Build. Int.* **2011**, *3*, 67–86. [CrossRef]
101. Ahmad, M.W.; Mourshed, M.; Mundow, D.; Sisinni, M.; Rezgui, Y. Building energy metering and environmental monitoring—A state-of-the-art review and directions for future research. *Energy Build.* **2016**, *120*, 85–102. [CrossRef]
102. Morán, A.J.; Profaizer, P.; Zapater, M.H.; Valdavida, M.A.; Bribián, I.Z. Information and Communications Technologies (ICTs) for energy efficiency in buildings: Review and analysis of results from EU pilot projects. *Energy Build.* **2016**, *127*, 128–137. [CrossRef]
103. Dubois, M.-C.; Blomsterberg, Å. Energy saving potential and strategies for electric lighting in future North European, low energy office buildings: A literature review. *Energy Build.* **2011**, *43*, 2572–2582. [CrossRef]
104. Jakica, N. State-of-the-art review of solar design tools and methods for assessing daylighting and solar potential for building-integrated photovoltaics. *Renew. Sustain. Energy Rev.* **2018**, *81*, 1296–1328. [CrossRef]

105. Davidavičienė, V. *Research Methodology: An Introduction*; Springer: Singapore, 2018; pp. 1–23.
106. Espuny, M.; Neto, A.F.; Reis, J.S.D.M.; Neto, S.T.D.S.; Nunhes, T.V.; de Oliveira, O.J. Building new paths for responsible solid waste management. *Environ. Monit. Assess.* **2021**, *193*, 1–20. [CrossRef] [PubMed]
107. *Scopus—The Database with an Eye on Global Research*; Elsevier: Amsterdam, The Netherlands, 2019; Available online: https://www.elsevier.com/__data/assets/pdf_file/0012/81120/Scopus-A4-Poster-HI-RES.pdf (accessed on 15 December 2019).
108. Cristino, T.M.; Neto, A.F.; Costa, A.F.B. Energy efficiency in buildings: Analysis of scientific literature and identification of data analysis techniques from a bibliometric study. *Science* **2017**, *114*, 1275–1326. [CrossRef]
109. Price, D.J.D.S. *Little Science, Big Science*; Columbia University Press: New York, NY, USA, 1963.
110. Gilbert, R.O. *Statistical Methods for Environmental Pollution Monitoring*, 1st ed.; John Wiley & Sons: Hoboken, NJ, USA, 1987.
111. Sun, Y.; Latora, V. The evolution of knowledge within and across fields in modern physics. *Sci. Rep.* **2020**, *10*, 1–9. [CrossRef]
112. Robinson, J. A software system for laboratory experiments in image processing. *IEEE Trans. Educ.* **2000**, *43*, 455–459. [CrossRef]
113. Dabi, Y.; Darrigues, L.; Katsahian, S.; Azoulay, D.; De Antonio, M.; Lazzati, A. Publication Trends in Bariatric Surgery: A Bibliometric Study. *Obes. Surg.* **2016**, *26*, 2691–2699. [CrossRef]
114. Rasure, J.; Argiro, D.; Sauer, T.; Williams, C. Visual language and software development environment for image processing. *Int. J. Imaging Syst. Technol.* **1990**, *2*, 183–199. [CrossRef]
115. Cristino, T.; Lotufo, F.; Delinchant, B.; Wurtz, F.; Neto, A.F. A comprehensive review of obstacles and drivers to building energy-saving technologies and their association with research themes, types of buildings, and geographic regions. *Renew. Sustain. Energy Rev.* **2021**, *135*, 110191. [CrossRef]
116. Cristino, T.; Neto, A.F.; Wurtz, F.; Delinchant, B. Barriers to the adoption of energy-efficient technologies in the building sector: A survey of Brazil. *Energy Build.* **2021**, *252*, 111452. [CrossRef]
117. Lobato, C.G.; Cristino, T.M.; Neto, A.F.; Costa, A.F.B. Lean System: Analysis of scientific literature and identification of barriers for implementation from a bibliometric study. *Gestão Produção* **2021**, *28*, 4769. [CrossRef]
118. Johnson, R.A.; Wichern, D.W. *Applied Multivariate Statistical Analysis*, 6th ed.; Pearson Education: New Jersey, NJ, USA, 2007.
119. Manly, B.F.J.; Navarro-Alberto, J.A. *Multivariate Statistical Methods: A Primer*, 4th ed.; CRC Press: Boca Raton, FL, USA, 2016.

energies

MDPI

Article

Optimized EMS and a Comparative Study of Hybrid Hydrogen Fuel Cell/Battery Vehicles

Elkhatib Kamal [1,2,*] and Lounis Adouane [3]

1 Ecole Centrale Nantes—LS2N (Laboratoire des Sciences du Numérique de Nantes), 1 Rue de la Noë, CEDEX 3, 44000 Nantes, France
2 Department of Industrial Electronics and Control Engineering, Faculty of Electronic Engineering, Menoufia University, Menouf 32952, Menoufia, Egypt
3 Electrical Engineering Department, Université de Technologie de Compiègne, Heudiasyc CNRS, 60200 Compiègne, France; lounis.adouane@hds.utc.fr
* Correspondence: elkhatib.ibrahim@ec-nantes.fr or elkhateb.ibrahim@el-eng.menofia.edu.eg

Abstract: This paper presents a new Fuel Cell Fuel Consumption Minimization Strategy (FCFCMS) for Hybrid Electric Vehicles (HEVs) powered by a fuel cell and an energy storage system, in order to minimize as much as possible the consumption of hydrogen while maintaining the State Of Charge (SOC) of the battery. Compared to existing Energy Management Strategies (EMSs) (such as the well-known State Machine Strategy (SMC), Fuzzy Logic Control (FLC), Frequency Decoupling and FLC (FDFLC), and the Equivalent Consumption Minimization Strategy (ECMS)), the proposed strategy increases the overall vehicle energy efficiency and, therefore, minimizes the total hydrogen consumption while respecting the constraints of each energy and power element. A model of a hybrid vehicle has been built using the TruckMaker/MATLAB software. Using the Urban Dynamometer Driving Schedule (UDDS), which includes several stops and accelerations, the performance of the proposed strategy has been compared with these different approaches (SMC, FLC, FDFLC, and ECMS) through several simulations.

Keywords: HEV; electromobility; hybrid drive; fuel cell; energy consumption; optimization process

Citation: Kamal, E.; Adouane, L. Optimized EMS and a Comparative Study of Hybrid Hydrogen Fuel Cell/Battery Vehicles. *Energies* **2022**, *15*, 738. https://doi.org/10.3390/en15030738

Academic Editors: Michał Jasinski, Zbigniew Leonowicz and Arsalan Najafi

Received: 4 November 2021
Accepted: 21 December 2021
Published: 20 January 2022

1. Introduction

Currently, as the levels of air pollution caused by the consumption of fossil fuels reach alarming levels, a less polluting fuel source is being considered. Hydrogen has good characteristics to become the fuel of the future such as a high calorific value and clean combustion without producing pollutants, but hydrogen fuel cell technology has not yet been mastered. In 2014, although hydrogen-powered cars and buses began to hit the streets, the technology is not yet fully controlled [1].

A Hydrogen-powered Fuel Cell (HFC) is a non-polluting energy source that generates electricity through the chemical reaction of hydrogen and oxygen. These must be continuously fed to the Fuel Cell (FC) so that it can provide the electricity requirements of the load. In on-board applications, the hydrogen is usually stored in the system, while the oxygen is obtained from the atmosphere by a compressor. Due to the mechanical time constant of the compressor, the HFC system is characterized by a slow response time to load changes, and an auxiliary energy source, such as batteries or supercapacitors, is used to support it in the energy requirements of the load [2,3].

In the last few decades, there has been much research on new transport solutions due to the emission reduction objectives, among which Hybrid Electric Vehicles (HEVs) based on FCs are becoming an attractive technology. Energy management in these vehicles allows for improved fuel economy (hydrogen in this case), which is a very promising solution due to its high mass and volume energy density compared to other polluting sources such as gasoline and diesel. Hydrogen represents a non-toxic, non-polluting fuel with zero emissions, where the combustion only releases water [4].

The future of sustainable transportation is closely linked to the development of these vehicles. Indeed, these vehicles are much quieter, non-polluting, and more efficient than vehicles based on Internal Combustion Engines (ICEs) [5]. However, the use of hydrogen in a running vehicle poses problems: the need for a storage system, the power electronics converters, the choice of the traction motor, and finally, the management of the energy flows. This latter corresponds to the subject of study proposed in this paper, which aims to improve the consumption of the fuel while respecting the constraints imposed by the sources. The FCs are high-current and low-voltage sources. In order to use them in electric vehicle powertrains, it is necessary to use adapted static converters in order to increase the operating voltage. The global optimization of these powertrains requires the hybridization of the battery using an Energy Storage System (ESS) [6].

A control strategy, in the field of hybrid vehicles, is an algorithm whose objective is to regulate the power distribution coming from the different propulsion parts. The considered input data are the operating conditions of the vehicle such as the speed, acceleration, or torque requested by the driver. The output data recovered can be the activation or non-activation of certain components, the increase or reduction of the power output, or the modification of the operating ranges [7]. An energy management strategy can be implemented to satisfy different demands. The most common ones are to ensure the driver's power demand, to maintain the State Of Charge (SOC) of the battery, to reduce the number of starts, to optimize the efficiency of the drive train, or to reduce fuel consumption and pollutant emissions. In general, a compromise must be found to achieve several of these objectives simultaneously [8,9]. The strategy varies according to the type of vehicle and the type of engine. Indeed, if we consider Hybrid Electric Vehicles (HEVs), the objective is to have a final *SOC* equal to the initial SOC. If we consider a Plug-in-HEV (PHEV), it is preferable to recharge the battery from the grid rather than from the engine. Therefore, the final *SOC* should be as close as possible to the minimum tolerated threshold. A typical method is to run the vehicle in full electric mode until the minimum threshold is reached and then maintain the SOC. This method is called Charge Depleting/Charge Sustaining (CD/CS). However, it is not the most suitable method [10–12]. To obtain an optimal solution, the principle consists of progressively discharging the battery and reaching the minimum at the end of the trip [11]. This method is often found in the literature under the term blended mode. C. Silva et al. [13] studied the factors of PHEVs affecting the fuel consumption and emissions of this type of vehicle. Recently, with the development of on-board systems in cars, it is possible to access many other parameters. Technologies such as GPS, the Geographical Information System (GIS), and the Intelligent Transport System (ITS) make it possible to define a route and observe the traffic conditions in real time [11]. Several strategies need to know the cycle to work, but some can be used without prior knowledge. In general, these strategies have been parameterized on a predefined cycle, and their efficiency is better on this one. Based on this observation, several works have set up cycle recognition strategies, which use data from the present and the past to determine the appropriate strategy [14–16]. Some studies have also been interested in cycle prediction, using for example Markov chains [17–20]. In this case, the efficiency of the strategy will depend directly on its ability to predict future events. The integration of available information while using new technologies therefore appears to be an interesting way of reducing emissions and consumption [21,22]. Management strategies can be broken down into two main families, rule-based methods and optimization methods. There are many studies that have been conducted on each of these two families. Reference [23] provided an overview of the existing control strategies. This allows us to have an indication of the methods that have been studied. In [24,25], the EMS based on deterministic rules or based on fuzzy logic or Neural Networks (NNs) was detailed. Similarly, the optimization methods based on Model Predictive Control (MPC) to solve the energy management problem online were presented in [26]. An EMS based on the Pontryagin Minimum Principle (PMP) was introduced as an optimal control solution [27,28]. The global optimization requires the driving profile to be known a priori. Therefore, the results are only valid in the laboratory, but they can be used as a basis for comparison with other real-time strategies. A real-time optimization algorithm called the Equivalent Consumption Minimization Strategy (ECMS) was also developed in [29–35]. It introduces the concept of equivalent consumption. A MATLAB model was implemented for the study of this algorithm, which was subsequently developed using Stateflow on the hybrid vehicle model.

In this paper, a new Fuel Cell Fuel Consumption Minimization Strategy (FCFCMS) is proposed and compared with existing EMSs (such as the well-known State Machine Strategy (SMC), Fuzzy Logic Control (FLC), Frequency Decoupling and FLC (FDFLC), and the Equivalent Consumption Minimization Strategy (ECMS)). The proposed strategy is to minimize the hydrogen consumption of the fuel cell during a vehicle run, while respecting the constraints of the power of the fuel cell and the *SOC* of the ESS. The model of a hybrid vehicle and the performance of the proposed strategy is compared with these different approaches (SMC, FLC, FDFLC, and ECMS) through TruckMaker/MATLAB software

This paper is organized as follows: In Section 2, the problem formulation of multi-source energy management is presented. In Section 3, the modeling of the hydrogen fuel cell/battery hybrid vehicle is presented. Section 4 presents the proposed hybrid EMS. Finally, Section 5 presents the results on different driving cycles by applying the different methods. Conclusions and future prospects are presented in Section 6.

2. Formulation of the Optimization Problem

The problem of energy management consists of finding the best distribution of power among the energy sources of the system. The presence of the ESS introduces additional degrees of freedom in the supply of the required power. However, this distribution must satisfy the power demand of the Electric Motor (EM) and respect the operating constraints (power of the *FC*, define the *SOC* of the battery). The HEV considered in this paper has three energy sources, as illustrated in Figure 1. The energy management problem can be formulated as a dynamic optimization problem in which the system, represented as a dynamic Equation (1), is controlled in order to minimize a cost criterion (2) by respecting equality constraints (3) and inequality constraints (4) [36].

$$\dot{x}(t) = f(x(t), u(t), t) \tag{1}$$

$$\int_{t_o}^{t_f} \Psi(x(t), u(t), t)dt \tag{2}$$

$$\varphi(x(t), u(t), t) = 0 \tag{3}$$

$$\phi(x(t), u(t), t) \leq 0 \tag{4}$$

where $x(t)$ represents the state variables and $u(t)$ the control variables.

2.1. Overall Multi-Criteria Optimization Formulation

The overall objective of the developed algorithms is to optimize the energy distribution among the energy sources in order to minimize the fuel cell hydrogen consumption, while respecting the constraints of the power limits of the fuel cell and the *SOC* of the ESS. The most used criterion is the fuel consumption (fuel for thermal engines, hydrogen for FCs, etc.). This criterion is also called the cost function and is expressed as (2) [6]. In our study, we consider that the state variable is the *SOC* of the battery, and by choosing the power supplied by the *FC* as the control variable, the equation governing the dynamics of the system is in this case [37–39]:

$$S\dot{O}C = \frac{-i}{Q_{bat}}, \qquad i_{bat} = \frac{V_{bat} - \sqrt{V_{bat}^2 - 4R_{bat}P_{bat}}}{2R_{bat}}, \tag{5}$$

where Q_{bat}, P_{bat}, V_{bat}, and R_{bat} are the capacity, power, voltage, and resistance of the battery, respectively. On the other hand, the hybrid system must ensure the instantaneous power demand, which results in the following equality constraint [40]:

$$P_{demand} = P_{FC} + P_{bat} \tag{6}$$

where P_{demand} is the power demanded and P_{FC} is the fuel cell power.

2.2. The Constraints

The design of the hybrid system's components imposes maximum and minimum limits on the power exchanged and the energy levels that can be reached. These limits form the inequality constraints (4), expressed as follows [41]:

$$P_{FC,min} \leq P_{FC}(k) \leq P_{FC,max} \tag{7}$$

$$\Delta P_{FC,drop} \leq \frac{dP_{FC}(k)}{dt} \leq \Delta P_{FC,rise} \tag{8}$$

$$SOC_{min} \leq SOC \leq SOC_{max} \tag{9}$$

where $P_{FC,max}$ and $P_{FC,min}$ are respectively the maximum and minimum power supplied by the *FC* and SOC_{max} and SOC_{min} the maximum and minimum *SOC* that can be reached by the battery. Knowing that the response time of the *FC* is large compared to other energy sources, the HEV will not be able to withstand certain slopes of charge power (accelerations). To remedy this problem, the battery will provide a power whose maximum slope will not exceed a value to be determined (at the rise rate $\Delta P_{FC,rise}$ and at the drop rate $\Delta P_{FC,drop}$ of the *FC* power slope). Finally, an additional condition is imposed on the system in order to guarantee the maintenance of the battery's SOC. In this respect, we considered that the *SOC* of this element at the end of the studied time horizon is equal to its initial state [41]:

$$SOC(t_o) = SOC(t_f) \tag{10}$$

where $SOC(t_o)$ and $SOC(t_f)$ are the initial value and the final value of the SOC.

2.3. The Optimization Criteria

The objective of optimal control applied to energy management is the minimization of fuel consumption in a time interval $[t_o;t_f]$ on a given mission profile. This amounts to finding at each instant the power to be requested from the *FC*, in order to minimize the energy consumed from the fuel tank in this interval while verifying the constraints (7)–(10). The energy consumed from the fuel tank can be expressed as a function of the net power delivered by the cell and its total effort, so the cost function to be minimized is expressed by the following equation [42]:

$$\Psi(x(t), u(t), t) = \frac{P_{FC}}{\eta_{FC}(P_{FC})} + (SOC(t_o) - SOC(t_f))^2 \tag{11}$$

Thus:

$$J = \int_{t_o}^{t_f} \frac{P_{FC}}{\eta_{FC}(P_{FC})} + (SOC(t_o) - SOC(t_f))^2 \tag{12}$$

The *FC* efficiency (η_{FC}) can be determined by [41]:

$$\eta_{FC} = \frac{-2V_{bat}F}{N_{cell}\Delta h} \tag{13}$$

where F is the Faraday constant (in As/mol) and N_{cell} is the number of fuel cells. Let us introduce the Hamiltonian applied to the system defined by the following equation:

$$H(x(t), u(t), t) = \Psi(x(t), u(t), t) + \lambda(x(t), u(t), t) \tag{14}$$

with,

$$\dot{\lambda}(t) = \frac{\partial H(x^*(t), u^*(t), \lambda(t))}{\partial x} = -2(x(t) - x(t_o)) \tag{15}$$

where * means the optimal Hamiltonian solution.

$$\lambda(t) = \lambda_o - 2\int_{t_o}^{t_f} (x(t) - x(t_o))dt \tag{16}$$

where λ_o is the initial value of λ. Therefore, the optimization problem consists of finding the values of the power demanded at the *FC* that allows respecting the condition of maintaining the *SOC* with the minimization of the Hamiltonian function, as indicated by the following equation:

$$P_{FC}^{*} = arg\ minH(x^{*}(t), u^{*}(t), \lambda(t)) \tag{17}$$

For the implementation of the algorithm, the driving cycle is known a priori, which gives us the values of the final time, as well as the values of the initial and final load states; what remains to be found is the initial value of the adjoint state (the Lagrangian multiplier), since we have a unique value that is suitable for an optimal trajectory to maintain the *SOC* inside the required constraints. A search was performed by implementing a dichotomy algorithm using a graphical approach. Nevertheless, this value is not totally exact, and it changes with the driving cycle and the initial *SOC* of the storage element.

2.4. Hydrogen Consumption and Overall Efficiency

To present the optimization-based energy management strategies, first, the equation that defines the hydrogen consumption must be determined. The energy consumed from the fuel tank can be expressed as a function of the net power delivered by the cell (P_{FC}) and the total efficiency of the generator set ($\eta_{overall}$), according to the following equation.

$$W_{H_2[g/s]} = \frac{P_{FC}}{\eta_{FC}\Delta H} \tag{18}$$

where the calorific value Δ is a property of the fuels whose value depends on the state of the water produced in the chemical reaction (liquid or gas). Then, if we perform the numerical integration, the hydrogen consumption results are given by:

$$Cons_{H_2[g/s]} = \int_{t_o}^{t_f} \frac{P_{FC}}{\eta_{FC}\Delta H} \tag{19}$$

The overall efficiency ($\eta_{overall}$) is given by [43]:

$$\eta_{overall} = \frac{P_{demand}}{P_{FC}^{in} + P_{bat}^{in} + P_{SC}^{in}} \tag{20}$$

where P_{FC}^{in}, P_{bat}^{in}, and P_{SC}^{in} are the fuel cell power (input of the DC/DC converter), battery power (input of the DC/DC converters), and supercapacitor power, respectively.

2.5. Global Optimization

If the driving profile is known a priori, a global optimization method can be used to determine the optimum of (20). Thus, the minimization of the consumption is equivalent to determining the power profile of the *FC* that produces the minimum hydrogen consumption to achieve the driving profile. The objective function to be optimized is therefore:

$$Cons_{H_2[min]} = Min \sum_{0}^{t_f} P_{FC}(t_k)\Delta t_k = Min(P_{FC}(t_0)\Delta t_0 + \cdots + P_{FC}(t_f)\Delta t_f) \tag{21}$$

with the constraints given in (7) and (9) and:

$$\Delta SOC = SOC(t_f) - SOC(t_o) = \sum_{0}^{t_f} P_{SC}(t_k).\Delta t_k = 0 \tag{22}$$

Therefore, Equation (6) gives the power balance, but (7) and (9) limit the power of the *FC* and the *SOC* of the battery, while (22) indicates that the energy balance of the ES must be zero at the end of the drive cycle to ensure that the final *SOC* of the battery is equal to the initial SOC.

3. Modeling of the Hydrogen Fuel Cell/Battery Hybrid Vehicle

The studied vehicle structure (cf. Figure 1) consists of Proton Exchange Membrane Fuel Cells (PEMFC) as the main energy source connected to a step-up DC/DC converter and the ESS connected to a bi-directional step-down DC/DC converter, a DC bus connected via a DC/AC converter that supplies the mechanical traction. The power converters of the chopper type or direct current (DC/DC) and inverter or DC/AC converter types are used to connect the vehicle's electrical power devices, namely the electrical machinery, the *FC*, and the ES. The vehicle's electrical architecture specifies how the connection is made. The DC/AC converter transforms the power in the DC buses into AC power to the Electric Motor (EM) and controls the traction torque of the motor. In addition, it must be reversible to allow energy recovery during braking. The DC/DC converters control the power distribution between the *FC* and the ES. They allow adapting the voltages of the *FC* and the ES to the DC bus and limiting their currents. The structure of the studied hybrid fuel cell vehicle is shown in Figure 1. SimPowerSystems includes already built models of the *FC*, battery, and supercapacitor, and the formulas determined for these components help in understanding them.

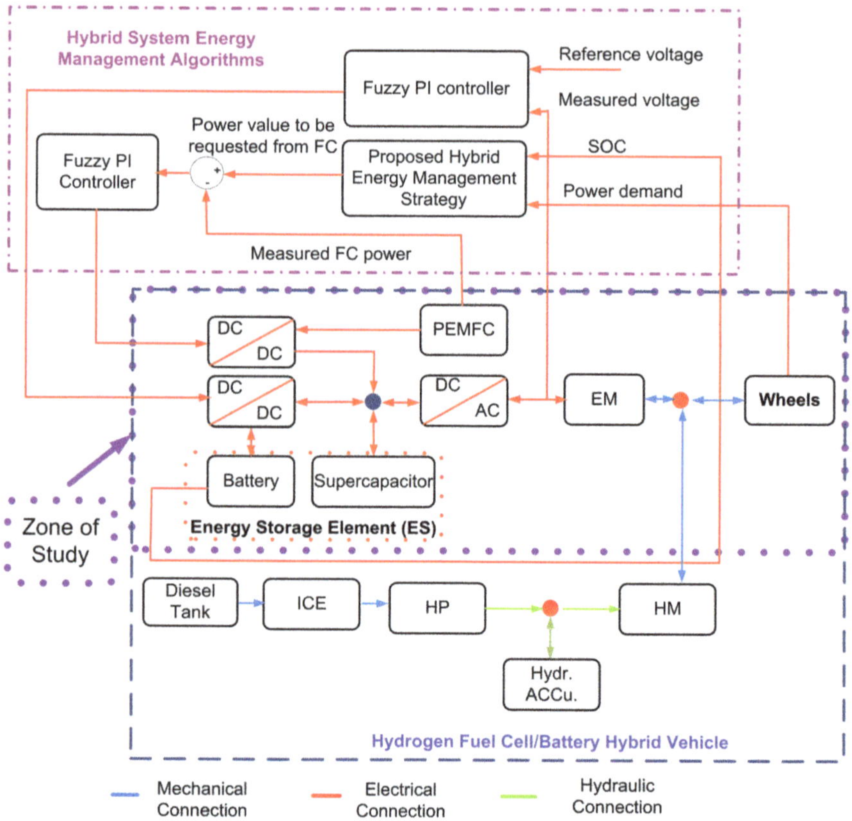

Figure 1. Hybrid system configuration and the powertrain power flows (ICE: Internal Combustion Engine, HP: Hydraulic Pump, HM: Hydraulic Motor, EM: Electric Motor).

3.1. The Static Model of PEMFC

Many works, such as [44–49], have proposed a static model describing the polarization curve of the PEM cell (cf. Figure 2) [50] as the sum of four terms: the theoretical open-circuit voltage E, the activation surge voltage V_{FC} (or activation drop: Region 1 in Figure 2), the ohmic surge voltage Vohm (or ohmic drop: Region 2), and the concentration surge voltage (or concentration drop: Region 3) [48].

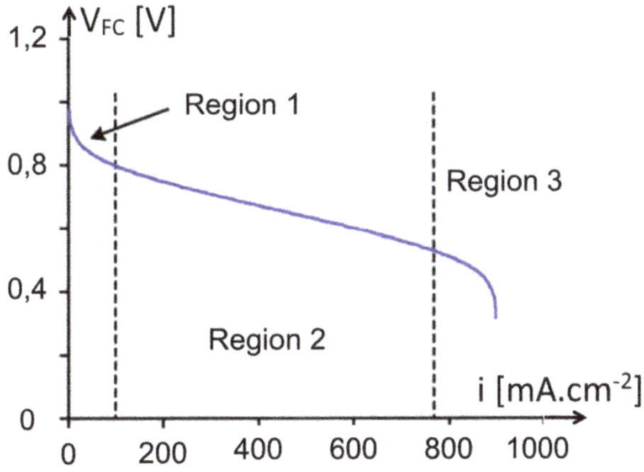

Figure 2. A typical PEMFC voltage–current curve.

This voltage–current characteristic fuel cell comportment can be defined as follows [51]:

$$V_{FC} = E - A\ log(\frac{I_{FC}+i_n}{i_o}) - R_m(I_{FC} + i_n) + B\ log(1 - \frac{I_{FC}+i_n}{i_L}) \tag{23}$$

where E is the reversible no loss voltage of the fuel cell; I_{FC} the delivered current; i_o the exchange current; A the slope of the Tafel line; i_L the limiting current; B the constant in the mass transfer term; i_n the internal current; R_m the membrane and contact resistances.

3.2. The Energy Storage Element

Two types of ESs are considered in the hybridization of vehicles: batteries and supercapacitors. Figure 3 shows the differences in the power and energy of some types of capacitors and batteries. The specific energy represents how much electrical energy per unit of mass can be stored by an energy source, while the specific power represents how much power per unit mass can be supplied by an energy source. The specific power also represents the ability of the source to provide or recover energy. The higher the specific power, the faster the source will supply or recover energy.

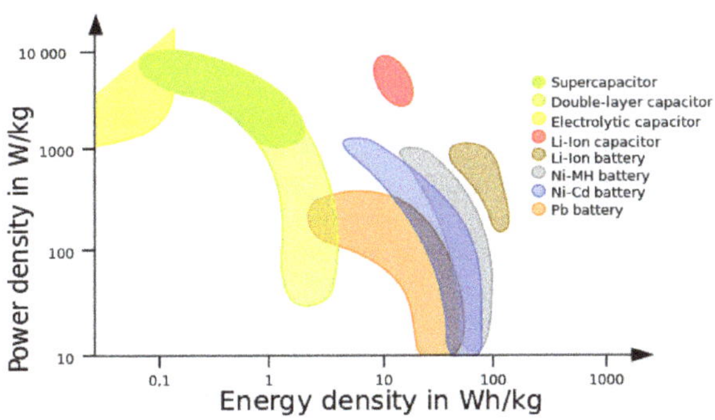

Figure 3. Power vs. energy density [52].

3.2.1. The Battery Model

Several battery models found in the literature use a simple model (cf. Figure 4a). This includes an electromotive force modeling the open circuit voltage of the battery, a capacitor modeling the internal capacity of the battery, and an internal resistance.

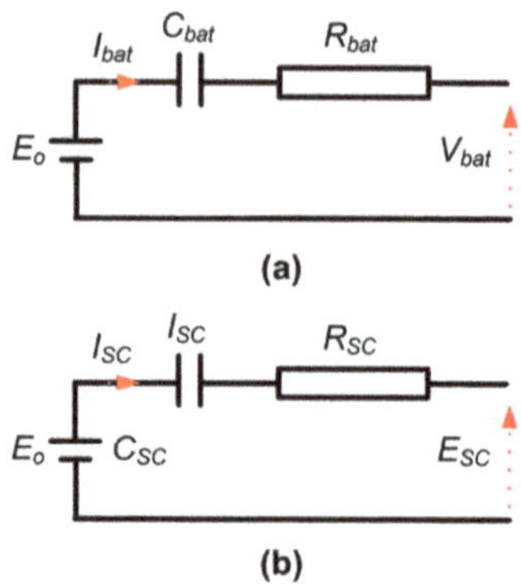

Figure 4. Electrical model of: (**a**) the battery, (**b**) the supercapacitor.

Thus, we have:

$$V_{bat} = E_o - R_{bat}I_{bat} - V_c \quad (24)$$

The SOC is defined as the ratio of the charge stored in the battery to the maximum charge capacity Q_{bat}:

$$\frac{d(SOC)}{dt} = -\frac{I_{bat}}{Q_{bat}} \quad (25)$$

In order to express I_{bat}, it should be noted that the instantaneous power delivered by the battery at load is:

$$P_{bat} = I_{bat}V_{bat} \quad (26)$$

We also determine the SOC of the battery by:

$$SOC = SOC(0) - \frac{1}{C_{bat}}\int I_{bat}dt \quad (27)$$

The SOC can be expressed according to the following equation:

$$\frac{d(SOC)}{dt} = -\frac{V_{bat} - \sqrt{V_{bat}^2 - 4R_{bat}P_{bat}}}{2R_{bat}Q_{bat}} \quad (28)$$

3.2.2. The Supercapacitor

Unlike the battery, the SC is mainly a power source with a low energy capacity. The energy stored in the SC is given by the following equation:

$$E_{SC} = \frac{1}{2}C_{SC}E_o^2 \quad (29)$$

where E_o is the voltage behind the impedance R_{SC} in the electrical diagram of the SC (cf. Figure 4b) and C_{SC} is the capacitance of the SC in Farads (F).

3.3. DC/AC and DC/DC Converter Models

The power converters allow adapting the currents and voltages between two electrical devices. The different types of converters are classified according to the type (AC or DC) and the characteristics of the electrical energy. Moreover, depending on the direction of the currents and voltages, the converters can have one or four according to the required current and voltage reversibilities. Finally, the converters have a local control, which generally integrates the current, voltage, and/or power current limitations, thus ensuring partial protection of the devices to which it is connected. The DC/DC converter and DC/DC converter are represented by an average-value model.

4. Hybrid System Energy Management Algorithms

This section presents the strategies and corresponding algorithms for the energy management of an HEV. The goal of power management in a hybrid FC HEV (FCHEV) is to determine the optimal power flow between the *FC* generator and the storage element in order to provide the power demanded by the load within the operating constraints and low hydrogen consumption, as well as providing high overall system efficiency. The energy management algorithms or strategies for a hybrid vehicle can be classified as given in [41]. Thus, we can divide the algorithms based on deterministic rules, based on fuzzy rules, real-time online optimization, and offline global optimization.

In this section, we develop rule-based algorithms, as well as an EMS based on optimization. The desired characteristic of the algorithms is the minimization of the hydrogen consumption during a given cycle. Moreover, the final *SOC* of the battery and supercapacitor should be equal to its initial *SOC*, which means that the balance of energy that the ES has gained or lost during the driving cycle must be zero at the end of the driving cycle. Thus, each of the algorithms to be performed will determine the reference current of the *FC* converter in the vehicle diagram shown in Figure 1, which minimizes the fuel consumption while respecting the operating constraints. In the case that the algorithm calculates a reference power, this will be transformed into a reference current. The proposed FC/battery/supercapacitor hybrid vehicular system shown in Figure 1 was implemented in MATLAB and SimPowerSystems software packages.

4.1. EMS Based on the State Machine Strategy

An EMS based on the State Machine Strategy (SMC) is based on [24] to distribute the required power between the fuel cell and the battery in order to maximize the efficiency of the system. It is a deterministic method based on rules that can contain many operating states to control the flow of energy between the different components of a hybrid fuel cell system [53]. Its implementation consists of eight states, as presented in Table 1. These rules are derived using the approach proposed in [54] and are based on the operational limits of the fuel cell and the battery in the system, the power demanded (P_{demand}) by the vehicle, and the *SOC* of the battery, where $P_{FC,max}$, $P_{FC,min}$, $P_{FC[req]]}$, and $P_{FC[opt]}$ are respectively the maximum, minimum, requested, and optimal power supplied by the *FC* and $P_{bat,min}$ is the minimum power of the battery.

Table 1. Power distribution among the different sources of the system for an EMS based on an SMS.

SOC	P_{demand}	$P_{FC[req]}$
High	$P_{demand} < P_{FC,min}$	$P_{FC[req]} = P_{FC,min}$
High	$P_{demand} \in [P_{FC,min}, P_{FC,max}]$	$P_{FC[req]} = P_{demand}$
High	$P_{demand} \geq P_{FC,max}$	$P_{FC[req]} = P_{FC,max}$
Normal	$P_{demand} < P_{FC[opt]}$	$P_{FC[req]} = P_{FC[opt]}$
Normal	$P_{demand} \in [P_{FC[opt]}, P_{FC,max}]$	$P_{FC[req]} = P_{demand}$
Normal	$P_{demand} \geq P_{FC,max}$	$P_{FC[req]} = P_{FC,max}$
Low	$P_{demand} < P_{FC,max}$	$P_{FC[req]} = P_{demand} - P_{bat,min}$
Low	$P_{demand} \geq P_{FC,max}$	$P_{FC[req]} = P_{FC,max}$

The power of the fuel cell is determined based on the *SOC* range of the battery and the load power. The implementation scheme is shown in Figure 5. Among the main drawbacks of this method is the need to take into account the control of the hysteresis (cf. Figure 6) during the switch between states, which affects the response of the EMS to change it in power demand. As shown in Figure 5, the output of the SMC algorithm is the reference power of the fuel cell, which is obtained by dividing the power determined by the algorithm by the efficiency of the converter, and the inputs are the *SOC* of the battery and the power demand. From all these data, the hydrogen consumption of the system is evaluated, and and this can be seen by the efficiency curve of the *FC*. The purpose of this implementation is to verify the guidelines set in the development of the strategy. The aim of the SMC is to decide the *FC* reference power with the state change. According to the hysteresis cycles for the *SOC* levels of battery and SCs, as shown in Figure 6, four states designed by the SMC were defined to obtain *FC* reference power $P_{FC[req]}$.

The simulation and validation results are shown in Section 5.

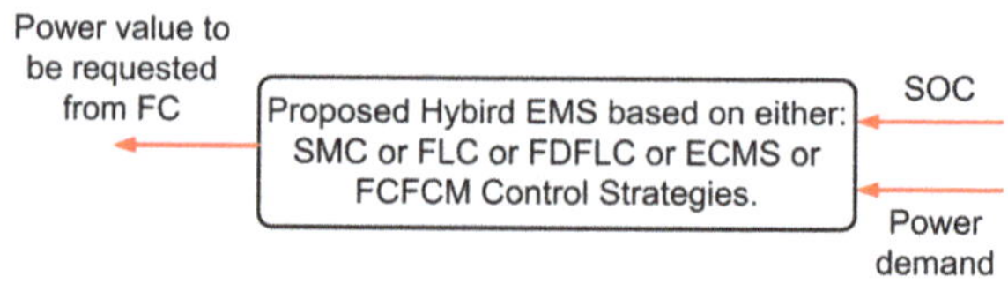

Figure 5. Representation of the implementation of the energy management strategy (SMC: State Machine Strategy, FLC: Fuzzy Logic Control, FDFLC: Frequency Decoupling and FLC, ECMS: Equivalent Consumption Minimization Strategy, FCFCMS: *FC* Fuel Consumption Minimization Strategy).

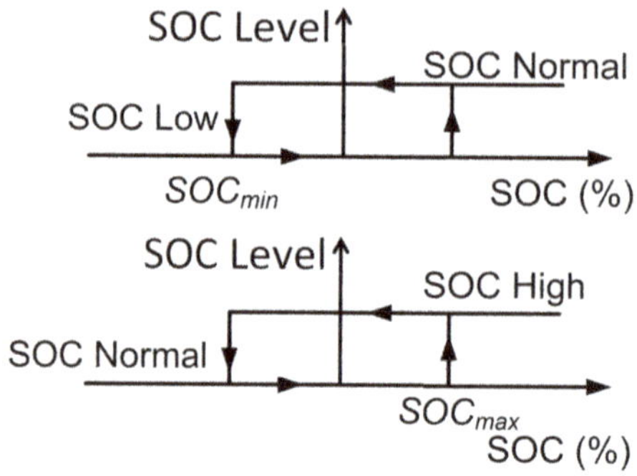

Figure 6. Hysteresis control.

4.2. EMS Based on Fuzzy Logic Rules

Based on the Fuzzy Logic Control (FLC) strategy presented in [25], an EMS was developed. This strategy presents a fast response to changes in the power demand compared to the SMC strategy due to the optimization procedure by adjusting the variation range of membership functions in order to reduce the consumption of hydrogen. The power of the fuel cell was obtained from the membership functions of the requested power and the *SOC* of the battery, as well as from the set of "if–then" rules. The diagram of EMS based on the FLC strategy is presented in Figure 5.

Trapezoidal membership functions were used, as shown in Figure 7, to design this approach. The rules derived from the decisions of the state machine are shown in Table 2, where H is High, M is Medium, L is Low, and VL is Very Low. Mamdani's fuzzy inference approach was used with the centroid method for defuzzification [55].

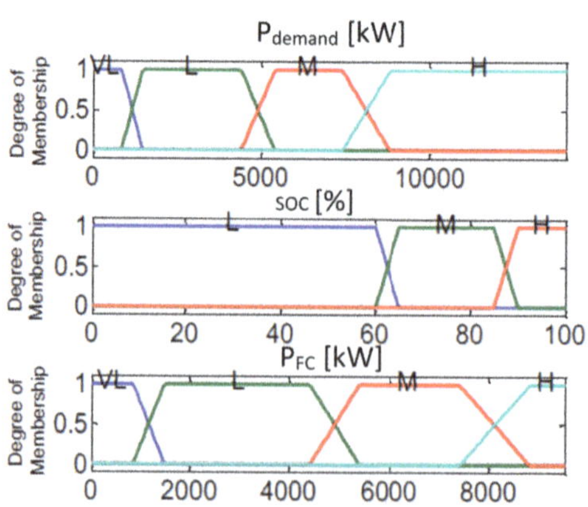

Figure 7. Membership functions for power demand, *SOC*, and battery power.

Table 2. Fuzzy logic rules assigned to the stack.

SOC	P_{demand}	$P_{FC[req]}$
H	VL	VL
H	L	L
H	M	ML
H	H	H
M	VL	VL
M	L	L
M	M	M
M	H	H
L	VL	L
L	L	M
L	M	H
L	H	H

4.3. Strategy Based on Frequency Decoupling and Fuzzy Logic Control

Based on the wavelet transform-FLC strategy [56], the Frequency Decoupling (FD) and FLC (FDFLC) is proposed for power splitting in the studied hybrid HEV. The FDFLC scheme was designed as the rule-based FLC, with the exception of a low-pass filter to cutout the high-frequency component of the power demand. The schematic of this strategy is shown in Figure 5. The FLC was developed based on [25]. This strategy presents better results than the FLC strategy (cf. Section 4.2), because here, the *FC* is not constantly subjected to variations coming from the power demand. This is due to the low-pass filter, which plays the role of rejecting all the high-frequency signals that could disturb the *FC*. This strategy, as the FLC strategy (cf. Section 4.2), is very useful in maintaining the *SOC*, but also ensures that the *FC* is moderately loaded. When evaluating the hydrogen consumption, both strategies consume almost the same amount of hydrogen.

4.4. Strategy Based in the Minimization of the Equivalent Consumption

This strategy is one of the real-time optimization approach control methods that is based on cost functions and used by many authors [29–34]. The objective is to reduce the hydrogen consumption by minimizing the hydrogen consumed by the fuel cell and the equivalent energy required for the final *SOC* to be equal to the initial *SOC* of the battery. The equivalent scheme is presented in Figure 5. The optimization problem to determine the equivalent hydrogen consumption can be formulated as follows:

Find the optimal solution:

$$x = [P_{FC}, \alpha_p, P_{bat}] \tag{30}$$

where P_{bat} and P_{FC} are the power of the battery and the *FC*, respectively. α_p is the penalty coefficient and is given in (32), which minimizes:

$$F = [P_{FC} + \alpha_p P_{bat}]\Delta T \tag{31}$$

where ΔT is the sampling time, under the constraints of the equalities (6), with:

$$\alpha_p = 1 - 2\mu\frac{(SOC - 0.5(SOC_{max} + SOC_{min}))}{SOC_{max} + SOC_{min}} \tag{32}$$

where μ is the *SOC* balance coefficient, with the limitations (7) and (9),

$$0 \leq \alpha_p \leq 100 \tag{33}$$

4.5. Proposed FC *Fuel Consumption Minimization Strategy*

In what follows, an online EMS is proposed. It aims at optimizing the equivalent hydrogen consumption in real time. To do so, we used a real-time optimization technique based on the ECMS. Still having the same objective, which is to minimize the hydrogen consumption in the system, a new optimization concept is introduced based on the maximization of the battery and supercapacitor energies at any given instant, while keeping the battery *SOC* and DC bus voltage (or supercapacitor SOC) within their operating limits, instead of minimizing the fuel consumption, which requires the evaluation of the equivalent fuel consumption. This strategy is called the *FC* Fuel Consumption Minimization Strategy (FCFCMS). As shown in Figure 8, the outputs of the FCFCMS algorithm are the battery reference power and the supercapacitor charge/discharge voltage.

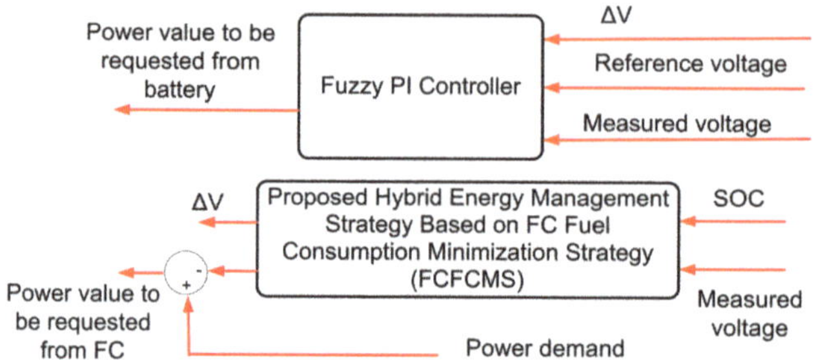

Figure 8. FC Fuel Consumption Minimization Strategy (FCFCMS).

The optimization problem to determine the equivalent hydrogen consumption can be formulated as follows:

Find the optimal solution:

$$x = [P_{bat}, \Delta V] \tag{34}$$

that minimizes:

$$F = -(P_{bat}\Delta T + \tfrac{1}{2}C_{SC}\Delta V^2) \tag{35}$$

where ΔV is the supercapacitor charge/discharge voltage and C_{SC} is the rated capacitance of the supercapacitor, with the following constraints:

$$P_{bat}\Delta T \leq (SOC - SOC_{min})V_{bat}Q \tag{36}$$

within the boundary conditions:

$$\begin{array}{l} P_{bat,min} \leq P_{bat} \leq P_{bat,max} \\ V_{DC,min} - V_{DC} \leq \Delta V \leq V_{DC,max} - V_{DC} \end{array} \tag{37}$$

where $V_{DC,min}$ and $V_{DC,max}$ are the minimum and maximum DC bus voltage (V_{DC}) and $P_{bat,max}$ and $P_{bat,min}$ are maximum and minimum power of the battery.

4.6. FC Fuel Consumption Minimization Based on Offline Optimization

If the driving profile is known, (16) must be optimized at each time to achieve the minimum fuel consumption. The optimization problem is defined as follows: The optimal solution is given by:

$$x = [P_{FC}(1), P_{FC}(2), P_{FC}(3), \ldots, P_{FC}(n)] \tag{38}$$

with $(k = 1, 2, 3, \ldots, n)$ and n is the number of samples ($n = T_p/\Delta T$), with T_p being the load profile duration. This minimizes the total fuel cell energy required for the whole power demand profile (F_p):

$$F_p = \sum_{k=1}^{n} P_{FC}(k)\Delta T \tag{39}$$

Minimizing F_p means minimizing the net fuel cell capacity (in Ampere-hours), hence H_2 consumption, with the following constraints:

$$y(k+1) \leq (SOC(t_o) - SOC_{min})V_{bat}Q \quad (40)$$

$$\sum_{k=1}^{n} P_{FC}(k) \geq n \times P_{FC,min} \quad (41)$$

with:

$$y(k+1) = y(k) + (P_{demand}(k) - P_{FC}(k))\Delta T \quad (42)$$

within the boundary conditions (7), where $y(k)$ is the battery energy after k samples. The minimum fuel consumption is obtained from the nominal fuel consumption ($Cons_{H_2[nom]}$) as:

$$Cons_{H_2[opt]} = \frac{F^{[opt]} Cons_{H_{2[nom]}}}{\sum_{k=1}^{n} P_{FC[nom]} \Delta T} \quad (43)$$

where $P_{FC[nom]}$ is the nominal fuel cell power.

5. Simulation and Validation Results

To evaluate the developed strategy, a simulation using MATLAB software for the simulation of hybrid vehicles on standard cycles was implemented. For comparison purposes, all the EMSs were designed based on the same requirements given in Table 3 with the same initial conditions (SOC initial value is $SOC(t_o)$ = [65%], battery, supercapacitor, and FC temperature are 30 °C, 25 °C and 40 °C, respectively). The supercapacitor and FC voltage are 270 V and 52 V, respectively, and were maintained for both cases. In this section, a precise comparative study of the performance of each EMS presented in Section 4 (namely the State Machine Control (SMC), EMS based on FLC, EMS based on Frequency Decoupling and Fuzzy Logic (FDFLC), EMS based on ECMS, and EMS based on FCFCMS), is made. The evaluation of each of the above EMSs was made on an appropriate and long standard driving profile. It corresponds to the Urban Dynamometer Driving Schedule (UDDS), which includes several stops and accelerations. Figure 9 shows this driving profile.

Figures 10–14 show the FC power, battery power, and supercapacitor power on the UDDS profile for the EMS based on SMC, FLC, FDFLC, ECMS, and FCFCMS, respectively. The SOC, H_2 consumption, and overall efficiency on the UDDS profile achieved with the HEV are presented in Figures 15–19 for the EMS based on SMC, FLC, FDFLC, ECMS, and FCFCMS, respectively.

Table 3. Energy management design requirements.

Design Requirements	Value
FC power $[P_{FC,min} - P_{FC,max}]$ (kW)	[1–10]
Energy power $[P_{bat,min} - P_{bat,max}]$ (kW)	[−1.2–4]
Energy SOC $[SOC_{min} - SOC_{max}]$ [%]	[60–90]
DC bus voltage $[V_{DC,min} - V_{DC,max}]$ (kW)	[250–280]

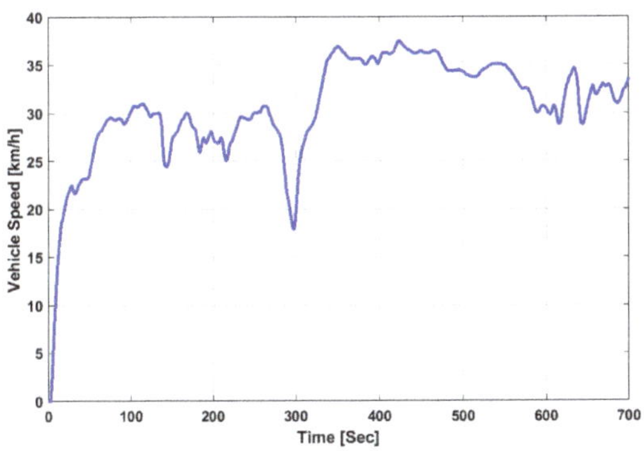

Figure 9. Speed profile of the UDDS standard velocity profile.

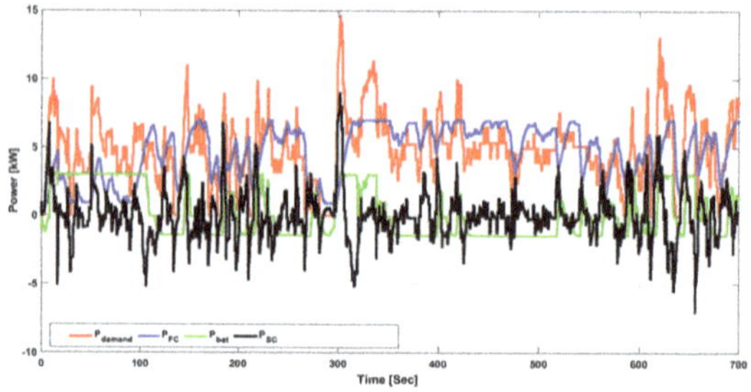

Figure 10. Power demand (P_{demand}), FC power (P_{FC}), battery power (P_{bat}), and supercapacitor power (P_{SC}) for the EMS based on SMC.

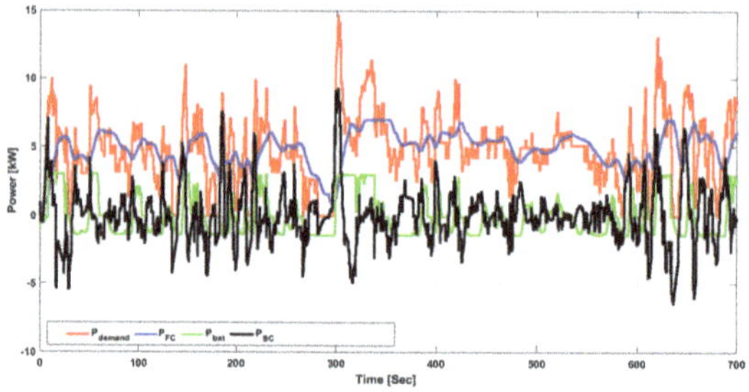

Figure 11. Power demand (P_{demand}), FC power (P_{FC}), battery power (P_{bat}), and supercapacitor power (P_{SC}) for the EMS based on FLC.

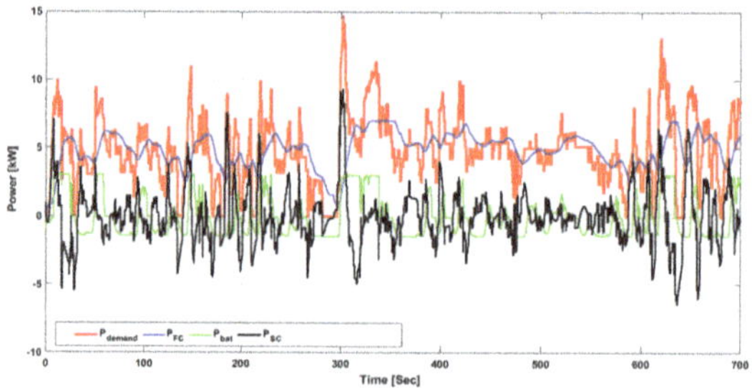

Figure 12. Power demand (P_{demand}), FC power (P_{FC}), battery power (P_{bat}), and supercapacitor power (P_{SC}) for the EMS based on FDFLC.

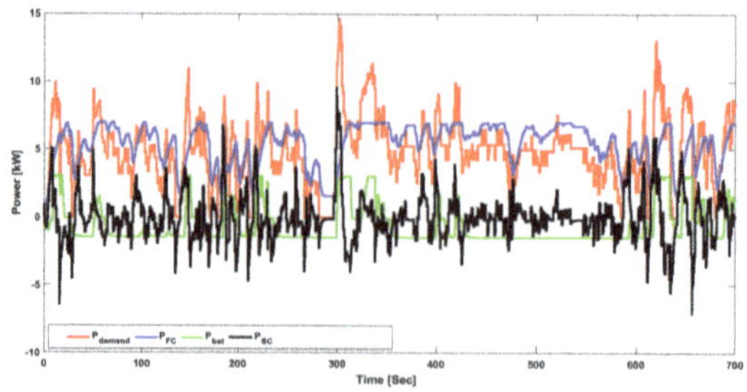

Figure 13. Power demand (P_{demand}), FC power (P_{FC}), battery power (P_{bat}), and supercapacitor power (P_{SC}) for the EMS based on the ECMS.

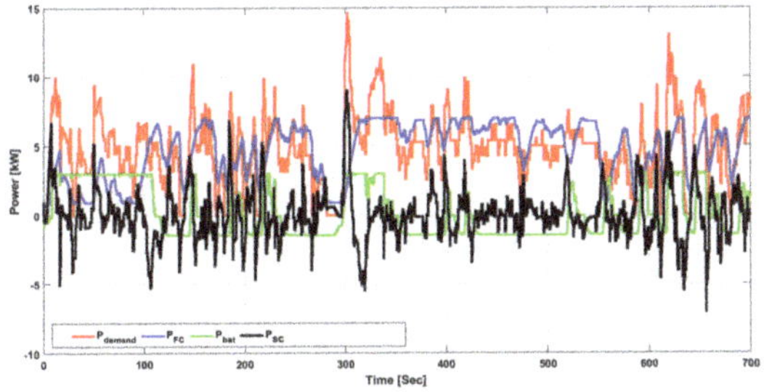

Figure 14. Power demand (P_{demand}), FC power (P_{FC}), battery power (P_{bat}), and supercapacitor power (P_{SC}) for the EMS based on the FCFCMS.

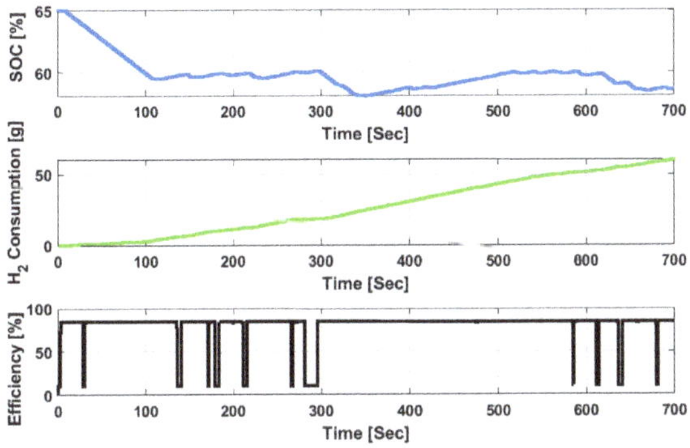

Figure 15. *SOC*, H_2 consumption, and overall efficiency for the EMS based on SMC.

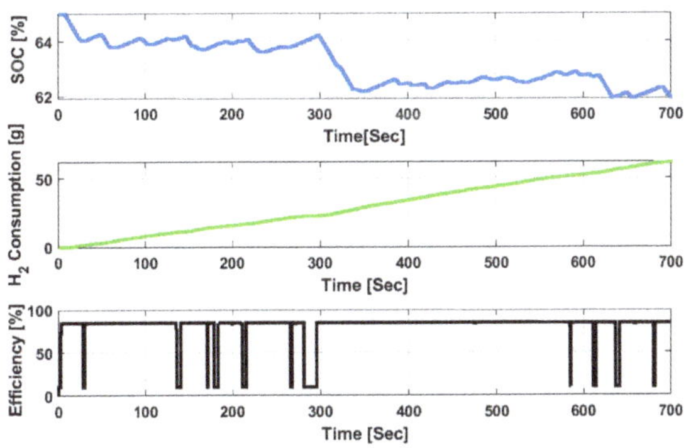

Figure 16. *SOC*, H_2 consumption, and overall efficiency for the EMS based on FLC.

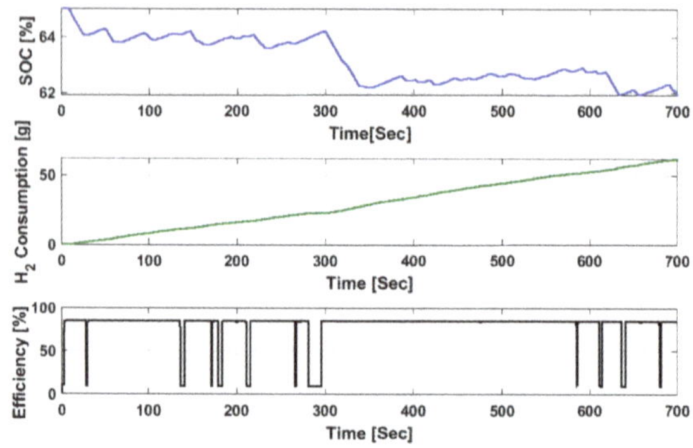

Figure 17. *SOC*, H_2 consumption, and overall efficiency for the EMS based on FDFLC.

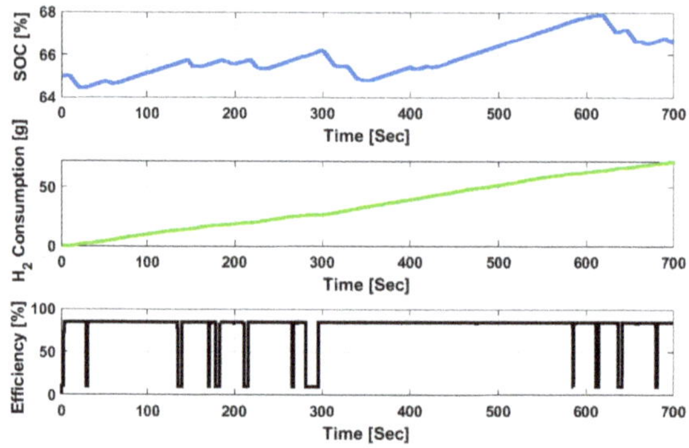

Figure 18. *SOC*, H_2 consumption, and overall efficiency for the EMS based on the ECMS.

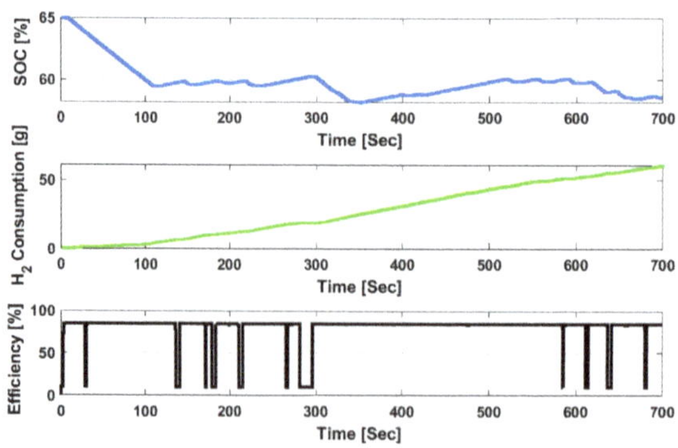

Figure 19. *SOC*, H_2 consumption, and overall efficiency for the EMS based on the FCFCMS.

Discussion

This paper tested the performance of the energy management strategies (ECMS and FCFCMS) based on optimal control compared to three other strategies (SMC, FLC, and FDFLC) while highlighting the results w.r.t. the fuel consumption, *SOC*, and overall efficiency. These strategies were examined on the UDDS driving profile. The results of this comparison are shown in Table 4.

Table 4. Overall performance obtained for the different studied energy management strategies. SMC: State Machine Strategy, FLC: Fuzzy Logic Control, FDFLC: Frequency Decoupling and FLC, ECMS: Equivalent Consumption Minimization Strategy, FCFCMS: *FC* Fuel Consumption Minimization Strategy; the initial value is $SOC(t_o)$ = [65%].

EMS Strategies	$SOC(t_f)$ (%)	H_2 Consumption (g)	Overall Efficiency (%)
SMC	59.10	59.10	85.20
FLC	62.01	60.92	84.93
FDFLC	62.50	61.50	84.09
ECMS	66.65	61.01	84.88
ECMS (offline)	66.65	58.20	...
FCFCMS	58.57	58.40	85.01
FCFCMS (offline)	58.57	57.90	...

The main criteria for the performance comparison were: H_2 consumption (g), the *SOC* (%) of the batteries/supercapacitors, and the overall efficiency (%). From the obtained results, it can be seen that energy management based on optimal control led to a good reduction of the hydrogen consumption while respecting the limits imposed by the sources, a good control of the *SOC*, and the stability of the *FC* during its operation. The FLC strategy presented a fast response to changes in the power demand compared to the SMC strategy. FLC provided a quite suitable structure compared to conventional control methods, especially for the systems composed of nonlinear behaviors, where an overall mathematical model is difficult to obtain. As expected, the lowest use of the battery energy was achieved with the frequency decoupling and fuzzy logic scheme, but at the expense of more fuel consumption and lower overall efficiency. In all the considered cases, optimal control-based management had the best performance and did not require a large amount of computation time. The FCFCMS performed slightly better compared to the ECMS in terms of efficiency and fuel consumption. The fact that it is an offline management strategy makes it less feasible except in applications where the driving path is known a priori such as street cars and high-speed roads. The following Table 5 compares the characteristics of the EMS algorithms, where H is High, DM is Difficulty Medium, L is Low, Imp is Implementation, ER is Easy to Realize, Comp is Complicated, and ET is the Execution time.

Table 5. Comparative table of the characteristics of the SMC, FLC, FDFLC, ECMS, and FCFCMS algorithms.

EMS Strategies	Imp	Setting	ET	$SOC(t_f)$	H_2 Consumption
SMC	simple	ER	L	not satisfied	M
FLC	DM	Comp	M	satisfied	M
FDFLC	DM	Comp	M	satisfied	The largest
ECMS	complex	DM	H	satisfied	M
FCFCMS	complex	DM	H	satisfied	The least

6. Conclusions

The paper presented a new Fuel Cell Fuel Consumption Minimization Strategy (FCFCMS) for Hybrid Electric Vehicles (HEVs). This strategy depends on the driving path and takes into account several performance criteria such as the slow dynamics of the *FC*, the reduction of fuel consumption, and good control of the storage element. In order to properly carry out this task, the work was divided into the following steps: (i) the definition of the main formulas that govern the operation of the system components, namely: the *FC*, the ES (the supercapacitor and the battery), the vehicle and its powertrain; (ii) the modeling of the hybrid vehicle; (iii) the implementation of the control strategies. A model of a hybrid vehicle was built using the TruckMaker/MATLAB software. Using the Urban Dynamometer Driving Schedule (UDDS), which includes several stops and accelerations, the performance of the proposed strategy was compared with these different approaches (SMC, FLC, FDFLC, and ECMS) through simulations. The results of this paper support that the proposed strategy: (i) is simple and more robust to changes of the power demand; (ii) increases the overall vehicle energy efficiency; (iii) minimizes the total hydrogen consumption and respects the constraints of each energy and power elements compared to the strategies based on SMC, FLC, FDFLC, and the ECMS. In a future work, it will be possible to set up an intelligent energy management strategy which will determine in a first step the priority of operation of the batteries and, in a second step, the type of strategies to adopt. This strategy will be further developed around a learning system in order to allow the system to make decisions according to the actual behavior of the system.

Author Contributions: Conceptualization, E.K. and L.A.; methodology, E.K. and L.A.; software, E.K. and L.A.; validation, E.K. and L.A.; formal analysis, E.K. and L.A.; investigation, E.K. and L.A.; resources, E.K. and L.A.; data curation, E.K. and L.A.; writing—original draft preparation, E.K. and L.A.; writing—review and editing, E.K. and L.A.; visualization, E.K. and L.A.; supervision, E.K. and L.A.; project administration, E.K. and L.A.; funding acquisition, E.K. and L.A. All authors have read and agreed to the published version of the manuscript.

Funding: This research received no external funding.

Institutional Review Board Statement: Not applicable.

Informed Consent Statement: Not applicable.

Data Availability Statement: Not applicable.

Conflicts of Interest: The authors declare no conflict of interest.

Abbreviations

HFC	Hydrogen-powered Fuel Cell
FC	Fuel Cell
HEVs	Hybrid Electric Vehicles
ICE	Internal Combustion Engine
ESS	Energy Storage System
SOC	State Of Charge
CD/CS	Charge Depleting/Charge Sustaining
GIS	Geographical Information System
EMS	Energy Management Strategy
NN	Neural Network
MPC	Model Predictive Control
PMP	Pontryagin Minimum Principle
ECMS	Equivalent Fuel Consumption Minimization
SE	Storage Element
FCS	Fuel Cell System
EM	Electric Motor
PEMFC	Proton Exchange Membrane Fuel Cells
HP	Hydraulic Pump
HM	Hydraulic Motor
SMC	State Machine Strategy
FLC	Fuzzy Logic Control
FDFLC	Frequency Decoupling (FD) and FLC
FCFCMS	FC Fuel Consumption Minimization Strategy
UDDS	Urban Dynamometer Driving Schedule

References

1. Hwang, H.-Y. Developing Equivalent Consumption Minimization Strategy for Advanced Hybrid System-II Electric Vehicles. *Energies* **2020**, *13*, 2033. [CrossRef]
2. Khan, M.J.; Iqbal, M.T. Dynamic Modelling and Simulation of a Fuel Cell Generator, Fuel Cells Special Issue: Modelling of Fuel Cell Systems. *Fuel Cells* **2005**, *5*, 97–104. [CrossRef]

3. He, Y.; Miao, C.; Wu, J.; Zheng, X.; Liu, X.; Han, F. Research on the Power Distribution Method for Hybrid Power System in the Fuel Cell Vehicle. *Energies* **2021**, *14*, 734. [CrossRef]
4. Fueling the Future Hydrogen Fuel Cell Vehicles in the 21st Century. Available online: http://serc.berkeley (accessed on 1 January 2021).
5. Chan, C.C.; Bouscayrol, A.; Chen, C.K. Electric, hybrid and fuel cell vehicles: architectures and modelling. *IEEE Trans. Veh. Technol.* **2010**, *59*, 589–598. [CrossRef]
6. Neffati, A. Stratégies de Gestion de l. Énergie Électrique d'un Système Multi-Source: Décision Optimisée pour véHicule éLectrique Hybride. Ph.D. Thesis, Université de Toulouse III Paul Sabatier, Toulouse, France, 2013.
7. Hajimiri, M.H.; Salmasi, F.R. A Fuzzy Energy Management Strategy for Series Hybrid Electric Vehicle with Predictive Control and Durability Extension of the Battery. In Proceedings of the IEEE Conference on Electric and Hybrid Vehicles, Pune, India, 18–20 December 2006; pp. 1–5.
8. Stockar, S.; Marano, V.; Canova, M.; Rizzoni, G.; Guzzella, L. Energy-optimal control of plug-in hybrid electric vehicles for real-world driving cycles. *IEEE Trans. Veh. Technol.* **2011**, *60*, 2949–2962. [CrossRef]
9. Wai, R.-J.; Jhung, S.-J.; Liaw, J.-J.; Chang, Y.-R. Intelligent Optimal Energy Management System for Hybrid Power Sources Including Fuel Cell and Battery. *IEEE Trans. Power Electron.* **2013**, *28*, 3231–3244. [CrossRef]
10. Gonder, J.; Markel, T. Energy Management Strategies for Plug-In Hybrid Electric Vehicles. In Proceedings of the SAE World Congress, Detroit, MI, USA, 16–19 April 2007; pp. 1–5.
11. Gong, Q.; Li, Y.; Peng, Z.-R. Trip-Based Optimal Power Management of Plug-in Hybrid Electric Vehicles. *IEEE Trans. Veh. Technol.* **2008**, *57*, 3393–3401. [CrossRef]
12. Wu, G.; Boriboonsomsin, K.; Barth, M.J. Development and Evaluation of an Intelligent Energy-Management Strategy for Plug-in Hybrid Electric Vehicles. *IEEE Trans. Intell. Transp. Syst.* **2014**, *15*, 1091–1100. [CrossRef]
13. Silva, C.; Ross, M.; Farias, T. Evaluation of energy consumption, emissions and cost of plug-in hybrid vehicles. *Energy Convers. Manag.* **2009**, *50*, 1635–1643. [CrossRef]
14. Lin, C.-C.; Peng, H.; Jeon, S.; Lee, J.M. Control of a Hybrid Electric Truck Based on Driving Pattern Recognition. In Proceedings of the 2002 Advanced Vehicle Control Conference, Hiroshima, Japan, 9–13 September 2002.
15. Ravey, A.; Blunier, B.; Lukic, S.; Miraoui, A. Control strategy of fuel cell hybrid electric vehicle based on driving cycle recognition. In Proceedings of the 2012 IEEE Transportation Electrification Conference and Expo (ITEC), Dearborn, MI, USA, 18–20 June 2012; pp. 1–5.
16. Recognition, R. Energy Management, and Plughybrid Electric Vehicles. Chalmers Publication Library Benefit of Route Recognition in Energy Management of Plug-In Hybrid Electric Vehicles. In Proceedings of the 2012 American Control Conference (ACC), Montréal, CA, USA, 27–29 June 2012.
17. Johannesson, L.; Asbogard, M.; Egardt, B. Assessing the Potential of Predictive Control for Hybrid Vehicle Powertrains Using Stochastic Dynamic Programming. *IEEE Trans. Intell. Transp. Syst.* **2007**, *8*, 71–83. [CrossRef]
18. Serrao, L.; Onori, S.; Rizzoni, G. A comparative analysis of energy management strategies for hybrid electric vehicles. *J. Dyn. Syst. Meas. Control.* **2011**, *133*, 031012. [CrossRef]
19. Souffran, F.; Miegeville, L.; Guerin, P. Simulation of RealWorld Vehicle Missions Using a Stochastic Marcov Model for Optimal Design Purposes. In Proceedings of the IEEE Vehicular Power and Propulsion Conference, Chicago, IL, USA, 6–9 September 2011; pp. 1–6.
20. Martinez, C.M.; Hu, X.; Cao, D.; Velenis, E.; Gao, B.; Wellers, M. Energy Management in Plug-in Hybrid Electric Vehicles: Recent Progress and a Connected Vehicles Perspective. *IEEE Trans. Veh. Technol.* **2016**, *66*, 4534–4549. [CrossRef]
21. Manzie, C.; Watson, H.; Halgamuge, S. Fuel economy improvements for urban driving: Hybrid vs. intelligent vehicles. *Transp. Res. Part Emerg. Technol.* **2007**, *15*, 1–16. [CrossRef]
22. Murphey, Y.L. Intelligent Vehicle Power Management: An Overview. In *Computational Intelligence in Automotive Applications*; Springer: Berlin/Heidelberg, Germany, 2008; Volume 190, pp. 169–190.
23. Zhang, P.; Yan, F.; Du, C. A comprehensive analysis of energy management strategies for hybrid electric vehicles based on bibliometrics. *Renew. Sustain. Energy Rev.* **2015**, *48*, 88–104. [CrossRef]
24. Kamal, E.; Adouane, L. Intelligent Energy Management Strategy Based on Artificial Neural Fuzzy for Hybrid Vehicle. *IEEE Trans. Intell. Veh.* **2017**, *3*, 112–125. [CrossRef]
25. Kamal, E.; Adouane, L. Hierarchical energy optimization strategy and its integrated reliable battery fault management for hybrid hydraulicelectric vehicle. *IEEE Trans. Veh. Technol.* **2018**, *67*, 3740–3754. [CrossRef]
26. Philippe, C.; Adouane, L.; Thuilot, B.; Tsourdos, A.; Shin, H.-S. Safe and online MPC for managing safety and comfort of autonomous vehicles in urban environment. In Proceedings of the 21st International Conference on Intelligent Transportation Systems (ITSC), Maui, HI, USA, 4–7 November 2018; pp. 300–306.
27. Ouddah, N.; Adouane, L.; Abdrakhmanov, R.; Kamal, E. Optimal energy management strategy of plug-in hybrid electric bus in urban conditions. In Proceedings of the 14th International Conference on Informatics in Control, Automation and Robotics, Madrid, Spain, 26–28 July 2017; pp. 304–311.
28. Kim, K.; Jeong, J.; Min, S.; Yang, H.; Vijayagopal, R.; Rousseau, A.; Cha, S.W. A Component-Sizing Methodology for a Hybrid Electric Vehicle Using an Optimization Algorithm. *Energies* **2021**, *14*, 3147. [CrossRef]
29. Ouddah, N.; Adouane, L.; Abdrakhmanov, R. From offline to adaptive online energy management strategy of hybrid vehicle using Pontryagin's minimum principle. *Int. J. Autom. Technol.* **2018**, *19*, 571–584. [CrossRef]

30. Volkan, S.; Metin, G.; Seta, B. A novel ECMS and combined cost map approach for high-efficiency series hybrid electric vehicles. *IEEE Trans. Veh. Technol.* **2011**, *60*, 3557–3570.
31. García, P.; Torreglosa, J.P.; Fernandez, L.M.; Jurado, F. Viability study of a FC-battery-SC tramway controlled by equivalent consumption minimization strategy. *Int. J. Hydrogen Energy* **2012**, *37*, 9368–9382. [CrossRef]
32. Torreglosa, J.P.; Jurado, F.; García, P.; Fernández, L.M. Hybrid fuel cell and battery tramway control based on an equivalent consumption minimization strategy. *Control Eng. Pract.* **2011**, *19*, 1182–1194. [CrossRef]
33. Rodatz, P.; Paganelli, G.; Sciarretta, A.; Guzzella, L. Optimal power management of an experimental fuel cell/supercapacitor-powered hybrid vehicle. *Control Eng. Pract.* **2005**, *13*, 41–53. [CrossRef]
34. Kwon, L.; Cho, D.-S.; Ahn, C. Degradation-Conscious Equivalent Consumption Minimization Strategy for a Fuel Cell Hybrid System . *Energies* **2021**, *14*, 3810. [CrossRef]
35. Abdrakhmanov, R.; Adouane, L. Dynamic programming resolution and database knowledge for online predictive energy management of hybrid vehicles. In Proceedings of the 14th International Conference on Informatics in Control, Automation and Robotics, Madrid, Spain, 26–28 July 2017; pp. 132–143.
36. Govardhan, O.M. Fundamentals and Classification of Hybrid Electric Vehicles. *Int. J. Eng. Tech.* **2017**, *3*, 194–198.
37. Zheng, C.H.; Oh, C.E.; Park, Y.I.; Lim, W.S.; Xu, G. PMP-Based Power Management Strategy of Fuel Cell Hybrid Vehicles Considering Multi-Objective Optimization. *Int. J. Precis. Eng. Manuf.* **2013**, *14*, 845–853. [CrossRef]
38. Li, C.Y.; Liu, G.P. Optimal fuzzy power control and management of fuel cell/battery hybrid vehicles. *J. Power Sources* **2009**, *192*, 525–533. [CrossRef]
39. Zheng, C.H.; Kim, N.W.; Park, Y.I.; Lim, W.S.; Xu, G. The effect of battery temperature on total fuel consumption of fuel cell hybrid vehicles. *Int. J. Hydrogen Energy* **2013**, *38*, 5192–5200. [CrossRef]
40. Feroldi, D.; Serra, M.; Riera, J. Energy Management Strategies based on efficiency map for Fuel Cell Hybrid Vehicles. *J. Power Sources* **2009**, *190*, 387–401. [CrossRef]
41. Hankache, W. Gestion Optimisée de l'Énergie Électrique d'un Groupe Électrogène Hybride à Pile à Combustible. Ph.D. Thesis, Université de Toulouse, Toulouse, France, 2008; pp. 36–41.
42. Ouddah, N.; Boukhnifer, M.; Raisemche, A. Two Control Energy Management Schemes for Electrical Hybrid Vehicle. Presented at the SSD Systems, Signals & Devices, Hammamet, Tunisia, 18–21 March 2013.
43. Wróblewski, P.; Drożdż, W.; Lewicki, W.; Miążek, P. Methodology for Assessing the Impact of Aperiodic Phenomena on the Energy Balance of Propulsion Engines in Vehicle Electromobility Systems for Given Areas. *Energies* **2021**, *14*, 2314. [CrossRef]
44. Khan, N.J.; Iqbal, M.T. Dynamic modeling and simulation of a small wind-fuel cell hybrid energy system. *Renew. Energy* **2005**, *30*, 421–439. [CrossRef]
45. Kishor, N.; Mohanty, S.R. Fuzzy modeling of fuel cell based on mutual information between variables. *Int. J. Hydrogen Energy* **2010**, *35*, 3620–3631. [CrossRef]
46. Shan, Y.; Choe, S.Y. Modeling and simulation of a PEM fuel cell stack considering temperature effects. *J. Power Sources* **2006**, *158*, 274–286. [CrossRef]
47. Kamal, E.; Aitouche, A. Fuzzy Observer-Based Fault Tolerant Control Against Sensor Faults for Proton Exchange Membrane Fuel Cells. *Int. J. Hydrogen Energy* **2020**, *45*, 11220–11232. [CrossRef]
48. Becherif, M.; Saadi, A.; Hissel, D.; Aboubou, A.; Ayad, M.Y. Static and dynamic proton exchange membrane fuel cell models. *J. Hydrocarb. Mines Environ. Res.* **2011**, *2*, 1–9.
49. Ham, S.-W.; Jo, S.-Y.; Dong, H.-W.; Jeongn, J.-W. A simplified PEM fuel cell model for building cogeneration applications. *Energy Build.* **2015**, *107*, 213–225. [CrossRef]
50. Zhang, C.; Li, W.; Hu, M.; Cheng, X.; He, K.; Mao, L. A Comparative Study of Using Polarization Curve Models in Proton Exchange Membrane Fuel Cell Degradation Analysis. *Energies* **2020**, *13*, 3759. [CrossRef]
51. Larminie, J.; Dicks, A. *Fuel Cell Systems Explained*, 2nd ed.; John Wiley & Sons: Hoboken, NJ, USA, 2003.
52. Available online: https://commons.wikimedia.org/wiki/File:Energiespeicher3.svg (accessed on 1 October 2021).
53. Fernandez, L.M.; Garcia, P.; Garcia, C.A.; Jurado, F. Comparison of control schemes for a fuel cell hybrid tramway integrating two DC/DC converters. *Int. J. Hydrogen Energy* **2010**, *35*, 5731–5744. [CrossRef]
54. Fernandez, L.M.; Garcia, P.; Garcia, C A.; Jurado, F. Hybrid electric system based on fuel cell and battery and integrating a single DC/DC converter for a tramway. *Energy Convers. Manag.* **2011**, *52*, 2183–2192. [CrossRef]
55. Kamal, E.; Adouane, L.; Abdrakhmanov, R.; Ouddah, N. Hierarchical and adaptive neuro-fuzzy control for intelligent energy management in hybrid electric vehicles. *IFAC-PapersOnLine* **2017**, *50*, 3014– 3021. [CrossRef]
56. Vural, B.; Boynuegri, A.R.; Nakirn, I.; Erdincn, O.; Balikci, A.; Uzunoglu, M.; Gorgun, H.; Dusmez, S. Fuel cell and ultra-capacitor hybridization: A prototype testbench based analysis of different energy management strategies for vehicular applications. *Int. J. Hydrogen Energy* **2010**, *35*, 11161–11171. [CrossRef]

energies

MDPI

Article

Long-Term Hydrogen Storage—A Case Study Exploring Pathways and Investments

Ciara O'Dwyer [1,2], Jody Dillon [2] and Terence O'Donnell [1,*]

1 School of Electrical and Electronic Engineering, University College Dublin, Belfield, D04 Dublin, Ireland; ciara.odwyer@energyreform.ie
2 Energy Reform, A98 Bray, Ireland; jody.dillon@energyreform.ie
* Correspondence: terence.odonnell@ucd.ie

Abstract: Future low-carbon systems with very high shares of variable renewable generation require complex models to optimise investments and operations, which must capture high degrees of sector coupling, contain high levels of operational and temporal detail, and when considering seasonal storage, be able to optimise both investments and operations over long durations. Standard energy system models often do not adequately address all these issues, which are of great importance when considering investments in emerging energy carriers such as Hydrogen. An advanced energy system model of the Irish power system is built in SpineOpt, which considers a number of future scenarios and explores different pathways to the wide-scale adoption of Hydrogen as a low-carbon energy carrier. The model contains a high degree of both temporal and operational detail, sector coupling, via Hydrogen, is captured and the optimisation of both investments in and operation of large-scale underground Hydrogen storage is demonstrated. The results highlight the importance of model detail and demonstrate how over-investment in renewables occur when the flexibility needs of the system are not adequately captured. The case study shows that in 2030, investments in Hydrogen technologies are limited to scenarios with high fuel and carbon costs, high levels of Hydrogen demand (in this case driven by heating demand facilitated by large Hydrogen networks) or when a breakthrough in electrolyser capital costs and efficiencies occurs. However high levels of investments in Hydrogen technologies occur by 2040 across all considered scenarios. As with the 2030 results, the highest level of investments occur when demand for Hydrogen is high, albeit at a significantly higher level than 2030 with increases in investments of large-scale electrolysers of 538%. Hydrogen fuelled compressed air energy storage emerges as a strong investment candidate across all scenarios, facilitating cost effective power-to-Hydrogen-to-power conversions.

Keywords: Hydrogen; renewable energy; investment planning; long-term storage

Citation: O'Dwyer, C.; Dillon, J.; O'Donnell, T. Long-Term Hydrogen Storage—A Case Study Exploring Pathways and Investments. *Energies* **2022**, *15*, 869. https://doi.org/10.3390/en15030869

Academic Editors: Zbigniew Leonowicz, Arsalan Najafi and Michał Jasinski

Received: 22 December 2021
Accepted: 19 January 2022
Published: 25 January 2022

1. Introduction

The decarbonisation of the energy system is a central pillar of climate change mitigation policies and is advancing at a great pace. Electrification of the heating and transport sectors coupled with greatly increased use of renewable generation are generally seen as critical to this decarbonisation effort. The EU has recently proposed to increase its target of energy use from renewable generation to 40% (from a previous target of 32%) as part of a package of measures to put the EU on track for a 55% reduction in carbon emissions by 2030 compared to 1990 levels, and zero net emissions by 2050 [1]. In Ireland, in 2020, 43% of electricity consumed was generated by renewable sources and a target has been set to increase the share of renewable generation on the grid up to 80% by 2030 [2]. At these high levels of variable renewable generation, maintaining the supply/demand balance becomes increasingly challenging as balancing challenges occur at vastly different time-scales, from minutes to seasons. Long periods of low renewable generation and high demand can occur at certain periods leading to under supply, while over supply, leading to high levels of curtailment become an issue for other periods. While many potential solutions exist, including storages of various durations, flexible generation and demand response, higher degrees of sector coupling can also facilitate more efficient solutions. However, finding optimal solutions is complex, and models must be able to capture the sector coupling, contain sufficient temporal and operational detail and optimise both short duration and seasonal storage over long durations, which is particularly challenging for typical investment models.

Hydrogen is gaining attention as an energy carrier of increasing importance with the potential to provide solutions to supply demand balancing in future high renewable energy systems. It has a wide range of potential applications and pathways [3,4] and can potentially facilitate high levels of sector coupling. "Green Hydrogen", produced by electrolysers using renewable electricity, can be used to avoid curtailment and is a potential fuel source for sectors which are difficult to decarbonise, with applications in industry, and significant potential as a transport fuel, particularly for heavy freight [5]. Hydrogen can also offer zero emissions options in the heating sector, with both the co-generation of heat and electricity via fuel cells, and the deployment of Hydrogen gas boilers emerging as likely candidates to meet future heating needs [6,7]. Hydrogen also has power-to-gas-to-power applications where it has significant potential to provide a clean fuel source for centrally dispatchable generation plants which will still remain important in high renewable systems to provide valuable system services and back up generation when renewable generation is not available. As regards Hydrogen-based generation plant, co-generation of heat and electricity via large-scale fuel cells and Hydrogen fuelled gas turbines are both advancing. However, the efficiency of the combustion process compared to the electro-chemical process is poor [7], and the more efficient co-generation process relies on a local heat demand to take advantage of the increased efficiencies. An alternative option for electricity generation is Hydrogen fuelled compressed air energy storage (CAES), which offers efficiency gains compared to Power-to-Hydrogen-to-power options [8]. While Hydrogen as an energy carrier has a wide range of applications, and offers many advantages as a potentially zero-emission fuel, some of the technologies are still relatively immature and costs are high. It is estimated that billions of euro are required to develop the necessary infrastructure, and support the research and development required to advance emerging technologies and achieve the necessary economies of scale for cost effective generation, storage and transport of the fuel [9]. Despite the enormous costs involved, Hydrogen is receiving very significant interest as the push to reach net zero emissions by 2050 continues, partly due to its potential to store vast amounts of energy at relatively low cost per MWh which is significantly advantageous for seasonal storage [10]. The question then arises, given current cost projections, as to when Hydrogen-based technologies for generation and storage are likely to become commercially advantageous in the energy system. Answering this question requires consideration of the interplay between future projected generation and demand scenarios including increased demand from electrification of heating and transport, and cost projections for Hydrogen-based technologies and their more conventional alternatives. In this work we attempt to answer this question using an optimal investment planning approach which determines the set of technologies which minimize system costs including investment and operational costs.

Modelling the future development of the energy system to achieve decarbonisation goals is challenging for a number of reasons. The potential roles of variable renewables, Hydrogen and long term underground storage give rise to increased interactions across timescales, energy sectors and regions, resulting in a number of modelling challenges. While Hydrogen generation and storage has significant potential value in the future system to facilitate the electrification of sectors such as industry, heating and transport, and can play an important role in system balancing across different time-scales (from minutes to seasons), adequate system models grow in scale and complexity. To effectively capture this potential value, long term investment models must include more operational detail and more energy sectors making them more difficult to solve. Additionally, large shares of variable renewable technologies give rise to increased need for flexibility. Adequate temporal and operational detail is therefore essential to capture these flexibility needs and the value flexible technologies can have in meeting these. For example system reserve and inertia which are required to manage short term variations in renewable generation and which should influence longer-term technology investments. In addition, there is a need to optimise long term storage investments and utilisation where energy can be arbitraged across seasons. Long-term investment models thus need to simultaneously allow short term operational issues to influence long term investments while also optimising long term storage investments.This is a challenging proposition because the resulting optimisation models must cover multiple energy sectors over long periods of time at relatively high temporal resolution resulting in a very large problem that is not easily solved.

A number of recent studies have presented energy system investment models, with a focus on Hydrogen technologies. For example, Sgobbi et al. [11] use a TIMES model of the EU28 energy system to assess the role of Hydrogen in a future decarbonised Europe. Yue et al. [12] explore optimal pathways to a 100% renewable energy system in Ireland, again using a TIMES model and a whole energy system approach. While the TIMES model used for both studies contains a high level of detail in terms of end uses and technologies, just 12 time slices are used. The coarse temporal resolution will not be capable of addressing short-term flexibility needs and curtailment will be underestimated. Optimisation of seasonal storage is also not addressed. Indeed underground Hydrogen storage is not included as an investment option. He et al. [13] develop a generalised modelling framework for co-optimising energy system investments and operations across the power and transport sectors and supply chains of electricity and Hydrogen. Representative weeks with an hourly resolution are used, allowing for the short-term flexibility needs of the system to be addressed. However, once again this approach does not allow for the optimisation of long-duration seasonal storage, and large-scale underground Hydrogen storage is not included as an investment option. Power-to-gas-power offers power systems a high degree of flexibility, but the poor round trip efficiency reduces the viability of such investments. However, Hydrogen fuelled CAES improves the efficiency. To the authors knowledge, no previous studies have included Hydrogen fuelled compressed air energy storage (CAES) as an investment option in low carbon energy system investment models. Standard energy system models are not capable of addressing all the challenges of future system operations with very high shares of variable renewable generation.

This paper presents a case study exploring optimal investments in Hydrogen generation and underground storage alongside other conventional generation and storage technologies in an energy system with a very high share of renewable generation (Ireland). A range of different technologies are considered as investment options. For the conversion of electricity to Hydrogen, electrolysers are considered and for the conversion of Hydrogen to electricity Hydrogen-based Open Cycle Gas Turbines and Hydrogen-based Compressed Air Energy Storage plants are considered. Investment in technologies will of course be influenced by various factors such as technology costs, the costs of alternatives such as conventional fossil fuels, and the costs of renewable generation (as the source of green Hydrogen). The influence of these factors on final investments are explored through the use of a set of scenarios which consider the impact of Hydrogen technology costs and efficiency, Hydrogen demand and fuel and carbon costs, on the uptake and demand for "green Hydrogen" generation and storage investments. A base level of demand for Hydrogen in the transport and industrial sectors is assumed, but as investment is likely to be influenced by potential increased demand in other sectors, one of scenarios specifically looks at how an increased demand for Hydrogen through its use in the heating sector would influence investments. The problem of optimising both investments in and the operation of seasonal storage is dealt with by using enhanced representative periods, which are mapped to equivalent days across the year, the trajectory of the large-scale Hydrogen storage can be optimised as part of the overall solution. The contribution of this paper is the exploration of alternative pathways for "green Hydrogen" investments, examining, through a case study, different drivers for investment decisions using an energy system model with high operational and temporal detail. The case study also demonstrates the optimisation of long-term storage investments and operations. The importance of the operational detail and the synergistic impact on multiple investment decisions according to the level of detail included is also highlighted. The results also highlight some of the complementarities and competition which exists between different solutions such as between conventional generation, battery storage and Hydrogen storage and renewable generation. The overall objective of the study is to explore important drivers for the introduction of large-scale Hydrogen investments in an energy system which targets a high levels of variable renewable generation. Although this is explored in the context of a specific case study based on Ireland, the authors believe that the insights gained have relevance to other systems which target very high levels of variable renewable generation.

The remainder of the paper is organised as follows: Section 2 describes the methodology, outlining the scenarios explored, the test system and provides details of the Hydrogen conversion and storage modelling. Section 3 summarises the results of the simulations while Section 4 discusses the implications of these results, along with the limitations of the model and recommendations for future work. Section 5 concludes.

2. Materials and Methods

This Section outlines the methodology used to optimise the investment decisions and storage trajectories outlined in Section 3. The objective was to demonstrate alternative pathways for large-scale Hydrogen investments, including long-term storage. An investment model is solved for two study years (2030 and 2040) considering annualised capital costs with an assumed discount rate of 6%. A high level of operational detail facilitates a detailed representation of operational costs, including O&M costs, start-up costs, fuel costs, taxes and penalties. First, a description of the scenarios explored is provided which focus on future pathways where Hydrogen plays a significant role in the drive towards a net-zero energy system. A description of the test system along with the implementation of the investment model in the SpineOpt tool [14] is provided, along with some more specific modelling details of the Hydrogen conversion processes and long-term storage.

2.1. Scenarios

A total of six scenarios are explored for two future years (2030 and 2040). In addition, a further set of simulations are completed for the six scenarios (2030 only) with a lower level of operational detail, in order to explore the impact and highlight the importance of considering a high level of operational detail in investment models exploring the future role of Hydrogen.

The ENTSO-E TYNDP 2020 [15] Global Ambition scenario (GA) was chosen as a base scenario for this work. Wind, solar and load time series, as well as Hydrogen demand and fuel and carbon prices are all based on the TYNDP Global Ambition scenario. Generation capacities for technologies which are not included as investment options in this work (e.g., hydro, waste, biomass) are also taken from this scenario. The remaining five scenarios facilitate the exploration of alternative pathways for Hydrogen adoption and the integration of large-scale Hydrogen storage within future energy systems. The six modelled scenarios are described below:

Global Ambition (GA): The Global Ambition scenario is one of three scenarios considered in ENTSO-E TYNDP 2020 and is in line with COP21 targets. In this scenario, there is a focus on a centralised approach to the energy transition. This scenario has been chosen as the base case for this work.

High Fuel Price (HFP): The objective of this scenario is to explore the impact of fossil fuel prices on investment decisions. In the High Fuel Price scenario, the assumed fuel prices, including for "blue Hydrogen" (i.e., Hydrogen from methane reforming combined with carbon capture and storage) are increased by 30% from those assumed in the GA scenario.

Carbon prices are increased to 50/100 EUR /tonne in 2030/2040, respectively, compared to 35/80 €/tonne in the GA scenario.
Hydrogen Network (HN): This scenario assesses the impact of increased Hydrogen demand on technology investments. Here, Hydrogen is assumed to meet a portion of the country's heating demand, in addition to the predominant demand from transport and industrial Hydrogen assumed in the GA scenario. Reassigning natural gas pipelines is one of many options being considered for the bulk transport of Hydrogen [16]. A portion of the natural gas network is assumed to be reassigned to carry Hydrogen, facilitating a relatively low cost (for the end-user) conversion to Hydrogen space heating. The HN Hydrogen demand time-series includes space and water heating demand for 100,000 dwellings in 2030, and 500,000 by 2040. It is also assumed that heat pump uptake is reduced when an alternative low carbon heating solution is available to householders, with the electricity demand in the HN scenario updated to reflect this.
Technology Breakthrough (TB): In the Technology Breakthrough scenario, the impact of uncertain electrolyser prices and efficiencies are explored. Large cost reductions for electolysers are anticipated in the coming decades, in addition to improved efficiencies driven by the ongoing research and development and anticipated economies of scale. The TB scenario assumes investment costs at the lower end of projections [17,18] for both 2030 and 2040 (700/300 EUR /kW, respectively, compared to 1000/600 EUR /kW, respectively, assumed for the GA scenario). Modest efficiency improvements are also assumed.
Variable Renewable Energy (VRE): For "green Hydrogen" to compete with other low carbon solutions and to be adopted at scale, wide-scale infrastructure will be required, as well as low cost and efficient electrolysers. Cost effective "green Hydrogen" will also be reliant on low cost renewable energy. In the Variable Renewable Energy scenario, lower investment costs for both wind and solar generation are assumed, exploring the synergistic relationship between low cost renewable generation and Hydrogen generation and storage.
Restricted CAES (RC): Hydrogen fuelled compressed air energy storage offers a flexible potential investment for future high renewables energy systems, providing valuable peaking capacity and energy storage across different time scales, using both the compressed and stored air, and the stored Hydrogen as a fuel source. However, the locating of large-scale CAES plants is geographically restricted, relying on suitable underground storage (e.g., underground salt formations). In the Restricted CAES scenario, limits are imposed for CAES investments, exploring both the impact on alternative generation and storage solutions and the impact on overall Hydrogen investments when this investment option is limited.

2.2. Test System and Investment Model

The test system used for this case study is based on the All-Island power system of Ireland. The input data [19], along with the model [20], are both openly available. Existing power plants which are expected to be still operational are included in the base model. Additional capacities are also included for technologies which are not included as investment options, e.g., waste and biomass plant, which have capacities fixed at the levels assumed for the GA scenario. For variable renewable generation, installed capacities in the base model are based on GA levels. However, additional investments are also possible, allowing total installed capacities to increase, depending on the modelled scenario. Table 1 shows the capacities included in the base model (before investments are considered).

Table 2 shows all considered investment options and the capacity considered for each investment decision. Note that for energy storage investments, decisions in increments of 1 MWh are considered, and for renewable generation and batteries, investment decisions are made in increments of 1 MW. For plants with more complex efficiency curves, investments are less granular and standard sizing is assumed, with the conventional plant aligning with the ENTSO-E data. With a focus on very large-scale Hydrogen generation, a plant size of 100 MW has been selected for the electrolyser. For the OCGT and CCGT plant, efficiencies and costs are all based on those assumed for the ENTSO-E TYNDP 2020 Global Ambition scenario. Both power and energy capacity investments can be made independently for the batteries, with costs based on [21]. CCS is modelled as a post combustion carbon capture and storage unit. The plant is represented in Spine as two separate units with independent investment variables and associated annualised costs. The plant performance, in terms of fuel use, electricity output and emissions is captured for all operating points at an hourly resolution using the user constraint (see Section 2.3). Costs and performance are modelled as per [22] and plant operation is co-optimised as part of the overall problem, in order to minimise costs, with bypassing of the CCS unit possible. More details are provided for the electrolysers and CAES plant in Section 2.3.

Table 1. Starting installed capacities in the base model.

Plant Type	Installed Capacity (MW)
Gas	2327
Distillate	324
Biomass & Waste	587
Hydro	238
PHES	292
Wind (Onshore)	9607
Wind (Offshore)	3430
PV	860

Table 2. Investment options and capacity available for investment.

Plant Type	Capacity (MW)
CCGT	400
CCGT-CCS	400
OCGT	80
H-OCGT	80
H-CAES	160
Wind (Onshore)	1
Wind (Offshore)	1
PV	1
Battery (charge/discharge)	1
Electrolyser	100
Energy Storage	**Capacity (MWh)**
Battery (energy)	1
Compressed Air (energy)	1
Hydrogen (energy)	1

The investment model is run for 2030 and 2040 (with the 2040 base portfolio updated based on 2030 results and anticipated retirements) using the SpineOpt co-optimised operations and investments model. SpineOpt is an energy system modelling framework, implemented in Julia [20] and developed specifically for detailed operational and planning studies for future energy systems with high shares of variable renewable generation and complex cross-sectoral interactions. SpineOpt's generic structure consisting of nodes, units and connections allows SpineOpt models to be extended easily to include any number of sectors, commodities and energy conversion units. The flexible spatial, temporal and stochastic structures allow the model detail to be carefully tailored for each sector and region of interest, ensuring meaningful results while managing the computational burden. SpineOpt is open source and the complete code and documentation is available online [23].

The SpineOpt modelling framework implements enhanced representative days with ordering and weighting using the SpinePeriods companion model [24] which allows for the reduction of the model size while capturing arbitrage across the full model optimisation horizon. Each period of the model horizon (which can be flexibly defined by the user, e.g., day or week) is mapped to a corresponding representative period. Most problem variables such as unit flows and unit online statuses exist only for the representative periods, thus reducing the size of the overall optimisation problem. However, the state variables of long term storage nodes exist for every real (non-representative) interval over the full model horizon. For each real interval, the storage state variables interacts with the other problem variables from the corresponding mapped representative intervals. This allows the state of charge of long term storage to be optimised across the full optimisation horizon and co-optimised with short term operations.

The objective function is shown in Equation (1), which considers investment costs, O&M costs, start-up costs, fuel costs, taxes and penalties which are associated with the slack variables of the demand balance and reserve constraints. The Mixed Integer Programming (MIP) optimisation is solved using CPLEX 12.9 [25] and an optimality gap of 1%. It should be noted here, that SpineOpt is a flexible modelling framework that allows specification of a wide variety of energy systems in a very flexible way using only nodes, units and connections. The problem formulation in its most general form is presented in the Mathematical Formulation section of the online documentation [23]. Here we present the formulation of the specific case of the model implemented for this work.

$$\begin{aligned} \min obj = {} & v_{unit\_investment\_costs} + v_{storage\_investment\_costs} \\ & + v_{fixed\_om\_costs} + v_{variable\_om\_costs} + v_{fuel\_costs} \\ & + v_{start\_up\_costs} + v_{taxes} + v_{objective\_penalties} \end{aligned} \tag{1}$$

Unit and storage investment decision variables ($v_{units\_invested}$,$v_{storages\_invested}$) are included for all units with a defined $p_{unit\_investment\_cost}$ and for all nodes with a defined $p_{storage\_investment\_costs}$, which are included in this model as annualised investment costs with an assumed discount rate of 6%. The total unit and storage investment costs are shown below in Equation (2) and Equation (3).

$$v_{unit\_investment\_costs} = \sum_{(u,t)} v_{units\_invested}(u,t) \times p_{unit\_investment\_cost}(u,t) \tag{2}$$

$$v_{storage\_investment\_costs} = \sum_{(n,t)} v_{storages\_invested}(n,t) \times p_{storage\_investment\_cost}(n,t) \tag{3}$$

In SpineOpt, the temporal resolution of energy flows, unit online decisions and investment decisions can all be defined independently and can change by look-ahead time using temporal block objects. The investment temporal block has a resolution of 1 year while the remaining decision variables have a resolution of 1 hour, using 12 weighted representative days generated using SpinePeriods [24]. A further temporal block is used to define the mapping of the non-representative days to a representative day, which allows the trajectory of long-term storages to be considered and the storage investments to be optimised. This will be described in more detail in Section 2.3.

Time series for demand, wind and solar generation are all taken from ENTSO-E TYNDP 2020 Global Ambition scenario, using the 1984 climate year. Total annual Hydrogen demand is also taken from the Global Ambition scenario, which comes predominantly from the industrial and transport sectors. A weekly profile for the transport related Hydrogen demand based on [26] is applied to the annual estimate for Hydrogen transport demand for Ireland from the Global Ambition scenario. Unit constraints include minimum generation levels and minimum up and down times and start-up costs are included. System constraints include an inertia floor and primary and tertiary operating reserve requirements. In addition to being met by generating units, demand can also be met by demand side response (DSM) with an assumed variable operation and maintenance (VOM) cost of €500/MWh. 10% of the DSM capacity can also provide system operating reserves. In addition, 2150 MW of DC interconnectors are included −1450 MW to GB and 700 MW to France. GB and France are each represented as a single generating unit with a time varying VOM cost, representing the marginal price, with an average value matching the TYNDP 2020 prices. The VOM varies with net load as per the country specific matched times series for TYNDP 2020 Global Ambition. This allows the flows on the interconnectors to be approximated and reserve provision is also facilitated. Future work will use a full European model to estimate the country specific marginal prices.

2.3. Hydrogen Conversion and Storage

SpineOpt has been designed as a generic energy system modelling framework and it does not assume specific types of energy carriers or sectors. A wide variety of energy systems, technologies and transport physics can be implemented using the fundamental elements of nodes (representing balance, storage and demand), connections (representing transport) and units (representing conversions). Any number of sectors can be included and co-optimised within the model and arbitrary energy conversion units can be added. SpineOpt is a powerful tool when considering a high degree of sector coupling and when modelling emerging technologies, such as electrolysers and Hydrogen-fueled CAES. This Section provides more details of the Hydrogen technologies included in the model. These include electrolysers for the conversion of electricity to Hydrogen, CAES and Hydrogen turbines for electricity generation and both the underground Hydrogen and compressed air storage. Figure 1 shows a simplified version of the Hydrogen node (labelled "H2")as implemented in the SpineOpt model. In the diagram the SpineOpt objects of units are in red and nodes are in purple, while the black lines represent the relationships between the various objects on which the various model parameters are defined. The red arrows indicate the direction of flow. The Hydrogen node has an associated time varying demand, with the time series depending on the year and the scenario. Storage can be added to the node in SpineOpt by giving the node a state, and for existing storage, defining a node state capacity *node_state_cap*. For nodes with storage investments enabled, *node_state_cap* represents the storage capacity per storage investment, which is set at 1 MWh in this model. An importer unit (labelled "Importer_H2") allows "blue Hydrogen" to be imported to the Hydrogen node based on cost estimates for the relevant years. "Blue Hydrogen" is assumed to be an important transitional fuel while "green Hydrogen" scales up. As such, the importer capacity is sized to meet the Hydrogen demand in the 2030 GA scenario. Additional Hydrogen demand, including increases assumed for 2040 scenarios, must be met by the generation of "green Hydrogen". The Hydrogen node, and any invested storage capacity at the Hydrogen node, i.e., underground Hydrogen storage, is connected to the electricity node (labelled "ELEC_IE") via three different unit types: electrolysers, Hydrogen CAES and as an alternative, a Hydrogen gas turbine.

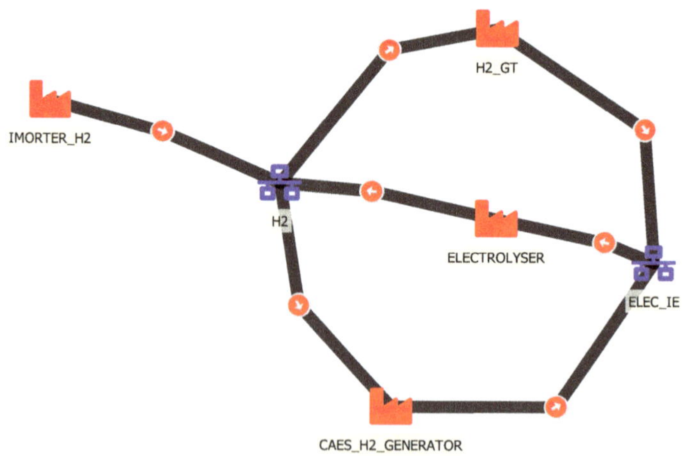

Figure 1. Simplified SpineOpt implementation of the Hydrogen network.

Units in SpineOpt may have any number of input flows from nodes and any number of output flows to nodes. Arbitrary affine constraints can be defined involving any or all of these flows to represent conversion processes. Electricity can be converted to Hydrogen via the electrolyser units which are included as an investment option. PEM electrolysers are assumed and a detailed operational model is included, with minimum and maximum load levels and efficiency which varies with input electrical energy as outlined in [27]. Here, the electrolyser efficiency curve is approximated using one of SpineOpt's generic conversion constraints. The *fix_ratio_out_in_unit_flow* constraint in its simplest form allows a linear relationship to be defined between the outgoing flow and the incoming flow from and to a unit (for the elctrolyser the flow of electricity from the electricity node, and the flow of Hydrogen to the Hydrogen node), using the parameter $p_{fix\_ratio\_out\_in\_unit\_flow}$. By including the $p_{fix\_units\_on\_coefficient\_out\_in}$ parameter, the varying efficiency of the electrolyser is captured (see Equation (4)). In SpineOpt, more complex efficiency curves can also be represented by defining an array of operating points for the unit, facilitating the decomposition of the flow variable in to multiple segments (i.e., incremental heat rates). Full details are provided in the documentation [23].

$$
\begin{aligned}
&\sum_{(u,n,d,t_{out}) \in (u,n_{out},:to\_node,t)} v_{unit\_flow}(u,n,d,t_{out}) \\
&== p_{fix\_ratio\_out\_in\_unit\_flow}(u,n_{out},n_{in},t) \\
&\times \sum_{(u,n,d,t_{in}) \in (u,n_{in},:from\_node,t)} v_{unit\_flow}(u,n,d,t_{in}) \\
&+ p_{fix\_units\_on\_coefficient\_out\_in}(u,n_{out},n_{in},t) \\
&\times \sum_{(u,t_{units\_on}) \in (u,t)} v_{units\_on}(u,t_{units\_on}) \\
&\forall (u,n_{out},n_{in}) \in ind(p_{fix\_ratio\_out\_in\_unit\_flow}), \\
&\forall t \in time\_slices
\end{aligned}
\tag{4}
$$

where v_{unit_flow} and v_{units_on} represent the flow and units online variables and the coefficients applied to the variables are the parameters $p_{fix\_ratio\_out\_in\_unit\_flow}$ and $p_{fix\_units\_on\_coefficient\_out\_in}$. The indices represent the units (u), nodes (n), direction of flow (d) and time-slice (t). Equation (4) is applied to all *unit, node, node* tuples which have a $p_{fix\_ratio\_out\_in\_unit\_flow}$ defined—i.e., it can also be applied to conventional generating units, such as the Hydrogen gas turbine, defining the relationship between the flow of Hydrogen to the gas turbine and the flow of electricity to the electricity node.

The CAES plant uses similar generic constraints to describe its operation. However, the CAES plant is slightly more complex, and is modelled as 3 separate units connected to 3 different nodes. Figure 2 shows a simplified implementation of the CAES plant in SpineOpt (for clarity, temporal blocks and stochastic structures are omitted). As per Figure 1, the units are shown in red, and the nodes are in purple. In addition, the yellow symbol represents user defined constraints. The black and grey lines represent various relationships between the model objects, with the black lines also representing flows, with the arrows indicating the direction of flow. Equation (4) is applied to the air compressor unit (*CAES_COMPRESSOR* in Figure 2) which defines the relationship between the flow of electricity from the electricity node and the flow of air to the compressed air storage node. On the generation side, the CAES plant is modelled as two additional units, *CAES_EXPANDER* and *CAES_H2_GENERATOR* in Figure 2, and once again, Equation (4) determines the relationship between the flow of compressed air and Hydrogen from their respective nodes and the flow of electricity from the two units, the combined flow being the power output of the plant. As the two generating units do not operate independently, a further user constraint is applied linking the output of the two units. SpineOpt's user constraint allows the user to define arbitrary linear constraints involving most of the problem variables. Equation (5) shows an instance of the generic user constraint with all the relevant parameters, which allows the relationships between the various flows of the CAES generating units to be captured, with the *uc* index representing the user constraints. In summary, the CAES plant is represented in Spine as three different units, with independent investment variables and associated annualised costs, and a compressed air node, which also has an associated investment variable and annualised investment cost. Equation (4) is used to quantify the efficiency of each unit component of the plant and the user constraint, Equation (5), links the operation of the CAES generating units (air expander and H2 generator). This methodology allows for a detailed representation of the CAES plant at an hourly resolution in terms of fuel use (both Hydrogen and compressed air) and electricity generation. The state of the compressed air node (i.e., energy content of the compressed air cavern) is also modelled at an hourly resolution and the plant operation is co-optimised as part of the overall investment problem.

$$+ \sum_{u,n \in unit\_\_node\_\_user\_constraint(uc),t} v_{unit\_flow}(u,n,d,t) \times p_{unit\_flow\_coefficient}(u,n,uc,t)$$
$$+ \sum_{u \in unit\_\_user\_constraint(uc),t} v_{units\_on}(u,t) \times p_{units\_on\_coefficient}(u,uc,t) == 0 \quad (5)$$
$$\forall uc, t \in constraint\_user\_constraint\_indices$$

As described previously, storage can be added to a node in SpineOpt by giving the node a state, and, for existing storages, defining a node state capacity *node_state_cap*. When investments in storage capacity at a given node are enabled, *node_state_cap* represents the storage capacity per storage investment, which is set at 1 MWh in this Model. However, optimising investments in long-term storage is challenging in typical investment models, which rely on reduced temporal representations to maintain tractability for very large problems. Long time horizons can be considered at a low resolution (e.g., one year at a daily resolution) which allows requirements for seasonal storage to be captured [28]. However, such a low resolution does not capture the flexibility needs of systems with high shares of variable generation, for which a high level of temporal detail is essential [29]. While it is possible to capture the systems short term flexibility needs with suitable selected representative periods, in order to simultaneously capture seasonal storage requirements, more advanced methodologies are required [30]. In this work, SpinePeriods [24] generates and orders representative days using an optimisation approach which approximates the annual duration curves [31]. The remaining, non-representative days are each mapped to a representative day, which is used to model the state of charge of a storage node over the full horizon, allowing the consideration of the arbitrage which takes place between the represented days. Thus an energy trajectory of the storage node can be generated and the energy capacity of the storage node optimised. A cyclic condition for the node state is also enforced, which ensures the node state at the end of the optimisation is at least as high as the initial value at the beginning of the optimisation. In SpineOpt, a map containing a representative day for each day in the horizon is included in a third temporal block (along with the investment temporal block, used for the investment decisions, and the representative days which are used for the operational decisions, see Section 2.2) which applies appropriate constraints to the energy level of the long term storage node for each day of the year. The full formulation is described in [30].

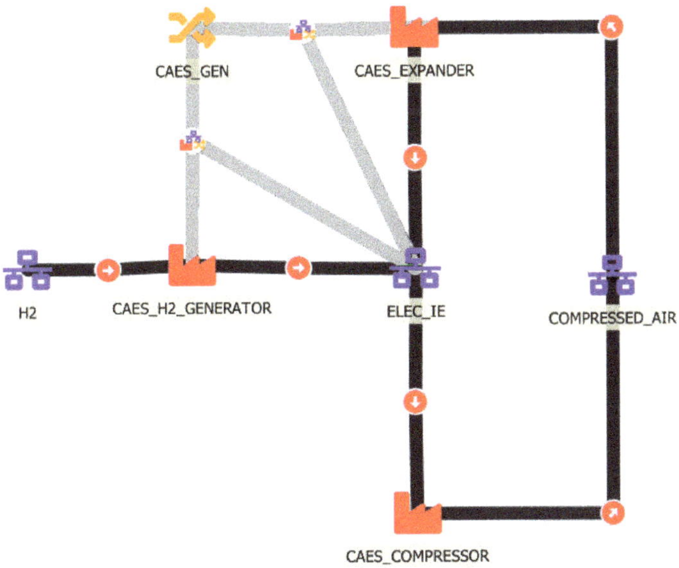

Figure 2. Simplified SpineOpt implementation of the CAES plant.

3. Results

The Results of the investment model are shown in this section, firstly for 2030, secondly for 2040 and finally a second set of 2030 results are presented with a lower level of operational detail, highlighting the impact such details have the results of long–term investment models.

3.1. 2030 Results

Figure 3 shows the capacities of dispatchable generation selected for investment by the 2030 investment models under the different scenarios. A high level of investments in electrolysers occur in the HFP and HN scenarios, with a lower level of investments occurring in the TB scenario triggered by the lower costs and higher efficiencies. However in the other three scenarios no investment in Electrolysers occur. Electrolyser investments are coupled with strong investments in Hydrogen CAES plant, and for the HN scenario, the flexibility and load shifting provided by the electrolysers and CAES suppresses battery investments.

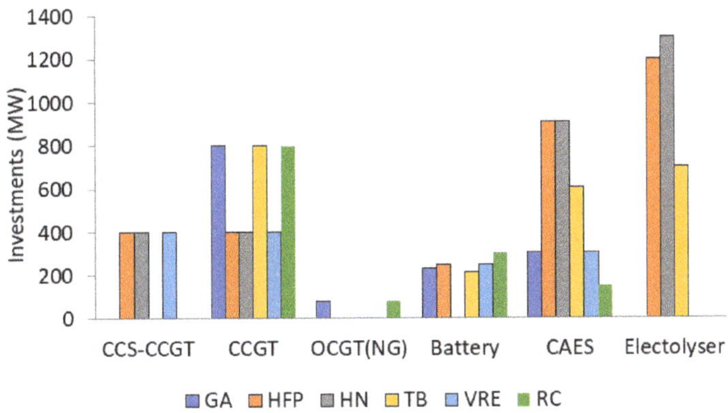

Figure 3. 2030 investments including dispatchable generation capacities and load (electrolysers). OCGT(NG) refers to natural gas powered OCGT.

In all modelled scenarios, high levels of wind generation are already included (as per the GA scenario). As shown in Figure 4a in 2030, variable renewable investments predominantly occur for PV, despite the relatively poor capacity factor for PV in Ireland. The installed capacity of PV in 2030 is modest (see Table 1) and a combination of favourable capital costs and a relatively high capacity factor in the summer months (when wind availability is lower) triggers a high level of investments. Interestingly, the HN scenario sees the lowest level of PV investments despite the additional Hydrogen demand. However, the additional Hydrogen demand included for this scenario is concentrated in the winter months (heating) when PV availability is poor, although alternative Hydrogen demand assumptions (e.g., higher transport or industry demand) would impact on this result.

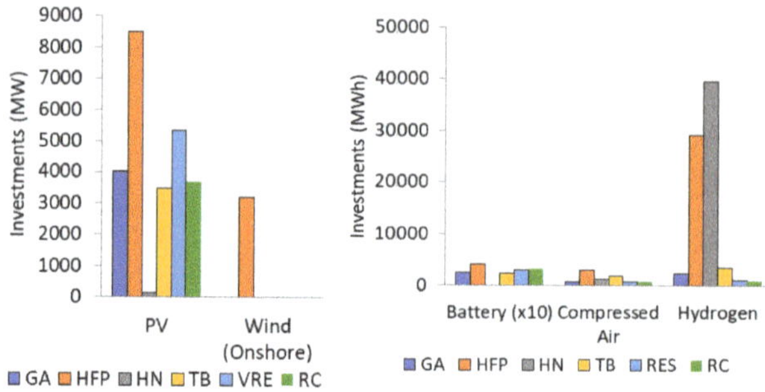

(a) (b)

Figure 4. (**a**) 2030 investments in variable renewable generation. (**b**) 2030 investments in energy storage capacity.

The energy storage investment capacities can be seen in Figure 4b. Hydrogen CAES emerges as principal technology selected for short/medium term balancing. However, only the scenarios with highest levels of electrolyser investments (HFP & HN) see large investments in the underground Hydrogen storage. The stored Hydrogen (primarily generated by the electrolysers) can be used both to meet the defined external Hydrogen demand and also by the dispatchable plant (i.e., CAES).

3.2. 2040 Results

Figure 5 shows the combined capacities of dispatchable generation selected by the 2030 and 2040 investment models. By 2040, lower costs for electrolysers are assumed as well as increased efficiencies. Fuel costs, and particularly carbon prices are also expected to rise which all favour "green Hydrogen" production. A very high level of investments in electrolysers occur in all scenarios. As with the 2030 scenarios, electrolyser investments are coupled with strong investments in Hydrogen CAES plant, and restricting investments in CAES (RC scenario) has a large impact on the remaining investment options. Investments in electrolysers and renewable generation decrease, and more gas plant is required (both OCGT and CCGT). While a share of "blue Hydrogen" import is still possible (i.e., 2030 demand levels), the imports are largely displaced by "green Hydrogen" production in most scenarios as an economically competitive alternative.

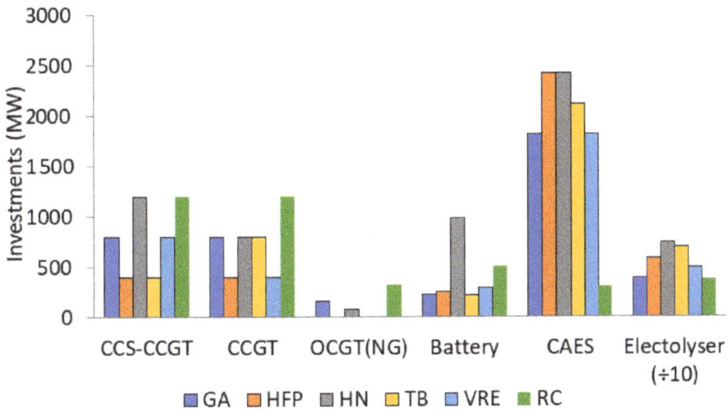

Figure 5. 2040 investments including dispatchable generation capacities and load (electrolysers).

As shown in Figure 6a, as with the 2030 scenarios, PV investments dominate, to complement the already high level of installed wind generation, although wind generation investments also occur for 5 of the 6 scenarios, the exception being the RC scenario. In the 2040 scenarios, with the large capacities of electrolysers, and the high demand for Hydrogen, we now see large investments in the underground Hydrogen storage to facilitate arbitrage across the year for all scenarios (see Figure 6b).

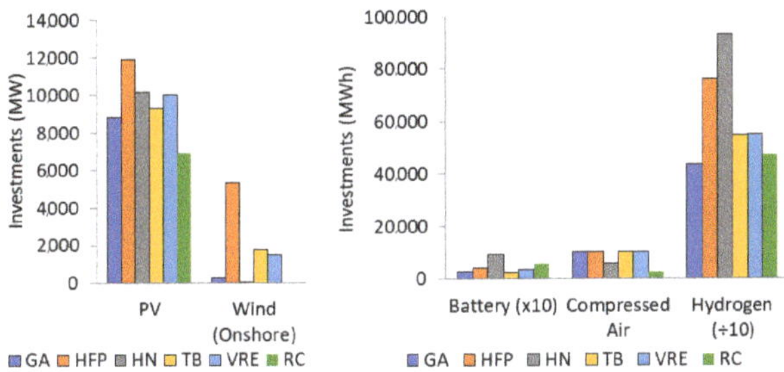

(**a**) (**b**)

Figure 6. (**a**) 2040 investments in variable renewable generation. (**b**) 2040 investments in energy storage capacity.

Figure 7 shows the state of charge of the underground Hydrogen storage across the full year in 2040 for three of the 2040 scenarios. GA and HFP follow a similar trajectory, albeit with HFP using a larger volume of storage, with a higher reliance on Hydrogen due to the high fuel prices. All three trajectories demonstrate charging when excess renewable generation occurs and discharging when demand for Hydrogen is high and net load is also high (as Hydrogen is used by the CAES plant). Differences in variable renewable investments will drive changes in the shape of the trajectory. The HN trajectory is markedly different from the other two, due to the high Hydrogen demand assumed for heating in the winter months.

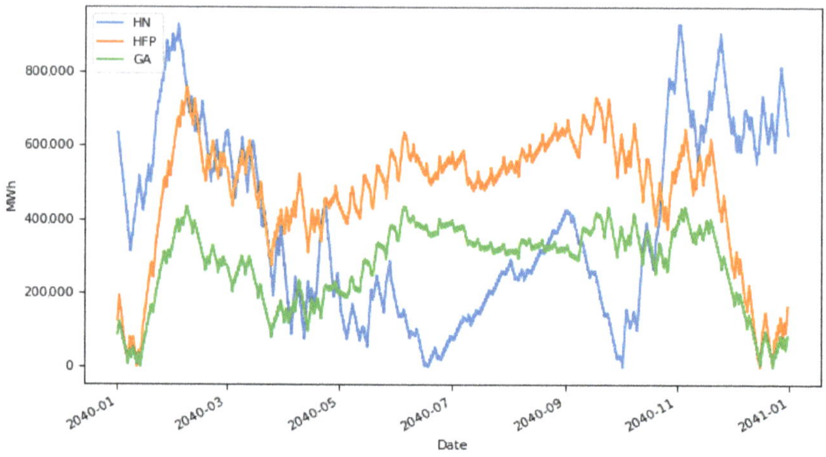

Figure 7. Trajectories of the state of charge of the underground Hydrogen storage for the 2040 scenarios—GA (green), HFP (orange) and HN (blue).

3.3. 2030 Results with Low Level of Operational Detail

To explore the impact of operational detail on the investment decisions, an additional set of simulations was completed for the 2030 scenarios, without considering either reserve requirements or the inertia floor. Figures 8 and 9 show the investment decisions for these runs. Compared to the original 2030 results, variable renewable generation is consistently over invested in. Neglecting reserve requirements leads to an underestimation of curtailment levels and renewable integration costs. Conventional plants, particularly the larger CCGT, play an important role in meeting the systems inertia requirements. When this constraint is neglected, lower investment levels occur for conventional dispatchable plants, particularly the CCGT (with or without CCS). Despite the higher levels of of renewable generation, as curtailment is underestimated, additional electrolyser investments do not result. Large investments in Hydrogen storage occur for the same two scenarios (HFP, HN), but at significantly higher levels. These differences highlight the importance of co-optimising short term operations with long term investments and storage operation for planning the future energy system, with the ability to include sufficient detail and also the ability to capture the interactions between the different sectors which are becoming increasingly integrated.

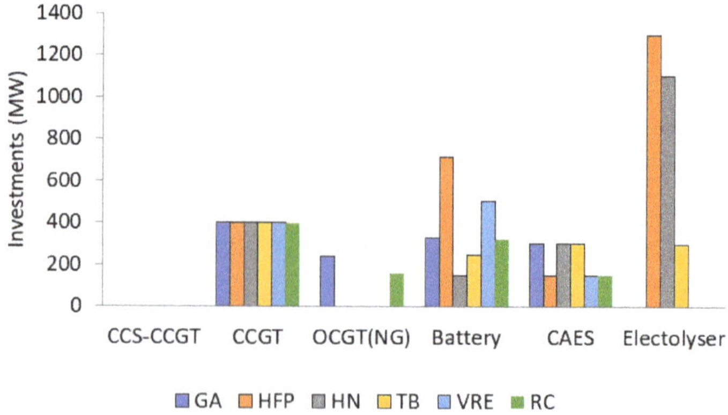

Figure 8. 2030 investments including dispatchable generation capacities and load (electrolysers).

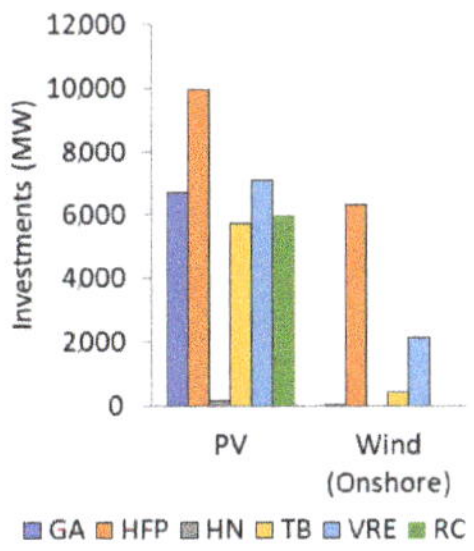

(a)

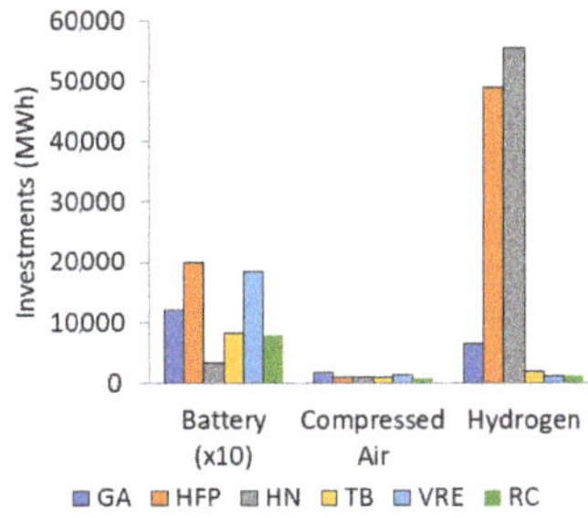

(b)

Figure 9. (**a**) 2030 investments in variable renewable generation with low operational detail. (**b**) 2030 investments in energy storage capacity with low operational detail.

4. Discussion

Important insights can be gained from the results of this work regarding the potential for the future transition to a Hydrogen economy. "Green Hydrogen" is seen to play a limited role in the 2030 results. In 3 of the 6 scenarios, no investments in large-scale electrolysers occur, and a maximum investment of 1300 MW seen for the remaining 3 scenarios. Limits on "blue Hydrogen" import trigger electrolyser investments in the HN scenario, driven by the increased Hydrogen demand, which represents "green Hydrogen" adoption under a climate of strong policy measures, i.e., strong demand for Hydrogen coupled with limits on "blue Hydrogen" production.Limits on the role of "blue Hydrogen" ties in with the vision of its use as a stepping stone to the wide-scale adoption of Hydrogen as an energy carrier, facilitating increasing demands as "green Hydrogen" scales up and becomes more economically viable. A large uptake of electrolysers is seen for the HFP scenario with investments of 1200 MW occurring in 2030. High fuel prices coupled with a high carbon price triggers a large increase in "green Hydrogen" production. This result highlights the crucial role that carbon prices play in promoting the adoption of costly alternative low-carbon energy solutions. Capital costs of both electroysers and renewable generation influence the uptake in Hydrogen production. While in 2030, capital cost reductions in the VRE scenario are not sufficient to trigger green Hydrogen production, electrolyser investments of 700 MW are seen for the TB scenario. By 2040 the synergistic relationship between variable renewable generation and electrolysers can be seen, with reduced capital costs for either triggering increased investments in both. Indeed, 2040 scenarios provide a much more favourable climate for Hydrogen investments. Again, the HN scenario prompts the highest level of investments with a 634% increase in electrolyser investments compared to 2030. By 2040 high levels of electrolyser investments are seen across all scenarios, including the base GA scenario, which still sees 60% of the investments in the HN scenario

The results presented in this paper provide insights into some of the drivers of "green Hydrogen" investments, with the representation of the long-term trajectory of the seasonal storage within the investment model forming an essential part of the optimisation. However, further advances are still required in long-term storage optimisation. Compromises need to be made between model detail and model accuracy, and the selection of an appropriate level of temporal and operational detail is highly system specific [28,30]. Decomposition techniques could be used to facilitate increases levels of operational and temporal detail.

Hydrogen gas turbines are not selected as investments in any of the considered scenarios. The round trip efficiency for power-to-Hydrogen-to-power with a gas turbine used for generation mean the losses are too high for this option to be favoured. An alternative Hydrogen fuelled power generation option is fuel cells which can achieve greater efficiencies when used for the co-generation of heat and electricity. Regions with high levels of district heating anticipate a significant role for fuel cells in power generation. As district heating is very limited in Ireland, and due to the large costs and disruptive nature of the addition of district heating, fuel cells were not considered in the six scenarios. Fuel cell uptake in the industrial sector is assumed, represented by the modelled Hydrogen demand. CAES is widely selected in each of the scenarios while, due to the unfavourable economics, Hydrogen turbines saw little investment. Due to geographic limitations, CAES may not be an option in all cases and alternative Hydrogen generation technologies should also be considered in future work.

5. Conclusions

Future low-carbon energy systems require radical changes to the structure and operation of the system. Planning for such a future requires complex modelling solutions which must be capable of adequately capturing a high degree of sector coupling at high levels of operational and temporal detail. Insufficient levels of model detail leads to misleading solutions, for example the under estimation of renewables integration costs and curtailment levels, and subsequently over investments in renewable generation. Hydrogen technologies can provide solutions on both the supply and demand side. However it is essential that these investments are not considered in isolation. Hydrogen electrolysers and generators such Hydrogen fuelled CAES can have a synergistic impact on renewable generation investments where investments in one would not be economically viable without the other. Seasonal storage, such as underground Hydrogen storage, can play a vital role in the long-term balancing of the energy system. However, it is difficult to optimise as part of a standard investment model. New methodologies are emerging which allow both investments in and operation of seasonal storage to be optimised, such as using ordered representative periods, mapped to equivalent periods across the year.

The case study explored in this paper demonstrates the use of an energy system model with a high level of sector coupling through the extensive use of Hydrogen as an energy carrier. A high level of operational and temporal detail are included and a methodology which optimises long-term storage is applied. The results show a large role for Hydrogen technologies in Ireland by 2040. The role of Hydrogen in 2030 is more limited, but moderate levels of investment in both electrolysers and Hydrogen fuelled generation (CAES) can still be seen in 2030, when for example, fuel and carbon prices are high, or high levels of Hydrogen demand occur, in this case study prompted by Hydrogen demand in the heating sector. Indeed, high investments in Hydrogen fuelled CAES are seen across all scenarios in this case study, which facilitates increased round trip efficiencies for power-to-Hydrogen-to-power.

Future work will use benders decomposition to iterate between the master investment problem and the operations sub-problem, allowing additional operational and temporal detail to be included, while maintaining tractability. Benders decomposition has already been implemented in SpineOpt, and work on including the long term storage trajectory is progressing. In addition, in future work, alternative generation technologies, including fuel cells will be considered, as well as a more detailed representation of Hydrogen end-uses and alternative storage solutions. A more thorough exploration of alternative pathways to a zero carbon future will also be completed, including different levels of electrification and biomass utilisation.

Author Contributions: Conceptualization, C.O. and J.D.; methodology, C.O. and J.D.; validation, J.D.; formal analysis, C.O.; investigation, C.O. and J.D.; data curation, C.O. and J.D.; writing—original draft preparation, C.O.; writing—review and editing, C.O., J.D. and T.O.; visualization, C.O.; supervision, T.O.; project administration, T.O.; funding acquisition, T.O. All authors have read and agreed to the published version of the manuscript.

Funding: This project has received funding from the European Union's Horizon 2020 research and innovation programme under grant agreement No. 774629.

Institutional Review Board Statement: Not applicable.

Informed Consent Statement: Not applicable.

Data Availability Statement: The data presented in this study are openly available in: https://github.com/Spine-project/spine-cs-c3, accessed on 21 January 2022.

Conflicts of Interest: The authors declare no conflict of interest.

Abbreviations

The following abbreviations are used in this manuscript:

CAES	Compressed Air Energy Storage
CCGT	Closed Cycle Gas Turbine
CCS	Carbon Capture and Storage
CCGT-CCS	Closed Cycle Gas Turbine with Carbon Capture and Storage
DSM	Demand Side Management
ENTSO-E	European association for the cooperation of transmission system operators (TSOs) for electricity
GA	Global Ambition scenario
H-CAES	Hydrogen Compressed Air Energy Storage
H-OCGT	Hydrogen Open Cycle Gas Turbine
HFP	High Fuel Price scenario
HN	Hydrogen Network scenario
MIP	Mixed Integer Programming
OCGT	Open Cycle Gas Turbine
O&M	Operation and Maintenance
RC	Restricted CAES scenario
TB	Technology Breakthrough scenario
TYNDP	Ten Year Network Development Plan
VRE	Variable Renewable Energy scenario
VOM	Variable Operation and Maintenance

References

1. European Commission. *Proposal for a Directive of the European Parliament and of the Council Amending Directive (EU) 2018/2001 and Directive 98/70/EC as Regards the Promotion of Energy from Renewable Sources, and Repealing Council Directive (EU) 2015/65*; Technical Report; European Commission: Brussels, Belgium, 2021.
2. Department of the Environment. Climate Action Plan 2021. Available online: https://www.gov.ie/en/publication/6223e-climate-action-plan-2021/ (accessed on 12 November 2021).
3. Evangelopoulou, S.; Vita, A.D.; Zazias, G.; Capros, P. Energy System Modelling of Carbon-Neutral Hydrogen as an Enabler of Sectoral Integration within a Decarbonization Pathway. *Energies* **2019**, *12*, 2551. [CrossRef]
4. Walker, S.; Al-zakwani, S.; Maroufmashat, A.; Fowler, M.; Elkamel, A. Multi-Criteria Examination of Power-to-Gas Pathways under Stochastic Preferences. *Energies* **2020**, *13*, 3151. [CrossRef]
5. Moriarty, P.; Honnery, D. Prospects for Hydrogen as a transport fuel. *Int. J. Hydrogen Energy* **2019**, *44*, 16029–16037. [CrossRef]
6. Baldino, C.; Searle, S.; Zhou, Y.; Christensen, A. *Hydrogen for heating? Decarbonization Options for Households in the United Kingdom in 2050*; Technical Report December; The International Council on Clean Transportation: Washington, DC, USA, 2020.
7. Staffell, I.; Scamman, D.; Velazquez Abad, A.; Balcombe, P.; Dodds, P.E.; Ekins, P.; Shah, N.; Ward, K.R. The role of Hydrogen and fuel cells in the global energy system. *Energy Environ. Sci.* **2019**, *12*, 463–491. [CrossRef]
8. Bartela, Ł. A hybrid energy storage system using compressed air and Hydrogen as the energy carrier. *Energy* **2020**, *196*, 117088. [CrossRef]
9. Erbach, G.; Jensen, L. *BRIEFING Towards Climate Neutrality*; Technical Report April; European Union: Brussels, Belgium, 2021.
10. Mongird, K.; Viswanathan, V.; Alam, J.; Vartanian, C.; Sprenkle, V.; Baxter, R. *2020 Grid Energy Storage Technology Cost and Performance Assessment*; Technical Report December; Pacific Northwest National Laboratory: Richland, WA, USA, 2020.
11. Sgobbi, A.; Nijs, W.; Miglio, R.D.; Chiodi, A.; Gargiulo, M.; Thiel, C. How far away is Hydrogen? Its role in the medium and long-term decarbonisation of the European energy system. *Int. J. Hydrogen Energy* **2016**, *41*, 19–35. [CrossRef]
12. Yue, X.; Patankar, N.; Decarolis, J.; Chiodi, A.; Rogan, F.; Deane, J.P.; O'Gallachoir, B. Least cost energy system pathways towards 100% renewable energy in Ireland by 2050. *Energy* **2020**, *207*, 118264. [CrossRef] [PubMed]
13. He, G.; Mallapragada, D.S.; Bose, A.; Heuberger, C.F.; Gençer, E. Sector coupling via hydrogen to lower the cost of energy system decarbonization. *arXiv* **2021**, arXiv:2103.03442.
14. Ihlemann, M.; Kouveliotis-Lysikatos, I.; Huang, J.; Dillon, J.; O'Dwyer, C.; Rasku, T.; Marin, M.; Poncelet, K.; Kiviluoma, J. SpineOpt: A flexible open-source energy system modelling framework. *Submitted for review.*
15. ENTSO-E. *TYNDP 2020 Main Report*; Technical Report January; ENTSO-E: Brussels Belgium, 2021.

16. Cerniauskas, S.; Jose Chavez Junco, A.; Grube, T.; Robinius, M.; Stolten, D. Options of natural gas pipeline reassignment for Hydrogen: Cost assessment for a Germany case study. *Int. J. Hydrogen Energy* **2020**, *45*, 12095–12107. [CrossRef]
17. Schmidt, O.; Gambhir, A.; Staffell, I.; Hawkes, A.; Nelson, J.; Few, S. Future cost and performance of water electrolysis: An expert elicitation study. *Int. J. Hydrogen Energy* **2017**, *42*, 30470–30492. [CrossRef]
18. IRENA. *Green Hydrogen Cost Reduction: Scaling Up Electrolysers to Meet the 1.5degC Climate Goal*; Technical Report; IRENA: Abu Dhabi, United Arab Emirates, 2020.
19. SpineProject. Case Study C3 Input Data. Available online: https://github.com/Spine-project/spine-cs-c3 (accessed on 12 November 2021).
20. SpineProject. SpineOpt Repository. Available online: https://github.com/Spine-project/SpineOpt.jl (accessed on 12 November 2021).
21. Cole, W.; Frazier, A.W.; Augistine, C. *Cost Projections for Utility-Scale Battery Storage: 2021 Update*; Technical Report June; National Renewable Energy Laboratory: Golden, CA, USA, 2021.
22. O'Dwyer, C.; Vorushylo, I.; Huang, Y.; Hewitt, N.; Flynn, D. Enabling flexible operation of CCS plant within a high renewables power system. In Proceedings of the International Conference on Renewable Power Generation (RPG), online, 14–15 October 2021; pp. 1–6.
23. SpineProject. SpineOpt Documentation. Available online: https://spine-project.github.io/SpineOpt.jl/latest/index.html (accessed on 12 November 2021).
24. SpineProject. Spine Periods Repository. Available online: https://github.com/Spine-project/SpinePeriods.jl (accessed on 12 November 2021).
25. IBM. ILOG CPLEX Optimization Studio 12.9.0. Available online: https://www.ibm.com/docs/en/icos/12.9.0 (accessed on 12 November 2021).
26. Welder, L.; Ryberg, D.S.; Kotzur, L.; Grube, T.; Robinius, M.; Stolten, D. Spatio-temporal optimization of a future energy system for power-to-Hydrogen applications in Germany. *Energy* **2018**, *158*, 1130–1149. [CrossRef]
27. Rabiee, A.; Keane, A.; Soroudi, A. Green Hydrogen: A new flexibility source for security constrained scheduling of power systems with renewable energies. *Int. J. Hydrogen Energy* **2021**, *46*, 19270–19284. [CrossRef]
28. Helist́o, N.; Kiviluoma, J.; Morales-España, G.; O'Dwyer, C. Impact of operational details and temporal representations on investment planning in energy systems dominated by wind and solar. *Appl. Energy* **2021**, *290*, 116712. [CrossRef]
29. Poncelet, K.; Delarue, E.; Six, D.; Duerinck, J.; D'haeseleer, W. Impact of the level of temporal and operational detail in energy-system planning models. *Appl. Energy* **2016**, *162*, 631–643. [CrossRef]
30. Gonzato, S.; Bruninx, K.; Delarue, E. Long term storage in generation expansion planning models with a reduced temporal scope. *Appl. Energy* **2021**, *298*, 117168. [CrossRef]
31. Poncelet, K.; Hanspeter, H.; Delarue, E.; Virag, A.; D'haeseleer, W. Selecting representative days for capturing the implications of integrating intermittent renewables in generation expansion planning problems. *IEEE Trans. Power Syst.* **2017**, *32*, 1936–1948. [CrossRef]

designs

Article

Identification of Citronella Oil Fractions as Efficient Bio-Additive for Diesel Engine Fuel

Noor Fitri [1,*], Rahmat Riza [2], Muhammad Kurnia Akbari [1], Nada Khonitah [1], Rifaldi Lutfi Fahmi [1] and Is Fatimah [1]

[1] Chemistry Department, Universitas Islam Indonesia, Kampus Terpadu UII, Jl. Kaliurang Km 14, Sleman, Yogyakarta 55584, Indonesia; 16612076@alumni.uii.ac.id (M.K.A.); 16612094@alumni.uii.ac.id (N.K.); 16612089@alumni.uii.ac.id (R.L.F.); isfatimah@uii.ac.id (I.F.)

[2] Department of Mechanical Engineering, Universitas Islam Indonesia, Kampus Terpadu UII, Jl. Kaliurang Km 14, Sleman, Yogyakarta 55584, Indonesia; rahmatriza@uii.ac.id

* Correspondence: noor.fitri@uii.ac.id; Tel.: +62-27-489-6439

Abstract: Escalation fuel consumption occurs in various regions of the world. However, world oil reserves decline from year to year so that it becomes scarce and causes oil prices to surge up. This problem can be solved by saving fuel consumption. One method of saving fuel is adding bio-additives from citronella oil as a sustainable resource to diesel fuels. Citronellal, citronellol and geraniol are the main components of citronella oil which can be used as fuel additives. This study aimed to evaluate the effect of citronella oil fractions as bio-additives to the performance of diesel engine. The research stages include: extraction of citronella oil, vacuum fractionation of citronella oil, physical chemical characterization of citronella oil and its fractions, formulation of bio-additive -fuel blending, characterization of blending, and evaluation of fuel efficiency. The effect of concentration of the bio-additives was examined towards three diesel fuels; dexlite, pertamina-dex, and biosolar. The results showed two main fractions of citronella oil; citronellal dominant component (FA) and citronellol-geraniol dominant components (FB). The concentration variation of bio-additives was 0.1–0.5%. Fuel consumption efficiency was tested using diesel engine at an engine speed of 2000 rpm and a load increment of 1000, 2000 and 3000 psi with 7 min running time. The fractions represented the different tendencies to enhance the fuel efficiency up to 46%, influenced by the mixture's concentration. Generally, citronella oil and the fractions showed the potency as bio-additive to diesel fuels.

Keywords: bio-additive; citronella oil; fuel additive; diesel fuel

Citation: Fitri, N.; Riza, R.; Akbari, M.K.; Khonitah, N.; Fahmi, R.L.; Fatimah, I. Identification of Citronella Oil Fractions as Efficient Bio-Additive for Diesel Engine Fuel. *Designs* **2022**, *6*, 15. https://doi.org/10.3390/designs6010015

Academic Editors: Zbigniew Leonowicz, Arsalan Najafi and Michał Jasinski

Received: 11 January 2022
Accepted: 8 February 2022
Published: 14 February 2022

1. Introduction

Since the beginning of the industrial revolution, fossil fuels have been primary energy sources and industrial chemicals. For decades, the benchmark for a country's development has been linked to the level of fossil fuel consumption, which is increasingly elevated. For example, in Indonesia, fuel consumption for transportation is by 8.6% per year, higher than power plant and household demands with 4.6 and 3.7%, respectively [1]. The International Energy Agency (IEA) estimates that the world will reach maximum world oil production between 1996 and 2035. Along with increasing fuel consumption, diesel and biodiesel demand is also escalated. The limited nature of fuel, coupled with the challenges of skyrocketing costs of conventional oil, global warming issues, and other environmental pollution problems, has led to in-depth research on the use of renewable and sustainable alternative fuels [2].

Various researchers have offered several solutions, such as biodiesel production [3,4] and additives for diesel engine fuel [5,6]. The diesel fuel additive for compressing fuel consumption is interestingly developed because it is closely related to reducing emissions [2,7]. Several compounds consisting of glycerol-based compounds and furfural-based compounds were synthesized for those purposes. Miscibility, renewability of the resource, and energy-efficiency in production are the requirements for further industrial developments [8]. In addition, the use of natural product with simple processes highlighted the use of essential oil as a potential and renewable resource for diesel fuel additives. Previous studies revealed the enhanced ignition efficiency and combustion quality of diesel fuel by clove oil [9]. The improving brake thermal efficiency (BTE) and reduced brake-specific fuel consumption were expressed by combining some essential oils. In addition, a significant reduction in particulate matter (PM) emissions have been reported [9,10]. Other investigation also reported the role of essential oils as bio-additives, such as, pine wood oil [11], cinnamon oil [12], sweet orange oil [13], patchouli oil [14], lemongrass oil [15], and clove oil [9,16].

Based on the chemical composition and chemical structures, the chemical interaction between the additive and the fuel is the basic idea for the essential oil utilization in fuel. The characteristics consist of low vapor point, solubility, and stability in mixture with diesel fuel are of the consideration [15]. In addition, essential oils contain various kinds of chemical compounds containing oxygen atoms so that they are able to complete the combustion system in diesel engines [11]. This reduces harmful emissions such as hydrocarbons (HC), particulate matter, CO_2, and NOx emissions when added to fuel. In addition, it is used to increase the viscosity, anti-knock, octane, cetane, and cold flow properties of fuels as well as improve thermal stability, cleanliness and prevent corrosion of engines and engine parts [17]. Additionally, the metal-free feature of essential oils can better replace conventional additives such as tetra-methyl-lead (TML) and tetra-ethyl-lead (TEL). TML and TEL contain Pb, which produce toxic gases and are harmful to humans and the environment.

Due to the abundance source and ease of production, citronella oil is one of the important essential oils in several countries in Asia [18,19]. Besides enormous potential in pharmaceutical, food, and other chemical industries, the main compounds of citronella oil, citronellal, citronellol and geraniol, have properties to be potentially bound with interactions similar to reported essential oil for bio-additive applications. Practically, the composition of citronella oil depends on the fraction within the distillation process. Different compositions and combinations of citronellal, citronellol and geraniol as the main ingredient in the citronella oil reflected the different oxygen content that influenced diesel fuel combustion performance.

The novelty of this research is the utilization of fractions of citronella oil as the bio-additives. To our knowledge, there was no report on the utilization of citronella oil fractions to improve efficiency. Our hypothesis is that the more oxygenate content in the fraction, the more effective the combustion process in the engine, which means a more efficient the use of fuel. The distillation setting itself, which is related to energy consumption, needs to be optimized for investigating the optimum condition of citronella oil fractionation for bio-additives application. Based on these backgrounds, a study on citronella oil composition to the bio-additive performance toward diesel (DI) fuels was investigated. The study focused on the effect of fractions as the bio-additives. The significance of this research is related to the use of sustainable natural resource for minimizing energy, which is the main issue for sustainable energy in the future perspective.

2. Materials and Methods

The Flowchart of this research work is shown in Figure 1. The experimental section includes extraction of citronella oil, vacuum fractionation of citronella oil, physical chemical characterization of citronella oil and its fractions, formulation of bio-additive -fuel blending, characterization of blending, and evaluation of fuel efficiency.

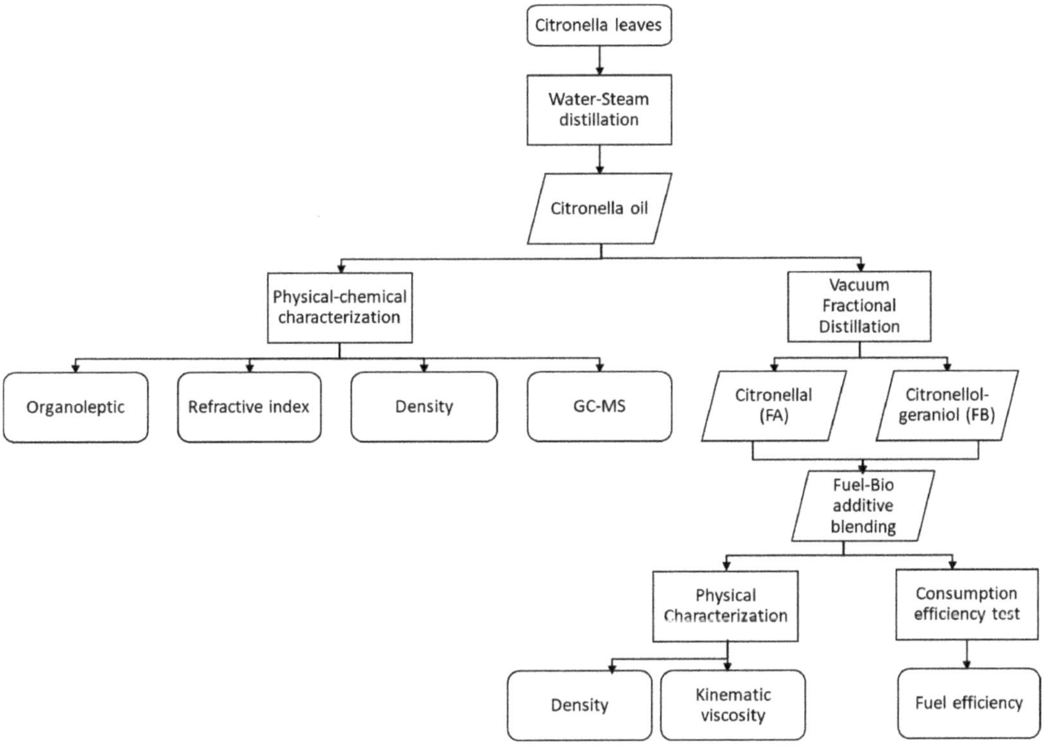

Figure 1. Flow chart of research.

2.1. Materials

The materials were citronella leaves, anhydrous sodium sulfate (Na_2SO_4), biodiesel, dexlite (light diesel), and pertamina-Dex (premium diesel). The equipments were laboratory glassware, refractometer, Oswald viscometer, steam distillation, vacuum fractional distillation equipment (B/R Instrument-Spinning Band Distillation System Model 36-100), Gas Chromatography-Mass Spectroscopy (GC-MS QP2010S Shimadzu, Tokyo, Japan), diesel engine (IN-DO R 180 iDi), dynotest hydraulic system, exhaust emission test equipment (StarGas 898).

2.2. Distillation of Citronella Oil

A total of 25 kg of citronella leaves were cut into small pieces, then put into a water-steamed distillation kettle. The chopped citronella leaves were distilled for 3–5 h with medium heat and turned on the cooling water circulation system for five repetitions. Every 30 min, the water-steam distillation system was checked. After that, the citronella oil was separated from hydrosol using a separator. Citronella oil was purified by adding anhydrous sodium sulfate (Na_2SO_4) until the water in the citronella oil separated. Finally, citronella oil was decanted and filtered to separate the Na_2SO_4 precipitate.

2.3. Vacuum Fractional Distillation

Citronella oil was further fractionated by fractional vacuum distillation equipment (B/R Instrument-Spinning Band Distillation System Model 36–100) at a pressure of 30 mmHg with a reflux ratio of 20:1 for 30 h. Two fractions containing citronellal (FA) and citronellol-geraniol (FB) were obtained by distilling citronella oil at the different conditions of fractional vacuum distillation, as shown in Table 1.

Table 1. The Operational conditions of fractional Vacuum distillation.

Parameter	Reflux Ratio 20:1	
	FA	FB
Open cut (°C)	200	225
Close cut (°C)	225	240
Heat rate (%)	20	20
Pressure (mmHg)	30	30
Condenser temp. (°C)	30–300	30–300

2.4. Physical and Chemical Characterization

The physical properties of distillates were performed using density (pycnometer method), refractive index (using Abbe refractometer), and color-odor that organoleptically refers to the Indonesian national standard method (SNI 06-3953-1995).

Chemical properties of the citronella oil were examined using gas chromatography-mass spectroscopy (GC-MS) using RTX-5MS column and helium gas as the mobile phase. We set injector temperatures of 80.0 and 300.0 °C, pressure of 42.3 kPa, and a total flow rate of 0.74 mL/minute. The Operational conditions of GC-MS shown in Table 2.

Table 2. Operational conditions of GC-MS.

Specification	Value
Model	GC-2010 Simadzu
Engine type	QP2010 SE
Colomn temperature	80.0 °C
Injector temperature	300.00 °C
Injectore mode	Separated
Flow control mode	Pressure
Pressure	42.3 kPa
Flow total	117.5 mL/min
Colomn Flow	0.74 mL/min
Linear velocity	31.8 cm/s
Removal flow	3.0 mL/min
Rate	10.00
Temperature	80.0–320.0
Source ion Temperature	25.00 °C
Interface Temperature	300.00 °C
Enhanced detector mode	Relative
Gain Detector	0.98 kV + 0.00 kV
Scan Speed	1250

2.5. Fuel Efficiency Test

The diesel bio-additive blending testing was performed onto three DI fuels: dexlite, pertamina-dex and biosolar. The tested bio-additive samples of citronella oil, FA and FB were at the varied concentration of 1.0; 1.5; 2.0; and 5.0%. For each test, the bio-additive:fuel volume ratio of 1:1 was utilized, and the stirring for fuel-citronella oil was conducted for 30 min before testing.

Determination of blending consumption and exhaust gas emissions was carried out. A total of 1000 mL blending analyzed for 7 min with variations in engine speed of 1500, 2000 and 4000 rpm, a load of 0, 25, 50, 25, 50, 75 and 100 W at constant torque using 125 cc diesel engine. The Diesel engine specification is shown in Table 3.

Table 3. Diesel engine specification.

Specification	Value
Model	R 180 iDi
Maximum power (kW)	5.7
Working power (kW)	5.0
Diameter (mm)	80 × 80
Volume (L)	0.402
Weight (kg)	72
Fuel consumption	205
Type	4 stroke horizontal engine
Cylinder	1
Start system	Crank
Cooling system	hopper/radiator
Lubricant system	pressure combination system and distribution
Combustion system	Antechamber

3. Results and Discussion

3.1. Characterization of Citronella Oil and the Fractions

Fractional distillation under vacuum conditions was performed to obtain the fractions of citronella oil, FA and FB, based on an immediate difference in boiling point in a vacuum at 30 mmHg pressure with a reflux ratio of 20:1. It referred to the optimum condition presented in previous work highlighting the pressure of 30 mmHg, which shows that the best reflux ratio for separating citronella oil is 20:1 [20]. A quick separation without any other chemicals required in the process is beneficial for fractionation [21]. In addition, the reduced pressure of the system to be less than 1 atm (760 mmHg) gives the reducing vapor point of the distillate without any excessive chemical change. Calculation of the vapor point of the solution in this condition can be completed through the Clausius–Clapeyron equation at a pressure of 30 mmHg. The enthalpy of vaporization of citronellal is 44.22 kJ/mol, citronellol 63.50 kJ/mol, and geraniol 54.61 kJ/mol. The results of calculations using the Clausius–Clapeyron equation obtained that the boiling points of citronellal, citronellol, and geraniol at a pressure of 30 mmHg were 97.86, 138.91, and 129.57 °C, respectively.

Physico-chemical characteristics of citronella oil and the fractionation results (FA and FB) were carried out based on SNI-06-3953-1995. Color is one of the parameters for beginning fractionation, and its measurement was carried out organoleptically or by direct eye observation at a distance of 30 cm. The results presented in Table 4 represent that the color of citronella oil and the fractions are pale yellow, which met the SNI standard stating the parameter of "no color difference", namely pale yellow-brown. Another organoleptic parameter is odor. Odor testing was carried out with the sense of smell at a distance of 5 cm. The test results show that citronella oil has a fresh smell typical of lemongrass. FA fraction has a high citronellal content and smells of pungent lemon. FA fraction has a strong odor because it has more citronellal content compared to citronella oil. All odor test results follow SNI of citronella oil.

Table 4. Physical Characteristics of Citronella Oil and Its Fractions.

Parameter	Unit	Citronella Oil	FA	FB	SNI-06-3953-1995
Color	-	Pale yellow	Pale yellow	Pale yellow	Pale yellow–brown
Smell	-	Fresh typical citronella oil	Strong typical citronella oil	medium typical citronella oil	Fresh typical citronella oil
Density	g/mL	0.882	0.867	0.878	0.880–0.922
Refractive Index	nD	1.475	1.449	1.467	1.466–1.475

Density characteristics were carried out using a pycnometer. The results showed that the densities of citronella oil, FA, and FA, respectively, were 0.882; 0.867; 0.878 g/mL. In its pure state, citronellal ($C_{10}H_{18}O$ BM 154.25 g/mol) has a density of 0.855 g/mL, citronellol

($C_{10}H_{20}O$ BM 156.27 g/mol) is 0.855 g/mL, and geraniol ($C_{10}H_{18}O$ BM 154.25 g/mol) is 0.889 g/mL. Compared to the SNI standard, the results are valued as good results within the range of 0.880 to 0.922 g/mL. Refractive index measurement was conducted on a refractometer, giving the values of 1.475; 1.449; 1,467 for citronella oil, FA, and FB, respectively. They are represented to be qualified within the acceptable range of refractive index of 1.466–1.375. Conclusively, citronella oil and its fractions are met following SNI standard from the four parameters.

The chemical components of citronella oil and their fractions were determined by GC-MS, and the compared chromatograms are presented in Figure 2. The data in Table 5 represent the identification results of citronellal, citronellol and geraniol. Citronellal is identified at 6.38 min of retention time with the percentage of peak area of 19.01%, citronellol is identified at 7.46 min and a peak area of 20.48%, and at 7.86 min geraniol was identified at 18.81%. Other minor components were identified at a percentage below 5%. It can be concluded that in terpenoid groups (monoterpenoid C_{10}) are found many essential oils, and they are composed as secondary metabolites and are characteristic of each essential oil [22,23]. The percentage is acceptable as it is similar to what was reported by [24], reporting the component of citronellal (36.11%), geraniol (20.07%) and citronellol (20.82%). Meanwhile, regarding citronella oil (Javanese type), ISO 3848:2016 states the range of citronellal content (31.00–40.00%), citronellol content (8.00–14.00%) and geraniol content (20.00–25.00%).

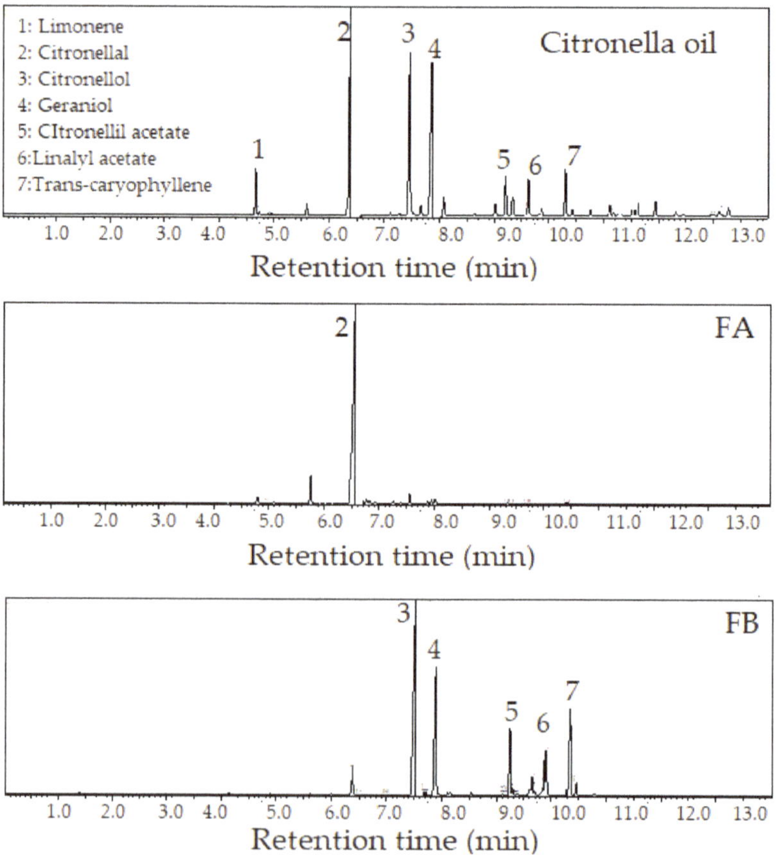

Figure 2. Chromatograms of Citronella Oil, FA and FB.

Table 5. Main Compounds of Citronella Oil.

Retention Time (minute)	Content (%)	Compound
4.670	3.840	Limonene
6.380	19.010	citronellal
7.460	20.480	Citronellol
7.860	18.810	Geraniol
9.210	3.720	Citronelyl acetate
9.620	3.850	Linalyl acetate
10.300	5.030	Trans-caryophyllene

Citronellal, citronellol and geraniol compounds tend to be semipolar compounds. In separation, the interaction of the three compounds with the GC column (nonpolar) is only affected by the boiling point of the compound. The three main compounds have similar polarity and interact weakly with the GC column so that the three compounds have relatively short retention times. Citronellal has a lower boiling point than citronellol and geraniol. The -OH group (alcohol) in citronellol and geraniol can form strong hydrogen bonds so that the boiling point becomes higher. Citronellal is a type of monoterpenoid aldehyde (-CHO). Therefore, citronellal has a faster retention time than citronellol and geraniol.

Physically, each fraction has a different color; FA is pale yellow, and FB 4 is yellow-brown. The increasing citronellal content was achieved in FA, as it is about 89.37% compared to citronella oil which is 19.01%. Likewise, FB contains more citronellol and geraniol—in the range of 27.36–31.71%—compared to citronella oil (18–20%). According to previous work [24], isolation of citronella oil by vacuum fractionation distillation was able to increase the levels of citronellal up to 90%, citronellol up to 30% and geraniol up to 45%. The density of citronella oil and its fraction is shown in Table 6 and the refractive index in Table 7.

Table 6. Density of Fractions of Citronella Oil.

Sample	Result (kg/m^3)	Standard (SNI) (kg/m^3)
Citronella Oil	882.6	880–922
Citronellal fraction (F2)	867.1	850–860
Citronellol-geraniol fraction (F3)	878.4	870–899

Table 7. The refractive index of Fractions of Citronella Oil.

Sample	Result	SNI Standard
Citronella Oil	1.4751	1.4660–1.4750
Citronellal fraction (FA)	1.4486	1.4440–1.4540
Citronellol-geraniol fraction (FB)	1.4668	1.4660–1.4770

Referring to [9], a good bio-additive must contain oxygenate (O atoms), which can increase the oxygen content in the fuel so that efficient combustion occurs in diesel engines. FA and FB have potential to be diesel fuel bio-additives. Both citronellol and geraniol have similar physicochemical properties, such as boiling points of 225 °C and 226 °C, respectively. It was not easy to separate them through fractional vacuum distillation (physics) which relies on differences in the boiling points of the compounds. In addition, citronellol is a functional group isomer of geraniol.

3.2. Utilization of Citronella Oil Fractions as Bio-Additives

The use of citronella oil fractions as bio-additives was tested based on previous research, which mentioned that the optimum percentage of at 0.1–1.0% can save fuel up to 50% [15,25]. There were 36 formulas used, with three different types of diesel fuel, two types of bio-additives, and four variations of concentration of bio-additives.

The physical characteristics of blending were determined to see the changes in physical properties after the addition of bio-additives. Density characteristics were carried out using a pycnometer and kinematic viscosity was carried out using an Oswald viscometer according to SNI 8220:2017. Viscosity indicates the ability of a fluid to flow through an area per unit of time. This is important related to the mechanism of fuel atomization shortly after leaving the nozzle into the combustion chamber [15].

The Effect of adding citronella oil and the fractions to density and viscosity of fuel is represented by the data in Tables 8 and 9. Based on the Indonesian standard (SNI 8220:2017) regarding the specifications for type 48 diesel fuel, the density of at least 815 kg/m^3 is a maximum of 860 kg/m^3 and the viscosity is a minimum of 2.0 m^2/s, a maximum of 4.5 m^2/s. This shows that after the addition of bio-additives, there is no significant change in the physical properties of diesel fuel. The maintaining viscosity values in the addition of the bio-additives are related to the similar response of the volume restricting the movement of the lon- chain hydrocarbon molecules. The viscosity of DI fuel is also related to temperature and pressure dependency, which are important for fuel performance [20].

Table 8. Density of fuel-bio additive blending.

Tested Bio-Additive	Concentration	Density (kg/m^3)		
		Biosolar	**Dexlite**	**Pertamina-Dex**
Without additive		853.133	839.873	808.819
FA	0.1	851.239	832.702	812.741
	0.15	854.030	836.888	809.214
	0.2	851.139	837.444	809.713
	0.5	853.432	835.985	811.708
FB	0.1	850.640	831.713	813.136
	0.15	851.538	841.851	808.801
	0.2	851.937	835.455	809.524
	0.5	853.432	833.166	810.092
Citronella oil	0.1	851.239	832.327	809.111
	0.15	853.332	837.319	808.457
	0.2	851.637	837.083	808.767
	0.5	851.837	834.160	809.696

Table 9. Viscosity of fuel-bio additive blending.

Tested Bio-Additive	Concentration (%)	Viscosity (m^2/s)		
		Biosolar	**Dexlite**	**Pertamina-Dex**
Without additive		4.58	4.19	2.47
		4.53	4.02	2.42
FA	0.1	4.54	3.83	2.48
	0.15	4.37	3.63	2.53
	0.2	4.42	3.86	2.59
	0.5	4.59	3.63	2.62
FB	0.1	4.34	3.70	2.47
	0.15	4.52	3.72	2.59
	0.2	4.39	3.68	2.52
	0.5	4.49	3.78	2.60
Citronella Oil	0.1	4.60	3.81	2.43
	0.15	4.51	3.96	2.50
	0.2	4.41	3.60	2.54
	0.5			

3.3. Characterization of Citronella Oil and the Fractions

The Effect of bio-additive on fuel consumption is the most crucial parameter in this research. Fuel consumption occurs in the combustion process due to air compression in the engine combustion chamber. The amount of fuel consumed is measured in weight units per unit of time. Mathematically, the fuel consumption is shown in the following equation (Equation (1)):

$$fc = \frac{b}{t} \times \gamma_f \frac{3600}{1,000,000} \tag{1}$$

where f_c is fuel consumption expressed in kg/h, b is the volume of fuel consumption expressed in milliliters (mL), t is the time used, which is expressed in seconds (s), and f is the density of fuel expressed in kg/m^3.

The efficiency of diesel fuel consumption is carried out volumetrically, namely by calculating the fuel economy consumption when the engine produces power within a certain time frame. The purpose of determining consumption efficiency is to determine the optimum formula to improve fuel quality. The diesel engine used is a one-cylinder compression ignition (CI) type with a maximum speed of 2500 rpm for 7 min. Measurements were made by comparing the consumption volume of blending with the control. The calculation of fuel consumption results shows a decrease in fuel consumption when bio-additives are added. The decrease in fuel consumption is due to the compounds in bio-additives acting as oxygen providers, also called oxygenate compounds. The more oxygen content is contained in fuel; the easier and more completely the combustion will occur [26,27].

The results of experiments are presented in Figure 3. It is seen that at all varied concentrations, citronella oil and the fractions significantly affect the efficiencies towards tested fuels; dexlite, pertamina-dex, and biosolar. In more detail, the increasing efficiency of biosolar is most intensive compared to pertamina-dex and dexlite. In addition, the effect of concentration is not linear with the efficiency for all tested fuels, particularly for biosolar; the trends are increasing efficiency at increasing concentration. A specific pattern was expressed by the use of FA and FB in the diesel fuel which showed the optimum concentration of 0.2%. Still, the efficiency decreased at the increasing concentration, representing the possible chemical or oxidative effect of the bio-additives.

Generally, by comparing FA, FB and citronella oil, it was found that FA had the most influence. This phenomenon is closely related to the greater abundance of oxygen in the fraction. This is also associated with more oxygen in biosolar, which presented better efficiency. The alcohol (-OH) and aldehyde (-CHO) groups (containing oxygen) in blending will react with CO gas and charcoal (C) to form CO_2, which causes fewer CO and green emissions [28]. In the combustion chamber, there are three main reactions: initiation, propagation, and termination. The availability of oxygen is important to produce an efficient, constant chemical reaction in the combustion chamber.

$$4\,H_{26}C_{12(l)} + 74\,O_{2(g)} \rightarrow 48\,CO_{2(g)} + 52\,H_2O_{(aq)} \tag{2}$$

$$2\,H_{26}C_{12(l)} + 31\,O_{2(g)} \rightarrow 12\,CO_{(g)} + 12\,CO_{2(s)} + 26\,H_2O_{(aq)} \tag{3}$$

When burning in a diesel engine, two possible combustion reactions will occur, a complete combustion reaction (Equation (2)) or an incomplete combustion (Equation (3)). Complete combustion occurs when there is sufficient oxygen in the engine combustion chamber. Incomplete combustion occurs when there are insufficient oxygen molecules to burn one complex hydrocarbon molecule in diesel fuels completely. In this study, diesel fuel, pertamina-dex, dexlite, and biosolar were mixed with bio-additives from fractions FA and FB of citronella oil. The diesel fuels-bio-additive mixture is expected to reduce emissions from CO gas and charcoal (soot). This is because the FA and FB fractions contain citronellal and citronellol-geraniol, which have oxygenated aldehyde (-CHO) and alcohol (-OH) functional groups to contribute to the availability of oxygen in the combustion chamber when the diesel engine is running. According to previous research, it was found that a mixture of citronella-diesel oil was able to reduce CO gas emissions by 23–30%, NOx 31–36%, SOx 12–22%, and particulates 30–33% compared to that without the addition of bio-additives [11,12].

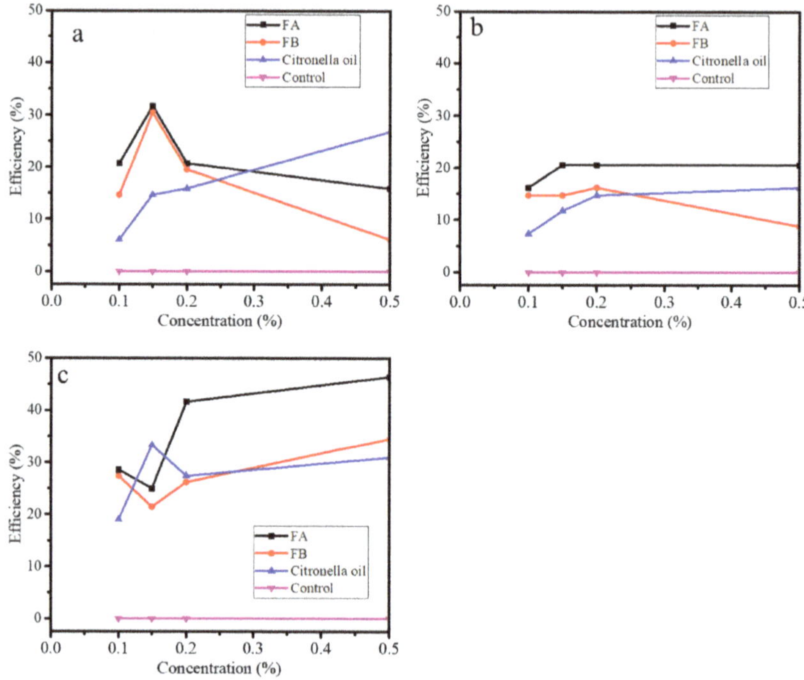

Figure 3. Effect of bio-additive concentration on fuel consumption efficiency for. (**a**) Pertamina-dex (**b**) Dexlite (**c**) Biosolar.

The bulk structure of the compounds contained in essential oils can reduce the strength of the van der Waals bonds of the compounds that make up the fuel. The bond between the fuel molecules is a fragile dispersion force [9,14]. Meanwhile, citronellal, citronellol, and geraniol compounds have dipole–dipole interactions between their molecules. Dipole–dipole interactions, which are stronger than dispersion forces, can facilitate the breaking of intermolecular bonds. Therefore, this causes the fuel molecules to break up easily to achieve more efficient combustion process. The interactions of compounds in diesel fuel after adding bio-additives are shown in Figures 4–6. The interactions that occur are induced dipole forces, namely interactions between compounds with permanent dipole moments that influence non-polar compounds. The presence of oxygen and the volatility of the citronellal, citronellol and geraniol supplied easier oxidation for faster compression in DI fuel. Oxidation reactivity of DI fuel is supplied by the combination of chemical interactions including the van der Waals and hydrogen bonding in the fuel mixture with additive. Similar results were reported by the influence of additive and oxygen-rich fuel [29,30].

Based on the volume of fuel consumption obtained, the fuel consumption efficiency can be calculated. Consumption efficiency shows an increase in fuel quality after adding bio-additives. Based on the fuel consumption test, FA with citronellal as the main component has a higher consumption efficiency compared to FB (citronellol and geraniol as the main components) and citronella oil. The lessened flash point is probably the main reason for this as citronellal, citronellol, and geraniol have flash points of 86 °C, 99 °C, and 108 °C, respectively. The lessened flash point represents the ease of being burnt and oxidized.

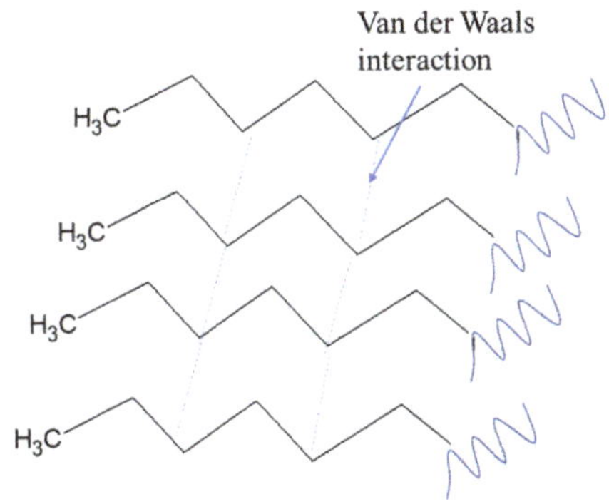

Figure 4. Hypothesis of Compounds interaction in diesel fuel [9].

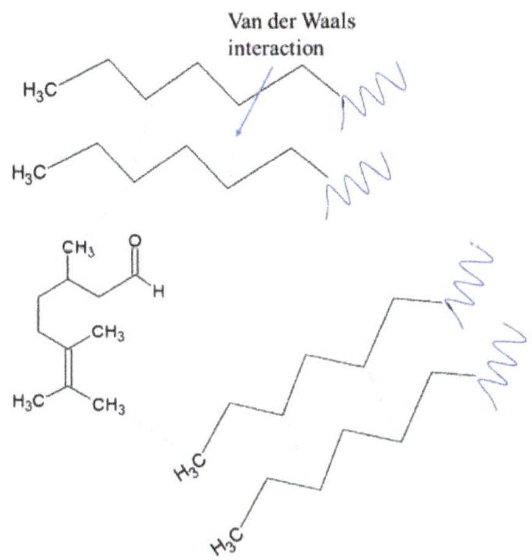

Figure 5. Hypothesis of diesel fuel interaction with FA (citronellal as main component) [9].

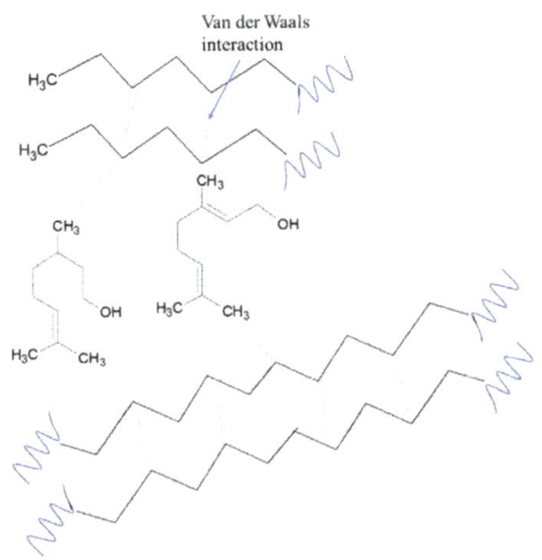

Figure 6. Hypothesis of diesel fuel interaction with FB (citronellol and geraniol as main components) [9].

Another factor is the viscosity, which affects liquid fuel properties. Viscosity is related to the flow rate and characteristics of the spray or mist of fuel into the combustion chamber [20]. The higher the viscosity, the more difficult the fuel is to move and spray or atomize into the combustion chamber, so the combustion process is not optimal. Low viscosity helps in atomization, evaporation, and diffusion, and increases the interaction of fuel with air. Based on the viscosity test of fuel that has been mixed with bio-additives, the fuel that has been mixed with the citronellal fraction (FA) gives the lowest viscosity value compared to the citronellol-geraniol (FB) fraction and citronella oil.

Generally speaking, the results from this work show the potential of the use of essential oil as a sustainable bio-additive in minimizing energy consumption, especially DI fuel. More exploration for other fuel as well as techno-economic studies are required.

4. Conclusions

This study initiated the examination of fractionation to citronella oil and its use as bio-additive to diesel fuel. The results showed two main fractions of citronella: dominant containing fraction (FA) and citronellol-geraniol dominant containing fraction (FB) obtained at different vacuum fractionation conditions. The fractions and citronella oil exhibited the capability to be bio-additives to diesel fuel as shown by the acceptable density and viscosity in tested varied concentrations (0.1–0.5%). In addition, the fuel consumption examination represented the significant ability of the tested sample to reduce fuel consumption to 46% depending on the concentration of the bio-additive. Generally, at the tested concentration range, the increasing concentration affects the reduction of fuel consumption. The next work to be carried out is emission testing to investigate whether the emission quality is improved using bio-additives.

Author Contributions: Conceptualisation, N.F., R.R.; methodology, N.F., M.K.A., R.L.F.; valida-tion, I.F.; formal analysis, N.F., R.R., M.K.A., N.K.; investigation, R.R., M.K.A., N.K., R.L.F.; re-sources, N.F.; data curation, N.F., R.L.F.; writing—original draft preparation, N.F.; writing—review and editing, I.F.; visualisation, I.F.; supervision, N.F.; project administration, R.L.F.; funding acquisition, N.F. All authors have read and agreed to the published version of the manuscript.

Funding: This research was funded by Directorate of Research and Community Services Universitas Islam Indonesia (DPPM UII) with grant number 004/Dir/DPPM/70/Pen.Unggulan/PIIII/XII/2018.

Institutional Review Board Statement: Not applicable.

Informed Consent Statement: Not applicable.

Data Availability Statement: This study did not report any data.

Acknowledgments: The authors would like to express appreciation for the financial support from the Directorate of Research and Community Services Universitas Islam Indonesia (DPPM UII), Department of Chemistry, Universitas Islam Indonesia, for the laboratory support, and also Muhammad Idris and Nur Muhammad Syafi'i for technical assistance.

Conflicts of Interest: The authors declare no conflict of interest.

References

1. Akhmad, A. Amir Study of fuel oil supply and consumption in indonesia. *Int. J. Energy Econ. Policy* **2018**, *8*, 13–20.
2. Manisalidis, I.; Stavropoulou, E.; Stavropoulos, A.; Bezirtzoglou, E. Environmental and Health Impacts of Air Pollution: A Review. *Front. Public Health* **2020**, *8*, 14. [CrossRef] [PubMed]
3. Dębowski, M.; Michalski, R.; Zieliński, M.; Kazimierowicz, J. A comparative analysis of emissions from a compression–ignition engine powered by diesel, rapeseed biodiesel, and biodiesel from chlorella protothecoides biomass cultured under different conditions. *Atmosphere* **2021**, *12*, 1099. [CrossRef]
4. Samuel, O.D.; Boye, T.G.E.; Enweremadu, C.C. Financial and parametric study of biodiesel production from hemp and tobacco seed oils in modified fruit blender and prediction models of their fuel properties with diesel fuel. *Bioresour. Technol. Rep.* **2020**, *12*, 100599. [CrossRef]
5. Permanasari, A.A.; Mauludi, M.N.; Sukarni, S.; Puspitasari, P.; Zaine, S.N.A.; Wahyunengsih, W. The Potential of Waste Cooking Oil B20 Biodiesel Fuel with Lemon Essential Oil Bioadditive: Physicochemical Properties, Molecular Bonding, and Fuel Consumption. *Bull. Chem. React. Eng. Catal.* **2021**, *16*, 555–564. [CrossRef]
6. Senthur Prabu, S.; Asokan, M.A.; Prathiba, S.; Ahmed, S.; Puthean, G. Effect of additives on performance, combustion and emission behavior of preheated palm oil/diesel blends in DI diesel engine. *Renew. Energy* **2018**, *122*, 196–205. [CrossRef]

7. Ge, S.; Ganesan, R.; Sekar, M.; Xia, C.; Shanmugam, S.; Alsehli, M.; Brindhadevi, K. Blending and emission characteristics of biogasoline produced using CaO/SBA-15 catalyst by cracking used cooking oil. *Fuel* **2022**, *307*, 121861. [CrossRef]
8. Akinnawo, C.A.; Mosia, L.; Alimi, O.A.; Oseghale, C.O.; Fapojuwo, D.P.; Bingwa, N.; Meijboom, R. Eco-friendly synthesis of valuable fuel bio-additives from glycerol. *Catal. Commun.* **2021**, *152*, 106287. [CrossRef]
9. Kadarohman, A.; Hernani; Khoerunisa, F.; Astuti, R.M. A potential study on clove oil, eugenol and eugenyl acetate as diesel fuel bio-additives and their performance on one cylinder engine. *Transport* **2010**, *25*, 66–76. [CrossRef]
10. Nair, J.N.; Singh, T.S.; Raju, V.D. Effect of addition of bio-additive clove oil to ternary fuel blends (diesel-biodiesel-ethanol) on compression ignition engine. *J. Phys. Conf. Ser.* **2021**, *2070*, 012212. [CrossRef]
11. Lawan, I.; Zhou, W.; Garba, Z.N.; Zhang, M.; Yuan, Z.; Chen, L. Critical insights into the effects of bio-based additives on biodiesels properties. *Renew. Sustain. Energy Rev.* **2019**, *102*, 83–95. [CrossRef]
12. Lawan, I.; Zhou, W.; Idris, A.L.; Jiang, Y.; Zhang, M.; Wang, L.; Yuan, Z. Synthesis, properties and effects of a multi-functional biodiesel fuel additive. *Fuel Process. Technol.* **2020**, *198*, 106228. [CrossRef]
13. Rahman, S.M.A.; Hossain, F.M.; Van Chu, T.; Dowell, A.; Islam, M.A.; Rainey, T.J.; Ristovski, Z.D.; Brown, R.J. Comparative evaluation of the effect of sweet orange oil-diesel blend on performance and emissions of a multi-cylinder compression ignition engine. *AIP Conf. Proc.* **2017**, *1851*, 020007. [CrossRef]
14. Alfian, D.G.C.; Prahmana, R.A.; Silitonga, D.J.; Muhyi, A.; Supriyadi, D. Performance Characterization of Gasoline Engine with Patchouli Oil as Bio-Additive for Gasoline with an Octane Number 90. *IOP Conf. Ser. Earth Environ. Sci.* **2019**, *258*, 012010. [CrossRef]
15. Kotaiah, K.; Periyasamy, P.; Prabhahar, M. Performance and emission characteristics of small agricultural diesel engine using Lemongrass oil and its diesel blends. *Mater. Today Proc.* **2020**, *33*, 658–662. [CrossRef]
16. Mallesham, B.; Govinda Rao, B.; Reddy, B.M. Production of biofuel additives by esterification and acetalization of bioglycerol. *Comptes Rendus Chim.* **2016**, *19*, 1194–1202. [CrossRef]
17. Cornejo, A.; Barrio, I.; Campoy, M.; Lázaro, J.; Navarrete, B. Oxygenated fuel additives from glycerol valorization. Main production pathways and effects on fuel properties and engine performance: A critical review. *Renew. Sustain. Energy Rev.* **2017**, *79*, 1400–1413. [CrossRef]
18. Vanisha, S. Nambiar and Hema Matela Potential Functions of Lemon Grass (*Cymbopogon citratus*) in Health and Disease. *Int. J. Pharm. Biol. Arch.* **2016**, 3, 1035–1043.
19. Boukhatem, M.N.; Ferhat, M.A.; Kameli, A.; Saidi, F.; Kebir, H.T. Lemon grass (*Cymbopogon citratus*) essential oil as a potent anti-inflammatory and antifungal drugs. *Libyan J. Med.* **2014**, *1*, 1–10. [CrossRef]
20. Schaschke, C.; Fletcher, I.; Glen, N. Density and viscosity measurement of diesel fuels at combined high pressure and elevated temperature. *Processes* **2013**, *1*, 30–48. [CrossRef]
21. Anwar, Y.; Siringoringo, V.S. Fractionation of Citronella Oil and Identification of Compounds by Gas Chromatography-Mass Spectrometry. *Pharm. Sci. Res.* **2020**, *7*, 138–144. [CrossRef]
22. Maurya, A.; Prasad, J.; Das, S.; Dwivedy, A.K. Essential Oils and Their Application in Food Safety. *Front. Sustain. Food Syst.* **2021**, *5*, 133. [CrossRef]
23. Dhifi, W.; Bellili, S.; Jazi, S.; Bahloul, N.; Mnif, W. Essential Oils' Chemical Characterization and Investigation of Some Biological Activities: A Critical Review. *Medicines* **2016**, 3, 25. [CrossRef] [PubMed]
24. Harianingsih, H.; Wulandari, R.; Harliyanto, C.; Andiani, C.N. Identifikasi GC-MS Ekstrak Minyak Atsiri dari Sereh Wangi (*Cymbopogon winterianus*) Menggunakan Pelarut Metanol. *Techno* **2017**, *18*, 23–27.
25. Tracy, N.I.; Chen, D.; Crunkleton, D.W.; Price, G.L. Hydrogenated monoterpenes as diesel fuel additives. *Fuel* **2009**, *88*, 2238–2240. [CrossRef]
26. Pattanaik, B.P.; Jena, J.; Misra, R.D. The effect of oxygen content in soapnut biodiesel-diesel blends on performance of a diesel engine. *Int. J. Automot. Mech. Eng.* **2017**, *14*, 4574–4588. [CrossRef]
27. Veza, I.; Said, M.F.M.; Latiff, Z.A. Progress of acetone-butanol-ethanol (ABE) as biofuel in gasoline and diesel engine: A review. *Fuel Process. Technol.* **2019**, *196*, 1–56. [CrossRef]
28. Niculescu, R.; Clenci, A.; Iorga-Siman, V. Review on the Use of Diesel–Biodiesel–Alcohol Blends in Compression Ignition Engines. *Energies* **2019**, *12*, 1194. [CrossRef]
29. Hernández, J.J.; Sanz-Argent, J.; Carot, J.M.; Jabaloyes, J.M. Ignition delay time correlations for a diesel fuel with application to engine combustion modelling. *Int. J. Engine Res.* **2010**, *11*, 199–206. [CrossRef]
30. Wei, J.; Lu, W.; Zeng, Y.; Huang, H.; Pan, M.; Liu, Y. Physicochemical properties and oxidation reactivity of exhaust soot from a modern diesel engine: Effect of oxyfuel type. *Combust. Flame* **2022**, *238*, 111940. [CrossRef]

designs

MDPI

Article

Optimal Cascade Non-Integer Controller for Shunt Active Power Filter: Real-Time Implementation

Hoda Nikkhah Kashani [1], Reza Rouhi Ardeshiri [2], Meysam Gheisarnejad [3] and Mohammad-Hassan Khooban [3,*]

1 Department of Electrical Engineering, Islamic Azad University of Iran, Garmsar Branch, Garmsar 91775-1111, Iran; nikkhah_hoda@yahoo.com
2 Department of Electrical Engineering, Shanghai Jiao Tong University, Shanghai 200240, China; r.rouhi.a@sjtu.edu.cn
3 Department of Electrical and Computer Engineering, Aarhus University, 8000 Aarhus, Denmark; m.gheisar@ece.au.dk
* Correspondence: khooban@ece.au.dk

Abstract: Active power filters (APFs) are used to mitigate the harmonics generated by nonlinear loads in distribution networks. Therefore, due to the increase of nonlinear loads in power systems, it is necessary to reduce current harmonics. One typical method is utilizing Shunt Active Power Filters (SAPFs). This paper proposes an outstanding controller to improve the performance of the three-phase 25-kVA SAPF. This controller can reduce the current total harmonic distortion (THD), and is called fractional order PI-fractional order PD (FOPI-FOPD) cascade controller. In this study, another qualified controller was applied, called multistage fractional order PID controller, to show the superiority of the FOPI-FOPD cascade controller to the multistage FOPID controller. Both controllers were designed based on a non-dominated sorting genetic algorithm (NSGA-II). The obtained results demonstrate that the steady-state response and transient characteristics achieved by the FO (PI + PD) cascade controller are superior to the ones obtained by the multistage FOPID controller. The proposed controller was able to significantly reduce the source current THD to less than 2%, which is about a 52% reduction compared to the previous work in the introduction. Finally, the studied SAPF system with the proposed cascade controller was developed in the hardware-In-the Loop (HiL) simulation for real-time examinations.

Keywords: three-phase shunt active power filter; repetitive controller; fractional-order (PI + PD) cascade controller; multistage fractional-order PID Controller

Citation: Nikkhah Kashani, H.; Rouhi Ardeshiri, R.; Gheisarnejad, M.; Khooban, M.-H. Optimal Cascade Non-Integer Controller for Shunt Active Power Filter: Real-Time Implementation. *Designs* **2022**, *6*, 32. https://doi.org/10.3390/designs6020032

Academic Editors: Zbigniew Leonowicz, Arsalan Najafi and Michał Jasinski

Received: 15 February 2022
Accepted: 21 March 2022
Published: 1 April 2022

1. Introduction

At present, developments in power electronic technology have led to a major increase in the usage of power electronic converters in the power grid while also increasing the use of electrical energy. However, power electronic converters generate reactive power and harmonics, which pollute the power system [1]. Therefore, the optimal compensation of nonlinear loads' harmonics is an important issue in power networks. Current harmonics boost losses, destroy the quality of the voltage sine waveform, cause metering devices to malfunction, and may lead to resonances and interferences [2]. As a result, distortions in current and voltage sine waveforms are not only a source of technical problems, but also have economic effects [3]. There are several popular devices such as active power filters (APFs), which may be of a series, shunt or hybrid type [4–6], static compensator, and unified power quality controller. These utilities are widely used to decrease power quality problems [7] that affect the distribution side [8].

From the viewpoint of circuit topology, Reference [9] has a more comprehensive taxonomy of available APFs, which are divided into parallel/series/hybrid type and other types. The active power filter is an effective inhibition device of active compensation harmonics that can efficiently omit harmonic contamination and improve the power factor compared with the classic passive filter [10]. Although series and/or shunt APFs are generally used to eliminate power quality problems, shunt APFs are used more often than series APFs due to their excellent performance [11,12]. SAPF is an especially efficient solution for power quality issues [13], and can reduce harmonic pollution [14–19] and compensate reactive power generated by linear/nonlinear loads in distribution networks. Therefore, SAPFs play an increasingly essential role in power distribution and delivery [20]. These filters are connected in parallel with the nonlinear loads to remove undesired current harmonics. A traditional PI controller or other control techniques, combined with the repetitive controller, can modify the dynamic response time of the repetitive controller [21].

In most cases, some characteristics should be complied with, such as low total harmonic distortion (THD) for the compensated currents and a fast transient response. To ensure an excellent design, it is believed that compromises must be found between some different necessities. Therefore, applying a multi-objective optimization approach to achieve a set of desirable objective functions based on the APFs' specifications is essential. SAPFs equipped with repetitive controllers promise an excellent compensation at steady-state with a slow, transient response [22]. In Reference [23], a traditional PI controller and FOPI controller were used to promote the performance of a 25-kVA parallel active filter based on the NSGA-II optimization approach. The optimization results proved that the obtained results using the FOPI controller were more acceptable than the achieved results by the traditional PI controller. The minimum obtained value of THD was about 3.8% in this article. In Reference [24], research was carried out looking at the superiority of FOPID controller compared to the integer-order PID controller. In fact, it is believed that the fractional-order PID/PI controller has a better performance than the integer-order PID/PI controller. In a different work, researchers presented the multistage PID controller for the automatic generation control of power systems. Hence, we were inspired by the multistage PID controller to devise a novel method, called a multistage fractional-order PID controller [25]. In the present research, a novel optimal fractional-order controller is proposed to achieve a better performance from the 25-kVA parallel active power filter, called a fractional-order (PI + PD) cascade controller, which was recently presented in Reference [26]. In addition, another fractional-order controller is designed for comparison with our proposed method. To the best of our knowledge, this controller was applied for the first time in this case study. It is called a multistage fractional-order PID controller.

As mentioned before, two different controllers were designed to optimize the performance of the shunt active power filter with the high-performance repetitive controller called the fractional order PI-fractional order PD cascade controller and multistage fractional-order PID controller. These controllers were designed based on the NSGA-II optimization method. This optimization approach is still a powerful multi-objective optimization technique to minimize the objective functions that the other researchers are using in several fields [27–39]. According to this optimization method, there are two result categories: one of them is related to the variables selected to design each controller, called the Pareto Optimal Set (POS), and another set of results is concerned with two objective functions, called the Pareto Optimal Front (POF). It is mandatory to choose an appropriate range for each variable in order to reach an excellent POF.

An acceptable performance means achieving both fast transient/settling time to obtain an appropriate transient response and a low THD to obtain a proper steady-state response [22,23]. It should be mentioned that transient time and THD are taken as two objective functions that must be simultaneously minimized, as well as settling time and THD. In fact, there is a compromise between two objective functions: the smaller the value of one, the higher the value of the other, and vice versa. In this research, first, the proposed controller is applied to acquire some transient/settling time and THDs, which are the same POF; therefore, the obtained results show the efficiency of this controller. Secondly, a multistage FOPID controller is used; different results are obtained for this, which include transient/settling time and THDs similar to the proposed controller. Eventually, the obtained results from both controllers are compared to show the better performance of the proposed method for a three-phase shunt active power filter. The key contributions of the present study are summarized as follows:

1. Fractional controller is applied due to its having more tunable parameters, which allow for more flexibility to achieve a high accuracy.
2. The cascade controller is able to rapidly reject disturbance before it leaks to the other parts of the system.
3. Multi-objective NSGA-II algorithm offers optimal solutions to multidimensional objective functions, which minimize the THD.

The rest of this paper is structured as follows: Section 2 outlines the system under study, which is a three-phase shunt active power filter with a high-performance repetitive controller. In Section 3, the proposed controllers are implemented. Section 4 describes the NSGA-II optimization method, objective functions, case studies, and design parameters. The real-time results are discussed in Section 5, and, finally, Section 6 summarizes the conclusions.

2. Shunt Active Power Filter and Repetitive Controller

Some devices, such as passive, active, or hybrid power filters and operation strategies, have been developed for the local correction of power-quality problems [40–43]. Since the performance of SAPFs is more dependent on the current control method, many current-control schemes have been proposed in the research [44–46]. However, in this research, a 25-kVA parallel active power Filter (Figure 1) with a high-performance repetitive controller (Figure 2) is optimized [22].

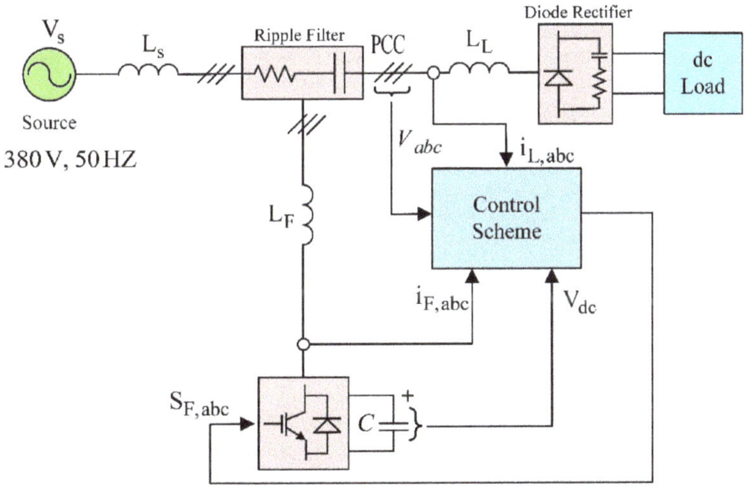

Figure 1. Structure of the 25-kVA SAPF [22,46].

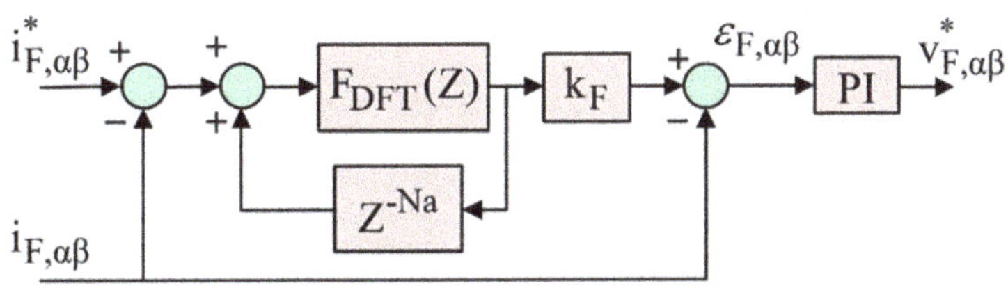

Figure 2. Structure of repetitive controller [22,46].

As seen in Figure 1, with these specifications Vs = 380 v, fs = 50 Hz and Is = 80 A [22], the functioning idea is based upon the injection of a compensating current into the network, which provides the basic reactive component and the harmonic currents due to the distorting load operation. Hence, a reference waveform for the current to be injected in the alternative current (AC) network should be provided by the control unit, so that the inverter is required to produce a current that is as close as possible to the reference. In Figure 1, L_S is the equivalent supply inductance, as seen by the bus where the active filter and the distorting load are connected; L_L is the equivalent inductance of the line supplying the load, while L_F is the inductance of the series inductor filter [47]. In 1981, the repetitive control notion was initially developed [48–50]. The primary motivations and representative examples include the rejection of periodic disturbances in a power supply control application [48,50] and the tracking of periodic reference inputs in a motion control application [49,50]. The repetitive controller is mainly used in continuous processes to track or reject periodic exogenous signals [50]. Although this controller has a high tracking operation, its operation is inherently slow. This controller is inserted in series with a used controller, which is a PI controller in this figure, as shown in Figure 2, and a discrete Fourier transform (DFT) is used. This DFT has a frequency response that almost equals the frequency response used to track the harmonic reference (Figure 2) [51]. Equation (1) gives us the discrete transfer function of the mentioned DFT.

$$F_{\mathrm{DFT}}(Z) = \frac{2}{N}\sum_{i=0}^{N-1}\left(\sum_{k\in N_h} cos\left[\frac{2\pi}{N}h(i+N_a)\right]\right)Z^{-i} \tag{1}$$

Here, N is the number of the coefficients; N_h is the set of selected harmonic frequencies, and N_a is the number of leading steps that are essential to guarantee the stability of the system. In fact, (1) can be considered a finite-impulse response (FIR) band pass filter of N taps with a unity gain at all selected harmonics h, and is also called a discrete cosine transform (DCT) filter [51].

3. Fractional Controllers

3.1. Fractional-Order PID Controller (FOPID Controller)

The traditional PID controllers are basic, robust, impressive, and easily implementable control techniques [25]. The transfer function of the PID controller is as follows:

$$T_{\mathrm{PID}}(s) = K_p + K_i S^{-1} + K_d S \tag{2}$$

In recent years, one of the best possibilities for improving the quality and robustness of PID controllers is to apply fractional-order controllers with non-integer derivation and integration parts [52,53]. The $PI^{\alpha}D^{\beta}$ controller generalizes the PID controller including an integrator of order α and a differentiator of order β.

The transfer function of the FOPID controller is acquired using the Laplace transformation, as given below:

$$T_{\mathrm{FOPID}}(s) = K_p + K_i S^{-\alpha} + K_d S^{\beta} \tag{3}$$

To design a FOPID controller, three parameters (K_p, K_i, K_d) and two non-integer orders (α, β) should be optimally determined.

3.2. Fractional-Order (PI + PD) Cascade Controller

As far as we know, it is difficult to achieve an excellent performance in terms of transient/steady-state response using a conventional PID controller. In this study, we applied a FOPI-FOPD cascade controller and a multistage FOPID controller instead of the traditional PI controller, as seen in Figure 2. Therefore, the FO (PI + PD) cascade controller is our proposed controller. It includes two controllers, which were connected in cascade, as shown in Figure 3. One of them is the FOPI controller and the other one is the FOPD controller. When the FOPI receives the ACE signal, the fractional-order PI controller produces a signal, which also operates as the input of another controller. The output of the FO (PI + PD) cascade controller is the reference power setting or control input.

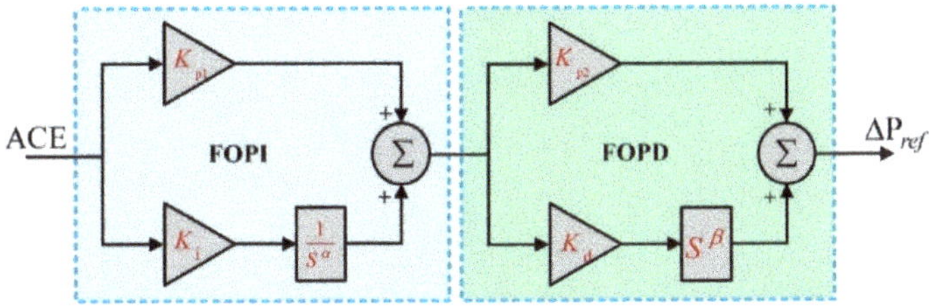

Figure 3. The structure of FO (PI + PD) controller.

ΔP_{ref} for the electric power systems to be controlled, as mathematically given by Equation (4):

$$\Delta P_{ref} = ACE \times \left(K_{p1} + K_i s^{-\alpha}\right) \times \left(K_{p2} + K_d s^{\beta}\right) \tag{4}$$

For $\alpha = 1$ and $\beta = 1$ the FOPI-FOPD cascade controller is transformed to a simpler form of conventional PI − PD cascade controller, i.e., K_{p1}, K_i, K_{p2}, K_d, α, and β are six variable parameters that must be optimized.

Three objectives contribute to the design of the FO (PI + PD) cascade controller. First of all, it should be economical, straightforward, and easy to apply and develop. As a result, its operation is comparable to that of a PID controller. Second, PI and PD controllers are cascaded, i.e., PI − PD, to combine the benefits of their distinct specifications and capabilities. On the other hand, a cascade controller has more adjustable parameters than a non-cascade controller, and it is obvious that if there are more adjustable parameters, the controller will provide a better system performance. Furthermore, the cascade controller is attractive because it can rapidly reject disturbances, before they reach the rest of the system. To comply with the third goal, a non-integer integrator/derivative order is considered, i.e., FOPI-FOPD, to enhance its freedom to design and promote PI – PD cascade controller performance [26].

3.3. Multistage Fractional-Order PID Controller

As stated before, it is difficult to obtain an excellent performance when applying a classic PID controller. According to Equation (2), increasing the integral gain to eliminate the steady-state error worsens the system's transient response. The existence of integral gain affects the speed and stability of the system during transient conditions, which leads to decreases in these parameters. To improve the transient response, the integrator must be disabled during the transient part [25]. A two-stage FOPD-FOPI controller with a first-stage fractional-order PD controller and a second-stage fractional-order PI controller can accomplish this. Sensors generate noise in an automated control system. This noise usually has a high frequency. Sometimes, the tie-line telemetry system generates noise. Due to this noise, if the derivative term is used, the plant input becomes excessively big. As a result, it can be removed by applying a first-order derivative filter that reduces the high-frequency noise. Figure 4 depicts the structure of the presented multistage FOPID controller. The transfer function of the multistage FOPID controller is represented by:

$$T_{\text{multistage-FOPID}}(s) = \left[K_p + K_d\left[\frac{N}{N + S^{\beta}}\right]\right] \times \left[1 + K_{pp} + \frac{K_i}{S^{\alpha}}\right] \tag{5}$$

In the controller scheme shown in Figure 4, K_p, K_d, β, K_i, α, K_{pp} and N are proportional, derivative, non-integer derivative, integral, non-integer integral, proportional gain, and filter coefficient, respectively. The input of the controller is Area Control Error (ACE), as well as output of the controller is (ΔF), which produces a control signal through these two stages. Afterward, this enters the power system. It is worth noting that the frequency deviation (ΔF) is the ACE in the case of a single-area system.

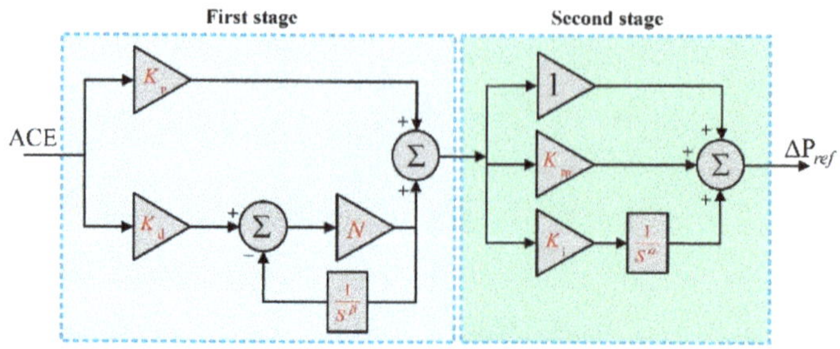

Figure 4. The structure of multi-stage FOPID controller.

4. NSGA-II Optimization Method and Objective Functions

4.1. NSGA-II: An Overview

NSGA-II is a popular multi-objective-optimization algorithm, which has three particular specifications: a speedy non-dominated sorting approach, prompt crowded distance estimate method and simple crowded comparison operator [54]. Typically, NSGA-II is described in detail as follows:

1. Population initialization:

The population must be initialized based upon the range of the problem and its limitations.

2. Non-dominated sorting process based upon non-domination criteria of the population that was initialized.
3. Crowding distance:

When the sorting is complete, the value of the crowding distance is determined in advance. The individuals in the population are chosen based on crowding distance and rating.

4. Selection:

Individuals are selected by applying a binary contest election with a crowded-comparison operator.

5. Genetic Operators:

Actual coded GA is achieved by applying simulated polynomial mutation and binary crossover.

6. Recombination and selection:

Population of children and population of the current generation are combined. The next generation is set by election. The new generation is filled by each front until the size of the population exceeds the current population size [55]. Figure 5 shows the NSGA-II procedure.

4.2. Objective Functions

The purpose of this work is to minimize the transient/steady-state response as two objective functions by the proposed controller based on the NSGA-II optimization technique. It offers optimal solutions to multidimensional objective functions [23]. Three objective functions have been chosen, which must be minimized in two case studies, as follows:

1. Steady-State Response (THD (up to the 50th harmonic) of the source current)

Steady-State Response: In electronics, steady-state is an equilibrium condition of a circuit that occurs when the effects of transients are no longer important. Steady-State determination is an important issue because many design features of electronic systems are given in terms of their steady-state characteristics. The periodic steady-state solution is also a prerequisite for small-signal dynamic modeling. The steady-state analysis is, therefore, an essential component of the design process.

Total harmonic distortion (THD) is a widely occupied concept when defining the level of harmonic content in alternating signals, which is measured in percentages.

2. Transient Response (Transient/Settling Time): In electrical engineering, transient response is the response of a system to changes from the equilibrium. The impulse response and step response are transient responses to a specific input (an impulse and a step, respectively).

Rise time or transient time (t_r) refers to the time required for a signal to alter from a specified low value to a specified high value. Usually, these values are 10% and 90% of the step height. Settling time (t_s) is the time needed for a response to become steady. This is defined as the time needed by the response to reach and remain within the determined range of from 2% to 5% of its final value. Therefore, the following two case studies were considered to be synchronously minimized:

Case study 1: THD (up to the 50th harmonic) and Transient (Rise) Time must be synchronously minimized.

Case study 2: THD (up to the 50th harmonic) and Settling Time must be synchronously minimized.

The set of designing parameters used to minimize the objective functions is presented in the next section.

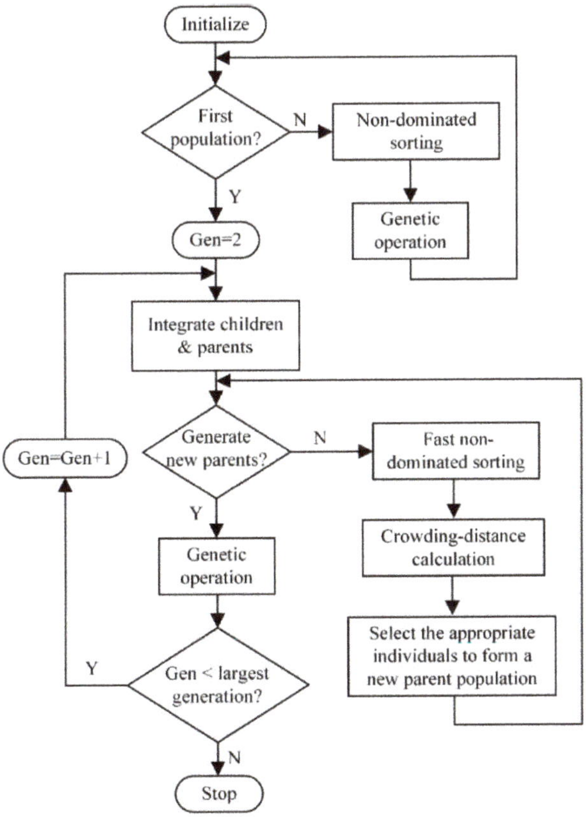

Figure 5. Flowchart of non-dominated sorting genetic algorithm (NSGA-II) [56].

4.3. Design Parameters

DC bus voltage (V_{dc}) and the FO (PI + PD) cascade controller parameters, which are K_{p1}, K_i, K_{p2}, K_d, α and β, in Figure 3, as well as V_{dc} and multistage FOPID controller parameters, which are K_p, K_d, β, K_i, α, K_{pp} and N in Figure 4, are determined based on the NSGA-II optimization technique. V_{dc} affects the transient response and the steady-state response in the shunt active power filter. In fact, it acts an important role to decrease current harmonics, i.e., THD. Hence, it was chosen as a design variable. In this optimization approach, the mentioned parameters are experimentally limited. These limitations dramatically reduce the computational time [41]. Therefore, it is said that K_{p1}, K_i, K_{p2}, K_d, α, β, V_{dc} are POS members for the FO (PI + PD) controller, and the POS for multistage FOPID controller has the following parameters:

K_p, K_i, K_{pp}, N, K_d, α, β, V_{dc}. As mentioned before, the different obtained values from POS members are called POF, and concern the values of the objective functions.

The general multi-objective optimization problem is considered as the following, with x as the design:

$$\begin{gathered}\text{Minimize}\\ g = f(x) = (f_1(x), \ldots f_i(x), \ \ldots f_k(x))\\ \text{Subject To :}\\ x = (x_1, x_2, \ldots x_n) \in X\end{gathered} \tag{6}$$

where k is the number of objective functions, n is the number of inequality constraints, x is a vector of design variables, and $f(x)$ is a vector of the objective functions to be minimized. Figure 6 depicts a Pareto front block diagram.

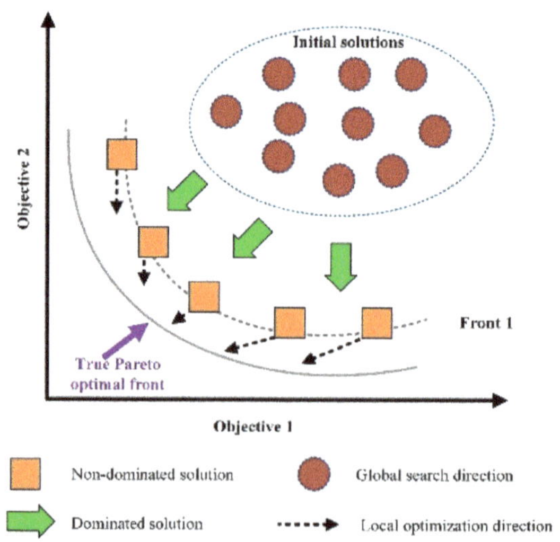

Figure 6. A Pareto front of the multi-objective problem.

Therefore, in this research, for the first case study, the goal is as follows:

$$\begin{gathered}\text{Minimize}\\ g = f(x) = (\text{THD(up to the 50th harmonic)}, t_r)\\ \text{Subject To :}\\ x = \left(K_{p1},\ K_i,\ K_{p2},\ K_d,\ \alpha,\ \beta,\ V_{dc}\right) \in X \textit{ for } \text{FO (PI + PD)controller},\\ x = \left(K_p,\ K_i,\ K_{pp},\ N,\ K_d,\ \alpha,\ \beta,\ V_{dc}\right) \in X \textit{ for } \text{multistage FOPID controller}\end{gathered} \tag{7}$$

For the second case study, the goal is as follows:

$$\begin{gathered}\text{Minimize}\\ g = f(x) = (\text{THD(up to the 50th harmonic)}, t_s)\\ \textit{Subject To}:\\ x = \left(K_{p1},\ K_i,\ K_{p2},\ K_d,\ \alpha,\ \beta,\ V_{dc}\right) \in X\ for\ \text{FO (PI + PD)controller},\\ x = \left(K_p,\ K_i,\ K_{pp},\ N,\ K_d,\ \alpha,\ \beta,\ V_{dc}\right) \in X\ for\ \text{multistage FOPID controller}\end{gathered} \tag{8}$$

5. Real-Time Simulation Results

In this research, the 25-kVA parallel APF in Figure was developed in the hardware-In-the Loop (HiL) to verify the efficiency of the proposed control scheme in the real-time framework. The HiL set-up based on the OPAL-RT simulator was adopted to consider the effects of the control errors and computation delays on the SAPF system (see Figure 7) [57]. The compensator was a three-phase PWM inverter with a switching frequency of 10 kHz. A 3.3 us dead time was also considered for the inverter's switches [22]. The NSGA-II algorithm in the aforementioned case studies was implemented for 20 generations. Each generation includes 30 individuals. For all the sections of the case studies, the remaining parameters for the repetitive controller were: $N = 200$, $N_a = 3$, and $k_f = 1$ [51]. For the FO (PI + PD) cascade controller and multistage FOPID controller, the Crone approximation with order 5 and frequency range equals [0.01; 1000] rad/s has been considered. According to the above description, all tables and figures related to POF show the optimization results after 20 generations based on the NSGA-II multi-objective optimization method.

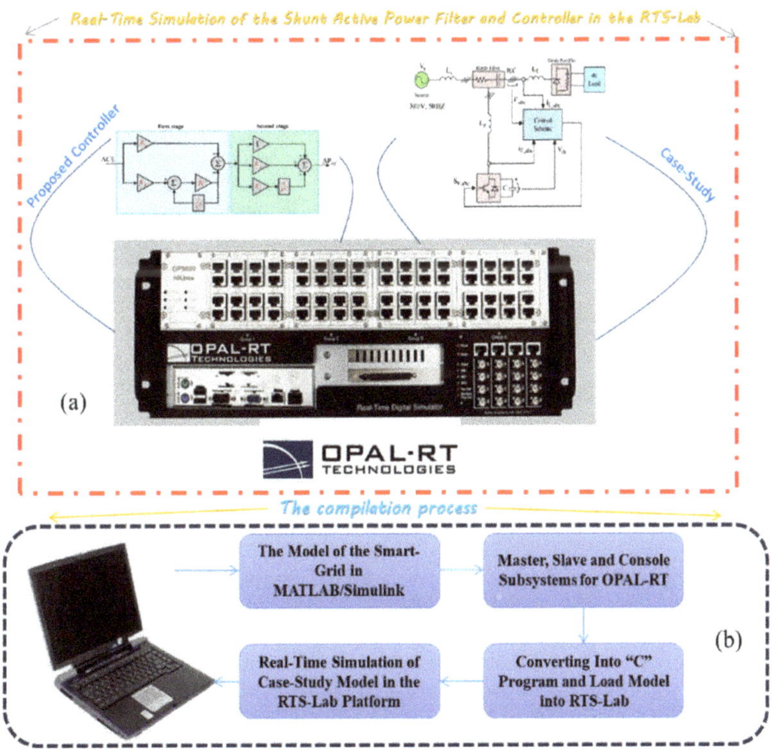

Figure 7. Process of HiL setup based on OPAL-RT (**a**) illustration of real-time simulation, (**b**) compilation process.

5.1. Case Study 1: THD (up to the 50th Harmonic) and Transient (Rise) Time Must Synchronously Be Minimized

Transient time and settling time can determine the transient response. In this case optimization, the rise time is the quantity that must be optimized with the THD. This time is defined as the time gap between the beginning of the compensation and the time when the THD starts to be lower than 5% [22]. Here, THD (up to the 50th harmonic) of the source current and rise time were synchronously minimized. The system outcomes achieved using an FO (PI + PD) cascade controller and multistage FOPID controller are as follows.

5.1.1. First Section of the First Case Study: Applying FO (PI + PD) Cascade Controller

In this work, the POS for FO (PI + PD) cascade controller included (K_{p1}, K_i, K_{p2}, K_d, α, β) and V_{dc}. The obtained results from POS members are known as POF; the POF is related to THD values and rise time. All obtained results are optimal, but the designer can pick one of them based on any other issues posed by the technical, economical, or managerial benefits requirements. The THD range and transient time for the FO (PI + PD) cascade controller are important from the technical viewpoint. According to Table 1, the lowest THD and the highest rise time are related to row no. 6; row no. 9 is concerned with the highest THD and the lowest rise time, as shown in Figure 8:

Table 1. POS and POF for FO (PI + PD) cascade controller.

No	K_{p1}	K_i	V_{dc}	K_{p2}	K_d	α	β	THD (%)	t_r(s)
1	2.4154	10.4602	815.0000	3.0000	0.4257	0.7481	0.3224	1.9673	0.0369
2	2.0321	10.4680	815.0000	2.5807	0.4287	0.7496	0.3224	1.9048	0.0387
3	1.9759	12.4680	815.0000	3.0000	0.4257	0.7481	0.3224	2.0416	0.0368
4	2.2689	12.4055	815.0000	3.0000	0.4287	0.7481	0.3224	2.0654	0.0367
5	1.9759	12.4680	815.0000	2.5807	0.4287	0.7481	0.3216	1.8360	0.0388
6	2.0321	12.4679	814.9904	2.5807	0.4287	0.7481	0.3224	1.8068	0.0388
7	2.2689	12.4055	814.9904	2.5807	0.4287	0.7481	0.3224	1.9122	0.0387
8	2.4154	10.4602	815.0000	2.9902	0.4703	0.7481	0.3302	2.1114	0.0366
9	2.2970	12.4680	814.6924	3.0000	0.4257	0.7481	0.3224	2.1513	0.0366
10	2.4154	10.4602	814.9994	2.9902	0.4257	0.7481	0.3224	2.1024	0.0366
11	2.4154	10.4602	815.0000	2.9207	0.4287	0.7481	0.3224	1.9616	0.0370
12	2.4154	10.4602	815.0000	2.5807	0.4287	0.7480	0.3212	1.9345	0.0385

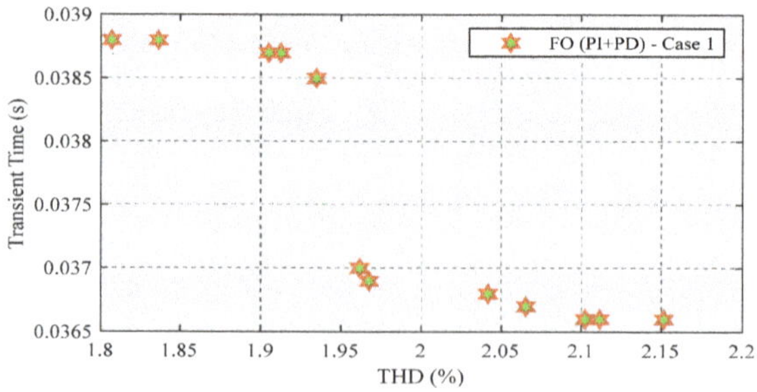

Figure 8. POF for FO (PI + PD) cascade controller.

The compensated source current (Figure 9) and THD (Figure 10) diagrams are related to row no. 6, also Is (Figure 11) and THD (Figure 12) diagrams are associated with row no. 9 as below:

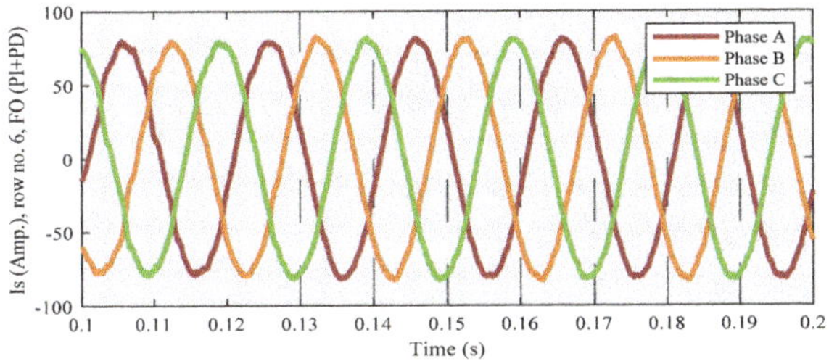

Figure 9. Compensated source current for row no. 6 (Lowest THD) using FO (PI + PD) cascade controller.

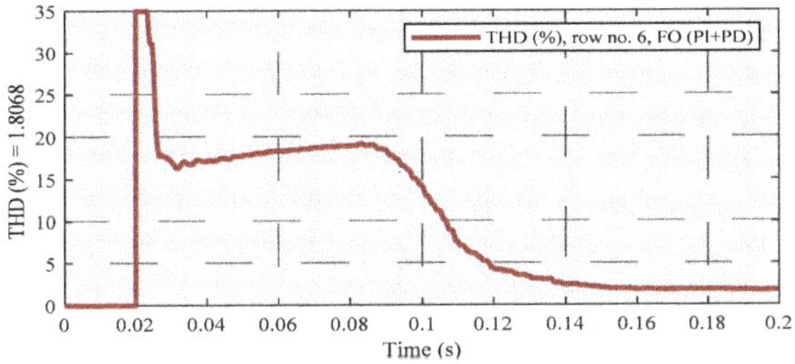

Figure 10. Row no. 6 (with the lowest THD at steady-state) using FO (PI + PD) cascade controller.

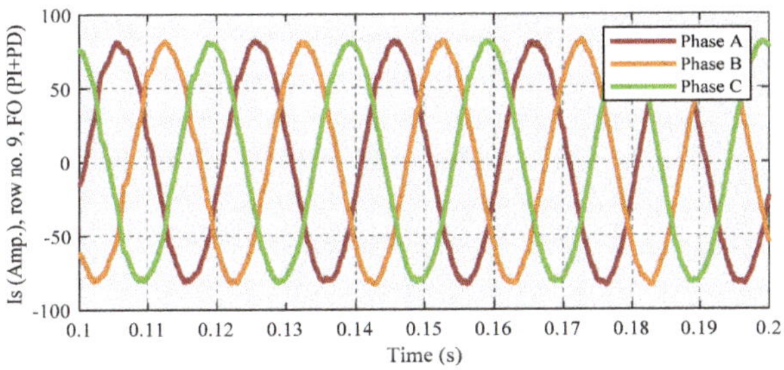

Figure 11. Compensated source current for row no. 9 (Highest THD) using FO (PI + PD) cascade controller.

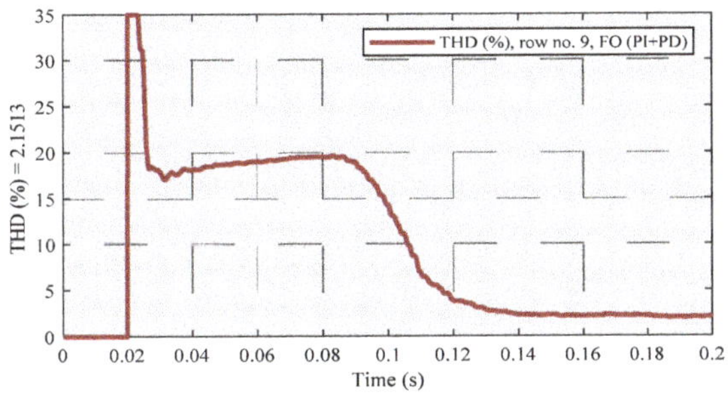

Figure 12. Row no. 9 (with the highest THD at steady-state) using FO (PI + PD) cascade controller.

Figure 10 (row no. 6) and Figure 12 (row no. 9) show the variations in THD using the FO (PI + PD) cascade controller. By looking at these two figures, we can see that the value of THD in Figure 10 is lower than the THD value in Figure 12. Additionally, this claim is valid for subsequent sections with similar conditions.

5.1.2. Second Section of the First Case Study: Applying Multistage FOPID Controller

According to Table 2, (K_p, K_i, V_{dc}, K_{pp}, N, K_d, α, β) are members of POS. THD and t_r are concerned with POF. In the case of variables for the multistage FOPID controller and V_{dc} has already been discussed. This table shows that row no. 2 is related to the lowest THD and the highest rise time; row no. 7 is associated with the highest THD and the lowest rise time, as shown in Figure 13.

Table 2. POS and POF for multistage FOPID controller.

No	K_p	K_i	V_{dc}	K_{pp}	N	K_d	α	β	THD (%)	t_r(s)
1	2.5400	12.4261	719.0607	11.2424	1.7693	0.9098	0.9000	0.0287	2.0274	0.0386
2	1.8393	10.7355	573.2281	11.7506	1.6734	0.1000	0.9408	0.0904	1.9219	0.0393
3	2.6000	13.2472	601.7378	11.7506	1.6734	0.6063	0.9041	0.0287	2.0992	0.0370
4	2.6000	10.0000	812.8484	11.7428	1.0054	0.2728	0.9683	0.0760	2.0621	0.0386
5	2.5315	13.8570	812.8484	11.7424	1.7693	0.2728	0.9683	0.0535	1.9946	0.0388
6	2.4969	13.8570	812.8484	11.6879	1.7547	0.9098	0.9683	0.0287	1.9299	0.0389
7	2.5764	11.1837	529.0360	11.3453	1.6186	0.4805	0.9335	0.0801	2.8242	0.0369
8	2.5764	11.5726	686.4316	11.9324	1.5213	0.6709	0.9335	0.0960	2.2386	0.0369
9	2.5764	11.5745	686.4515	11.9324	1.5213	0.6709	0.9336	0.0967	2.1409	0.0369
10	2.5764	11.5764	686.4515	11.9324	1.5213	0.6709	0.9336	0.0967	2.2158	0.0369

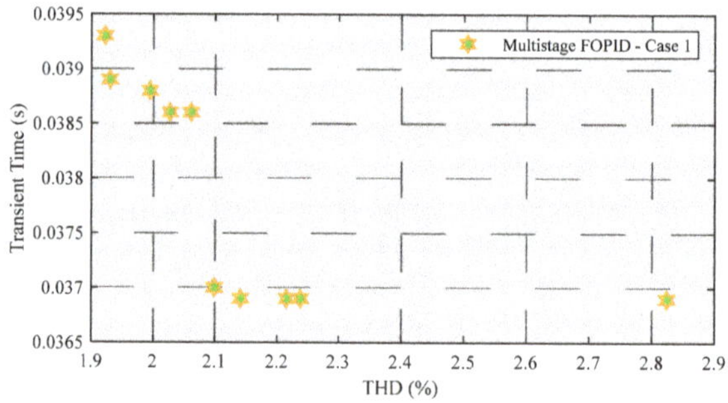

Figure 13. POF for multistage FOPID.

Is (Figure 14) and THD (Figure 15) diagrams are concerned with row no. 2, also the compensated source current (Figure 16) and THD (Figure 17) diagrams are related to row no. 7 as follows:

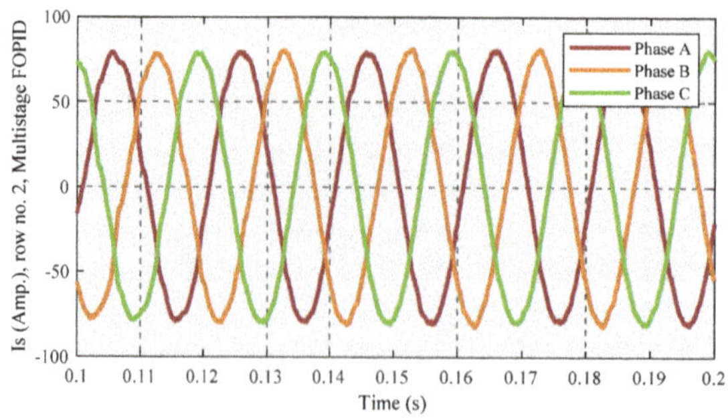

Figure 14. Compensated source current for row no. 2 (lowest THD) using the multistage FOPID controller.

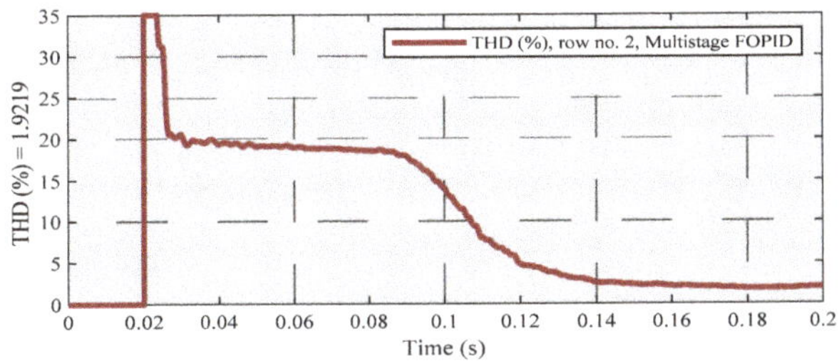

Figure 15. Row no. 2 (with the lowest THD at steady-state) using a multistage FOPID controller.

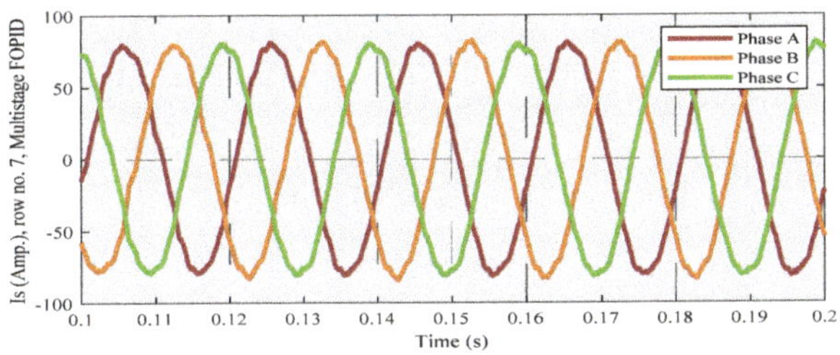

Figure 16. Compensated source current for row no. 7 (highest THD) using a multistage FOPID controller.

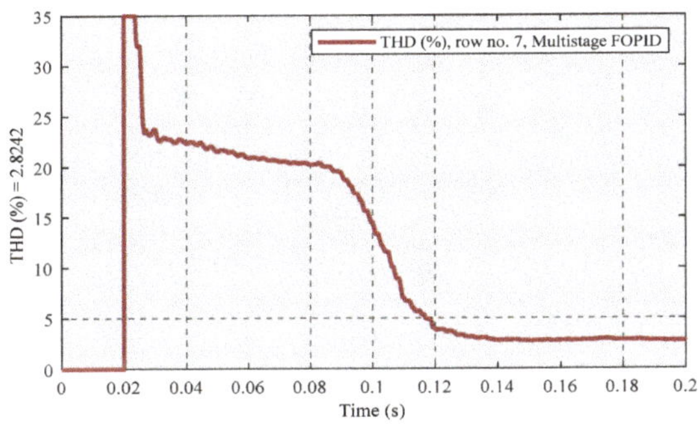

Figure 17. Row no. 7 (with the highest THD at steady-state) using a multistage FOPID controller.

Figure 15 (row no. 2) and Figure 17 (row no. 7) illustrate the changes in source current THD using a multistage FOPID controller. Therefore, by comparing Figures 15 and 17, it is obvious that the THD value in Figure 17 is higher than the THD value in Figure 15.

5.1.3. Third Section of the First Case Study: Comparison between FO (PI + PD) Cascade Controller and Multistage FOPID Controller

In this section, the real-time results of two controllers are compared with each other. Figure 18 depicts a comparison between the obtained POF values of FO (PI + PD) cascade controller and the acquired POF values of a multistage FOPID controller. This figure proves that the values obtained by the FO (PI + PD) cascade controller dominate all the values using a multistage FOPID controller. Figure 19, which has two parts, shows a comparison between the values of current THD that were obtained using these two controllers. The magnified part precisely demonstrates that the FO (PI + PD) cascade controller has better behavior than the multistage FOPID controller. Additionally, this figure shows that the THD (around 1.8068%) related to our proposed controller reached steady-state earlier than the THD (around 1.9219%), which is related to the other controller.

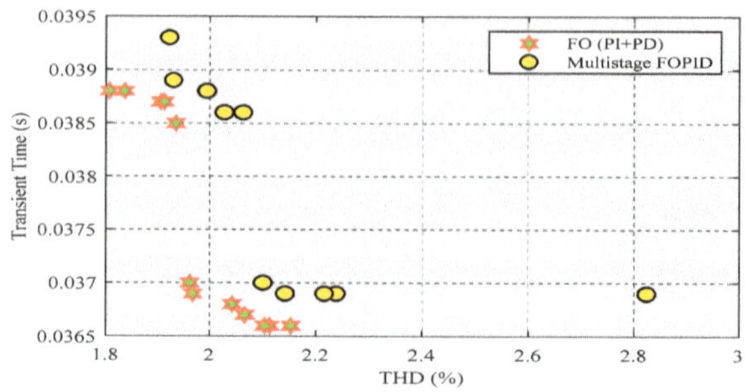

Figure 18. POF of FO (PI + PD) cascade and multistage FOPID controller—a comparison.

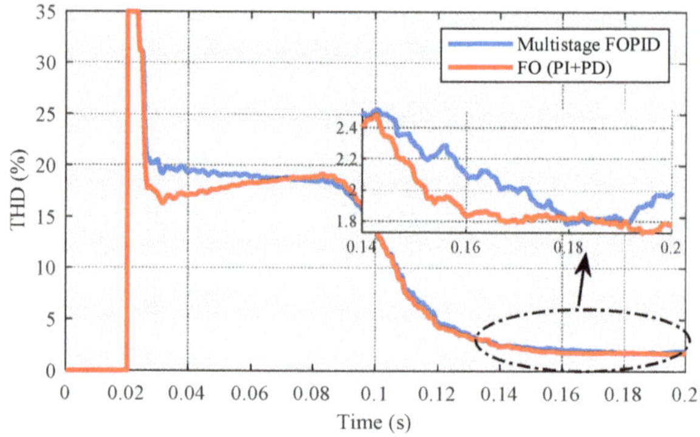

Figure 19. THD of "row no. 6 in Table 1" and "row no. 2 in Table 2"—a comparison.

In the following, it is worth noting that Figure 20 shows that the THD (around 2.1513%), which is related to row no. 9 in Table 1, is much lower than the THD (around 2.8242%), which is seen in row no. 7 in Table 2. As a result, this figure also confirms the superiority of the FO (PI +PD) cascade controller compared to the multistage FOPID controller in this research.

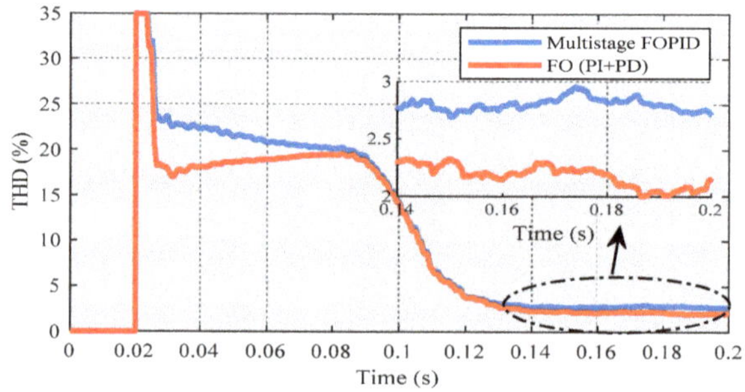

Figure 20. THD of "row no. 9 in Table 1" and "row no. 7 in Table 2"—a comparison.

5.2. Case Study 2: THD (up to the 50th Harmonic) and Settling Time Must Synchronously Be Minimized

In the second case study, the THD (up to the 50th harmonic) of the source current and settling time were chosen to be minimized at the same time. Therefore, a low THD is necessary, and the transient response of the compensator is momentous, especially when quick and frequentative variations occur in the load [22]. In this part, the settling time was computed based on the time needed for the source current THD to reach and stay inside a ±2% error band near its steady-state value. The system results that were obtained by means of FO (PI + PD) cascade controller and multistage FOPID controller are as follows.

5.2.1. First Section of the Second Case Study: Applying FO (PI + PD) Cascade Controller

According to Table 3, as stated, (K_{p1}, K_i, K_{p2}, K_d, α, β) and V_{dc} are related to POS. THD and t_s are associated with POF. This table indicates that the lowest THD and the highest settling time are related to row no. 3. Row no. 7 is concerned with the highest THD and the lowest settling time, as shown in Figure 21.

Table 3. POS and POF for FO (PI + PD) cascade controller.

NO	K_{p1}	K_i	V_{dc}	K_{p2}	K_d	α	β	THD (%)	t_s(s)
1	1.2000	13.9922	816.6667	2.1886	1.2999	0.8416	0.3667	1.8429	0.1130
2	1.2000	13.9922	816.6667	2.1618	0.0100	0.7333	0.6349	1.8836	0.1013
3	2.4667	13.4800	816.6667	2.2449	1.2999	0.8416	0.3667	1.7717	0.1197
4	3.0667	14.2904	552.2070	3.0000	1.3000	0.6000	0.3667	2.7445	0.0643
5	2.6000	10.0000	616.6667	2.1667	0.8700	0.6000	0.1000	2.0936	0.0720
6	2.6000	10.0000	616.6667	2.1618	0.8700	0.9999	0.1000	2.0609	0.0763
7	2.6000	10.0000	500.0000	2.1618	1.3000	0.6002	0.1000	2.7926	0.0590

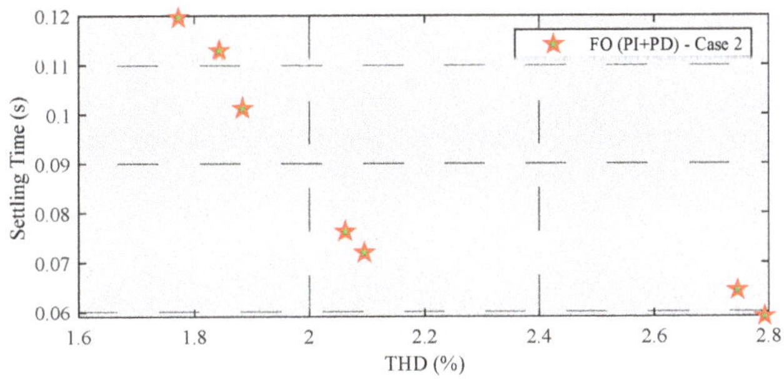

Figure 21. POF for FO (PI + PD) cascade controller.

For a fair comparison in the third section, "row no. 1" should be selected as an example, instead of "row no. 3", because the POF values related to the first row can thoroughly dominate the values of the corresponding POF in the next table. Therefore, the compensated source current (Figure 22) and THD (Figure 23) diagrams are associated with row no. 1, also Is (Figure 24) and THD (Figure 25) diagrams are concerned with row no. 7, as shown below:

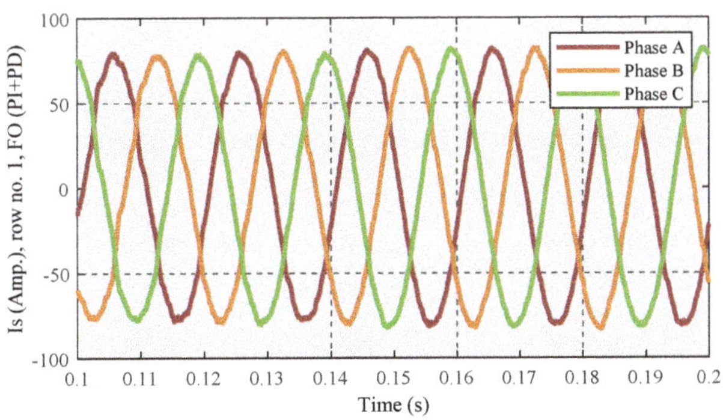

Figure 22. Compensated source current for row no. 1 (THD = 1.8429) using FO (PI + PD) cascade controller.

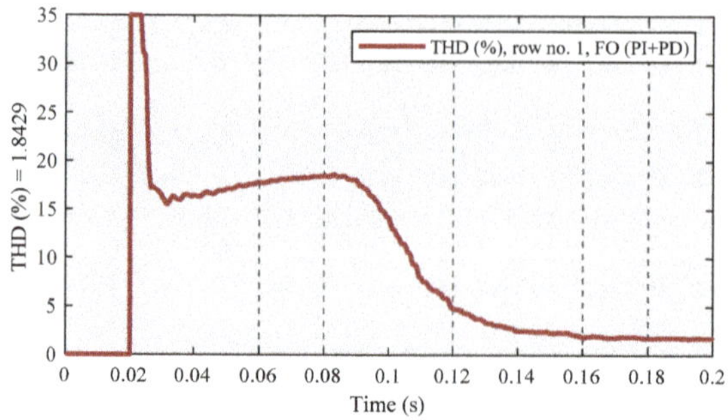

Figure 23. Row no. 1 (with THD = 1.8429 at steady-state) using FO (PI + PD) cascade controller.

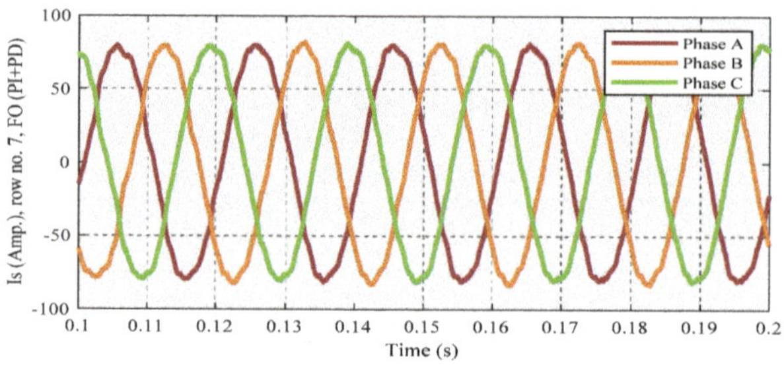

Figure 24. Compensated source current for row no. 7 (highest THD) using FO (PI + PD) cascade controller.

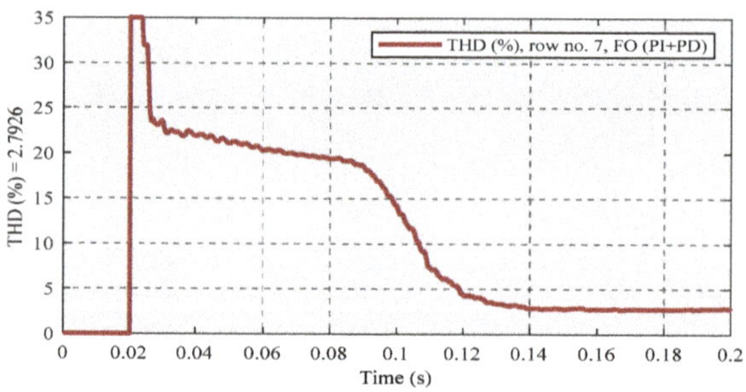

Figure 25. Row no. 7 (with the highest THD at steady-state) using FO (PI + PD) cascade controller.

Figure 23 (row no. 1) and Figure 25 (row no. 7) show the changes in THD using the FO (PI + PD) cascade controller.

Based on the above-mentioned results, we can recognize that Figure 25, which has a higher THD value, reached steady-state earlier than Figure 23. In other words, Figure 25 has a lower settling time than Figure 23.

5.2.2. Second Section of the Second Case Study: Applying Multistage FOPID Controller

According to Table 4, as previously mentioned, (K_p, K_i, V_{dc}, K_{pp}, N, K_d, α, β) are members of POS. THD and t_s are related to POF. This table shows that the lowest THD and the highest settling time are shown in row no. 1. The highest THD and the lowest settling time are shown in row no. 3, as seen in Figure 26:

Table 4. POS and POF for multistage FOPID controller.

NO	K_p	K_i	V_{dc}	K_{pp}	N	K_d	α	β	THD (%)	t_s(s)
1	2.5434	11.0508	794.7464	11.6264	1.3662	0.8016	0.9822	0.0984	1.8948	0.1180
2	2.5432	11.8920	763.7147	11.9952	1.2958	0.8016	0.9399	0.0321	2.1301	0.0829
3	2.4183	11.7547	490.0000	11.9461	0.1583	1.0000	0.9169	0.0886	2.9171	0.0597
4	2.1557	11.8764	763.7147	11.2523	0.2918	0.4703	0.9822	0.0984	1.9191	0.1100
5	2.1426	11.8998	763.7147	11.2523	0.2754	0.4703	0.9688	0.0984	1.9363	0.1065
6	2.1426	11.8920	763.7147	11.2523	0.2918	0.8016	0.9688	0.0666	1.9108	0.1153
7	2.3213	10.6331	547.4531	11.0189	1.2924	0.2266	0.9267	0.0921	2.2803	0.0762

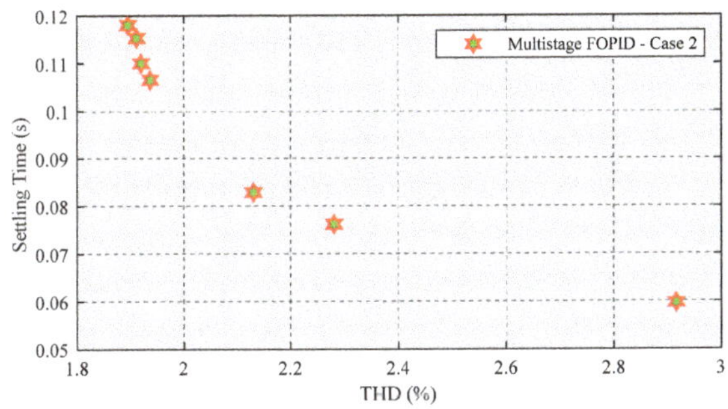

Figure 26. POF for multistage FOPID controller.

Is (Figure 27) and THD (Figure 28) diagrams are related to row no. 1, also Is (Figure 29) and THD (Figure 30) diagrams are associated with row no. 3, as follows:

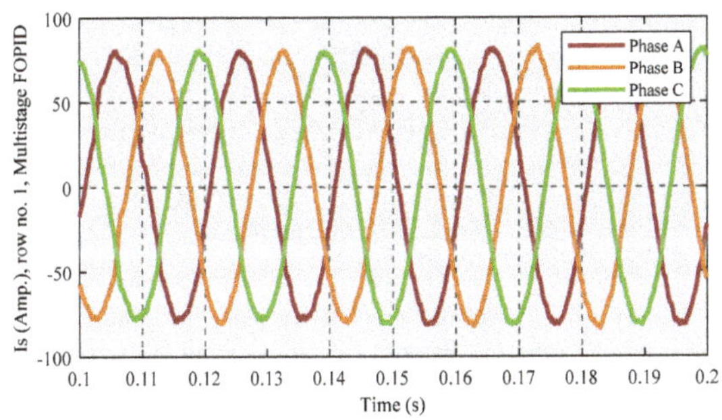

Figure 27. Compensated source current for row no. 1 (lowest THD) using multistage FOPID controller.

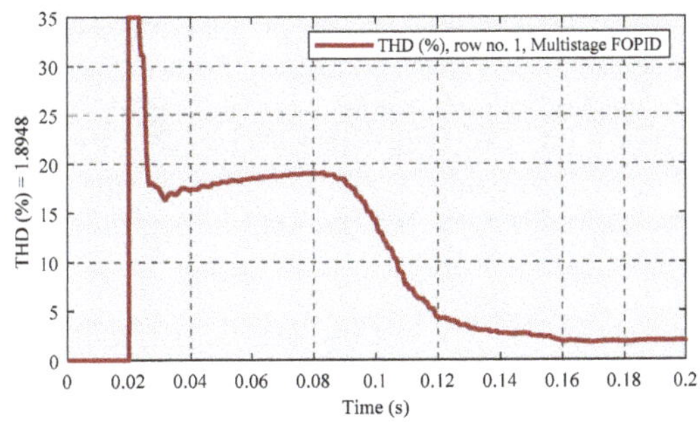

Figure 28. Row no. 1 (with the lowest THD at steady-state) using multistage FOPID controller.

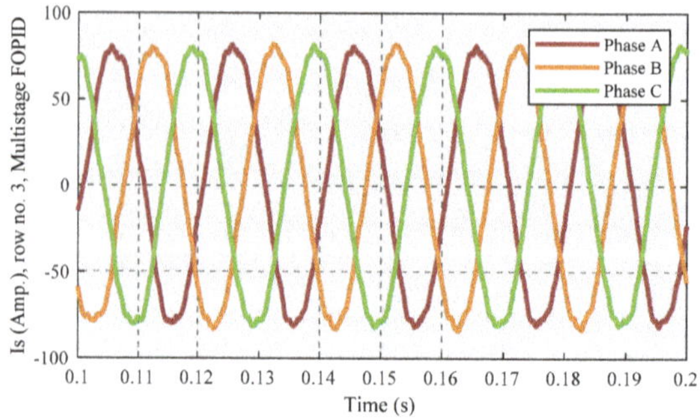

Figure 29. Compensated source current for row no. 3 (highest THD) using multistage FOPID controller.

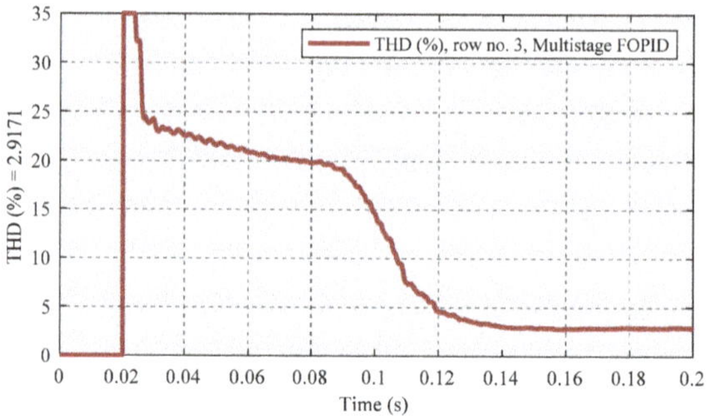

Figure 30. Row no. 3 (with the highest THD at steady-state) using multistage FOPID controller.

Figure 28 (row no. 1) and Figure 30 (row no. 3) indicate the variations in THD using a multistage FOPID controller. As in the two aforementioned figures, the difference between the values of THD and settling time is quite explicit.

5.2.3. Third Section of the Second Case Study: Comparison between FO (PI + PD) Cascade Controller and Multistage FOPID Controller

In this section, Figure 31 shows a comparison between Figures 21 and 26. As discussed earlier, Figure 21 is related to POF values, which were obtained using FO (PI + PD) cascade controller, and Figure 26 is concerned with POF that was achieved using the multistage FOPID controller. This figure confirms that the values acquired using the FO (PI + PD) cascade controller dominate the values by means of the multistage FOPID controller. Figure 32 shows a comparison between the THD values obtained by the mentioned controllers. The magnified part of this figure affirms that the multistage FOPID controller was dominated by the proposed controller. Moreover, this figure demonstrates that the THD (around 1.8429%), which is associated with the FO (PI + PD) controller, reached steady-state sooner than the THD (around 1.8948%) related to the multistage FOPID controller.

Finally, Figure 33 demonstrates that the THD (around 2.7926%), which is related to row no. 7 in Table 3, is lower than the THD (around 2.9171%), which is seen in row no. 3 in Table 4. Hence, this figure also affirms that the performance of the proposed controller is better than the multistage FOPID controller in this research.

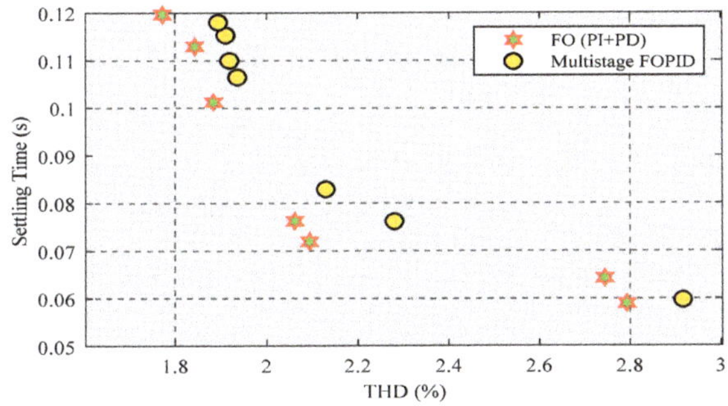

Figure 31. POF of FO (PI + PD) cascade and multistage FOPID controller—a comparison.

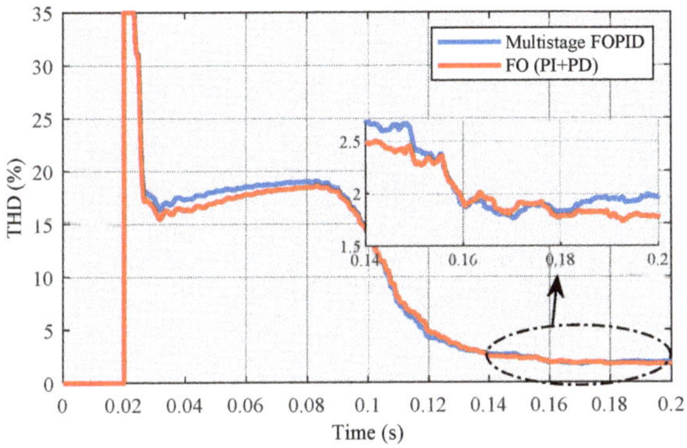

Figure 32. THD of "row no. 1 in Table 3" and "row no. 1 in Table 4"—a comparison.

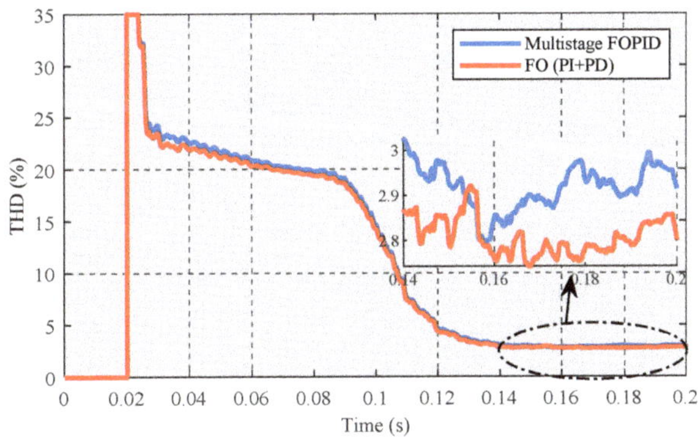

Figure 33. THD of "row no. 7 in Table 3" and "row no. 3 in Table 4"—A comparison.

5.3. Summary

In this section, this work and the obtained results are stated. The steps were as follows:

- We designed two different controllers to improve the performance of a shunt active power filter based on the NSGA-II optimization approach.
- The mentioned controllers were the FO (PI + PD) cascade controller and multistage FOPID controller.
- For the first time, we devised a multistage FOPID controller using the inspired multi-stage PID.
- FO (PI + PD) cascade controller was our proposed controller, which was compared with the other controller.

- The obtained results demonstrate that the first controller is superior to the other one. Table 5 shows the compared THDs with their corresponding t_r and t_s.

Table 5. POF for FO (PI + PD) cascade/ Multistage FOPID controller—a comparison.

NO	Controller	THD (%)	t_r(s)	THD (%)	t_s(s)
1	FO (PI + PD) cascade	1.8068	0.0388	1.8429	0.1130
2	Multistage FOPID	1.9219	0.0393	1.8948	0.1180
3	FO (PI + PD) cascade	2.1513	0.0366	2.7926	0.0590
4	Multistage FOPID	2.8242	0.0369	2.9171	0.0597

Row no. 1/2 and row no. 3/4 show the lowest/highest THD for each controller, respectively. The obtained POF demonstrates that the lowest values belong to the FO (PI + PD) cascade controller.

6. Conclusions

In this paper, two new controllers, called the FO (PI + PD) cascade and multistage FOPID controller, were employed to promote the performance of a 25-kVA parallel active power filter with a repetitive controller. They were devised based on the NSGA-II optimization method, and each controller was applied instead of the classic PI controller in the repetitive controller. Although both are powerful and practical, the FO (PI + PD) cascade controller was the proposed compensator in this study. It should be mentioned that the cascade controller can rapidly reject disturbance before it leaks to the other parts of the system. Eventually, real-time results based on the HiL setup proved that the intended controller has better behavior than the multistage FOPID controller in terms of its steady-state/transient response. Despite the successful performance of the proposed scheme, it suffers from a lack of adaptivity, because the control gains are adjusted in an offline manner. Therefore, the future work can be directed towards the development of robust control design using the training ability of neural networks.

Author Contributions: Investigation, H.N.K.; methodology, R.R.A., hardware, M.G.; supervision, M.-H.K. All authors have read and agreed to the published version of the manuscript.

Funding: This research received no external funding.

Institutional Review Board Statement: Not applicable.

Informed Consent Statement: Not applicable.

Conflicts of Interest: The authors declare no conflict of interest.

References

1. Chen, D.; Xiao, L.; Yan, W.; Li, Y.; Guo, Y. A harmonics detection method based on triangle orthogonal principle for shunt active power filter. *Energy Rep.* **2021**, *7*, 98–104. [CrossRef]
2. Xie, C.; Zhao, X.; Savaghebi, M.; Meng, L.; Guerrero, J.M.; Vasquez, J.C. Multirate fractional-order repetitive control of shunt active power filter suitable for microgrid applications. *IEEE J. Emerg. Sel. Top. Power Electron.* **2016**, *5*, 809–819. [CrossRef]
3. Buła, D.; Grabowski, D.; Maciążek, M. A Review on Optimization of Active Power Filter Placement and Sizing Methods. *Energies* **2022**, *15*, 1175. [CrossRef]
4. Karuppanan, P.; Mahapatra, K.K. PI and fuzzy logic controllers for shunt active power filter—A report. *ISA Trans.* **2012**, *51*, 163–169. [CrossRef]
5. Chebabhi, A.; Fellah, M.K.; Kessal, A.; Benkhoris, M.F. A new balancing three level three dimensional space vector modulation strategy for three level neutral point clamped four leg inverter based shunt active power filter controlling by nonlinear back stepping controllers. *ISA Trans.* **2016**, *63*, 328–342. [CrossRef]
6. Toumi, T.; Allali, A.; Meftouhi, A.; Abdelkhalek, O.; Benabdelkader, A.; Denai, M. Robust control of series active power filters for power quality enhancement in distribution grids: Simulation and experimental validation. *ISA Trans.* **2020**, *107*, 350–359. [CrossRef]
7. Pandove, G.; Singh, M. Robust repetitive control design for a three-phase four wire shunt active power filter. *IEEE Trans. Ind. Inform.* **2018**, *15*, 2810–2818. [CrossRef]

8. Sugavanam, K.R.; Sundaram, K.M.; Jeyabharath, R.; Veena, P. Convolutional Neural Network-based harmonic mitigation technique for an adaptive shunt active power filter. *Automatika* **2021**, *62*, 471–485. [CrossRef]
9. Li, D.; Wang, T.; Pan, W.; Ding, X.; Gong, J. A comprehensive review of improving power quality using active power filters. *Electr. Power Syst. Res.* **2021**, *199*, 107389. [CrossRef]
10. Zhang, H.; Liu, Y. Adaptive RBF neural network based on sliding mode controller for active power filter. *Int. J. Power Electron.* **2020**, *11*, 460–481. [CrossRef]
11. Sanjan, P.S.; Gowtham, N.; Bhaskar, M.S.; Subramaniam, U.; Almakhles, D.J.; Padmanaban, S.; Yamini, N.G. Enhancement of Power Quality in Domestic Loads Using Harmonic Filters. *IEEE Access* **2020**, *8*, 197730–197744. [CrossRef]
12. Morales-Caporal, R. Optimal indirect model predictive control for single-phase two-level shunt active power filters. *J. Power Electron.* **2022**, *22*, 84–93. [CrossRef]
13. França, B.W.; Aredes, M.; Da Silva, L.F.; Gontijo, G.F.; Tricarico, T.C.; Posada, J. An enhanced Shunt Active Filter based on Synchronverter concept. *IEEE J. Emerg. Sel. Top. Power Electron.* **2021**, *10*, 494–505. [CrossRef]
14. Hoon, Y.; Radzi, M.A.M.; Zainuri, M.A.A.M.; Zawawi, M.A.M. Shunt active power filter: A review on phase synchronization control techniques. *Electronics* **2019**, *8*, 791. [CrossRef]
15. Jauhari, M.; Widarsono, K.; Kurdianto, A.A. Shunt active power filter for harmonic mitigation based on PQ theory. In Proceedings of the International Conference on Electrical, Electronics and Information Engineering (ICEEIE), Denpasar, Indonesia, 3–4 October 2019; pp. 11–14.
16. Balasubramaniam, P.M.; Sudhakar, S.; Krishnamoorthy, S.; Sriram, V.P.; Dhanaraj, S.; Subramaniyaswamy, V.; Rajesh, T. An efficient control strategy of shunt active power filter for asymmetrical load condition using time domain approach. *J. Discret. Math. Sci. Cryptogr.* **2021**, *24*, 19–34. [CrossRef]
17. Upadhyay, S.; Singh, S. Shunt Active Power Filter (SAPF) Design and Analysis of Harmonics Mitigation in Three-Phase Three-Wire Distribution System. In *Computing Algorithms with Applications in Engineering*; Springer: Singapore, 2020; pp. 201–217.
18. Imam, A.A.; Kumar, R.S.; Al-Turki, Y.A. Modeling and Simulation of a PI Controlled Shunt Active Power Filter for Power Quality Enhancement Based on P-Q Theory. *Electronics* **2020**, *9*, 637. [CrossRef]
19. Frifita, K.; Boussak, M. A novel strategy for high performance fault tolerant control of shunt active power filter. *Electr. Eng.* **2022**, preview. [CrossRef]
20. Çelik, D. Lyapunov based harmonic compensation and charging with three phase shunt active power filter in electrical vehicle applications. *Int. J. Electr. Power Energy Syst.* **2022**, *136*, 107564. [CrossRef]
21. Bessa, R.J.; Trindade, A.; Miranda, V. Spatial-temporal solar power forecasting for smart grids. *IEEE Trans. Ind. Inf.* **2015**, *11*, 232–241. [CrossRef]
22. Rafiei, S.M.R.; Limongi, L.; Griva, G.; Bojoi, R. Optimal Design of High Performance Repetitive Controller for Shunt Active Filters Using Strength Pareto Multi-Objective Optimization Approach. In Proceedings of the 13th European Conference on Power Electronics and Applications, Barcelona, Spain, 8–10 September 2009.
23. Kashani, H.N.; Rafiei SM, R. Optimal control of active power filters using fractional order controllers based on NSGA-II optimization method. *Int. J. Electr. Power Energy Syst.* **2014**, *63*, 1008–1014. [CrossRef]
24. Tehrani, K.; Amirahmadi, A.; Rafiei, S.M.R.; Griva, G.; Barrandon, L.; Hamzaoui, M.; Rasoanarivo, I.; Sargos, F.M. Design of fractional order PID controller for boost converter based on Multi-Objective optimization. In Proceedings of the 14th International Power Electronics and Motion Control Conference, EPE-PEMC, Ohrid, Macedonia, 6–8 September 2010; pp. T3-179–T3-185. [CrossRef]
25. Sivalingam, R.; Chinnamuthu, S.; Dash, S.S. A hybrid stochastic fractal search and local unimodal sampling based multistage PDF plus (1+ PI) controller for automatic generation control of power systems. *J. Frankl. Inst.* **2017**, *354*, 4762–4783. [CrossRef]
26. Çelik, E. Design of new fractional order PI–fractional order PD cascade controller through dragonfly search algorithm for advanced load frequency control of power systems. *Soft Comput.* **2021**, *25*, 1193–1217. [CrossRef]
27. Zhou, X.; Shi, Y.; Zhu, J.; Zhao, L.; Zhu, Z. Structural multi-objective optimization on a MUAV-based pan–tilt for aerial remote sensing applications. *ISA Trans.* **2020**, *100*, 405–421. [CrossRef] [PubMed]
28. Ghogare, M.G.; Patil, S.L.; Patil, C.Y. Experimental validation of optimized fast terminal sliding mode control for level system. *ISA Trans.* 2021, in press. [CrossRef]
29. Xu, T.; Ren, Y.; Guo, L.; Wang, X.; Liang, L.; Wu, Y. Multi-objective robust optimization of active distribution networks considering uncertainties of photovoltaic. *Int. J. Electr. Power Energy Syst.* **2021**, *133*, 10719. [CrossRef]
30. Sadeghi, A.; Daneshvar, A.; Zaj, M.M. Combined ensemble multi-class SVM and fuzzy NSGA-II for trend forecasting and trading in Forex markets. *Expert Syst. Appl.* **2021**, *185*, 115566. [CrossRef]
31. Ghaderian, M.; Veysi, F. Multi-objective optimization of energy efficiency and thermal comfort in an existing office building using NSGA-II with fitness approximation: A case study. *J. Build. Eng.* **2021**, *41*, 102440. [CrossRef]
32. Jain, K.; Gupta, S.; Kumar, D. Multi-objective power distribution optimization using NSGA-II. *Int. J. Comput. Methods Eng. Sci. Mech.* **2021**, *22*, 235–243. [CrossRef]
33. Hu, H.X.; Shao, L.H.; Hu, Q.; Zhang, Y.; Hu, Z.Y. Multi-objective Reservoir Optimal Operation Based on GCN and NSGA-II Algorithm. In Proceedings of the 4th International Conference on Advanced Electronic Materials, Computers and Software Engineering (AEMCSE), Changsha, China, 26–28 March 2021; pp. 546–551.

34. Rego, M.F.; Pinto, J.C.E.; Cota, L.P.; Souza, M.J. A mathematical formulation and an NSGA-II algorithm for minimizing the makespan and energy cost under time-of-use electricity price in an unrelated parallel machine scheduling. *PeerJ Comput. Sci.* **2022**, *8*, e844. [CrossRef]
35. Ahmed, F.; Zhu, S.; Yu, G.; Luo, E. A potent numerical model coupled with multi-objective NSGA-II algorithm for the optimal design of Stirling engine. *Energy* **2022**, *247*, 123468. [CrossRef]
36. Suman, G.K.; Guerrero, J.M.; Roy, O.P. Stability of microgrid cluster with Diverse Energy Sources: A multi-objective solution using NSGA-II based controller. *Sustain. Energy Technol. Assess.* **2022**, *50*, 101834. [CrossRef]
37. Bagheri-Esfeh, H.; Dehghan, M.R. Multi-objective optimization of setpoint temperature of thermostats in residential buildings. *Energy Build.* **2022**, *261*, 111955. [CrossRef]
38. Bao, L.; Zheng, M.; Zhou, Q.; Gao, P.; Xu, Y.; Jiang, H. Multi-objective optimization of partition temperature of steel sheet by NSGA-II using response surface methodology. *Case Stud. Therm. Eng.* **2022**, *31*, 101818. [CrossRef]
39. Zhang, X.; Fan, X.; Yu, S.; Shan, A.; Fan, S.; Xiao, Y.; Dang, F. Intersection Signal Timing Optimization: A Multi-Objective Evolutionary Algorithm. *Sustainability* **2022**, *14*, 1506. [CrossRef]
40. Ghosh, A.; Ledwich, G. A unified power quality conditioner (UPQC) for simultaneous voltage and current compensation. *Electr. Power Syst. Res.* **2001**, *59*, 55–63. [CrossRef]
41. Rafiei, M.R.; Toliyat, H.A.; Ghazi, R.; Gopalarathanam, T. An optimal and flexible control strategy for active filtering and power factor correction under nonsinusoidal line voltages. *IEEE Trans. Power Delivery* **2001**, *16*, 297–305. [CrossRef]
42. Milanés, M.I.; Romero, E.; Barrero, F. Comparison of control strategies for shunt active power filters in three-phase four-wire systems. *IEEE Trans. Power Electron.* **2007**, *2*, 229–236.
43. da Silva, C.H.; Pereira, R.R.; da Silva LE, B.; Torres, G.L.; Pinto, J.O.P. Unified hybrid power quality conditioner (UHPQC). In Proceedings of the 2010 IEEE International Symposium on Industrial Electronics, Bari, Italy, 4–7 July 2010; pp. 1149–1153.
44. Buso, S.; Fasolo, S.; Malesani, L.; Mattavelli, P. A dead-beat adaptive hysteresis current control. *IEEE Trans. Ind. Appl.* **2000**, *36*, 1174–1180. [CrossRef]
45. Garcia-Cerrada, A.; Pinzon-Ardila, O.; Feliu-Batlle, V.; Roncero-Sanchez, P.; Garcia-Gonzalez, P. Application of a Repetitive Controller for a Three-Phase Active Power Filter. *IEEE Trans. Power Electron.* **2007**, *22*, 237–246. [CrossRef]
46. Limongi, L.R.; Bojoi, R.; Griva, G.; Tenconi, A. Performance comparison of DSP-based current controllers for three-phase active power filters. In Proceedings of the IEEE International Symposium on Industrial Electronics, Cambridge, UK, 30 June–2 July 2008; pp. 136–141.
47. Cavallini, A.; Montanari, G. Compensation strategies for shunt active-filter control. *IEEE Trans. Power Electron.* **1994**, *9*, 587–593. [CrossRef]
48. Inoue, T.; Nakano, M.; Kubo, T.; Matsumoto, S.; Baba, H. High Accuracy Control of a Proton Synchrotron Magnet Power Supply. *IFAC Proc. Vol.* **1981**, *14*, 3137–3142. [CrossRef]
49. Inoue, T.; Nakano, M.; Iwai, S. High accuracy control of servomechanism for repeated contouring. In Proceedings of the 10th Annual Symposium on Incremental Motion Control Systems and Devices, Rosemont, IL, USA, 1–4 June 1981; pp. 258–292.
50. Wang, Y.; Gao, F.; Doyle, F.J., III. Survey on iterative learning control, repetitive control, and run-to-run control. *J. Process Control.* **2009**, *19*, 1589–1600. [CrossRef]
51. Mattavelli, P.; Marafão, F. Repetitive-Based Control for Selective Harmonic Compensation in Active Power Filters. *IEEE Trans. Ind. Electron.* **2004**, *51*, 1018–1024. [CrossRef]
52. Khooban, M.H.; Gheisarnejad, M. Islanded microgrid frequency regulations concerning the integration of tidal power units: Real-time implementation. *IEEE Trans. Circuits Syst. II: Express Briefs* **2019**, *67*, 1099–1103. [CrossRef]
53. Gheisarnejad, M.; Khooban, M.H. An intelligent non-integer PID controller-based deep reinforcement learning: Implementation and experimental results. *IEEE Trans. Ind.Electron.* **2020**, *68*, 3609–3618. [CrossRef]
54. Deb, K.; Pratap, A.; Agarwal, S.; Meyarivan, T.A.M.T. A fast and elitist multiobjective genetic algorithm: NSGA-II. *IEEE Trans. Evol. Comput.* **2002**, *6*, 182–197. [CrossRef]
55. Yusoff, Y.; Ngadiman, M.S.; Zain, A.M. Overview of NSGA-II for Optimizing Machining Process Parameters. *Procedia Eng.* **2011**, *15*, 3978–3983. [CrossRef]
56. Liu, M.; Liu, R.; Zhu, Z.; Chu, C.; Man, X. A Bi-Objective Green Closed Loop Supply Chain Design Problem with Uncertain Demand. *Sustainability* **2018**, *10*, 967. [CrossRef]
57. Esfahani, Z.; Roohi, M.; Gheisarnejad, M.; Dragičević, T.; Khooban, M.-H. Optimal non-integer sliding mode control for frequency regulation in stand-alone modern power grids. *Appl. Sci.* **2019**, *9*, 3411. [CrossRef]

energies

Article

Nested Decomposition Approach for Dispatch Optimization of Large-Scale, Integrated Electricity, Methane and Hydrogen Infrastructures

Lukas Löhr *, Raphael Houben, Carolin Guntermann and Albert Moser

Institute of High Voltage Equipment and Grids, Digitalization and Energy Economics (IAEW), RWTH Aachen University, Schinkelstraße 6, 52062 Aachen, Germany; r.houben@iaew.rwth-aachen.de (R.H.); c.guntermann@iaew.rwth-aachen.de (C.G.); a.moser@iaew.rwth-aachen.de (A.M.)

* Correspondence: l.loehr@iaew.rwth-aachen.de; Tel.: +49-241-80-97651

Abstract: Energy system integration enables raising operational synergies by coupling the energy infrastructures for electricity, methane, and hydrogen. However, this coupling reinforces the infrastructure interdependencies, increasing the need for integrated modeling of these infrastructures. To analyze the cost-efficient, sustainable, and secure dispatch of applied, large-scale energy infrastructures, an extensive and non-linear optimization problem needs to be solved. This paper introduces a nested decomposition approach with three stages. The method enables an integrated and full-year consideration of large-scale multi-energy systems in hourly resolution, taking into account physical laws of power flows in electricity and gas transmission systems as boundary conditions. For this purpose, a zooming technique successively reduces the temporal scope while first increasing the spatial and last the technical resolution. A use case proves the applicability of the presented approach to large-scale energy systems. To this end, the model is applied to an integrated European energy system model with a detailed focus on Germany in a challenging transport situation. The use case demonstrates the temporal, regional, and cross-sectoral interdependencies in the dispatch of integrated energy infrastructures and thus the benefits of the introduced approach.

Keywords: multi-energy systems; optimal power and gas flow; dispatch optimization; hydrogen infrastructure; large-scale optimization; decomposition

Citation: Löhr, L.; Houben, R.; Guntermann, C.; Moser, A. Nested Decomposition Approach for Dispatch Optimization of Large-Scale, Integrated Electricity, Methane and Hydrogen Infrastructures. *Energies* **2022**, *15*, 2716. https://doi.org/10.3390/en15082716

Academic Editors: Zbigniew Leonowicz, Michał Jasinski and Arsalan Najafi

Received: 14 March 2022
Accepted: 5 April 2022
Published: 7 April 2022

1. Introduction

1.1. Motivation

In order to achieve greenhouse gas neutrality, energy policies like the Green Deal of the European Commission aim for energy system integration [1]. Besides high energy efficiency, integrated energy systems are characterized by a versatile energy mix that includes molecule-based energy carriers in addition to electricity [2]. These include natural gas transitionally and hydrogen, as well as biogenic and synthetic methane in the long term. Moreover, a coordinated and cross-sectoral operation of the energy infrastructures, hereinafter referred to as integrated energy infrastructures (IEI), is an important property of integrated energy systems [2].

In renewable energy systems, the electricity infrastructure needs to integrate large amounts of intermittent renewable energy sources (RES). This results in high demand for spatial and temporal flexibility in terms of transport as well as short-term and seasonal storage options [3]. A bidirectional coupling with the existing gas infrastructure by gas-fired power plants and power-to-gas plants can provide this flexibility [4]. In addition to the existing natural gas infrastructure, dedicated hydrogen infrastructures are supposed to integrate RES and supply a future hydrogen economy [5].

In order to derive design principles for the future system, a comprehensive understanding of interactions between these infrastructures in operation is required. For example, cost-benefit analyses can be applied to analyze, evaluate, and compare different concepts of IEI. Such analyses require a dispatch simulation in order to determine key indicators such as costs, emissions or energy not served [6,7].

For modeling the dispatch of IEI, different requirements arise. IEI enable raising system-wide synergies to operate the energy system cost-efficiently, sustainably, and securely. However, interdependencies between the coupled infrastructures increase. To adequately address these, an integrated modeling approach rather than a co-simulation approach is necessary [8]. Furthermore, the provision of temporal flexibility by IEI requires the modeling of a full year in at least hourly resolution [9]. Besides temporal resolution, the choice of spatial resolution has a strong impact on the results of energy system modeling [10]. To sufficiently model spatial flexibility, its transmission losses, and operating limits, the physical laws determining the power flows must be considered [8,11]. Finally, the modeling should be applicable to real interconnected energy systems, such as the European energy system to draw application-related conclusions. Thus, the following criteria serve as requirements for this paper:

- Integrated modeling of multi-energy-infrastructures
- Consideration of a full year in at least hourly resolution
- Consideration of transmission systems in nodal resolution and the physical laws determining their power flows (electricity and gas flows)
- Application on real, large-scale energy systems

1.2. Literature Review on Dispatch Models for IEI

In the following, selected models are analyzed with respect to the raised requirements. They either explicitly or implicitly simulate and optimize the dispatch of energy infrastructures. Table 1 provides an overview of selected models. The discussed models consider at least two coupled infrastructures and pursue an integrated modeling approach. This list does not claim to be complete but is intended to provide a broad overview of the literature.

Integrated electricity and gas market models represent a model class of dispatch models that focus on simulating the dispatch of power plants (Unit Commitment and Economic Dispatch) and gas supply [12–16]. Due to present dependencies on district heating systems, these are mostly modeled as additional boundary conditions for dispatch. Market simulations have a high application orientation and are therefore usually applied to real energy systems such as the European or American energy markets [12,14]. These models focus on a full-year consideration with high temporal resolution (1 h or 15 min) as well as a high level of technical detail in the modeling of power plants.

In contrast, the modeling of transmission networks for electricity is often simplified by considering exchange capacities between bidding zones, following the zonal electricity market design [12,13,16]. Transport within a bidding zone is then assumed to be free of congestion. Alternatively, market simulations can model the nodal electricity market design such as the commercial software *PLEXOS* [16]. *PLEXOS* applies the DC power flow approximation. In market simulations, the transmission networks for gas are usually modeled with a network flow algorithm with linear transfer capacities neglecting fluid mechanics. Market simulations often apply decomposition approaches such as Lagrange relaxation [17] to solve the resulting linear (LP) or mixed-integer problem (MIP) [13,16].

Investment models have similar qualities to market simulations since this model class needs to model the system dispatch to adequately derive investment decisions. In contrast to market simulations, investment models inherently consider the energy system as a whole, so that interactions between all relevant energy carriers are taken into account. In addition to aggregation in the spatial dimension, investment models such as *DIMENSION+* [18], *REMod-D* [19], *IKARUS* [20], and *PRIMES* [21] often aggregate in the temporal dimension to extrapolate the operating costs from type days or weeks. Other investment models like *TIMES* [22], *REMix* [23,24], and *PyPSA* [25] increase their spatial resolution by modeling smaller regions or even network nodes and applying a DC power flow (see next paragraph) between its interconnectors. Nemec-Begluk [26] develops a nested decomposition approach that allows modeling the transmission grid in nodal resolution with a DC power flow. Subsequent to the investment decisions, the total period is sliced into smaller sections. Again, gas infrastructures are not physically modeled in these approaches. Nonetheless, *REMix* uses a more detailed representation of the gas infrastructure compared to other investment models. It considers additional operational aspects of gas infrastructure dispatch such as estimations for driving energy of compressors [24].

Optimal power and gas flow models (OPGF) represent a class of dispatch models, which model the physical laws of power and gas flows in detail. Thus, they consider the physical potentials voltage and pressure. Since the integrated operation of the networks also includes the dispatch of conversion plants such as power-to-gas plants or gas-fired power plants, these models often optimize the dispatch of these plants in addition to network optimization. For these models, modeling the non-linear physical laws resulting from power and gas flows is the main challenge. The AC power flow equations describe the trigonometric and quadratic dependence of active and reactive power flow from voltage magnitude and phase angle. By assuming flat voltage profiles, small voltage angle differences and small resistance to reactance ratios (R/X), linear dependencies arise for modeling the active power flow [27]. This so-called DC power flow approximation is a permissible and common simplification for planning issues at transmission grid level [11]. Gas flows are determined by fluid mechanics and described by three differential equations for mass, momentum, and energy conservation [8]. Thus, transient modeling of dynamic gas flows requires high spatial and temporal resolution. Complexity can be reduced by assuming steady-state conditions, which is a common simplification for planning issues of gas transmission networks [28]. However, this still results in a non-linear system of equations [29]. Under so called quasi-steady-state conditions, slow dynamics of mass conservation of gases and linepack flexibility can be simplified considered in hourly resolution [29–32].

The commercial software *SAInt* [8,33] provides an integrated modeling approach with AC power flows and transient gas flows. Source [30,34] also model the physical flows in detail. Other approaches like [31,35–39] apply the DC power flow approximation as well as steady-state gas flow assumptions to reduce the complexity of the OPGF problem. Sources [29–32] consider simplified gas dynamics by using a quasi-steady-state formulation. The non-linear optimization problem is often solved by applying piecewise linearization approaches like in [35,38,39] or non-linear solvers like in [30,34,37]. The resulting problem is often applied to small test systems and periods of usually 24 h since these approaches are difficult to scale up for large problem sizes [29]. Chaudry et al. [36] solve the OPGF problem for large-scale systems and 24 h using a commercial solver based on successive linear programming (SLP). However, they also address problems with further scalability. Löhr et al. [29] introduce a SLP-based approach showing good scaling properties. It is applied to a power and gas transmission system with over 500 nodes each for a 24 h period.

The OPGF problem commonly considers the electricity and natural gas infrastructure. A bidirectional coupling of electricity and gas infrastructure is a comparatively new aspect. Therefore, power-to-gas plants are only modeled in [29,30,33]. Schwele et al. [40] additionally consider heat infrastructures with physical thermal power flows. Hydrogen transport grids are a new research topic in energy system analysis and are not considered explicitly in the presented literature. The *Energy Hub Concept* by Geidl [41,42] basically allows modeling any number of infrastructures and conversion processes between different energy carriers as well as AC power and steady-state gas flows.

Therefore, this literature review shows a research gap. On the one hand, there are energy system models, that can be applied for long periods and large-scale systems but have a low spatial and technical resolution when modeling transmission infrastructures. On the other hand, there are models with high spatial and technical detail, but can only be applied to short periods and small systems. Thus, to the best of the author's knowledge, no model that meets all listed requirements for dispatch simulation of IEI exists.

Table 1. Overview of selected existing dispatch models for integrated energy infrastructures.

References	Name	Model Class	Energy Carrier	Physics Power/Gas	Spatial Scope/Resolution	Temporal Scope/Resolution
[12]	Riepin et al.	Market	electricity, gas	network flow	Europe/country	year/1 h
[13]	Baumann	Market	electricity, gas	network flow	Europe/country	year/1 h (1 d)
[14,15]	Quelhas et al.	Market	electricity, gas, coal	network flow	USA/US regions	year/1 h
[16]	*PLEXOS*	Market	electricity, gas, heat	DC/network flow	USA/Europe/nodal	year/($\geq$1 min)
[21]	*PRIMES*	Invest	total energy system	DC/network flow	Europe /country	years/1 h (type)
[23,24]	*REMix*	Invest	total energy system	DC/network flow	Germany/regions	years/1 h
[25]	*PyPSA*	Invest	total energy system	DC/network flow	Europe/nodal	years/1 h
[26]	Nemec-Begluk	Invest	electricity, gas, heat	DC/network flow	Austria/nodal	year/1 h
[41,42]	Geidl	OPGF	all (modular)	AC/steady state	3×3 test/nodal	day/1 h
[38]	Correa-P. et al.	OPGF	electricity, gas	DC/steady state	24×20 test/nodal	day/1 h
[30]	Sun et al.	OPGF	electricity, gas	AC/q.-steady state	24×20 test/nodal	day/1 h
[40]	Schwele et al.	OPGF	electricity, gas, heat	DC/steady state	$24 \times 12 \times 3$ test/nodal	day/1 h
[8,33]	*SAInt*, Pambour	OPGF	electricity, gas	AC/transient	30×25 test/nodal	day/1 h
[36]	Chaudry et al.	OPGF	electricity, gas	DC/steady state	16×168 UK/nodal	day/1 h
[29]	Löhr et al.	OPGF	electricity, gas	DC/q.-steady state	542×524 GER/nodal	day/1 h

1.3. Contribution and Paper Organization

Hence, the purpose of this paper is to introduce a method that enables dispatch modeling for IEI meeting the requirements listed in Section 1.1. The novelty of this method is the combined capability of modeling non-linear physical power and gas flows while allowing applicability to large-scale systems and long periods.

The introduced model is based on an integrated optimization approach that models electricity, methane, and hydrogen infrastructures "as a whole" integrated energy infrastructure. A DC-power flow and a quasi-steady-state gas flow formulation allow detailed analyses of IEI on grid node level. Thus, network bottlenecks and transport losses can be determined and the temporal flexibility of gas infrastructures through linepacking can be considered. The basic optimization problem describing the dispatch problem for IEI is formulated in Appendix A.

These specifications result in a complex mathematical problem, that cannot be solved in a closed-loop optimization with currently available solvers and resources. To enable this level of detail, various model reduction and decomposition techniques are applied. The approach of this paper builds on the SLP-based OPGF model introduced in [29]. This model is integrated into a three-staged nested heuristic. The nested decomposition approach applies successive zooming techniques that focus first on the temporal, then on the spatial, and finally on the technical dimension. Complexity is handled by model reductions of the other dimensions in each stage, which enables scalability to an entire year in hourly resolution and large-scale systems in high technical detail. The main contribution of this paper is to demonstrate the combined application of several model reduction and decomposition techniques to handle application-oriented, large-scale problems. Therefore, it focuses on the methodology presented in Section 2.

The closing investigations in Section 3 are intended to prove the applicability of the approach to large-scale systems and illustrate the temporal, spatial, and cross-sectoral interdependencies in IEI and therefore the benefits of the introduced approach. For this purpose, a use case of the future interconnected European energy system in 2040 with a focused analysis of the dispatch in Germany is considered. Application-oriented analyses of European energy infrastructure with the presented spatial, temporal, and technical scope also represent a novelty. Section 4 concludes the main findings of this paper.

2. Nested Decomposition Approach

2.1. Integrated Dispatch Optimization Problem

The nested decomposition approach is based on an integrated dispatch optimization problem. For reasons of clarity, this is only briefly characterized in this section and is presented mathematically in Appendix A.

The dispatch optimization problem considers the energy infrastructure in an integrated manner as a coherent graph for the energy carrier electricity (AC and DC), hydrogen, and methane. The nodes of the IEI graph describe busbars and gas stations. Branches either connect two nodes within an infrastructure or connect two infrastructures with each other (conversion plants). In the power system, branches represent AC lines, DC lines, and (phase-shifting) transformers. In the infrastructures for gases, branches model pipelines, compressors, and pressure regulating valves. The considered conversion plants are electrolyzers, gas-fired power plants, fuel cells, steam methane reforming, and methanation plants. Power and gas storage, as well as other feed-in and feed-out plants, are connected to nodes. Degrees of freedom of other feed-in and feed-out plants is the dispatch of power plants, gas imports, RES curtailment, or demand-side response (DSR).

The objective (A25) of the optimization is to minimize dispatch costs. These are in particular fuel costs, costs for DSR, and costs for loss of load (energy not served). Perfect foresight is assumed. The following constraints must be considered when dispatching the system:

- Node balance equation in each infrastructure (A1)
- DC power flow Equations (A3) and (A5)
- Electrical transmission loss Equation (non-linear) (A6)
- Steady-state pressure loss Equation (non-linear) (A7)
- Linepack Equations (non-linear) (A8)–(A11)
- Horsepower equations for losses in compression (non-linear) (A16)–(A18)
- Conversion continuity Equation (A19)
- Storage continuity Equation (A20)
- Technical operation limits in Equations (A2), (A4), (A12)–(A15), (A21)–(A24)

The resulting OPGF problem thus uses the DC load flow approximation including transmission losses and a quasi-steady state gas flow formulation. The formulated optimization problem has a linear objective function, linear and non-linear constraints, and no integer variables.

2.2. Analysis of Complexity Drivers

In energy system modeling, complexity can result due to various drivers in technological, temporal, and spatial dimension [43]. One driver is the pure size of the problem, which results from considering multiple infrastructures, large-scale systems, or large periods [44]. Linking variables and constraints, which increase dependencies between different technologies, regions, and time steps, make it difficult to decompose the problem [43]. Another complexity driver results from non-linear and integer relations, which make the problem harder and complicate the application of efficient algorithms [43,45].

The formulated dispatch optimization problem already avoids some complexity drivers such as integer decisions and non-linear objective functions. Nevertheless, non-linear constraints remain to model the hydraulic gas flows, linepack, compressors, as well as electrical losses. Moreover, the problem is extensive for large network sizes and periods. If the stated problem is applied to the European scenario in Section 3, this results in nearly 400 million variables and over 140 million constraints. Among them are several linking constraints. The population of the coefficient matrix is suitable to identify and visualize the structure of the optimization problem and its linking constraints [43,44]. Figure 1 shows the population of the coefficient matrix of the linearized integrated dispatch problem in full technical detail (see Section 2.7) when applied to the European scenario.

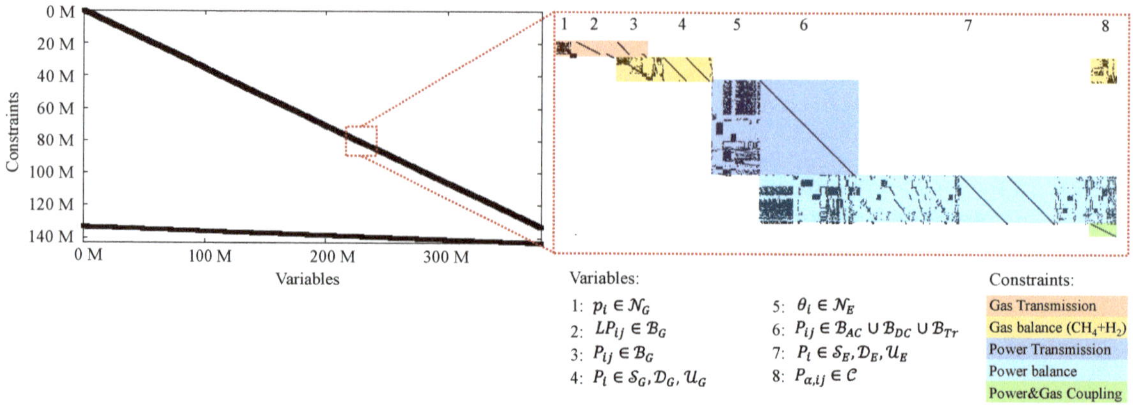

Figure 1. Structure of coefficient matrix (model input).

The coefficient matrix shows at the top a diagonal block structure, which is typical for energy system models [44]. A block represents a discrete time step $t \in \mathcal{T}$, that consists of constraints for nodal balances and physical transport for electricity and gases, respectively. In addition, a block contains coupling constraints between the infrastructures. Electrical or thermal power flows on the lines is included in both the transport constraints and the nodal balance. In contrast, physical potentials and linepack variables are only incorporated into the transport equations, feed-ins, and feed-outs only into the nodal balance. Conversion plants are an exception as they are additionally present in coupling constraints. These constraints represent spatial and technological linking constraints. Temporal linking constraints, which result from storages and linepack, form a second diagonal at the bottom of the matrix. These connect variables of different blocks with each other.

2.3. Overview on Model Reduction and Decomposition Techniques

Various techniques exist for reducing modeling complexity. First, model reduction techniques are briefly introduced. According to [46], model reduction techniques can be subdivided into techniques of slicing and aggregation in the dimensions of time, space, and technology. Slicing approaches consider a subproblem of the original problem, reducing its scope and neglecting interactions between subproblems. Slicing is especially effective to decompose linking constraints and variables. In contrast, aggregation approaches reduce the model detail while keeping the full scope of the problem. Among others, aggregation can be useful to relax non-linear or integer relations. Table 2 presents an overview of common model reduction techniques.

Table 2. Model reduction techniques based on [46].

Dimension	Slicing	Aggregation
Time	• Division of periods in subperiods • Analysis of type-hours/days/weeks	• Down sampling temporal resolution
Space	• Neglection of peripheral systems • Decoupling of interconnected systems	• Clustering of network nodes • Network reduction algorithms
Technology	• Neglection of technologies and infrastructures	• Merging single plants to classes • Simplified models (e.g., linearization)

Accordingly, slicing narrows the scope of the problem, for example by considering subperiods or subregions and neglecting whole infrastructures or technologies. Aggregation reduces temporal or spatial granularity and simplifies technological models. Network reduction algorithms such as the Ward-method for power systems represent an example of spatial aggregation [47]. This method aims to reduce network nodes in peripheral areas while still representing the physical power flows within the focus area. For this purpose, suitable virtual lines and feed-in/outs that correspond to the electrical behavior of the peripheral area are determined. The simplified physical power and gas flow formulations introduced in Section 1.2 represent technical aggregation methods.

According to the principle "divide and conquer", decomposition techniques intend to make model size and complexity manageable by dividing the problem. A distinction can be made between mathematically exact methods and heuristic approaches [46]. Mathematically exact methods reformulate the problem and solve it in an iterative process. Examples are the Dantzig-Wolfe—or the Benders Decomposition, which divides the problem into one main and (various) sub-problems [48,49]. Lagrange Relaxation transfers linking constraints into the objective function [17]. Lagrange Relaxation and Dantzig-Wolfe Decomposition are commonly used to deal with linking constraints, whereas Benders Decomposition is more promising to reduce computation time dealing with linking variables. While these methods maintain the guarantee for optimality, they are also not applicable to large-scale problems with heterogeneous complexity drivers, such as those resulting from the stated requirements. In contrast, heuristic approaches such as nested decomposition approaches can be specifically designed to find a near-optimal, but not guaranteed optimal, solution for an individual problem [46]. However, such a solution may be sufficient for the purpose of energy system analysis. In nested decomposition approaches, the problem is sequentially and coordinately solved multiple times by applying different model reduction techniques in each stage [46]. The solutions of each stage can serve as boundary conditions for the subsequent stages. Zooming techniques represent an example of nested decomposition approaches. Thereby, wide scopes are first modeled in low resolution to afterward model subsections in greater detail. The interactions between the subsections are transferred from the previous stages. Zooming techniques are often applied in temporal dimensions [44] but can also be applied in spatial or technological dimensions. For a detailed description of reduction and decomposition techniques, please refer to [43,44,46].

2.4. Applied Model Reduction and Decomposition Techniques

The developed nested approach primarily decomposes the optimization problem in the temporal dimension due to the shown structure of the coefficient matrix. It applies a zooming technique, where first the total period is considered by performing model reductions in technical and spatial dimension. In two following stages, the technical and spatial level of detail is increased successively by slicing the total period into smaller subperiods. This approach enables parallel computing, resulting in advantages in computation time. Table 3 summarizes the model reduction techniques of the three stages applied. Figure 2 gives a schematic overview of the nested decomposition approach. In the following, the three stages are described in detail.

Table 3. Applied model reduction techniques.

Dimension	Stage 1	Stage 2	Stage 3
Time		• Slicing to subperiods	• Further slicing in subperiods
Space	• Aggregation on country level		• Aggregation by network reduction
Technology	• Aggregation to plant classes • Aggregation to network flow (power and gas)	• Aggregation to network flow (gas)	

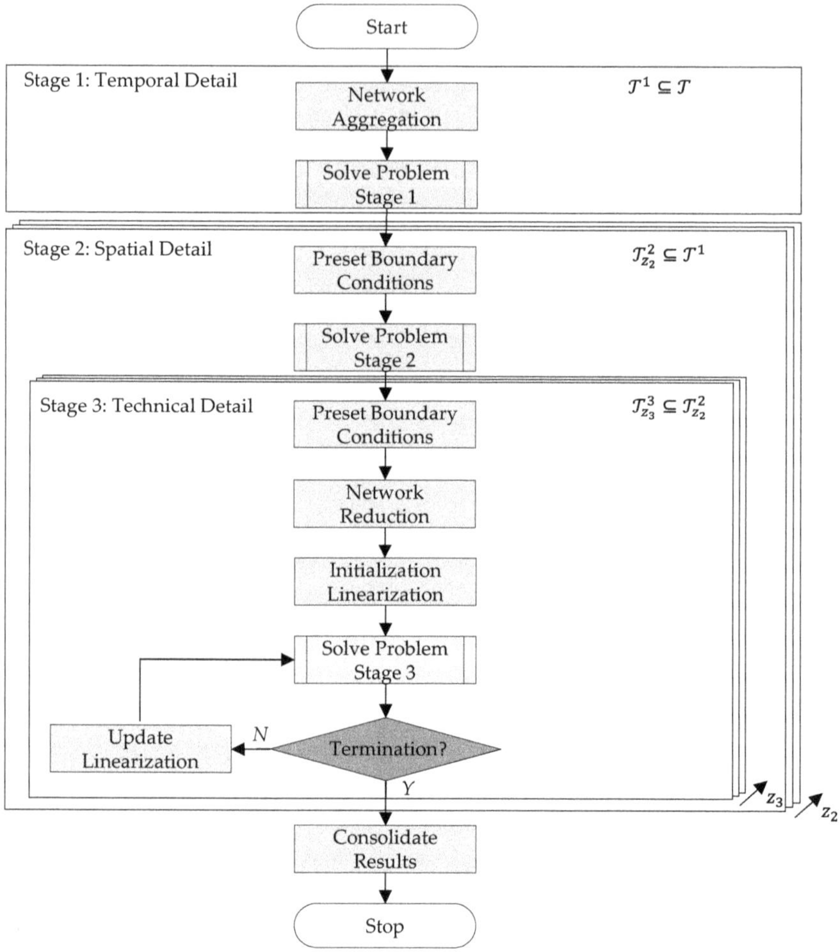

Figure 2. Flowchart of the nested decomposition approach.

2.5. Stage 1: Storage Level Optimization in Full Temporal Detail

The first stage aims to adequately model the dispatch of seasonal storage for electricity and gases. Thus, the focus of this stage is on the temporal dimension. It considers a full year $\mathcal{T}_1 = \mathcal{T}$ in hourly resolution. To solve this large-scale optimization problem, nodes are aggregated in the spatial dimension to defined regions $\mathcal{R}$ ("Network Aggregation"). In this paper, regions are defined at the country level. This aggregation is done for all electricity nodes and gas nodes of each gas type in one region. Further model reductions are applied in the technical dimension. On the one hand, physical laws for power and gas flows are neglected. Instead, the exchange between regions is modeled using a network flow algorithm and network losses are roughly estimated as additional feed-out depending on the residual load of each region. Thus, Equations (A3)–(A18) are not considered in this stage. Exchange capacities result from aggregating interconnector capacities. In the power system, only 70% of the (already reduced) transport capacity of each line is considered as exchange capacity to account for loop flows. Capacities in gas networks are estimated depending on the pipe geometry (pipe length l and diameter d), maximum velocity w_{max} and heating value h_u of the gas. The thermodynamic state of the gas such as the compressibility factor K is estimated based on nominal quantities such as the rated pressure p_N of the pipeline.

$$P_{G,ij}^{max} = \frac{p_N}{p_n} \cdot \frac{T_n}{T_m} \cdot \frac{1}{K_m} \cdot w_{max} \cdot \frac{\pi}{4} \cdot d^2 \cdot h_u \quad (1)$$

On the other hand, plants such as power plants or conversion plants are aggregated into plant classes to further reduce complexity. Storages are an exception to this principle. All model reductions result in a linear optimization problem that can be solved in manageable computing time ("Solve Problem Stage 1").

2.6. Stage 2: Simplified Dispatch Optimization in Full Spatial Detail

The objective of the second stage is still a linear but more detailed estimation of the dispatch of IEI compared to stage 1. Stage 2 derives further constraints for the subsequent final dispatch optimization. Full complexity in spatial and higher complexity in technical dimensions is enabled by slicing the total period $\mathcal{T}$ into subperiods $\mathcal{T}_z^2$. An applicable period duration in stage 2 is 168 h (1 week).

$$\bigcup_{z_2 \in \mathcal{Z}^2} \mathcal{T}_{z_2}^2 = \mathcal{T}, \quad \bigcap_{z_2 \in \mathcal{Z}^2} \mathcal{T}_{z_2}^2 = \varnothing \quad (2)$$

The target storage level at the beginning and end of a sub-period is transferred from the first stage solution ("Preset Boundary Conditions"). This maintains the information about the total period dispatch of storage. In stage 2, all plants are modeled individually, and the transport networks are considered on a nodal level in all defined regions $\mathcal{R}$. The DC power flow approximation is applied in the electricity network. In the gas network, a network flow under consideration of (1) is still assumed. Again, electric network losses are estimated as additional feed-outs. Thus, Equations (A6)–(A18) are not considered in this stage. The resulting linear optimization problem is solved ("Solve Problem Stage 2").

2.7. Stage 3: Dispatch Optimization in Full Technical Detail

The third stage identifies the final dispatch considering the full detail of the formulated optimization problem in Appendix A. To allow this level of detail, model reductions must be made in other dimensions. In the temporal dimension, the subperiods of the second stage are further divided. An applicable period duration in stage 3 is 24 h (1 day).

$$\bigcup_{z_3 \in \mathcal{Z}^3} \mathcal{T}_{z_3}^3 = \mathcal{T}, \quad \bigcap_{z_3 \in \mathcal{Z}^3} \mathcal{T}_{z_3}^3 = \varnothing \quad (3)$$

In spatial dimension, the considered regions are divided into focus regions $\mathcal{R}_{focus} \subseteq \mathcal{R}$ and external regions $\mathcal{R}_{ex} \subseteq \mathcal{R}$. The dispatch of plants in external regions is captured from the second stage. Thus, the exchange with the focus area serves as a boundary condition. In addition, storage levels at the beginning and end of a sub-period are transferred from stage 2 solution ("Preset Boundary Conditions"). A network reduction is performed around the focus area ("Network Reduction"). In the electricity network, the Ward method is applied to reduce external nodes but maintain the electric behavior of the network in the focus area. In contrast, hydraulic interdependencies with the external system are neglected in the gas network, since gas flows can be dispatched well by controllers. The full technical model scope is considered in the focus region. This includes in particular the consideration of physical gas flows, electric network losses, and linepack in addition to the DC power flow. Thus, the conservation of mass in the gas network is considered within the periods. Since the time-coupled constraints of the linepack are of short-term nature, dependencies between subperiods are neglected to reduce complexity. The resulting non-linear optimization problem is solved using the successive linearization approach introduced and validated in [29]. In the focus region, the results from stage 2 serve as starting solution for the SLP algorithm ("Initialize Linearization"). The problem is solved successively ("Solve Problem Stage 3"). In each iteration, the linearization is updated, and the intervals of variable bounds are reduced to achieve convergence of the SLP algorithm ("Update Linearization"). The SLP algorithm is terminated when the linearization error of the pressure losses and electrical losses becomes marginal, and the value of the objective function no longer changes significantly. The results of the nested decomposition approach are consolidated by updating the results from stage 2 by the results from stage 3 in the focus regions ("Consolidate Results").

3. Application of the Approach

The introduced nested decomposition approach is applied to an integrated European energy system scenario for the year 2040 to demonstrate its applicability to large-scale systems. The European states are defined as regions. In the third stage, Germany is considered a focus region to analyze the operational restrictions for long-distance transport in detail. The problem is calculated for a full year in hourly resolution. Therefore, stage 1 considers 8760 h time coupled. Stage 2 slices the year into weekly periods (~52 × 168 h), stage 3 in daily periods (365 × 24 h). Parallel computing on the HPC infrastructure of RWTH Aachen University enables a total calculation time of about 3 h. Stage 1 requires about half of the time (100 min). Subsequently, the subperiods in stage 2 and seven successive periods in stage 3 are calculated in one instance. A total of 53 instances are calculated in parallel. The average computation time of an instance is about 90 min since a subperiod of stage 2 requires 15 min, a subperiod of stage 3 requires 10 min on average. It usually takes 10 iterations until the convergence of the SLP algorithm. The approach is implemented in C++ using Gurobi 9.1.1 (Gurobi Optimization, LLC., Beaverton, OR, USA).

3.1. Scenario Description

The considered scenario is based on the Global Ambition 2040 scenario of TYNDP 2022 (draft) [50]. This scenario assumes ambitious RES expansion targets that enable large-scale hydrogen production and transport in Europe. Tables 4–6 show the framework data of the scenario. Table 4 shows the commodity and CO_2 prices. In Table 5, demand data for Europe and the focus area Germany are given as well as installed capacities of RES and flexibility options such as power plants, storages, electrolyzers or DSR. In addition, domestic biomethane and conventional natural gas potentials are given. Table 6 provides import potentials for natural gas and green hydrogen in Europe. The import potentials of [50] are scaled upward according to the additional hydrogen demand, due to a larger spatial scope in this scenario. Figure 3 presents the applied European infrastructure models. These include the electric transmission grid and its power plant fleet, the natural gas transmission grid, and a visionary hydrogen grid. RES capacities from photovoltaic (PV), wind, biomass and run of river are not shown for clarity. The infrastructure models build on databases and models of IAEW at RWTH Aachen University. The electric transmission grid model and power plant fleet are based on publicly available sources, in particular the ENTSO-E Grid map [51] and decentralized data research. Expansion projects up to 2040 are integrated and remaining structural bottlenecks are eliminated by further line expansion. The natural gas transmission grid is also built on several publicly available sources, most notably ENTSOG Grid Map [52] and Rövekamp [53]. The hydrogen network is based on current designs of the European Hydrogen backbone [5].

Table 4. Costs assumption (modified based on [50,54]).

Type	Costs
Nuclear	1.7 €/MWh_{th}
Hard Coal	6.9 €/MWh_{th}
Oil	34.6 €/MWh_{th}
Biomethane	61.0 €/MWh_{th}
Natural Gas imports	14.7 €/MWh_{th}
Green Hydrogen Imports	57.9 €/MWh_{th}
Demand Side Response	1000 €/MWh_{el}
Value of Lost Load Electricity	5000 €/MWh_{el}
Value of Lost Load Gases	1000 €/MWh_{th}
CO_2	123 €/tCO_2

Table 5. Scenario data (modified based on [50]).

Scenario 2040	Europe	Germany
Final Energy Demand Electricity	3994 TWh/a	718 TWh/a
Final Energy Demand Methane	2439 TWh/a	435 TWh/a
Final Energy Demand Hydrogen	1391 TWh/a	345 TWh/a
Installed Capacity Nuclear	120 GW	0 GW
Installed Capacity Hard Coal	3 GW	0 GW
Installed Capacity Methane	233 GW	35 GW
Installed Capacity Hydrogen	0.5 GW	0 GW
Installed Capacity Oil	8 GW	3 GW
Installed Capacity Other Non-RES	38 GW	9 GW
Installed Capacity PV	828 GW	124 GW
Installed Capacity Wind Onshore	630 GW	117 GW
Installed Capacity Wind Offshore	341 GW	52 GW
Installed Capacity Run of River	58 GW	5 GW
Installed Capacity Hydro Storage	190 GW	10 GW
Installed Capacity Other RES	35 GW	9 GW
Installed Capacity Battery	40 GW	7 GW
Installed Capacity Electrolysis	160 GW	20 GW
Installed Capacity Methanation	10 GW	3 GW
Installed Capacity DSR	18 GW	3 GW
Biomethane Potential	749 TWh/a	51 TWh/a
Conventional Methane Production	106 TWh/a	0 TWh/a

Table 6. Import potentials in Europe (modified based on [50]).

Source	Natural Gas	Hydrogen
Russia	2044 TWh/a	216 TWh/a
Norway	818 TWh/a	244 TWh/a
North Africa	337 TWh/a	291 TWh/a
South East	514 TWh/a	0 TWh/a
LNG	1515 TWh/a	-

3.2. Analysis of Dispatch Costs

The dispatch costs represent the value of the objective function (A25) of the integrated dispatch optimization problem. It is composed of the dispatch costs of stage 2 of the nested optimization process updated by the costs of stage 3 for the focus region. Table 7 shows the different types of dispatch costs for all considered regions.

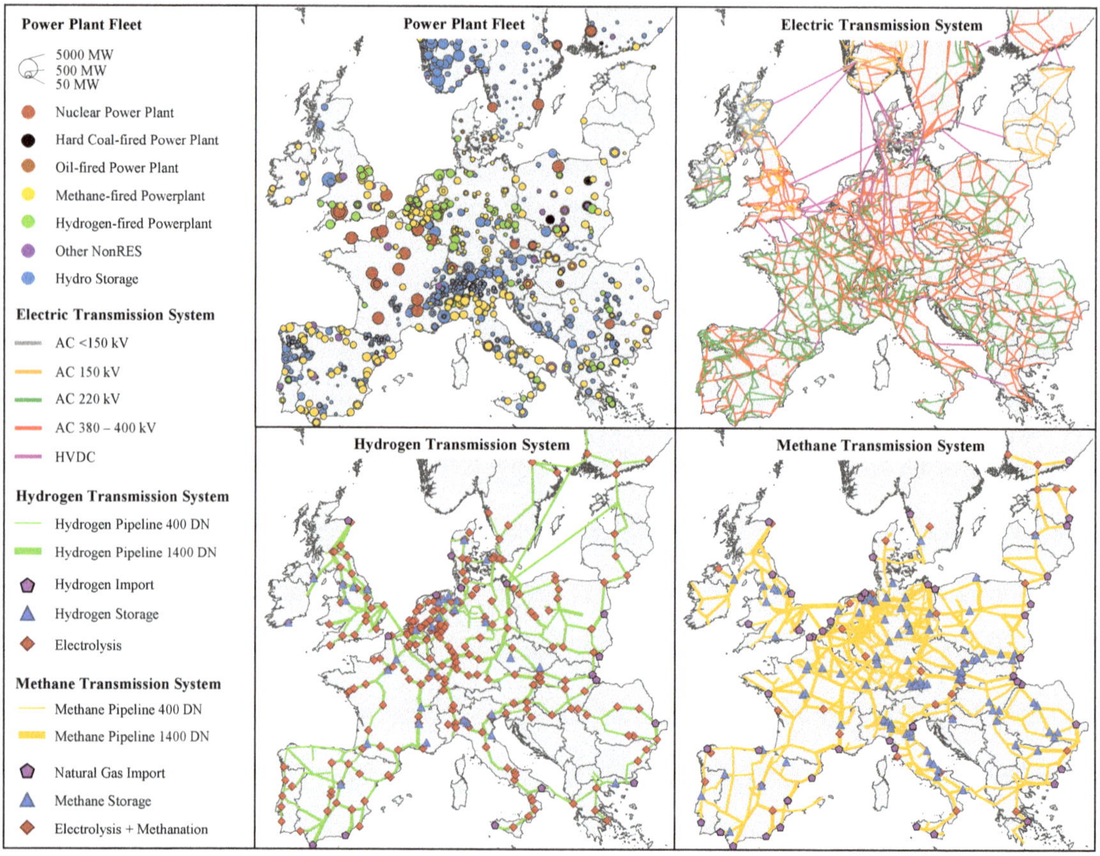

Figure 3. Considered infrastructure models: power plant fleet (**top left**), electric transmission system (**top right**), hydrogen transmission system (**bottom left**), natural gas (methane) transmission system (**bottom right**).

Table 7. European dispatch costs (model result).

Type	Costs
Nuclear Power Plants	3.07 bn €/a
Hard Coal-fired Power Plants [1]	0.06 bn €/a
Oil Power Plants [1]	0.02 bn €/a
Biomass Power Plants	9.90 bn €/a
Natural Gas Imports [1]	73.12 bn €/a
Natural Gas Conventional Production [1]	4.19 bn €/a
Biomethane Production	45.60 bn €/a
Green Hydrogen Imports	37.35 bn €/a
Demand Side Response	0.66 bn €/a
Value of Lost Load Electricity	0.98 bn €/a
Value of Lost Load Gases	0.00 bn €/a
SUM	147.97 bn €/a

[1] including CO_2 costs.

Power generation from power plants excluding gas-fired power plants accounts for only 7.5% of the system's total dispatch costs. This is remarkable since electricity demand constitutes more than 50% of the total final energy demand considered. This can be explained by the high share of RES generation without marginal costs and high nuclear power generation with low fuel costs (see Section 3.4). In contrast, methane supply accounts for 70% of the total system costs with a final demand of only 30%. The high costs result from fuel and CO_2 costs for natural gas imports and conventional production as well as biomethane production. Unlike hydrogen supply, renewable methane production at a low marginal cost is small. Low-cost electricity is mainly used for hydrogen production due to higher hydrogen import costs and better efficiency at conversion. Therefore, hydrogen supply accounts for 21% of the total dispatch costs by 18% of the total final energy demand. Additional dispatch costs arise from DSR with scheduled load shedding. However, these are low at below 1% of the total system costs. Moreover, unscheduled load shedding (ENS) is necessary for the power system. A structural bottleneck in Poland's electricity transmission system results in dispatch costs of 1 bn. €. There is no ENS in the methane and hydrogen system.

The analysis of the total system dispatch costs, e.g., on a European scale, thus represents an output of the presented method.

3.3. Analysis of Seasonal Dispatch

The introduced nested decomposition approach allows for examination of an entire year in hourly resolution. Thus, the annual dispatch of seasonal storage can be optimized. This represents the added value of stage 1. Its result is updated in stage 2 and stage 3. These provide additional information on the optimal storage dispatch resulting from more detailed modeling of spatial and technical scope within the time slices. It must be stated, that due to this decomposition there is a loss of guarantee of optimality. However, it can be assumed that the seasonal storage levels are largely independent of the detailed modeling of grids and plants. Figure 4 shows the annual storage level of representative storages of different technologies as a result of all three stages of the nested decomposition approach.

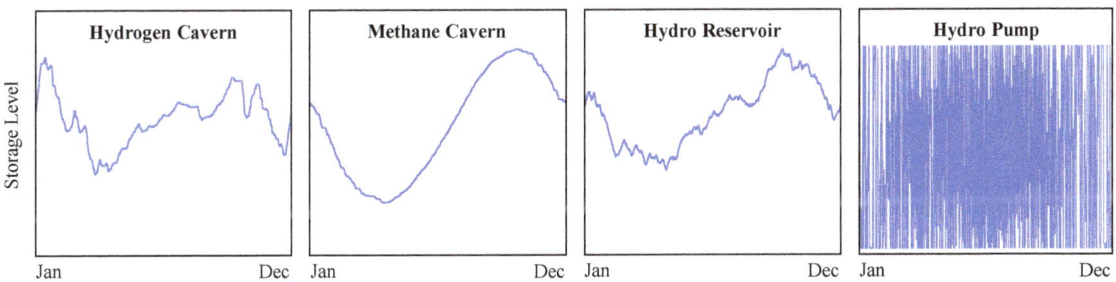

Figure 4. Annual storage levels for different storage technologies (model result).

The storage level of hydrogen, methane and reservoir storages show a seasonal pattern, where the storage tends to be discharged in winter and charged in summer. However, since hydrogen demand, unlike methane demand, does not have a seasonal pattern in this scenario, this behavior is determined by the hydrogen supply side, especially wind energy. The cavern is filled in situations of high hydrogen production from electrolysis and is discharged in situations of low hydrogen production. Methane storages show a typical pattern to supply seasonal heat demand in winter. Hydro reservoirs have similar patterns to hydrogen caverns, as the dispatch of electrolysis and electric storage both are highly dependent on RES generation. Moreover, snowmelt and a slightly seasonal electric load affect the hydro reservoir level. In contrast, pump storages and batteries have a daily pattern due to their relatively low capacity to power ratio.

The analysis of the storage dispatch shows that due to the highly seasonal pattern of hydro and gas storage, optimization with a full-year horizon is necessary. The nested approach can use this information in the following detailed spatial and technical analyses. It must be noted that the optimality guarantee is lost due to the nested optimization of storage levels. Detailed spatial and technical information such as network congestion close to the storage cannot be included in the seasonal pattern. However, in the subsequent iterations, there is the chance to adjust the storage levels within a time slice using the more detailed spatial and technical information.

3.4. European Infrastructure Analysis

The nested decomposition method enables the analysis of large-scale energy systems such as the European electricity, methane, and hydrogen infrastructures. The annual dispatch of IEI includes both the supply of the mentioned energy sources by different plants and the modeling of the transport networks on-grid node and unit level. This corresponds to the resolution of Figure 3. For reasons of clarity, Figure 5 shows the electricity, methane, and hydrogen supply as well as energy exchange aggregated on a regional level. The energy flows reflect mainly the result of stage 2, however, it is updated by the result of stage 3 for the focus region. The results are available in hourly resolution. The electricity system is characterized by a high share of domestic generation. Exchanges are low compared to gas systems due to limited exchange capacities. Electricity generation is dominated by intermittent generation from PV and wind. In Scandinavia and the Alpine region, hydropower also accounts for a large share of electricity generation. This results in shares between 51% and 100% of renewable power generation in the European countries. In addition, nuclear power accounts for up to 49% of electricity generation, especially in France, the UK, and Eastern Europe. Electricity generation from hydrogen and natural gas is low in energy terms, due to high fuel costs. However, these power plants are required as secured capacity during peak loads and "Dunkelflauten". Fossil baseload power plants fired by lignite and hard coal have largely been phased out. Therefore, CO_2-intensive power generation is below 13% in all and close to 0% in many considered countries. Compared to historical conditions, this shifts the net positions of countries that had a high coal-fired generation in the past. For example, Germany imports 88 TWh_{el} (net), mainly from offshore wind from the north and nuclear power from the west. Italy is the largest importer with 125 TWh_{el} (net). In contrast, France is the biggest exporter of nuclear and RES with a net export of 152 TWh_{el} (net), which supplies all neighboring countries.

The hydrogen system has a balanced mix of domestic production (54%, 764 TWh_{th}) and imports (46%, 647 TWh_{th}). Electrolysis takes a share of 83% of the domestic generation. Hydrogen imports enter Europe from all directions: from the north from Norway (208 TWh_{th}), from the south from North Africa (223 TWh_{th}), and from the east from Russia (216 TWh_{th}). The import potential is thus largely exploited. Germany is the central sink of hydrogen flows. It is the largest net importer with 243 TWh_{th} (net) and produces 66 TWh_{th} from electrolysis and 37 TWh_{th} from steam reforming. Denmark (12 TWh_{th} (net)) and France (8 TWh_{th} (net)) are besides Norway the only net exporters. France has the largest domestic hydrogen production of 148 TWh_{th}. The European hydrogen demand of the conversion sector is only 10 TWh_{th}.

The natural gas system continues to be supplied primarily by imports (68%). Domestic production from biomethane has a share of 28%, conventional production 4%. The share of methanation (0.5%) is small, since the use of hydrogen in the hydrogen system is favored in terms of efficiency. Pipeline imports enter Europe mainly from the northeast from Russia (592 TWh_{th}) and the north from Norway (728 TWh_{th}). Other imports are provided from the south from North Africa (83 TWh_{th}) and from the southeast (106 TWh_{th}). Moreover, LNG imports (224 TWh_{th}) enter the system at the coasts mainly in western Europe. Due to lower fuel costs, power generation from natural gas is preferred (122 TWh_{th}). The methane demand of the conversion sector is 279 TWh_{th} in total.

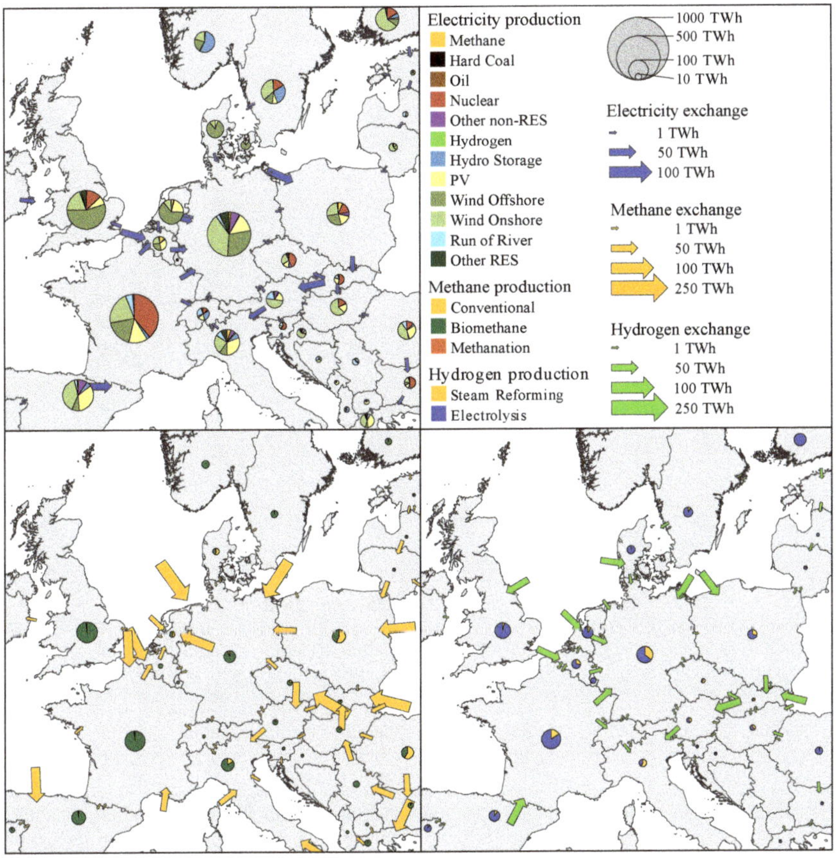

Figure 5. European production and exchange of electricity (**top**), methane (**bottom left**) and hydrogen (**bottom right**), (model result).

The analysis of the European dispatch shows high energy exchanges between the heterogeneous regions. Thus, these interactions must be considered in the dispatch analysis of individual countries. Due to the nested approach, exchanges can be used as boundary conditions in the detailed dispatch analysis in the focus region. By this approach there is again no guarantee for optimality, since the optimal energy exchanges could change due to additional information from the detailed technical analysis in stage 3. For example, bottlenecks in the gas network can be detected by modeling pressures or previous linear loss estimations can have errors, leading to a different optimal exchange. However, the influence of these changed constraints is small compared to the overall dispatch of the system and can be minimized by suitable linear estimations in stage 2 (see Section 2.6).

3.5. Snapshot Analysis for Germany

In the focus region Germany, the nested decomposition approach allows to update the annual dispatch on grid node and unit level from stage 2 to a higher level of technical detail. Network losses in the power and gas system as well as hydraulic gas flows and linepack can be modeled. In order to exemplify the results, a particularly stressful situation with a high need for transportation is shown in Figure 6. The following analyses can be performed for all 8760 h of the year.

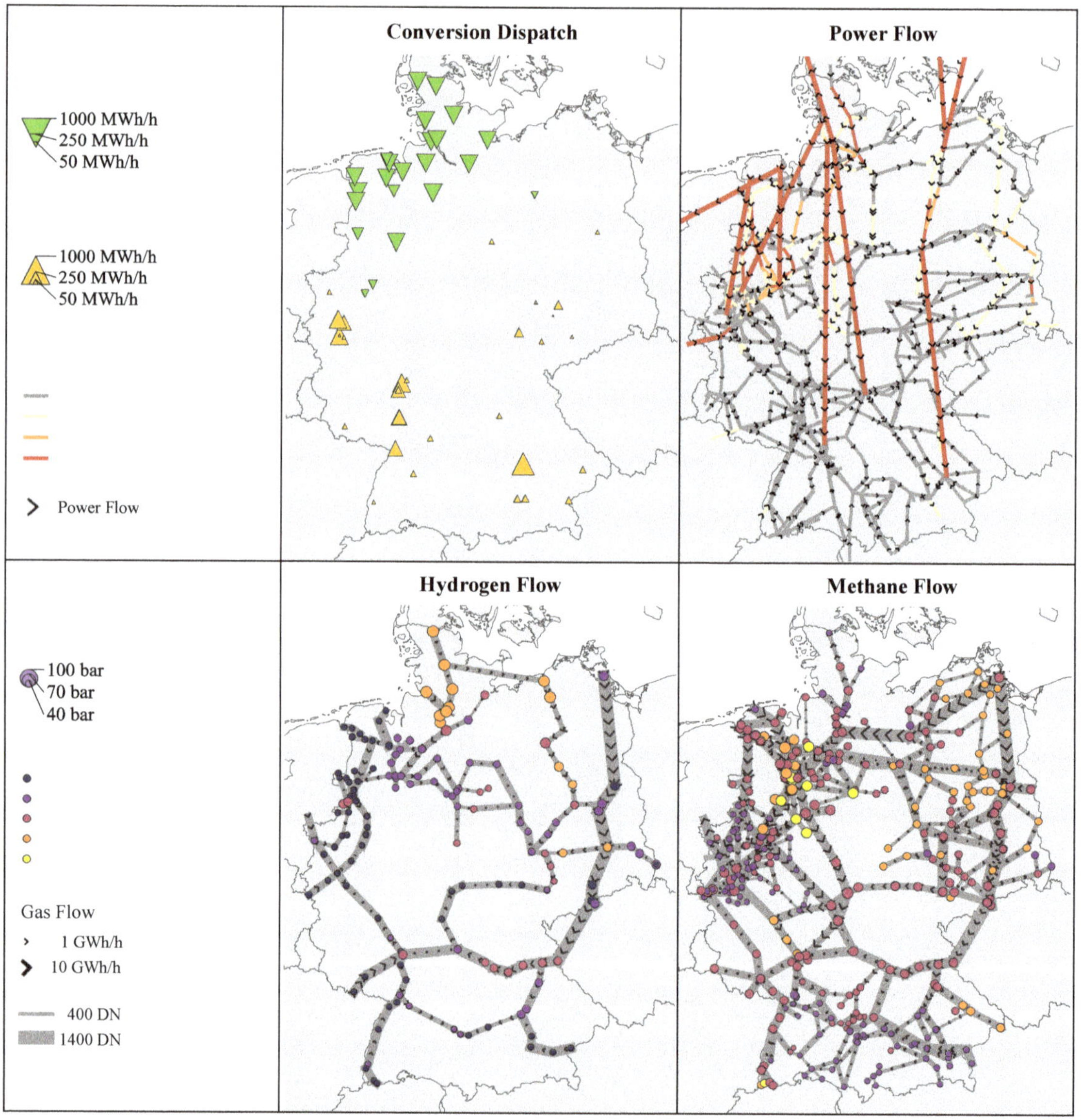

Figure 6. Detailed dispatch results for stressful snapshot in Germany: conversion dispatch (**top left**), electric power flow (**top right**), hydrogen gas flow (**bottom left**), natural gas (methane) gas flow (**bottom right**), (model result).

This snapshot represents an hour on a winter evening. There is simultaneously high generation from wind power (70.0 GWh_{el}), especially in the north, and high demand (87.1 GWh_{el}), especially in the load centers in the southwest. Figure 6 shows the electrical and thermal power flows in the system for the three infrastructures considered. The utilization of the power system is illustrated by showing the loading of the lines. In the gas systems, pressures at nodes are presented absolutely by their thickness and in relation to their pressure limits by color. Moreover, sector coupling is illustrated as triangles by feed-in or feed-out of gas-fired power plants and power-to-gas plants.

The supply task of the described snapshot results in high transport demand from the north to the south in the power system. Thus, all DC lines and several AC lines in north to south direction must be operated at their operating limits. This results in grid losses of 2.8 GWh_{el}. Overloads are not feasible due to the modeling. Instead, there is a partial shift of the transport task from the electricity grid to the gas grid. This is indicated by the simultaneous dispatch of power-to-gas plants (12.2 GWh_{el}) and gas-fired power plants (6.2 GWh_{el}), which show a clear north-south separation between the bottlenecks in the power system in Figure 6. Simultaneous dispatch of these conversion plants should be avoided from a cost and energy efficiency perspective. However, it is necessary in this snapshot to avoid load shedding due to congestion in the electricity transmission system. Although a temporal shifting of the transportation task is more efficient than sectoral shifting, electrical storage can only supply 2.2 GWh_{el}. This is because the remaining stored energy is needed in adjacent hours. In addition, a part of the transportation task is avoided by curtailing 5.0 GWh_{el} of electricity from wind power in the north. Other power plants contribute 15.0 GWh_{el}. The remaining electricity demand is mainly imported from France.

From a hydrogen infrastructure perspective, the IEI dispatch results in a feed-in of 8.1 GW_{th} of hydrogen from electrolysis in the north. In this snapshot, there is no demand for power generation from hydrogen. 4.6 GW_{th} is supplied by steam methane reforming. The majority of the 38.5 GWh_{th} final energy demand must therefore be covered by imports. These are provided mainly by Russia in the northeast of Germany (7.9 GWh_{th}), followed by France (6.4 GWh_{th}) Belgium (3.0 GWh_{th}) and the Netherlands (2.6 GWh_{th}) in the west (1.7 GWh_{th}). In addition, Denmark (2.0 GWh_{th}) and Norway (1.7 GWh_{th}) export from the north. The rest is imported in the south, supplied by southwestern and southeastern Europe. The linepack supplies 1.2 GWh_{th}, hydrogen storages demand 0.8 GWh_{th} (net). This results in energy flows mainly from north to southwest. These can be visualized by the flow directions as well as the hydraulic potentials (pressures) in Figure 6. To keep the pressures within the technical and contractual operating limits, 151 MWh_{th} and 40 MWh_{el} of driving energy is necessary from gas-fired and electric compressors, respectively. Only a few of the assumed compressor stations are operated in this snapshot, indicating that the hydrogen network is not fully utilized. Despite this, some pressures are operated at their minimal limits to minimize driving energy. Relatively high pressures are required in the north to transport electrolysis production.

Last, the methane infrastructure is discussed. The demand of 57.3 GWh_{th} is relatively high compared to the scenario year, but a significant decrease in absolute terms compared to current levels [51]. Power plants demand 10.8 GWh_{th}. In total, about 65.0 GWh_{th} are imported mainly from Norway (33.6 GWh_{th}) and Russia (19.6 GWh_{th}). 15.0 GWh_{th} are exported mainly to the Netherlands. 7.6 GWh_{th} are provided as biomethane, 14.7 GWh_{th} by storages and 3.9 GWh_{th} by linepack. This results in a northeast to southwest flow. However, the transmission grid for methane is underutilized as well. In particular, the transit pipelines with large diameters are operated at relatively low pressures. The methane network is mainly operated close to its lower limits to minimize driving energy. This results in 90 MWh_{th} and 47 MWh_{el} driving losses for compression. Due to the thermodynamic properties of methane, the driving losses are lower than in the hydrogen network despite higher demand.

The snapshot analysis illustrates the cross-sectoral interactions in dispatch of IEI. The added value of modeling physical power and gas flow is particularly evident in the dispatch of conversion plants, which can avoid grid congestion if necessary. All in all, the application of the nested decomposition approach thus successfully demonstrates its applicability to large-scale IEI and its potential use cases. These represent dispatch costs analysis, analysis of annual dispatch of storage and other plants in hourly resolution, as well as physical flows and losses in the electricity and gas grids.

4. Conclusions

Energy system integration enables to raise synergies, but also increases interdependencies between electricity, methane, and hydrogen infrastructures. As IEI are increasingly based on RES, which are often located far from demand, requirements for adequately modeling of long-distance transports in electricity and gas systems are rising. For these reasons, this paper introduced a dispatch model, that combines the following features:

- Integrated energy system modeling
- Scalability to full-year modeling in hourly resolution
- Applicability to large-scale energy system model in nodal resolution
- Detailed modeling of non-linear physical laws:
    - Electric power flows with DC approximation and electric losses
    - Quasi-steady state methane and hydrogen flows and compression losses

The novelty of this method is therefore the combined capability of considering non-linear physics in dispatch optimization, while allowing applicability to large-scale systems and long periods. Since a mathematically exact optimization is not manageable with current computing capabilities, a three-stage nested decomposition approach is developed. It successively applies different model reduction techniques based on aggregation and slicing in each stage.

However, the nested decomposition approach results in the following limitations:

- Loss of optimality guarantee due to successive decomposition, but a near-optimal solution may be sufficient for the purpose of energy system analysis. The optimality gap cannot be quantified due to the complexity of the problem. However, this gap can be rationally estimated as small as important interactions are taken into account by the applied problem-specific decomposition.
- Reduced spatial scope for high detailed technical analysis, however, the achievable regional scope can still be classified as large-scale compared to other applications in literature.
- Remaining model simplifications like DC power flow, quasi-steady-state gas flow, a rough estimation of contingencies and the neglect of non-linear operating constraints, but they appear to be largely justified in long-term system planning.
- No inherent consideration of uncertainties, but this can be addressed through scenario analysis.
- The presented approach will have the following use cases and enhancements:
- Annual dispatch analysis of IEI including cost and emissions analysis taking into account technical constraints and losses of energy transmission and conversion.
- Application in cost-benefit analyses to assess system design principles or single infrastructure projects such as HVDC-links, pipelines or electrolyzers. For this purpose, a delta consideration of two scenarios can be performed.
- Planned model enhancements include the integration of district heating infrastructures and additional linearized dispatch constraints of plants.

In addition to the applicability to large-scale energy systems, the use case demonstrates the temporal, regional and cross-sectoral interdependencies in the dispatch of IEI. This shows that the requirements for energy system models stated in this paper are justified.

Author Contributions: Conceptualization, L.L. and A.M.; methodology, L.L. and R.H.; software, L.L. and R.H.; validation, L.L. and C.G.; formal analysis, L.L.; investigation, L.L. and C.G.; data curation, L.L.; writing—original draft preparation, L.L.; writing—review and editing, C.G. and R.H.; visualization, L.L., R.H. and C.G.; supervision, A.M.; project administration, L.L. All authors have read and agreed to the published version of the manuscript.

Funding: This research received no external funding.

Data Availability Statement: Not applicable.

Acknowledgments: Simulations were performed with computing resources granted by RWTH Aachen University.

Conflicts of Interest: The authors declare no conflict of interest.

Nomenclature

Quantities			
$\mathcal{A}$	Incident branches	$\mathcal{T}$	Time steps
$\mathcal{B}$	Branches	$\mathcal{U}$	Consumers
$\mathcal{D}$	Storages	$\mathcal{R}$	Regions
$\mathcal{E}$	Energy carrier	$\mathcal{Z}$	Subperiods
$\mathcal{N}$	Nodes	δ	Incident feed-in and feed-out
$\mathcal{S}$	Supplies		
Variables			
LP	Linepack	θ	Phase angle
P	Electrical or thermal power flow	ϑ	Phase angle shift
W	Storage level	π	Squared pressure
Parameters			
a	Annuity factor	T	Temperature
c	Cost	U_b	Operational voltage
d	Pipeline diameter	X	Impedance
e	Specific emissions	η	Efficiency
h_u	Lower heating value	κ	Isentropic exponent
K	Compressibility factor	λ	Pipe friction coefficient
l	Pipeline length	ρ	Density
R	Resistance	γ	Maximal pressure reduction ratio
R_s	Specific gas constant	Γ	Maximal pressure rise ration
Indexes			
AC	Alternating current	in	Inflow
CH_4	Methane	$inst$	Installed capacity
CO_2	Carbon dioxide	$loss$	Losses
Com	Compressor	m	Medium
DC	Direct current	max	Maximum
dis	Dispatchable	min	Minimum
dr	Driving	n	Standard condition
DSR	Demand side response	out	Outflow
E	Electricity	PST	Phase-shifting transformer ($PST \subseteq Tr$)
ENS	Energy not served	Reg	Pressure regulator
$fuel$	Fuel costs	t	Index for time steps
G	Gas (pipelines)	Tr	Transformer
H_2	Hydrogen	z	Index for sub periods
i	Index for nodes and plants	$\alpha,\ \beta,\ \omega$	Indexes for energy carrier
ij	Index for branches from node i to and node j	$+$	In-feeding
int	Intermittent	$-$	Out-feeding
is	Isentropic		

Appendix A. Integrated Dispatch Optimization Problem

This appendix formulates the IEI dispatch as an optimization problem, that serves as the basis for the developed nested decomposition approach. The Nomenclature is given at the end of the paper. The energy infrastructure is described in an integrated manner as a coherent graph for different energy carriers $\alpha, \beta, \ldots, \omega \in \mathcal{E}$. The infrastructures of the following energy carriers are considered:

- Electricity E with subsets AC and DC,
- Gases G with subsets hydrogen H_2 and methane CH_4

The nodes $\mathcal{N}$ of the IEI graph describe busbars and gas stations for hydrogen and methane. Kirchhoff's first law applies to all power flows P (electrical or thermal) into or out of nodes from incidental branches $\mathcal{A}$ and feed-ins/feed-outs δ of all infrastructures:

$$\sum_{ij\in\mathcal{A}^+(i)} P_{ij,t} + \sum_{i\in\delta^+(i)} P_{i,t} - \sum_{ij\in\mathcal{A}^-(i)} P_{ij,t} - \sum_{i\in\delta^-(i)} P_{i,t} = 0 \quad \forall i \in \mathcal{N}, t \in \mathcal{T} \tag{A1}$$

Two nodes are connected by branches. Branches either connect two nodes within an infrastructure or convert energy when connected to two nodes of different infrastructures. Thus, conversion plants also represent branches. AC and DC power lines as well as (phase-shifting) transformers represent branches in the power grid. Gas pipelines, pressure regulating valves and compressor stations are branches in the gas grid. In addition to branches, other feed-in or feed-out such as consumers, generating units, import stations and storages are attached to nodes. Power flows cannot be set arbitrarily, but are subject to physical and technical constraints. These as well as the objective function of the dispatch optimization are introduced in the following.

Appendix A.1. Power System

To describe the physical laws of power flows on AC power lines and transformers, this model applies the DC power flow approximation. Therefore, the power flow is determined by the difference of the phase angles θ and the impedance X of the branch assuming a constant operational voltage U. The power flow is limited by the thermal capacity of the equipment. For AC equipment, only 70% of the nominal thermal capacity is assumed to account for contingencies [55].

$$0 \le P_{ij,t} \le P_{ij}^{max} \qquad \forall ij \in \mathcal{B}_{AC} \cup \mathcal{B}_{DC} \cup \mathcal{B}_{Tr}, t \in \mathcal{T} \tag{A2}$$

$$P_{ij,t} = \frac{U^2}{X_{ij}} \cdot (\theta_{i,t} - \theta_{j,t}) \quad \forall\, ij \in \mathcal{B}_{AC} \cup \mathcal{B}_{Tr},\ t \in \mathcal{T} \tag{A3}$$

In contrast to AC power lines and transformers, HVDC-converters can actively dispatch the power flow over HVDC-lines within their operating limits. In addition, phase-shifting transformers (PST) can control the power flow over AC lines by injecting a supplementary voltage. PST are modeled in a simplified way by an injected voltage angle ϑ with a continuous operating range.

$$\vartheta_{ij}^{min} \le \vartheta_{ij,t} \le \vartheta_{ij}^{max} \quad \forall ij \in \mathcal{B}_{PST},\ t \in \mathcal{T} \tag{A4}$$

$$P_{ij,t} = \frac{U_b^2}{X_{ij}} \cdot (\theta_{i,t} - \theta_{j,t} + \vartheta_{ij,t}) \quad \forall ij \in \mathcal{B}_{PST}, t \in \mathcal{T} \tag{A5}$$

Transmission losses of all electrical equipment are modeled as feed-out, which is equally divided to the inlet and outlet nodes as a subset of $\delta^-(i)$. They depend linearly on the ohmic resistance R and quadratically on the current.

$$P_{i,t}^{loss} = \frac{1}{2} \sum_{ij\in\mathcal{A}(i)} R_{ij} \cdot \frac{P_{ij,t}^2}{U^2} \quad \forall\, ij \in \mathcal{B}_{AC} \,\cup\, \mathcal{B}_{DC} \,\cup \mathcal{B}_{Tr}, t \in \mathcal{T} \tag{A6}$$

Appendix A.2. Gas Systems

Physical gas flows are modeled applying a quasi-steady-state formulation that model gas dynamics without considering transients. To describe pressure losses on pipelines, the applied variant of the integrated Darcy-Weisbach equation links gas flow and pressures. It assumes steady-state gas flows, isothermal gas flow ($T_m = const$) and horizontal pipelines. This still results in the non-linear, non-convex relation in Equation (A7). Pressure losses depend on the gas flow, pipeline geometry (pipe length l and diameter d) and the thermodynamic state of the gas. Compressibility K_m and friction coefficient λ_m are themselves dependent on the thermodynamic state and the considered gas, which is modeled by the formulas of Papay and Zanke for methane and empirical approximations for hydrogen. Squared pressures π are considered instead of pressures p to reduce complexity in the pressure loss equation. The formula is related to the thermal power flow via the calorific value h_u.

$$|P_{ij,t}| \cdot P_{ij,t} = \frac{\pi^2\, d^5\, T_n}{16\, \lambda_m\, \rho_n\, p_n\, T_m\, K_m\, l\, h_u^2} \cdot (\pi_{i,t} - \pi_{j,t}), \quad \forall ij \in \mathcal{B}_G, t \in \mathcal{T} \tag{A7}$$

Due to the compressibility of gases, the gas network represents an inherent storage that provides flexibility to itself. To model linepack, the inflow P^{in} of a pipeline can differ from the outflow P^{out}. Both flows are linked to the average gas flow used in the pressure-loss equation. Linepack LP depends on the pipeline geometry and the state of the gas. Mass conservation for each pipeline between each time step and the considered total period is modeled by Equations (A10) and (A11). The reader could refer to [29,31,32] for further detail.

$$P_{ij,t} = \frac{P_{ij,t}^{in} + P_{ij,t}^{out}}{2} \quad \forall ij \in \mathcal{B}_G, t \in \mathcal{T} \tag{A8}$$

$$LP_t = \frac{\pi\, d^2\, l\, T_n}{4\, T_m\, p_n\, K_m} \cdot \sqrt{\frac{\pi_{i,t} + \pi_{j,t}}{2}} \quad \forall ij \in \mathcal{B}_G, t \in \mathcal{T} \tag{A9}$$

$$LP_{ij,\,t} = LP_{ij,\,t-1} + (P_{ij,t}^{in} - P_{ij,t}^{out})/h_u \quad \forall ij \in \mathcal{B}_G, t \in \mathcal{T} \tag{A10}$$

$$LP_{ij,0} = LP_{ij,T} \qquad \forall ij \in \mathcal{B} \tag{A11}$$

Pressures must be maintained within their technical and contractual limits during the operation of gas transmission systems. By dispatching compressors, the pressure at their outlets can be increased. Pressure regulating valves can decrease the pressure at their outlets. The maximum gas flow and maximum pressure ratio (Γ and γ) must be respected.

$$\pi_i^{min} \le \pi_{i,t} \le \pi_i^{max} \quad \forall i \in \mathcal{N}_G, t \in \mathcal{T} \tag{A12}$$

$$\gamma_{ij} \cdot \pi_{i,t} \le \pi_{j,t} \le \pi_{i,t} \quad \forall ij \in \mathcal{B}^{Reg}, t \in \mathcal{T} \tag{A13}$$

$$\pi_{i,t} \le \pi_{j,t} \le \Gamma_{ij} \cdot \pi_{i,t} \quad \forall ij \in \mathcal{B}^{Com}, t \in \mathcal{T} \tag{A14}$$

$$0 \le P_{ij,t} \le P_{ij}^{inst} \quad \forall ij \in \mathcal{B}^{Com}, t \in \mathcal{T} \tag{A15}$$

The horsepower equation determines the required work for compression P_{Gas}. It is dependent from the thermodynamic state, the considered gas properties, and the isentropic efficiency η_{ij}^{is}. Driving power P^{dr} considering drive efficiency η^{dr} can be taken either directly from the inlet node by a gas turbine or from the power system using an electric motor. Other operating restrictions are neglected.

$$P_{Gas} = \frac{\rho_1 \cdot P_{ij,t}}{h_u \cdot \eta_{ij}^{is}} \cdot \frac{\kappa}{\kappa - 1} \cdot Z_1 \cdot R_s \cdot T_1 \cdot \left(\left[\frac{\pi_2}{\pi_1} \right]^{\frac{\kappa-1}{2\,\kappa}} - 1 \right) \ \forall\, ij \in \mathcal{B}^{Com}, t \in \mathcal{T} \tag{A16}$$

$$P_{Gas} = \frac{P_{G,ij,t}^{dr}}{\eta_{G,ij}^{dr}} + \frac{P_{E,ij,t}^{dr}}{\eta_{E,ij}^{dr}} \quad \forall ij \in \mathcal{B}^{Com}, t \in \mathcal{T} \tag{A17}$$

$$0 \le P_{\alpha,ij,t}^{dr} \le P_{\alpha,ij}^{max} \quad \forall ij \in \mathcal{B}^{Com}, t \in \mathcal{T} \tag{A18}$$

Appendix A.3. Conversion Plants

Conversion plants connect at least two nodes from different infrastructures by converting energy. Following the energy hub concept [42], a conversion plant is modeled as a black box that converts input energy carriers P_α^{in} into output energy carriers P_β^{out}. The conversion processes and their efficiency are described by the conversion coefficient η.

$$P_\beta^{out} = \eta_{\alpha\beta,i} \cdot P_\alpha^{in} \quad \forall\, ij \in \mathcal{C},\ t \in \mathcal{T} \tag{A19}$$

Equation (A19) can model several technologies like electrolyzers, methanation plants, gas-fired power plants, fuel cells or steam methane reforming. The rating of the conversion plant limits P_α^{in} and P_α^{out} . Other operating restrictions such as minimum power, power gradients, minimum operating times and downtimes are neglected.

Appendix A.4. Storages

Energy storages for electricity and gases can decouple energy supply and demand in time. The continuity equation ensures conservation of energy connecting injections P^{in} and withdrawals P^{out}, storage inflows from the environment $\overline{W}$ and the storage levels W between two points in time. In addition, injection η^{in} and withdrawal losses η^{out} must be considered.

$$W_{i,t} = W_{i,t-1} + \overline{W}_{i,t} + \left(\frac{1}{\eta_i^{in}}\, P_{i,t}^{in} - \eta_i^{out}\, P_{i,t}^{out}\right) \cdot 1h \;\forall\, i\ \in \mathcal{D},\ t \in \mathcal{T} \tag{A20}$$

Again, the rating of injection and withdrawal power, as well as storage capacity, limits the storage's operation. The start level of the storage must correspond to the level at the end of the considered period, in order not to add energy to the system.

Appendix A.5. Other Feed-In and Feed-Out

In addition to conversion plants and storage facilities, all infrastructures can be supplied by other feed-in plants $\mathcal{S}_\alpha$. These do not gain energy from explicitly modeled infrastructures but energy is added externally to the system. Dispatchable plants can feed-in up to their installed capacity. These include power plants and gas imports.

$$P_{i,\,\alpha}^{min} \leq P_{\alpha,i,\,t} \leq P_{i,\,\alpha}^{max} \qquad \forall i \in \mathcal{S}_\alpha^{Dis}, t \in \mathcal{T} \tag{A21}$$

In contrast, the maximum feed-in of intermittent RES is externally determined. For these, curtailment represents their degree of freedom in operation.

$$0 \leq P_{\alpha,i,\,t} \leq P_{\alpha,i}^{max} \qquad \forall i \in \mathcal{S}_\alpha^{Int}, t \in \mathcal{T} \tag{A22}$$

The feed-out of final consumers and distribution networks is also externally determined. However, flexibility can be provided through demand side response (DSR), which is implemented in this paper as scheduled load shedding.

$$0 \leq P_{\alpha,i,t}^{DSR} \leq P_{\alpha,i}^{max} \qquad \forall i \in \mathcal{U}_\alpha, t \in \mathcal{T} \tag{A23}$$

In addition, unscheduled load shedding (energy not served) is considered to ensure feasibility.

$$0 \leq P_{\alpha,i,t}^{ENS} \leq P_{\alpha,i}^{max} \qquad \forall i \in \mathcal{U}_\alpha, t \in \mathcal{T} \tag{A24}$$

Appendix A.6. Objective

The objective function (A25) aims to minimize the operating costs of IEI. The most significant terms of the objective function are fuel costs including CO_2 costs $c_{\alpha,i}^{fuel}$ from other feed-in plants such gas imports or other power plants. Note that fuel costs of conversion plants and storages must not be taken into account explicitly, as they are already considered beforehand when supplying the input energy carrier. The second component of the objective function are costs for DSR $c_{\alpha,i}^{DSR}$. Last, macroeconomic costs for unscheduled load shedding $c_{\alpha,i}^{ENS}$ (value of lost load) are considered, which are generally higher in the power system than in the gas system [6,7]. Other operating costs, e.g., variable operation and maintenance costs for plants and grids, are not shown for clarity.

$$\min o = \sum_{t=\mathcal{T}} \sum_{\alpha \in \mathcal{E}} \left(\sum_{i \in \mathcal{S}_{\alpha}^{Dis}} c_{\alpha,i}^{fuel} \cdot P_{\alpha,i,t} + \sum_{i \in \mathcal{U}_{\alpha}} c_{\alpha,i}^{DSR} \cdot P_{\alpha,i,t}^{DSR} + \sum_{i \in \mathcal{U}_{\alpha}} c_{\alpha,i}^{ENS} \cdot P_{\alpha,i,t}^{ENS} \right) \tag{A25}$$

References

1. European Commission. The European Green Deal. Brussels. 2019. Available online: https://eur-lex.europa.eu/legal-content/EN/TXT/?qid=1576150542719&uri=COM%3A2019%3A640%3AFIN (accessed on 20 February 2022).
2. European Commission. Powering a Climate-Neutral Economy: An EU Strategy for Energy System Integration. Brussels. 2020. Available online: https://ec.europa.eu/commission/presscorner/detail/en/ip_20_1259 (accessed on 20 February 2022).
3. Kondziella, H.; Bruckner, T. Flexibility requirements of renewable energy based electricity systems—A review of research results and methodologies. *Renew. Sustain. Energy Rev.* **2016**, *53*, 10–22. [CrossRef]
4. Sterner, M. *Bioenergy and Renewable Power Methane in Integrated 100% Renewable Energy Systems*; Kassel University Press: Kassel, Germany, 2009; Volume 14.
5. Guidehouse. Extending the European Hydrogen Backbone—A European Hydrogen Infrastructure Vision Covering 21 Countries. Utrecht, The Netherlands. 2021. Available online: https://gasforclimate2050.eu/sdm_downloads/extending-the-european-hydrogen-backbone/ (accessed on 20 February 2022).
6. ENTSO-E. 2nd ENTSO-E Guideline for Cost Benefit Analysis of Grid Development Projects. 2018. Available online: https://tyndp.entsoe.eu/cba (accessed on 20 February 2022).
7. ENTSOG. 2nd ENTSOG Methodology for Cost-Benefit Analysis of Gas Infrastructure Projects. 2018. Available online: https://www.entsog.eu/methodologies-and-modelling#2nd-cba-methodology (accessed on 20 February 2022).
8. Pambour, K.A. Modelling, Simulation and Analysis of Security of Supply Scenarios in Integrated Gas and Electricity Transmission Networks. Ph.D. Thesis, University of Groningen, Groningen, The Netherlands, 2018.
9. Pfenninger, S.; Hawkes, A.; Keirstead, J. Energy systems modeling for twenty-first century energy challenges. *Renew. Sustain. Energy Rev.* **2014**, *33*, 74–86. [CrossRef]
10. Frysztacki, M.M.; Hörsch, J.; Hagenmeyer, V.; Brown, T. The strong effect of network resolution on electricity system models with high shares of wind and solar. *Appl. Energy* **2021**, *291*, 116726. [CrossRef]
11. Nolden, C.; Schönfelder, M.; Eßer-Frey, A.; Bertsch, V.; Fichtner, W. Network constraints in techno-economic energy system models: Towards more accurate modeling of power flows in long-term energy system models. *Energy Syst.* **2013**, *4*, 267–287. [CrossRef]
12. Riepin, I.; Möbius, T.; Müsgens, F. Integrated Electricity and Gas Market Modeling—Effects of Gas Demand Uncertainty. In Proceedings of the 15th International Conference on the European Energy Market (EEM), Lodz, Poland, 27–29 June 2018. [CrossRef]
13. Baumann, C. Simulation der europäischen Märkte für Erdgas und Strom. Aachener Beiträge zur Energieversorgung. Ph.D. Thesis, RWTH Aachen University, Aachen, Germany, 2015.
14. Quelhas, A.; Gil, E.; McCalley, J.D.; Ryan, S.M. A Multiperiod Generalized Network Flow Model of the U.S. Integrated Energy System: Part I—Model Description. *IEEE Trans. Power Syst.* **2007**, *22*, 829–836. [CrossRef]
15. Quelhas, A.; Gil, E.; McCalley, J.D.; Ryan, S.M. A Multiperiod Generalized Network Flow Model of the U.S. Integrated Energy System: Part II—Simulation Results. *IEEE Trans. Power Syst.* **2007**, *22*, 837–844. [CrossRef]
16. Deane, J.P.; Ciaráin, M.Ó.; Gallachóir, B.P.Ó. An integrated gas and electricity model of the EU energy system to examine supply interruptions. *Appl. Energy* **2017**, *193*, 479–490. [CrossRef]
17. Lemaréchal, C. Lagrangian Relaxation, Computational Combinatorial Optimization, Optimal or Provably Near-Optimal Solutions. In *Lecture Notes in Computer Science*; Springer: Berlin, Germany, 2000; pp. 112–156.
18. Deutsche Energie-Agentur GmbH (DENA). DENA-Leitstudie Integrierte Energiewende: Impulse für die Gestaltung des Energiesystems bis 2050. 2018. Available online: https://www.dena.de/integrierte-energiewende/ (accessed on 20 February 2022).
19. Henning, H.-M.; Palzer, A. *Energiesystem Deutschland 2050: Sektor- und Energieträgerübergreifende, modellbasierte, ganzheitliche Untersuchung zur langfristigen Reduktion energie-bedingter CO_2-Emissionen durch Energieeffizienz und den Einsatz Erneuerbarer Energien*; Fraunhofer ISE: Freiburg, Germany, 2013.
20. Markewitz, P.; Stein, G. *Das IKARUS-Projekt: Energietechnische Perspektiven für Deutschland: Abschlussbericht des Projektes IKARUS*; Forschungszentrum Jülich: Jülich, Germany, 2003.
21. E3MLab. *Primes Model Version 6 2016–2017, Detailed Model Description*; E3MLab: Athens, Grecee, 2017.
22. Loulou, R.; Lehtilä, A.; Kanudia, A.; Remme, U.; Goldstein, G. *Documentation for the TIMES Model*; ETSAP: Stuttgart, Germany, 2021.
23. Gils, H.C.; Scholz, Y.; Pregger, T.; Luca de Tena, D.; Heide, D. Integrated modelling of variable renewable energy-based power supply in Europe. *Energy* **2017**, *123*, 173–188. [CrossRef]

24. Fette, M.; Brandstätt, C.; Gils, H.C.; Gardian, H.; Pregger, T.; Schaffert, J.; Tali, E.; Brücken, N. Multi Sektor Kopplung: Modellbasierte Analyse der Integration erneuerbarer Stromerzeugung durch die Kopplung der Stromversorgung mit dem Wärme, Gas- und Verkehrssektor—Endbericht. 2020. Available online: https://elib.dlr.de/135971/ (accessed on 20 February 2022).
25. Brown, T.; Hörsch, J.; Schlachtenberger, D. PyPSA: Python for Power System Analysis. *J. Open Res. Softw.* **2018**, *6*, 4. [CrossRef]
26. Nemec-Begluk, S. Modeling and Optimization of Multi-Energy Systems and the Future Role of Storage and Coupling Technologies. Ph.D. Thesis, TU Wien, Vienna, Austria, 2020.
27. Purchala, K. Modeling and Analysis of Techno-Economic Interactions in Meshed High Voltage Grids Exhibiting Congestion. Ph.D. Thesis, University of Leuven, Leuven, Belgium, 2005.
28. Mischner, J.; Fasold, H.G.; Heymer, J. *gas2energy.net. Systemplanerische Grundlagen der Gasversorgung*; Deutscher Industrieverlag: München, Germany, 2015.
29. Löhr, L.; Houben, R.; Moser, A. Optimal power and gas flow for large-scale transmission systems. *Electr. Power Syst. Res.* **2020**, *189*, 106724. [CrossRef]
30. Sun, G.; Chen, S.; Wei, Z.; Chen, S. Multi-period integrated natural gas and electric power system probabilistic optimal power flow incorporating power-to-gas units. *J. Mod. Power Syst. Clean Energy* **2017**, *5*, 412–423. [CrossRef]
31. Ordoudis, C.; Pinson, P.; Morales, J.M. An integrated market for electricity and natural gas systems with stochastic power producers. *Eur. J. Oper. Res.* **2019**, *272*, 642–654. [CrossRef]
32. Schwele, A.; Ordoudis, C.; Kazempour, J.; Pinson, P. Coordination of Power and Natural Gas Systems: Convexification Approaches for Linepack Modeling. In Proceedings of the IEEE PowerTech, Milan, Italy, 23–27 June 2019. [CrossRef]
33. Pambour, K.A.; Erdener, B.C.; Bolado-Lavin, R.; Dijkema, G.P.J. SAInt—A novel quasi-dynamic model for assessing security of supply in coupled gas and electricity transmission networks. *Appl. Energy* **2017**, *203*, 829–857. [CrossRef]
34. An, S.; Li, Q.; Gedra, T.W. Natural gas and electricity optimal power flow. In Proceedings of the IEEE/PES Transmission and Distribution Conference and Exposition, Dallas, TX, USA, 7–12 September 2003. [CrossRef]
35. Urbina, M.; Li, Z. A Combined Model for Analyzing the Interdependency of Electrical and Gas Systems. In Proceedings of the IEEE-39, North American Power Symposium, Las Cruces, NM, USA, 30 September–2 November 2007. [CrossRef]
36. Chaudry, M.; Jenkins, N.; Strbac, G. Multi-time period combined gas and electricity network optimization. *Electr. Power Syst. Res.* **2008**, *78*, 1265–1279. [CrossRef]
37. Chiang, N.-Y.; Zavala, V.M. Large-scale optimal control of interconnected natural gas and electrical transmission systems. *Appl. Energy* **2016**, *168*, 226–235. [CrossRef]
38. Correa-Posada, C.; Sanchez-Martin, P.; Lumbreras, S. Security-constrained model for integrated power and natural-gas system. *J. Mod. Power Syst. Clean Energy* **2017**, *5*, 326–336. [CrossRef]
39. Ojeda-Esteybar, D.M.; Rubio-Barros, R.G.; Vargas, A. Integrated operational planning of hydrothermal power and natural gas systems with large scale storage. *J. Mod. Power Syst. Clean Energy* **2017**, *5*, 299–313. [CrossRef]
40. Schwele, A.; Arrigo, A.; Vervaeren, C.; Kazempour, J.; Vallée, F. Coordination of Electricity, Heat, and Natural Gas Systems Accounting for Network Flexibility. *Electr. Power Syst. Res.* **2020**, *189*, 106776. [CrossRef]
41. Geidl, M.; Koeppel, G.; Favre-Perrod, P.; Klöckl, B.; Andersson, G.; Fröhlich, K. The Energy Hub—A Powerful Concept for Future Energy Systems. In Proceedings of the Third Annual Carnegie Mellon Conference on the Electricity Industry, Pittsburgh, PA, USA, 13–14 March 2007.
42. Geidl, M. Integrated Modeling and Optimization of Multi-Carrier Energy Systems. Ph.D. Thesis, ETH Zürich, Zürich, Switzerland, 2007. [CrossRef]
43. Kotzur, L.; Nolting, L.; Hoffmann, M.; Groß, T.; Smolenko, A.; Priesmann, J.; Busing, H.; Beer, R.; Kullmann, F.; Singh, B.; et al. A modeler's guide to handle complexity in energy systems optimization. *Adv. Appl. Energy* **2021**, *4*, 100063. [CrossRef]
44. Scholz, Y.; Fuchs, B.; Borggrefe, F.; Cao, K.K.; Wetzel, M.; von Krbek, K.; Cebulla, F.; Gils, H.C.; Fiand, F.; Bussieck, M.; et al. Speeding up Energy System Models—A Best Practice Guide. 2020. Available online: https://elib.dlr.de/135507/ (accessed on 20 February 2022).
45. Traub, J.F.; Woniakowski, H. Complexity of linear programming. *Oper. Res. Lett.* **1982**, *1*, 59–62. [CrossRef]
46. Cao, K.; von Krbel, K.; Wetzel, M.; Cebulla, F.; Schreck, S. Classification and Evaluation of Concepts for Improving the Performance of Applied Energy System Optimization Models. *Energies* **2019**, *12*, 4656. [CrossRef]
47. Ward, J.B. Equivalent Circuits for Power Flow Studies. *AIEE Trans. Power Appl. Syst.* **1949**, *68*, 373–382. [CrossRef]
48. Dantzig, G.B.; Wolfe, P. Decomposition Principle for Linear Programs. *Oper. Res.* **1960**, *8*, 101–111. [CrossRef]
49. Benders, J.F. Partitioning Procedures for Solving Mixed-Variable Programming Problems. *Numer. Math.* **1962**, *4*, 238–252. [CrossRef]
50. ENTSO-E; ENTSOG. TYNDP 2022 Draft Scenario Report, Brussels. 2021. Available online: https://2022.entsos-tyndp-scenarios.eu/ (accessed on 20 February 2022).
51. ENTSO-E. Grid Map. 2021. Available online: https://www.entsoe.eu/data/map/ (accessed on 20 February 2022).
52. ENTOG. Transparency Platform. 2021. Available online: https://transparency.entsog.eu/#/map (accessed on 20 February 2022).
53. Rövekamp, J. Transportnetzberechnung zur Feststellung der Erdgasversorgungssicherheit in Deutschland unter Regulatorischem Einfluss. Ph.D. Thesis, TU Clausthal, Clausthal, Germany, 2014.

54. Gils, H.C. Balancing of Intermittent Renewable Power Generation by Demand Response and Thermal Energy Storage. Ph.D. Thesis, Universität Stuttgart, Stuttgart, Germany, 2015. [CrossRef]
55. Hörsch, J.; Ronellenfitsch, H.; Witthaut, D.; Brown, T. Linear optimal power flow using cycle flows. *Electr. Power Syst. Res.* **2018**, *158*, 126–135. [CrossRef]

energies

MDPI

Article

Application of Generator-Electric Motor System for Emergency Propulsion of a Vessel in the Event of Loss of the Full Serviceability of the Diesel Main Engine

Zbigniew Łosiewicz [1,*], Waldemar Mironiuk [2], Witold Cioch [3], Ewelina Sendek-Matysiak [4] and Wojciech Homik [5]

1 Faculty of Maritime Technology, West Pomeranian University of Technology in Szczecin, al. Piastów 41, 71-065 Szczecin, Poland
2 Faculty of Navigation and Naval Weapons, Polish Naval Academy of the Heroes of Westerplatte, Śmidowicza Street 69, 81-127 Gdynia, Poland; w.mironiuk@amw.gdynia.pl
3 Faculty of Mechanical Engineering and Robotics, AGH University of Science and Technology, al. Mickiewicza 30, 30-059 Krakow, Poland; cioch@agh.edu.pl
4 Faculty of Management and Computer Modeling, Kielce University of Technology, al. Tysiąclecia Państwa Polskiego 7, 25-314 Kielce, Poland; esendek@tu.kielce.pl
5 Faculty of Mechanical Engineering and Aeronautics, Rzeszow University of Technology, al. Powstańców Warszawy, 35-959 Rzeszow, Poland; whomik@prz.edu.pl
* Correspondence: zbigniew.losiewicz@zut.edu.pl

Abstract: Oil tanker disasters have been a cause of major environmental disasters, with multi-generational impacts. One of the greatest hazards is damage to the propulsion system that causes the ship to turn sideways to a wave and lose stability, which in storm conditions usually leads to capsizing and sinking Despite the perceived consequences of maritime disasters in the current solutions for the propulsion of oil tankers, there are no legal or real solutions for independent emergency main propulsion in this type of ship. Stressing that the reliability of the propulsion system has a significant impact on the ship's safety at sea, the authors presented a new solution in the form of a power take-off/power take-in (PTO/PTI) system. This is the emergency use of a shaft generator as the main electric motor, operating in parallel in a situation when the main engine (ME), (the main engine of the ship's direct high-power propulsion system that is slow-speed) loses the operational capability to propel the ship. Since one cause of wear, or failure, of the main engines is improper operational decisions, the paper shows the wear mechanism in relation to the accuracy of operational decisions. Using classical reliability theory, it also shows that the use of the proposed system results in an increase in the reliability of the propulsion system. The main topic of the paper was the use of an electrical system called PTO/PTI as an emergency propulsion system on the largest commercial vessels, such as bulk carriers and crude oil tankers, which has not been used before in maritime technical solutions. Semi-Markov processes, continuous in time, discrete in states, and which are used in technology, were also proposed as a tool describing the process of the operation of such a ship propulsion system, and they are useful to support operational decisions affecting the state of the technical condition of the engine. There are two ship operation strategies that can be adopted: the four-state model, for normal operation, and the three-state model, which operates with the occurrence of failure. For these types of models, their limiting distributions were defined in the form of probabilities. It was also demonstrated that faster than expected engine wear and the occurrence of inoperability of the main engine can be caused by wrong operational decisions made by the shipowner or crew. Using this type of main engine operating methodology, it is possible to support the decision of the engineer to stop the main engine and to subject it to the process of restoration to an acceptable state of technical condition (before the failure or during the failure in severe storm conditions), with the parallel use of the proposed electric propulsion (PTO/PTI) as an emergency propulsion, giving the crew a chance to maintain the steering necessary to maintain safe lateral stability.

Keywords: shaft generator–electric motor (PTO/PTI); reliability of the ship's main propulsion; model of exploitation process; technical states of the engine; semi-Markov process

Citation: Łosiewicz, Z.; Mironiuk, W.; Cioch, W.; Sendek-Matysiak, E.; Homik, W. Application of Generator-Electric Motor System for Emergency Propulsion of a Vessel in the Event of Loss of the Full Serviceability of the Diesel Main Engine. *Energies* **2022**, *15*, 2833. https://doi.org/10.3390/en15082833

Academic Editor: Theodoros Zannis

Received: 30 December 2021
Accepted: 6 April 2022
Published: 13 April 2022

1. Introduction

Ships such as bulk carriers of the "Capesize" class, with a deadweight of more than 150,000 t, and tankers of the very large crude carrier (VLCC) and ultra large crude carrier (ULCC) classes are among the largest types of maritime transport. ULCC tankers are ships of up to about 360 m in length, and with a carrying capacity of up to 400,000 t. Their cargo tanks can hold up to 500,000 t of crude oil, and their fuel tanks up to 10,000 t of toxic heavy fuel oil (HFO).

Such large amounts of fuel are needed to drive large power propulsion units, such as steam turbines or compression ignition reciprocating internal combustion engines (e.g., marine low-speed, long/super long stroke two-stroke main engines, with power outputs of up to 80,080 kW [1]). These engines and equipment are necessary to ensure the movement of the ship, propelled by the main engine and the other main propulsion equipment of the ship, and to drive its auxiliary machinery, powered mainly by the diesel electric generators.

The safety of a ship performing transportation tasks in difficult, unpredictable sea conditions depends on many factors. One of the main requirements, apart from hull integrity and lateral stability [2–5], is to maintain the ship's maneuverability, which is possible only when the positive speed of the ship is relative to the water. This requires an adequate reliability of the main propulsion system [6], in particular the main engines (ME) [7,8]. A failure of the main propulsion system or steering device, or damage to the hull plating, often leads to shipwreck [3,9–12], resulting in environmental disasters. In the search for alternatives to propulsion powered by internal combustion engines fueled by hydrocarbon fuels, research is being conducted on internal combustion engines fueled by gaseous fuels, alcohol, and biofuel, in addition to investigations into propulsion powered by electricity. Due to the subject of this paper, research related to electric drives was analyzed.

In the paper [13], one of the largest marine engine manufacturers, MAN Energy Solutions, proposed the solution of using a generator/electric motor solution, i.e., a power take-off/power take-in (PTO/PTI) system. This solution concerns using the generator as a motor, but only as a support for the main combustion engine when fully operational and working at full load. Such a solution is aimed at increasing the efficiency or lowering the exhaust gas emission.

A thematically similar study on the cooperation of a shaft alternator and an internal combustion engine is presented in [14]. The results of the study on the use of the IHIMU-CRP hybrid electric fuel propulsion system in coastal, short-range ships are shown. A method for optimizing this propulsion, thereby reducing fuel consumption and exhaust emissions, is presented.

Authors of the paper [15] analyzed the probability of main engine failure during severe weather conditions at sea, and effective ways of dealing with such a situation by the shipboard crew. They pointed out that the described bulk carriers, carrying up to 400,000 t (dwt) and powered by engines of up to 80,000 kW, may fail during severe weather conditions. They also suggested that this work should be adopted as a guideline by seafarers to assist in ship risk management.

The work [16] presented a new decision-making process based on reducing multiple evaluation criteria (sometimes unquantifiable) to an evaluation against a single quantifiable criterion—financial values. This methodology was used to compare the performance of a ship with diesel-electric hybrid propulsion against a ship with conventional propulsion. The analysis is presented using the example of a selected vessel.

Wartsila, a manufacturer of marine engines and propulsion systems, proposed hybrid solutions [17,18], using medium-speed, medium-power engines as the main internal combustion engines driving the electric generators that power the electric motors of the ship's main propulsion.

In paper [19], the key application during the use of multiple electric generators was the use of high-power current rectifiers, and paper [20] presents assumptions for the design of electrical systems and overload protection.

In paper [21], technical solutions for environmental protection and alternative propulsion systems, including electric propulsion, were described. Their advantages and disadvantages were then described.

The application of electric drives on warships is presented in [22]. This refers to ships of special military application.

The literature on alternative solutions powered by electric energy also impact the solutions for inland waterway vessels.

In publication [23], the concept of using alternative configurations of propulsion systems for inland waterway vessels in order to reduce their carbon footprint was presented, as well as models for assessing emissions and related costs over the lifetime (life cycle assessments—LCAs) of a propulsion system using an internal combustion engine, and of electricity powered engines in various configurations (including batteries and photovoltaics). The economic viability of both solutions, according to the life cycle cost assessment (LCCA), was compared using the GREET 2020 program.

Due to the specificity of river navigation, the authors proposed the concept of diesel-hydraulic and hybrid propulsion system for inland waterway vessels. The solution of the hybrid design with a pumping system driven by a battery bank in regard to the aspect of energy efficiency was also presented. The results of experimental investigations carried out on a natural scale parallel hybrid and a diesel-electric drive controlled by a smart propulsion system are presented in [24].

Based on the measurements, the authors analyzed the fuel consumption and investment costs of four alternative propulsion systems. A simplified method of cost and savings analysis was presented. A solution of "green propulsion" on a passenger ship in the "green shipping" concept was presented in [25].

An analysis of the available literature shows that the use of battery electric propulsion is very limited to low-power, short-range craft, often used in inland passenger shipping. The biggest problem is the availability of battery charging stations, as well as the battery capacity. Analogies can be seen with the problems encountered in automotive transport [26,27]. In contrast, hybrid systems, in which the main engines are high-power electric motors, depend on the power of the internal combustion (emission) engines driving the electric current generators, which supply these electric motors.

So far, no research has been found on the issue of the electric main propulsion of very large merchant ships, especially during a failure of the internal combustion slow-speed, two-stroke main engine.

Ship constructors and engine manufacturers proposed the solution of using the shaft generator as the engine (PTI) only to support the main propulsion of the ship at full load e.g., in difficult sailing conditions with the excess energy coming from generator sets [28–30].

However, it should be noted main engine failure often occurs due to improper engine operation. Therefore, the safe operation of ships requires making proper, and rational operational decisions concerning the use and operation of the main engines (ME) in particular [7,31]. Damage in such large ships, especially sinking, can cause disasters, and difficult to estimate losses due to the contamination of the marine environment that is not only harmful to flora and fauna, but also to many generations of the population living on polluted coasts [32]. Therefore, in addition to equipping main engines (MEs) with appropriate diagnostic systems (SDGs) to warn of engine wear (deterioration) or failure, the use of diesel-electric propulsion (referred to as PTO/PTI)is proposed, using an electric motor supplied back from the ship's mains.

The solution proposed by the authors is to use a shaft generator, working in PTI mode, as an emergency main propulsion engine. The configuration will be calculated by the constructor to ensure the ship's minimum maneuvering speed in difficult sailing conditions, taking into account the adoption of the most safety-compliant emergency course.

However, apart from the use of PTI as an emergency solution, the biggest problem for mechanics is making the rational decision of at what point emergency propulsion needs to be used, as this requires significant changes in the structure of this type of propulsion system.

Many researchers have described operational uncertainty states using various sophisticated research methods, such as in publications [33,34].

The authors assumed that in order to rationally make such decisions (not allowing the complete destruction of the internal combustion engine or the possibility of explosion), it is necessary to assess the technical condition of the main engine on an ongoing basis, using methods presented in [7,8,31,35]. It has been shown that the wrong operational decisions lead to permanent degradation of the engine structure, causing faster wear of the engine than it was designed for, which was the basis for determining the operational time between scheduled overhauls.

The authors adopted a design solution that enabled a significant improvement of shipping safety, by confirming the correctness of such actions using the general theory of reliability of the ME presented in publications [36,37]. The design solution also considers the protection of the marine environment, and other forms of protection resulting from rational engine operation are presented in publications [38,39].

In Figure 1, an illustrative diagram of the ship's main propulsion is shown, using a low-speed, two-stroke piston engine as the main engine with a shaft generator, and a PTO/PTI system (generator (G)) driving the common propeller shaft of the ship.

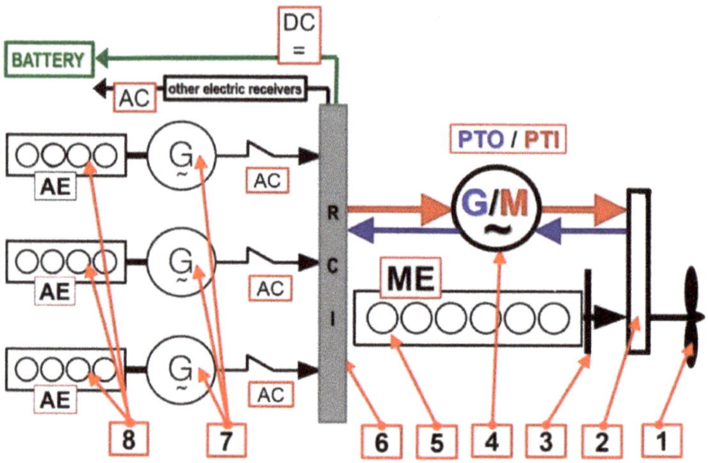

Figure 1. Example of the main propulsion system of the vessel: 1. Unregulated pitch drive screw propeller (fixed pitch propellers—FPP); 2. Electricity generator gearbox; 3. Clutch disconnector; 4. Shafted alternator (PTO/PTI); 5. Marine low-speed, two-stroke main engine (ME); 6. RCI main power unit (AC/DC rectifier, AC/AC converter, DC/AC inverter); 7. Auxiliary engine (AE)—driven generators (G); 8. Auxiliary engine (AE) driving generators of ship power plants.

Further consideration of the use of marine diesel-electric propulsion systems (Figure 1) ensured the safety of the ship in emergency operating situations (the scenarios studied and presented in publications [3,10,40]) in probabilistic terms. The bases of this method are developed in publications [41–43], and are made on the example of the propulsion system of an oil tanker adapted to transport crude oil. Particular attention was paid to the considerations of reliability, one of the most important features of such a system.

2. General Description of the Reliability of the Propulsion System of an Oil Tanker

An oil tanker adapted for the transport of crude oil is characterized by its very large size. It enables the use of ship propellers with a non-adjustable (fixed) pitch, which are driven by means of marine low-speed, two-stroke piston, internal combustion engines [1].

A drive system with such a screw and engine often consists of the following devices [30]:

- marine low-speed, two-stroke, internal combustion piston crosshead engine;
- disconnectable or inseparable coupling;
- intermediate shaft with bearings;
- screw shaft with stern tube;
- drive screw with non-adjustable stroke (unregulated).

Such a system is called a direct propeller system, where the engine speed is equal to the propeller speed. The devices of such a system are, in terms of reliability, elements connected in series, which makes its reliability structure a serial one. Since the failure of one of the elements of the serial structure causes the failure of the whole system, they require the correct exploitation decisions. The dependence of correct exploitation decisions on the exploitation conditions is presented in publications [7,44–46]. A schematic diagram of the structure of the aforementioned drive system is shown in Figure 2. From this diagram, it is clear that the reliability structure of such a drive train is a serial structure.

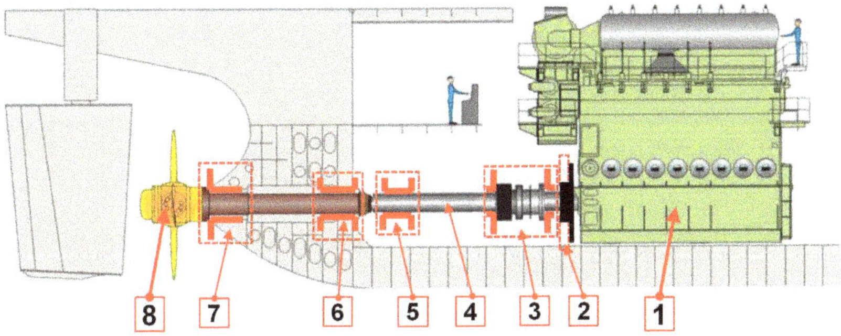

Figure 2. Sample diagram of the main propulsion system of the ship: 1—marine low-speed, two-stroke internal combustion main engine (ME), winged and slow-running; 2—non-disconnectable hydraulic coupling; 3—thrust bearing; 4—intermediate shaft; 5—bearing of intermediate shaft; 6—bearing of screw shaft; 7—scabbard of screw shaft; 8—fixed pitch propeller (own elaboration using MAN BW drive drawing [30]).

The reliability of any such drive system, as a system with a serial structure, is closely linked to the reliability of all its individual components [33,34,36,37,47]. This is because the inability of any of the elements of this system causes failure. The reliability of such a drive system is unambiguously characterized by the reliability function, which can be presented in the form of dependency (1):

$$R(t) = P\{T_1 > t,\ T_2 > t,\ \ldots\ T_8 > t\} = P\{T_1 > t\}P\{T_2\}\ldots P\{T_8 > t\} = \prod_{i=1}^{8} R_i(t) \quad (1)$$

From the diagram shown in Figure 2, it follows that the elements (in terms of reliability) are: 1. marine low-speed, two-stroke diesel engine—main engine (ME); 2. hydraulic separable coupling; 3. thrust bearing; 4. intermediate shaft; 5. bearing of intermediate shaft; 6. support bearing of screw shaft; 7. main bearing of screw shaft; 8. stern tube; and 9. fixed pitch drive screw. This diagram also shows that the reliability of the vessel's propulsion system can be increased if a shaft generator (4) driven by the main engine (1), through a gearbox (3), is used. A diagram of such an arrangement is shown in Figure 3.

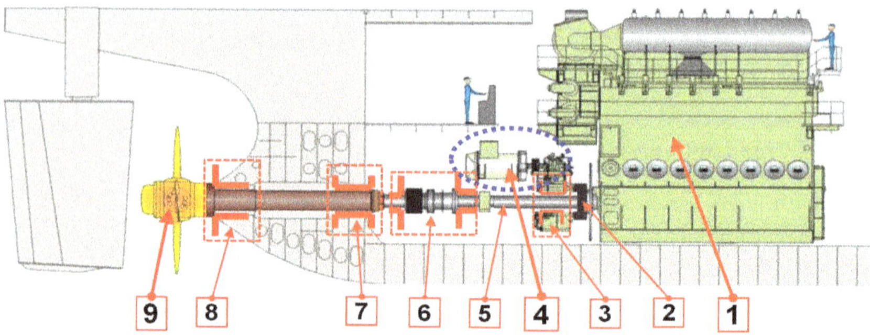

Figure 3. Example of ship's main propulsion with shaft generator and PTO/PTI system: 1—marine low-speed, two-stroke diesel engine—main engine (ME); 2—hydraulic separable coupling; 3—shaft generator gear; 4—shaft generator (PTO/PTI); 5—intermediate shaft; 6—thrust bearing of intermediate shaft; 7—bearing of screw shaft; 8—stern tube; 9—fixed pitch propeller (own elaboration using MAN drawing [30]).

The use of a shaft generator in the ship's propulsion system allows one to drive the ship's propeller, in case of damage to the main engine. In such a case, it operates as an electric engine, powered by electricity generated by the vessel's power plant generating sets. As a result, a parallel reliability structure of the system can be obtained, as shown in Figure 4. The reliability of such a system $R_s(t)$, consisting of shaft generator systems (*SGS*): generator set (4) into gearbox (3) with distribution $F_1(t)$; and main engine set (*MES*): main engine (1) into clutch (2) with distribution $F_2(t)$, capable of driving a common intermediate shaft (5), together or separately, may be described by Formula (2):

$$R_S(t) = 1 - F_1(t)F_2(t) \tag{2}$$

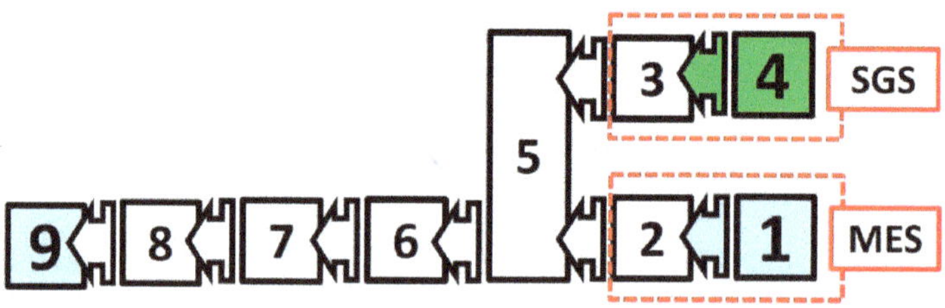

Figure 4. Scheme of main propulsion of the ship with shaft generator and PTO/PTI system: 1—two-stroke main motor; 2—hydraulic coupling; 3—shaft generator gear; 4—shaft generator (PTO/PTI); 5—intermediate shaft; 6—thrust bearing of intermediate shaft; 7—bearing of screw shaft; 8—stern tube; 9.—fixed pitch propeller; *SGS*—shaft generator systems; *MES*—main engine set (own elaboration).

The reliability function of the propulsion system with the reliability structure is presented in Figure 4, and including the PTO/PTI system, can be presented as Formula (3):

$$R_S(t) = \left(1 - \prod_{j=1}^{4} F_j(t)\right) \prod_{5}^{9} R_j(t) \tag{3}$$

where:

$$F_j(t) = (1 - R_1(t)R_2(t))$$

and:

$F_j(t)$—the distribution function (distributor), the probability of failure of the jth *MES* unit consisting of main engine (ME), and clutch or *SGS* consisting of a generator and gearbox (PTO/PTI);

$j = 1, 2, 3, 4$;

$R_i(t)$—the reliability of the ith component of the propulsion system, other than the component of *MES* and *SGS* units, and *SGS*, $i = 4, 5, 6, 7, 8, 9$;

T—time of correct operation of *MES* and *SGS*.

The approach presented to determine the reliability of the ship's propulsion system with the diagram shown in Figure 4 is the result of the adoption of an alternative classification of its reliability states of the type: fit/unfit. In the operational practice of sea-going vessels, other states of their propulsion equipment are also important, especially the main engines (ME), such as the fully effective and ecological condition [43,44], the fully effective condition, the partial condition, and the unfit condition. For this reason, there is a need for a clear interpretation of these states.

3. Change in Engine Condition Due to Engine Operation

The mechanical energy generated by the engine, under strictly defined conditions, is considered as a measure of its ability to perform We. For this purpose, the formulas developed in [6,7] determining the action of A_M are used. Action is a concept that has been defined differently. In this paper it is taken as represented in [6,7], as a physical quantity with a unit of measure called *"joulesecond"* (product joule times second). Thus, the action defined always results in energy consumption E, and requires time t, and the less efficient it is, the higher the energy consumption. Compression ignition internal combustion engines produce and transmit energy to consumers (e.g., the ship's propeller) in the entire load range for which they are designed and manufactured. The basic figure that clearly defines the motor load is the torque M_0. Torque is closely related to the average dose of fuel G_p, injected successively into each engine cylinder, and thus, to the average effective pressure p_e. The dependence of M_o on the dose G_p and pressure p_e can be presented as follows [7]:

$$M_o = C_1 G_p = C_2 p_e \tag{4}$$

Therefore, it was assumed that it is most convenient to measure the torque M_0 with a torsiometer. This allows the measurement the angle φ of the crankshaft torsion, or the tangential stresses τ created in the shaft due to its torsion, which depend on the engine's M_o torque, according to the dependence [7]:

$$M_0 = s\varphi \; lub \; M_0 = \tau W_0 \tag{5}$$

Therefore, engine action can be determined by Equation (6):

$$A_\varphi = \int_{t_0}^{t_n} s\varphi\omega t dt \tag{6}$$

or

$$A_\tau = \int_{t_0}^{t_n} \tau W_0 \omega t dt \tag{7}$$

where:

s—the torsional stiffness of the shaft;

W_0—the indicator of the torsional strength of the tested section of the shaft, where:

$$W_0 = \frac{\pi d^3}{16}, \quad d - shaft\ diameter$$

Taking the pressure p_e of the Formula (8):

$$p_e = \frac{W_d p_0 \eta_i}{V_{A0} R T_0 \lambda} \eta_v \eta_m \tag{8}$$

where:
W_d—gas calorific value;
P_0—ambient pressure;
V_{A0}—amount of air theoretically necessary for combustion of 1 kg of fuel;
R—gas constant;
T_0—ambient temperature;
η_i—indicated efficiency;
η_v—filling efficiency;
η_m—mechanical efficiency;
λ—excess air ratio.

Therefore, Formula (6) can take the form:

$$A_M = \int_{t_0}^{t_n} C_2 \frac{W_d p_0 \eta_i}{V_{A0} R T_0 \lambda} \eta_v \eta_m \, \omega t dt \quad (9)$$

Instead of p_e, a momentary pressure $p_M = f(\omega t)$ was used to allow the engine to generate a momentary torque. Formula (6) can be presented as follows:

$$A_M = \int_{t_0}^{t_n} C_2 p_M \, \omega t dt \quad (10)$$

Using Poisson's process, it is possible to present a physical interpretation of the process of reducing W_e using a fixed value of e [7]. Assuming that from the moment the engine starts operating (assuming that this is the moment $t_0 = 0$) until the moment when the measuring device registers for the first time, event A consisting of reducing the operation of W_e by a value of $\Delta W_e = e$, any value of W_e's operation (including at the maximum load of the engine) can be performed in particular periods of the engine's operation. As time passes, further wear and tear of the engine causes further drops in W_e's operating value by another equal e value, recorded by the measuring device. Therefore, if the accumulated number of B_t of events A, described by a homogeneous Poisson process, is recorded up to t, a total reduction of W_e by ΔW_e up to t, can be presented with the following relation:

$$\Delta W_e = e B_t \quad (11)$$

The random variable B_t has a distribution [6]:

$$P(B_t = k) = \frac{(\lambda t)^k}{k} \exp(-\lambda t); k = 1, 2, \ldots, n \quad (12)$$

where:
λ—the constant quantity (λ + idem), interpreted as the intensity of the reduction of W_e by equal e values, recorded by the measuring device during tests, $\lambda > 0$.

The expected value of $E(B_t)$ and the variance of the process of the increase in the number of events A, i.e., the decrease in W_e's operation by successive values e recorded by the measuring device, can be presented as follows:

$$E(B_t) = \lambda t; \quad D^2(B_t) = \lambda t \quad (13)$$

Therefore, the expected value and the standard deviation of the reduction of W_e's work performed by the motor until t, can be expressed by formulas:

$$E[\Delta W_e(t)] = eE(B_t) = e\lambda t \qquad \sigma_l(t) = e\sqrt{D^2(B_t)} = e\sqrt{\lambda t} \quad (14)$$

Assuming that a new engine (when t = 0) performs the greatest work, i.e., that $W_e(0) = W_{emax}$, can be expressed by a mathematical relation describing the reduction of work with time t, with the formula:

$$W_e(t) = \begin{cases} W_{emax} & for\ dla\ \ t = 0 \\ W_{emax} - e\left(\lambda t \pm e\sqrt{\lambda t}\right) & for\ dla\ \ t > 0 \end{cases} \tag{15}$$

The graphical interpretation of the relationship recorded in Formula (15) is shown in Figure 5.

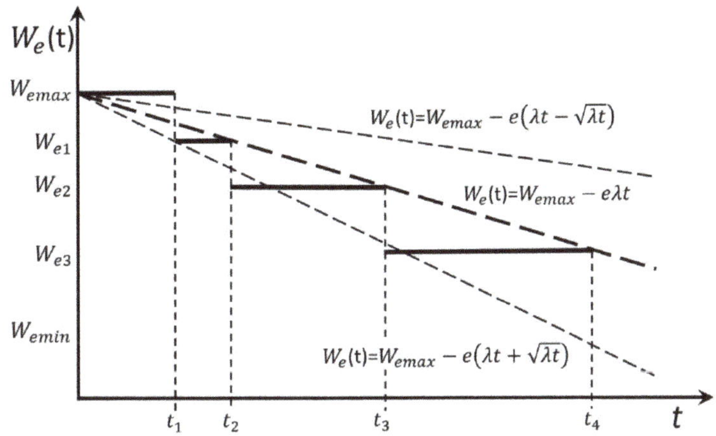

Figure 5. Graphic interpretation of the example realization of the reduction of the useful work of the engine: W_e—useful work, e—quantum by which the work of W_e is changed [6,7].

From Formula (12), it follows that for any moment t, the work We performed by the engine can be determined, and that it is possible to determine the probability of the appearance of such a reduction in the work We of the engine wear and tear that will make it impossible to perform a given task in the operation process. Thus, the probability $P(B_t = k$, $k = 1, 2, \ldots, n)$ as determined by the Formula (12) is considered as an indicator of engine reliability, or is taken as an indicator of engine operating safety in case it concerns such a reduction of We performance that it may lead, e.g., to a shipping accident.

The above presented relations between the action of a human being and the engine they are operating, in a certain period of time, causing the drop of the work done W_e as a result of the wear and tear of the device. This allows the introduction of the theory of the vector of the accuracy of decisions (decision accuracy vector) and the concept of the curve of the engine technical condition potential (C_{ETCP}) [6].

Introducing the curve of the engine technical condition potential (CETCP) is a necessary element in the analysis of the processes. It is presented as a set of vectors of decision accuracy $\overline{V_{DAi}}$ where i = [1, . . . , 6], with different values of the consequences of these decisions other than the value in point A. It was assumed that as a result of the exploiter's action, the use of the engine, being in a technical condition with a specific potential, causes adequate wear and tear of the engine, lowering the technical condition potential, which has an impact on lowering the level of ability to perform an exploitation task, and thus, lowering its life span. It was assumed on the basis of expert knowledge that the lower the potential of the engine's technical condition, the lower the quality of processes taking place in the engine (e.g., combustion, charging, and lubrication due to degradation of oil properties), and consequently, the quicker the process of engine wear [8,31].

In addition to the processes that are a natural consequence of using the engine with different potential, it is important to note, for example, that the desire of users to perform a transportation task even/or especially when the potential of the engine's technical condition is reduced, results in additional negative events that accelerate engine wear. An incorrect operating decision may cause the set of technical condition potential values of the engine, presented as the C_{ETCP} engine condition potential curve (wear and tear during operation), to decrease by an appropriate value ΔS_{ETC} (S_{ETC}—state of technical condition of the engine). This phenomenon has been shown as the curve of the state of the engine technical condition, with C_{SETCn}, C_{SETC2}, C_{SETC3}, C_{SETCP4}, and C_{SETC5} presented as consequences of adequate decisions, shown in Figure 6 as decision accuracy vector $\overline{V_{DAi}}$ where $i = \{1, \ldots, 6\}$.

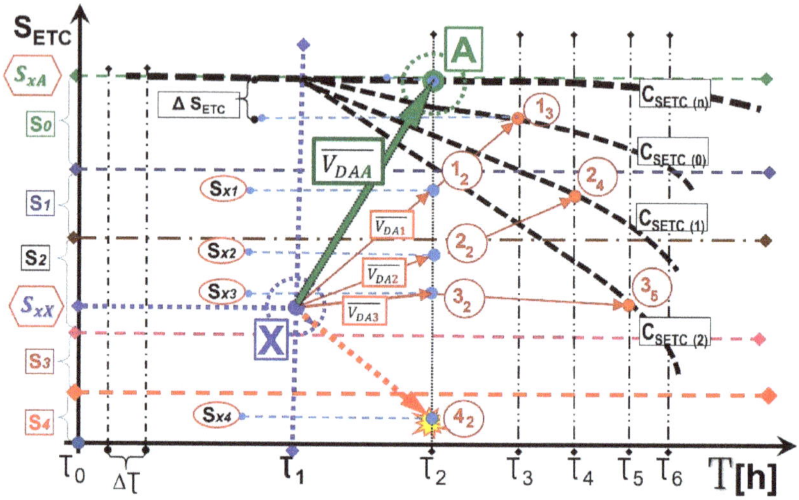

Figure 6. An overview drawing presenting graphically the vector concept of the decision accuracy vector $\overline{V_{DA}}$ where: S_{ETC} axis—the axis of the state of the engine's technical condition; T axis—the axis of the engine's lifetime; τ_W axis—the axis of the relative lifetime of the engine; "X"—the point where values of the state of the engine's technical condition have been assigned to the corresponding technical condition S_x in time τ_1; "A"—the point where values of the state of the engine's technical condition have been assigned to the corresponding technical condition S_{xA} in time τ_2; $\overline{V_{DAA}}$—the decision accuracy vector in the area "A"; C_{STCP}—curve of the state engine technical condition; C_{STCPn}—curve of the state of the engine's technical condition nominal (n), i.e., predicted during the design process (author's own elaboration).

4. Engine Technical Conditions

From operational experience, it is important to maintain a high level of reliability for an oil tanker when performing transportation tasks in difficult, unpredictable sea conditions, especially with regard to the main engine (ME). This means that it is necessary to rationally control both the use of the motor and its operation [46]. In order to act rationally in this respect, diagnostic tests should be carried out to determine the technical condition of the engine [41].

The technical condition of an engine is defined by the set of technical characteristics of its structure that enable it to operate reliably, and carry out its operational tasks in accordance with the intended use for which it was designed, manufactured, and assembled. As presented in works [6,7], this state, at any moment t of operation depends not only on this moment, but also on the technical condition of the engine at the initial moment to $<t$, any changes of the engine load in the time interval $[t_0, t]$, and the course of control of the engine in this interval. This control has a major impact on the change in engine condition. The state of the engine at the end of the manufacturing process depends on many factors, which are of a random nature (these dependencies are presented in publications [7,31,37,41,42,44]). This means that the process of changing the condition of each engine is stochastic, continuous in states, and over time. Assuming that the criterion

for creating a set of states for the engine's suitability to perform a transport task while maintaining ecological standards, the following classes of technical states are distinguished, called directly the states (determination of technical states of machines, including marine engines, has been presented in publications [7,31,37,41,42,44]):

- *The state of full effective and ecological operability of the engine* s_1, which allows the engine to be used in the whole load range for which it was designed;
- *The state of full effective operability of the engine* s_2, which allows the engine to be used in the whole load range for which it was designed, but with significantly lower overall efficiency, i.e., significantly higher fuel consumption and not meeting environmental protection standards;
- *The state of partial operability of the engine* s_3, which allows the engine to be used in a limited load range, smaller than that for which it was designed, does not meet environmental protection standards, and where the levels of wear and tear of the engine components, exceeding the alarm parameters of the media, indicate the need to perform maintenance allowing the engine to be renewed (at least) to s_2 in order to take the ship to a place of refuge;
- *Engine inoperability state* s_4, which prevents the engine from being used as intended (e.g., due to damage, prevention work, etc.).

The set elements $S = \{s_i; i = 1, 2, 3, 4\}$ are the process values of $\{W(t): t \geq 0\}$, which are consecutive states of $s_i \in S$, known to be causal.

It is important to distinguish between states $s_i \in S$ ($i = 1, 2, 3, 4$) because it is crucial to use the motors when they are in state s_1, or possibly in state s_2. When engines are in state s_2, and s_3, they are used in the shortest possible time due to the occurrence of intense degradation processes.

Shipboard internal combustion piston engines in *full operability* s_1 can be used at any time under different loads and ecological standards. In the s_2 condition, they can be freely loaded, but they cannot meet ecological standards. Partially operability s_3 can be used or operated depending on the decision situation, i.e., when the operating conditions (high wave, strong wind, proximity to land, too low depth, saturation of the body of water with other objects, etc.) are not sufficient. This does not put the engine out of action and take up service, but this condition gives a chance to reach the place of shelter [3].

An engine in the *inoperability state* s_4 must always be operated, especially at sea (which gives a chance of survival to the vessel) or in port, if this is still economically justified. During states s_1, s_2, and s_3, the engines can operate and thus send out diagnostic signals, which makes it possible to recognize the elementary states classified into the listed state classes. The changes in these states are influenced by the operating conditions of the engines. Knowledge of these conditions enables rational engine control [45].

If diagnostic information (parameters, change trends) and information about expected engine operating conditions (e.g., expected weather conditions) are available, the operator (engine user) may risk undertaking the task when the engine is in state s_2, or even risk undertaking some tasks in state s_3.

However, the problem is that to analyze the data displayed on the screens of units processing engine parameters requires knowledge exceeding the knowledge and skills of about 70% of mechanics, especially those who do not have an academic education (IMO requires secondary education). Secondary education does not require knowledge of the so-called higher mathematics. However, knowledge of the basics of so-called higher mathematics is necessary to correctly read the graphs displayed on the screens of computers processing engine parameters. Diagrams appearing on the ship's computer screens are shown in Figure 7.

In state s_3, when the operating parameters of the engine indicate significant degradation of the engine structure and further operation can lead to serious damage, called failure, the only rational decision is to stop using the engine and start maintenance. This involves putting the engine out of service, dismantling its components, renewing those components that need to be renewed, or replacing them with new ones. Such a decision taken during the voyage involves stopping the ship, which may result in a loss of maneuverability and the ship's sideways alignment.

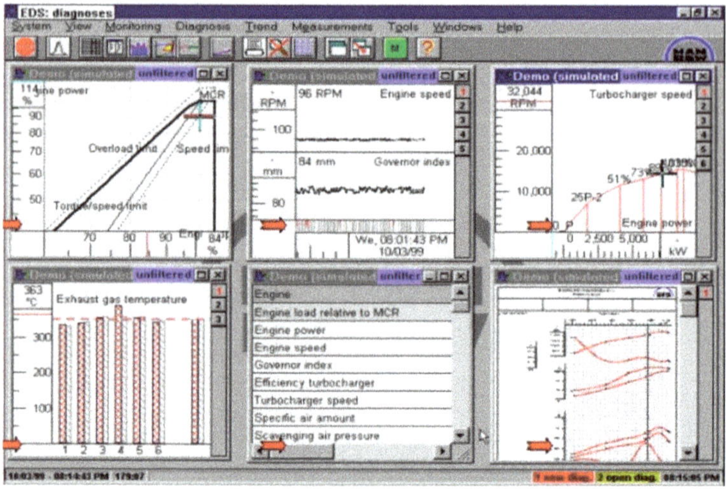

Figure 7. Complicated graphs of mathematical dependencies (at the level of higher mathematics, e.g., nonlinear functions) showing trends in changes of engine condition, on the basis of analysis of data provided by sensors placed on the engine, requiring [48].

On the other hand, if the weather conditions are good, the decision to start a service of the engine causes the voyage to be stopped and too long a stoppage means the possibility of suffering the economic consequences of not entering the port on time. In unfavorable weather conditions, or only unfavorable sea waves, during lateral heeling the ship may lose its lateral stability and drown. Therefore, during the voyage of each oil tanker, it is necessary that the propulsion system of that tanker is in the condition s1 to maintain the ship's maneuverability. This state of the ship's propulsion system is provided by the PTO/PTI electrical unit. This is a solution proposed by engine manufacturers. The selection of the shaft alternator (PTO) should be determined by the electricity demand necessary to maintain the safe movement of the vessel when operating the main propulsion system, which the shaft alternator (PTO/PTI) produces electricity according to the electrical power demand of the auxiliary equipment.

At the planned load of the ship's propulsion system, the transmitted output of the shaft generator (PTO/PTI) is about 10–15% of the power of this main engine [29].

Where it is not relevant in the operating strategy adopted for these engines to distinguish between states s_1 and s_2, a simpler process of $\{W^*(t): t \geq 0\}$ changes in the technical states of the engines may be considered, namely a model with a set of states $S = \{s_1, s_2, s_3\}$ [42], with the following interpretation of these states:

- The *full operability state* s_1, which allows the engine to be used under all the conditions and load ranges for which it is designed and manufactured (including eco-sufficiency);
- The *partial operability state* s_2, which allows the engine to be used under restricted conditions and load ranges smaller than those for which it is designed and manufactured (e.g., "green" and "economic"), but with less favorable economic coefficients;
- The *engine inoperability state* s_3, the failure condition which prevents the engine from being used as intended (e.g., due to damage to components or subassemblies, preventive work on subassemblies, or an inability to provide adequate parameters for the operating media, etc.).

Thus, the process consists of three stages with continuous execution over time. The elements of the set $S = \{s_1, s_2, s_3\}$ are also considered as the values of the said process $\{W^*(t) : t \geq 0\}$ following one another during engine operation. This process is characterized by the fact that, if states s_2 or s_3 do not occur, the internal combustion engine is in state s_1, and that the operating time of the internal combustion engines is not a good measure of the wear of their structural structure. This means that a change in the technical condition of this type of engine is poorly correlated with the wear and tear of the expected operating hours. This was justified in the works [7,31]. It follows from the above consideration that the commonly used strategy of carrying out preventive maintenance of internal combustion engines (including ship engines) after the expiry of the designated, fixed periods of their proper operation (time reservations) is ineffective, and therefore, unreasonable. Hence, the conclusion that preventive maintenance of engines should be carried out depending on the results of wear tests, which is possible if the appropriate technical diagnostics are used. The application of such diagnostics requires the development of a diagnostic model for a given type of internal combustion engine, and the application of an appropriate diagnostic system adapted to that model, which is adopted to identify the technical condition of those engines [8,38].

5. Change of Engine States Model

For each marine diesel engine, the change process in engine states is random. The scientific works [31,37] show that the models of this process can be a stochastic process $\{W(t)$: $t \geq 0\}$, with a discrete set of states and continuous during the distinguished technical states of these engines. The development of a model of the process of changing the technical states of diesel engines in the form of a stochastic process requires the establishment of a finite set of S states of these engines. Taking as a criterion for distinguishing such states of suitability of compression ignition engines for performing operational tasks, a set of technical states (significant in operational practice) can be defined in the form of relation:

$$S = \{s_i; i = 1, 2, 3, 4\}, \tag{16}$$

with the interpretation presented earlier applicable here, i.e., s_1—full effective and environmental *operability*; s_2—full effective *operability*; s_3—partial *operability*; s_4—*inoperability*.

The set elements $S = \{s_i;\ i = 1, 2, 3, 4\}$ are the values of the process $\{W(t) :\ t \geq 0\}$, which are consecutive states of $s_i \in S$, and causal to each other. The distinction between $s_i \in S$ ($i = 1, 2, 3, 4$) states for a ship's main engines is extremely important as it is vital to use these engines when they are in state s_1 or, if necessary, for the shortest possible time (after which they should be renewed) when they are in state s_2.

This process is fully specified if its functional matrix is known:

$$Q(t) = [Q_{ij}(t)], \tag{17}$$

the non-zero elements of which have the following interpretation:

$$Q_{ij}(t) = P\{W(\tau_{n+1}) = s_j,\ \tau_{n+1} - \tau_n < t \mid W(\tau_n) = s_i\};\ s_i, s_j \in S;\ i, j = 1, 2, 3, 4;\ i \neq j$$

When the initial distribution is given:

$$p_i = P\{W(0) = s_i\},\ s_i \in S;\ i = 1, 2, 3, 4 \tag{18}$$

It was assumed that the changes in state of the ship's main two-stroke, slow-speed main engine used for the propulsion of sea-going vessels, such as cargo and oil tankers, take place according to the transition graph, which is shown in Figure 8 [31].

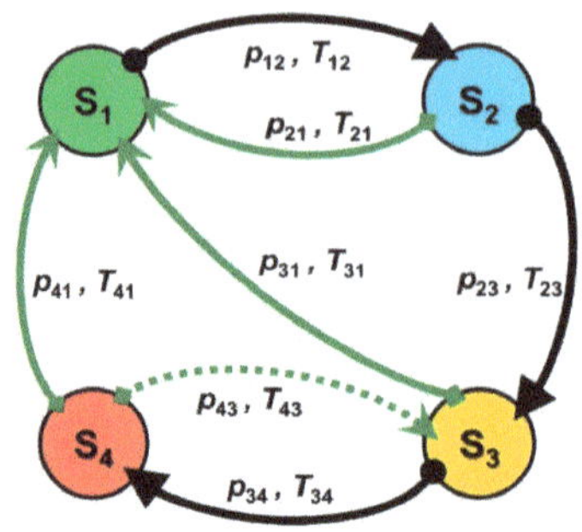

Figure 8. The graph of state changes $s_i \in S(i = 1, 2, 3, 4)$ of the process $\{W(t) : t \geq 0\}$ [4].

In Figure 8 a transition is marked, with an arc depicted by a dashed line, from the state s_4 to the state s_3, which in rational control can only take place in exceptional situations, and therefore, as it is unreasonable, is not included in the developed model under consideration of the process of changes in states of the ship's main engine. Therefore, the process limit distribution $\{W(t)$: $t \geq 0\}$ can be presented in the formulas [7]:

$$P_1 = E(T_1)M^{-1},\ P_2 = E(T_2)M^{-1},\ P_3 = p_{23}E(T_3)M^{-1},\ P_4 = p_{23}p_{34}E(T_4)M^{-1} \quad (19)$$

whereby:

$$M = E(T_1) + E(T_2) + p_{23}p_{34}E(T_4) \quad (20)$$

where:

$E(T_j)$—expected duration of the state $s_j \in S(j = 1, 2, 3, 4)$,

P_{ij}—probability of the process passing $\{W(t)$: $t \geq 0\}$ from state s_j to state s_j (s_i, $s_j \in S$; $i, j = 1, 2, 3, 4; i \neq j$).

The individual probabilities Pj ($j = 1, 2, 3, 4$), as defined by Formulae (19), are interpreted as follows:

$$P_1 = \lim_{t\to\infty} P\{W(t) = s_1\},\ P_2 = \lim_{t\to\infty} P\{W(t) = s_2\}, P_3 = \lim_{t\to\infty} P\{W(t) = s_3\}, P_4 = \lim_{t\to\infty} P\{W(t) = s_4\}$$

In the presented model of state changes s_i ($i = 1, 2, 3, 4$), there are situations when the operator may decide to use the PTO/PTI system: at state s_2 of the engine at time $\Delta\tau_{12}$; at state s_3 of the main engine at a time presented as $\Delta\tau_{45}$ or $\Delta\tau_{89}$, and in good weather conditions, the PTO/PTI system should be used and operated with ME; and at state s_4 at time $\Delta\tau_{56}$, the propulsion with PTO/PTI is necessary and gives a chance for the survival of the vessel (Figure 9).

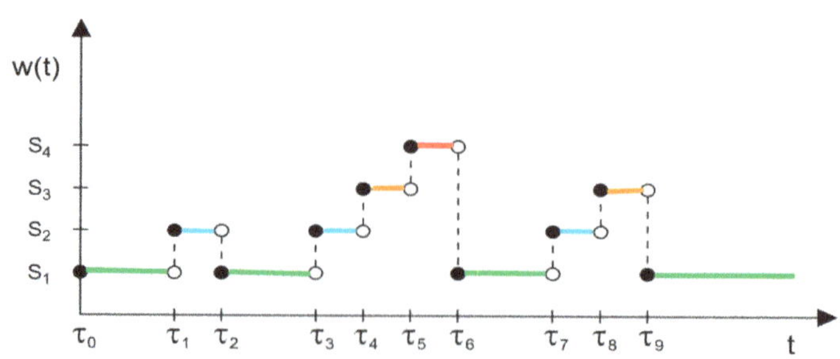

Figure 9. Example of the implementation of the $\{W(t)$: $t \in T\}$ process of a compression ignition engine: $\{W(t)$: $t \in T\}$—process of state changes; t—operating time; s_1—full serviceability; s_2—incomplete serviceability; s_3—partial serviceability; s_4—unsuitable state; periods of time during which the use of a PTO/PTI system is justified.

An important feature of the graph shown in Figure 9 is that the relationship between states reflecting the execution of full recovery to state s_1 is taken into account if the engine is in state s_2 or s_3. Therefore, this variant of $s_i \in S$ condition means changes should be taken into account in creating a simpler (tri-state) graph to show the technical condition changes of a ship's main engines and a related diagnostic model of this type of engine (Figure 10).

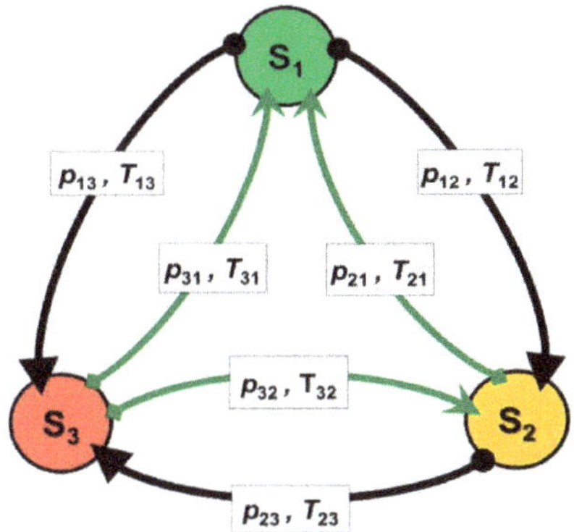

Figure 10. Ship's main engine state change graph: s_1—full effective operability, s_2—partial operability, s_3—inoperability. p_{ij}—probability of process transition from s_i to s_j, T_{ij}—duration of s_i provided that process transition to s_j, i, j = 1, 2, 3.

In the strategy adopted for the operation of marine diesel main engines, it is not relevant to distinguish between states s_1 and s_2, it can be considered a simpler process of $\{U(t) : t \in T\}$ changes in the technical states of these engines, namely a process with a set of states, can be considered:

$$S = \{s_1, s_2, s_3\} \tag{21}$$

for interpretation: s_1—full effective *operability*, s_2—partial *operability*, s_3—*inoperability*.

The graph of state changes s_i (i = 1, 2, 3) of the process $\{u(t) : t \in T\}$ of these engines is shown in Figure 7.

This process is also a three state process (like the process $\{W(t) : t \geq 0\}$) with continuous execution over time. It is assumed that if either of the states s_2 or s_3 do not occur, the internal combustion engine is in state s_1.

Therefore, the set of technical states $S = \{s_1, s_2, s_3\}$ can be considered as a set of stochastic process values $\{U(t) : t \in T\}$, with fixed compartments and right-hand continuous execution (Figure 11).

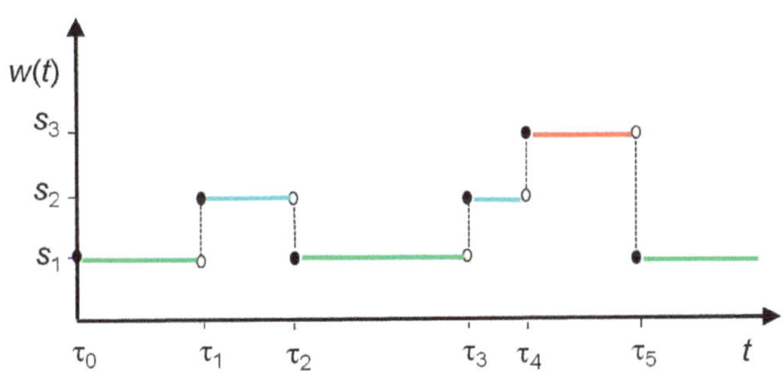

Figure 11. Example of the implementation of the process: $\{U(t) : t \in T\}$ of a marine diesel main engine; (ME): $\{U(t) : t \in T\}$—process of state of repair; t—service life; s_1—full effective *operability*; s_2—partial *operability*; s_3—*inoperability*. Periods of time during which the use of the PTO/PTI system is justified.

The initial distribution of the process under consideration (Figure 11) about the transition graph (Figure 10) is defined by a formula:

$$P_i = P\{W(0) = s_i\} = \begin{cases} 1\ for\ i = 1 \\ 0\ for\ i = 2,3 \end{cases} \quad (22)$$

The function matrix, on the other hand, is as follows, when the function $Q_{32}(t)$ is non-zero, or $Q_{32}(t) \neq 0$:

$$\mathbf{Q(t)} = \begin{bmatrix} 0 & Q_{12}(t) & Q_{13}(t) \\ Q_{21}(t) & 0 & Q_{23}(t) \\ Q_{31}(t) & Q_{32}(t) & 0 \end{bmatrix} \quad (23)$$

When the function $Q_{32}(t)$ is zero, i.e., $Q_{32}(t) = 0$, the matrix (23) will take the form:

$$\mathbf{Q(t)} = \begin{bmatrix} 0 & Q_{12}(t) & Q_{13}(t) \\ Q_{21}(t) & 0 & Q_{23}(t) \\ Q_{31}(t) & 0 & 0 \end{bmatrix} \quad (24)$$

Therefore, for the presented process $\{U(t){:}\ t \in T\}$ with the functional matrix defined by the Formula (24), the following limit distribution can be determined:

$$P_1 = \frac{\pi_1 E(T_1)}{H}, P_2 = \frac{\pi_2 E(T_2)}{H}, P_3 = \frac{\pi_3 E(T_3)}{H}, \quad (25)$$

whereby:

$$\pi_1 = \frac{1}{2 + p_{12}p_{23}}, \pi_2 = \frac{p_{12}}{2 + p_{12}p_{23}}, \pi_3 = \frac{1 - p_{12}p_{21}}{2 + p_{12}p_{23}},$$
$$H = \pi_1 E(T_1) + \pi_2 E(T_2) + \pi_3 E(T_3),$$

where:

P_1, P_2, P_3—the probability that the compression ignition engine is in the states s_1, s_2, s_3 respectively;

π_j—limit probability of the process $\{U(t) : t \in T\}$ of the Markov string describing the possibility of the state s_j appearing, $j = 1, 2, 3$;

p_{ij}—probability of the process passing $\{W(t) : t \in T\}$ from the state s_i to the state s_j;

$E(T_j)$—expected duration value of the state s_j.

The presented technical conditions of the marine diesel main engine are related to the respective operating conditions of these engines. These technical states and the associated operating states (states of use and operation) are mutually exclusive. Therefore, taking into account these types of conditions when making operational decisions, it is reasonable to control the operation of the ship's propulsion system by taking into account the application of the PTO/PTI system, in operational situations where it is reasonable to exclude the ME from service in case of malfunction, or when the ME is in an unsuitable state.

A properly designed computer program based on the proposed models could work out unambiguous conclusions from the analysis of trends in changes of engine states, and unambiguously suggest to the mechanic-operator to change the type of drive to the emergency one, i.e., to use the PTI system. However, this requires the involvement of the engine manufacturer, who avoid such unambiguous solutions and leave such decisions to the mechanics. This is related to complicated post-accident procedures conducted by insurance companies in order to determine responsibility for failures.

6. Conclusions

The paper presented the concept of a technical solution of a ship propulsion system, consisting of a two-stroke, low-speed main diesel engine (ME) and a shaft generator that can be used as an emergency electric motor when it is necessary to put the ME out of service.

This is an innovative solution, because it is not currently used on very large ships where there is such a large disproportion between the main and the emergency power.

The main problems to be solved in the research process identified by the research team can be divided into direct and indirect problems.

Indirect issues are that with this type of main engine propulsion system, and the use of a fixed pitch propeller, the shaft generator cannot be used to power the ship's power plant while the ship is stationary at the loading terminal, or used as a power source for the cargo pumps. Therefore, the power of the shaft generator is designed adequately for the energy demand for powering the equipment ensuring the movement of the ship at sea. At the same time, when controlling the power of such an engine by changing the revolutions, shaft generators are used only when the hydrometeorological conditions make operation possible. This means a limited range of revolutions and use only during a sea voyage.

The direct problems consist of the fact that in order to achieve the minimum maneuvering speed, generators with a power corresponding to at least about 30–40% of the nominal power of the two-stroke diesel engine (depending on the state of loading of the ship) are needed, with standard generators of about 6–10% of the nominal power of the ME.

Therefore, engine manufacturers will not use over-powered generators, associated with an increase in investment and operating costs, and with a decrease in the efficiency of the generator when only partially loaded during the voyage.

Based on operational experience, it was suggested to change the thinking from "economic" to "safe", which requires additional costs, but a similar cost increase occurs with "safe" to "ecological" thinking.

Therefore, in justifying the proposed solution, the paper shows that it is possible to increase the reliability of the main propulsion system of crude oil tankers, and bulk cargo ships adapted to carry bulk cargo, by using a shaft generator as an emergency engine (PTO/PTI) when the main engine loses its fully operational condition.

Since one of the biggest exploitation problems leading to failure is rational decision making, it was shown that depending on the adopted exploitation strategy, one of the two exploitation process models of marine main engines can be applied using the semi-Markov process, i.e., four-state, when making decisions in order to protect the environment [16], or three-state, when the main engine must be stopped before it reaches the state of inoperability. For both of these models, their limiting distributions were determined, which formed the probabilities of the engine staying in their respective states.

It was shown that such a model was useful in the operational practice of the ships under consideration, especially in the process of diagnosing the engine technical condition and facilitating the operational decision at the right moment.

In conclusion, on the basis of the above-mentioned research, it seems reasonable that in order to avoid both serious breakdown and a loss of maneuverability of the ship, the PTO/PTI system should be applied at the appearance of the main engine partial operational state (s_2 in the three-state model). This is a prerequisite for maintaining the ship's maneuverability and safely performing engine maintenance. It was also found that when a ship's propulsion engine is in an unserviceable condition (s_3 in three-state model), it is absolutely necessary to use the shaft generator (PTO/PTI) as an electric motor of emergency main propulsion.

Author Contributions: Conceptualization, Z.Ł., W.M., W.C., E.S.-M. and W.H.; methodology, Z.Ł., W.M. and W.C.; software, E.S.-M. and W.H.; validation, Z.Ł., W.M., W.C., E.S.-M. and W.H; formal analysis, W.C., E.S.-M. and W.H.; investigation, Z.Ł., W.M., W.C.; resources, Z.Ł., W.M., W.C., E.S.-M. and W.H.; data curation, Z.Ł.; writing—original draft preparation, Z.Ł.; writing—review and editing, Z.Ł., W.M. and W.C.; visualization, Z.Ł.; supervision, Z.Ł.; project administration, Z.Ł.; funding acquisition, W.M. All authors have read and agreed to the published version of the manuscript.

Funding: Polish Naval Academy of the Heroes of Westerplatte, Śmidowicza Street 69, PL 586-010-46-93, Gdynia 81-127, Poland.

Institutional Review Board Statement: Not applicable.

Informed Consent Statement: Not applicable.

Data Availability Statement: Not applicable.

Conflicts of Interest: The authors declare no conflict of interest.

References

1. Schmitter, T. *Project Guide, Very Large Crude Oil Carrier, Sulzer RTA84T-B Main Engine, 280000/300000 Dwt—2 Million Barrel Tanker;* Wartsila NSD Switzzerland Ltd.: Winterthur, Switzerland, 1998.
2. Mironiuk, W.; Łosiewicz, Z. *Experimental Studies of Pitching Training Ship Model in Terms of Maritime Transport Security*, 6th ed.; CD 3; Logistyka: Gdynia, Poland, 2014; pp. 7534–7539.
3. Łosiewicz, Z.; Mironiuk, W. *Ocena Bezpieczeństwa Statków Handlowych Różnych Typów w Warunkach Morskich—Wg Przyjętych Kryteriów (Assessment of the Safety of Merchant Ships of Different Types in Sea Conditions—According to Adopted Criteria);* Nr 12/2015 s.2012-2015; TTS Technika Transportu Szynowego: Radom, Poland, 2015.
4. Mironiuk, W. Model-based investigations on dynamic ship heels in relation to maritime transport safety. Badania modelowe dynamicznych przechyłów statków w odniesieniu do bezpieczeństwa transportu morskiego. *Arch. Transp.* **2015**, *33*, 69–80. [CrossRef]
5. Szozda, Z. *Stateczność Statku Morskiego (Stability of a Seagoing Ship);* Wydawnictwo Naukowe Akademii Morskiej: Szczecin, Poland, 2009.
6. Łosiewicz, Z. *Bezpieczeństwo Żeglugi w Aspekcie Eksploatacji Okrętowych Silników Spalinowych Tłokowych Napędu Głównego Statku (Safety of Navigation in Terms of Operation of Marine Piston Main Propulsion Diesel Engines);* Oficyna Wydawnicza Politechniki Warszawskiej: Warszawa, Poland, 2019; ISBN 978-83-7814-870-8.
7. Girtler, J.; Kuszmider, S.; Plewiński, L. *Wybrane Zagadnienia Eksploatacji Statków Morskich w Aspekcie Bezpieczeństwa Żeglugi (Selected Issues of Operation of Seagoing Vessels in the Aspect of Safety of Navigation);* WSM w Szczecinie: Szczecin, Poland, 2003.
8. Łosiewicz, Z. Probabilistyczny Model Diagnostyczny Silnika Napędu Głównego Statku (A Probabilistic Diagnostic Model of the Ship's Main Propulsion Engine). Ph.D. Thesis, Faculty of Ocean Engineering and Shipbuilding, Gdańsk University of Technology, Gdańsk, Poland, 2008.
9. Mironiuk, W. Comparative analysis of the flooding time main engine ship type 888 definited from calculations and model research. *J. KONES Powertrain Transp.* **2010**, *17*, 345–352.
10. Mironiuk, W.; Pawlędzio, A. *Modelling Studies of the Roll and the Pitch Training Ship*; Maritime Transport & Shiping: London, UK, 2013.
11. Pawlędzio, A.; Mironiuk, W. Analiza wpływu ujemnej wysokości metacentrycznej na położenie pontonu prostopadłościennego (Analysis of the effect of negative metacentric height on the position of the perpendicular pontoon). *Sci. J. Pol. Nav. Acad. Gdyn.* **2006**, *47*, 61–75.
12. Szubrycht, T. Marine accidents as potential crisis situations on the Baltic Sea. *Arch. Transp.* **2020**, *54*, 125–135. [CrossRef]
13. MAN Energy Solutions. *Shaft Generators for Low Speed Main Engines;* MAN Energy Solutions SE: Copenhagen, Denmark, 2021.
14. Yamada, H.; Miyabe, H.; Saeki, A. Energy Saving Technology of the Diesel-Electric Propulsion—System for Japanese Coastal Vessels. *IHI Eng. Rev.* **2011**, *44*, 12–16.
15. Anantharaman, M.; Islam, R.; Khan, F.; Vikram Garaniya, V.; Lewarn, B. *Emergency Preparedness for Management of Main Propulsion Engine Failure on a Bulker during Harsh Weather at Sea*; Springer Nature Switzerland AG, Safety in Extreme Environments: Cham, Switzerland, 2020.
16. Jeong, B.; Oguz, E.; Wang, H.; Peilin, W.; Zhou, P. Multi-criteria decision-making for marine propulsion: Hybrid, diesel electric and diesel mechanical systems from cost-environment-risk perspectives. *Appl. Energy* **2018**, *230*, 1065–1081. [CrossRef]
17. WÄRTSILÄ Electrical & Automation. *Diesel-Electric Propulsion Systems;* Wärtsilä Corporation: Helsinki, Finland, 2016.
18. Wärtsilä, H.Y. (Ed.) *WÄRTSILÄ Hybrid Solutions;* Wärtsilä Corporation: Helsinki, Finland, 2018.
19. Jeon, H.; Kim, J.; Kyoungkuk Yoon, K. Large-Scale Electric Propulsion Systems in Ships Using an Active Front-End Rectifier. *J. Mar. Sci. Eng.* **2019**, *7*, 168. [CrossRef]
20. Institute of Electrical and Electronics Engineers, Inc. *IEEE Recommended Practice and Requirements for Harmonic Control in Electric Power Systems;* IEEE Std 519-2014; IEEE: New York, NY, USA, 2014; pp. 5–9.
21. Krakowski, R. Technical and ecological analysis of different solutions of the power supply of a marine diesel engine. *J. KONES Powertrain Transp.* **2019**, *26*, 4. [CrossRef]
22. Dumitrescu, M. Overview on Efficient Naval Power Architecture. In Proceedings of the 14th International Conference on Interdisciplinarity in Engineering—INTER-ENG 2020, Târgu Mures, Romania, 8–9 October 2020; MDPI: Basel, Switzerland, 2020; Volume 63, p. 22.
23. Percic, M.; Vladimir, N.; Korican, M. Electrification of Inland Waterway Ships Considering Power System Lifetime Emissions and Costs. *Energies* **2021**, *14*, 7046. [CrossRef]
24. Wojciech Litwin, W.; Wojciech Leśniewski, W.; Piątek, D.; Niklas, K. Experimental Research on the Energy Efficiency of a Parallel Hybrid Drive for an Inland Ship. *Energies* **2019**, *12*, 1675. [CrossRef]
25. Kunicka, M.; Litwin, W. Energy Demand of Short-Range Inland Ferry with Series Hybrid Propulsion Depending on the Navigation Strategy. *Energies* **2019**, *12*, 3499. [CrossRef]

26. Sendek-Matysiak, E.; Pyza, D. The Assignment of Vehicle Assessment Based on Multi Criteria Decision Making. In *Archives of Transport 2018*; Warsaw University of Technology: Warsaw, Poland, 2018; Volume 48.
27. Sendek-Matysiak, E.; Łosiewicz, Z. Analysis of the Development of the Electromobility Market in Poland in the Context of the Implemented Subsidies. *Energies* **2021**, *14*, 222. [CrossRef]
28. MAN Diesel & Turbo. *Hybrid Propulsion Flexibility and Maximum Efficiency Optimally Combined*; MAN Energy Solutions: Augsburg, Germany, 2017.
29. MAN Diesel & Turbo. *S70MC-C Mk 7 Project Guide Two-stroke Engines*; MAN Diesel & Turbo—Copenhagen SV: Kopenhaga, Denmark, 2005.
30. MAN Diesel & Turbo. *Shaft Generators for Low Speed Main Engines*; MAN Diesel & Turbo—Copenhagen SV: Kopenhaga, Denmark, 2015.
31. Girtler, J.; Łosiewicz, Z. Analysis of Properties of the Diagnostic model of engine crankshaft-piston systems, adjusted to needs of rational control over the operation of these engines in the context of ship safety and protection of marine environment. In *Interim Report on the Execution of the Author's Research Project No N N509 494638 Financed by the Ministry of Science and Higher Education and Entitled: "Decisionmaking Control over the Operation of Crankshaft-Piston Systems of Main Propulsion Engines on Sea-Going Ships with Application of Technical Diagnostics and Taking into Account Safety and Protection of the Environment"*; Prace Badawcze No. 06//10/PB; Gdansk University of Technology, Faculty of Ocean Engineering and Ship Technology: Gdansk, Poland, 2010. (In Polish)
32. Mambra, S. *The Complete Story of the Exxon Valdez Oil Spill*; 10/2020; Marine Insight: Bangalore, India, 2020.
33. Jacyna, M.; Semenov, I. *Models of Vehicle Service System Supply under Information Uncertainty, Eksploatacja i Niezawodność—Maintenance and Reliability*; Polish Maintenance Society: Warsav, Poland, 2020; Volume 22.
34. Jacyna, M.; Wasiak, M.; Lewczuk, K.; Kłodawski, M. Simulation model of transport system of Poland as a tool for developing sustainable transport. *Arch. Transp.* **2014**, *31*, 3. [CrossRef]
35. Łosiewicz, Z. *Zbiory Parametrów Diagnostycznych do Identyfikacji Stanów Technicznych Okrętowego Tłokowego Silnika Spalinowego (Sets of Diagnostic Parameters for the Identification of Technical States of a Marine Reciprocating Internal Combustion Engine)*; Nr 6/2016, s. 983-985; Autobusy, Technika, Eksploatacja, Systemy Transportowe: Radom, Poland, 2016.
36. Czajgucki, J.Z. *Niezawodność Spalinowych Siłowni Okrętowych (Reliability of Marine Diesel Engines)*; Wydawnictwo Morskie: Gdańsk, Poland, 1984.
37. Czajgucki, J.Z. Principles of Modelling of Processes and Examples of Interpretation of These Principles with Reference to Technology. *Syst. Sci.* **2009**, *35*, 15–21.
38. Łosiewicz, Z. Identifying the issue of reducing the emission of harmful compounds in the exhaust gas from marine main engines and description of the emission process of these compounds in probabilistic approach. *J. Pol. Marit. Res.* **2017**, *24*, 89–95. [CrossRef]
39. Rudnicki, J.; Zadrąg, R. Problems of modelling toxic compounds emitted by a marine internal combustion engine in unsteady states. *J. Pol. Marit. Res.* **2014**, *21*, 57–65. [CrossRef]
40. Dyrcz, C. Analysis of sea accidents. *Annual of Navigation. Annu. Navig.* **2016**, *23*, 151–157. [CrossRef]
41. Girtler, J. Sami-Markovian model of the process of technical state changes of technical objects. *Pol. Marit. Res.* **2004**, *11*, 3–7.
42. Girtler, J. Physical aspects of application and usefulness of semi-Markovian processes for modeling the processes occurring in operational phase of technical objects. *Pol. Marit. Res.* **2004**, *11*, 25–30.
43. Łosiewicz, Z. Application of probabilistic diagnostic models in diagnostic service decisions-aiding systems for ship main propulsion engine. *J. Pol. CIMAC* **2008**, *3*, 9–13.
44. Łosiewicz, Z. Analysis of the actual M.E. diagnostic processes on vessels—Usefulness of selected systems diagnosing the ship engines and of the operating procedures on example of the actual operating events. *J. Pol. CIMAC* **2009**, *4*, 199–205.
45. Łosiewicz, Z. Use assessment of modern diagnostic systems of internal marine combustion engines at the support phase of exploitation decisions in the aspect of economic criteria. *J. Pol. CIMAC* **2010**, *5*, 121–130.
46. Łosiewicz, Z. *Wpływ Czynnika Ludzkiego na Bezpieczną Eksploatację Statku w Aspekcie Różnych Faz Życia Statków (Impact of Human Factors in the Safe Operation of Ships in Terms of Different Phases of Life Ships)*; Nr 12/2015 s.2307-2309; TTS Technika Transportu Szynowego: Radom, Poland, 2015.
47. Grabski, F. *Semi-Markowskie Modele Niezawodności i Eksploatacji. (Semi-Markov Reliability and Operation Models) Wyd*; Seria BADANIA SYSTEMOWE, t.30; IBS PAN: Warszawa, Poland, 2002.
48. MAN Energy Solutions. *Strategic-Expertise of Efficiency-Control*; MAN Energy Solutions SE: Copenhagen, Denmark, 2021.

energies

Article

Modelling and Simulation/Optimization of Austria's National Multi-Energy System with a High Degree of Spatial and Temporal Resolution

Matthias Greiml *, Florian Fritz, Josef Steinegger, Theresa Schlömicher, Nicholas Wolf Williams, Negar Zaghi and Thomas Kienberger

Energy Network Technology, Montanuniversitaet of Leoben, 8700 Leoben, Austria
* Correspondence: matthias.greiml@unileoben.ac.at

Abstract: The European Union and the Austrian government have set ambitious plans to expand renewable energy sources and lower carbon dioxide emissions. However, the expansion of volatile renewable energy sources may affect today's energy system. To investigate future challenges in Austria's energy system, a suitable simulation methodology, temporal and spatially resolved generation and consumption data and energy grid depiction, is necessary. In this paper, we introduce a flexible multi-energy simulation framework with optimization capabilities that can be applied to a broad range of use cases. Furthermore, it is shown how a spatially and temporally resolved multi-energy system model can be set up on a national scale. To consider actual infrastructure properties, a detailed energy grid depiction is considered. Three scenarios assess the potential future energy system of Austria, focusing on the power grid, based on the government's renewable energy sources expansion targets in the year 2030. Results show that the overwhelming majority of line overloads accrue in Austria's power distribution grid. Furthermore, the mode of operation of flexible consumer and generation also affects the number of line overloads as well.

Keywords: 100% renewable energy sources (RESs); multi-energy system (MES) modelling; multi-energy system (MES) simulation; hybrid grid; national multi-energy system (MES)

Citation: Greiml, M.; Fritz, F.; Steinegger, J.; Schlömicher, T.; Wolf Williams, N.; Zaghi, N.; Kienberger, T. Modelling and Simulation/Optimization of Austria's National Multi-Energy System with a High Degree of Spatial and Temporal Resolution. *Energies* **2022**, *15*, 3581. https://doi.org/10.3390/en15103581

Academic Editors: Zbigniew Leonowicz, Michał Jasiński and Arsalan Najafi

Received: 20 April 2022
Accepted: 10 May 2022
Published: 13 May 2022

1. Introduction

Climate change is seen as a serious problem by ninety-three per cent of Europeans. According to the European Commission, the same number of Europeans have taken at least one action to tackle climate change. By setting up the ambitious "European Green Deal" program in December 2019, the European Commission aims to achieve a climate-neutral European Union by 2050 [1,2]. Concretizing the path towards achieving "European Green Deal" targets, the European Commission set up the "Fit for 55" program as an interim goal in July 2021. This program aims to reduce the European Union's carbon dioxide emissions by fifty-five per cent by the year 1990 [3].

As a member of the European Union, the Austrian government has set even more ambitious targets, aiming to achieve net CO_2 neutrality by 2040 [4]. Furthermore, the Austrian #mission2030 aims to achieve a hundred per cent renewable power generation net-balanced over one year until the year 2030. To achieve this target, renewable energy sources (RES), mainly volatile wind and photovoltaics, have to be expanded significantly [5].

The enhanced usage of RESs presents challenges for both the energy system and its operators since RESs are decentralized, hardly predictable, and introduce volatility into energy grids [6]. Achieving a hundred per cent RES might require:

- A spatial and timely compensation of energy;
- An increase in flexibility for both demand and generation in an energy system;
- The ability to cope with high instantaneous penetration of RES;
- Curtailment of RES as ultima-ratio.

In order to integrate a high share of RESs into existing energy systems and to avoid previously described issues, new approaches are necessary. In recent years, research focused on a cross-sectoral approach to consider the energy carriers' individual advantages in an energy system. This approach allows for the implementation of power, natural gas, district heating, hydrogen, and carbon dioxide grids, combined with storage and sector coupling (SC) options [7].

A need to address previously described challenges can be derived by looking at the distribution of RES in Austria. Sejkora et al. [8] provide a comprehensive overview of Austria's spatial distribution of technical exergy potentials of RES, which can be directly converted into RES energy potential [8]. Referring to Figure 1, it can be seen that renewable potentials are widely spread all over Austria, fluctuating in both the type of RES and the quantity in each district. However, it can be seen that wind potentials are mainly to be found in eastern Austria, whereas hydropower potentials are located in the western parts of Austria. Biomass and photovoltaics can be considered as more evenly distributed across Austria.

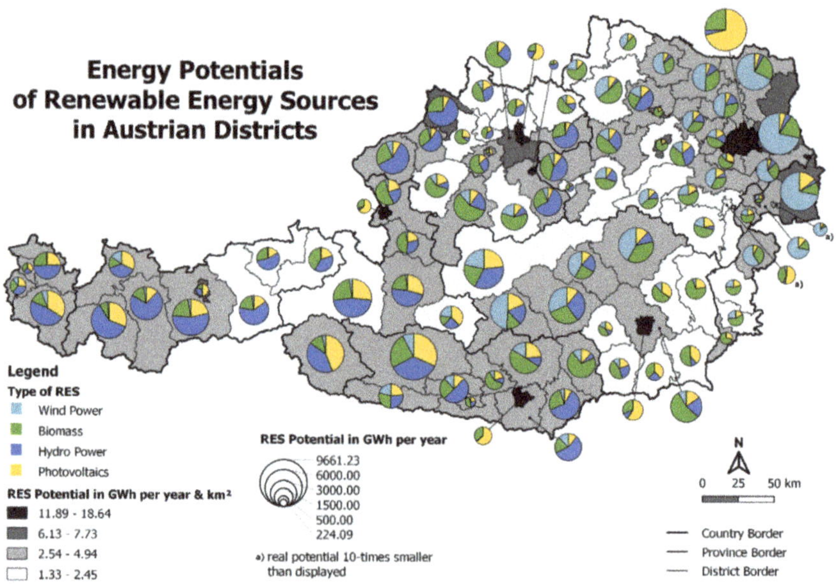

Figure 1. Technical energy potentials of RES per Austrian district, derived from [8,9].

Integrating further RES into the current energy system might lead to issues as previously disclosed. To address them, we introduce an updated multi-energy-system (MES) simulation framework, HyFlow, and discuss simulation results based on potential scenarios of the Austrian energy system in the year 2030.

1.1. Literature Overview and Research Need

As there are numerous publications on the topic of MES models, this section aims to display the current state of research regarding MES simulation and optimization approaches. Furthermore, we introduce research focusing on Austria's national multi-energy system models.

1.1.1. MES Simulation and Optimization Approaches

According to Klemm and Vennemann [10], energy system models can be methodologically categorized in optimization, forecasting/simulation, and back-casting. Depending on the defined objective function, optimization is capable of determining an optimal solution or scenario. Forecasting or Simulation models show the system's behavior according to the selected input parameter. This scenario-based approach likely doesn't represent an optimal solution with regard to the selected boundary conditions. In back-casting models, an envisioned future state or properties are defined. Based on the future state, the back-casting model develops paths to these future conditions. Further categorization criteria could be assessment criteria, as well as analytical or mathematical approaches and challenges. The structural and technological details of MES models can include geographic coverage, spatial and temporal resolution, time horizon, sectoral coverage, and demand sectors [10].

Several pieces of research provide a comprehensive overview and comparison of existing MES assessment approaches, such as [10–14]. As can be seen in the before-mentioned sources, the predominant modelling approach for MES is optimization followed by simulation. Since the methodology described in this paper can be categorized as an MES simulation model, the following section focuses on MES simulation models. Still, it will also display differences compared to optimization models.

Bottechia et al. [12] introduce the modular, multi-energy carrier, and multi-nodal multi-energy system simulator (MESS). The framework is designed for urban areas; however, wider spatial coverage is also possible. The authors compare and investigate the cause of differences of MESS results with Calliope (MES optimization framework), based on a simple MES. The outcome of both methods tends to be similar yet different due to the individual model's target function and mode of operation. One further advantage of MESS over Calliope is a much faster computation time. A main disadvantage of MESS might be that only one grid level can be depicted [12].

A combination of Pandapower [15] and Pandapipes [13] is proposed by Lohmeier et al. [13] to create a multi-energy grid simulation framework with a focus on detailed energy grid depiction. The so-called multi-energy controller, similar to the energy hub concept, allows for the implementation of sector coupling or energy storage options. The authors demonstrate the capabilities of the MES simulation framework based on two use cases. Since detailed energy grid calculations, as well as multi-energy controllers, are computation time intensive tasks, one disadvantage of the proposed model is extensive computation time, when simulating a full year in 15 min time steps [13].

Böckl et al. [16] introduces a previous version of the MES simulation framework HyFlow, which is utilized for various research questions, such as [17–19]. By applying HyFlow, potential fields of improvements became visible, since the previous HyFlow version is not capable of addressing issues, such as:

- More than two individual network levels;
- No energy transfers across various network levels, since only step-by-step energy transfer via each network level is possible (energy must always flow via each network level without skipping network levels);
- A lack of selectable control strategies for both sector coupling and energy storage options;
- Implementation of further components of an energy system is only possible with high programming effort.

In [20,21], an MES optimization framework is proposed, consisting of individual energy hubs, interconnected with individual energy grids. Both models provide a two-stage optimization for the energy hub and the whole system. In dependence of the target function, individual research questions are addressed.

1.1.2. MES Investigations on National Level

This section aims to provide an overview of existing research on national MESs to demonstrate current research approaches.

Sejkora et al. [22] display how Austria's future energy system could be composed if exergy efficiency is defined as optimization criteria in a fully decarbonized energy system, where RES are expanded according to #mission2030 targets [5,22]. The research shows that with restricted RES expansion in Austria, significant imports of sustainable methane and hydrogen will be necessary in the future. This research can provide guidelines for future technologies in MES, but cannot address any spatial problems [22]. In comparison, the project ONE100 from Austrian Gas Grid Management shows an energy system where RES are expanded until their maximum potential. In this case, import demand is significantly reduced to four per cent of total energy consumption. This research aims to achieve an economically optimized energy system. The model contains a rough spatial resolution of Austria, dividing Austria into 19 interconnected regions [23].

In [24], a comprehensive overview of research in the field of optimizing national energy system models is provided. However, the spatial resolution of each model shown is quite low. A lack of subnational data availability is seen as the main reason for the low spatial resolution [24].

1.1.3. Research Need

Flexible MES simulation frameworks to cover a wide range of individual problems are not available yet. Furthermore, to the best of our knowledge, it has never been attempted before to set up a national MES simulation model with detailed spatial resolution and detailed infrastructure depiction.

In this paper, we aim to close previously described scientific gaps by presenting a new version of our self-developed MES simulation framework HyFlow and demonstrate its capabilities based on Austria's energy system in 2030. The following research questions are to be investigated in this paper:

- Based on the scientific gap, how should an MES simulation framework be designed to cope with a national MES and various other research questions with a high degree of both spatial and temporal resolution?
- What steps have to be taken to model Austria's national energy system with detailed spatial and temporal resolution?
- What are the effects on power infrastructure based on #mission2030 renewable energy sources expansion, considering different modes of flexibility operation and power load flow optimization?

To answer the research questions this paper is structured as follows. The following subchapter describes considered challenges modelling the Austrian energy system. In Section 2, the methodology to set up an MES simulation framework and national MES model is described. Investigated scenarios and their corresponding results are disclosed in Sections 3 and 4, respectively. Simulation results are discussed in Section 5, followed by a conclusion and an outlook for potential further research in Section 6.

1.2. Problem Description

To address previously mentioned research questions, various obstacles need to be addressed beforehand. As demonstrated in the literature research, MES simulation frameworks that are currently available can be significantly improved to address a wide and flexible range of research questions, especially in the following fields:

- The flexible depiction of various network levels of all energy carriers (power, gas, heat), independent of spatial resolution. This should enable a large range of spatial resolution to be able to depict various areas, from single consumers up to the state level.
- The possibility to assign various flexibility options, such as sector coupling technologies, storage options, demand-side management (DSM), and operation-flexible power plants. This should include the possibility of flexibly adding any further components to expand the MES framework's functionality. Flexibilities may operate, e.g., as load following units or with various optimization-based operation strategies, such as maximizing profits or maximizing the degree of self-sufficiency in a specified area.
- State of art load flow consideration via adequate load flow calculation for all considered energy carriers. Depending on the user's selection, power flow simulation or optimal power flow load flow calculations should be selectable for the power grid.

The updated HyFlow MES simulation framework can address the previously mentioned points to develop a generic and flexible MES simulation framework.

As outlined in [24], a lack of subnational data is a major challenge when modelling a national MES. This challenge is addressed twice in this paper:

- Suitable approaches and data to be found to achieve a detailed spatial resolution. This includes consumption and generation data for all energy carriers. If data are not available in low spatial resolution, a suitable approach must be found to distribute general data towards smaller entities.
- Currently, no models of Austria's power, natural gas, and district heating energy infrastructure are openly available. To allow for the consideration of real grid properties, an energy grid model must be developed, based on available data to depict the Austrian energy infrastructure.

To demonstrate the capabilities of HyFlow, three scenarios of the Austrian energy system in the year 2030 are simulated to show the effects of RES expansion and various modes of operation of flexibilities, such as heat pumps, electric vehicles, power storage, and gas-fired power plants.

2. Methodology

This section is split into two main parts to provide a methodological overview of the HyFlow MES simulation framework and all the necessary steps to create an MES model of Austria, to be assessed with HyFlow.

2.1. *HyFlow*

To provide a comprehensive overview of the HyFlow MES simulation framework, the general modelling structure, the input data, the calculation procedure, and the implementation of flexibility options are discussed in the following sub-chapters.

2.1.1. General Modelling Structure

In HyFlow, the examined area can be divided into several cells. In this work, so-called substation districts are used (refer to Section 2.2.5. Spatial Data Distribution). This approach is called the cellular approach; further details can be found in [16]. All entities within one cell are aggregated into its corresponding cell. Therefore, a cell represents the smallest spatially resolved area, resulting in a node. In Figure 2, an example of a single node is displayed. To implement consumption and generation in one single term, the term "residual load" (RL) is used and defined as per Equation (1).

$$P_{RL}[t] = P_{demand}[t] - P_{Generation}[t] \tag{1}$$

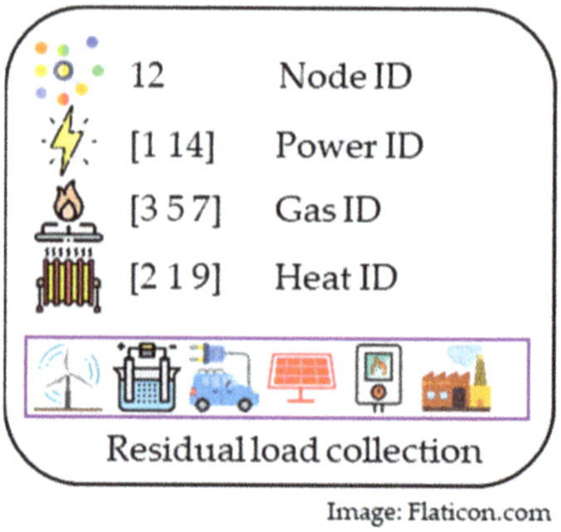

Figure 2. Example of a single node.

In Table 1, an overview of the node parameters is provided. Further parameters such as maximum and minimum voltage, pressure, and temperature can also be defined.

Table 1. Overview node class parameters.

Parameter	Type	Description
Node ID	scalar	One unique number is assigned to each node.
Power ID	1 by 2 vector	Vector at [1 1] is currently a spare parameter. Vector at [1 2] indicates the node's position in the power grid.
Gas ID	1 by 3 vector	Vector at [1 1] indicates the node's pressure level (e.g., 2 = 70 bar). Vector at [1 2] indicates the node's subgroup. Vector at [1 3] indicates the node's number within the subgroup.
Heat ID	1 by 3 vector	Please refer to Gas ID.
RL collection	array	This array contains all objects, and their behavior can be expressed in active and reactive power, gas, or heat RL. Power RL is defined as per Equation (1), valid for all other energy carriers too. The RL collection in Figure 2 includes wind energy, electrolysis, electric car, photovoltaics, gas to heat (GtH), and an industrial consumer.

Nodes can be interconnected with other nodes. Depending on the availability of energy grids, a node-edge depiction is established. An example of several nodes with various connections of energy carriers (edges) is shown in Figure 3. It can be seen that all nodes are connected to the power grid. Nodes 12, 13, 14, and 327 represent one gas sub-grid. Node number 14 supplies gas to a lower pressure network (since the first vector position of gas ID is higher), consisting of nodes 26 and 27. As an example of RL collection objects, further implementable objects that can be added to a node's RL collection, such as consumer, producer, sector coupling technology, storage options, and electric vehicles, are shown adjacent to their corresponding node.

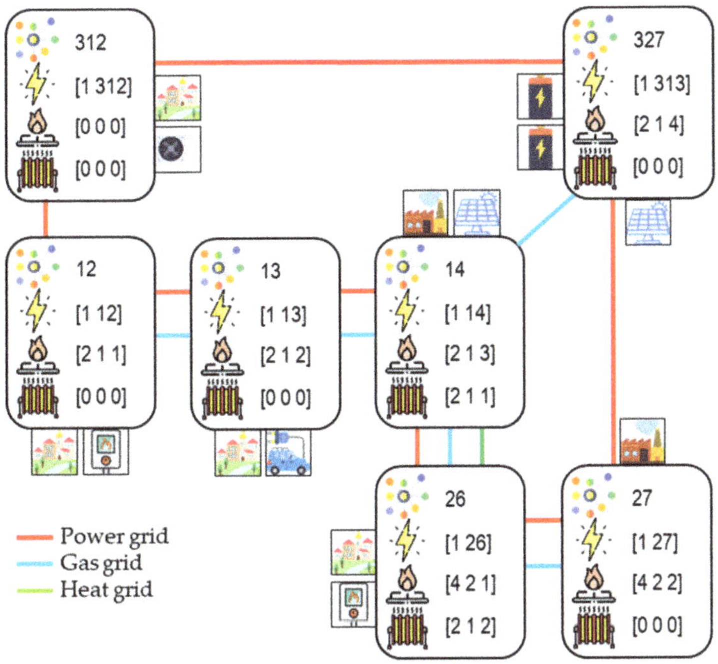

Cell connected to

cell	power	gas	heat
312	☑	☒	☒
327	☑	☑	☒
12	☑	☑	☒
13	☑	☑	☒
14	☑	☑	☑
26	☑	☑	☑
27	☑	☑	☒

Image: Flaticon.com

Figure 3. Energy infrastructure depiction.

To ensure that only the objects that are represented by an RL (refer to all classes hierarchically below "RL" in Figure 4) are addable to a RL collection, the programming principle of inheritance is used. A basic class "RL" is defined with simple properties (refer to Table 2 and functions. Based on the "RL" class, any derivative object can be developed and implemented by the user, with additional parameters to accommodate each object's individual need. Figure 4 displays available derivatives of the "RL" class. In each class, individual operating strategies can be implemented, depending on the user's need and addressed research question.

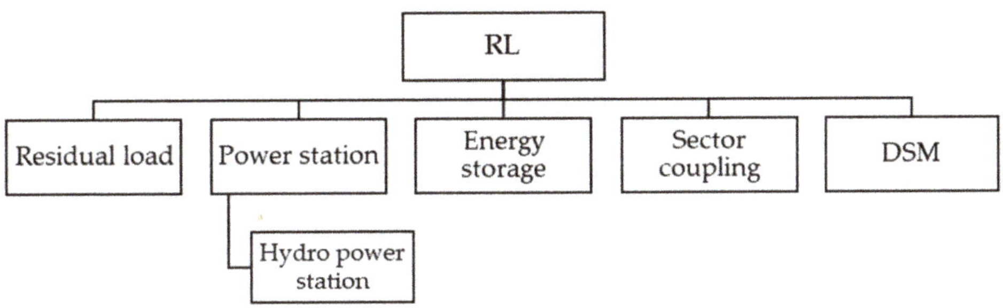

Figure 4. RL class and inherited derivatives.

Table 2. Overview of RL class properties.

Parameter	Description
type	Defines the type of underlying object. **Residual load or Power station** 1: Pre-defined residual load profile or power station with predefined temporally resolved profiles (e.g., residual load, water flowrate). **Sector coupling** 2: Power to gas and heat (PtGH). 3: Power to heat (PtH). 4: Heat to power (HtP). 5: Gas to heat (GtH). 6: Gas to power and heat (GtPH). **Energy storage** 7: Power storage. 8: Gas storage. 9: Heat storage. **DSM** 10: Electric vehicle. 11: Demand-side management.
RLgas	This vector contains the object's pre-set or calculated gas RL. The calculated gas RL depends on the object's operating strategy.
RLheat	Refer to RLgas.
RLpower	Like RLgas, except that active and reactive power RLs are considered.
RLgasFlex	These parameters contain the object's RL flexibility. The usage of these parameters depends on the object's operating strategy. The implementation of flexibility is explained in Section 2.1.4.
RLheatFlex	
RLpowerFlex	

2.1.2. Input Data

Before a simulation can be carried out in HyFlow, various input data need to be defined and read in for further processing. The input data are stored in individual objects, according to Table 2 and Figure 4. Node data must be defined, including parameters described in Table 1. Temporally resolved gas, heat, active, and reactive power RL data, including their associated node, can be defined. Properties for sector coupling options, storage, electric vehicle/DSM, and power stations (including their corresponding operating strategy) must be defined. Additionally, temporally resolved data for storage (e.g., water inflow in (pumped)-storage hydropower plant) are necessary. Properties include rated power, conversion or in-/output efficiencies, storage capacities, operating strategy, and further technology-specific properties.

Since the open-source power flow framework MATPOWER is used for power flow (PF) or optimal power flow (OPF) calculations, input data must reflect MATPOWER framework requirements. Therefore, tables for branch (=edge), bus (=node), generator, and generator cost data must be defined. The structure can be found in the MATPOWER documentation [25,26].

Gas and heat network properties follow the same principle scheme. For gas and heat, two tables need to be created. In the first sheet, connections between nodes at the same pressure level can be defined (e.g., between nodes 13 and 14 in Figure 3). The sheet number two contains connections between nodes at different pressure levels (e.g., between nodes 14 and 26 in Figure 3). Parameters are the gas or heat IDs of the connected nodes, length, diameter, roughness, and—in the case of heat—thermal conductivity of grid sections (edges).

2.1.3. Calculation Procedure and Grid Simulation

The calculation process of HyFlow is shown in Figure 5. The process of each dashed box will be explained further.

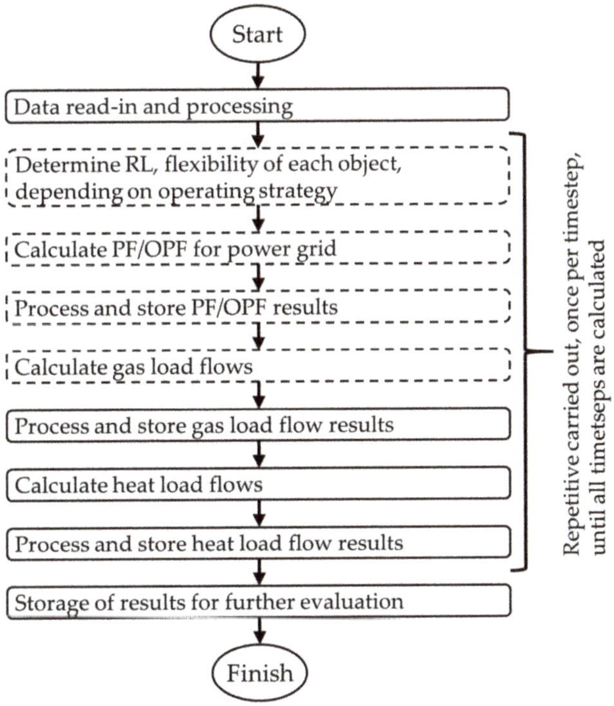

Figure 5. Overview of the calculation process.

Determination of RL

All objects in the RL collection of each specific node are assessed. Each object in the RL collection must provide an RL or flexibility based on the operating strategy of the object. The calculated summarized RLs and flexibilities of each node are transferred to the subsequent load flow calculations. The implementation and usage of flexibility are further explained in Section 2.1.4.

Power Grid PF or OPF

Depending on the users' selection MATPOWER PF or OPF, calculations can be performed for the power grid. Depending on the usage of PF or OPF, input data must be determined differently. In the case of PF simulation, all node residual loads and generator in-/outputs must be determined before a PF simulation can be performed. The PF simulation determines power load flows "as they physically are", without considering line restrictions or generation costs, for example. OPF mathematically optimizes load flows and generator dispatch, considering generation costs, line restrictions, and maximum/minimum generator power, based on the target function of minimum generation costs in the total power system [26]. Optimization restrictions, such as transmission line capacities or insufficient generation capacities, might cause an OPF to be incapable to converge. The advanced capabilities of OPF, compared to PF, come at the cost of higher complexity and increased likelihood of calculation failures. Further details regarding MATPOWER are provided in [26].

Before performing a PF or OPF calculation, bus (active and reactive power RL) and generator data (generation) must be updated in the MATPOWER data structure, based on the previous step's results (Determination of RL).

Gas and Heat Load Flow Calculation

Rüdiger adopts the node potential analysis for power grids in combination with Darcy's equation (refer to Equation (2)) to determine gas load flows [27].

$$\Delta p = \lambda \cdot \frac{8 \cdot \rho \cdot l \cdot \dot{V}^2}{d^5 \cdot \pi^2} \tag{2}$$

For heat load flows, Rüdiger's approach is extended by a second iteration loop to determine node temperatures and heat losses (refer to Equation (3) [28]) in both forward and return flow recursively.

$$T_{endnode} = (T_{startnode} - T_{ambient}) \cdot e^{\frac{-2 \cdot \pi \cdot k \cdot l}{c_p \cdot \rho \cdot \dot{V}}} + T_{ambient} \tag{3}$$

Both gas and heat load flow calculations can be characterized as steady-state load flow calculation approaches.

Process and Storage of Results

In the case of power OPF, MATPOWER determines each generator's generation based on minimum system generation costs. Therefore, the determined generation must be transferred to the corresponding object in the RL collection. The same procedure is necessary in case flexibilities are used. Depending on the energy carrier, further load flow calculation results such as load flows, voltage, angle, pressure, and temperature levels are stored.

2.1.4. Implementation and Usage of Flexibility Options

Power flexibility might generally accrue from sector coupling options, energy storage, or DSM, displayed by the yellow circles in Figure 6. Figure 6 also shows the general representation of each MES node, which is automatically adapted, depending on the properties of the node. A node optimization, similar to Chen et al. [29], is used to determine the node's flexibility band (green box in Figure 7) and usage of objects to provide flexibility (yellow box in Figure 7). Generally, node optimization aims to increase the profit of an MES (equal to minimizing costs), considering RL coverage, energy prices, technical properties of SC, and storage options [29]. This approach and its optimization target is adapted to determine the maximum/minimum power flexibility and the usage of objects providing flexibility, as explained further shortly.

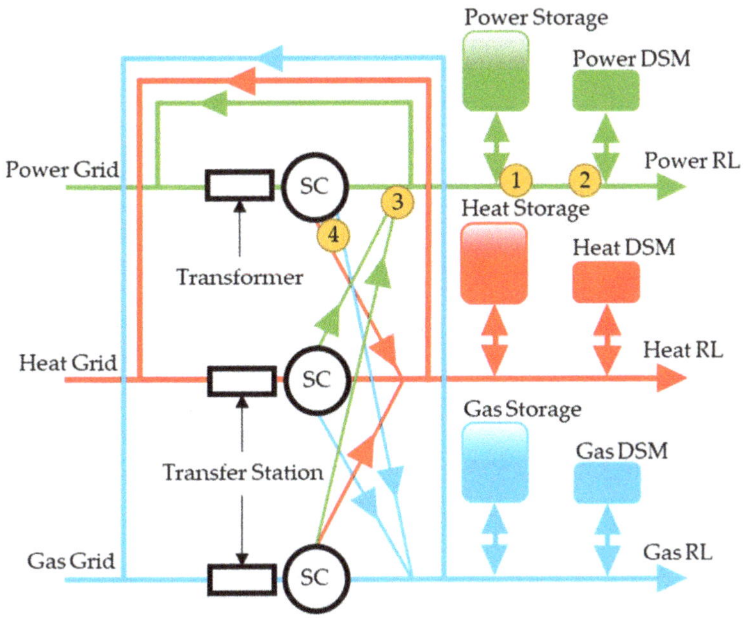

Figure 6. Generic node optimization representation.

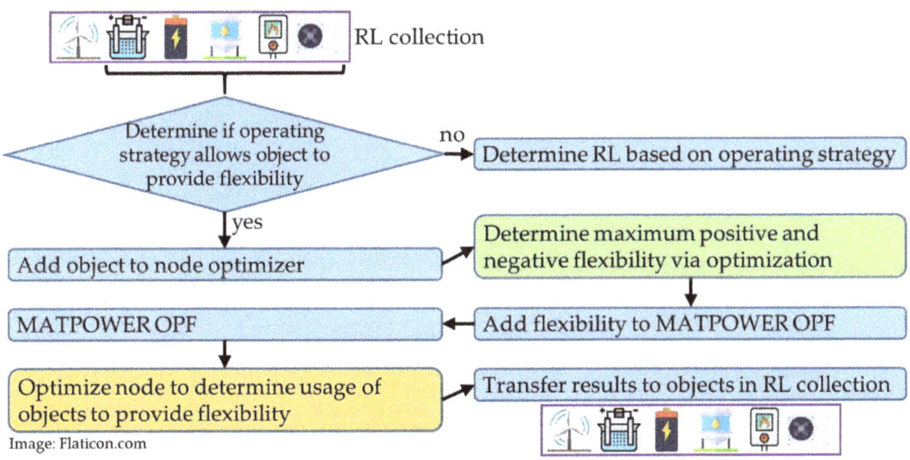

Figure 7. Process for optimized operating strategy, providing flexibility.

To consider flexibility in the HyFlow calculation process, the calculation process displayed in Figure 5 must be adapted. This adaption affects the upper three dashed boxes in Figure 5. The additional calculations to be performed are depicted in Figure 7.

If an object provides flexibility according to its operating strategy, it is considered for the following calculation procedure. If no flexibility is provided by the object, the RL is determined based on the objects operating strategy. To determine the total available flexibility per node, two optimizations are performed, aiming to determine the maximum possible positive and negative power RL (green box in Figure 7, target functions in Equations (A4) and (A5) in Appendix A), resulting in a flexibility band between both maximum and minimum power. For these optimizations, no energy prices are considered. The previously described calculation is carried out in the "Determine RL" section outlined in Figure 5.

Once a node's maximum positive and negative power flexibility is determined, it can be implemented in the MATPOWER framework as a generator at the corresponding node. The OPF, considering the whole depicted power system, determines the actual need for flexibility, ranging between the minimum and maximum possible flexibility of each node providing flexibility. So far, the need for flexibility at each node is determined; however, it is yet unknown which objects are used to what extent to provide the determined flexibility. To address this question, another node optimization is carried out (yellow box in Figure 7) to determine the actual usage of each object providing flexibility (target function in Equation (A1) in Appendix B). This optimization is carried out considering energy prices; therefore, the usage of the object is optimized economically. The determined optimal usage of each object providing flexibility is transferred to each corresponding object.

Yalmip [30] and Gurobi [31] are used to solve the optimization problems. Refer to "Appendix A. Node Optimization" for further mathematical details.

2.2. Austrian MES Modelling

The following modelling approaches are applied to develop an Austrian MES model. This includes an infrastructural depiction of power, natural gas, and heat grids, as well as timely and spatially resolved consumption and generation profiles.

2.2.1. Power Grid

The basis for the power grid model is a transmission grid plan. It shows the name of a substation's location and the transmission capacity of each line between substations for 110, 220, and 380 kV. However, the transmission grid plan only shows a past grid status. To determine a potential future power grid in 2030, potential grid expansion projects have to be included. The 220 and 380 kV transmission grid is operated by the Austrian Power Grid (APG), providing a grid development plan annually [32–37]. The 110 kV distribution grid is mainly owned and operated by nine local utilities in each federal state. Particularly for Upper Austria and Carinthia, detailed 110 kV grid expansion information is available [38,39]. Since the location of current and future substations and lines is roughly known, the geographic information system software QGIS [40], satellite images [41], and Open Street Map [42] are used to determine the exact location of substations and the course of power lines. MATPOWER requires line resistance, reactance, total line charging susceptance, and the maximum allowed apparent power flow [26]. APG provides detailed technical data for the transmission grid, which are used to parametrize the 220 and 380 kV grid [43]. The 110 kV grid is parametrized with literature values for resistance, reactance, and total line charging susceptance, as well as other already published projects [17,18,44,45], based on the maximum transmission current in the transmission grid plan.

2.2.2. Natural Gas Grid

To spatially depict Austria's natural gas infrastructure, we apply a similar approach compared to the power grid. Length and diameter for transnational pipelines and primary distribution system pipelines are available at Austrian Gas Grid Management (AGGM) [46]. The pipeline routing and length of national network level one (national transmission grid) and two (national distribution grid) can be derived from [47,48]. The diameter and pressure level are determined using statistical data [49], as well as information from utilities provided by request and previous projects [17]. Wall roughness is parameterized with the wall roughness of welded and seamless steel tubes [50].

2.2.3. Heat Grid

Currently, heat grids in Austria cover regional heat demand. Since the spatial resolution for an Austrian MES model is inadequate to depict regional heat grids, technical properties are assumed for interconnected regions, especially in urban areas. As a guideline, results from [51,52] are used to determine which areas of Austria are supplied with district heat.

2.2.4. Model of Austrian Natural Gas and Power Grid

Figure 8 displays the created model of Austria's power and natural gas infrastructure. It can be seen that the availability of Austria's energy infrastructure is much denser in urban and suburban areas in comparison to rural areas. Based on the infrastructure depiction in Figure 8 and Voronoi diagram methodology (refer to Section 2.2.5. Spatial Data Distribution), a corresponding node-edge model can be derived.

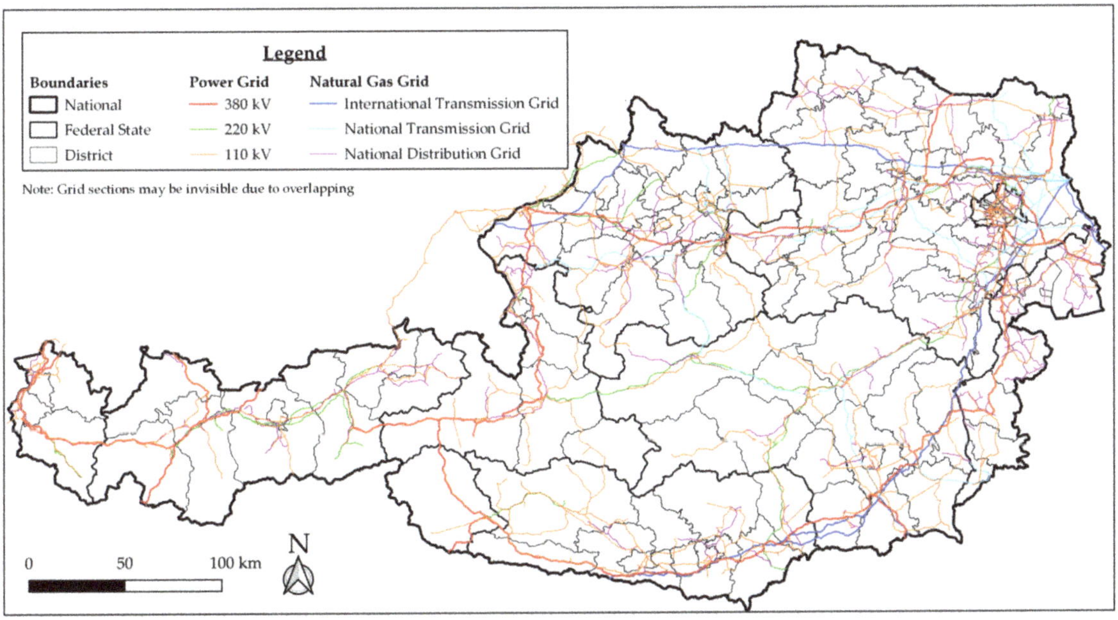

Figure 8. Model of Austria's power and natural gas infrastructure [9].

2.2.5. Spatial Data Distribution

To achieve a spatial resolution of Austria, a suitable methodology has to be selected to determine a spatial division of Austria. This is necessary to aggregate all objects (e.g., storage, power stations, RLs) within a spatial division into one node. A Voronoi diagram creates polygons starting from central points, dividing a layer into areas of equal nearest neighbors [53]. This approach is used with substations of the power grid as central points to determine single areas. The area covered by one substation is further referred to as a substation district (SSD). An example of the created SSDs within a selected area of Austria can be seen in Figure 13. In [8], the RES potentials of each Austrian community are determined. The RES potential data of each community are summed up to determine the SSD RES potential if a community is located within the boundaries of an SSD. Furthermore, power and natural gas (for both process and heating use) final energy consumption data from industrial, private, agricultural, and public and private service sectors are provided at a district level in [8]. To distribute final energy consumption data at the district level to a single community, the share of employees or households per community from *Statistik Austria*, in comparison to the district, is used [54]. Heat demand is modelled using the Austrian Heat Map [52]. Heat demand data are from 2012 but are quite stable till now [55]. Since heat demand data are available on a district level, and the same approach as for power and gas is used to distribute district demand to municipalities and then to SSDs. The useful energy analysis from *Statistik Austria* also provides information at the federal state level regarding the energy carrier used to provide heat [55].

2.2.6. Temporally Resolved Consumption Data

Annual energy consumption data of each SSD must be combined with temporally resolved load profiles to determine a temporally resolved RL profile for each SSD. For industrial power and gas demand, subsector specific load profiles are derived from [56]. For household, agriculture, and public and private service power consumption standardized load profiles (SLP) H0, L0, and G0 are used [57]. The reactive power is considered using literature and empiric values [58–60]. A cos(φ) of 0,98 is used. The temporal resolution of natural gas for non-heating purposes for households, agriculture, and service is determined based on the relatively steady cooking gas SigLinDe function [61]. The annual heating RL of each SSD is determined using the sectors corresponding SigLinDe function [61]. The temperature used as input data for the SigLinDe function is obtained for each SSDs substation from [62,63].

2.2.7. Temporally Resolved Generation Data

Oesterreichs Energie published a map of all power generation facilities in Austria with a power generation capacity greater than 10 MW [64,65]. In the following, we explain how each category of power station has been implemented into Austria's MES model.

Hydro Run-Off and Storage Power Station

The basic model of hydro run-off and storage power stations can be seen in Figure 9. Power generation is calculated using Equation (4) [66].

$$P = \eta \cdot \rho \cdot Q_{Turbine} \cdot g \cdot \Delta h \quad (4)$$

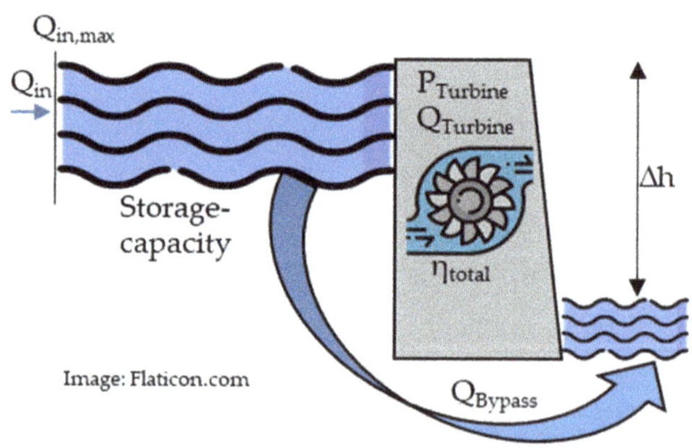

Figure 9. Hydro run-off and storage power plant model.

Each run-off and storage power station with a generation capacity greater than 10 MW is implemented into Austria's MES model in its corresponding SSD. Power station data are sourced from [67–79]. The temporally resolved generation is determined using run-off water measurements [80]. If the measurement point is different to the power station's location, interpolation is conducted between two measurement points. For hydropower stations with less than 10 MW, a different approach had to be used. *Kleinwasserkraft Österreich* [81] provides power and annual generation data for small-scale hydropower stations. These data are used together with hydropower potentials from [8] to determine small-scale hydropower in each SSD. Since no sufficient run-off measurements are available for small rivers, a standardized load profile based on measured data from small rivers (refer to Appendix B for measurement points) [80] is created, presented in Figure 10. A polynomial trend curve is used to smoothen the curve.

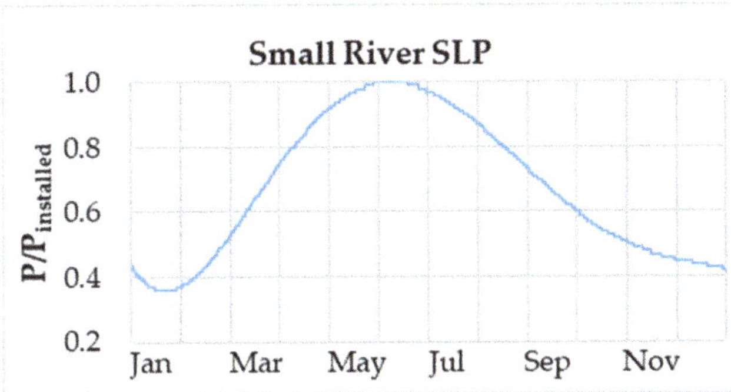

Figure 10. SLP small river, for small-scale hydro run-off power plants.

To cope with an additional 5 TWh of power, based on the government's RES expansion target [4], an increase in generation is carried out according to hydropower potentials from [82]. The small river SLP is used as a temporally resolved generation profile for these power stations.

(Pumped)-Storage Hydropower Plant

(Pumped)-storage hydropower plants are modelled as a simplified, flexible cascade of reservoirs, interconnected with pumps and turbines (refer to Figure 11).

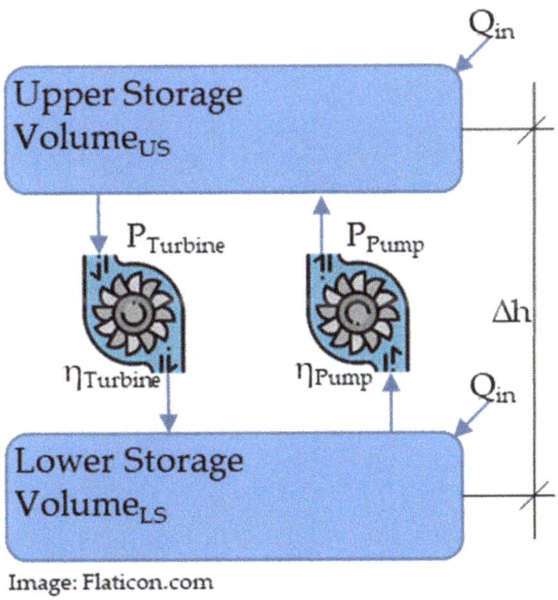

Figure 11. (Pumped)-storage hydropower plant model.

Technical properties such as storage capacity (Volume), annual inflow from natural sources (Q_{in}), and pump and turbine power ($P_{Turbine}$, P_{Pump}) are sourced from [68–70,73–76,79,82–84]. Furthermore, future projects such as [85,86] are considered as well. The pump (η_{Pump}) and turbine ($\eta_{Turbine}$) efficiency is set to 0.88 [87]. Reservoirs are naturally fed by water from glacier or snow melt. To determine a temporally resolved water inflow, suitable measurement data from [80] are used to derive an annual water inflow characteristic, displayed in Figure 12 (refer to Appendix C for measurement points). It can be seen that the majority of natural inflow occurs during the summer months; in contrast, hardly any inflow can be expected in the winter months.

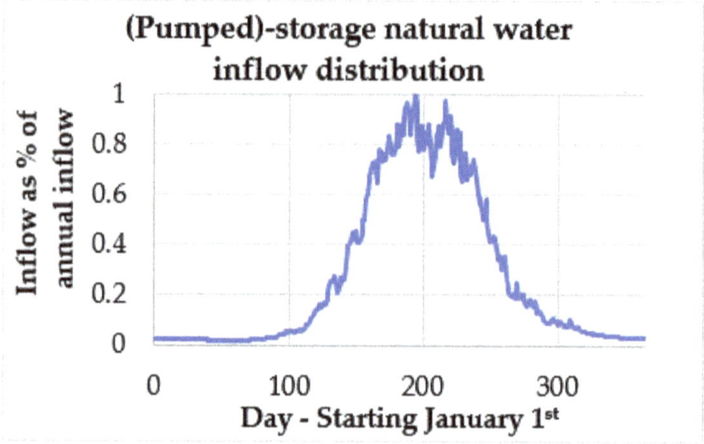

Figure 12. Annual (pumped)-storage hydropower plant natural inflow distribution.

The scenario-dependent generation and consumption profile for (pumped)-storage hydropower plants is described in the following chapter.

Biomass Combined Heat and Power (CHP) and Biogas Power Plants

Biomass CHP and biogas power plants are sourced from [88,89] and temporally resolved via SLP (E0) [57].

Photovoltaics

The installed photovoltaic power for each federal state is sourced from [90] and evenly distributed to each SSD via PV potentials from [8]. To reach the national goal of 11 TWh photovoltaics [4], photovoltaics are expanded, according to potentials in [8], by applying a split between rooftop and open area potentials of nine to one. Temporally resolved generation profiles are considered for each SSD, sourced from [62,63,91].

Wind

The locations of each wind park and their corresponding power levels are sourced from [92] and aggregated to the installed wind power of each SSD. To reach the national goal of 10 TWh wind power addition [4], the power at each SSD is evenly expanded according to potentials in [8]. Temporally resolved generation profiles are considered for each SSD, sourced from [62,63,93].

Thermal Generation

Technical data, such as power and efficiencies, of Austria's (combined cycle) gas turbine and large-scale CHP power plants are based on operators' publications [67,72,94–97]. If no efficiency data are available, an estimation based on comparable power plants is applied. The scenario-depended generation profile is described in the following chapter.

2.2.8. Power Exchange with Neighboring Countries

Power exchange with neighboring countries of Austria is considered based on data from *ENTSO-E*'s transparency platform for the year 2019 [98].

2.2.9. Example of Energy Infrastructure Depiction

An example of Austria's energy infrastructure and power plants can be seen in Figure 13. Exemplarily, an SSD is highlighted in yellow. The highlighted SSD contains several biogas plants and is connected to the power and natural gas grid. It can be seen that substations are concentrated in urban areas, whereas the substation density is lower in rural areas. More substations, compared to assignable municipalities, may especially occur in urban areas. In this case, suitable substations are manually selected and merged to create the Voronoi diagram. Hydropower plants are concentrated along large rivers, whereas wind, biomass CHP, and biogas are spread all over the area.

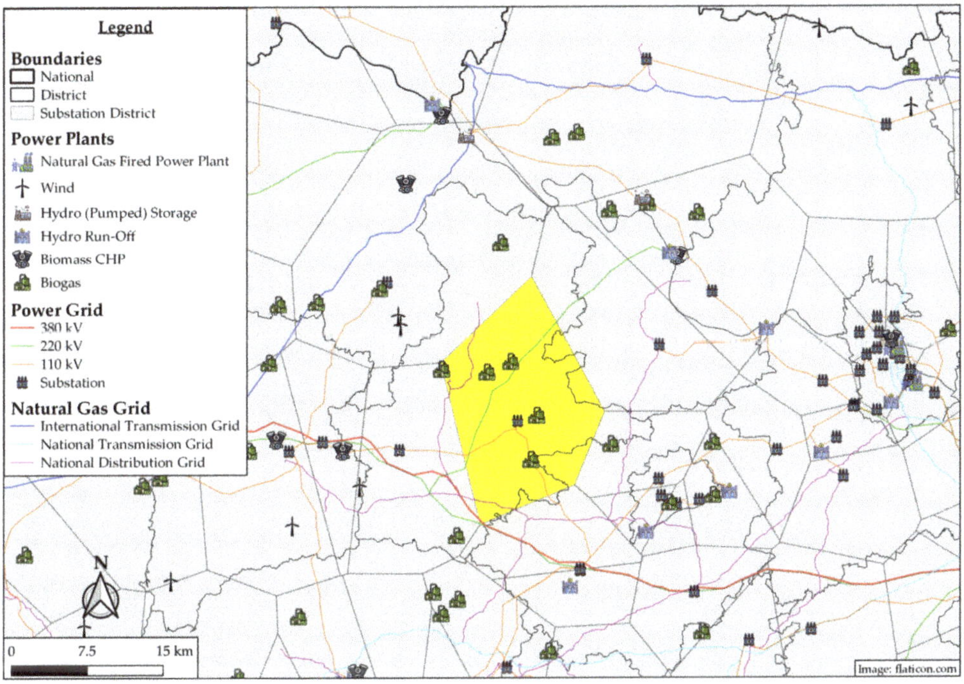

Figure 13. Example of Austria's energy infrastructure and power plants [9].

3. Scenarios

In this chapter, potential developments of the Austrian energy system are investigated based on three scenarios for the year 2030. Therefore, we apply the methodologies shown in Section 2. The following points are considered for each scenario:

- Since power, gas, and heat consumption are based on past data, sufficient studies need to be found to estimate the energy consumption in 2030. Austria's *Umweltbundesamt* (UBA) [99] estimates energy consumption in the years 2020, 2030, and 2050, based on the year 2015. Power demand is expected to remain stable between 2015 and 2020 and then increase between seven and thirty per cent, depending on the scenario. Electric vehicles, heat pumps, and electrolysis are seen as major drivers of power consumption growth. Since these consumers are additionally considered in each following scenario, we assume that the power demand will stay constant without. Depending on the scenario, a slight increase or decrease in natural gas consumption is assumed by *UBA*; therefore, we assume constant consumption [99]. Based on #mission2030 targets, thermal renovation of existing buildings should be doubled to two per cent per year, from current levels of around one per cent [5]. If a 50% heat demand reduction through a thermal renovation is assumed, heat demand might drop by thirteen per cent, until 2030. The assumption of 50% heat demand reduction seems reasonable since subsidiaries are granted if more than 40% heat demand reduction is achieved [100]. The 13% heat reduction is between both *UBA* scenarios (WAM, WEM) for the final energy consumption of buildings [99].

- The number of electric vehicles in each SSD is determined based on vehicle registration data and trends in each federal state [101]. The share of electric vehicles in 2030 is expected to be 20%, based on scenarios in [102]. In Austria, a car is used for an average of 13,700 km per year. Based on the average consumption of 20 kWh/100 km, an annual electric energy demand per car of about 2750 kWh can be expected [103,104]. The temporal charging characteristic is scenario-dependent and can be explained in the following subchapters. All electric vehicles account for approximately 3 TWh of additional power consumption.
- The share of heat pumps for each SSD is determined based on the share of ambient heat usage for heating purpose divided by an assumed coefficient of performance of three [55]. Heat pump usage might increase by six-fold until 2030 based on [105]. The mode of operation depends on the scenario. All heat pumps account for approximately 1.9 TWh of additional power consumption.
- For every fourth household, battery energy storage with a storage capacity of 8 kWh, charging–discharging power of 2 kW, and an input–output efficiency of 90% [106] is implemented with a scenario-dependent mode of operation.
- Renewable energy sources are expanded according to plans of the federal government, shown in Table 3.

Table 3. Expansion of RES until 2030 [4,107].

RES	Generation 2018	Expansion Until 2030
Hydro	37.6 TWh	+5 TWh
Wind	6.0 TWh	+10 TWh
Photovoltaics	1.5 TWh	+11 TWh
Biomass	4.9 TWh	+1 TWh
Total	50 TWh	+27 TWh

Table 4 below provides an overview of the differences between each scenario. The modes of operation are explained in each scenario description.

Table 4. Scenario parameters.

	Scenario 1	Scenario 2	Scenario 3
Thermal generation and (pumped)-storage hydro	ENTSO-E	ENTSO-E	Flexibility
Electric vehicle	SLP	Optimized	Optimized
Battery storage	Greedy	Optimized	Optimized
Heat pump	Load following	Optimized with storage	Optimized with storage
Power grid calculation	PF	PF	OPF

3.1. Scenario 1—BAU

In this scenario, certain elements of the energy system are operated in the business-as-usual (BAU) mode. Electric vehicles are charged according to the SLP, derived from [108], with 3.7 kW charging power. Since the number of electric vehicles is above 1000 for the vast majority of substation districts, a low coincidence factor can be applied [108,109]. Heat pumps are operated as heat demand occurs, without a storage option. Temporal battery storage behavior is determined as follows. The energy demand of an average household per SSD is coupled with an SLP (H0 [57]) and a 5 kW photovoltaic generation capacity, considering the SSDs' individual PV generation profile. One-quarter of households are equipped with battery storage. The battery storages operate according to the greedy algorithm to minimize the household energy demand from the power grid. Examples of the application of the greedy algorithm can be found in [110,111]. (Pumped)-storage hydropower and thermal power plants are operated according to *ENTSO-E* generation data from 2019 [98]. The resulting power RLs are added to each SSDs RL.

3.2. Scenario 2—Demand Optimization

Demand optimization is applied in this scenario to operate certain elements economically. An optimization concept similar to energy hubs is used to determine an economically optimized mode of operation of energy storage, heat pumps, and electric vehicles [29]. To enable a flexible operation of heat pumps, each heat pump is equipped with a thermal storage capacity of 50 kWh and a charging–discharging capacity of 10 kW. Electric vehicles charge their average daily consumption of about 7.5 kWh with 3.7 kW charging power. Peak demand times (6:00–9:00 and 17:00–20:00) are excluded from charging. The optimization is carried out, using power price data from 2019 [112] and each node's RL. As a result, price-optimal RLs of heat pumps, power storage, and electric vehicles are determined and added to each node's RLs.

3.3. Scenario 3—Demand Optimization and Flexibility

Heat pumps, electric vehicles, and battery power storage are operated, like in Scenario 2. Thermal power plants and (pumped)-storage hydropower plants are operated as additional flexibility (refer to Section 2.1.4). Since generation costs are considered in OPF for generator dispatch, the generation costs of each power source are set as follows:

- Subsidized forms of power generation, such as biogas, biomass CHP, wind, photovoltaics, and small-scale hydropower with 10 EUR/MWh;
- Large-scale hydropower 50 EUR/MWh;
- Flexibilities (gas turbine and CHP and (pumped)-storage hydropower) and import/export with 100 EUR/MWh.

4. Results

In all three discussed MES scenarios, natural gas and district heat grids show no critical overloads. Therefore, power grid results are discussed in detail. In Table 5, a comparison of overloaded distribution power grid (DG) and transmission power grid (TG) lines are displayed. Scenario 2 shows that the number of time steps, as well as affected power grid lines, increases compared to Scenario 1. This can be explained by the price-optimized mode of operation, since demand increases disproportionately in time steps with cheaper power, leading to RL peaks.

Table 5. Comparison of scenario results.

	Scenario 1	Scenario 2	Scenario 3
Overload time DG	182,205 time steps	206,592 time steps	81,991 time steps
Average DG line overload	38.8%	42.5%	35.5%
Count of overloaded DG lines	39/480	57/480	40/480
Overload time TG	3904 time steps	8131 time steps	144 time steps
Average TG line overload	16.0%	16.3	30.1%
Count of overloaded TG lines	5/104	7/104	6/104

To evaluate the degree of power line overloads, two different overloads are evaluated. As displayed in Figure 14, the average line overload and the top five per cent (based on the number of overloaded time steps) of line overloads are determined for each power grid line.

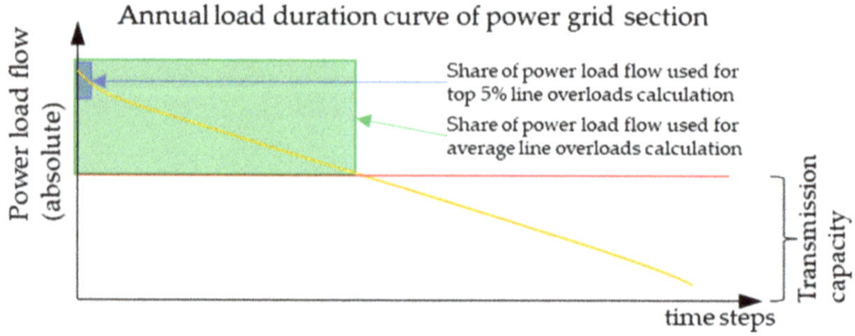

Figure 14. General determination of line overloads (an average and the top five per cent).

Subsequently, the worst (Scenario 2) and best (Scenario 3) case scenarios in terms of line overloads are displayed. In Figure 15, the average line loadings of Austria's power grid are displayed for Scenario 2. The thickness of each line qualitatively indicates the maximum transmission capacity of each line. Green lines indicate grid sections that are not affected by overloads. Orange to red lines indicate the average degree of overloads of the affected line section. Exemplarily, some grid sections are marked with a purple circle or ellipse-shaped indicators, allowing for the differentiation of the following overloads types:

- Continuous purple line—sections with low transmission capacity, e.g., single three-wire system;
- Dotted purple line—overloaded lines in urban areas;
- Dashed purple line—branch line with low transmission capacity in combination with either high potential of RES generation or demand;
- Dashed-dotted purple line—high potential of RES expansion.

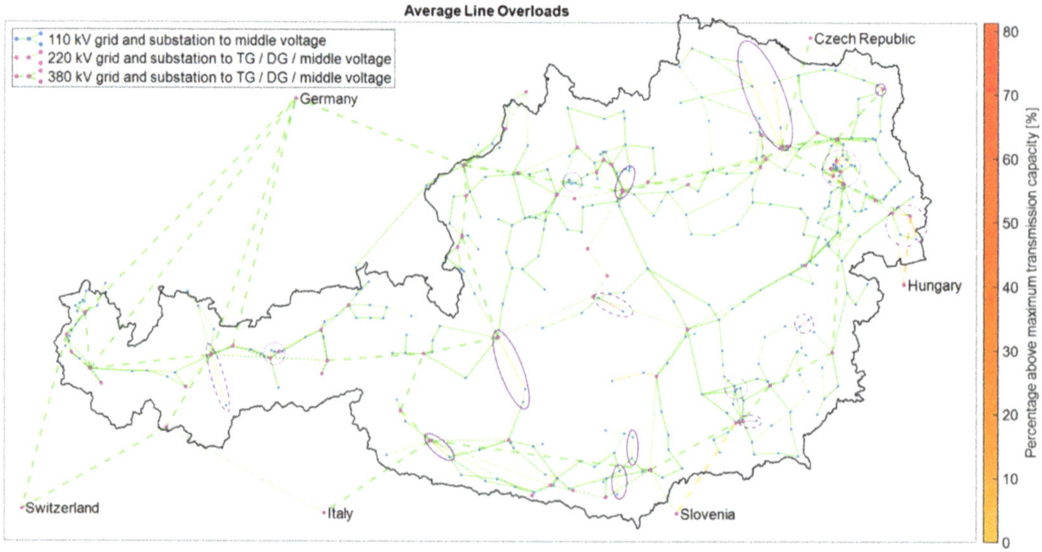

Figure 15. Average line overloads of Austria's power grid—Scenario 2.

In Figure 16, the top five per cent overloads of Austria's electricity grid are displayed for Scenario 2. It can be seen that most overloaded transmission and distribution grid lines are overloaded by a rather small degree.

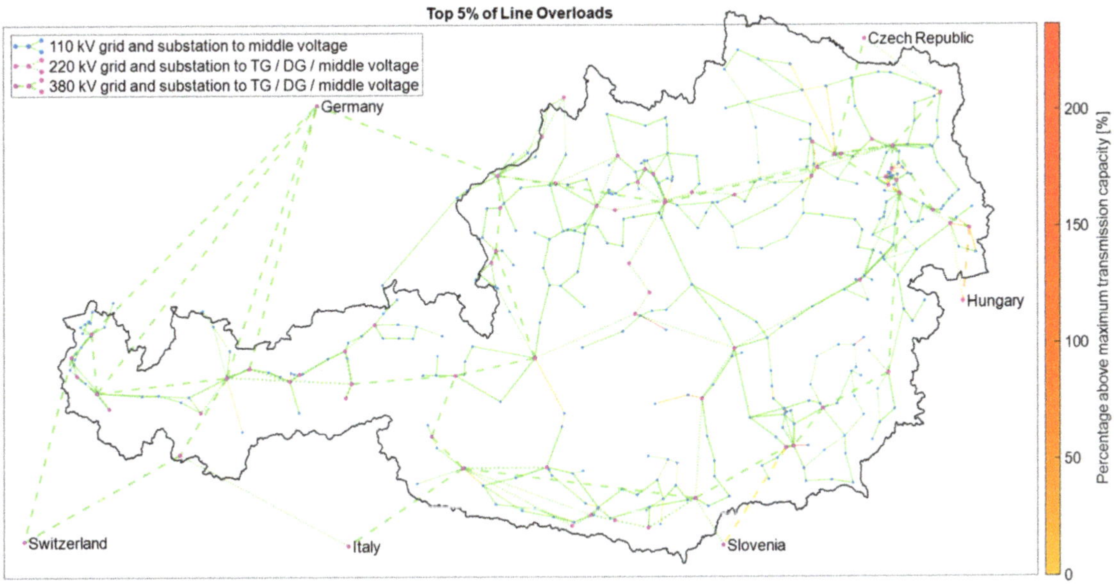

Figure 16. The top 5% line overloads of Austria's power grid—Scenario 2.

In following Figure 17 the annual load curve for a highly overloaded power branch line is displayed. It can be seen that the maximum transmission capacity is exceeded in both positive as well as negative direction. A positive and negative sign is related to the direction of power flow. This means that branch line overloads are caused by both consumptions at the end of the branch line and excess generation flowing from the end of the branch line towards the distribution grid.

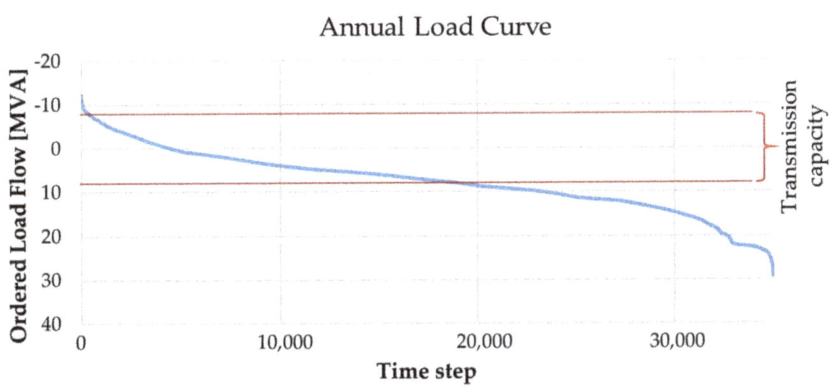

Figure 17. Annual load curve of a power branch line (ranked from min to max).

In Figure 18, the average power grid line overloads are displayed for Scenario 3. In comparison to Figure 15, the magnitude of the average overloads is significantly lower (refer to the scale magnitude). This observation is supported by line overload data displayed in Table 5, where line overloads in Scenario 3 are approximately halved in terms of count, compared to Scenarios 1 and 2. A similar context can be observed by comparing Figures 16 and 19 where the magnitude of the top five per cent of overloads are significantly lower in Scenario 3, compared to Scenario 2.

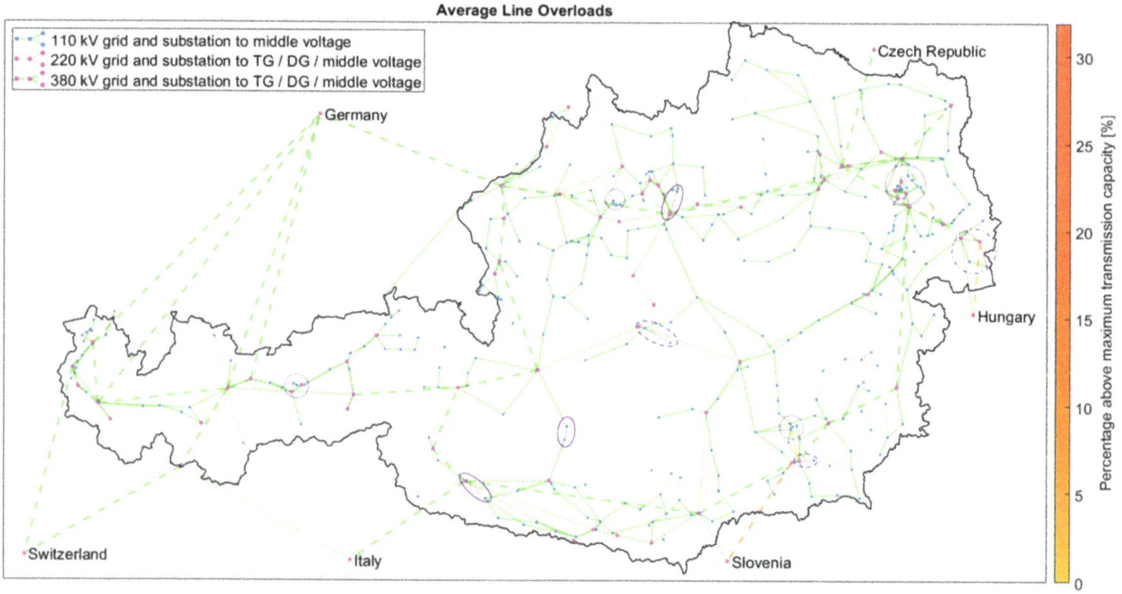

Figure 18. The average line overloads of Austria's power grid—Scenario 3.

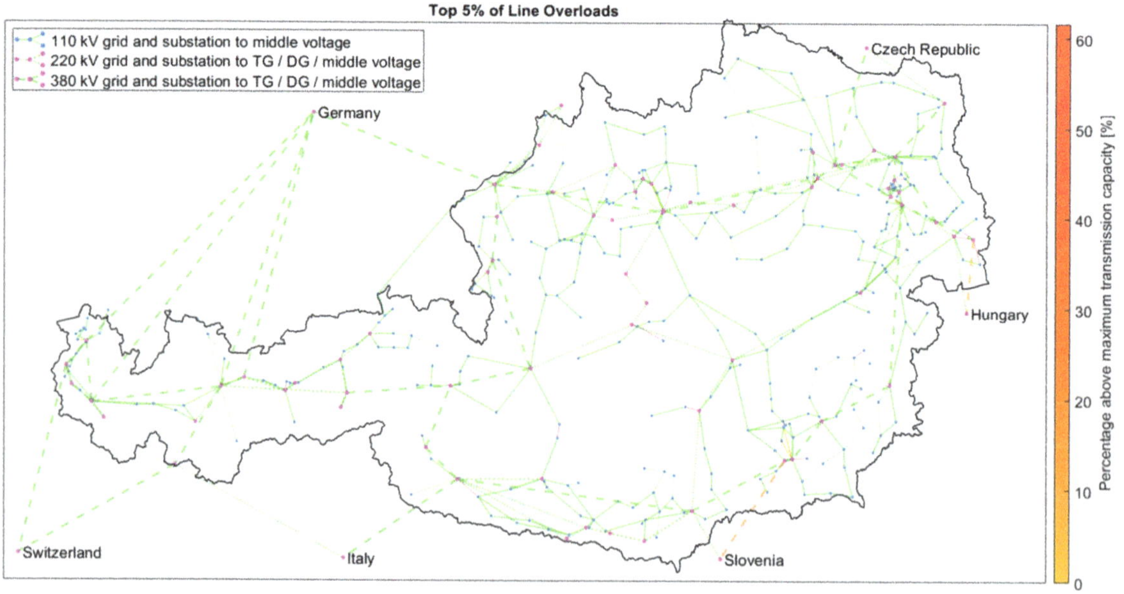

Figure 19. The top 5% line overloads of Austria's power grid—Scenario 3.

In Table 6, the power generation from each source is displayed for each scenario. It can be seen that the results for Scenarios 1 and 2 are identical, except for a small difference in imported and exported power. However, in Scenario 3, the generation from gas turbine and CHP and (pumped)-storage hydropower is reduced by about fifty per cent compared to generations from Scenarios 1 and 2. The lower generation from both before-mentioned sources reduces power exports by about one third compared to Scenarios 1 and 2. The curtailment of generation occurs in Scenario 3 only, since OPF is used instead of PF for power load flow calculation to avoid line overloads. Imports and exports are calculated based on power line load flows connecting Austria with neighboring countries; therefore, the calculated power energy imports and exports do not consider loop flows.

Table 6. Comparison of power generation for each scenario.

<table>
<tr><th></th><th>Scenario 1</th><th>Scenario 2</th><th>Scenario 3</th></tr>
<tr><td>Biomass</td><td>6.0 TWh</td><td>6.0 TWh</td><td>6.0 TWh</td></tr>
<tr><td>Photovoltaics</td><td>12.1 TWh</td><td>12.1 TWh</td><td>12.1 TWh</td></tr>
<tr><td>Wind</td><td>15.5 TWh</td><td>15.5 TWh</td><td>15.5 TWh</td></tr>
<tr><td>Gas turbine and CHP</td><td>10.8 TWh</td><td>10.8 TWh</td><td rowspan="2">Consumption: 0.83 TWh
Generation: 8.2 TWh</td></tr>
<tr><td>(Pumped)-storage hydropower</td><td colspan="2">Consumption: 1.5 TWh
Generation: 8.5 TWh</td></tr>
<tr><td>Hydropower >10 MW</td><td>27.9 TWh</td><td>27.9 TWh</td><td>27.9 TWh</td></tr>
<tr><td>Hydropower ≤10 MW</td><td>11.6 TWh</td><td>11.6 TWh</td><td>11.6 TWh</td></tr>
<tr><td>Import</td><td>8.1 TWh</td><td>10.5 TWh</td><td>3.7 TWh</td></tr>
<tr><td>Export</td><td>36.0 TWh</td><td>37.5 TWh</td><td>23.0 TWh</td></tr>
</table>

5. Discussion

In this section, we discuss the temporally and spatially resolved MES model of Austria, as well as the simulation results.

5.1. MES Model of Austria

Although Austria's #mission2030 aims for a net-balanced RES power supply over one year, depending on the applied scenario, significant power exports compared to imports are visible in Table 6 [5]. This gap can be explained exemplarily for Scenario 3 as follows. In Austria's MES model, power generation from company-owned CHP and power plants is not considered, since power is generated behind the meter and, therefore, company internally used. The internal generation reduces a company's power demand from the grid, as considered in the consumption data source [8]. This internal generation accounts for a total power generation of approximately 8 TWh [107]. The 7.5 TWh generation of the gas turbine and CHP has to be considered as well, since it is not a RES and therefore is not considered in the #mission2030 power generation target [5]. The remaining 3.8 TWh contain power grid losses, self-consumption of power plants, pumped-storage hydropower losses, variations of input data from [8], and others. This is in accordance with Austria's #mission2030.

The lack of subnational data available is seen as a main reason for a low spatial resolution in MES optimization projects [24]. These issues also present a main challenge within this work; however, based on experiences in [8], proven strategies are used to distribute aggregated data to more detailed granularity.

The Voronoi diagram is used because a more detailed approach may require further infrastructure data (e.g., roads) [113]. The Voronoi diagram does not take any local and geographical properties into account. Therefore, a community might be assigned to an SSD located across a mountain chain, for example. This case of misallocation is investigated manually since a small number of municipalities are affected. However, in this case, the municipalities are rather small in terms of energy consumption; therefore, the error of misallocation is considered to be negligible.

The usage of SLPs is valid if a number of several 100 consumers is aggregated [114]. This number is achieved for residential and to a smaller degree, for agricultural as well as public and private services consumers. The number of industrial consumers is significantly lower compared to the residential and service sectors. The quality of temporal consumption data can be further improved in the industrial sector, provided that more accurate industrial load profiles are available.

A temporal resolution of minutes to hours and days is common for MES frameworks which cover local levels up to regional and national levels [115]. This is important since the availability of data defines the achievable temporal resolution. For example, SLPs are available for 15 min (residential, agricultural, public, and private services) or, in the case of industrial SLPs, one-hour intervals. In comparison, wind and photovoltaic generation profiles are temporally resolved over one hour, whereas water flow rates, used to calculate hydropower plants generation, are available as daily averages. Although the simulation is carried out in 15 min interval time steps, a one-hour time step might be considered in the future to decrease computation time. Generally, simulating an MES system of the displayed size for a full year in 15 min's interval time steps takes approximately 2 days of calculation time. If node optimization is used additionally (Scenario 3), the computation time increases further.

5.2. Simulation Results

As displayed in Table 5, line overloads can be significantly reduced by more than fifty per cent in Scenario 3 compared to Scenarios 1 and 2, showing a positive effect of OPF and flexibility usage. Transmission line capacities represent a constraint using OPF applied in Scenario 3, therefore a number of zero line overloads should be expected. However, based on simulation results from Scenarios 1 and 2, the capacity of overloaded power grid sections is increased to enable the OPF to converge, since any unsolvable violation of transmission capacity would result in a termination of an OPF calculation. The count of the overload time for Scenario 3 is carried out using the original transmission capacity used in Scenario 1 and 2, considering the load flow occurring with increased line capacity.

Depending on the scenario, more than ninety-five per cent of line overloads occur in the distribution grid. Overloads can be caused by various reasons such as high RES potentials, low transmission capacity or in urban areas. No clear reason could be identified for overloads in urban areas. Potential issues might arise from the data source (consumption data) or loss of precision due to the need for grid simplification in urban areas.

In Scenario 3, the export of power is reduced significantly in comparison to Scenarios 1 and 2, since the power generation from natural gas CHP, gas turbine and especially (pumped)-storage hydropower plants is reduced. This is achieved by operating gas turbine and CHP, and (pumped)-storage hydropower plants as dispatchable flexibility. Since (pumped)-storage hydropower plants are located in western Austria and gas CHP and turbine are close to cities (=high power consumption), natural gas CHP and turbine are more likely to be activated due to lower transmission losses. This point can be further addressed through different flexibility pricing to favour (pumped)-storage hydropower usage over natural gas turbine and CHP power plants. However, this might have effects on the west–east power transmission in Austria's power grid.

6. Conclusions and Future Outlook

Within this work, we introduce a unique MES simulation framework and investigate the effects of the expansion of renewable energy sources on Austria's energy infrastructure based on a created MES model within this work.

The literature review has shown that current available multi-energy system simulation and optimization frameworks are not capable of depicting a national MES with a high degree of both high spatial and timely resolution. To overcome this scientific gap, the updated MES simulation framework HyFlow is introduced. The proposed MES simulation framework is capable of implementing the energy carrier power, natural gas and district heat (and their corresponding energy grid infrastructure), a broad range of individual consumers and producers, as well as storage and sector coupling options. Due to the flexible MES depiction approach, a wide range of research questions can be addressed. We believe that the HyFlow framework is unique in both existing and potential expansion capabilities and should be used to address various further research questions. This may include the addition of further capabilities to be added to the existing MES framework, such as further modes of operation, new objects, or improved gas and heat load flow calculations.

To depict a national MES, three main points must be addressed. First, detailed energy infrastructure models must be available. Due to the unavailability of Austria's energy grid infrastructure models, sufficient available sources and data from existing research are used to create a detailed model of Austria's energy infrastructure. Second and third, the examined area must be fed with both spatial and time-resolved consumption and generation data. To spatially resolute Austria, Voronoi diagrams based on power grid substations are used to divide Austria into so-called substation districts. To time-resolve the energy demands and generations of each substation district, a combination of SLPs and real-measured data is used. The created MES model of Austria may serve as a foundation for any further assessments of the Austrian MES. Potential fields could be the implementation of further flexibilities (e.g., storage and sector coupling) or assessment of other energy grids (gas, heat). If detailed RES expansion plans are available, the spatial distribution of RES expansion can be updated and its effects on energy grid infrastructure can be further investigated. New energy grid projects can be added to increase the transmission capacity between SSDs. We believe that this demonstrated approach can be a useful guideline to create a spatially and temporally resolved model of any national or regional MES.

Based on the created MES model of Austria and the presented MES simulation framework HyFlow, three scenarios are examined. The scenarios investigate the effects of Austrian government targets to achieve a one hundred per cent renewable power generation, net-balanced over one year. Renewable generation is expanded (mainly volatile wind and photovoltaics) by the aimed amount of the Austrian government. Additionally, electric vehicles, battery storage, and heat pumps are implemented into the MES simulation to an expectable future degree. Each scenario considers the same renewable expansion but differentiates between modes of flexibilities operation, such as (pumped)-storage hydropower, gas-fired power plants, heat pumps, electric vehicles, and battery storage. Results show that the mode of operation of flexibilities and the power load flow calculation methodology (PF and OPF) can lead to significantly different results, in terms of power line overload counts. By optimizing the consumption and generation of electric vehicles, battery storage and heat pumps based on the power price timeline (=market oriented), short demand peaks can occur, leading to the highest count of power grid overloads of all three investigated scenarios. In contrast, using OPF in combination with the flexible dispatch of natural gas-fired and (pumped)-storage hydropower plants line overloads are reduced by more than fifty per cent. The usage of OPF is therefore advantageous in contrast to PF, in terms of flexibility usage. It can be concluded that a solely price-optimized operation (market oriented) leads to grid overloads due to the neglectance of the power grids' transmission capacities. Therefore, both market and energy grid transmission capacity should be considered. A high degree of flexibilities, in a grid-supporting operation, are favorable to mitigate power grid overloads. Potentially, the addition of further flexibilities might have further positive impacts on the power grid.

Author Contributions: Conceptualization: M.G. and T.K.; methodology, software, validation, and formal analysis: M.G., F.F., J.S., N.Z. and N.W.W.; data curation: M.G., F.F., J.S., N.W.W. and T.S.; writing—original draft preparation: M.G.; writing—review and editing: N.W.W. and T.K.; visualization, M.G. and N.W.W.; supervision, T.K. All authors have read and agreed to the published version of the manuscript.

Funding: This research received no external funding.

Conflicts of Interest: The authors declare no conflict of interest.

Abbreviations

APG	Austrian Power Grid
CHP	combined heat and power
DG	distribution grid
EC	energy carrier
GtH	gas to heat
GtPH	gas to power and heat
HtP	heat to power
MES	multi-energy system
MESS	multi-energy system simulator
OPF	optimal power flow
PF	power flow
PtGH	power to gas and heat
PtH	power to heat
RES	renewable energy source
RL	residual load
SC	sector coupling
SLP	standardized Load Profile
SSD	sub-station district
TG	transmission grid
UBA	Umweltbundesamt (Federal Environment Agency)

Appendix A. Node Optimization

The node optimization and flexibility determination can be depicted in a four-stage process, displayed in Figure A1. Main parameters are explained in the following subchapters.

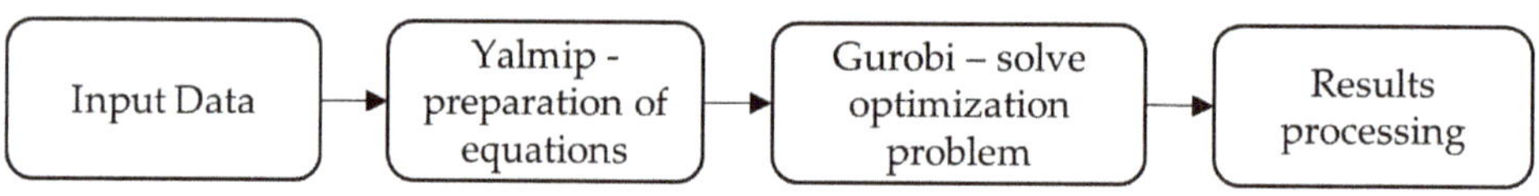

Figure A1. Optimization process.

Appendix A.1. Input Data

In Table A1, the input data for node optimization and flexibility determination are explained. Depending on the optimization problem, different input data are required. The optimization is adapted based on [29].

Table A1. Optimization input data.

Parameter	Type	Description
P_RL_Set	scalar	Residual load setting for the current time step.
t	scalar	Duration of one time step [h].
Converter	matrix	Defines properties of each converter. Each row represents one converter. [*FP*, *FH*, *FG*, *TP*, *TH*, *TG*, *MinP*, *MaxP*, *Ramp*, *PPrevPeriod*] *FP*—convert from power (1 or 0). *FH*—convert from heat (1 or 0). *FG*—convert from gas (1 or 0). *TP*—convert to power (η in [1]). *TH*—convert to heat (η in [1]). *TG*—convert to gas (η in [1]). *MinP*—minimum power [W]. *MaxP*—maximum power [W]. *Ramp*—ramp rate based on MaxP. *PPrevPeriod*—power of previous period [W].
P_Storage *H_Storage* *G_Storage*	matrix	Defines properties of each storage. Each row represents one storage. For power, heat, and gas, individual matrices have to be set up with the following structure. [*LP*, *MinSL*, *MaxSL*, *ISL*, *ηIn*, *MinIn*, *MaxIn*, *ηOut*, *MinOut*, *MaxOut*] *LP*—storage loss per period [1/h]. *MinSL*—minimum allowed storage level [Wh]. *MaxSL*—maximum allowed storage level [Wh]. *ISL*—initial storage level [Wh]. *ηIn*—input efficiency [1]. *MinIn*—minimum input power [W]. *MaxIn*—maximum input power [W]. *ηOut*—output efficiency [1]. *MinOut*—minimum output power [W]. *MaxOut*—maximum output power [W].
P_DSM *H_DSM* *G_DSM*	matrix	Defines properties of each DSM. Each row represents one DSM. For power, heat, and gas, individual matrices have to be set up with the following structure. [*MinP*, *MaxP*] *MinP*—minimum feed-in (demand) or maximum feed-out (generation) power. *MaxP*—maximum feed-in (demand) or minimum feed-out (generation) power.
eVehicle	vector	Defines parameters for electric vehicles. [*EC*, *CP*, *NoP*, *SB_1*, *EB_1*, *SB_2*, *EB_2*] *EC*—energy to be charged within *NoP* [Wh]. *CP*—charging power [W]. *NoP*—number of time steps before charging of EC must be finished (e.g., one day). *SB_1*, *SB_2*—the start of the charging break period. *EB_1*, *EB_2*—the end of the charging break period.
P_Price *H_Price* *G_Price*	vector	The number of rows is equivalent to the forecasting period. A separate vector must be defined for each energy carrier's price.
P_RL *H_RL* *G_RL*	vector	The number of rows is equivalent to the forecasting period. A separate vector must be defined for each energy carrier's residual load [W].
G_Connect *H_Connect*	scalar	Indicates if the node is connected (variable = 1) to gas/heat grid or not (variable = 0).
ops	vector	Contains settings for the Gurobi optimizer.

Appendix A.2. Node Optimization Target Function (TF)

$$TF = min(\sum_{EC}^{P,H,G} \left(\left(P_{EC} - P_{EC}^{€} \right) \cdot EP_{EC} \cdot t \right)) \tag{A1}$$

P_{EC}—consumed power for each energy carrier (EC).
$P_{EC}^{€}$—feed-in (sold) power for each EC.
EP_{EC}—price of EC.

Appendix A.3. Flexibility Determination Target Function

Positive and negative flexibility is determined according to (A2) and (A3).

$$TF_{Flex,\ neg} = \max(P_{C,out}) \tag{A2}$$

$P_{C,out}$—power converter power output.

$$TF_{Flex,\ pos} = \max(P_{C,in}) \tag{A3}$$

$P_{C,in}$—power converter power input.

To determine the total positive (A4) and negative flexibility (A5), storage and DSM options also have to be considered.

$$Flex_{pos} = TF_{Flex,\ pos} + \min\left(MaxIn, \frac{MaxSL - ISL}{t \cdot \eta IN} \right) + DSM_{MaxP} \tag{A4}$$

$$Flex_{neg} = TF_{Flex,\ neg} + \min\left(MaxOut, \frac{ISL \cdot \eta Out}{t} \right) + DSM_{MinP} \tag{A5}$$

Appendix A.4. Results

The main optimization outputs are time-resolved RLs for each converter, storage, and DSM, as well as specific results such as storage level. Furthermore, an additional parameter indicates if the optimization problem can be solved successfully.

Appendix B. Measurement Points for Small Hydropower Plant SLP

For each federal state, a random small river measurement point from [80] is used to create an SLP for small-scale hydropower plants, as displayed in Table A2.

Table A2. Measurement points used for small hydropower plant SLP [80].

Federal State	Measurement Point #	Name
Vorarlberg	200105	Garsella
Tyrol	230706	In der Au
Salzburg	203265	Schweighofbrücke
Carinthia	213389	Kaunz
Styria	211029	Anger
Upper Austria	204784	Riedau
Lower Austria	208041	Hollenstein
Burgenland	210039	Piringsdorf (Pfarrbrücke)
Vienna	None	None

Appendix C. Measurement Points for Natural Inflow Curve for (Pumped)-Storage Hydropower Plants

Natural water inflow into (pumped)-storage hydropower plants originates from water sources at high altitudes, such as snow and glacier melt. Therefore, measurement points from [80] at high elevation are selected to determine an annual inflow characteristic for (pumped)-storage hydropower plants, as displayed in Table A3.

Table A3. Measurement points used for annual (pumped)-storage hydropower plant inflow SLP [80].

Measurement Point Name	Measurement Point #	Elevation [m]
Gepatschalm	230300	1895
Vent (oberhalb Niedertalbach)	201350	1891
Obergurgl	201376	1879
Neukaser	201996	1786
Innergschlöß	212068	1686
Kees	203893	2040

References

1. European Commission. The European Green Deal, Brussels, 2019. Available online: https://eur-lex.europa.eu/resource.html?uri=cellar:b828d165-1c22-11ea-8c1f-01aa75ed71a1.0002.02/DOC_1&format=PDF (accessed on 12 January 2022).
2. European Commission. What is the European Green Deal? Brussels, 2019. Available online: https://ec.europa.eu/commission/presscorner/detail/en/fs_19_6714 (accessed on 12 November 2021).
3. European Commission. Fit for 55: Delivering the EU's 2030 Climate Target on the Way to Climate Neutrality, Brussels, 2021. Available online: https://eur-lex.europa.eu/legal-content/EN/TXT/PDF/?uri=CELEX:52021DC0550&from=DE (accessed on 12 November 2021).
4. Bundeskanzleramt. Aus Verantwortung für Österreich: Regierungsprogramm 2020–2024, Wien, 2020. Available online: https://www.bundeskanzleramt.gv.at/dam/jcr:7b9e6755-2115-440c-b2ec-cbf64a931aa8/RegProgramm-lang.pdf (accessed on 11 November 2021).
5. Bundesministerium Nachhaltigkeit und Tourismus; Bundesministerium Verkehr Innovation und Technologie. #Mission2030: Die Österreichische Klima- und Energiestrategie, Wien, 2018. Available online: https://www.bundeskanzleramt.gv.at/dam/jcr:903d5cf5-c3ac-47b6-871c-c83eae34b273/20_18_beilagen_nb.pdf (accessed on 12 January 2022).
6. Mancarella, P. MES (multi-energy systems): An overview of concepts and evaluation models. *Energy* **2014**, *65*, 1–17. [CrossRef]
7. Hansen, K.; Breyer, C.; Lund, H. Status and perspectives on 100% renewable energy systems. *Energy* **2019**, *175*, 471–480. [CrossRef]
8. Sejkora, C.; Kühberger, L.; Radner, F.; Trattner, A.; Kienberger, T. Exergy as Criteria for Efficient Energy Systems—A Spatially Resolved Comparison of the Current Exergy Consumption, the Current Useful Exergy Demand and Renewable Exergy Potential. *Energies* **2020**, *13*, 843. [CrossRef]
9. Bundesamt für Eich- und Vermessungswesen. *Katalog Verwaltungsgrenzen (VGD)—Stichtagsdaten 1:5000*; Bundesamt für Eich- und Vermessungswesen: Wien, Austria; Available online: https://www.data.gv.at/katalog/dataset/bev_verwaltungsgrenzenstichtagsdaten150000 (accessed on 15 February 2022).
10. Klemm, C.; Vennemann, P. Modeling and optimization of multi-energy systems in mixed-use districts: A review of existing methods and approaches. *Renew. Sustain. Energy Rev.* **2021**, *135*, 110206. [CrossRef]
11. Pfenninger, S.; Hawkes, A.; Keirstead, J. Energy systems modeling for twenty-first century energy challenges. *Renew. Sustain. Energy Rev.* **2014**, *33*, 74–86. [CrossRef]
12. Bottecchia, L.; Lubello, P.; Zambelli, P.; Carcasci, C.; Kranzl, L. The Potential of Simulating Energy Systems: The Multi Energy Systems Simulator Model. *Energies* **2021**, *14*, 5724. [CrossRef]
13. Lohmeier, D.; Cronbach, D.; Drauz, S.R.; Braun, M.; Kneiske, T.M. Pandapipes: An Open-Source Piping Grid Calculation Package for Multi-Energy Grid Simulations. *Sustainability* **2020**, *12*, 9899. [CrossRef]
14. Weinand, J.M.; Scheller, F.; McKenna, R. Reviewing energy system modelling of decentralized energy autonomy. *Energy* **2020**, *203*, 117817. [CrossRef]
15. Thurner, L.; Scheidler, A.; Schafer, F.; Menke, J.-H.; Dollichon, J.; Meier, F.; Meinecke, S.; Braun, M. Pandapower—An Open-Source Python Tool for Convenient Modeling, Analysis, and Optimization of Electric Power Systems. *IEEE Trans. Power Syst.* **2018**, *33*, 6510–6521. [CrossRef]
16. Böckl, B.; Greiml, M.; Leitner, L.; Pichler, P.; Kriechbaum, L.; Kienberger, T. HyFlow—A Hybrid Load Flow-Modelling Framework to Evaluate the Effects of Energy Storage and Sector Coupling on the Electrical Load Flows. *Energies* **2019**, *12*, 956. [CrossRef]
17. Kienberger, T.; Traupmann, A.; Sejkora, C.; Kriechbaum, L.; Greiml, M.; Böckl, B. Modelling, designing and operation of grid-based multi-energy systems. *Int. J. Sustain. Energy Plan. Manag.* **2020**, *29*, 7–24. [CrossRef]

18. Greiml, M.; Fritz, F.; Kienberger, T. Increasing installable photovoltaic power by implementing power-to-gas as electricity grid relief—A techno-economic assessment. *Energy* **2021**, *235*, 121307. [CrossRef]
19. FFG. SBM_Ind: Smart Business Models for Industry. Available online: https://projekte.ffg.at/projekt/3093356 (accessed on 13 November 2021).
20. Farshidian, B.; Ghahnavieh, A.R. A comprehensive framework for optimal planning of competing energy hubs based on the game theory. *Sustain. Energy Grids Netw.* **2021**, *27*, 100513. [CrossRef]
21. Cheng, Y.; Zhang, P.; Liu, X. Collaborative Autonomous Optimization of Interconnected Multi-Energy Systems with Two-Stage Transactive Control Framework. *Energies* **2020**, *13*, 171. [CrossRef]
22. Sejkora, C.; Kühberger, L.; Radner, F.; Trattner, A.; Kienberger, T. Exergy as criteria for efficient energy systems—Maximising energy efficiency from resource to energy service, an Austrian case study. *Energy* **2022**, *239*, 122173. [CrossRef]
23. Wagner & Elbling GmbH. ONE100: Österreichs Nachhaltiges Energiesystem—100% Dekarbonisiert, Wien, 2021. Available online: https://www.aggm.at/files/get/b2327a5f19bc53c7fa0f2ae1dcd4edb1/ONE100_Kurzfassung.pdf (accessed on 14 November 2021).
24. Aryanpur, V.; O'Gallachoir, B.; Dai, H.; Chen, W.; Glynn, J. A review of spatial resolution and regionalisation in national-scale energy systems optimisation models. *Energy Strategy Rev.* **2021**, *37*, 100702. [CrossRef]
25. Zimmerman, R.D.; Murillo-Sanchez, C.E.; Thomas, R.J. MATPOWER: Steady-State Operations, Planning, and Analysis Tools for Power Systems Research and Education. *IEEE Trans. Power Syst.* **2011**, *26*, 12–19. [CrossRef]
26. Zimmerman, R.D.; Murillo-Sánchez, C.E. MATPOWER User's Manual: Version 7.1. 2020. Available online: https://matpower.org/docs/MATPOWER-manual.pdf (accessed on 15 December 2021).
27. Rüdiger, J. Enhancements of the numerical simulation algorithm for natural gas networks based on node potential analysis. *IFAC-Pap.* **2020**, *53*, 13119–13124. [CrossRef]
28. Langeheinecke, K.; Jany, P.; Thieleke, G.; Langeheinecke, K.; Kaufmann, A. *Thermodynamik für Ingenieure*; Springer Fachmedien Wiesbaden: Wiesbaden, Germany, 2013; ISBN 978-3-658-03168-8.
29. Chen, Z.; Zhang, Y.; Tang, W.; Lin, X.; Li, Q. Generic modelling and optimal day-ahead dispatch of micro-energy system considering the price-based integrated demand response. *Energy* **2019**, *176*, 171–183. [CrossRef]
30. 2004 IEEE International Conference on Robotics and Automation (IEEE Cat. No.04CH37508). In Proceedings of the 2004 IEEE International Conference on Robotics and Automation (IEEE Cat. No.04CH37508), New Orleans, LA, USA, 26 April–1 May 2004.
31. Gurobi Optimization. Gurobi Optimizer Reference Manual 2021. 2021. Available online: https://www.gurobi.com/documentation/9.5/refman/index.html (accessed on 6 December 2021).
32. Austrian Power Grid AG. Netzentwicklungsplan 2015: Für das Übertragungsnetz der Austrian Power Grid AG (APG), Wien, 2015. Available online: https://www.apg.at/~{]/media/009493CEFD824A85962F65CEAA6521C0.pdf (accessed on 18 November 2021).
33. Austrian Power Grid AG. Netzentwicklungsplan 2016: Für das Übertragungsnetz der Austrian Power Grid AG (APG), Wien, 2016. Available online: https://www.apg.at/~{]/media/2A8B8AC633414A359DB163BBE5104AD8.pdf (accessed on 18 November 2021).
34. Austrian Power Grid AG. Netzentwicklungsplan 2017: Für das Übertragungsnetz der Austrian Power Grid AG (APG), Wien, 2017. Available online: https://www.apg.at/~{]/media/6B16E721BF8D45F49A907C11A7C095EC.pdf (accessed on 18 November 2021).
35. Austrian Power Grid AG. Netzentwicklungsplan 2018: Für das Übertragungsnetz der Austrian Power Grid AG (APG), Wien, 2018. Available online: https://www.apg.at/-/media/B1CF20C3D97B4496AE3E06AF5B351AB7.pdf (accessed on 18 November 2021).
36. Austrian Power Grid AG. Netzentwicklungsplan 2019: Für das Übertragungsnetz der Austrian Power Grid AG (APG), Wien, 2019. Available online: https://www.apg.at/api/sitecore/projectmedia/download?id=bd6645e4-f83d-456a-a9d4-5757b5098a70 (accessed on 18 November 2021).
37. Austrian Power Grid AG. Netzentwicklungsplan 2020: Für das Übertragungsnetz der Austrian Power Grid AG (APG), Wien, 2020. Available online: https://www.apg.at/de/Stromnetz/Netzentwicklung#download (accessed on 18 November 2021).
38. Austrian Power Grid AG; LINZ NETZ GmbH; Netz Oberösterreich GmbH. Stromnetz-Masterplan Oberösterreich 2028: Ausbau des Hochspannungs-Stromnetzes (≥110 kV) in Oberösterreich. Planungszeitraum 2018–2028. 2018. Available online: https://www.land-oberoesterreich.gv.at/Mediendateien/Formulare/Dokumente%20UWD%20Abt_US/us-en_Stromnetz-Masterplan_Oberoesterreich_2028.pdf (accessed on 18 November 2021).
39. Land Kärnten. Energie Masterplan Kärnten, Klagenfurt. Available online: https://www.ktn.gv.at/DE/repos/files/ktn.gv.at/Abteilungen/Abt8/Dateien/energie/energiemasterplan%5fkaernten?exp=478252&fps=cbe8bb636710ede50d5a94df838d40cbaebea6d1 (accessed on 18 November 2021).
40. QGIS Development Team. QGIS Geographic Information System; Open Source Geospatial Foundation Project. 2021. Available online: http://qgis.osgeo.org (accessed on 18 November 2021).
41. Google. Map Data: Google, Terrametrics, Kartendaten (C) 2021. Available online: maps.google.at (accessed on 18 November 2021).
42. OpenStreetMap-Mitwirkende. OpenStreetMap. Available online: www.openstreetmap.org/copyright (accessed on 18 November 2021).
43. Austrian Power Grid AG. Statistische Netzdaten. Available online: https://www.apg.at/api/sitecore/projectmedia/download?id=703efbb9-bd69-49db-b2f6-bb676cac466b (accessed on 16 September 2020).
44. Heuck, K.; Dettmann, K.-D.; Schulz, D. (Eds.) *Elektrische Energieversorgung*; Springer Fachmedien Wiesbaden: Wiesbaden, Germany, 2013; ISBN 978-3-8348-1699-3.
45. Heuck, K.; Dettmann, K.-D.; Schulz, D. Anhang. In *Elektrische Energieversorgung*; Heuck, K., Dettmann, K.-D., Schulz, D., Eds.; Springer Fachmedien Wiesbaden: Wiesbaden, Germany, 2013; pp. 742–753.

46. AGGM Austrian Gas Grid Management AG. Unser Netz im Detail. Available online: https://www.gasconnect.at/netzinformationen/unser-netz-im-detail (accessed on 19 November 2021).
47. Austrian Gas Grid Management AGGM. *Erdgasinfrastruktur—Österreich: Georeferenzierte Darstellung nach Netzebenen*; Austrian Gas Grid Management AGGM: Wien, Austria, 2018.
48. Gas Connect Austria. *Erdgasleitungen & Erdgaslagerstätten in Österreich*; Gas Connect Austria: Wien, Austria, 2017.
49. E-Control. Erdgas—Bestandsstatistik: Leitungslängen zum 31. Dezember–Jahresreihen. Leitungslängen von Fern- und Verteilleitungen zum Jahresende. Available online: https://www.e-control.at/de/statistik/gas/bestandsstatistik (accessed on 19 November 2021).
50. Cerbe, G. *Grundlagen der Gastechnik: Gasbeschaffung, Gasverteilung, Gasverwendung*; Hanser: München, Germany, 2004; ISBN 978-3-446-22803-0.
51. Büchele, R.; Haas, R.; Hartner, M.; Hirner, R.; Hummel, M.; Kranzl, L.; Müller, A.; Ponweiser, K.; Bons, M.; Grave, K.; et al. Bewertung des Potentials für den Einsatz der hocheffizienten KWK und effizienter Fernwärme- und Fernkälteversorgung, Wien. 2015. Available online: https://ec.europa.eu/energy/sites/ener/files/documents/Austria_MNE%282016%2950514.pdf (accessed on 21 January 2021).
52. Bundesministerium Klimaschutz, Umwelt, Energie, Mobilität, Innovatoin und Technologie. Austrian Heat Map: Fernwärme und Kraft-Wärme-Kopplung in Österreich. Available online: http://www.austrian-heatmap.gv.at/das-projekt/ (accessed on 19 November 2021).
53. Navarro, A.; Rudnick, H. Large-Scale Distribution Planning—Part II: Macro-Optimization With Voronoi's Diagram and Tabu Search. IEEE Trans. *Power Syst.* **2009**, *24*, 752–758. [CrossRef]
54. Statistik Austria. STATatlas. Available online: https://www.statistik.at/atlas/ (accessed on 20 November 2021).
55. Statistik Austria. Nutzenergieanalyse. Available online: https://www.statistik.at/web_de/statistiken/energie_umwelt_innovation_mobilitaet/energie_und_umwelt/energie/nutzenergieanalyse/index.html (accessed on 20 November 2021).
56. Gobmaier, T.; Mauch, W.; Beer, M.; von Roon, S.; Schmid, T.; Mezger, T.; Habermann, J.; Hohlenburger, S. Simulationsgestütze Prognose des elektrischen Lastverhaltens, München. 2012. Available online: https://docplayer.org/5749688-Simulationsgestuetzte-prognose-des-elektrischen-lastverhaltens.html (accessed on 20 November 2021).
57. Austrian Power Clearing and Settlement. Synthetische Lastprofile: Prognose von Verbrauchswerten mittels Lastprofilen, Wien. 2019. Available online: https://www.apcs.at/de/clearing/technisches-clearing/lastprofile (accessed on 20 November 2021).
58. Dock, J.; Janz, D.; Weiss, J.; Marschnig, A.; Kienberger, T. Time- and component-resolved energy system model of an electric steel mill. *Clean. Eng. Technol.* **2021**, *4*, 100223. [CrossRef]
59. Groiss, C.; Grubinger, D.; Schwalbe, R. Blindleistungsbilanz im Salzburger Verteilnetz. In *EnInnov 2018: 15. Symphosium Energieinnovation*; Institut für Elektrizitätswirtschaft und Energieinnovation, Ed.; Verlag der Technischen Universität Graz: Graz, Austria, 2018; ISBN 978-3-85125-586-7.
60. Thormann, B.; Purgstaller, W.; Kienberger, T. Evaluating the potential of future e-mobility use cases for providing grid ancillary services. In Proceedings of the 26th International Conference and Exhibition on Electricity Distribution, Online, 20–23 September 2021.
61. BDEW; VKU; GEODE. BDEW/VKU/GEODE-Leitfaden: Abwicklung von Standardlastprofilen Gas, Berlin. 2016. Available online: https://www.bdew.de/media/documents/Leitfaden_20160630_Abwicklung-Standardlastprofile-Gas.pdf (accessed on 20 November 2021).
62. Pfenninger, S.; Staffell, I. Renewables.ninja. Available online: https://www.renewables.ninja/ (accessed on 20 November 2021).
63. Rienecker, M.M.; Suarez, M.J.; Gelaro, R.; Todling, R.; Bacmeister, J.; Liu, E.; Bosilovich, M.G.; Schubert, S.D.; Takacs, L.; Kim, G.-K.; et al. MERRA: NASA's Modern-Era Retrospective Analysis for Research and Applications. *J. Clim.* **2011**, *24*, 3624–3648. [CrossRef]
64. Oesterreichs Energie. Kraftwerkskarte Österreich. Available online: https://oesterreichsenergie.at/kraftwerkskarte-1 (accessed on 21 November 2021).
65. Oesterreichs Energie. *Stromerzeugung in Österreich: Kraftwerke der Österreichischen E-Wirtschaft*; Oesterreichs Energie: Vienna, Austria, 2019.
66. Crastan, V. *Elektrische Energieversorgung 2*; Springer: Berlin/Heidelberg, Germany, 2008; ISBN 978-3-540-70877-3.
67. Verbund. Unsere Kraftwerke. Available online: https://www.verbund.com/de-at/ueber-verbund/kraftwerke/unsere-kraftwerke (accessed on 21 November 2021).
68. Wikipedia. Liste von Wasserkraftwerken in Österreich. Available online: https://de.wikipedia.org/wiki/Liste_von_Wasserkraftwerken_in_%C3%96sterreich (accessed on 21 November 2021).
69. KELAG AG. Kraftwerke. Available online: https://www.kelag.at/kraftwerke (accessed on 21 November 2021).
70. TIWAG. Kraftwerkspark: Unsere Kraftwerke im Überblick. Available online: https://www.tiwag.at/ueber-die-tiwag/kraftwerke/bestehende-kraftwerke/kraftwerkspark/ (accessed on 21 November 2021).
71. Energie AG. Die Wasserkraftwerke der Energie AG Oberösterreich. Available online: https://www.energieag.at/Themen/Energie-fuer-Sie/Kraftwerke/Wasserkraftwerke (accessed on 21 November 2021).
72. Salzburg AG. Unsere Erzeugungsanlagen. Available online: https://www.salzburg-ag.at/ueber-die-salzburg-ag/unternehmen/erzeugung/erzeugungsanlagen.html (accessed on 21 November 2021).

73. VERBUND Hydro Power AG. *Strom aus den Hohen Tauern: Die Wasserkraftwerke in Salzburg*; VERBUND Hydro Power AG: Vienna, Austria, 2013.
74. VERBUND Hydro Power AG. *Strom aus den Zillertaler Alpen: Die Wasserkraftwerke in Tirol*; VERBUND Hydro Power AG: Vienna, Austria, 2013.
75. VERBUND Hydro Power AG. *Strom aus den Hohen Tauern und aus der Drau: Die Wasserkraftwerke in Kärnten*; VERBUND Hydro Power AG: Vienna, Austria, 2013.
76. Illwerke VKW. Kraftwerksanlagen der Illwerke VKW. Available online: https://www.illwerkevkw.at/kraftwerke_uebersicht.htm (accessed on 21 November 2021).
77. Energie Steiermark. Wasserkraft. Available online: https://www.e-steiermark.com/ueber-uns/energieerzeugung/wasserkraft (accessed on 21 November 2021).
78. EVN Naturkraft. Wasserkraft. Available online: http://www.evn-naturkraft.at/Oekostrom/Wasser.aspx (accessed on 21 November 2021).
79. Geodatenstellen des Landes Tirol. *TIRIS*; Amt der Tiroler Landesregierung: Innsbruck, Austria; Available online: https://maps.tirol.gv.at (accessed on 22 November 2021).
80. Bundesministerium für Landwirtschaft, Regionen und Tourismus. eHYD. Available online: https://ehyd.gv.at/ (accessed on 16 November 2021).
81. Kleinwasserkraft Österreich. Nutzen der Kleinwasserkraft. Available online: https://www.kleinwasserkraft.at/en/fakten/ (accessed on 21 November 2021).
82. Pöyry. Wasserkraftpotenzialstudie Österreich. 2018. Available online: https://oesterreichsenergie.at/fileadmin/user_upload/Oesterreichs_Energie/Publikationsdatenbank/Studien/2018/WasserkraftpotenzialOesterreich2018.pdf (accessed on 21 November 2021).
83. TIWAG. *Kraftwerksgruppe Sellrain-Silz*; Kraftwerksgruppe Sellrain-Silz. TIWAG: Innsbruck, Austria; Available online: https://www.tiwag.at/unternehmen/unsere-kraftwerke/kraftwerk/kraftwerksgruppe-sellrain-silz/ (accessed on 21 November 2021).
84. KELAG AG. Pumpspeicherkraftwerke. Available online: https://www.kelag.at/corporate/pumpspeicherkraftwerke-852.htm (accessed on 21 November 2021).
85. TIWAG. Unsere Kraftwerksprojekte. Available online: https://www.tiwag.at/ueber-die-tiwag/kraftwerke/wasserkraftausbau/unsere-kraftwerksprojekte/ (accessed on 21 November 2021).
86. Verbund. Pumpspeicherkraftwerk Limberg 3. Available online: https://www.verbund.com/de-at/ueber-verbund/kraftwerke/unsere-kraftwerke/kaprun-oberstufe-limberg-3 (accessed on 21 November 2021).
87. Zahoransky, R. *Energietechnik*; Springer: Wiesbaden, Germany, 2019; ISBN 978-3-658-21846-1.
88. Österreichischer Biomasse-Verband. *Bioenergie Atlas Österreich 2019*, 2nd ed.; Österreichischer Biomasse-Verband: Wien, Austria, 2019; ISBN 978-3-9504380-3-1.
89. ÖVGW. Montanuniversität Leoben, WU Wien, DBI Gas- und Umwelttechnik GmbH, TU Wien, JKU Linz, ERIG. Greening the Gas: Forschungsbericht 2019, Vienna. 2020. Available online: https://www.ovgw.at/media/medialibrary/2020/03/OVGW_JB_forschung19_hi_corr2.pdf (accessed on 21 November 2021).
90. Photovoltaik Austria. Die österreichische Photovoltaik & Speicher-Branche in Zahlen, Wien. 2020. Available online: https://www.pvaustria.at/wp-content/uploads/2020_07_05_Fact_Sheet_PV_Branche.pdf (accessed on 22 November 2021).
91. Pfenninger, S.; Staffell, I. Long-term patterns of European PV output using 30 years of validated hourly reanalysis and satellite data. *Energy* **2016**, *114*, 1251–1265. [CrossRef]
92. The Wind Power. Windparks—Österreich. Available online: https://www.thewindpower.net/windfarms_list_de.php?country=AT (accessed on 10 December 2020).
93. Staffell, I.; Pfenninger, S. Using bias-corrected reanalysis to simulate current and future wind power output. *Energy* **2016**, *114*, 1224–1239. [CrossRef]
94. LINZ AG. Die Kraftwerke der LINZ AG: Effiziente, umweltschonende Energieerzeugung, Linz. 2018. Available online: https://www.linzag.at//media/dokumente/linzag/folder-kraftwerke.pdf (accessed on 22 November 2021).
95. EVN. Thermische Erzeugung. Available online: https://www.evn.at/EVN-Group/Energie-Zukunft/Energie-aus-Niederosterreich/Gas-und-Kohle.aspx (accessed on 22 November 2021).
96. Wien Energie. Energie ist unsere Verantwortung: Konsolidierte Umwelterklärung 2021 der Strom- und Wärmeerzeugungsanlagen der Wien Energie GmbH gemäß EMAS-Verordnung, Wien. 2021. Available online: https://dokumente.wienenergie.at/wp-content/uploads/umwelterklaerung-2021.pdf (accessed on 22 November 2021).
97. Energie AG. Übersicht thermische Kraftwerke. Available online: https://www.energieag.at/Themen/Energie-fuer-Sie/Kraftwerke/Thermische-Kraftwerke (accessed on 22 November 2021).
98. ENTSO-E. Transparancy Platform. Available online: https://transparency.entsoe.eu (accessed on 22 November 2021).
99. Krutzler, T.; Zechmeister, A.; Stranner, G.; Wiesenberger, H.; Gallauner, T.; Gössl, M.; Heller, C.; Heinfellner, H.; Ibesich, N.; Lichtblau, G.; et al. *Energie- und Treibhausgas-Szenarien im Hinblick auf 2030 und 2050: Synthesebericht 2017*; Umweltbundesamt: Wien, Austria, 2017; ISBN 978-3-99004-445-2.
100. Österreich.gv.at. Sanierungsoffensive 2021/2022. Available online: https://www.oesterreich.gv.at/themen/bauen_wohnen_und_umwelt/energie_sparen/1/sanierungsoffensive.html (accessed on 23 November 2021).

101. Statistik Austria. Kraftfahrzeuge—Bestand. Available online: https://www.statistik.at/web_de/statistiken/energie_umwelt_innovation_mobilitaet/verkehr/strasse/kraftfahrzeuge_-_bestand/index.html (accessed on 23 November 2021).
102. ÖAMTC; ARBÖ. Expertenbericht Mobilität & Klimaschutz 2030, Wien. 2018. Available online: https://www.oeamtc.at/%C3%96AMTC+Expertenbericht+Mobilit%C3%A4t+%26+Klimaschutz+2030+Web.pdf/25.789.593 (accessed on 23 November 2021).
103. Umweltbundesamt. Verkehrsmittel in Österreich. Available online: https://www.umweltbundesamt.at/fileadmin/site/themen/mobilitaet/daten/ekz_doku_verkehrsmittel.pdf (accessed on 23 November 2021).
104. ADAC. Elektroautos im Test: So hoch ist der Stromverbrauch. Available online: https://www.adac.de/rund-ums-fahrzeug/tests/elektromobilitaet/stromverbrauch-elektroautos-adac-test/ (accessed on 23 November 2021).
105. Hartl, M.; Biermayr, P.; Schneeberger, A.; Schöfmann, P. Österreichische Technologie-Roadmap für Wärmepumpen, Wien. 2016. Available online: https://nachhaltigwirtschaften.at/resources/nw_pdf/1608_endbericht_oesterreichische_technologieroadmap_fuer_waermepumpen.pdf?m=1469661515& (accessed on 23 November 2021).
106. Weniger, J.; Orth, N.; Lawaczeck, I.; Meissner, L.; Quaschning, V. Energy Storage Inspection 2021, Berlin, 2021. Available online: https://pvspeicher.htw-berlin.de/wp-content/uploads/Energy-Storage-Inspection-2021.pdf (accessed on 23 November 2021).
107. Statistik Austria. Energiebilanzen. Available online: http://www.statistik.at/web_de/statistiken/energie_umwelt_innovation_mobilitaet/energie_und_umwelt/energie/energiebilanzen/index.html (accessed on 20 November 2021).
108. Thormann, B.; Kienberger, T. Evaluation of Grid Capacities for Integrating Future E-Mobility and Heat Pumps into Low-Voltage Grids. *Energies* **2020**, *13*, 5083. [CrossRef]
109. Thormann, B.; Kienberger, T. Estimation of Grid Reinforcement Costs Triggered by Future Grid Customers: Influence of the Quantification Method (Scaling vs. Large-Scale Simulation) and Coincidence Factors (Single vs. Multiple Application). *Energies* **2022**, *15*, 1383. [CrossRef]
110. Shi, H.; Blaauwbroek, N.; Nguyen, P.H.; Kamphuis, R. Energy management in Multi-Commodity Smart Energy Systems with a greedy approach. *Appl. Energy* **2016**, *167*, 385–396. [CrossRef]
111. Xie, Y.; Ueda, Y.; Sugiyama, M. Greedy energy management strategy and sizing method for a stand-alone microgrid with hydrogen storage. *J. Energy Storage* **2021**, *44*, 103406. [CrossRef]
112. Energy Exchange Austria. Historische Marktdaten. Available online: https://www.exaa.at/marktdaten/historische-marktdaten/ (accessed on 15 January 2020).
113. Kays, J.; Seack, A.; Smirek, T.; Westkamp, F.; Rehtanz, C. The Generation of Distribution Grid Models on the Basis of Public Available Data. *IEEE Trans. Power Syst.* **2017**, *32*, 2346–2353. [CrossRef]
114. Esslinger, P.; Witzmann, R. Entwicklung und verifikation eines stochastischen Verbraucherlastmodells für Haushalte. In *12. Symposium Energieinnovation: Alternativen für die Energiezukunft Europas*; Elektrizitatswirtschaft und Energieinnovation, Ed.; Verlag der Technischen Universitat Graz: Graz, Austria, 2012; ISBN 978-3-85125-200-2.
115. Ringkjøb, H.-K.; Haugan, P.M.; Solbrekke, I.M. A review of modelling tools for energy and electricity systems with large shares of variable renewables. *Renew. Sustain. Energy Rev.* **2018**, *96*, 440–459. [CrossRef]

energies
MDPI

Article

Source-Load Coordinated Low-Carbon Economic Dispatch of Electric-Gas Integrated Energy System Based on Carbon Emission Flow Theory

Jieran Feng [1], Junpei Nan [1], Chao Wang [1], Ke Sun [2,3], Xu Deng [4] and Hao Zhou [1,*]

1 College of Electrical Engineering, Zhejiang University, Hangzhou 310027, China; jieran_feng@zju.edu.cn (J.F.); junpei_nan@zju.edu.cn (J.N.); 21910121@zju.edu.cn (C.W.)
2 State Grid Zhejiang Electric Power Co., Ltd., Hangzhou 310063, China; sun_ke@zj.sgcc.com.cn
3 Zhejiang Huayun Electric Power Engineering Design Consulting Co., Ltd., Hangzhou 310014, China
4 The People's Government of Guangzhou Municipality, Guangzhou 510032, China; dengxu926@163.com
* Correspondence: zhouhao_ee@zju.edu.cn

Abstract: The development of emerging technologies has enhanced the demand response (DR) capability of conventional loads. To study the effect of DR on the reduction in carbon emissions in an integrated energy system (IES), a two-stage low-carbon economic dispatch model based on the carbon emission flow (CEF) theory was proposed in this study. In the first stage, the energy supply cost was taken as the objective function for economic dispatch, and the actual carbon emissions of each energy hub (EH) were calculated based on the CEF theory. In the second stage, a low-carbon DR optimization was performed with the objective function of the load-side carbon trading cost. Then, based on the modified IEEE 39-bus power system/Belgian 20-node natural gas system, MATLAB/Gurobi was used for the simulation analysis in three scenarios. The results showed that the proposed model could effectively promote the system to reduce the load peak-to-valley difference, enhance the ability to consume wind power, and reduce the carbon emissions and carbon trading cost. Furthermore, as the wind power penetration rate increased from 20% to 80%, the carbon reduction effect basically remained stable. Therefore, with the growth of renewable energy, the proposed model can still effectively reduce carbon emissions.

Keywords: carbon emission flow; demand response; integrated energy system; ladder-type carbon price; low-carbon economic dispatch; Shapley value

Citation: Feng, J.; Nan, J.; Wang, C.; Sun, K.; Deng, X.; Zhou, H. Source-Load Coordinated Low-Carbon Economic Dispatch of Electric-Gas Integrated Energy System Based on Carbon Emission Flow Theory. *Energies* **2022**, *15*, 3641. https://doi.org/10.3390/en15103641

Academic Editors: Zbigniew Leonowicz, Michał Jasiński and Arsalan Najafi

Received: 25 April 2022
Accepted: 14 May 2022
Published: 16 May 2022

1. Introduction

Emissions of greenhouse gases such as carbon dioxide produced by the development of human society have exceeded the capacity of the Earth, causing the greenhouse effect to become increasingly apparent [1,2]. Energy systems are a major source of carbon emissions [3]. Under the Sustainable Development Goals (SDGs), energy systems are in urgent need of low-carbon development [4–6]. Integrated energy systems (IESs) have a prominent low-carbon emission potential, attracting a large number of domestic and foreign scholars to conduct relevant research [7,8]. The research on low-carbon IESs has become a hotspot in the international energy field [3,9].

So, how can we effectively mitigate the carbon emissions in IESs? The recent research [10] systematically combs through carbon emission mitigation strategies from the aspects of policies, sector specific technologies and initiatives, and general societal initiatives. Specifically for integrated energy systems, carbon reduction strategies can be roughly divided into two categories: system internal strategies and policy incentive strategies. The internal strategies include developing renewable energy power generation technologies to replace fossil fuel power generation technology, developing energy storage technology to promote renewable energy consumption, optimizing operation strategies to improve energy utilization, developing carbon capture and utilization storage technology, and so on [10,11]. Policy incentive strategies include the use of a carbon tax, carbon trading, time-of-use energy pricing, and other policies to stimulate the energy supply side and consumption side to change toward a direction that is conducive to mitigating carbon emissions [10,12].

Research focusing on the system internal strategies of IESs to reduce carbon emissions have been widely conducted. The study by [13] proposed a day-ahead energy trading strategy for a regional integrated energy system (RIES) that considered energy cascade utilization to improve the energy utilization efficiency. In [14], the concept of a sharing economy was introduced into the energy interaction process of an IES and proposed a distributed electrical–gas–thermal energy sharing mechanism to improve the energy efficiency and promote the optimal resource allocation. The use of power-to-ammonia in high-renewable multi-energy systems is superior to that of regular batteries and power-to-gas storage for system operational economy and renewable energy accommodation [15].

In terms of policy incentive strategies, some studies have focused on the carbon reduction effects of a carbon tax and carbon trading policies on the energy supply side. A carbon tax was introduced into the objective function of the economic dispatch model to improve the system economy and low-carbon performance in [16]. In [17], an economic dispatch model for RIESs was proposed using the ladder carbon prices. The reward and punishment ladder-type carbon trading mechanism was used to calculate the carbon trading cost of an IES considering the carbon capture technology in [9].

On the other hand, some studies have mainly focused on the policy impact on the energy consumption side. Stimulating demand response (DR) through policies is an effective way to increase the renewable energy consumption and reduce the system carbon emissions [18]. The study in [19] used the time-of-use electricity and gas prices to drive the integrated demand response to reduce the system carbon emissions. In [20], they studied the effect of the time-of-use electricity pricing policy on smart home participation in the power demand response. The authors in [21] studied the impact of the dynamic electricity tariffs on the household's electricity demand response. Demand response guided by real-time electricity prices has also been studied [22,23]. These studies [19–23] were all from the perspective of energy price policy to stimulate the demand response.

However, in the context of decarbonizing the energy system, the demand response motivated by energy price policies is not straightforward enough. Energy demand is the root cause of carbon emissions in the energy system. The responsibility of the energy demand side for system carbon emissions cannot be ignored. Therefore, it is more direct to guide the demand response through carbon price policies. At present, the demand response under the incentive of user-side carbon trading has not been fully studied, and its carbon reduction effectiveness and advantages remain to be discussed.

To study this problem, the first challenge is to calculate the actual carbon responsibility on the user side. The carbon emission flow (CEF) theory first proposed in [24] solves this problem well. The carbon emission flow theory considering power network losses has also been studied [25]. Furthermore, CEF theory has also been adopted in IESs [3,26], which can allocate the actual carbon emission responsibility from the energy supply side to the demand side.

To sum up, the demand response carbon reduction strategy under the CEF-based user-side carbon trading incentive is worth studying. To study the carbon reduction effect of the strategy, a two-stage low-carbon economic dispatch model was proposed in this paper. The main contributions of this paper are as follows:

1. This paper proposes a two-stage low-carbon economic dispatch model of the IES based on the CEF theory. On the basis of considering the load-side carbon responsibility and demand response, the source and load are coordinately optimized to realize the low-carbon economic operation of the IES;

2. Based on the Shapley value method, a method for the division of the carbon emission responsibility grades of each energy hub is proposed, and a ladder-type carbon trading mechanism is formulated;
3. Through the analysis of three scenarios, the effectiveness of the model proposed in this paper is verified in the IESs with different renewable energy penetration rates, and the mechanism of the demand response is elaborated. Furthermore, the proposed method is proven to have superiority over the other two related existing methods in carbon reduction.

2. Load-Side Carbon Responsibility Allocation Method Based on Carbon Emission Flow Theory and Shapley Value Method

The structure of an electric-gas IES is shown in Figure 1. The solid blue lines represent the power flow from the power source to the load through the power grid. The solid green lines indicate that natural gas flows from the gas source to the gas load through the natural gas network. Electricity and natural gas loads form energy hubs [27]. The blue and green dotted lines in Figure 1 represent the carbon emission flows of electricity and natural gas, respectively. To clarify the carbon emission responsibility that each EH should undertake in the process of using power and natural gas, calculations and analyses can be carried out based on the CEF theory [26]. Subsequently, based on the Shapley value method, the carbon emission responsibility for each EH can be divided into several grades. As a result, a ladder-type carbon-trading mechanism can be formed.

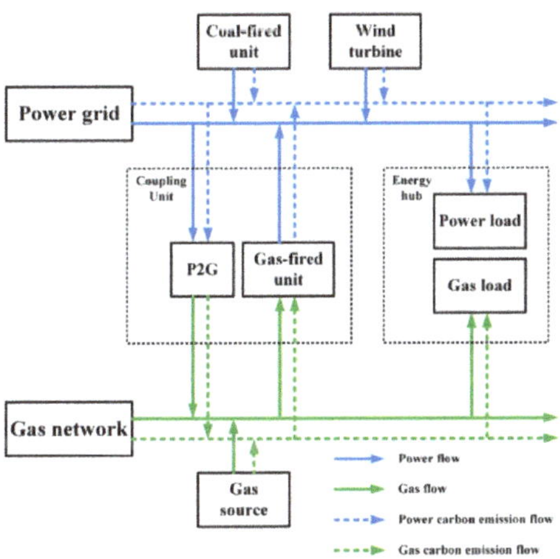

Figure 1. The structure of an electric-gas IES.

2.1. CEF in Power System Considering Grid Losses

The power system grid loss rate can reach 7–9% [28], and therefore, the carbon emissions caused by grid losses cannot be ignored. Therefore, the carbon emission responsibility caused by grid losses must be attributed to the load side by the CEF considering the grid losses. According to the method introduced in [29], the power flux of bus i is defined as

$$P_{Bi} = \sum_{j \in i^+} P_{ji} + P_{Gi} \tag{1}$$

where P_{ji} is the active power at the end of branch $j - i$, and the positive direction is from j to i; i^+ represents the set of start buses of the branches where the power flows injected into bus i are located; P_{Gi} is the generator output at bus i. Equation (2) can be derived from Equation (1):

$$\boldsymbol{P}_B = (\boldsymbol{E}^{gross} - \boldsymbol{A}^{gross})^{-1} \boldsymbol{P}_G \tag{2}$$

where P_B is an n-dimensional column vector representing the power fluxes of buses for a power grid with n buses; E^{gross} is an n-order identity matrix; A^{gross} is an $n \times n$ coefficient matrix, each element of which is defined as

$$A_{ij}^{gross} = \begin{cases} P_{ji}^{gross} / P_{Bj} \; j \in i^+ \\ 0 \; else \end{cases} \tag{3}$$

where P_{ij}^{gross} is the active power at the start of branch $j - i$, which is defined as $P_{ij}^{gross} = P_{ji} + P_{ij}^{loss}$; P_{ij}^{loss} is the grid loss on branch $j - i$. Correspondingly, it can be obtained as:

$$P_L^{gross} = B^{gross} P_B \tag{4}$$

where P_L^{gross} is an n-dimensional column vector, which denotes the equivalent electrical load value after allocating the grid losses to each load, B^{gross} is an n-order diagonal coefficient matrix, each element of which is defined as

$$B_{ij}^{gross} = \begin{cases} P_{Li} / P_{Bj} \; i = j \\ 0 \; i \neq j \end{cases} \tag{5}$$

From Equations (2) and (5), it can be obtained:

$$P_L^{gross} = B^{gross} (E^{gross} - A^{gross})^{-1} P_G = T^{gross} P_G \tag{6}$$

where T^{gross} is the distribution matrix from the source to the load, which can be calculated from the direct current (DC) optimal power flow results considering the grid losses. The element T_{ij}^{gross} represents the percentage of generator output at bus j supplied to the load at bus i, and therefore the sum of elements in each column of T^{gross} is 1.

Because the carbon emission flow is a virtual flow attached to the active power flow, analogous to Equation (2), it can be obtained as:

$$R_{ele,B} = (E^{gross} - A^{gross})^{-1} R_G \tag{7}$$

where R_G is an n-dimensional column vector representing the carbon flow rate of the generators in tCO_2/h. The calculation method for the elements is

$$R_{Gi} = e_{Gi} P_{Gi} \tag{8}$$

where e_{Gi} is the carbon emission intensity of the generator at bus i, in tCO_2/MWh.

Analogous to Equation (6), this is:

$$R_{ele,L}^{gross} = T^{gross} R_G \tag{9}$$

where $R_{ele,L}^{gross}$ is an n-dimensional column vector that denotes the load-side carbon emission responsibility after the source-side carbon emissions are attributed to the load side considering grid losses. On this basis, the bus carbon intensity e_{ele}^{gross} can be calculated, the unit of which is tCO_2/MWh. The calculation method for the elements is as follows:

$$e_{ele,i}^{gross} = R_{D,ele,i}^{gross} / P_{Bi} \tag{10}$$

2.2. CEF in a Natural Gas System

The CEF of an isolated lossless gas system is completely determined by the mass flow. Assuming that the carbon emission intensities of all gas sources are the same, the carbon emission responsibility of the load side can only be calculated directly according to the load value. However, as the carbon intensity of the power-to-gas (P2G) node changes, the carbon intensity of each gas node is no longer constant. Therefore, it is necessary to adopt the CEF theory to calculate the carbon emission responsibility on the gas load side.

In this study, the steady-state modeling of a natural gas system neglecting pipe storage and pipeline losses was adopted; thus, the CEF theory without considering the network losses can be applied. Because the CEF without grid losses is a special case of the CEF in Section 2.1, the following parts can be obtained.

The mass-flow flux F_{Bi} of node i is:

$$F_{Bi} = \sum_{j \in i^+} F_{ji} + F_{Si} \tag{11}$$

where F_{ji} is the gas mass flow rate of pipeline $j-i$, and the positive direction is from j to i; i^+ represents the set of start nodes of the pipelines where the gas flows injected into node i is located; F_{si} is the mass flow rate of the gas source at node i. By analogy, it can be obtained as:

$$\boldsymbol{F_B} = (\boldsymbol{E} - \boldsymbol{A})^{-1}\boldsymbol{F_s} \tag{12}$$

where $\boldsymbol{F_B}$ is an m-dimensional column vector representing the node mass flow flux for a gas network with m nodes; $\boldsymbol{E}$ is an m-order identity matrix; $\boldsymbol{A}$ is an $m \times m$ coefficient matrix, each element of which is defined as

$$\boldsymbol{A}_{ij} = \begin{cases} F_{ji}/F_{Bj} \; j \in i^+ \\ 0 \; else \end{cases} \tag{13}$$

where F_{ji} is the gas mass flow rate of pipeline $j-i$; correspondingly, which can be obtained as follows:

$$\boldsymbol{F_L} = \boldsymbol{B}(\boldsymbol{E} - \boldsymbol{A})^{-1}\boldsymbol{F_s} = \boldsymbol{TF_s} \tag{14}$$

where $\boldsymbol{F_L}$ is an m-dimensional column vector representing the gas load and $\boldsymbol{B}$ is an m-order diagonal coefficient matrix. Each element is defined as follows:

$$\boldsymbol{B}_{ij} = \begin{cases} F_{Di}/F_{Bj} \; i = j \\ 0 \; i \neq j \end{cases} \tag{15}$$

where $\boldsymbol{T}$ is the distribution matrix of the gas network from the source to load. For the carbon emission flow rate in the gas network,

$$\boldsymbol{R}_{gas,B} = (\boldsymbol{E} - \boldsymbol{A})^{-1}\boldsymbol{R}_s \tag{16}$$

where $\boldsymbol{R}_s$ is an m-dimensional column vector in tCO_2/h, representing the carbon flow rate of the gas source. The calculation method of the elements is

$$R_{si} = e_{si}F_{si} \tag{17}$$

where e_{si} is the carbon emission intensity of the gas source at node i, in $tCO_2/MBtu$.

$$\boldsymbol{R}_{gas,L} = \boldsymbol{TR_S} \tag{18}$$

where $\boldsymbol{R}_{gas,L}$ is an m-dimensional column vector, which represents the carbon emission responsibility on the load side. Correspondingly, the node carbon intensity of the gas network $\boldsymbol{e}_{gas}$, in $tCO_2/MBtu$, can be calculated as follows:

$$e_{gas,i} = R_{gas,B,i}/F_{Bi} \tag{19}$$

2.3. Allocation of Carbon Emission Responsibility Based on the Shapley Value Method

The energy hubs in an IES form a natural cooperative game alliance. The total carbon emissions of the IES are a joint responsibility for all EHs. Therefore, the carbon emission responsibility allocation can be regarded as a classic cost allocation problem. Many methods can be used to solve the cost allocation problem. Among them, the Shapley value (SV) method and generalized nucleolus (GN) method are the most widely used because of their unique solutions and good properties. Compared to the GN method, the SV method is superior in terms of equivalence (the mutual influence between any two members is the same) [30]. Hence, this study adopted the SV method to allocate the carbon emission responsibility among the EHs.

The SV method was proposed by Lloyd Shapley in 1953 and emphasizes the marginal effect of each member for different alliances. According to the definition of the Shapley value, the carbon emission responsibility shared by each EH should be the weighted average of all of its marginal effects, which can be expressed as:

$$X_i = \sum_{S \subseteq N \setminus \{i\}} \frac{n_s!(n_N - n_s - 1)!}{n_N!} [C(S \cup \{i\}) - C(S)] \tag{20}$$

where n_N represents the number of members in the entire alliance N; S represents any sub-alliance without the member i; n_s represents the number of members in the sub-alliance S; $n_s!(n_N - n_s - 1)!/n_N!$ represents the probability of the occurrence of sub-alliance S; $C(S)$ represents the carbon emission responsibility of the sub-alliance S; $S \cup \{i\}$ represents a new alliance formed by incorporating the alliance member i into alliance S; $C(S \cup \{i\}) - C(S)$ represents the carbon emission responsibility marginal effects of member i on sub-alliance S.

For each member in an alliance with n_N members, it has $2^{n_N - 1}$ marginal effects. Hence, the minimum and maximum marginal effects of member i can be defined as $X_{i,min}$, $X_{i,max}$.

$$X_{i,min} = min\{C(S \cup \{i\}) - C(S)\} \tag{21}$$

$$X_{i,max} = max\{C(S \cup \{i\}) - C(S)\} \tag{22}$$

The Shapley value X_i is the weighted average of all marginal effects and therefore $X_{i,min} < X_i < X_{i,max}$. The Shapley value of the member i is defined as $X_{i,mid}$.

$$X_{i,mid} = X_i \tag{23}$$

3. A Two-Stage Low-Carbon Economic Dispatch Model Considering Demand Response

3.1. The Two-Stage Model Overview

The two-stage low-carbon economic dispatch model considering the DR was employed to illustrate the source–load interaction coordination. The detailed process of the two-stage model is presented in Figure 2.

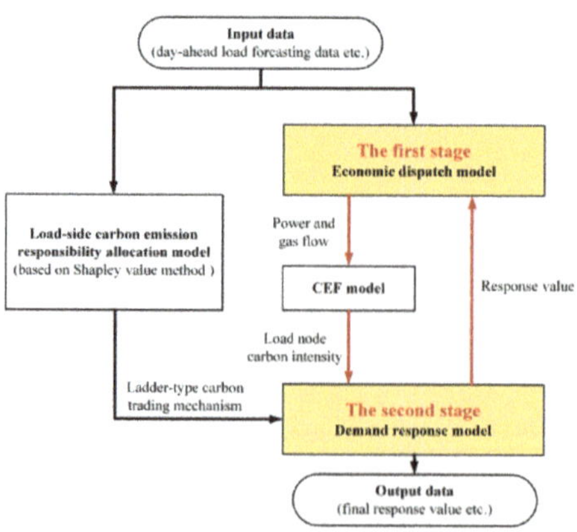

Figure 2. The framework of the two-stage mathematical model.

This can be divided into four parts:

1. Preparatory work: Based on the method proposed in Section 2, the carbon emission responsibility grades of each EH are calculated using the input data. Thus, a ladder-type carbon trading mechanism is formed to calculate the load-side carbon trading cost in the second stage;
2. The first stage: An economic dispatch optimization is performed with the objective function of minimizing the energy supply cost. The flow data of the IES are obtained by optimization. These are then passed to the CEF model to calculate the carbon intensity of each load node;
3. The second stage: Based on the first stage, this stage is optimized with the objective function of minimizing the carbon trading cost of the load side. The optimal demand response values are obtained after the optimization;
4. Loop solution and output: The response values optimized by the second stage cause load changes, and therefore, it is necessary to use the new load values to perform the first-stage economic dispatch calculation. Finally, a two-stage loop calculation is performed until the objective function values of the two stages tend to be stable.

The modeling of the IES and demand-side response in this study was based on the following hypothesis:

- Without considering the impact of the reactive power flow on the system carbon emissions, the DC power flow modeling was adopted for the power system;
- Without considering the pipeline inventory, the steady-state pipeline modeling was adopted for the gas network;
- The demand-side response was based on the premise that the total energy demand for a day remained unchanged;
- Without considering the power or gas load types, the load response ranges are set to constrain the load response capability.

3.2. *The First Stage: Economic Dispatch Model*

3.2.1. Objective Function

The objective function is established as flows:

$$min \sum_{t=0}^{T} \left(\sum_{CFU=1}^{N_{CFU}} c_{CFU} P_{CFU,t} + \sum_{wind=1}^{N_{wind}} c_{wind} P_{wind,t} + \sum_{gas=1}^{N_{gas}} c_{gas} F_{gas,t} \right) \tag{24}$$

where c_{CFU} represents the power generation cost coefficient of coal-fired units, which is determined by the fuel cost, power generation efficiency, and so on; c_{wind} is the power generation cost coefficient of the wind turbine, which is determined by the operation cost, maintenance cost, and so on; c_{gas} represents the cost coefficient of natural gas; $P_{CFU,t}$ and $P_{wind,t}$ represents the outputs of the coal-fired units and wind turbines at time t, respectively; $F_{gas,t}$ denotes the output mass flow rate of the natural gas source at time t.

3.2.2. Constraints

(1) Power system model

To consider the speed and accuracy of the calculation, this study adopted the DC optimal power flow model considering branch losses [31]. The system has the following constraints.

(a) Unit constraints:

$$P_{CFU,min} \leq P_{CFU,t} \leq P_{CFU,max} \tag{25}$$

$$P_{GFU,min} \leq P_{GFU,t} \leq P_{GFU,max} \tag{26}$$

$$P_{wind,min} \leq P_{wind,t} \leq P_{wind,max} \tag{27}$$

$$P_{GFU,i} = \eta_{GFU} F_{GFU,m} \tag{28}$$

$$Ramp_{CFU,min} \leq P_{CFU,t} - P_{CFU,t-1} \leq Ramp_{CFU,max} \tag{29}$$

$$Ramp_{GFU,min} \leq P_{GFU,t} - P_{GFU,t-1} \leq Ramp_{GFU,max} \tag{30}$$

where $P_{CFU,max}$ and $P_{CFU,min}$ represent the upper and lower output limits of the coal-fired units; $P_{GFU,max}$ and $P_{GFU,min}$ represent the upper and lower output limits of the gas-fired units, respectively; $P_{wind,max}$ and $P_{wind,min}$ represent the upper and lower output limits of the wind turbines; $P_{CFU,t}$, $P_{GFU,t}$, and $P_{wind,t}$ are the actual outputs of the coal-fired units, gas-fired units, and wind turbines at time t, respectively; η_{GFU} is the power generation efficiency of the gas-fired units; $F_{GFU,m}$ is the mass flow rate of natural gas consumed by the gas-fired unit, respectively; $Ramp_{CFU,max}$ and $Ramp_{CFU,min}$ represent the upper and lower output limits of the coal-fired units; $Ramp_{GFU,max}$ and $Ramp_{GFU,min}$ represent the upper and lower output limits of the gas-fired units, respectively.

(b) Branch constraints:

$$P_{ij,t} = \frac{\theta_{ij,t}}{x_{ij}} \tag{31}$$

$$P_{ij,t}^{loss} = g_{ij}\theta_{ij,t}^2 \tag{32}$$

$$P_{ij,min} \leq P_{ij} \leq P_{ij,max} \tag{33}$$

where $P_{ij,t}$ and $P_{ij,t}^{loss}$ represent the power flow and branch losses of branch $i-j$ at time t, respectively; $\theta_{ij,t}$ is the phase angle difference between the two ends of branch $i-j$ at time t; x_{ij} and g_{ij} are the reactance and conductance of branch $i-j$; $P_{ij,max}$ and $P_{ij,min}$ are the upper and lower power transmission limits of branch $i-j$.

(c) Bus constraints

$$P_{CFU,i} + P_{GFU,i} + P_{wind,i} = \sum_{j\in\Omega_i} P_{ij} + \sum_{j\in\Omega_i} \frac{1}{2}P_{loss,ij} + P_{L,i} + P_{P2G,i} \tag{34}$$

$$F_{P2G,m} = \eta_{P2G}P_{P2G,i} \tag{35}$$

$$\theta_{ij,min} \leq \theta_{ij,t} \leq \theta_{ij,max} \tag{36}$$

$$\theta_{ref,t} = 0 \tag{37}$$

where $P_{CFU,i}$, $P_{GFU,i}$ and $P_{wind,i}$ represent the power injected into bus i from the coal-fired units, gas-fired units, and wind turbines, respectively; $P_{L,i}$ represents the load on bus i; Ω_i represents the set of all buses around bus i; $P_{P2G,i}$ represents the power consumed by the P2G equipment on bus i; η_{P2G} is the energy conversion efficiency of P2G; $F_{P2G,m}$ represents the gas mass flow rate supplied by P2G to node m; $\theta_{ij,max}$ and $\theta_{ij,min}$ represent the upper and lower phase difference limits of branch $i-j$; $\theta_{ref,t}$ is the phase angle of the slack bus at time t.

(2) Natural gas system model

The natural gas system adopts the steady-state modeling based on the Weymouth function [32]. The constraints of the natural gas system are as follows.

(a) Natural gas source constraints:

$$F_{S,m}^{min} \leq F_{S,m} \leq F_{S,m}^{max} \tag{38}$$

where $F_{S,m}$ represents the mass flow rate of the gas source at node m; $F_{S,m}^{max}$ and $F_{S,m}^{min}$ represent the upper and lower mass flow rate limits of the gas source at node m.

(b) Pipeline constraints:

$$F_{mn}|F_{mn}| = k_{mn}\left(\pi_m^2 - \pi_n^2\right) \tag{39}$$

$$F_{mn}^{min} \leq F_{mn,t} \leq F_{mn}^{max} \tag{40}$$

where $F_{mn,t}$ represents the gas mass flow rate of pipeline $m-n$ at time t. k_{mn} is a constant that depends on the length, diameter, and absolute rugosity of the pipe and the gas composition [32]; π_m and π_n denote the gas pressure at nodes m and n, respectively; F_{mn}^{max} and F_{mn}^{min} are the upper and lower mass flow rate limits of pipeline $m-n$.

(c) Node constraints:

$$F_{S,m} + F_{P2G,m} = \sum_{n\in\Omega_m} F_{mn} + F_{L,m} + F_{GFU,m} \tag{41}$$

$$\pi_m^{min} \leq \pi_m \leq \pi_m^{max} \tag{42}$$

where $F_{L,m}$ represents the gas load at node m; π_m^{max} and π_m^{min} represent the upper and lower gas pressure limits of node m.

3.3. The Second Stage: Demand Response Model

In the second stage, the optimization focus is shifted from the energy supply side to the energy demand side. The demand response is the behavior of the energy demand side actively changing the demand under market incentives to coordinate with the energy supply. To promote low-carbon energy consumption, a demand-response low-carbon optimization model with a ladder-type carbon price is established in the second stage. The ladder-type carbon price was adopted as the incentive signal, and the minimum demand-side carbon trading cost was set as the goal to optimize the response value.

3.3.1. Objective Function

The objective function of the second stage is to minimize the load-side carbon trading cost.

$$min \sum_{t=0}^{T} \sum_{i=1}^{N_{EH}} C_{i,t}^{CT} \tag{43}$$

where $C_{i,t}^{CT}$ is the carbon trading cost of the EH_i at time t; N_{EH} represents the total number of energy hubs; T represents the time period of an optimization, which is 24 h in this paper.

3.3.2. Constraints

(1) Carbon trading cost constraints:

If EHs equally share the carbon emission responsibility of the entire IES, it is bound to be an unfair and unreasonable solution. To distribute the carbon emission responsibility on the load side fairly, it is necessary to determine the emission responsibility according to the load value. The carbon emission responsibility of each EH at time t should be in a reasonable range, neither greater than the maximum value of the member's marginal effect nor less than the minimum value of the marginal effect at time t (i.e., the interval $[X_{i,t,min}, X_{i,t,max}]$). Therefore, based on the Shapley value method, $X_{i,t,min}$, $X_{i,t,mid}$, and $X_{i,t,max}$ can be adopted as the carbon emission responsibility boundaries of each EH_i. According to Equations (20)–(23), $X_{i,t,min}$, $X_{i,t,mid}$, and $X_{i,t,max}$ can be calculated. For every hour, each EH has corresponding marginal effects, and therefore the $X_{i,t,min}$, $X_{i,t,mid}$, and $X_{i,t,max}$ of EH_i have corresponding 24 different values in a day.

However, in practical engineering applications, it is impractical and difficult for each EH to update the carbon emission responsibility boundaries hourly. Therefore, in this study, the average values of the carbon emission responsibility boundaries of each EH for 24 h were taken as the carbon emission responsibility boundaries for the whole day, as shown in Equations (44)–(46).

$$X_{i,minavg} = \sum_{t=1}^{T} X_{i,t,min} \tag{44}$$

$$X_{i,midavg} = \sum_{t=1}^{T} X_{i,t,mid} \quad (45)$$

$$X_{i,maxavg} = \sum_{t=1}^{T} X_{i,t,max} \quad (46)$$

where $X_{i,t,min}$, $X_{i,t,mid}$, and $X_{i,t,max}$ represent the minimum, medium, and maximum carbon emission responsibility marginal effects of EH_i at time t, respectively; $X_{i,minavg}$, $X_{i,midavg}$, $X_{i,maxavg}$ represent the 24-h average values of $X_{i,t,min}$, $X_{i,t,mid}$, $X_{i,t,max}$, respectively.

Based on the above, the ladder-type carbon trading cost is formed as Equation (47). Figure 3 shows the schematic diagram of the ladder-type carbon price model.

$$C_{i,t}^{CT} = \begin{cases} \lambda_1 (X_{i,minavg} - E_{i,t}) & 0 \le E_{i,t} < X_{i,minavg} \\ \lambda_2 (E_{i,t} - X_{i,minavg}) & X_{i,minavg} \le E_{i,t} < X_{i,midavg} \\ \lambda_2 \left(X_{i,midavg} - X_{i,minavg}\right) + \lambda_3 \left(E_{i,t} - X_{i,midavg}\right) & X_{i,midavg} \le E_{i,t} < X_{i,maxavg} \\ \lambda_2 \left(X_{i,midavg} - X_{i,minavg}\right) + \lambda_3 \left(X_{i,maxavg} - X_{i,midavg}\right) + \lambda_4 (E_{i,t} - X_{i,maxavg}) & E_{i,t} \ge X_{i,maxavg} \end{cases} \quad (47)$$

where $E_{i,t}$ is the carbon emission responsibility of the EH_i at time t; $\lambda_1 - \lambda_4$ are the carbon prices of the four grades.

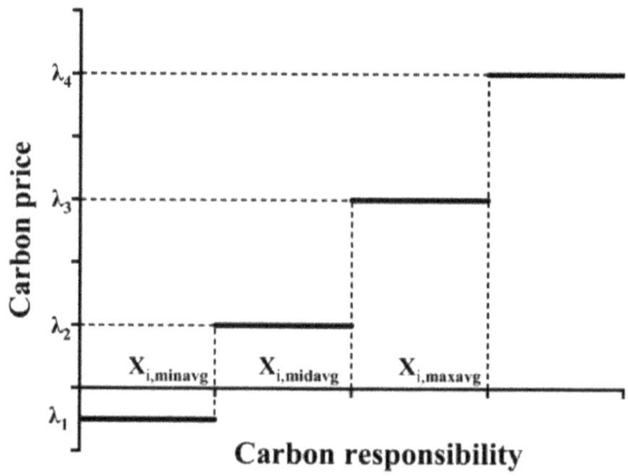

Figure 3. The schematic diagram of the ladder-type carbon price.

(2) Carbon emission constraints:

Based on the CEF theory, the actual carbon emissions of each EH can be calculated according to the actual energy consumption on the demand side. The carbon emission responsibility of EH_i at time t consists of the carbon responsibility of the original energy demand and the carbon responsibility of the response value.

$$E_{i,t} = R_{ele,L,i,t}^{gross} + e_{ele,i,t}^{gross} D_{i,t}^{ele} + R_{gas,L,i,t} + e_{gas,i,t} D_{i,t}^{gas} \quad (48)$$

where $R_{ele,L,i,t}^{gross}$ and $R_{gas,L,i,t}$ denote the EH_i carbon emission responsibility at time t; $e_{ele,i,t}^{gross}$ and $e_{gas,i,t}$ denote the carbon intensity of the electricity bus and gas node in EH_i at time t;

$D_{i,t}^{ele}$ and $D_{i,t}^{gas}$ represent the response values of the electric load and gas load of the EH_i at time t.

(3) Demand response constraints:

Demand response values are the state variables of the second-stage optimization model. In this paper, the demand response was modeled according to its actual characteristics. The ranges of the electric- and gas-demand response values are denoted as Equations (49) and (50). This paper assumes that the total load remained constant after the response, that is, the sum of all response values in time period T was zero, denoted as Equations (51) and (52). Equations (53) and (54) represent the response change constraints of the electric and gas loads between two adjacent moments, which characterize the flexibility of the demand response.

$$k_{ele}^{min} P_{L,i,t} \leq D_{i,t}^{ele} \leq k_{ele}^{max} P_{L,i,t} \tag{49}$$

$$k_{gas}^{min} F_{L,i,t} \leq D_{i,t}^{gas} \leq k_{gas}^{max} F_{L,i,t} \tag{50}$$

$$\sum_{t=1}^{T} D_{i,t}^{ele} = 0 \tag{51}$$

$$\sum_{t=1}^{T} D_{i,t}^{gas} = 0 \tag{52}$$

$$ramp_{ele,i}^{min} \leq D_{i,t}^{ele} - D_{i,t-1}^{ele} \leq ramp_{ele,i}^{max} \tag{53}$$

$$ramp_{gas,i}^{min} \leq D_{i,t}^{gas} - D_{i,t-1}^{gas} \leq ramp_{gas,i}^{max} \tag{54}$$

where $D_{i,t}^{ele}$ and $D_{i,t}^{gas}$ represent the response values of the power load and gas load of the EH_i at time t, respectively; $P_{L,i,t}$ and $F_{L,i,t}$ represent the real-time power load and gas load of the EH_i at time t, respectively; k_{ele}^{max} and k_{ele}^{min} are the ratios of the upper and lower limits of the power load response value; k_{gas}^{max} and k_{gas}^{min} are the ratios of the upper and lower limits of the gas load response value; $ramp_{ele,i}^{max}$ and $ramp_{ele,i}^{min}$ represent the upper and lower limits of the power load response change of EH_i; $ramp_{gas,i}^{max}$ and $ramp_{gas,i}^{min}$ represent the upper and lower limits of the gas load response change of EH_i.

4. Case Study and Discussion

A modified IEEE 39-bus power system/Belgian 20-node natural gas system was employed to demonstrate the effectiveness of the proposed model. All case studies were implemented using MATLAB/Gurobi on a PC with an Intel Core i7-11th processor and 16 GB of RAM. The economic dispatch period was 24 h, and the time step was 1 h.

4.1. A. Modified IEEE 39-Bus Power System/Belgian 20-Node Natural Gas System

4.1.1. Basic Data of the System

The modified electric-gas IES is shown in Figure 4. The power system includes four coal-fired units G1–G4, a wind turbine unit G5, and two gas-fired units G6–G7. The parameters of the thermal power units are listed in Table 1 [33,34]. Typical forecast data of wind turbine output were directly employed, as in other studies [27,35]. Wind turbine G5 had a cost coefficient of 15 $/MW and carbon emission intensity of 0.006 t CO_2/MW. The natural gas system contained five gas sources, whose parameters are listed in Table 2 [32]. The power grid and gas network were coupled by the P2G equipment with a capacity of 50 MW. Five power/gas loads were paired to form five energy hubs: $\boldsymbol{EH_A}$–$\boldsymbol{EH_E}$. The detailed data of the power load and gas load of each EH are listed in Table 3. The per-unit values of the 24 h maximum wind power outputs, power demands, and gas demands are shown in Figure 5. The base values of the wind power output, power load, and gas load were 658.8 MW, 3197.6 MW, and 2334.2 MBtu/h, respectively. The relevant data of the ladder-type carbon prices are shown in Table 4.

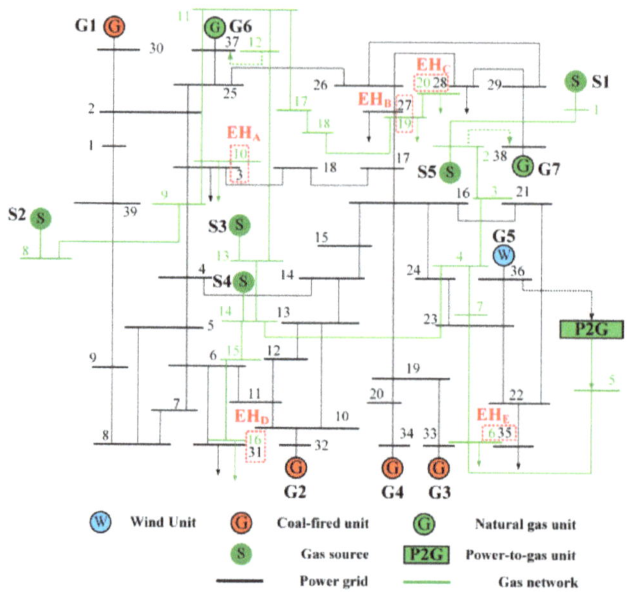

Figure 4. The modified IEEE 39-bus power system and the Belgian 20-node gas system.

Table 1. Parameters of the thermal power units.

Unit	Type	Capacity /(MW)	Cost Coefficient /($/MW)	Emission Intensity /(t CO_2/MW)
G1	Coal-fired	1040	35	1.280
G2	Coal-fired	725	35	1.300
G3	Coal-fired	652	35	1.290
G4	Coal-fired	508	35	1.270
G6	Gas-fired	564	\	0.564
G7	Gas-fired	865	\	0.550

Table 2. Parameters of the natural gas sources.

Capacity /(MBtu/h)	Cost Coefficient /($/MBtu)	Emission Intensity /(t CO_2/MBtu)	Capacity /(MBtu/h)	Cost Coefficient /($/MBtu)
S1	15,000	11.6	0.0566	1.300
S2	20,000	10.8	0.0566	1.290
S3	1000	12.0	0.0566	1.270
S4	250	10.2	0.0566	0.564
S5	250	10.0	0.0566	0.550

Table 3. Data of the energy hubs.

EH	Power Load/(MW)	Gas Load/(MBtu/h)
EH_A	913.6	526.2
EH_B	685.2	25.0
EH_C	685.2	158.6
EH_D	456.8	1290.9
EH_E	456.8	333.5

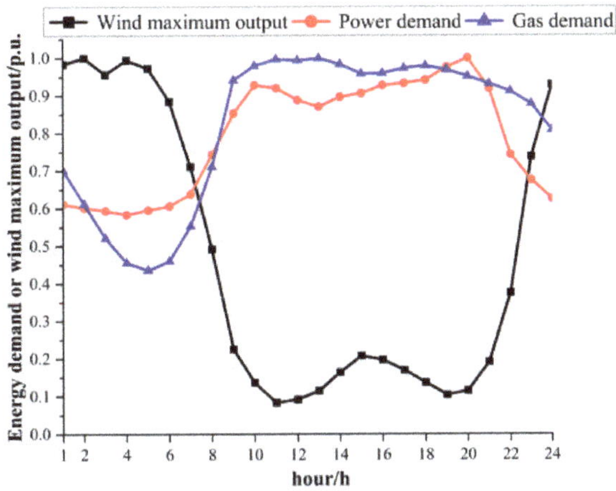

Figure 5. The modified IEEE 39-bus power system and the Belgian 20-node gas system.

Table 4. Parameters of the ladder-type carbon price.

Piecewise Interval	Carbon Price/($/t CO_2)
$0 \sim X_{i,minavg}$	$\lambda_1 = -5$
$X_{i,minavg} \sim X_{i,midavg}$	$\lambda_2 = 15$
$X_{i,midavg} \sim X_{i,maxavg}$	$\lambda_3 = 30$
$X_{i,maxavg} \sim \infty$	$\lambda_4 = 60$
$0 \sim X_{i,minavg}$	$\lambda_1 = -5$

4.1.2. Formation of the Ladder-Type Carbon Trading Mechanism for the Five EHs

It can be observed from Figure 4 that there are five EHs, denoted as A–E for convenience. Thus, the entire alliance is $N = \{A, B, C, D, E\}$, and there are 31 non-empty sub-alliances of N. Taking $t = 1$ as an example, the economic dispatch model of the IES is solved under the conditions of different sub-alliances, and the results for the system carbon emission responsibilities are shown in Table 5.

Table 5. The carbon emission responsibility of each sub-alliance.

Sub-Alliance	Carbon Emission Responsibility /(t CO_2)	Sub-Alliance	Carbon Emission Responsibility /(t CO_2)
{A}	574.70	\	\
{B}	375.29	{A, B}	1075.64
{C}	382.63	{A, C}	1087.93
{D}	48.77	{A, D}	1124.59
{E}	10.40	{A, E}	1089.93
{B, C}	1067.71	{A, B, C}	1480.67
{B, D}	929.60	{A, B, D}	1505.31
{B, E}	883.47	{A, B, E}	1476.86
{C, D}	939.98	{A, C, D}	1526.90
{C, E}	897.98	{A, C, E}	1489.16
{D, E}	574.60	{A, D, E}	1139.50
{B, C, D}	1120.59	{A, B, C, D}	1550.08
{B, C, E}	1082.62	{A, B, C, E}	1508.31
{B, D, E}	1120.91	{A, B, D, E}	1520.23
{C, D, E}	1130.49	{A, C, D, E}	1541.82
{B, C, D, E}	1520.13	{A, B, C, D, E}	2458.65

Taking EH_A as an example, it can be obtained from Equations (20)–(23):

$$X_{A,max} = max\{C(S \cup \{A\}) - C(S)\} = C(A, B, D, E) - C(B, D, E) = 399.32\ tCO_2 \quad (55)$$

$$X_{A,min} = min\{C(S \cup \{A\}) - C(S)\} = C(A, D) - C(D) = 1075.82\ tCO_2 \quad (56)$$

$$X_{A,mid} = \sum_{S} \frac{|S|!(5 - |S| - 1)!}{5!}[C(S \cap \{A\}) - C(S)] = 674.8\ tCO_2 \quad (57)$$

where S is any sub-alliance containing A, and $|S|$ is the number of members in the sub-alliance S.

In analogy with Equations (55)–(57), the carbon emission responsibility boundaries for the other 23 h can be obtained. Thus, the average value of the 24-h EH_A carbon responsibility boundaries is

$$X_{A,minavg} = 694.80\ tCO_2 \quad (58)$$

$$X_{A,midavg} = 904.32\ tCO_2 \quad (59)$$

$$X_{A,maxavg} = 1105.66\ tCO_2 \quad (60)$$

In summary, the ladder-type carbon trading cost of EH_A at time t is:

$$C_{A,t}^{CT} = \begin{cases} \lambda_1(694.80 - E_{A,t}) & 0 \leq E_{A,t} < 694.80 \\ \lambda_2(E_{A,t} - 694.80) & 694.80 \leq E_{A,t} < 904.32 \\ 209.52\lambda_2 + \lambda_3(E_{A,t} - 904.32) & 904.32 \leq E_{A,t} < 1105.66 \\ 209.52\lambda_2 + 201.34\lambda_3 + \lambda_4(E_{A,t} - 1105.66) & E_{A,t} \geq 1105.66 \end{cases} \quad (61)$$

On this basis, the 24-h carbon emission responsibility grades of the five EHs can be divided after the allocation of the carbon responsibility of the load side based on the Shapley value method. The results are shown in Figure 6. The carbon responsibility grades from low to high are represented in green, yellow, orange, and red, respectively. The three dashed horizontal lines represent the 24 h averages of the responsibility boundaries for different color grades. Based on these three averages, a one-day ladder-type carbon price can be formed for each EH, which will be used to calculate the carbon trading cost later.

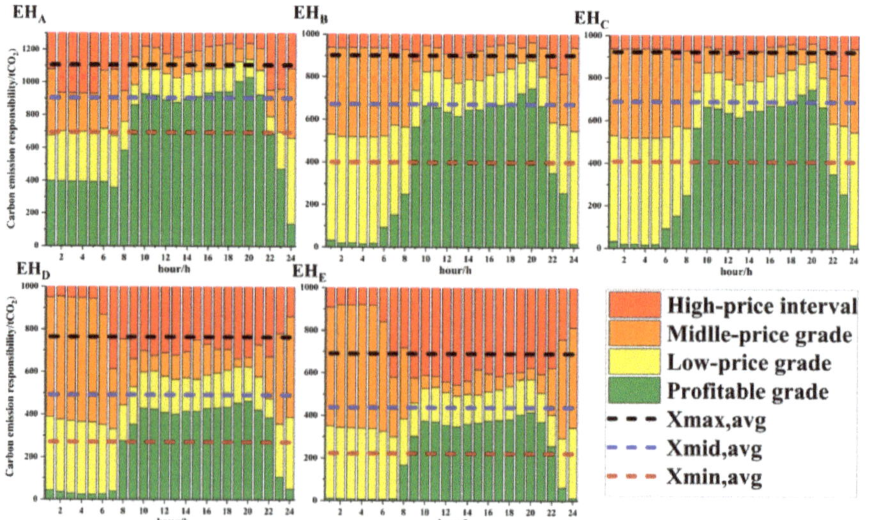

Figure 6. The carbon emission responsibility grades of each EH allocated by the Shapley value method in the scenario with wind power.

The calculation results of the carbon emission responsibility boundaries are closely related to the load value and location. In particular, the Shapley value X_{mid} (i.e., the upper boundary of the yellow grade) depends largely on the load value. Due to the low-carbon intensity of natural gas, the carbon emissions of each EH are mainly determined by the power load value. The power loads of EH_B and EH_C are equal, so it can be seen from Figure 6 that their carbon emission responsibility grades were basically similar, as were EH_D and EH_E. Due to the large power load of EH_A, its carbon emission responsibility boundaries were also obviously higher than those of the other four EHs. However, the carbon emission responsibility boundaries of EH_E were slightly lower than those of EH_D. This is because the location of EH_E was closer to the wind turbine, and a higher proportion of the power consumed came from wind power. From the above analysis, it can be seen that the Shapley value method can reasonably and effectively determine the carbon emission responsibility grades among the EHs.

4.2. Analysis of Scenarios with and without Wind Power

4.2.1. Scenario 1: Without Wind Power

In order to better study the impact of the two-stage model on the IES, the wind turbine G5 on bus 36 in the power system was replaced by a 300 MW coal-fired unit with a carbon emission intensity of 1.28 t CO_2/MWh as the blank control group. Based on the Shapley value method, the 24 h carbon emission responsibility boundaries of the five EHs in the IES were calculated again, and the 24-h average values were used to divide the ladder-type carbon price grades, as shown in Figure 7. Comparing Figures 6 and 7, it can be seen that the 24 h trend of the Shapley value X_{mid} (i.e., the upper boundary of the yellow grade) for the same EH in the two different scenarios was basically consistent and similar to the load curve, since the load values did not change. However, in Scenario 1, the profitable grades were significantly higher than those in the scenario with wind power, especially at night. This is because the carbon intensity of wind power was much lower than those of thermal power and natural gas. In Scenario 1, the EH consumes energy with a high carbon intensity, causing it to emit more carbon with the same load. Therefore, in the calculation of the Shapley value method, it is reasonable that the minimum carbon emission responsibility boundary (i.e., the upper boundary of the green grade) can be adjusted according to the actual carbon emissions of the EH.

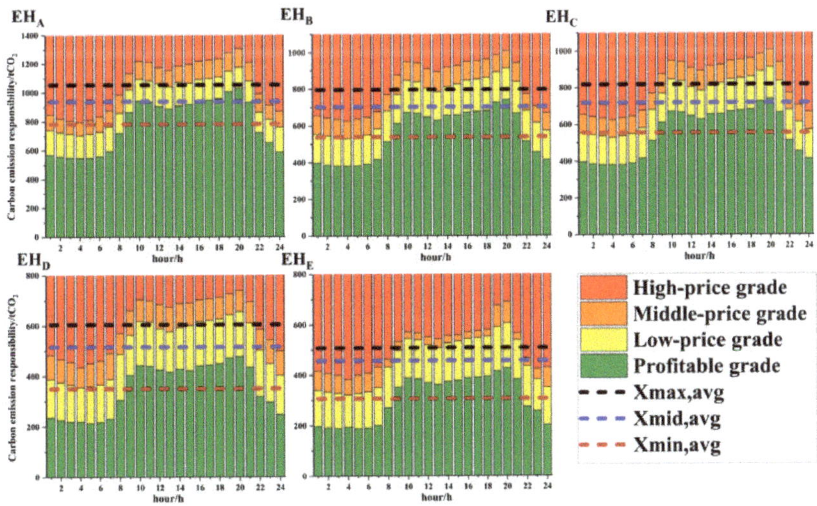

Figure 7. The carbon emission responsibility grades of each EH allocated by the Shapley value in Scenario 1.

After applying the two-stage model considering the DR, the total power/gas demand, energy supply cost, carbon trading cost, and the carbon emissions before and after the optimization are shown in Figures 8 and 9, respectively. As shown in Figure 8, in an IES without wind power, the model could effectively shave the peaks and fill valleys for the system load. Figure 9 shows that the energy supply cost and carbon emissions remained basically unchanged before and after the optimization because there was no low-carbon and low-cost energy injection. However, the carbon trading cost was visibly reduced by approximately 28.9%. This is because, after peak shaving and valley filling under the condition of the constant total load, when optimizing the carbon trading cost, the daytime load could jump from the relatively high-price carbon responsibility grade toward the lower price grade, and the night load increased as much as possible within the original price grade to achieve the lowest total carbon trading cost. From the above results, it can be seen that the model proposed in this paper can effectively guide the load to shave peaks and fill valleys, thus reducing the carbon trading cost.

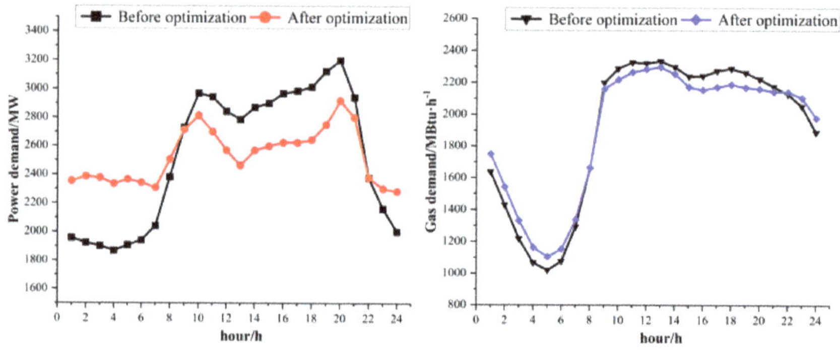

Figure 8. The total power/gas demand before and after the optimization in Scenario 1.

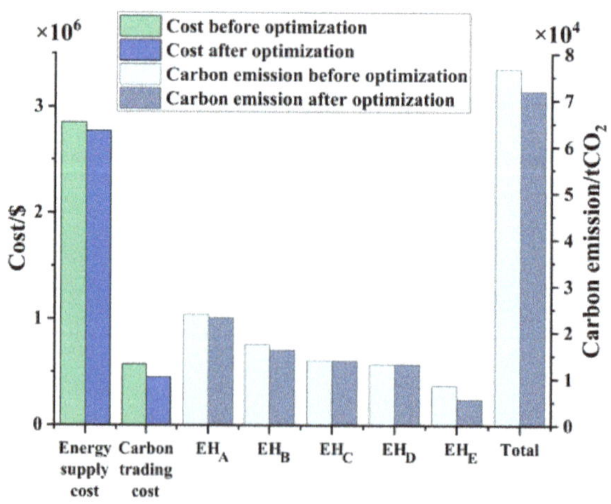

Figure 9. The energy supply cost, carbon trading cost, and carbon emissions of each EH before and after the optimization in Scenario 1.

4.2.2. Scenario 2: With Wind Power

In Scenario 2, the IES structure diagram and carbon emission responsibility grades are shown in Figures 4 and 6, respectively. Figure 10 shows that the two-stage model can still effectively promote load shaving and valley filling in Scenario 2. In Figure 11, the wind power consumption was greatly improved after the response at 1–8 h and 22–24 h.

Wind power has the characteristics of low carbon and low cost. Under the action of the two-stage model, the system load can respond in the direction of consuming as much wind power as possible. Before the optimization, except for 8–22 h, there was abandoned wind power for the rest of the time, and the wind power consumption rate throughout the day was only 43.2%. After the optimization, the night load actively participated in the response to consume excess wind power. Consequently, the wind-power consumption rate reached 93.0%.

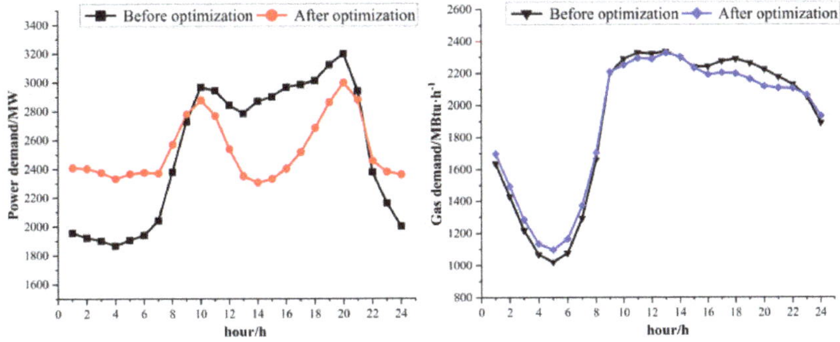

Figure 10. The total power/gas demand before and after the optimization in Scenario 2.

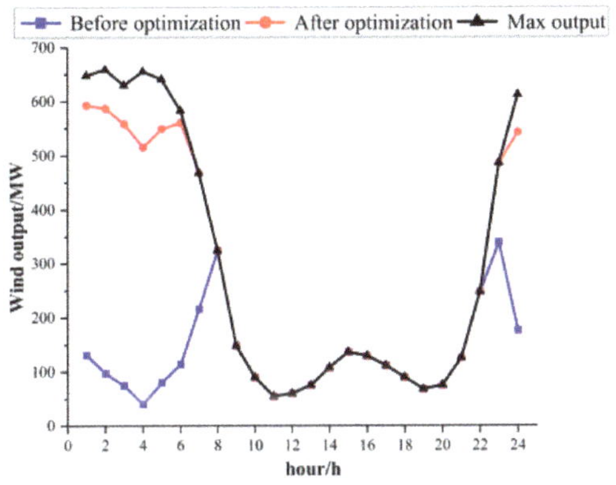

Figure 11. The total power/gas demand before and after the optimization in Scenario 2.

Figure 12 presents the energy supply cost, carbon trading cost, and carbon emissions before and after the optimization in Scenario 2. Compared with Figure 9, Figure 12 shows that the system energy supply cost and carbon emissions were significantly reduced after the optimization in the scenario including low-carbon and low-cost wind power. Specifically, the energy supply cost, carbon trading cost, and total carbon emissions were reduced by 2.9%, 21.7%, and 6.2%, respectively. From the above results of Scenario 2, it can be proven that the proposed model can not only effectively promote the load peak shaving and valley filling and reduce the load-side carbon trading cost, but also greatly improve the renewable energy consumption capacity of the IES, reducing the carbon emissions and energy supply cost.

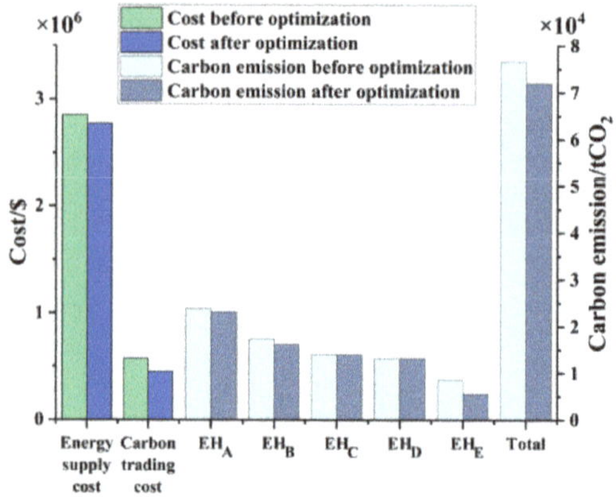

Figure 12. The energy supply cost, carbon trading cost, and carbon emissions of each EH before and after the optimization in Scenario 2.

4.2.3. Scenario 3: With Different Wind Penetration Rates

To investigate the effect of the two-stage model on the IES with different renewable energy penetration rates, the carbon emissions and percentages of carbon emission reduction after the optimization were studied with the wind power penetration rate (installed wind power capacity/system maximum system load) increasing from 20% to 80% in steps of 10%. As shown in Figure 13, an increase in the penetration rate led to the gradual reduction in the system carbon emissions, and the percentage of carbon reduction after the optimization remained basically stable, with a slight increase from 6.2% to 6.8%. Therefore, with the growth in renewable energy, the two-stage model considering the DR proposed in this paper can still effectively reduce carbon emissions.

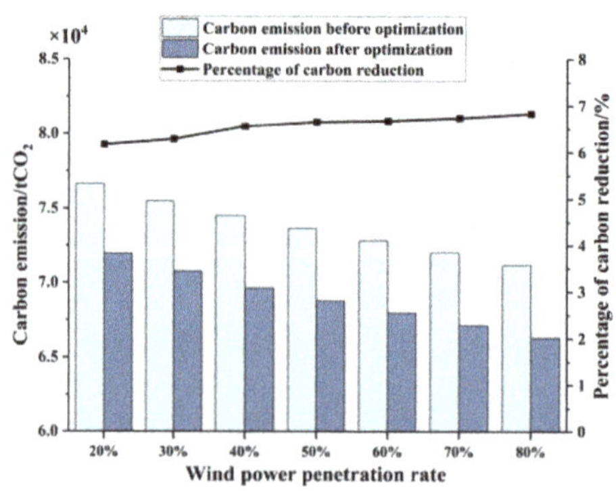

Figure 13. The carbon emissions and carbon reduction percentages before and after the optimization in different wind power penetration systems.

4.3. Discussion of the DR Mechanism

From the above scenario analysis considering the load-side carbon trading cost as the objective function, the system could achieve the load peak shaving and valley filling through the DR, thereby reducing the system carbon emission cost. To further explore the essential mechanism of the DR, the following discussion focused on two factors: carbon emissions and ladder-type carbon prices, which determine the objective function of the second stage.

4.3.1. Influence Mechanism of the Ladder-Type Carbon Prices on the DR

The influence mechanism of the ladder-type carbon price on the DR is shown in Figure 14. The direction of the DR is determined by the carbon price of the current load carbon emission responsibility grade. More specifically, the daytime load at a relatively high-price grade shifts the carbon emission responsibility toward a lower-price grade through a negative response. The night load at a low-price grade responds positively within the original carbon price grade as much as possible to maintain the total load conserved throughout the day. Thus, the carbon price gap can guide the DR to reduce the carbon trading cost on the load side.

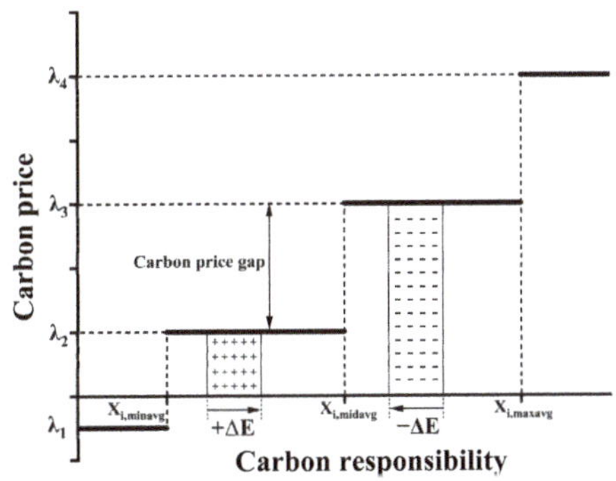

Figure 14. The influence mechanism of the ladder-type carbon price on the DR.

To demonstrate the promoting effect of the carbon price gap on the DR, two different carbon pricing cases were compared in Scenario 2.

- Case 1: Constant carbon pricing;
- Case 2: Ladder-type carbon pricing.

Case 2 adopted the ladder-type carbon price model proposed in this paper and its parameters are shown in Table 4. As the blank control group, Case 1 adopted a constant carbon price of 7.5 \$/t CO_2. Thus, Case 1 and Case 2 were equal in the carbon trading cost before the two-stage optimization. The results before and after the optimization of the two cases are shown in Figures 15 and 16. In Figure 15, the demand response after the optimization in Case 2 was more obvious than that in Case 1. In Figure 16, after the optimization, the carbon trading cost and carbon emissions in Case 2 were lower than those in Case 1. The carbon trading cost reduction percentages of Case 1 and Case 2 were 1.0% and 21.7%, respectively. The carbon emission reduction percentages of Case 1 and Case 2 were 1.9% and 6.2%, respectively. These results confirm that the carbon price gap can better guide the DR to reduce carbon emissions and the carbon trading cost.

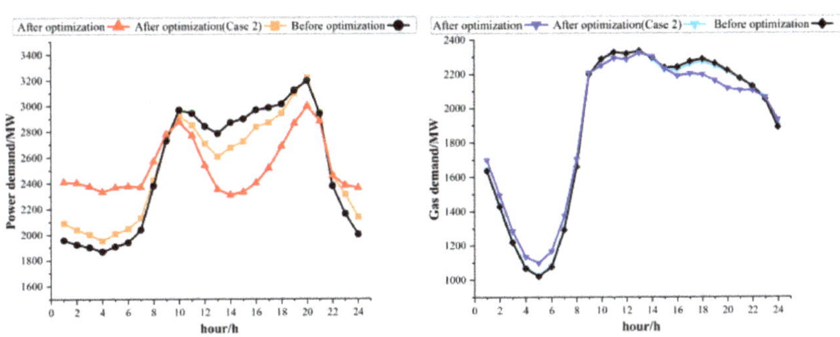

Figure 15. The power/gas demand before and after the optimization in Case 1 and Case 2.

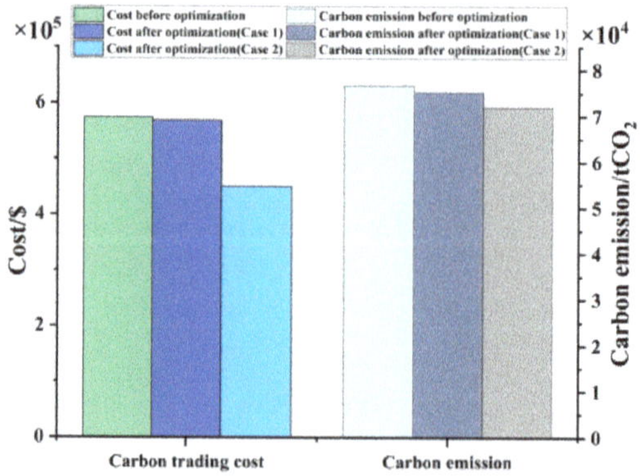

Figure 16. The carbon trading cost and carbon emissions before and after the optimization in Case 1 and Case 2.

4.3.2. Influence Mechanism of Carbon Emissions on the DR

The load carbon emissions depend on the load carbon intensity and load value. Therefore, the impact of carbon emissions on the DR can be further investigated by focusing on the carbon intensity. The following discussion takes the power demand response in Scenario 2 as an example. According to the principle of proportional sharing [29], the carbon intensity of each bus is determined by the power component injected into the bus, and its value is the weighted average of the carbon intensity of each power component. The power composition of the five power loads in Scenario 2 is shown in Figure 17. The corresponding results of the bus carbon intensities are indicated by the blue line in Figure 18.

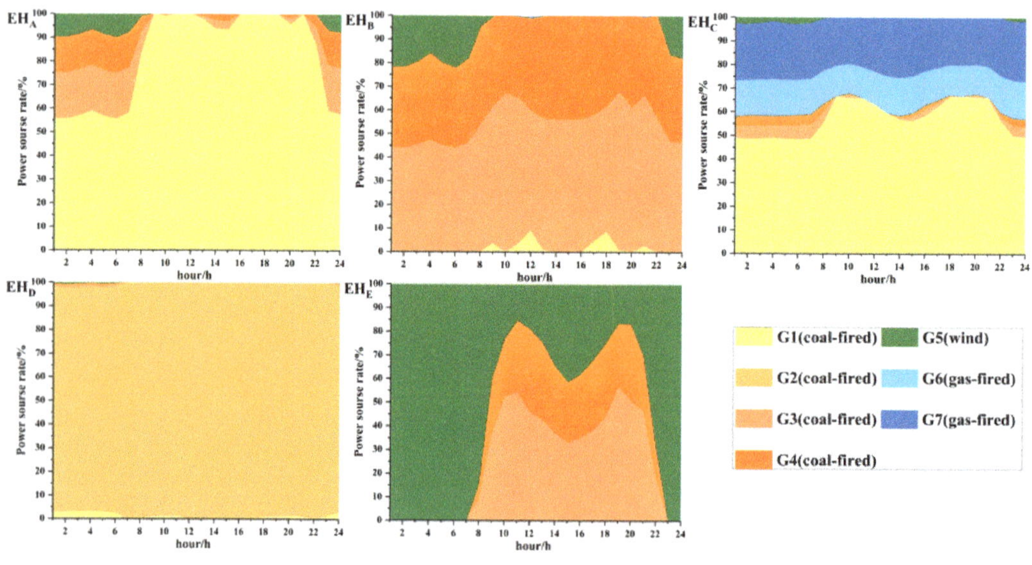

Figure 17. The power composition analysis diagram of each power load in Scenario 2.

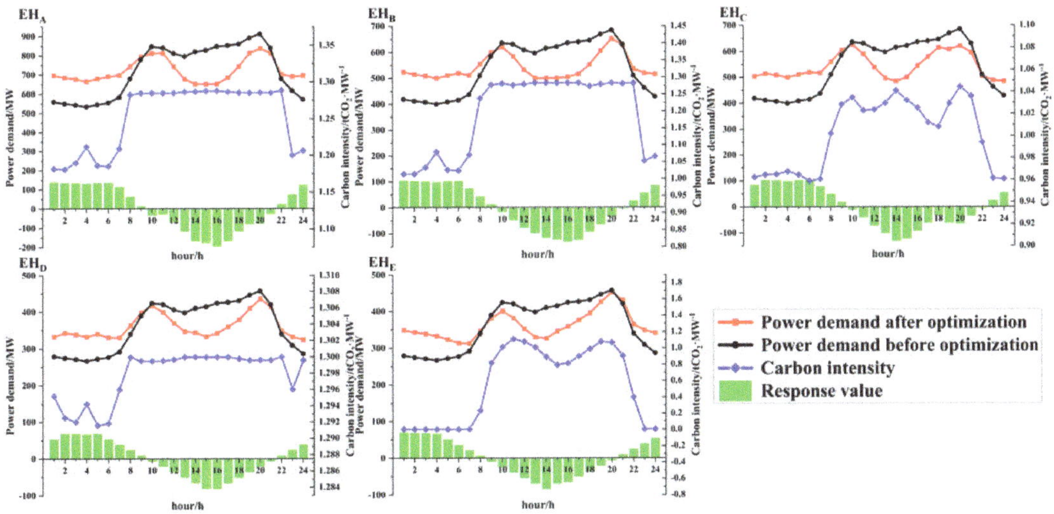

Figure 18. The response analysis diagram of each power demand in Scenario 2.

Figure 18 shows the relationship between the response value of each power load and bus carbon intensity. It can be seen from Figure 18 that in the period when the carbon intensity was relatively high, the load reduced the consumption of high-carbon power through a negative response. During the period when the carbon intensity was relatively low, the load increased the consumption of low-carbon power through a positive response to meet the conservation of total power demand. The response value is affected by the relative size of the carbon intensity, response range, response change ability, etc. Therefore, the carbon trading cost on the load side can be reduced by directly reducing the carbon emissions from energy consumption. Therefore, the carbon intensity difference caused by the renewable energy connected to the system can promote the DR to reduce the carbon trading cost on the load side.

4.4. Discussion of Three Carbon Reduction Methods

To study whether the proposed method had a superior carbon reduction in the IES, three carbon reduction methods in the existing related research and this paper were compared As shown in Table 6, Method 1, referenced from [9], is the low-carbon economic dispatch considering the source-side carbon trading. Method 2, referenced from [19], is the economic dispatch considering the DR driven by the time-of-use tariff. Method 3, which was proposed in this paper, is the economic dispatch considering the DR driven by the load-side carbon trading.

Table 6. Details of the three carbon reduction methods in the existing related research and this paper.

Method	Carbon Trading		DR	Reference
	Source-Side	Load-Side		
Method 1	✔			[9]
Method 2			✔	[19]
Method 3		✔	✔	This paper

The three methods were tested in the modified IEEE 39-bus power system/Belgian 20-node natural gas system of this paper, and the total carbon emissions and wind power consumption rates of the system before and after adopting the three methods were obtained, as shown in Figure 19. The carbon reduction effect of the proposed Method 3 was 281.8%

and 203.7% of Method 1 and Method 2, respectively. The wind power consumption rate of the proposed Method 3 was 178.3% and 135.3% of Method 1 and Method 2, respectively. Therefore, the proposed method had significant superiority in promoting the wind power consumption and reducing the system carbon emissions compared with Methods 1 and 2.

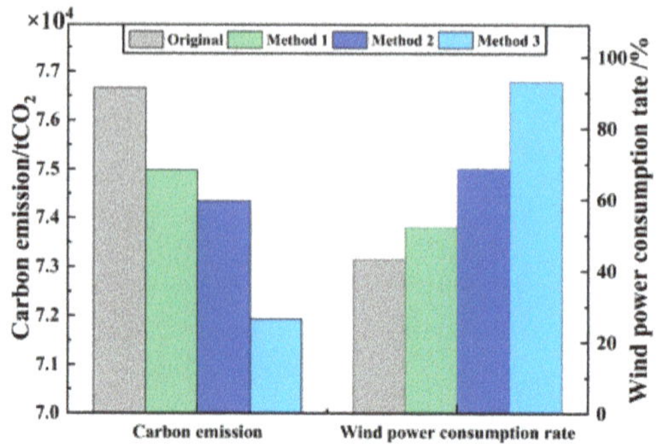

Figure 19. The system carbon emissions and wind power consumption rates with the different methods.

5. Conclusions

In this paper, a two-stage low-carbon economic dispatch model of an electric-gas integrated energy system considering the demand response was proposed. In the first stage, the economic dispatch of the integrated energy system was carried out with the objective of minimizing the energy supply cost, and the carbon emission responsibility of the load side was obtained based on the carbon emission flow theory. In the second stage, the low-carbon demand response optimization was carried out with the objective of minimizing the carbon trading cost on the load side. Additionally, a reward and punishment ladder-type carbon trading mechanism, which was used to calculate the carbon trading cost in the second stage, was formulated for each energy hub based on the Shapley value method. Cases based on a modified IEEE 39-bus power system/Belgian 20-node natural gas system were studied to demonstrate the effectiveness of the proposed model. By analyzing the all-thermal-power scenario, wind-included scenario, and scenario with varying wind power penetration rates, four conclusions can be drawn.

1. The two-stage model proposed in this paper could effectively reduce the peak-to-valley difference, enhance the ability of the system to consume wind power, and reduce the system carbon emissions by 6.2% and the carbon trading cost by 21.7%.
2. As the wind power penetration rate of the system increased from 20% to 80%, the carbon reduction effect of the model remained basically stable, with a slight increase from 6.2% to 6.8%. Therefore, with the growth in renewable energy, the two-stage model considering the demand response proposed in this paper can still effectively reduce carbon emissions.
3. Based on the Shapley value method, the user-side ladder-type carbon trading mechanism was thus formulated. The method proposed in this paper drives the demand response through user-side carbon trading. By comparing the proposed method with the other two related existing methods, the carbon emission reduction effect of the proposed method was 281.8% and 203.7% of the other two methods, respectively. Therefore, the proposed method was proven to have superiority over the other two methods in carbon reduction.

However, there were some limitations in this paper. For example, in the demand response, the loads can be further classified into important loads, shiftable loads, and adjustable loads. Furthermore, the impact of the load-side energy storage and distributed renewable energy was not considered. Based on this study, future research on the low-carbon demand response can be conducted by considering factors such as load-side energy storage, distributed renewable energy, and load types.

Author Contributions: Conceptualization, J.F. and J.N.; Methodology, J.F., J.N. and C.W.; Software, J.F. and C.W.; Validation, K.S., X.D. and H.Z.; Formal analysis, J.F. and J.N.; Investigation, X.D.; Resources, K.S. and X.D.; Data curation, J.F., J.N. and C.W.; Writing—original draft preparation, J.F. and J.N.; Writing—review and editing, J.F., J.N. and H.Z.; Visualization, J.F.; Supervision, H.Z.; Project administration, K.S., X.D. and H.Z.; Funding acquisition, K.S. All authors have read and agreed to the published version of the manuscript.

Funding: This research was funded by the Science and Technology Project of Zhejiang Huayun Electric Power Engineering Design Consulting Co., Ltd. under grant HYJL-2105035-01F.

Institutional Review Board Statement: Not applicable.

Informed Consent Statement: Not applicable.

Data Availability Statement: The data that support the findings of this study are available from the corresponding author upon reasonable request.

Conflicts of Interest: The authors declare that they have no known competing financial interests or personal relationships that could have appeared to influence the research reported in this paper.

References

1. Moore, F.C.; Lacasse, K.; Mach, K.J.; Shin, Y.A.; Gross, L.J.; Beckage, B. Determinants of emissions pathways in the coupled climate–social system. *Nature* **2022**, *603*, 103–111. [CrossRef] [PubMed]
2. Sognnaes, I.; Gambhir, A.; van de Ven, D.-J.; Nikas, A.; Anger-Kraavi, A.; Bui, H.; Campagnolo, L.; Delpiazzo, E.; Doukas, H.; Giarola, S.; et al. A multi-model analysis of long-term emissions and warming implications of current mitigation efforts. *Nat. Clim. Change* **2021**, *11*, 1055–1062. [CrossRef]
3. Huang, W.; Zhang, N.; Cheng, Y.; Yang, J.; Wang, Y.; Kang, C. Multienergy Networks Analytics: Standardized Modeling, Optimization, and Low Carbon Analysis. *Proc. IEEE* **2020**, *108*, 1411–1436. [CrossRef]
4. Wang, Y.; Wang, Y.; Huang, Y.; Yu, H.; Du, R.; Zhang, F.; Zhang, F.; Zhu, J. Optimal Scheduling of the Regional Integrated Energy System Considering Economy and Environment. *IEEE Trans. Sustain. Energy* **2019**, *10*, 1939–1949. [CrossRef]
5. Liu, N.; Wang, J.; Wang, L. Hybrid Energy Sharing for Multiple Microgrids in an Integrated Heat–Electricity Energy System. *IEEE Trans. Sustain. Energy* **2019**, *10*, 1139–1151. [CrossRef]
6. Zhu, X.; Yang, J.; Liu, Y.; Liu, C.; Miao, B.; Chen, L. Optimal Scheduling Method for a Regional Integrated Energy System Considering Joint Virtual Energy Storage. *IEEE Access* **2019**, *7*, 138260–138272. [CrossRef]
7. Zhang, Z.; Jing, R.; Lin, J.; Wang, X.; van Dam, K.H.; Wang, M.; Meng, C.; Xie, S.; Zhao, Y. Combining agent-based residential demand modeling with design optimization for integrated energy systems planning and operation. *Appl. Energy* **2020**, *263*, 114623. [CrossRef]
8. Fu, Y.; Sun, Q.; Wennersten, R. The effect of correlation of uncertainties on collaborative optimization of integrated energy system. *Energy Rep.* **2021**, *7*, 586–592. [CrossRef]
9. Wang, R.; Wen, X.; Wang, X.; Fu, Y.; Zhang, Y. Low carbon optimal operation of integrated energy system based on carbon capture technology, LCA carbon emissions and ladder-type carbon trading. *Appl. Energy* **2022**, *311*, 118664. [CrossRef]
10. Chen, L.; Msigwa, G.; Yang, M.; Osman, A.I.; Fawzy, S.; Rooney, D.W.; Yap, P.S. Strategies to achieve a carbon neutral society: A review. *Environ. Chem. Lett.* **2022**, 1–34. [CrossRef]
11. Huang, Y.; Ding, T.; Li, Y.; Li, L.; Chi, F.; Wang, K.; Wang, X.; Wang, X. Decarbonization Technologies and Inspirations for the Development of Novel Power Systems in the Context of Carbon Neutrality. *Proc. CSEE* **2021**, *41*, 28–51. [CrossRef]
12. Peñasco, C.; Anadón, L.D.; Verdolini, E. Systematic review of the outcomes and trade-offs of ten types of decarbonization policy instruments. *Nat. Clim. Change* **2021**, *11*, 257–265. [CrossRef]
13. Miao, B.; Lin, J.; Li, H.; Liu, C.; Li, B.; Zhu, X.; Yang, J. Day-Ahead Energy Trading Strategy of Regional Integrated Energy System Considering Energy Cascade Utilization. *IEEE Access* **2020**, *8*, 138021–138035. [CrossRef]
14. Peng, Q.; Wang, X.; Kuang, Y.; Wang, Y.; Zhao, H.; Wang, Z.; Lyu, J. Hybrid energy sharing mechanism for integrated energy systems based on the Stackelberg game. *CSEE J. Power Energy Syst.* **2021**, *7*, 911–921. [CrossRef]
15. Xu, D.; Zhou, B.; Wu, Q.; Chung, C.Y.; Li, C.; Huang, S.; Chen, S. Integrated Modelling and Enhanced Utilization of Power-to-Ammonia for High Renewable Penetrated Multi-Energy Systems. *IEEE Trans. Power Syst.* **2020**, *35*, 4769–4780. [CrossRef]
16. Gu, N.; Wang, H.; Zhang, J.; Wu, C. Bridging Chance-Constrained and Robust Optimization in an Emission-Aware Economic Dispatch With Energy Storage. *IEEE Trans. Power Syst.* **2022**, *37*, 1078–1090. [CrossRef]
17. Wang, Y.; Wang, Y.; Huang, Y.; Yang, J.; Ma, Y.; Yu, H.; Zeng, M.; Zhang, F.; Zhang, Y. Operation optimization of regional integrated energy system based on the modeling of electricity-thermal-natural gas network. *Appl. Energy* **2019**, *251*, 113410. [CrossRef]
18. Misconel, S.; Zöphel, C.; Möst, D. Assessing the value of demand response in a decarbonized energy system—A large-scale model application. *Appl. Energy* **2021**, *299*, 117326. [CrossRef]

19. Lv, G.; Cao, B.; Dexiang, J.; Wang, N.; Li, J.; Chen, G. Optimal scheduling of regional integrated energy system considering integrated demand response. *CSEE J. Power Energy Syst.* **2021**, 1–10. [CrossRef]
20. Yu, B.; Sun, F.; Chen, C.; Fu, G.; Hu, L. Power demand response in the context of smart home application. *Energy* **2022**, *240*, 122774. [CrossRef]
21. Guo, B.; Weeks, M. Dynamic tariffs, demand response, and regulation in retail electricity markets. *Energy Econ.* **2022**, *106*, 105774. [CrossRef]
22. Yuan, G.; Gao, Y.; Ye, B.; Liu, Z. A bilevel programming approach for real-time pricing strategy of smart grid considering multi-microgrids connection. *Int. J. Energy Res.* **2021**, *45*, 10572–10589. [CrossRef]
23. Yuan, G.; Gao, Y.; Ye, B. Optimal dispatching strategy and real-time pricing for multi-regional integrated energy systems based on demand response. *Renew. Energy* **2021**, *179*, 1424–1446. [CrossRef]
24. Zhou, T.; Kang, C.; Xu, Q.; Chen, Q. Preliminary Theoretical Investigation on Power System Carbon Emission Flow. *Autom. Electr. Power Syst.* **2012**, *36*, 38–43.
25. Kang, C.; Zhou, T.; Chen, Q.; Wang, J.; Sun, Y.; Xia, Q.; Yan, H. Carbon Emission Flow From Generation to Demand: A Network-Based Model. *IEEE Trans. Smart Grid* **2015**, *6*, 2386–2394. [CrossRef]
26. Cheng, Y.; Zhang, N.; Wang, Y.; Yang, J.; Kang, C.; Xia, Q. Modeling Carbon Emission Flow in Multiple Energy Systems. *IEEE Trans. Smart Grid* **2019**, *10*, 3562–3574. [CrossRef]
27. Cheng, Y.; Zhang, N.; Zhang, B.; Kang, C.; Xi, W.; Feng, M. Low-Carbon Operation of Multiple Energy Systems Based on Energy-Carbon Integrated Prices. *IEEE Trans. Smart Grid* **2020**, *11*, 1307–1318. [CrossRef]
28. Electric power transmission and distribution losses. *OECD/IEA* **2018**.
29. Bialek, J. Tracing the Flow of Electricity. *IEE Proc. Gener. Transm. Distrib.* **1996**, *143*, 313–320. [CrossRef]
30. Zhou, Q.; Feng, D.; Xu, C.; Feng, C.; Sun, T.; Ding, T. Methods for Allocating Carbon Obligation in Demand Side:a Comparative Study. *Autom. Electr. Power Syst.* **2015**, *7*, 153–159.
31. Yang, L.; Xu, Y.; Sun, H.; Zhao, X. Two-Stage Convexification-Based Optimal Electricity-Gas Flow. *IEEE Trans. Smart Grid* **2020**, *11*, 1465–1475. [CrossRef]
32. De Wolf, D.; Smeers, Y. The Gas Transmission Problem Solved by an Extension of the Simplex Algorithm. *Manag. Sci.* **2000**, *46*, 1454–1465. [CrossRef]
33. Zimmerman, R.D.; Murillo-Sánchez, C.E. *MATPOWER User's Manual*; Power Systems Engineering Research Center: Washington, DC, USA, 2011.
34. Wang, Y.; Zhou, S.; Wang, Y.; Qin, X.; Chen, F.; Ou, X. Comprehensive assessment of the environmental impact of China's nuclear and other power generation technologies. *J. Tsinghua Univ. (Sci. Technol.)* **2021**, *61*, 377–384. [CrossRef]
35. Cao, Y.; Wei, W.; Wu, L.; Mei, S.; Shahidehpour, M.; Li, Z. Decentralized Operation of Interdependent Power Distribution Network and District Heating Network: A Market-Driven Approach. *IEEE Trans. Smart Grid* **2019**, *10*, 5374–5385. [CrossRef]

mathematics

MDPI

Article

Solving Multi-Group Reflected Spherical Reactor System of Equations Using the Homotopy Perturbation Method

Mohammad Shqair [1,*], Emad A. M. Farrag [2] and Mohammed Al-Smadi [3,4]

[1] Physics Department, Faculty of Science and Humanities, Prince Sattam bin Abdulaziz University, Al-Kharj 11942, Saudi Arabia
[2] Physics Department, Faculty of Sciences and Arts, Jouf University, Tabarjal 75764, Saudi Arabia; eafarrag@ju.edu.sa
[3] College of Commerce and Business, Lusail University, Doha 122104, Qatar; mhm.smadi@bau.edu.jo
[4] Nonlinear Dynamics Research Center (NDRC), Ajman University, Ajman 20550, United Arab Emirates
* Correspondence: m.shqair@psau.edu.sa or shqeeeer@gmail.com; Tel.: +966-59-649-8633

Abstract: The solution of the complex neutron diffusion equations system of equations in a spherical nuclear reactor is presented using the homotopy perturbation method (HPM); the HPM is a remarkable approximation method that successfully solves different systems of diffusion equations, and in this work, the system is solved for the first time using the approximation method. The considered system of neutron diffusion equations consists of two consistent subsystems, where the first studies the reactor and the multi-group subsystem of equations in the reactor core, and the other studies the multi-group subsystem of equations in the reactor reflector; each subsystem can deal with any finite number of neutron energy groups. The system is simplified numerically to a one-group bare and reflected reactor, which is compared with the modified differential transform method; a two-group bare reactor, which is compared with the residual power series method; a two-group reflected reactor, which is compared with the classical method; and a four-group bare reactor compared with the residual power series.

Keywords: neutron diffusion; homotopy perturbation method; flux calculation; critical system; reflected reactor; multi-group

MSC: 82D75

Citation: Shqair, M.; Farrag, E.A.M.; Al-Smadi, M. Solving Multi-Group Reflected Spherical Reactor System of Equations Using the Homotopy Perturbation Method. *Mathematics* **2022**, *10*, 1784. https://doi.org/10.3390/math10101784

Academic Editors: Zbigniew Leonowicz, Arsalan Najafi and Michał Jasinski

Received: 6 March 2022
Accepted: 20 May 2022
Published: 23 May 2022

1. Introduction

The approximation methods have been used to solve system of differential equations; these methods can deal with complicated systems that never have been solved in classical methods, and HPM is one of the most important approximation methods.

The proposed system can be considered as two subsystems of neutron diffusion equations; each subsystem will be solved separately, where the core reflector boundary conditions are valid. This system represents two parts of a nuclear spherical reactor: the first part consists of the nuclear fuel called the core part, and the other has a material that reflects the neutrons to the core part, which improves the efficiency of the fission inside the reactor.

The general system of this work will be simplified as special cases to compare with other works [1,2] that have studied these cases using approximation and classical methods [3].

To solve the neutron diffusion equation, HPM gives the achieved result, and it is chosen because it has succeeded in solving simpler cases of neutron diffusion equations [4–7], and it is hoped to solve this general system. Solving nuclear reactor equations using other methods is also presented [1,2], and many works that solve different cases of the neutron diffusion equation and some other works [8–14] where another related approximation method studying diffusion equation is accomplished [15] are cited.

HPM was created firstly by He JH in 1999 [16] and performed a success in solving different fields [17–19], and soon, many works were accomplished using it [6,7]; the creator of HPM still depends on it in dealing with new physical problems [20,21].

The methodology of HPM is constructed on the combination of the topology concept (homotopy) and perturbation theory; this method depends on continuously changing the original difficult problem to a simple one that can be easily solved as the embedding parameter changes from unity to zero.

The full description of HPM, which is described in detail previously [6,16], will be clarified as we study this system.

The theoretical study of this work will be studied in Section 2, while the special case studies and their numerical examples are given in Section 3.

2. Theory

The basic concept of the neutron diffusion equation comes from applying Fick's law to simplify the transport neutron equation; this simplification is fair where the behavior of the neutrons in the reactor is still reasonable. The simple neutron diffusion equation is used to express the neutrons moving in one velocity, and this is known as one-group case; for more reality, the neutrons will be divided in many velocities; this is known as the multi-group case. This study presents two multi-group subsystems where the reflector neutron diffusion equations subsystem is added to improve the work. This system consists of the fuel surrounded by a reflector, which saves the core and improves the fission inside it [22–25]. The mathematical manipulation for this system is studied next.

2.1. The Reactor Core Part

Here, the multi-group neutrons diffusion equations subsystem in the reactor core will be studied. Buckling in the core part is not unique, such as occurs in the bare reactor, there are two buckling named principal and alternate buckling, and each neutron flux is a linear combination of both principal and alternate buckling fluxes.

In HPM, fluxes depend on intersecting between constants connecting the fluxes, while the classical method depends on the buckling calculation.

Buckling is one of the most essential concepts in the nuclear reactor theory, as it represents the leaks of the neutrons in the reactor, which means it has an essential role in the stability of the nuclear reactor [24].

The multi-group neutron diffusion equations for principal buckling in the reactor core are:

$$\begin{aligned} &\nabla^2\varnothing_1(r) + N_{11}\ \varnothing_1(r) + N_{12}\varnothing_2(r) + N_{13}\varnothing_3(r) + \ldots + N_{1n}\varnothing_n(r) = 0, \\ &\nabla^2\varnothing_2(r) + N_{21}\varnothing_1(r) + N_{22}\varnothing_2(r) + N_{23}\varnothing_3(r) + \ldots + N_{2n}\varnothing_n(r) = 0, \\ &\nabla^2\varnothing_3(r) + N_{31}\varnothing_1(r) + N_{32}\varnothing_2(r) + N_{33}\varnothing_3(r) + \ldots + N_{3n}\varnothing_n(r) = 0, \\ &\vdots \\ &\nabla^2\varnothing_n(r) + N_{n1}\varnothing_1(r) + N_{n2}\varnothing_2(r) + N_{n3}\varnothing_3(r) + \ldots + N_{nn}\varnothing_n(r) = 0, \end{aligned} \tag{1}$$

where D_i is the ith group diffusion coefficient, and N_{ij} is a constant that connects between fluxes in different energy groups of neutrons [1], which is defined as follows:

$$\begin{cases} N_{ii} = \frac{\chi_i \nu_i \Sigma_{fi} - (\Sigma_{\gamma i} + \sum \Sigma_{sij})}{D_i}, \\ N_{ij} = \frac{\sum \Sigma_{sji} + \chi_i \nu_j \Sigma_{fj}}{D_i}, \\ D_i = \frac{1}{3(\Sigma_{fi} + \Sigma_{sii} + \sum \Sigma_{sij} + \Sigma_{\gamma i})}. \end{cases} \tag{2}$$

Constants in Equation (2) have been defined in terms of different macroscopic cross-sections; the number of neutrons produced per fission for each group $\nu_{i,}$, and the fraction of fission neutrons emitted with energies in the ith group is χ_i.

The time-independent diffusion system of the multi-group at the core of the spherical reactor, after substituting the Laplacian in the radial part dependent on spherical coordinates, can be written as:

$$\begin{aligned} r\varnothing_1''(r) + 2\varnothing_1(r) + r(N_{11}\ \varnothing_1(r) + N_{12}\varnothing_2(r) + N_{13}\varnothing_3(r) + \ldots + N_{1n}\varnothing_n(r)) = 0, \\ r\varnothing_2''(r) + 2\varnothing_2(r) + r(N_{21}\varnothing_1(r) + N_{22}\varnothing_2(r) + N_{23}\varnothing_3(r) + \ldots + N_{2n}\varnothing_n(r)) = 0, \\ r\varnothing_3''(r) + 2\varnothing_3(r) + r(N_{31}\varnothing_1(r) + N_{32}\varnothing_2(r) + N_{33}\varnothing_3(r) + \ldots + N_{3n}\varnothing_n(r)) = 0, \\ \vdots \\ r\varnothing_n''(r) + \varnothing_n(r) + r(N_{n1}\varnothing_1(r) + N_{n2}\varnothing_2(r) + N_{n3}\varnothing_3(r) + \ldots + N_{nn}\varnothing_n(r)) = 0. \end{aligned} \tag{3}$$

This system of equations describes the behavior of the neutrons in the nuclear reactor where each flux $\varnothing_i$ expresses the neutron flux with a specific energy. Each flux has a maximum value at the center of the reactor, while its derivative vanishes there, so the initial conditions can be written in the mathematical form as:

$$\varnothing_i(0) = I, \quad \varnothing_i(0) = 0, \quad i = 1, 2, \ldots, n. \tag{4}$$

In order to solve these equations using HPM, we construct the homotopy [6,16] as:

$$\begin{aligned} H(\phi_1(r),\ \phi_2(r), \phi_3(r) \ldots \phi_n(r), p) = \ (1-p)\ F_1(\phi_1(r),\ \phi_2(r), \phi_3(r) \ldots \phi_n(r)) + \\ pL_1(\phi_1(r),\ \phi_2(r), \phi_3(r) \ldots \phi_n(r)). \end{aligned} \tag{5}$$

where H is the homotopy, L_i is the original problem, and F_i is the simple problem.

Thus,

$$\begin{aligned} &H_1(\phi_1(r), \phi_2(r), \phi_3(r) \ldots \phi_n(r), p) = r^2\phi_1''(r) + 2r\phi_1(r) + \\ &pr^2(N_{11}\ \varnothing_1(r) + N_{12}\varnothing_2(r) + N_{13}\varnothing_3(r) + \ldots + N_{1n}\varnothing_n(r)) = 0, \\ &H_2(\phi_1(r), \phi_2(r), \phi_3(r) \ldots \phi_n(r), p) = r^2\phi_2''(r) + 2r\phi_2(r) + \\ &pr^2(N_{21}\ \varnothing_1(r) + N_{22}\varnothing_2(r) + N_{23}\varnothing_3(r) + \ldots + N_{2n}\varnothing_n(r)) = 0, \\ &H_3(\phi_1(r), \phi_2(r), \phi_3(r) \ldots \phi_n(r), p) = r^2\phi_3''(r) + 2r\phi_3(r) + \\ &pr^2(N_{31}\ \varnothing_1(r) + N_{32}\varnothing_2(r) + N_{33}\varnothing_3(r) + \ldots + N_{3n}\varnothing_n(r)) = 0, \\ &\ldots \\ &H_1(\phi_1(r), \phi_2(r), \phi_3(r) \ldots \phi_n(r), p) = r^2\phi_n''(r) + 2r\phi_n(r) + \\ &pr^2(N_{n1}\ \varnothing_1(r) + N_{n2}\varnothing_2(r) + N_{n3}\varnothing_3(r) + \ldots + N_{nn}\varnothing_n(r)) = 0. \end{aligned} \tag{6}$$

while the identical powers of p are:

$$\begin{aligned} &p^0: r^2\phi_{i0}''(r) + 2r\phi_{i0}'(r) = 0, \phi_{i0}\ (0)\text{:finite}, \\ &p^1: r^2\phi_{i1}'';1(r)\ + 2r\phi_{i1}'\ (r)\ =\ -r^2\ N_{11}\ \varnothing_{i0}(r) + N_{12}\varnothing_{i2}(r) + N_{i3}\varnothing_3(r) + \ldots + N_{1,n}\varnothing_n(r)),\ \phi_{i,1}\ (0)\ =\ 0, \\ &p^2: r^2\phi_{i2}''\ (r) + 2r\phi_{i2}'\ (r) = -r^2\ N_{11}\varnothing_{i1}(r) + N_{22}\varnothing_{i2}(r) + N_{i3}\varnothing_{i3}(r) + \ldots + N_{2n}\varnothing_n(r)),\ \phi_{i2}\ (0)\ =\ 0, \\ &\ldots \\ &p^k: r^2\phi_{ik}''\ (r)\ +\ 2r\phi_{ik}''\ (r)\ =\ -r^2\ N_{k1}\varnothing_1(r) + N_{k2}\varnothing_2(r) + \ldots + N_{kk}\varnothing_k(r) + \ldots + N_{kn}\varnothing_n(r)),\ \phi_{ik}\ (0)\ =\ 0 \end{aligned} \tag{7}$$

Then,

$$\begin{aligned} &T_{i,0} = I_i, \\ &T_{i,n} = N_{ii}\ T_{i,n-2} + N_{i1}T_{1,n-2} + N_{i2}T_{2,n-2} + + N_{i3}T_{3,n-2} + + \ldots + N_{in}T_{n,n-2}. \end{aligned} \tag{8}$$

The solution of the first component of Equation (7) is:

$$\phi_{i,0}\ (r) = I_i, \tag{9}$$

Similarly, we obtain:

$$\begin{aligned} \phi_{i,1}\ (r)\ &=\ -\tfrac{T_{i,2}}{3!}r^2\ , \\ \phi_{i,2}\ (r)\ &=\ \tfrac{T_{i,4}}{5!}r^4\ , \\ &\ldots \\ \phi_{i,k}\ (r) &= -(-1)^k\tfrac{T_{i,2k}}{(2k+1)!}r^{2k}, \end{aligned} \tag{10}$$

In summary, the fluxes, in this case, are given by:

$$\phi_i\,(r) = \sum_{k=0}^{\infty}(-1)^k \frac{T_{i.2k}}{(2k+1)!} r^{2k} \tag{11}$$

Now, multi-group neutron diffusion equations for alternate buckling in the reactor core part will be:

$$\begin{array}{l}
\nabla^2\varnothing_1(r) - L_{11}\,\varnothing_1(r) - L_{12}\varnothing_2(r) - L_{13}\varnothing_3(r) - \ldots - L_{1n}\varnothing_n(r) = 0, \\
\nabla^2\varnothing_2(r) - L_{21}\varnothing_1(r) - L_{22}\varnothing_2(r) - L_{23}\varnothing_3(r) - \ldots - L_{2n}\varnothing_n(r) = 0, \\
\nabla^2\varnothing_3(r) - L_{31}\varnothing_1(r) - L_{32}\varnothing_2(r) - L_{33}\varnothing_3(r) - \ldots - L_{3n}\varnothing_n(r) = 0, \\
\vdots \\
\nabla^2\varnothing_n(r) L_{n1}\varnothing_1(r) - L_{n2}\varnothing_2(r) - L_{n3}\varnothing_3(r) - \ldots - L_{nn}\varnothing_n(r) = 0,
\end{array} \tag{12}$$

where L_{ij} is a constant connects between fluxes in different energy groups of neutrons. Respectively, L_{ii}, L_{ij}, and D_i are defined as:

$$\begin{cases}
L_{ii} = \frac{\chi_i v_i \sum_{fi} - (\Sigma_{\gamma i} + \Sigma \Sigma_{sij})}{D_i}, \\
L_{ij} = \frac{\Sigma \Sigma_{sji} + \chi_i v_j \sum_{fj}}{D_i}, \\
D_i = \frac{1}{3(\Sigma_{fi} + \Sigma_{sii} + \Sigma \Sigma_{sij} + \Sigma_{\gamma i})}.
\end{cases} \tag{13}$$

Now, the diffusion system of multi energy groups of neutrons at spherical reactor can be written as:

$$\begin{array}{l}
r\varnothing_1''(r) + 2\varnothing_1'(r) + (-rL_{11}\,\varnothing_1(r) - rL_{12}\varnothing_2(r) - rL_{13}\varnothing_3(r) - \ldots - rL_{1n}\varnothing_n(r)) = 0, \\
r\varnothing_2''(r) + 2\varnothing_2'(r) + (-rL_{21}\varnothing_1(r) - rL_{22}\varnothing_2(r) - rL_{23}\varnothing_3(r) - \ldots - rL_{2n}\varnothing_n(r)) = 0, \\
r\varnothing_3''(r) + 2\varnothing_3'(r) + (-rL_{31}\varnothing_1(r) - rL_{32}\varnothing_2(r) - rL_{33}\varnothing_3(r) - \ldots - rL_{3n}\varnothing_n(r)) = 0, \\
\vdots \\
r\varnothing_n''(r) + \varnothing_n'(r) + (L_{n1}\varnothing_1(r) - rL_{n2}\varnothing_2(r) - rL_{n3}\varnothing_3(r) - \ldots - rL_{nn}\varnothing_n(r)) = 0.
\end{array} \tag{14}$$

One more time, the mathematical form of initial conditions can be written as:

$$\varnothing_i(0) = I, \quad \varnothing_i'(0) = 0, \quad i = 1, 2, \ldots, n. \tag{15}$$

In order to solve these equations using HPM, we construct the homotopy as:

$$\mathrm{p}L_1(\phi_1(r),\ \phi_2(r), \phi_3(r) \ldots \phi_n(r)). \tag{16}$$

To obtain a solution for this subsystem using HPM, the homotopy will be:
Thus,

$$\begin{array}{l}
H_1(\phi_1(r), \phi_2(r), \phi_3(r) \ldots \phi_n(r), p) = r^2\phi_1''(r) + 2r\phi_1(r) + \\
pr^2(-L_{11}\,\varnothing_1(r) - L_{12}\varnothing_2(r) - L_{13}\varnothing_3(r) - \ldots - L_{1n}\varnothing_n(r)) = 0, \\
H_2(\phi_1(r), \phi_2(r), \phi_3(r) \ldots \phi_n(r), p) = r^2\phi_2''(r) + 2r\phi_2(r) + \\
pr^2(-L_{21}\,\varnothing_1(r) - L_{22}\varnothing_2(r) - L_{23}\varnothing_3(r) - \ldots - L_{2n}\varnothing_n(r)) = 0, \\
H_3(\phi_1(r), \phi_2(r), \phi_3(r) \ldots \phi_n(r), p) = r^2\phi_3''(r) + 2r\phi_3(r) + \\
pr^2(-L_{31}\,\varnothing_1(r) - L_{32}\varnothing_2(r) - L_{33}\varnothing_3(r) - \ldots - L_{3n}\varnothing_n(r)) = 0, \\
\ldots \\
H_1(\phi_1(r), \phi_2(r), \phi_3(r) \ldots \phi_n(r), p) = r^2\phi_n''(r) + 2r\phi_n(r) + \\
pr^2(-L_{n1}\,\varnothing_1(r) - L_{n2}\varnothing_2(r) - L_{n3}\varnothing_3(r) - \ldots - L_{nn}\varnothing_n(r)) = 0.
\end{array} \tag{17}$$

Taking identical powers of p as:

$$
\begin{aligned}
&p^0: r^2\phi''_{i0}(r) + 2r\phi_{i0}(r),\ \phi_{i0}\ (0)\text{: finite,} \\
&p^1: r^2\phi''_{i1};1(r)\ +\ 2\mathrm{r}\phi_{i1}\ (\mathrm{r}) \\
&= r^2\ (L_{11}\ \varnothing_{i0}(r) + L_{12}\varnothing_{i2}(r) + L_{13}\varnothing_3(r) + \ldots + L_{1n}\varnothing_n(r)),\ \phi_{i1}\ (0)\ =\ 0, \\
&p^2: r^2\phi''_{i2}\ (r)\ +\ 2\mathrm{r}\phi_{i2}\ (\mathrm{r}) \\
&= r^2\ (L_{i0}\varnothing_{i1}(r) + L_{22}\varnothing_{i2}(r) + L_{i3}\varnothing_{i3}(r) + \ldots + L_{2n}\varnothing_n(r)),\ \phi_{i2}\ (0)\ =\ 0, \\
&\ldots \\
&p^k: r^2\phi''_{ik}\ (r)\ +\ 2\mathrm{r}\phi''_{ik}\ (\mathrm{r}) \\
&= r^2\ (L_{k1}\varnothing_1(r) + L_{k2}\varnothing_2(r) + \ldots + L_{kk}\ \varnothing_k(r) + \ldots + L_{kn}\varnothing_n(r)),\ \phi_{ik}\ (0)\ =\ 0,
\end{aligned}
\tag{18}
$$

Now,

$$
\begin{aligned}
&T_{i,0}= I_i, \\
&T_{i,n}= N_{ii}\ T_{i,n-2}+N_{i1}T_{1,n-2}+N_{i2}T_{2,n-2}+ + N_{i3}T_{3,n-2}+ + \ldots + N_{in}T_{n,n-2} + .,
\end{aligned}
\tag{19}
$$

The solution of first component of Equation (18) is given by:

$$\phi_{i,0}\ (r) = I_i, \tag{20}$$

The other components are:

$$
\begin{aligned}
&\phi_{i,1}\ (r) = \tfrac{T_{i,2}}{3!}r^2, \\
&\phi_{i,2}\ (r) = \tfrac{T_{i,4}}{5!}r^4, \\
&\ldots \\
&\phi_{i,k}\ (r) = \tfrac{T_{i,2k}}{(2k+1)!}r^{2k},
\end{aligned}
\tag{21}
$$

In summary, the fluxes of this part are:

$$\phi_i\ (r)= \sum\nolimits_{k=0}^{\infty} \frac{T_{i.2k}}{(2k+1)!}r^{2k} \tag{22}$$

As is mentioned, the total flux of any group in the core part is linear combination of the two cases as:

$$\phi_i\ (r) = U\sum\nolimits_{k=0}^{\infty} \frac{T_{i.2k}}{(2k+1)!}r^{2k} + \mathrm{V}\sum\nolimits_{k=0}^{\infty}(-1)^k \frac{T_{i.2k}}{(2k+1)!}r^{2k} \tag{23}$$

Clearly, all fluxes should satisfy the needed boundary conditions.

2.2. The Reactor Reflected Part

After the core part, we study the reactor reflector; the multi-group neutron diffusion equations of this subsystem [3] is:

$$
\begin{aligned}
&\nabla^2\varnothing_1(r) - M_{11}\ \varnothing_1(r) - M_{12}\varnothing_2(r) - M_{13}\varnothing_3(r) - \ldots - M_{1n}\varnothing_n(r) = 0, \\
&\nabla^2\varnothing_2(r) - M_{21}\varnothing_1(r) - M_{22}\varnothing_2(r) - M_{23}\varnothing_3(r) - \ldots - M_{2n}\varnothing_n(r) = 0, \\
&\nabla^2\varnothing_3(r) - M_{31}\varnothing_1(r) - M_{32}\varnothing_2(r) - M_{33}\varnothing_3(r) - \ldots - M_{3n}\varnothing_n(r) = 0, \\
&\vdots \\
&\nabla^2\varnothing_n(r)M_{n1}\varnothing_1(r) - M_{n2}\varnothing_2(r) - M_{n3}\varnothing_3(r) - \ldots - M_{nn}\varnothing_n(r) = 0,
\end{aligned}
\tag{24}
$$

Each $M_{i,j}$ is a real (positive, zero, or negative) constant:

$$
\begin{cases}
M_{i,i} = \dfrac{-(\Sigma_{\gamma i} + \Sigma\Sigma_{sij})}{D_i}, \\
M_{ij} = \dfrac{\Sigma\Sigma_{sji}}{D_i}, \\
D_i = \dfrac{1}{3(\Sigma_{sii} + \Sigma\Sigma_{sij} + \Sigma_{\gamma i})}.
\end{cases}
\tag{25}
$$

Applying Laplacian in spherical coordinates:

$$\begin{aligned}
r\varnothing_1''(r) + 2\varnothing_1(r) - r(M_{11}\varnothing_1(r) + M_{12}(r)\varnothing_2 + M_{13}(r)\varnothing_3 + \ldots + M_{1n}\varnothing_n(r)) = 0,\\
r\varnothing_2''(r) + 2\varnothing_2(r) - r(M_{21}\varnothing_1(r) + M_{22}\varnothing_2(r) + M_{23}\varnothing_3(r) + \ldots + M_{2n}\varnothing_n(r)) = 0,\\
r\varnothing_3''(r) + 2\varnothing_3(r) - r(M_{31}\varnothing_1(r) + M_{32}\varnothing_2(r) + M_{33}\varnothing_3(r) + \ldots + M_{3n}\varnothing_n(r)) = 0,\\
\vdots\\
r\varnothing_n''(r) + \varnothing_n(r) - r(M_{n1}\varnothing_1(r) + M_{n2}\varnothing_2(r) + M_{n3}\varnothing_3(r) + \ldots + M_{nn}\varnothing_n(r)) = 0.
\end{aligned} \tag{26}$$

Now, the homotopy [6,16] will be:

$$\begin{aligned}
&H_1(\phi_1(r),\ \phi_2(r), \phi_3(r)\ldots\phi_n(r), p) = r^2\phi_1''(r) + 2r\phi_1(r)\\
&+pr^2(-M_{11}\varnothing_1(r) - M_{12}\varnothing_2(r) - M_{13}\varnothing_3(r) - \ldots - M_{1n}\varnothing_n(r)) = 0,\\
&H_2(\phi_1(r),\ \phi_2(r), \phi_3(r)\ldots\phi_n(r), p) = r^2\phi_2''(r) + 2r\phi_1(r)\\
&+pr^2(-M_{21}\varnothing_1(r) - M_{22}\varnothing_2(r) - M_{23}\varnothing_3(r) - \ldots - M_{2n}\varnothing_n(r)) = 0,\\
&H_3(\phi_1(r),\ \phi_2(r), \phi_3(r)\ldots\phi_n(r), p) = r^2\phi_3''(r) + 2r\phi_3(r)\\
&+pr^2(-M_{31}\varnothing_1(r) - M_{32}\varnothing_2(r) - M_{33}\varnothing_3(r) - \ldots - M_{3n}\varnothing_n(r)) = 0,\\
&\ldots\\
&H_n(\phi_1(r),\ \phi_2(r), \phi_3(r)\ldots\phi_n(r), p) = r^2\phi_n''(r) + 2r\phi_n(r)\\
&+pr^2(-M_{n1}\varnothing_1(r) - M_{n2}\varnothing_2(r) - M_{n3}\varnothing_3(r) - \ldots - M_{nn}\varnothing_n(r)) = 0
\end{aligned} \tag{27}$$

Here, we obtain identical powers of p terms as a set of equations:

$$\begin{aligned}
&p^0 : r^2\phi_{i0}''(r) + 2r\phi_{i0}'(r),\\
&p^1 : r^2\phi_{i1}''(r) + 2r\phi_{i1}'(r) = -r^2\,(-M_{11}\,\varnothing_1(r) - M_{12}\varnothing_2(r) - M_{13}\varnothing_3(r) - \ldots - M_{1n}\varnothing_n(r)),\\
&p^2 : r^2\phi_{i2}''(r) + 2r\phi_{i2}'(r) = -r^2\,(-M_{21}\varnothing_1(r) - M_{22}\varnothing_2(r) - M_{23}\varnothing_3(r) - \ldots - M_{2n}\varnothing_n(r)),\\
&p^k : r^2\phi_{ik}''(r) + 2r\phi_{ik}''(r) = -r^2\,(M_{k1}\varnothing_1(r) + M_{k2}\varnothing_2(r) + \ldots + M_{kk}\,\varnothing_k(r) + \ldots + M_{kn}\,\varnothing_n(r)),
\end{aligned} \tag{28}$$

The solution of the first component of Equation (28) is:

$$\varphi_{i,0}\,(r) = A_{i,0} + \frac{B_{i,0}}{r}, \tag{29}$$

The following constants $(A_{i,n})$ depend on $(A_{i,0})$ as:

$$A_{i,k} = \sum\nolimits_{j=1}^{n} M_{ij}\,A_{j,k-1}, \tag{30}$$

In addition, $(B_{i,n})$ depends on $(B_{i,0})$ as:

$$B_{i,k} = \sum\nolimits_{j=1}^{n} M_{i,j}\,A_{j,k-1}, \tag{31}$$

Then,

$$\begin{aligned}
&\varphi_{i,0}\,(r) = A_{i,0} + B_{i,0}\,\tfrac{1}{r}\\
&\varphi_{i,1}\,(r) = A_{i,1}\tfrac{r^2}{3!} + B_{i,1}\tfrac{r}{2} + C_{i,0} + D_{i,0}\tfrac{1}{r}\\
&\varphi_{i,2}\,(r) = A_{i,2}\tfrac{r^4}{5!} + B_{i,2}\tfrac{r^3}{4!} + C_{i,1}\tfrac{r^2}{3!} + D_{i,1}\tfrac{r}{2} + E_{i,0} + F_{i,0}\tfrac{1}{r}\\
&\varphi_{i,3}\,(r) = A_{i,3}\tfrac{r^6}{7!} + B_{i,3}\tfrac{r^5}{6!} + C_{i,2}\tfrac{r^4}{5!} + D_{i,2}\tfrac{r^3}{4!} + E_{i,1}\tfrac{r^2}{3!} + F_{i,0}\tfrac{r}{2} + + G_{i,0} + H_{i,0}\tfrac{1}{r}
\end{aligned} \tag{32}$$

We can define $(C_{i,n}, D_{i,n}, F_{i,n}, \ldots)$ in the same way as $(A_{i,n}$, and $B_{i,n})$, so:

$$\begin{aligned}
\varphi_i(r) = \tfrac{1}{r}\,[(A_{i,0}\tfrac{r}{1!} + A_{i,1}\tfrac{r^3}{3!} + A_{i,2}\tfrac{r^5}{5!} + A_{i,3}\tfrac{r^7}{7!} + \ldots) + (B_{i,0} + B_{i,1}\tfrac{r^2}{2} + B_{i,2}\tfrac{r^4}{4!} +\\
B_{i,3}\tfrac{r^6}{6!} + \ldots) + (C_{i,0}\tfrac{r}{1!} + C_{i,1}\tfrac{r^3}{3!} + C_{i,2}\tfrac{r^5}{5!} + C\tfrac{r^7}{7!} + \ldots) + (D_{i,0} + D\tfrac{r^2}{2} + D_{i,2}\tfrac{r^4}{4!} +\\
D_{i,3}\tfrac{r^6}{6!} + \ldots) + \ldots]
\end{aligned} \tag{33}$$

On the other hand,

$$\varphi_i(r) = \frac{1}{r}\left[(\{A_{i,0}+C_{i,0}+E_{i,0}+\ldots\}\frac{r}{1!}+\{A_{i,1}+C_{i,1}+E_{i,1}+\ldots\}\frac{r^3}{3!}+\{A_{i,2}+C_{i,2}+E_{i,2}+\ldots\}\frac{r^5}{5!}+\{A_{i,3}+C_{i,3}+E_{i,3}+\ldots\}\frac{r^7}{7!}+\ldots)+(\{B_{i,0}+D_{i,0}+F_{i,0}+\ldots\}+\{B_{i,1}+D_{i,1}+F_{i,1}+\ldots\}\frac{r^2}{2}+\{B_{i,2}+D_{i,2}+F_{i,2}+\ldots\}\frac{r^4}{4!}+\{B_{i,3}+D_{i,3}+F_{i,3}+\ldots\}\frac{r^6}{6!}+\ldots)+\ldots\right] \tag{34}$$

Let

$$\begin{aligned}\alpha_{i,k} &= A_{i,k}+C_{i,k}+E_{i,k}+\ldots\\ \beta_{i,k} &= B_{i,k}+D_{i,k}+F_{i,k}+\ldots\end{aligned} \tag{35}$$

Its consequence is that:

$$\begin{aligned}\alpha_{i,k} &= \textstyle\sum_{j=1}^{n} M_{i,j}\,\alpha_{i,k-1},\\ \beta_{i,k} &= \textstyle\sum_{j=1}^{n} M_{i,j}\,\beta_{i,k-1}.\end{aligned} \tag{36}$$

The final solution of Equation (34) is given by:

$$\varphi_i(r) = \frac{1}{r}\left[\sum_{k=0}^{\infty}\alpha_{i,k}\frac{r^{2k+1}}{(2k+1)!}+\sum_{k=0}^{\infty}\beta_{i,k}\frac{r^{2k}}{(2k)!}\right] \tag{37}$$

2.3. *The Core-Reflector Boundary Conditions*

After finding the solution of neutron diffusion equations in the reactor core and reflector parts, it is essential to apply the boundary condition, the point R, on the surface between them.

The neutron fluxes ($\varphi_{ci}(r)$) must be continuous as well as their currents ($J_i\,(r)$) [24], which mathematically can be expressed as:

$$\varphi_{ci}\,(R) = \varphi_{ri}\,(R),\ J_{ci}\,(\mathrm{R}) = J_{ri}\,(R), \tag{38}$$

Neutron currents in Equation (38) will be defined as:

$$J_i\,(r)) = -D_i\varphi_i\,(r). \tag{39}$$

Therefore, after inserting the values of fluxes and currents in Equation (38), this system of equations can be written as:

$$\begin{aligned}&U\sum_{k=0}^{\infty}\frac{T_{i,2k}}{(2k+1)!}R^{2k}+\mathrm{V}\sum_{k=0}^{\infty}(-1)^k\frac{T_{i,2k}}{(2k+1)!}R^{2k} = \frac{1}{R}\left[\sum_{k=0}^{\infty}\alpha_{i,k}\frac{R^{2k+1}}{(2k+1)!}+\sum_{k=0}^{\infty}\beta_{i,k}\frac{R^{2k}}{(2k)!}\right],\\ &D_{ic}\frac{d}{dr}\left(U\sum_{k=0}^{\infty}\frac{T_{i,2k}}{(2k+1)!}R^{2k}+\mathrm{V}\sum_{k=0}^{\infty}(-1)^k\frac{T_{i,2k}}{(2k+1)!}R^{2k}\right) = D_{ir}\frac{d}{dr}\left(\frac{1}{R}\left[\sum_{k=0}^{\infty}\alpha_{i,k}\frac{R^{2k+1}}{(2k+1)!}+\sum_{k=0}^{\infty}\beta_{i,k}\frac{R^{2k}}{(2k)!}\right]\right).\end{aligned} \tag{40}$$

By solving this system of equations computationally, we can find R.

3. Special Cases Numerical Study

The theoretical results reached in Section 2 will be simplified to compare numerically with special cases that were obtained [1–3]; this comparison will assure the theory. Special cases start with one-group and two-group, and finally, the multi-group is taken as a general case.

3.1. *One-Group Nuclear Reactor*

Here, a one-group bare reactor (which has fuel only) is studied; then, we focus on the reflector reactor (core and reflected parts). The needed cross-section data are originally taken from [25], which clarified them also, and [2] used cross-section data only. Theoretical and numerical results for both bare and reflected reactors will be compared with modified differential transform method (MDTM).

3.1.1. One-Group Bare Nuclear Reactor

Now, the neutron diffusion equation of a one-group bare reactor [24] can be written as:

$$\nabla^2\varnothing\ (r) + B^2\varnothing\ (r) = 0, \tag{41}$$

where $B^2 = (\nu\Sigma f\ \ - (\Sigma f\ + \Sigma\gamma))/D$ is the buckling, and $D = 1/3(\Sigma f\ + \Sigma s + \Sigma\gamma)$ is the diffusion coefficient [24].

After applying HPM, the flux will be:

$$\varnothing\ (r) = \frac{A}{r}\sum_{k=0}^{\infty}\frac{(-1)^k(Br)^{2k+1}}{(2k+1)!} = \frac{A}{r}sin(Br) \tag{42}$$

One-group bare spherical reactor is taken numerically when 1-MeV neutrons diffuse in pure ^{235}U [2] the MDTM, which will be compared with this studied case.

At this point, necessary nuclear reactor data [2,25] are found in Table 1.

Table 1. One-group bare reactor cross-sections data.

ν	σ_f	σ_s	σ_γ
2.42	1.336	5.959	0.153
	$N = 0.0478 \times 10^{24}$ atoms cm^{-3}		

The essential critical radius result using Mathematica software is tabulated in Table 2.

Table 2. The critical radius of one-group bare reactor.

	HPM	**MDTM**
Critical radius	10.528	10.528

Critical radius calculation using HPM, where zero flux boundary condition is used, gives the same results as MDTM, with the flux behavior tabulated in Table 3 graphed in Figure 1.

Table 3. Normalized flux in one-group bare reactor.

Method	r/a_c	0.0	0.25	0.50	0.75	1.0
Classical	Flux	1.0000	0.9069	0.6563	0.3309	0.03109
RPSM	Flux	1.0000	0.9069	0.6563	0.3309	0.03109

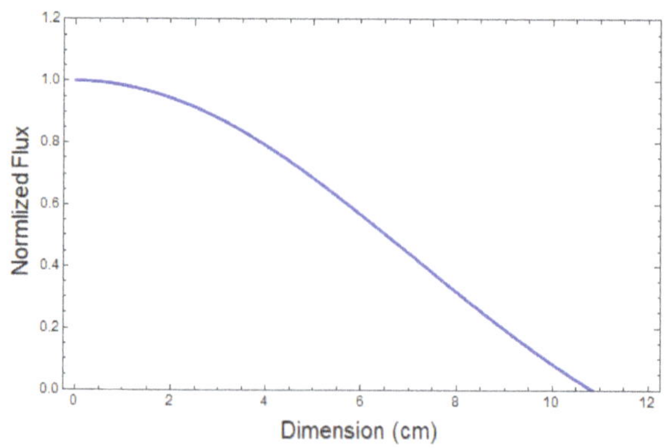

Figure 1. Flux distribution in one-group bare reactor.

After applying HPM, the flux has its normalization value at the sphere center, and it has its zero value at the critical radius, and this is the expected behavior; here, MDTM results are reproduced.

3.1.2. One-Group Reflected Nuclear Reactor

The one-group reflected reactor [6] is taken after a bare reactor, and the neutron diffusion equations for core and reflected parts are:

$$\begin{aligned} \nabla^2 \varnothing_c(r) + B^2 \varnothing_c(r) &= 0, \\ \nabla^2 \varnothing_r(r) - \tfrac{1}{L^2} \varnothing_r(r) &= 0, \end{aligned} \tag{43}$$

where B^2 is buckling, and L^2 is called diffusion area, and L is diffusion length [24] The core and reflected parts fluxes, after applying HPM, will be:

$$\begin{aligned} \varnothing_c(r) &= \frac{A_c}{r} \sum_{k=0}^{\infty} \frac{(-1)^k (Br)^{2k+1}}{(2k+1)!} = \frac{A_c}{r} sinBr, \\ \varnothing_r(r) &= \frac{A_r}{r} \sum_{k=0}^{\infty} \frac{\left(\frac{r}{L}\right)^{2k+1}}{(2k+1)!} = \frac{A_r}{r} sinh\frac{r}{L}. \end{aligned} \tag{44}$$

The solution of the neutron diffusion equation is obtained to study this numerical example, where ^{235}U is the core part, and H_2O is the reflector of this reactor. More needed cross-sections [2] are given in Table 4.

Table 4. One-group reflected reactor cross-sections data.

ν_c	Σ_{cf}	Σ_{cs}	$\Sigma_{c\gamma}$
2.7971	0.06528	0.24806	0.01306
ν_r	Σ_{rf}	Σ_{rs}	$\Sigma_{r\gamma}$
0.0000	0.0000	0.24937	0.03264

For large reflector, the flux $\varnothing_r(r)$ will be

$$\varnothing\,(r) = \frac{A}{r} e^{-\frac{r}{L}} \tag{45}$$

The critical radius of the reflected spherical reactor is compared with both MDTM and transport theory data, and the transport theory can be admitted as a benchmark where the diffusion theory is approximation when Fick's law is used; after using of Mathematica software, the result is tabulated in Table 5.

Table 5. Critical radius of one-group reflected reactor.

	HPM	**MDTM**	**Transport Theory Data**
Critical radius	6.5128	6.5128	6.1275

The critical radius of the one-group reflected reactor, when zero flux boundary condition is used, is reproduced as the results of MDTM and has a good agreement with transport theory data. After this, flux behavior will be studied in both core and reflected parts, which are described in Table 6 and Figure 2.

Table 6. Normalized flux in one-group reflected reactor.

Method	r/a_c	0.0	0.25	0.50	0.75	1.0
Classical	Flux	1.0000	0.9567	0.8301	0.6415	0.4199
RPSM	Flux	1.0000	0.9567	0.8301	0.6415	0.4199

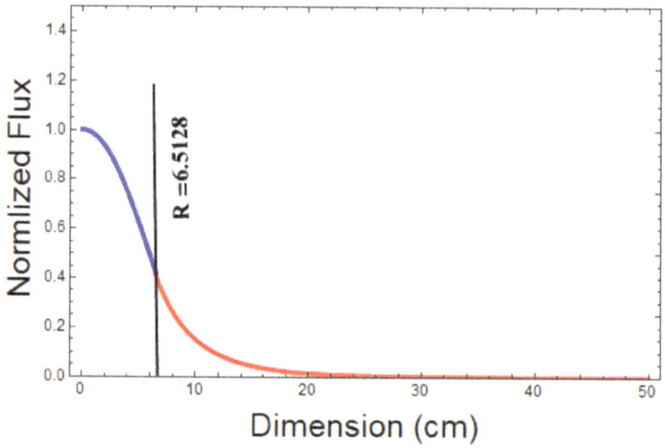

Figure 2. Flux distribution of one-group reflected reactor. Blue, core flux; Red, reflector flux.

The flux distribution using HPM, which has the same MDTM results, using core reflector boundary conditions, and it is clear that flux behavior is in good agreement with classical calculations [3,24].

3.2. Two-Group Nuclear Reactor

After one-group study, the two-group case for bare and reflected reactors are considered; neutrons are divided in two-groups: fast and thermal groups. Numerical results for the bare reactor are compared with the residual power series method (RPSM) [1], while the reflected reactor is compared with the classical method [3].

3.2.1. Two-Group Bare Nuclear Reactor

Now, a bare reactor will be obtained when neutrons move in fast and thermal velocities, and neutron diffusion equations [3] will be:

$$\begin{array}{c} \nabla^2 \varnothing_1(r) + N_{11}\varnothing_1(r) + N_{12}\varnothing_2(r) = 0, \\ \nabla^2 \varnothing_2(r) + N_{21}\varnothing_1(r) + N_{22}\varnothing_2(r) = 0. \end{array} \quad (46)$$

Fluxes, after using HPM, can be written as:

$$\varnothing_i(r) = \sum_{k=0}^{\infty} (-1)^k \frac{1}{(2k+1)!} T_{i,2k} r^{2k}, \ i = 1,2. \quad (47)$$

Computational determination of analytical results is useful in comparing them with RPSM [1]; where Mathematica software is used, the important cross-section is taken from [1], where neutrons undergo fast and thermal diffusion in uranium with enrichment ratio is 93%. Two-group bare reactor cross-sections data in Table 7 are taken from [25], and [1] used them, while [25] gave meaning to each.

Table 7. Two-group bare reactor cross-sections data.

$\Sigma_{f1} = 0.0010484\ \mathrm{cm}^{-1}$	$\Sigma_{\gamma 1} = 0.0010046\ \mathrm{cm}^{-1}$	$\Sigma_{S11} = 0.62568\ \mathrm{cm}^{-1}$
$\Sigma_{S12} = 0.029227\mathrm{cm}^{-1}$	$\nu_1 = 2.5$	$\chi_1 = 1.0$
$\Sigma_{f2} = 0.050632\ \mathrm{cm}^{-1}$	$\Sigma_{\gamma 2} = 0.025788\ \mathrm{cm}^{-1}$	$\Sigma_{S22} = 2.44383\ \mathrm{cm}^{-1}$
$\Sigma_{S21} = 0.0000\ \mathrm{cm}^{-1}$	$\nu_2 = 2.5$	$\chi_2 = 0.0$

Values of N_{11}, N_{12}, N_{21}, and N_{22} are determined depending on these cross-sections as shown in Table 8.

Table 8. Two-group bare reactor fluxes coefficients.

$N_{11} = -0.0564834$	$N_{12} = 0.249474$	$N_{21} = 0.220978$	$N_{22} = -0.577793$

For a two-group bare reactor, the critical radius is obtained using both zero flux (ZF) and extrapolated boundary conditions (EBC), and it is found that fluxes vanish before a certain distance, which is called extrapolated distance, where EBC is named; next, data are compared with RPSM, and transport theory data are taken as a benchmark.

Here, 93% enriched uranium is taken as a numerical example, and the critical radius is tabulated in Table 9.

Table 9. Critical radius of two-group bare reactor.

BC	HPM	RPSM	Transport Theory Date
ZF	17.120	17.120	-
EBC	16.251	16.251	16.049836

This critical radius of two-group bare reactor of HPM shows that when RPSM is used, both of them have logical results with the benchmark (Transport Theory).

The behavior of fluxes in the two-group bare reactor is found in Table 10 and Figure 3.

Table 10. Normalized thermal, fast, and total fluxes in two-group bare reactor.

Flux	Method	r/a_c	0.0	0.25	0.50	0.75	1.0
Fast	Classical	Flux	2.7671	2.5238	1.7802	1.0063	0.1835
	RPSM	Flux	2.7671	2.5238	1.7802	1.0063	0.1835
Thermal	Classical	Flux	1.0000	0.9121	0.6759	0.3637	0.0663
	RPSM	Flux	1.0000	0.9121	0.6759	0.3637	0.0663
Total	Classical	Flux	3.7671	3.4359	2.5460.	1.3699	0.2498
	RPSM	Flux	3.7671	3.4359	2.5460	1.3699	0.2498

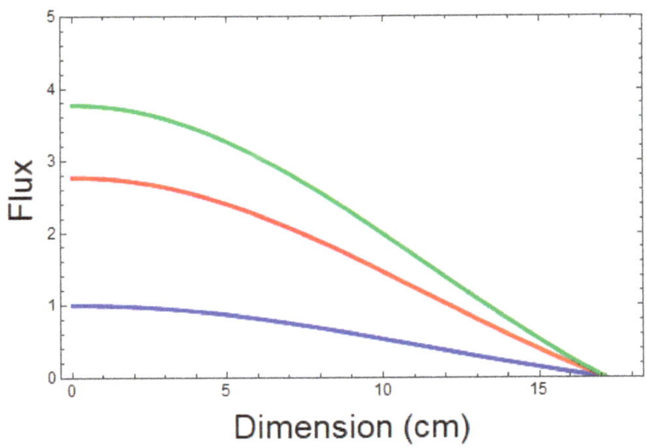

Figure 3. Fluxes distribution in two-group bare reactor. Blue, thermal flux; Red, fast flux; Green, total flux.

HPM has the same performance as RPSM in tabulated and graphical representation of the two-group bare reactor, where both fluxes and their sum converge at the same point.

As it is explained, the real critical dimension calculated by transport theory data is less than the HPM calculated point, and this is reasonable after applying Fick's law approximation.

3.2.2. Two-Group Reflected Nuclear Reactor

A two-group reflected reactor will be studied here; unlike the bare reactor, the buckling is not unique in its principal and alternate buckling for the core part of the reactor.

The neutron diffusion equations [3] corresponding to principal buckling are:

$$\begin{aligned} \nabla^2 \varnothing_1(r) + N_{11}\varnothing_1(r) + N_{12}\varnothing_2(r) = 0, \\ \nabla^2 \varnothing_2(r) + N_{21}\varnothing_1(r) + N_{22}\varnothing_2(r) = 0. \end{aligned} \tag{48}$$

Similarly, this for alternate buckling will be:

$$\begin{aligned} \nabla^2 \varnothing_1(r) - L_{11}\ \varnothing_1(r) - L_{12}\varnothing_2(r) = 0, \\ \nabla^2 \varnothing_2(r) - L_{21}\varnothing_1(r) - L_{22}\varnothing_2(r) = 0. \end{aligned} \tag{49}$$

After applying HPM for each case, the solution is a linear combination of both cases, which can be written as:

$$\phi_{ic}\ (r) = U \sum_{k=0}^{\infty} \frac{T_{i.2k}}{(2k+1)!} r^{2k} + V \sum_{k=0}^{\infty} (-1)^k \frac{T_{i.2k}}{(2k+1)!} r^{2k},\ i = 1, 2. \tag{50}$$

Furthermore, the reflector of neutron diffusion equations will be:

$$\begin{aligned} \nabla^2 \varnothing_1(r) - M_{11}\ \varnothing_1(r) - M_{12}\varnothing_2(r) = 0, \\ \nabla^2 \varnothing_2(r) - M_{21}\varnothing_1(r) - M_{22}\varnothing_2(r) = 0. \end{aligned} \tag{51}$$

After taking into consideration that there is no fission in the reflector and no upper scattering case, this is a known case that will be studied numerically: constant $M_{12} = 0$.

The fast-group flux will be:

$$\varphi_{1r}(r) = \frac{1}{r}\left[\sum_{k=0}^{\infty} \frac{r^{2k+1}}{(2k+1)!} \right]. \tag{52}$$

while thermal flux is:

$$\varphi_{2r}(r) = \frac{1}{r}\left[\sum_{k=0}^{\infty} \alpha_{i,k} \frac{r^{2k+1}}{(2k+1)!} + \sum_{k=0}^{\infty} \beta_{i,k} \frac{r^{2k}}{(2k)!} \right] \tag{53}$$

After finding the solution of neutron diffusion equations in core and reflector parts, it is essential to apply the boundary conditions in the surface between reactor core and reflector, where fluxes ($\varphi_1\ (x)$, $\varphi_2\ (x)$) and their currents ($J_1\ (r)$, $J_2\ (r)$) are continuous [3], which mathematically can be expressed as:

$$\begin{aligned} \varphi_{c1}\ (R) &= \varphi_{r1}\ (R),\ J_{c1}\ (R) = J_{r1}\ (R), \\ \varphi_{c2}\ (R) &= \varphi_{r2}\ (R),\ J_{c2}\ (R) = J_{r2}\ (R), \end{aligned} \tag{54}$$

Hence, the neutron currents are $J_1\ (r)) = -D_1\ \varphi_1\ (r)$, $J_2\ (r)) = -D_2\ \varphi_2\ (r)$.

The derived analytical formalism is computationally determined to verify the theory, while cross-section data are taken from [3] which given in Table 11.

Table 11. Two-group reflected reactor cross-sections data.

	Core data	
$\tau_C = 45.3$ cm^2 $p_C = 1.0$	$\Sigma_R^1 = 0.03179$ cm^{-1} $D_1 = 1.44$ cm	$\Sigma_a^2 = 0.1104$ cm^{-1} $D_2 = 0.229$ cm
	Reflector data	
$\tau_R = 40.0$ cm^2 $p_R = 1.0$	$\Sigma_a = 1.85$ cm^{-1} $D_1 = 1.85$ cm	$\Sigma_a^2 = 0.01226$cm^{-1} $D_2 = 0.204$ cm

Values of N_{ij} and L_{ij} $i, j = 1, 2$ are shown in Table 12.

Table 12. Two-group reflected reactor fluxes coefficients.

$N_{11} = -0.022076.$	$N_{12} = 0.100893$	$N_{21} = 0.1388210$	$N_{22} = -0.482096$
$L_{11} = -0.025000$	$L_{12} = 0.000000$	$L_{21} = 0.226716$	$L_{22} = -0.060098$

Classical calculations completely depend on finding the buckling, principal, and alternate buckling; for a more than two-group reactor, this calculation will be more complicated.

HPM studies the two-group reflected reactor and, in general, any number of group reactor uses more advance and accurate method for determination of critical radius, and clarification of the fluxes distribution depends on the combination between fluxes' constants; these constants depend on all groups' cross sections.

Necessary Mathematica software is used to numerically reach the critical radius for the two-group reflected reactor; critical radius of the reactor core can be found in Table 13.

Table 13. Critical radius of two-group core part of reflected reactor.

BC	HPM	Classical Method
Critical radius	30.713	31.9

After finding the critical radius for this system, fast and thermal fluxes and their sum are considered in Table 14 and Figure 4.

Table 14. Two-group reflected reactor fluxes and total flux.

Flux	Method	r/a_c	0.0	0.25	0.50	0.75	1.0
Group 1	HPM	Flux	1.0000	0.94312	0.76591	0.51492	0.23573
	Classical method	Flux	1.0000	0.93730	0.76360	0.51698	0.24140
Group 2	HPM	Flux	0.2860	0.26791	0.21757	0.14654	0.11478
	Classical method	Flux	0.2841	0.26638	0.21702	0.14717	0.11286
Total	HPM	Flux	1.2860	1.21103	0.98348	0.66146	0.35051
	Classical method	Flux	1.2841	1.20368	0.98062	0.66415	0.35426

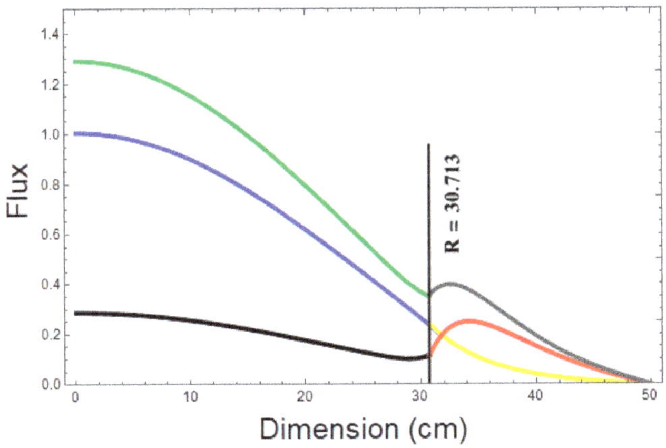

Figure 4. Flux distribution in the core and reflected parts of two-group reflected reactor. Black, $\varnothing_{1c}(r)$; Blue, $\varnothing_{2c}(r)$; Red, $\varnothing_{1r}(r)$; Yellow, $\varnothing_{2r}(r)$; Green, $\varnothing_{tot.c}(r)$; Gray, $\varnothing_{tot.r}(r)$.

Table 14 gives fast and thermal fluxes and their total flux values, and thus, it is obvious that total flux decreases when the reactor radius increases and vanishes at reflector radius.

As we compare HPM results and classical results, it is seen that the HPM critical dimension is less than that of the classical method, and the total flux converges faster, which reduces the fuel and improves the reactor fission. This can be one step forward, using HPM, in accuracy of critical dimension calculation and flux distribution determination.

3.3. Multi-Group Nuclear Reactor

The four-group of neutrons diffusion equations case [1], which is an example of a multi-group reactor, is discussed as one step forward, which is represented by the following system:

$$\begin{aligned} \nabla^2\varnothing_1(r) + N_{11}\varnothing_1(r) + N_{12}\varnothing_2(r) + N_{13}\varnothing_3(r) + N_{14}\varnothing_4(r) = 0, \\ \nabla^2\varnothing_2(r) + N_{21}\varnothing_1(r) + N_{22}\varnothing_2(r) + N_{23}\varnothing_3(r) + N_{24}\varnothing_4(r) = 0, \end{aligned} \quad (55)$$

with initial conditions:

$$\varnothing_i(0) = I_i,\ \varnothing_i(0) = 0, i = 1,2,3,4. \quad (56)$$

Hence, the solutions of ith four-group reactor flux according to HPM are given by

$$\varnothing_i(r) = \sum_{k=0}^{\infty}(-1)^k \frac{1}{(2k+1)!} T_{i,2k} r^{2k}. \quad (57)$$

Mathematica software is used in numerical solutions; the solution is obtained numerically for cross sections related to interactions of the four-group system. The following data obtained from [1] are correspondingly used in Table 15.

Table 15. Four-group reflected reactor cross-sections data.

Group 1 (1.35 Mev–10 Mev)	$\nu_1 \sum_{f1} = 0.0096\ \text{cm}^{-1}$ $\sum_{S13} = 0.00$	$\sum_a = 0.0049\ \text{cm}^{-1}$ $\sum_{S14} = 0.00$ $\chi_1 = 0.575$	$\sum_{S12} = 0.0831\ \text{cm}^{-1}$ $D_1 = 2.162$ cm
Group 2 (9.1 keV–1.35 Mev)	$\nu_2 \sum_{f2} = 0.0012\ \text{cm}^{-1}$ $\sum_{S23} = 0.0.0585\ cm^{-1}$	$\sum_{a2} = 0.0028\ \text{cm}^{-1}$ $\sum_{S24} = 0.00\ \text{cm}^{-1}$ $\chi_2 = 0.425$	$\sum_{S21} = 0.00\ \text{cm}^{-1}$ $D_2 = 1.087$ cm
Group 3 (0.4 ev–9.1 kev)	$\nu_3 \sum_{f3} =$ 0.0.0177 cm^{-1} $\sum_{S32} = 0.00\text{cm}^{-1}$	$\sum_{a3} = 0.0.0305\ \text{cm}^{-1}$ $\sum_{S34} = 0.0651\ \text{cm}^{-1}$ $\chi_3 = 0.0$	$\sum_{S31} = 0.00\ \text{cm}^{-1}$ $D_3 = 0.632$ cm
Group 4 (0.0 ev–0.4 ev)	$\nu_4 \sum_{f4} = 0.1851\ \text{cm}^{-1}$ $\sum_{S42} = 0.00\text{cm}^{-1}$	$\sum_{a4} = 0.1210\ \text{cm}^{-1}$ $\sum_{S43} = 0.00\text{cm}^{-1}$ $\chi_4 = 0.0$	$\sum_{S41} = 0.00\ \text{cm}^{-1}$ $D_4 = 0.354$ cm

Values of $N_{ij},\ i,j = 1,2,3,4$ are shown in Table 16 is derived from Table 15 data.

Table 16. Four-group fluxes coefficients.

$N_{11} = -0.038150$	$N_{12} = 0.000319$	$N_{13} = 0.004707$	$N_{14} = 0.049229$
$N_{21} = 0.080202$	$N_{22} = -0.055925$	$N_{23} = 0.083370$	$N_{24} = 0.148820$
$N_{31} = 0.092563$	$N_{32} = 0.092563$	$N_{33} = -0.151266$	$N_{34} = 0.092563$
$N_{41} = 0.183898$	$N_{42} = 0.183898$	$N_{43} = 0.183898$	$N_{44} = -0.341808$

The critical radius of a four-group reactor is calculated depending on ZF and EBC boundary conditions, and generated data are listed in Table 17.

Table 17. Critical radius of four-group reactor.

BC	HPM	RPSM
ZF	8.770	8.770
EBC	7.905	7.905

Fluxes values of the four-group reactor are given in Table 18 and Figure 5.

Table 18. Four-group fluxes and total flux reactor.

Flux	Method	r/a_c	0.0	0.25	0.50	0.75	1.0
Group 1	HPM	Flux	1.0000	0.918535	0.697804	0.400643	0.107646
	RPSM	Flux	1.0000	0.918535	0.697804	0.400643	0.107646
Group 2	HPM	Flux	4.1716	3.83172	2.91093	1.67131	0.449052
	RPSM	Flux	4.1716	3.83172	2.91093	1.67131	0.449052
Group 3	HPM	Flux	2.7361	2.5132	1.90926	1.0962	0.294531
	RPSM	Flux	2.7361	2.5132	1.90926	1.0962	0.294531
Group 4	HPM	Flux	3.0931	2.84115	2.1584	1.23924	0.332964
	RPSM	Flux	3.0931	2.84115	2.1584	1.23924	0.332964
Total	HPM	Flux	11.0008	10.1046	7.67639	4.40739	1.18419
	RPSM	Flux	11.0008	10.1046	7.67639	4.40739	1.18419

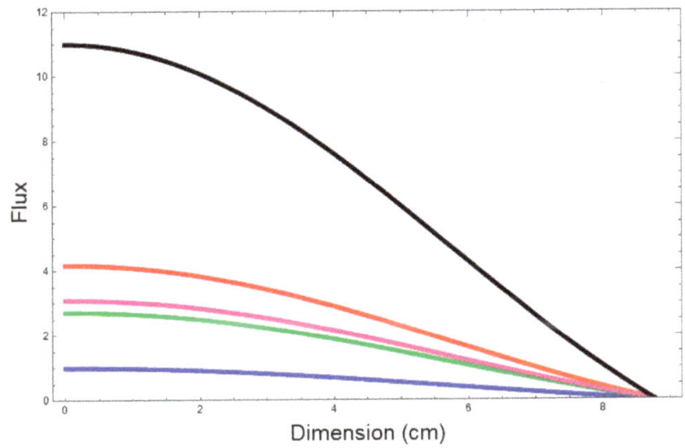

Figure 5. Four-group fluxes and total flux. Blue , $\varnothing_1(r)$; Red, $\varnothing_2(r)$; Green, $\varnothing_3(r)$; Pink, $\varnothing_4(r)$; Black, Total flux.

Both tabulated and graphical representation show that all fluxes decrease when the reactor radius increases and vanishes at the critical radius, and this expected behavior reproduces RPSM.

4. Conclusions

The application of HPM, as an approximation method, in the reflected spherical reactor to solve diffusion equations of a multi-group system is accomplished in this work. To assure the theory, the solutions are simplified and compared with RPSM, MDTM, and classical theory. The results can be easily reproduced with the approximation methods, while flux converged faster when compared with the classical calculations; this improves the efficiency of the reactor by reducing the critical mass and dimensions using HPM. We can assure that the utilities of HPM in this work and previous studies still has the capability to solve different branches of science problems.

Author Contributions: Conceptualization, M.S.; methodology, M.A.-S.; software, E.A.M.F.; validation, M.S., E.A.M.F. and M.A.-S.; investigation, M.S. and E.A.M.F., writing—original draft preparation, M.S.; writing—review and editing, M.S.; visualization, E.A.M.F.; supervision, M.A.-S.; project administration, M.S.; funding acquisition, M.S. All authors have read and agreed to the published version of the manuscript.

Funding: This research was funded by the Deanship of Scientific Research at Prince Sattam Bin Abdulaziz University, grant number 2021/01/17993 and The APC was funded by the Deanship of Scientific Research at Prince Sattam Bin Abdulaziz University.

Data Availability Statement: Not applicable.

Acknowledgments: The authors extend their appreciation to the Deputyship for Research and innovation, Ministry of Education in Saudi Arabia for funding this research work through the project number (IF-PSAU/2021/01/17993).

Conflicts of Interest: The authors declare no conflict of interest. The funders had no role in the design of the study; in the collection, analyses, or interpretation of data; in the writing of the manuscript, or in the decision to publish the results.

References

1. Shqair, M.; El-Ajou, A.; Nairat, M. Analytical Solution for Multi-Energy Groups of Neutron Diffusion Equations by a Residual Power Series Method. *Mathematics* **2019**, *7*, 633. [CrossRef]
2. Shqair, M.; Elzahar, E. Analytical solution of neutron diffusion equation in reflected reactors using modified differential transform method. In *Computational Mathematics and Applications*; Zeidan, D., Padhi, S., Burqan, A., Peer Ueberholz, P., Eds.; Springer: Singapore, 2020; pp. 129–145.
3. Meghreblian, R.; Holmes, D. *Reactor Analysis*; McGraw-Hill: New York, NY, USA, 1960.
4. Khasawneh, K.; Dababneh, S.; Odibat, Z. A solution of the neutron diffusion equation in hemispherical symmetry using the homotopy perturbation method. *Ann. Nucl. Energy* **2009**, *36*, 1711–1717. [CrossRef]
5. Dababneh, S.; Khasawneh, K.; Odibat, Z. An alternative solution of the neutron diffusion equation in cylindrical symmetry. *Ann. Nucl. Energy* **2010**, *38*, 1140–1143. [CrossRef]
6. Shqair, M. Solution of different geometries reflected reactors neutron diffusion equation using the homotopy perturbation method. *Results Phys.* **2018**, *12*, 61–66. [CrossRef]
7. Shqair, M. Developing a new approaching technique of homotopy perturbation method to solve two-group reflected cylindrical reactor. *Results Phys.* **2019**, *12*, 1880–1887. [CrossRef]
8. Cassell, J.; Williams, M. A solution of the neutron diffusion equation for a hemisphere with mixed boundary condition. *Ann. Nucl. Energy* **2004**, *31*, 1987–2004. [CrossRef]
9. Ray, S.; Patra, A. Application of homotopy analysis method and Adomian decomposition method for the solution of neutron diffusion equation in the hemisphere and cylindrical reactors. *J. Nucl. Eng. Technol.* **2011**, *1*, 1–14.
10. Martin, O. A homotopy perturbation method for solving a neutron transport equation. *Appl. Math. Comput.* **2011**, *217*, 8567–8574. [CrossRef]
11. Williams, M. Neutron transport in two interacting slabs II: The influence of transverse leakage. *Ann. Nucl. Energy* **2002**, *29*, 1551–1596. [CrossRef]
12. Nahla, A.; Al-Malki, A.; Rokaya, M. Numerical Techniques for the Neutron Difusion Equations in the Nuclear Reactors. *Adv. Stud. Theor. Phys.* **2012**, *6*, 649–664.
13. Aboanber, A.; Nahla, A.; El Mhlawy, A.; Maher, O. An efficient exponential representation for solving the two-energy group point telegraph kinetics model. *Ann. Nucl. Energy* **2022**, *166*, 108698. [CrossRef]
14. Aboanber, A.; Nahla, A.; Aljawazneh, S. Fractional two energy groups matrix representation for nuclear reactor dynamics with an external source. *Ann. Nucl. Energy* **2021**, *153*, 108062. [CrossRef]
15. Nairat, M.; Shqair, M.; Alhalholy, T. Cylindrically Symmetric Fractional Helmholtz Equation. *Appl. Math. E-Notes* **2019**, *19*, 708–717.
16. He, H. Homotopy perturbation technique. *Comp. Methods Appl. Mech. Eng.* **1999**, *178*, 257–262. [CrossRef]
17. He, H. A coupling method of a homotopy technique and perturbation technique for non-linear problems. *Int. J. Non-Linear Mech.* **2000**, *35*, 37–43. [CrossRef]
18. He, J.H. Homotopy perturbation method: A new nonlinear analytical technique. *Appl. Math. Comp.* **2003**, *135*, 73–79. [CrossRef]
19. He, H. Homotopy perturbation method for solving boundary value problems. *Phys. Lett. A* **2006**, *350*, 87–88. [CrossRef]
20. He, H. Some asymptotic methods for strongly nonlinear equations. *Int. J. Mod. Phys. B* **2006**, *20*, 1141–1199. [CrossRef]
21. He, H.; El-Dib, O. Homotopy perturbation method with three expansions. *J. Math. Chem.* **2021**, *59*, 1139–1150. [CrossRef]
22. Stacey, W.M. *Nuclear Reactor Physics*; John Wiley & Sons: Boston, MA, USA, 2001.
23. Duderstadt, J.J.; Hamilton, L.J. *Nuclear Reactor Analysis*; John Wiley & Sons: Boston, MA, USA, 1976.

24. Lamarsh, J.R. *Introduction to Nuclear Engineering*, 2nd ed.; Addison-Wesley: Boston, MA, USA, 1983.
25. Sood, A.; Forster, R.D.; Parsons, R. Analytical Benchmark Test Set for Criticality Code Verification. *Prog. Nucl. Energy* **2003**, *42*, 55–106. [CrossRef]

energies

MDPI

Article

Distributionally Robust Multi-Energy Dynamic Optimal Power Flow Considering Water Spillage with Wasserstein Metric

Gengli Song [1,2] and Hua Wei [1,2,*]

[1] School of Electrical Engineering, Guangxi University, Nanning 530004, China; glisong@outlook.com
[2] Guangxi Key Laboratory of Power System Optimization and Energy Technology, Guangxi University, Nanning 530004, China
* Correspondence: hweigxu@outlook.com

Abstract: This paper proposes a distributed robust multi-energy dynamic optimal power flow (DR-DOPF) model to overcome the uncertainty of new energy outputs and to reduce water spillage in hydropower plants. The proposed model uses an ambiguity set based on the Wasserstein metric to address the uncertainty of wind and solar power forecasting errors, rendering the model data-driven. With increasing sample size, the conservativeness of the ambiguity set was found to decrease. By deducing the worst-case expectation in the objective function and the distributed robust chance constraints, the exact equivalent form of the worst-case expectation and approximate equivalent form of the distributed robust chance constraints were obtained. The test results of the IEEE-118 and IEEE-300 node systems indicate that the proposed model could reduce water spillage by more than 85% and comprehensive operation cost by approximately 12%. With an increasing number of samples, the model could reduce conservativeness on the premise of satisfying the reliability of safety constraints.

Keywords: multi-energy system; water spillage; distributed robust optimization; Wasserstein; dynamic optimal power flow

Citation: Song, G.; Wei, H. Distributionally Robust Multi-Energy Dynamic Optimal Power Flow Considering Water Spillage with Wasserstein Metric. *Energies* **2022**, *15*, 3886. https://doi.org/10.3390/en15113886

Academic Editors: Zbigniew Leonowicz, Michał Jasiński and Arsalan Najafi

Received: 6 April 2022
Accepted: 12 May 2022
Published: 25 May 2022

1. Introduction

Renewable energy sources, such as hydropower, wind, and solar power, have attracted considerable attention worldwide. Still, the problem of renewable energy power consumption must be solved urgently. The problem of water, wind, and light spillage is very serious, among which water spillage is highly prevalent. Therefore, solving the water spillage problem has become a critical research topic [1]. In addition to poor consumption, the uncertainty of new energy outputs is also an important factor. On the one hand, when dealing with such uncertainty, the traditional dispatching method gives priority to hydropower regulation, which makes for a large regulation burden. On the other hand, the traditional dispatching method is based on the dispatcher's experience in allocating the power imbalance among hydropower plants. Due to the lack of reasonable and accurate planning, this method is prone to improper dispatching, which increases the risk of water spillage. Hence, to solve the water spillage problem, the uncertainty in wind and solar power output must be urgently addressed, and a multi-energy optimal power flow model is needed to fully utilize the characteristics of complementarity and coordination between power plants.

The optimal power flow (OPF) problem has been of concern since it was proposed, and new methods have been proposed in recent years. An enhanced quasi-reflection jellyfish optimization algorithm was proposed in [2] to solve the OPF problem, which performed well and showed resilience in the simulation. An algorithm of social network search optimizers was used for optimal power system operation in [3], and the study in this paper proved that the algorithm has significant stability. An improved heap-based optimization algorithm was proposed in [4] to address the OPF problem, and its effectiveness and robustness was demonstrated. A multi-objective quasi-reflected jellyfish search optimizer was proposed in [5] to solve the multi-dimensional OPF issue with diverse objectives. The above methods showed good performance in solving OPF problems; however, the uncertainty of new energy outputs was not considered in those models. Therefore, a new OPF model considering the uncertainty of wind and solar power output is needed to reduce water spillage in hydropower plants.

Many advancements have been made in uncertainty research, and effective methods, such as stochastic optimization (SO) [6,7] and robust optimization (RO) [8–10], have been proposed. SO assumes that the considered random variable follows a certain probability distribution, and the original problem is transformed into a deterministic problem through formula deduction [11]. SO requires complete knowledge of the probability distribution information of uncertain parameters, which is usually difficult to obtain in practice. In this case, assuming that the random variable follows a certain probability distribution appears to be too optimistic, which may lead to incorrect decisions. In contrast to SO, RO only needs a sample set of uncertain variables and does not require knowledge of a specific probability distribution. RO limits all possible scenarios of the considered random variables to an uncertain set and then transforms the original problem into a deterministic optimization problem under extreme scenarios. RO can ensure that the constraints are met in the case of absolute risk aversion, but the optimization results are often highly conservative, because this approach ignores the distribution information of uncertain variables.

Distributed robust optimization (DRO) [12] combines the advantages of SO and RO methods. DRO assumes that the real distribution is located within an ambiguity set. DRO does not require knowledge of the specific probability distribution and can effectively overcome the conservativeness of RO. At present, the moment-based ambiguity set is the most widely used option [13–15], i.e., the first- and second-order moments of the probability distribution are used to describe the uncertain set. However, when considering only moment information, the distribution of samples cannot be described in detail, and much probability information is omitted, especially when the number of samples is large. Therefore, this model cannot converge to the real distribution when the sample size approaches infinity.

To overcome this defect, an ambiguity set based on the Wasserstein metric [16,17] was developed. The Wasserstein ambiguity set is a data-driven set. With increasing sample size, the ambiguity set decreases and finally converges to the real distribution. Through research, scholars have made achievements in this area. A distributed robust chance-constrained approximate OPF based on the Wasserstein metric was proposed in [18]. This method employed the distributed robust optimization method of the Wasserstein metric to solve the OPF problem for the first time and used a decoupled linear power flow model to approximate the classical power flow equation. A distributed robust approximation framework of unit commitment based on the Wasserstein metric was developed in [19]. This method obtained an upper approximation of the original problem through mathematical deduction and transformed the original model into a mixed integer linear programming problem. A data-driven distributed robust chance-constrained real-time scheduling model was established in [20], which transformed the original problem into a linear programming problem by linearly reconstructing the secondary generation cost and distributed robust chance constraints. A two-stage distributed robust optimization method was proposed in [21] to address the uncertainty in wind power output in an integrated electric–gas thermal energy system. The above methods have been applied to solve problems of power systems, but research on multi-energy dynamic optimal power flow considering the water spillage of cascade hydropower stations under distributed robust opportunity constraints is lacking.

Accordingly, a distributed robust multi-source dynamic optimal power flow model is proposed in this paper considering abandoned water under the Wasserstein metric. The main contributions are as follows:

(1) Cascade hydropower stations coupled in time and space are introduced into the distributed robust dynamic OPF. The model uses the ambiguity set with the Wasserstein metric to address the uncertainty of wind and solar output. The water spillage cost of hydropower plants is also considered to solve the problem of a large amount of water spillage caused by wind and solar power output uncertainty.

(2) An exact equivalent form of the extreme distribution term in the objective function is obtained via dual reformulation and mathematical deduction. The equivalent form is an affine function of the control variable. The form is concise and the scale remains unchanged with increasing numbers of samples, so this form achieves satisfactory computational performance.

(3) An initial equivalent form is obtained by transforming the distributed robust chance constraint. Although the initial equivalent form is accurate, the number of constraints and variables rapidly increases with an increasing number of samples, resulting in a significant decline in operation efficiency. Therefore, an approximate equivalent model is proposed to overcome the defects. Finally, the original problem is transformed into a mixed-integer linear programming problem, which can be solved efficiently with a commercial solver.

2. Dynamic Optimal Power Flow with Distributed Robust Chance Constraints Based on Wasserstein Metric

2.1. Wasserstein Metric and Ambiguity Set

The ambiguity set based on the Wasserstein metric is constructed as follows:

According to a sampling set of historical data, the empirical distribution can be obtained as $\hat{\mathbb{P}}_N = \frac{1}{N}\sum_{n=1}^{N}\delta_{\hat{\omega}_i}$, where $\delta_{\hat{\omega}_i}$ is the Dirac measure of $\hat{\omega}_i$ and $\hat{\mathbb{P}}_N$ can be used as an estimate of the real distribution $\mathbb{P}$. To determine the deviation degree between the constant $\hat{\mathbb{P}}_N$ and the real distribution of $\mathbb{P}$, the Wasserstein metric is defined as follows:

Let $\mathbb{P}_1$ and $\mathbb{P}_2$ be two arbitrary probability distributions. Then, the Wasserstein metric $W:\ \mathcal{M}(\Xi)\times\mathcal{M}(\Xi)\rightarrow R_+$ is defined as [16]:

$$W(\mathbb{P}_1,\mathbb{P}_2)=\inf_{\Pi}\left\{\int_{\Xi^2}\|\omega_1-\omega_2\|\Pi(d\omega_1,d\omega_2)\right\} \tag{1}$$

where $\mathcal{M}(\Xi)$ denotes the set of all probability distributions whose support set is Ξ, Π denotes the joint distribution of ω_1 and ω_2, $\|\cdot\|$ denotes any norm on $\mathbb{R}^m$, and $\|\omega_1-\omega_2\|$ denotes the cost of moving an object of unit mass from distribution $\mathbb{P}_1$ to $\mathbb{P}_2$. Because of its tractability, the L_1-norm was adopted, and the ambiguity set is defined as:

$$\mathbb{P}_N=\{\mathbb{P}\in\mathcal{M}(\Xi)\,|W(\hat{\mathbb{P}}_N,\mathbb{P}_2)<\delta(N)\} \tag{2}$$

The above equation defines a Wasserstein ball with empirical distribution $\hat{\mathbb{P}}_N$ as the center and $\delta(N)$ as the radius, which can be calculated by using the following expression (3) [18]:

$$\delta(N)=C\sqrt{\frac{1}{N}\ln\left(\frac{1}{1-\beta}\right)} \tag{3}$$

where $1-\beta$ is the confidence level and C is a constant, which can be obtained by solving the following optimization problem:

$$C=2\inf_{\eta>0}\sqrt{\frac{1}{2\eta}\left\{1+\ln\left[\frac{1}{N}\sum_{k=1}^{N}\exp\left(\eta\|\hat{\omega}_i-\hat{\mu}\|_1^2\right)\right]\right\}} \tag{4}$$

where $\hat{\mu}$ denotes the sample mean value. This equation is a unimodal function of scalar η, which can be solved via the golden section or binary search method.

The ambiguity set based on the Wasserstein metric exhibits the following characteristics: It is a data-driven ambiguity set; with an increasing number of historical samples, the set decreases. When the sample size follows $N\rightarrow\infty$, the radius of the Wasserstein ball converges to 0 and the corresponding ambiguity set converges to the real distribution.

2.2. Power Flow Equation and Its Reformulation

The power flow equation and line power constraints are as follows:

$$\begin{cases} P_i = V_i \sum_{j\in i} V_j (G_{ij} cos\theta_{ij} + Bsin\theta_{ij}) \\ Q_i = V_i \sum_{j\in i} V_j (G_{ij} sin\theta_{ij} - Bcos\theta_{ij}) \\ P^l_{i,j} = -V_i^2 g_{ij} + V_i V_j (g_{ij} cos\theta_{ij} + b_{ij} sin\theta_{ij}) \end{cases} \quad (5)$$

To deal with fluctuations in wind and solar output, automatic generation control (AGC) is the most widely used scheme in real-world applications. This scheme ensures a real-time power balance of the system by distributing unbalanced wind and photovoltaic power and assigning the unbalanced power to each AGC unit in the form of an affine function. According to the scheme, the actual generating power of each generator unit is:

$$\begin{cases} \widetilde{P}^g_{i,t} = -\alpha^g_{i,t} e^T \omega_t + P^g_{i,t} \\ \widetilde{P}^h_{i,t} = -\alpha^h_{i,t} e^T \omega_t + P^h_{i,t} \\ \widetilde{P}^w_{i,t} = \omega_{i,t} + P^w_{i,t} \\ \widetilde{Q}_{i,t} = Q_{i,t} + \sigma\omega_{i,t} \end{cases} \quad (6)$$

where ω_t denotes the vector of the combined forecasting errors of wind and solar power, and $\alpha^g_{i,t}$ and $\alpha^h_{i,t}$ are the participation factors of thermal power units and hydropower units, respectively, satisfying $\sum_{i\in G} \alpha^g_{i,t} + \sum_{i\in H} \alpha^h_{i,t} = 1$, vector $e = [1, 1, \cdots, 1]$, and parameter $\sigma = \sin\varphi / \cos\varphi$, where $\cos\varphi$ is the power factor. As the power flow equation (Equation (5)) is nonlinear and cannot be addressed under distributed robust chance constraints, a decoupled linear power flow model [22] was adopted in this paper. According to this model, expressions of the true values of the voltage phase angle, node voltage, and line power can be obtained, as expressed in Equation (7) (the detailed process can be found in [18,22]):

$$\begin{cases} \widetilde{\theta}_{it} = -(e^T \omega_t) A^\theta_{i:} \left(\alpha^g_t + \alpha^h_t\right) + B^\theta_{i:} \omega_t + \theta_{i,t} \\ \widetilde{V}_{it} = -(e^T \omega_t) A^v_{i:} \left(\alpha^g_t + \alpha^h_t\right) + B^v_{i:} \omega_t + V_{i,t} \\ \widetilde{P}^l_{k,t} = -(e^T \omega_t) A^l_{i:} \left(\alpha^g_t + \alpha^h_t\right) + B^l_{i:} \omega_t + P^l_{k,t} \end{cases} \quad (7)$$

where A^θ, B^θ, A^v, B^v, A^l, and B^l denote the constant coefficient matrices determined by the network parameters, $A^\theta_{i:}$ denotes the vector composed of the elements in the ith row of matrix A^θ, and the other parameters are similar.

2.3. Constraints on Safe Operation and Cascade Hydropower Plants

The safe operation constraint of the power system adopts the following form of distributed robust chance constraints (DRCC):

$$\inf_{\mathbb{P}\in\mathbb{P}_N} \mathbb{P}\left\{ \underline{V}_i \le \widetilde{V}_{i,t} \le \overline{V}_i \right\} \ge 1 - \rho_v \quad (8)$$

$$\inf_{\mathbb{P}\in\mathbb{P}_N} \mathbb{P}\left\{ \underline{R}^g_i \le -\alpha^g_{i,t} e^T \omega_t \le \overline{R}^g_i \right\} \ge 1 - \rho_r \quad (9)$$

$$\inf_{\mathbb{P}\in\mathbb{P}_N} \mathbb{P}\left\{ \underline{R}^h_i \le -\alpha^h_{i,t} e^T \omega_t \le \overline{R}^h_i \right\} \ge 1 - \rho_r \quad (10)$$

$$\inf_{\mathbb{P}\in\mathbb{P}_N} \mathbb{P}\left\{ \underline{P}^g_i \le P^g_{i,t} - \alpha^g_{i,t} e^T \omega_t \le \overline{P}^g_i \right\} \ge 1 - \rho_p \quad (11)$$

$$\inf_{\mathbb{P}\in\mathbb{P}_N} \mathbb{P}\left\{ \underline{P}^h_i \le P^h_{i,t} - \alpha^h_{i,t} e^T \omega_t \le \overline{P}^h_i \right\} \ge 1 - \rho_p \quad (12)$$

$$\inf_{\mathbb{P}\in\mathbb{P}_N} \mathbb{P}\left\{ \underline{P}^l_i \le \widetilde{P}^l_{i,t} \le \overline{P}^l_i \right\} \ge 1 - \rho_l \quad (13)$$

The above expressions indicate that the probability that the variable matches at least $1-\rho$ satisfies the corresponding constraints, of which Equation (8) expresses the upper and lower bound constraints of all node voltages, Equations (9) and (10) are the reserve capacity constraints of the generator set, and Equations (11)–(13) are the upper and lower bound constraints of the generator set output and line power.

The constraints of cascade hydropower plants are expressed as follows:

$$\begin{cases} r_{i,t}^{h} = r_{i-1,t}^{h} - (q_{i,t} - s_{i,t} + j_{i,t})\Delta t + \widetilde{q}_{Ini}\Delta t \\ r_{i0}^{h} = r_{i}^{ini}, r_{i,T}^{h} = r_{i}^{fin}, \underline{r}_{i}^{h} \le r_{i,t}^{h} \le \overline{r}_{i,t}^{h} \\ q_{i,t} = \sum_{m=1}^{M} q_{i,m,t} + \underline{q}_{i,m} \\ \overline{q}_{i,m} u_{i,m,t} \le q_{i,m,t} \le \overline{q}_{i,m} \\ \overline{q}_{i,m} u_{i,m,t} \le q_{i,m,t} \le \overline{q}_{i,m,t} u_{i,m-1,t} \\ P_{i,t} = \sum_{m=1}^{M} k_{i,m} q_{i,m,t} + \underline{P}_{i}^{h} \end{cases} \tag{14}$$

where $r_{i,t}^{h}$, $q_{i,t}$, $s_{i,t}$, $\widetilde{q}_{Ini}$, and $j_{i,t}$ denote the storage capacity, power generation flow, water spillage flow, inflow, and natural inflow of reservoir i during period t, respectively, and r_{i}^{ini} and r_{i}^{fin} denote the initial and termination storage capacity, respectively. To improve the solution efficiency, the reservoir flow and generation power constraints are approximated with piecewise linear functions, where $q_{i,m,t}$ denotes the flow of the mth segment of hydropower unit i during period t, $k_{i,m}$ and $\overline{q}_{i,m}$ denote the generation power coefficient and upper limit of flow of the mth segment, respectively, $q_{i,t}$ and $P_{i,t}^{h}$ denote the generation flow and generation power of hydropower unit i during period t, respectively, and $u_{i,m,t}$ is a 0–1 variable. When the flow of hydropower station i during period t exceeds the second segment, the variable is 1; otherwise, it is 0. In addition, there are upper and lower bound constraints and reserve capacity constraints of hydropower stations, which can be expressed in the form of distributed robust opportunity constraints among the safe operation constraints of power systems (Equations (10) and (12)).

2.4. Objective Function and Distributed Robust Optimization Framework

The objective function of this model is to minimize the sum of the actual power generation cost, reserve cost, water spillage cost, and regulation cost of each AGC unit under extreme distribution:

$$\min\left[\sum_{i\in G}\sum_{t\in T}\left(FG_i\left(P_{i,t}^{g}\right) + FR_{i,t}\right) + FS\right] + \sup_{\mathbb{P}\in\mathbb{P}_N} \mathbb{E}_{\mathbb{P}}\left(\sum_{i\in G}\sum_{t\in T}\alpha_{i,t}^{g} d_i^{g} \left|e^{T}\omega_t\right|\right) \tag{15}$$

of which:

$$FR_{it} = \overline{c}_i^{g}\overline{R}_{i,t}^{g} + \underline{c}_i^{g}\underline{R}_i^{g} + \overline{c}_i^{h}\overline{R}_{i,t}^{h} + \underline{c}_i^{h}\underline{R}_i^{h} \tag{16}$$

$$FS = c_s \sum_{i\in H}\sum_{t\in T} s_{i,t} \tag{17}$$

where $FG_i\left(P_{i,t}^{g}\right)$, $FR_{i,t}$, and FS denote the power generation cost, reserve cost, and water spillage cost, respectively. Parameters c_i^{g}, c_i^{h}, and c_s denote the cost coefficient of thermal power, hydropower reserve capacity, and water spillage cost, respectively. Furthermore, d_i^{g} is the adjustment cost coefficient of the thermal power unit in response to wind and solar forecasting errors. Power generation cost $FG_i\left(P_{i,t}^{g}\right)$ is a nondecreasing quadratic function, which can be approximated with a piecewise linear function to improve calculation efficiency, i.e., $FG_i\left(P_{i,t}^{g}\right)$ can be replaced by decision variables with the following constraints [23]:

$$\Phi_{i,t} \ge k_i^{n} P_{i,t}^{g} + b_i^{n} \tag{18}$$

where k_i^{n} and b_i^{n}, respectively, denote the slope and intercept of the nth segment in the piecewise linear approximation of thermal power unit i, which can be determined via the piecewise interpolation method [24]. This objective function can improve not only the economy but also the feasibility of decisions, as the decisions under this objective function need less adjustment in practice.

The problems involved in this model can be expressed in the following distributed robust optimization framework:

$$\min_{x} c^T x + \sup_{\mathbb{P}\in\mathbb{P}_N} \mathbb{E}_{\mathbb{P}}\{F(x,\omega)\} \tag{19a}$$

$$s.t.\ h_i(x) = 0, \tag{19b}$$

$$g_j(x) \geq 0, \tag{19c}$$

$$\sup_{\mathbb{P}\in\mathbb{P}_N} \mathbb{P}\left\{ a_k^T(x)\omega + b_k(x) \leq 0 \right\} \geq 1-\rho \tag{19d}$$

Optimization problem (19) cannot be solved directly, because the objective function and constraints contain random variables. It must be reformulated and transformed into a deterministic problem with only control variables.

3. Model Reformulation

3.1. Reformulation of The Objective Function

The objective function contains the worst-case expectation. Note that $e^T\omega_t$ is a scalar and appears as a whole in the objective function. Therefore, $\xi = e^T\omega_t$ is preferable, and its supporting set is $\left[\underline{\xi}, \overline{\xi}\right]$. Let $(\hat{\xi}_1, \hat{\xi}_2, \cdots, \hat{\xi}_N)$ correspond to the sample data $(\hat{\omega}_1, \hat{\omega}_2, \cdots, \omega_N)$; then, the worst-case expectation is transformed as follows:

$$\sup_{\mathbb{P}\in\mathbb{P}_N} \mathbb{E}_{\mathbb{P}}\left\{ \sum_{i\in G} \sum_{t\in T} \alpha_{i,t}^g d_i^g |\xi| \right\} \tag{20}$$

According to [16], Equation (20) can be transformed into the following equation by using a strong duality:

$$\inf_{\kappa\geq 0} \left\{ \kappa\delta + \frac{1}{N}\sum_{i=1}^N \sup_{\underline{\xi}\leq\xi\leq\overline{\xi}} \left[\sum_{i\in \mathcal{G}} \sum_{t\in T} \alpha_{i,t}^g d_i^g |\xi| - \kappa\|\xi - \hat{\xi}_i\| \right] \right\} \tag{21}$$

Because problem (21) is an affine function with respect to variable ξ in the interval $\left[\underline{\xi}, \hat{\xi}_i\right]$ and $\left[\hat{\xi}_i, \overline{\xi}\right]$, the optimal solution must be obtained at the vertices of the feasible region, i.e., $\underline{\xi}$, $\hat{\xi}_i$, or $\overline{\xi}$. Therefore, the worst-case expectation is equivalent to the following problem:

$$\sup_{\mathbb{P}\in\mathbb{P}_N} \mathbb{E}_{\mathbb{P}}\left\{ \sum_{i\in G} \sum_{t\in T} \alpha_{i,t}^g d_i^g |\xi| \right\} = \begin{cases} \inf_{\kappa\geq 0} \kappa\delta + \frac{1}{N}\sum_{i=1}^N \eta_i \\ s.t.\ \eta_i \geq \sum_{i\in G} \sum_{t\in T} \alpha_{i,t}^g d_i^g \overline{\xi} - \kappa(\overline{\xi} - \hat{\xi}_i), \forall i \leq N \\ \quad \eta_i \geq \sum_{i\in G} \sum_{t\in T} \alpha_{i,t}^g d_i^g \underline{\xi} + \kappa\left(\underline{\xi} - \hat{\xi}_i\right), \forall i \leq N \\ \quad \eta_i \geq \sum_{i\in G} \sum_{t\in T} \alpha_{i,t}^g d_i^g \hat{\xi}_i, \forall i \leq N \end{cases} \tag{22}$$

Lemma 1. *The optimal value of problem (22) is equal to Equation (23):*

$$\sup_{\mathbb{P}\in\mathbb{P}_N} \mathbb{E}_{\mathbb{P}}\left\{ \sum_{i\in G} \sum_{t\in T} \alpha_{i,t}^g d_i^g |\xi| \right\} = b \sum_{i\in G} \sum_{t\in T} \alpha_{i,t}^g d_i^g \tag{23}$$

where $b = \min\left\{ max\left\{ \overline{\xi}, -\underline{\xi} \right\}, \delta + \sum_{i=1}^N |\hat{\xi}_k| \right\}$ *and* δ *is the Wasserstein radius. The proof is given in the Appendix A.*

Compared with the complex form of the worst-case expectation (Equation (22)), Equation (23) is concise and is only decided by the control variable α in affine form. Therefore, the number of variables and constraints will not increase with an increased number of samples.

3.2. Reformulation of The Distributed Robust Chance

Equations (8)–(13) express the joint distributed robust chance constraint (DRCC). According to [20,25], if the function in the distributed robust chance constraint is affine with respect to both control variable x and random variable ω, i.e., the function exhibits the form of Equation (19d), where $a_k^T(x)$ is an affine function of x; then, the DRCC can be transformed into a set as follows:

$$Z = \left\{ x \middle| \begin{array}{l} \delta\lambda_k - \rho\beta_k + \frac{1}{N}\sum_{i=1}^{N} z_i^k \leq 0 \\ a_k^T(x)\hat{\omega}_i + b_k(x) + \beta_k - z_i^k \leq 0 \\ \|a_k(x)\|_* \leq \lambda_k \\ \lambda_k, \beta_k, z_i^k \geq 0 \end{array} \right\} \tag{24}$$

Model (24) is an approximate equivalent form of the DRCC with high accuracy, but the number of inequalities and variables in the model can rapidly increase with the number of samples, resulting in a significant decrease in the calculation efficiency of this method when the sample size is large. To overcome this defect, the second equation in Model (24) is substituted into the first one, and the first equation is retained to obtain the approximate set Z_1 of set Z. As the L_1-norm is used when constructing the ambiguity set, the dual norm $\|\cdot\|_*$ takes the infinite norm L_∞, and the final approximate set Z_1 is expressed in (25). Compared with set Z, the number of inequalities and variables contained in set Z_1 does not increase with the sample size; therefore, it has better computational performance when dealing with a large sample size.

$$Z_1 = \left\{ x \middle| \begin{array}{l} \delta\lambda_k - \rho\beta_k + \frac{1}{N}\sum_{i=1}^{N} z_i^k \leq 0 \\ a_k^T(x)\frac{1}{N}\sum_{i=1}^{N} \hat{\omega}_i + b_k(x) \leq (\rho - 1)\beta_k - \delta\lambda_k \\ -\lambda_k \leq a_k(x) \leq \lambda_k \\ \lambda_k, \beta_k, z_i^k \geq 0 \end{array} \right\} \tag{25}$$

As such, the difficult part of the original problem is transformed into an easy-to-manage form. The original problem is transformed into a mixed-integer linear programming problem, which can be efficiently solved with a commercial solver.

4. Calculation Results and Discussion

This section presents case studies on modified IEEE-118 and 300 node systems. The 118-node system was used to verify the effectiveness of the proposed DR-DOPF method. The test of the 300-bus system focused on the characteristics of the DR-DOPF method by comparing it to other methods to address the uncertainty. The number of periods of all models was set to 24, and the parameters of wind and solar output data were retrieved from https://www.tennet.eu/, accessed on 24 July 2019. The parameters of hydropower stations came from three stations on the Yellow River in China: the Bapanxia, Yanguoxia, and Daxia Hydropower Stations. All the programs were run on a PC with an Intel Core i5 CPU and 8 GB RAM by calling CPLEX with MATLAB 2021a.

4.1. IEEE-118 Node System

4.1.1. Effectiveness Verification

The modified IEEE118 node system includes 16 thermal power plants, three cascade hydropower plants, three wind power plants with a capacity of 200 MW, and two solar power plants with a capacity of 150 MW. The specific parameters of wind and solar power plants are shown in Table 1. To verify the effectiveness of DR-DOPF in reducing the comprehensive power generation cost and water spillage, the scenario displayed in

Figure 1 is selected from the sample. Under this scenario, the real output of the sum wind and solar power considerably exceeds the forecasted output in periods 5–9 and 15–

Table 1. Parameters of wind and solar power.

	Number of Plants	Connection Nodes	Capacity	Power Factor
Wind farm	3	28, 59, 83	200 MW	0.95
Solar power plant	3	36, 74, 97	150 MW	0.95

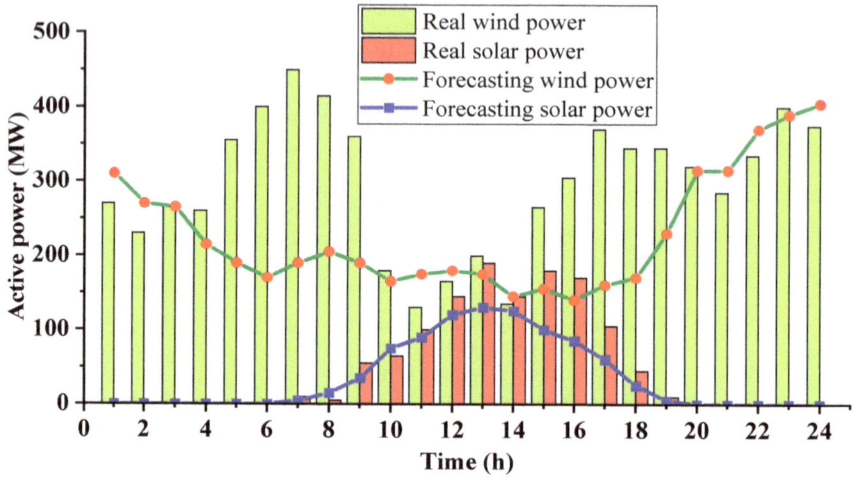

Figure 1. Output of wind and solar power.

For convenience of expression, the traditional dynamic optimal power flow without considering water spillage is abbreviated as TN-DOPF. The TN-DOPF model does not consider the uncertainty of wind and solar output, and the objective function does not include the cost of water spillage. Regarding the DR-DOPF model, the confidence level of the Wasserstein ambiguity set is set to 0.95, and the risk factor is set to $\rho_v = \rho_r = \rho_p = \rho_l = 0.05$.

The result calculated is the day-ahead generation scheduling. To obtain the operation curve and water spillage of hydropower plants in real-time dispatching, the two models are treated as follows.

Regarding the DR-DOPF model, the solution of the real-time output value is given by Equation (6). Specifically, at the real-time dispatching stage, the real output of wind and solar power and the forecasting errors are already known, and the participation factors $\alpha^g_{i,t}$, $\alpha^h_{i,t}$ and the unit planned output $P^g_{i,t}$, $P^h_{i,t}$ are calculated. Hence, the real output of each plant can be directly determined according to Equation (6). Actually, this method dynamically distributes the unbalanced power to each unit according to the participation factor. The flowchart of the DR-DOPF model is shown in Figure 2.

Regarding TN-DOPF, because this model does not involve a participation factor, the actual output cannot be directly determined, but this issue can be addressed according to the actual dispatching situation. To embody the principle by which the dispatcher prioritizes hydropower adjustment in practical operation, it is assumed that 90% of the wind and solar power forecast error is balanced by the hydropower plants and the remaining 10% by the thermal plants, and the power imbalance is assumed to be evenly distributed between the three hydropower plants.

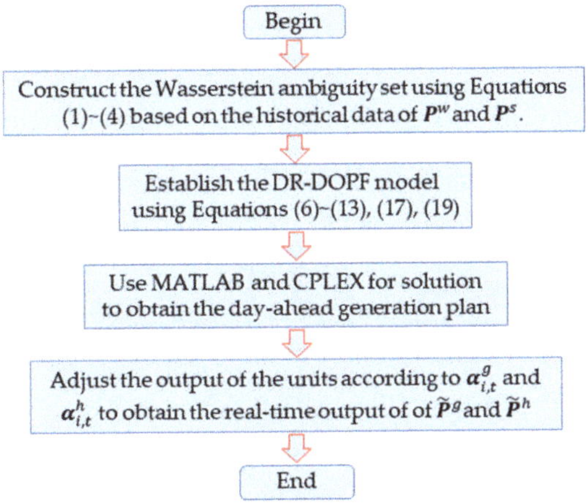

Figure 2. Flowchart of DR-DOPF model.

Figure 3 shows the optimal output of the day-ahead scheduling and the real-time output of the power plant under the DR-DOPF model when the sample size is 100. The real-time output of hydropower and thermal power is lower than that based on the generation plan, because the real output of wind and solar power exceeds the predicted value and the output curve of thermal power is relatively smooth in the previous generation plan but fluctuates in the real-time output curve, especially during periods 5–10 and 15–19. During these two periods, the actual output of wind and solar power is significantly higher than expected, which indicates that the thermal power unit under the DR-DOPF model greatly participates in system power regulation. Although this suggests that the thermal units will produce higher regulation cost, compared to the regulation mode mainly relying on hydropower, the joint optimal regulation mode of hydropower and thermal power can produce higher economic benefits. This point is discussed in detail later.

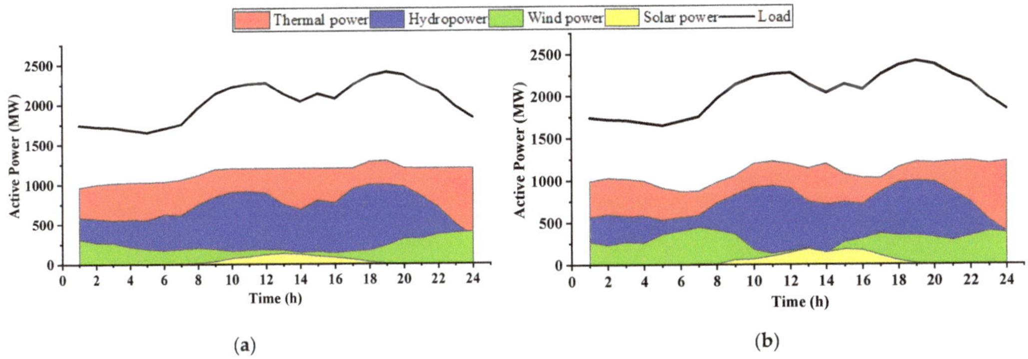

Figure 3. Output of various power plants. (**a**) Day-ahead scheduling; (**b**) output of real-time dispatch.

Figure 4 displays the day-ahead and real-time output of hydropower plant 1 under the DR-DOPF and TN-DOPF models; the other two hydropower plants are similar and are not described here. Figure 4a,c reveal that neither model produces water spillage in the day-ahead scheduling. However, at the real-time stage, both models produce water spillage (Figure 4b,d), which is considerably smaller for DR-DOPF than TN-DOPF. The reason is that when dealing with forecasting errors of wind and solar power output, the hydropower plants under these two models must bear a certain amount of the power imbalance. Because the real value of the wind power output during periods 5–9 and 15–19 greatly exceeds the predicted value, under real-time dispatching, to maintain the system power balance, hydropower plant 1 must greatly reduce the power generation based on the previous power generation plan, resulting in water spillage. In addition, Figure 5 shows that the adjusting power of hydropower plant 1 is lower under the DR-DOPF model than under the TN-DOPF model, especially during periods 5–9 and 15–19; consequently, the water spillage amount under the DR-DOPF model is much smaller.

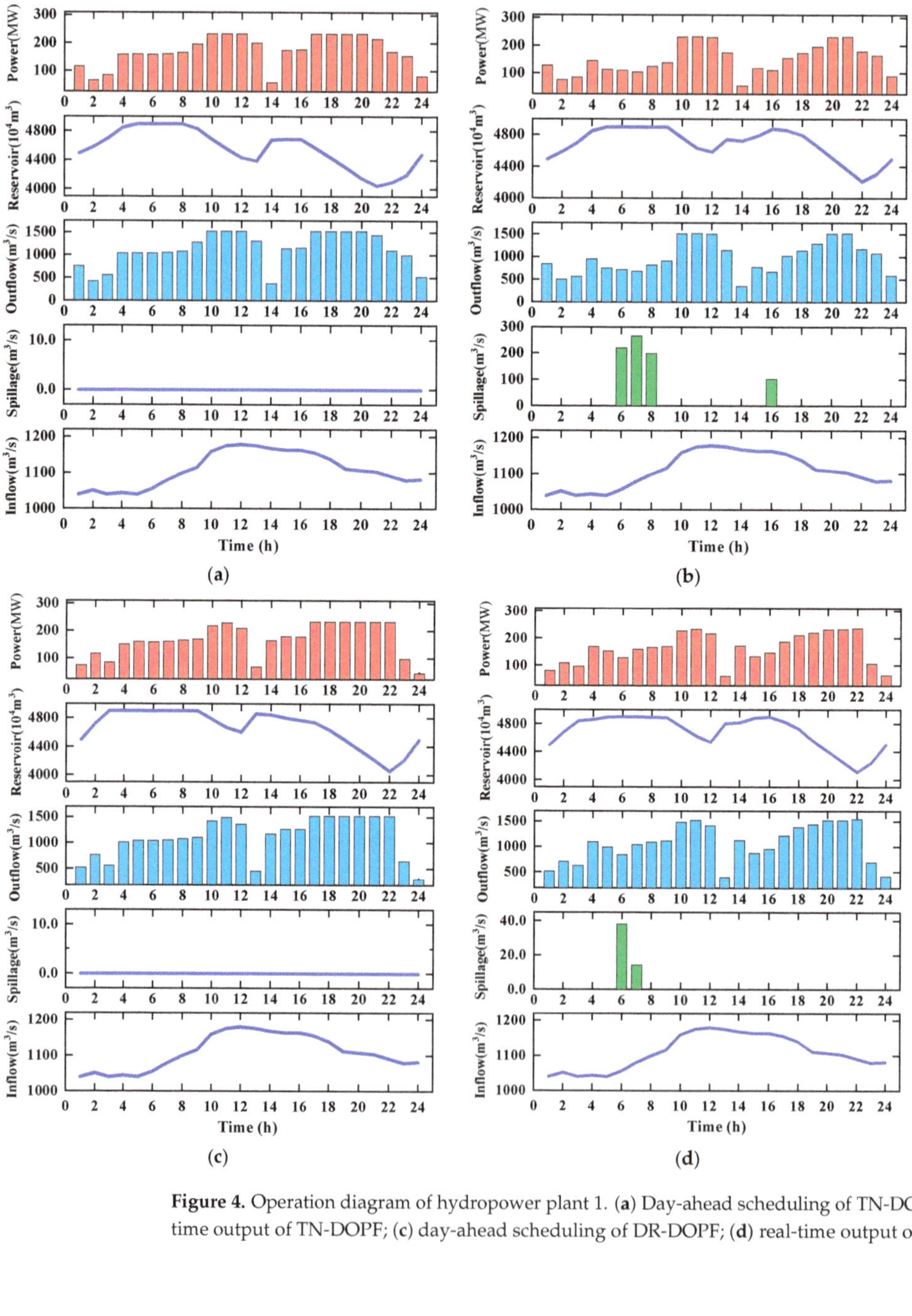

Figure 4. Operation diagram of hydropower plant 1. (**a**) Day-ahead scheduling of TN-DOPF; (**b**) real-time output of TN-DOPF; (**c**) day-ahead scheduling of DR-DOPF; (**d**) real-time output of DR-DOPF.

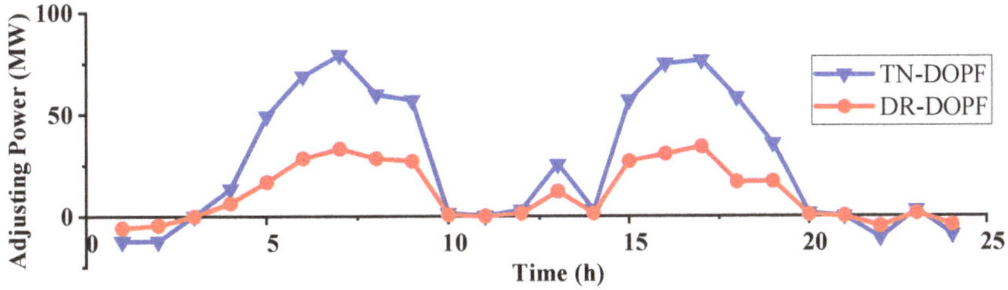

Figure 5. Adjusting power of hydropower plant 1 under two models.

Figure 6 displays the total unbalanced power and adjusting power of thermal power and hydropower under the two models. It can be seen that the adjusting thermal power under DR-DOPF is larger than that of TN-DOPF because it is determined by the participation factors calculated by the day-ahead scheduling. The day-ahead plan considers the uncertainty of wind and solar output, and the objective function contains the water spillage cost. To minimize the water spillage cost, this inevitably requires that the decisions made according to the participation factors can optimally assign the unbalanced power and fully utilize the regulation capacity of the thermal power units. Hence, the regulation burden of the hydropower units can be reduced, and consequently, the water spillage can be decreased. The NDOPF model is modeled on the actual dispatching situation, in which dispatchers tend to prioritize the hydropower plant-adjusting power, and thermal power units participate in regulation to a lesser degree. In this case, power regulation mainly depends on hydropower. Due to the high regulation burden of the hydropower units and the unreasonable distribution of regulation, when the real value of wind and solar power output greatly exceeds the expectation, water spillage easily occurs.

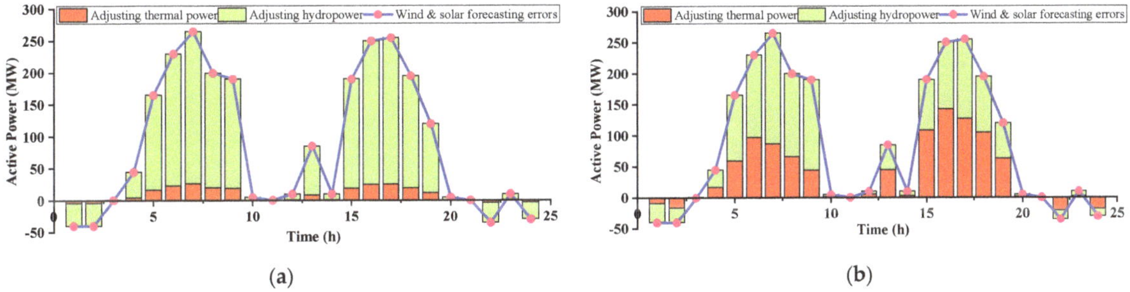

Figure 6. Total adjusting power of thermal power and hydropower with (**a**) TN-DOPF model and (**b**) DR-DOPF model.

Under the TN-DOPF model, the proportion of thermal power participating in regulation is low, which can reduce the regulation cost of thermal power units; however, this phenomenon results in large water spillage. Although the water spillage is not reflected in the cost function of TN-DOPF, it cannot be ignored. To fairly compare the comprehensive cost of TN-DOPF and DR-DOPF, the water spillage of TN-DOPF is converted into the water spillage cost according to Equation (17) and added to its objective function value to obtain the comprehensive cost.

Table 2 presents the comprehensive generation cost and total water spillage according to the two methods. Compared with the TN-DOPF model, the total water spillage of DR-DOPF decreases considerably, with a decrease ratio of more than 86%, and the comprehensive cost decreases by more than 12%, which indicates that this model can effectively reduce the total water spillage and comprehensive generation cost. This analysis confirms the previous conclusion that the joint optimal regulation of hydropower and thermal power can produce higher comprehensive economic benefits compared to regulation mainly relying on hydropower.

Table 2. Comparison of TN-DOPF and DR-DOPF results.

Model	Comprehensive Cost		Total Water Spillage	
	Value ($\times 10^5$ USD)	Reduction Ratio	Value ($\times 10^5$ m^3)	Reduction Ratio
NDR-DOPF	2.830	–	69.746	–
DR-DOPF	2.473	12.6%	9.275	86.7%

Table 3 summarizes the test results of DR-DOPF for various sample sizes. The Wasserstein radius exhibits a negative correlation with sample size, because when more sample data are used, more sufficient information can be obtained about the real probability distribution. Therefore, impossible distributions can be excluded from the Wasserstein ball, resulting in a narrower radius range, less conservativeness of the ambiguity set, and a lower comprehensive cost and water spillage. It can be seen that when the sample size increases from 20 to 5000, the comprehensive power generation cost of the system reduces from 2.613 $\times$ 10^5 USD to 2.409 $\times$ 10^5 USD, a reduction of 7.8%; and total water spillage reduces from 9.458 $\times$ 10^5 m^3 to 9.201 $\times$ 10^5 m^3, a reduction of 2.7%. This result verifies that with the increase in the sample size, the conservatism of the model lessens and the economy of the model improves. In addition, it should be noted that the number of constraints and variables of DR-DOPF do not increase with the sample size, and the operation time basically remains stable from 28–30 s, which indicates that the DR-DOPF model achieves good computational performance.

Table 3. Test results of DR-DOPF under various sample numbers.

Sample Size	$\delta(N)$	Number of Constraints	Number of Variables	CPU Time (s)	Comprehensive Cost ($\times 10^5$ USD)	Total Water Spillage ($\times 10^5$ m^3)
20	9.731	7341	34,806	29.2	2.613	9.458
50	3.406	7341	34,806	28.6	2.518	9.326
200	1.557	7341	34,806	28.3	2.448	9.238
1000	0.348	7341	34,806	29.7	2.418	9.217
2000	0.102	7341	34,806	29.1	2.409	9.201

4.1.2. Comparisons with RO and SO

In this part, the characteristics of the DR-DOPF method and RO and SO are evaluated. The RO model requires that safety constraints (8–13) must be satisfied in any case. The SO method assumes that samples obey a normal distribution [26,27]. In this case, constraints (8–13) can be transformed into second-order cone constraints [28]. The calculation results obtained using these methods are displayed in Figure 7.

Figure 7 shows that the RO method has the largest water spillage and comprehensive cost, and the SO method has the smallest. Regardless of the sample size, the water spillage and comprehensive cost of DR-DOPF always vary between RO and SO because RO completely ignores the distribution information of samples and is too conservative, and SP assumes that the sample follows a certain distribution, which is too aggressive, while DR-DOPF completely depends on the obtained samples. When limited samples are used, the corresponding Wasserstein radius is large. The extreme distribution in the ambiguity set deviates considerably from the real distribution, so the calculation result is relatively conservative, and is closer to RO at this stage. When the sample size increases, the ambiguity set shrinks, and the extreme distribution is closer to the real distribution. Therefore, the conservatism of the calculation result decreases, and the result is closer to SP. In summary, the order of conservatism from high to low is RO > DR-DOPF (20) > DR-DOPF (50) > DR-DOPF (200) > DR-DOPF (1000) > DR-DOPF (2000) > SO.

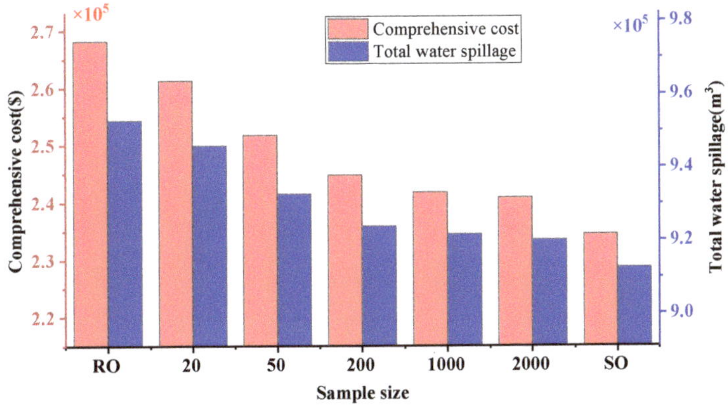

Figure 7. Test results under DR-DOPF, RO, and SO methods.

4.2. IEEE-300 Node System

The modified IEEE-300 system includes 30 thermal power plants, among which the parameters of cascade hydropower stations, wind farms, and photovoltaic power plants are the same as those of the 118-node system, and the scenario shown in Figure 1 is still selected as the scenario of wind and solar output. The wind farms are connected on bus 77, 186, and 235 respectively, and the solar power plants are connected on bus 49, 123, and 284, respectively. DR-DOPF, RO, and SO results are obtained on the same sample set. In addition, the Monte Carlo simulation method is applied, involving 10^4 samples, to evaluate the performance beyond the sample, i.e., the minimum reliability of all security constraints.

4.2.1. Effectiveness Verification

Similar to the method of node 118, in this part, we first compare the test results of DR-DOPF and TN-DOPF models to verify the effectiveness of the proposed DR-DOPF model in reducing the comprehensive power generation cost and water spillage. The results are shown in Table 4.

Table 4. Comparison of TN-DOPF and DR-DOPF results.

Model	Comprehensive Cost (USD)		Total Water Spillage (m^3)	
	Value ($\times 10^5$ USD)	Reduction Ratio	Value ($\times 10^5$ m^3)	Reduction Ratio
TN-DOPF	6.035	–	53.154	–
DR-DOPF	5.341	11.5%	6.219	88.3%

Table 4 presents the test results of TN-DOPF and DR-DOPF when the sample size is 100. Compared with TN-DOPF, the total water spillage of DR-DOPF decreases considerably, with a decrease ratio of more than 88%, and the comprehensive cost decreases by approximately 12%. The results of the IEEE-300 node system indicate that the proposed DR-DOPF model can reduce the comprehensive generation cost and water spillage effectively. This conclusion is similar to that for the 118-node system.

4.2.2. Comparisons with RO and SP

Table 5 summarizes the comparison test results of the DR-DOPF model with RO and SO methods using various sample sizes, where the confidence level of the Wasserstein ambiguity set is set to 0.95, the risk factor is set to $\rho_v = \rho_r = \rho_p = \rho_l = 0.05$, and the processing method of RO and SO is the same as that of the IEEE118 node.

Table 5. Comparison of the DR-DOPF, RO, and SP results.

Model	Sample Size	Comprehensive Cost ($\times 10^6$ USD)	Total Water Spillage ($\times 10^5$ m^3)	CPU Time (s)	Reliability (%)
DR-DOPF	20	5.468	6.338	47.2	99.98
	50	5.379	6.247	48.6	99.56
	200	5.323	6.195	47.7	99.18
	1000	5.291	6.162	46.2	98.86
	2000	5.276	6.147	48.5	98.52
RO	2000	5.540	6.416	18.3	100
SO	2000	5.198	6.039	27.9	93.14

As can be seen from Table 5, the results of the IEEE-300 node system are similar to those of the IEEE-118 system. Specifically, RO has the highest comprehensive generation cost and total water spillage, SO has the lowest, and the value of DR-DOPF is always between RO and SO and decreases with the sample size. In terms of the calculation time, DR-DOPF requires the longest time, followed by SO, and RO requires the shortest time. When the sample size increases, the calculation time of DR-DOPF basically remains unchanged because the number of constraints and variables do not change with the number of samples; therefore, the computational burden basically remains constant.

Table 5 also compares the reliability levels of the safety constraints of the three methods. RO exhibits the highest reliability of 100%, but obtaining high reliability has drawbacks; notably, the comprehensive cost and total water spillage of RO are the highest in all cases. In other words, the RO model is the most conservative. SO attains the lowest reliability and does not satisfy 95% of the minimum reliability requirements of the safety constraints. This occurs because SO assumes that the sample follows a certain distribution (normal distribution). However, the real distribution of the prediction error may not follow a normal distribution [29,30], which results in an over-aggressive strategy; therefore, it does not satisfy the minimum reliability requirements. The reliability of DR-DOPF lies between RO and SO, which can satisfy the minimum requirements of reliability. With an increasing number of samples, the reliability of the DR-DOPF model gradually decreases. This occurs because with increased samples, the Wasserstein ambiguity set decreases, which reduces the conservativeness of the DR-DOPF model and the comprehensive cost and reliability. Notably, on the premise of meeting the minimum reliability requirements of the safety constraints, with increasing sample size, the DR-DOPF model sacrifices reliability in exchange for higher economic benefits. It is easy to calculate that when the sample size is 2000, the comprehensive cost of DR-DOPF is only 1.5% higher and the water spillage is only 1.8% more compared to SO, which shows that when the sample is large enough, the DR-DOPF model can ensure reliability based on small economic loss. Compared with the over-conservatism of RO and the over-aggressiveness of SO, DR-DOPF can balance economy and conservatism well. The above analysis once again confirms the previous conclusion on the conservativeness ranking; i.e., in the IEEE-300 system, conservativeness is still ranked from high to low as: RO > DR-DOPF (20) > DR-DOPF (50) > DR-DOPF (200) > DR-DOPF (1000) > DR-DOPF (2000) > SO.

In addition, we also study the influence of the confidence level of the Wasserstein ambiguity set on the results. The sampling number is $N = 1000$, and the results of DR-DOPF at different confidence levels and of SO and RO are obtained. The results are displayed in Figure 8.

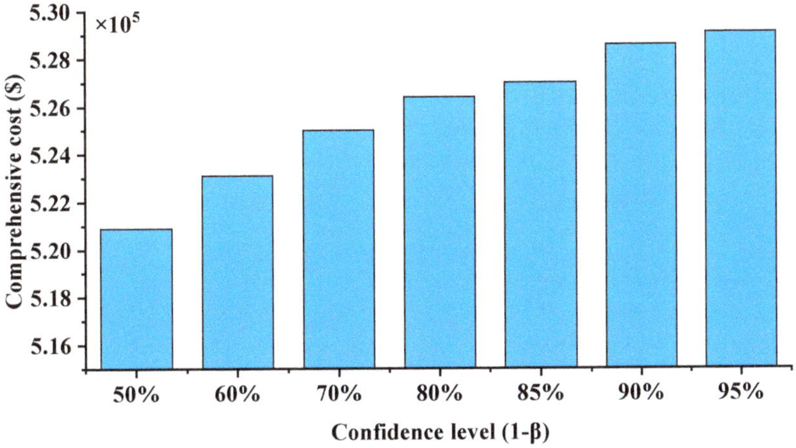

Figure 8. Comprehensive cost with confidence level.

Figure 8 shows that the comprehensive cost increases with the confidence level, because when the confidence level increases, to ensure that the ambiguity set contains the real distribution with a higher probability, the Wasserstein radius must become larger. Therefore, the extreme distribution contained in the ambiguity set deviates from the real distribution, and the corresponding result becomes more conservative; consequently, the comprehensive generation cost increases. Hence, controlling the confidence level can limit the conservativeness of the results and ensure a balance between conservativeness and economy in the model.

5. Conclusions

In this study, a robust multi-energy dynamic distribution optimal power flow model is proposed. The model uses the ambiguity set with the Wasserstein metric to address the uncertainty of wind and solar output, and also considers water spillage. The Wasserstein ambiguity set is data-driven and does not require an assumption of the distribution information in advance. Regarding the worst-case expectation in the objective function, its exact equivalent form is obtained through reformulation. The equivalent form is highly concise, and its scale does not change with an increasing number of samples; therefore, it exhibits excellent computational performance. Further, the distributed robust chance constraints are transformed into tractable reformulations. By the above means, a tractable DR-DOPF model is proposed. By minimizing the comprehensive power generation cost, the DR-DOPF model provides the factor coefficients of each power plant participating in the regulation process during real-time dispatching. The test results of IEEE-300 and other systems indicate that, if the power system operates according to the day-ahead scheduling and regulates based on participation factors given by the DR-DOPF model, the comprehensive operation cost and water spillage of hydropower stations can be effectively reduced. In addition, with an increased sample size, the DR-DOPF model can reduce conservatism and improve economy based on satisfying the reliability of safety constraints.

Author Contributions: Conceptualization, H.W.; methodology, G.S.; software, G.S.; validation, H.W.; formal analysis, H.W. and G.S.; investigation, H.W. and G.S.; resources, H.W.; data curation, G.S.; writing—original draft preparation, G.S.; writing—review and editing, H.W. and G.S.; visualization, G.S.; supervision, H.W.; project administration, H.W.; funding acquisition, H.W. All authors have read and agreed to the published version of the manuscript.

Funding: This research was funded by the National Natural Science Foundation of China, Research on minimum water spillage mechanism of multi-energy power system based on big data driven distributed robust optimization (Funding number: 51967002).

Institutional Review Board Statement: Not applicable.

Informed Consent Statement: Not applicable.

Data Availability Statement: Not applicable.

Conflicts of Interest: The authors declare no conflict of interest.

Appendix A

Proof of Lemma 1. The optimization variables in problem (22) are κ and η, and the control variable α is equivalent to a parameter here. Let $\psi = \sum_{i\in G}\sum_{t\in T}\alpha_{i,t}d_i$, and problem (22) can be rewritten as:

$$\inf_{\kappa\geq 0}\kappa\delta + \frac{1}{N}\sum_{i=1}^{N}\eta_i \quad \text{(A1a)}$$

$$s.t.\ \eta_i \geq \psi\overline{\xi} - \kappa\left(\overline{\xi} - \hat{\xi}_i\right), \forall i \leq N \quad \text{(A1b)}$$

$$\eta_i \geq -\psi\underline{\xi} + \kappa\left(\underline{\xi} - \hat{\xi}_i\right), \forall i \leq N \quad \text{(A1c)}$$

$$\eta_i \geq \psi\left|\hat{\xi}_i\right|, \forall i \leq N \quad \text{(A1d)}$$

Because κ is a scalar, the discussion of κ is simple. Specifically, the optimal solution of problem (A1) can be obtained either in $\kappa > \psi$ or $0 \leq \kappa \leq \psi$. First, the optimal solution of the problem must not be obtained at $\kappa > \psi$. The proof is as follows:

For $\kappa \geq \psi$, the right side of Equation (A1b) is equal to:

$$\psi\overline{\xi} - \kappa\left(\overline{\xi} - \hat{\xi}_i\right) = (\psi - \kappa)\left(\overline{\xi} - \hat{\xi}_i\right) + \psi\hat{\xi}_i \leq \psi\hat{\xi}_i \leq \psi\left|\hat{\xi}_i\right| \quad \text{(A2)}$$

The right side of Equation (A1c) minus $\psi\left|\hat{\xi}_i\right|$ equals the following:

$$-\psi\underline{\xi} + \kappa\left(\underline{\xi} - \hat{\xi}_i\right) - \psi\left|\hat{\xi}_i\right| = \begin{cases} (\kappa - \psi)\underline{\xi} - (\kappa + \psi)\hat{\xi}_i < 0, \hat{\xi}_i \geq 0 \\ (\kappa - \psi)\left(\underline{\xi} - \hat{\xi}_i\right) < 0, \hat{\xi}_i < 0 \end{cases} \quad \text{(A3)}$$

Therefore, the right side of Equation (A1c) is also less than $\psi\left|\hat{\xi}_i\right|$.

Hence, constraints (A1b–d) can be combined, which is equivalent to constraint (A1d). To minimize the objective function, η_i must take its minimum value $\psi\left|\hat{\xi}_i\right|$. After further analysis, for $\kappa \geq \psi$, the first term $\kappa\delta$ in the objective function increases with κ, while the second term $\frac{1}{N}\sum_{i=1}^{N}\eta_i = \frac{1}{N}\sum_{i=1}^{N}\psi\left|\hat{\xi}_i\right|$ and remains unchanged. Hence, the total objective function value increases with κ. Thus, when $\kappa > \psi$, the objective function is greater than when $\kappa = \psi$. In other words, the original problem cannot yield the optimal solution for $\kappa > \psi$. Accordingly, the optimal solution must be obtained on the complement of $\kappa > \psi$, namely, $0 \leq \kappa \leq \psi$.

Because problem (A1) is a linear programming problem about κ and η, the optimal solution must be at the vertices, i.e., $\kappa = 0$ or $\kappa = \psi$. This is analyzed in the two cases below.

For $\kappa = 0$, problem (A1) is simplified as:

$$\inf_{\eta}\left\{\frac{1}{N}\sum_{i=1}^{N}\eta_i\middle|\eta_i \geq \psi\overline{\xi}, \eta_i \geq -\psi\underline{\xi}, \eta_i \geq \psi\left|\hat{\xi}_i\right|\right\} = \max\left\{\psi\overline{\xi}, -\psi\underline{\xi}\right\} \quad \text{(A4)}$$

For $\kappa = \psi$, problem (A1) is simplified as:

$$\inf_{\eta}\left\{\psi\delta + \frac{1}{N}\sum_{i=1}^{N}\eta_i\middle|\eta_i \geq \psi\hat{\xi}_i, \eta_i \geq -\psi\hat{\xi}_i, \eta_i \geq \psi\left|\hat{\xi}_i\right|\right\} = \psi\left(\delta + \frac{1}{N}\sum_{i=1}^{N}\left|\hat{\xi}_i\right|\right) \quad \text{(A5)}$$

Therefore, the optimal value of the original problem is the smaller value of the above two cases, namely:

$$\min\left\{\max\left\{\psi\overline{\xi},-\psi\underline{\xi}\right\},\psi\left(\delta+\tfrac{1}{N}\sum_{i=1}^{N}\left|\hat{\xi}_k\right|\right)\right\}=\min\left\{\max\left\{\overline{\xi},-\underline{\xi}\right\},\delta+\tfrac{1}{N}\sum_{i=1}^{N}\left|\hat{\xi}_k\right|\right\}\cdot\psi=b\psi \tag{A6}$$

As such, the complete proof is provided. □

Appendix B

Table A1. Parameters of hydropower stations.

Hydropower Station	Connection Node	r_i^{ini} ($\times 10^4$ m^3)	r_i^{fin} ($\times 10^4$ m^3)	$\underline{r}_i^h$ ($\times 10^4$ m^3)	$\overline{r}_i^h$ ($\times 10^4$ m^3)	$\underline{P}_i^h$ (MW)	$\overline{P}_i^h$ (MW)
H1	78	4500	4500	4000	4900	45.90	232.56
H2	34	21,600	21,600	21,000	22,000	90.44	445.74
H3	92	8600	8600	5500	9000	58.80	318.30

H1: Bapanxia Hydropower Station; H2: Yanguoxia Hydropower Station; H3: Daxia Hydropower Station.

Table A2. Parameters of hydropower stations.

Hydropower Station	$k_{i,1}$(MW·s/m^3)	$k_{i,2}$ (MW·s/m^3)	$k_{i,3}$ (MW·s/m^3)	$k_{i,4}$ (MW·s/m^3)	$\underline{q}_{i,m}$ (m^3/s)	$\overline{q}_{i,m}$ (m^3/s)	$\overline{q}_i$(m^3/s)
H2	0.145	0.161	0.145	0.161	300	305	1520
H1	0.315	0.331	0.315	0.331	280	275	1380
H3	0.188	0.204	0.188	0.204	300	331	1624

Table A3. Parameters of wind and solar output data.

Time	P^w (MW)	P^s (MW)	$\widetilde{P}^w$ (MW)	$\widetilde{P}^s$ (MW)	Time	P^w (MW)	P^s (MW)	$\widetilde{P}^w$ (MW)	$\widetilde{P}^s$ (MW)
1	310.2	0	269.3	0	13	176.1	129.6	202.1	191.8
2	271.3	0	231.4	0	14	145.4	123.8	135.5	145.1
3	263.3	0	263.7	0	15	154	101.5	264.8	179.6
4	214.8	0	258.4	0	16	141.1	85.7	303.9	171.4
5	190.7	0	357.8	0	17	158.6	58.8	370	105
6	171.4	0	403	0	18	168.7	24.3	345.1	45.2
7	192.7	5.1	451.7	10.2	19	229.8	4.9	344.7	9.8
8	203.6	14.8	413.2	5	20	314.9	0	318.4	0
9	190.6	34.4	362.1	54.8	21	315.2	0	285.6	0
10	165	75.3	178.9	66.3	22	370.8	0	335.3	0
11	176.1	90.3	130.4	102.7	23	390	0	400.7	0
12	179.8	120.6	166	144.5	24	406.1	0	374.6	0

P^w: Forecasting wind power; P^s: Forecasting solar power; $\widetilde{P}^w$: Real wind power; $\widetilde{P}^s$: Real solar power.

References

1. National Energy Administration. *Implementation Plan for Solving the Problem of Spillage Water, Air and Light*; National Energy Administration: Beijing, China, 2017.
2. Shaheen, A.M.; Elsayed, A.M.; El-Sehiemy, R.A.; Ghoneim, S.S.; Alharthi, M.M.; Ginidi, A.R. Multi-dimensional energy management based on an optimal power flow model using an improved quasi-reflection jellyfish optimization algorithm. *Eng. Optim.* **2022**, 1–23. [CrossRef]
3. El-Sehiemy, R.; Elsayed, A.; Shaheen, A.; Elattar, E.; Ginidi, A. Scheduling of Generation Stations, OLTC Substation Transformers and VAR Sources for Sustainable Power System Operation Using SNS Optimizer. *Sustainability* **2021**, *13*, 11947. [CrossRef]
4. Shaheen, A.M.; El-Sehiemy, R.A.; Hasanien, H.M.; Ginidi, A.R. An improved heap optimization algorithm for efficient energy management based optimal power flow model. *Energy* **2022**, *250*, 123795. [CrossRef]
5. Shaheen, A.M.; El-Sehiemy, R.A.; Alharthi, M.M.; Ghoneim, S.S.; Ginidi, A.R. Multi-objective jellyfish search optimizer for efficient power system operation based on multi-dimensional OPF framework. *Energy* **2021**, *237*, 121478. [CrossRef]
6. Achiluzzi, E.; Kobikrishna, K.; Sivabalan, A.; Sabillon, C.; Venkatesh, B. Optimal asset planning for prosumers considering energy storage and photovoltaic (PV) units: A stochastic approach. *Energies* **2020**, *13*, 1813. [CrossRef]
7. Zheng, Q.P.; Wang, J.; Liu, A.L. Stochastic optimization for unit commitment—A review. *IEEE Trans. Power Syst.* **2014**, *30*, 1913–1924. [CrossRef]

8. Agabalaye-Rahvar, M.; Mansour-Saatloo, A.; Mirzaei, M.A.; Mohammadi-Ivatloo, B.; Zare, K.; Anvari-Moghaddam, A. Robust optimal operation strategy for a hybrid energy system based on gas-fired unit, power-to-gas facility and wind power in energy markets. *Energies* **2020**, *13*, 6131. [CrossRef]
9. Bai, X.; Qu, L.; Qiao, W. Robust AC optimal power flow for power networks with wind power generation. *IEEE Trans. Power Syst.* **2015**, *31*, 4163–4164. [CrossRef]
10. Lorca, A.; Sun, X.A. Multistage robust unit commitment with dynamic uncertainty sets and energy storage. *IEEE Trans. Power Syst.* **2016**, *32*, 1678–1688. [CrossRef]
11. Han, S.; Kim, H.-J.; Lee, D. A Long-Term Evaluation on Transmission Line Expansion Planning with Multistage Stochastic Programming. *Energies* **2020**, *13*, 1899. [CrossRef]
12. Ren, J.; Xu, Y.; Wang, S. A Distributed Robust Dispatch Approach for Interconnected Systems with a High Proportion of Wind Power Penetration. *Energies* **2018**, *11*, 835. [CrossRef]
13. Zhao, C.; Jiang, R. Distributionally robust contingency-constrained unit commitment. *IEEE Trans. Power Syst.* **2017**, *33*, 94–102. [CrossRef]
14. Jabr, R.A. Distributionally robust CVaR constraints for power flow optimization. *IEEE Trans. Power Syst.* **2020**, *35*, 3764–3773. [CrossRef]
15. Wang, Z.; Bian, Q.; Xin, H.; Gan, D. A distributionally robust co-ordinated reserve scheduling model considering CVaR-based wind power reserve requirements. *IEEE Trans. Sustain. Energy* **2015**, *7*, 625–636. [CrossRef]
16. Mohajerin Esfahani, P.; Kuhn, D. Data-driven distributionally robust optimization using the Wasserstein metric: Performance guarantees and tractable reformulations. *Math. Program.* **2018**, *171*, 115–166. [CrossRef]
17. Yang, I. Data-driven distributionally robust stochastic control of energy storage for wind power ramp management using the Wasserstein metric. *Energies* **2019**, *12*, 4577. [CrossRef]
18. Duan, C.; Fang, W.; Jiang, L.; Yao, L.; Liu, J. Distributionally robust chance-constrained approximate AC-OPF with Wasserstein metric. *IEEE Trans. Power Syst.* **2018**, *33*, 4924–4936. [CrossRef]
19. Zhu, R.; Wei, H.; Bai, X. Wasserstein metric based distributionally robust approximate framework for unit commitment. *IEEE Trans. Power Syst.* **2019**, *34*, 2991–3001. [CrossRef]
20. Zhou, A.; Yang, M.; Wang, M.; Zhang, Y. A linear programming approximation of distributionally robust chance-constrained dispatch with Wasserstein distance. *IEEE Trans. Power Syst.* **2020**, *35*, 3366–3377. [CrossRef]
21. Liu, H.; Fan, Z.; Xie, H.; Wang, N. Distributionally Robust Joint Chance-Constrained Dispatch for Electricity-Gas-Heat Integrated Energy System Considering Wind Uncertainty. *Energies* **2022**, *15*, 1796. [CrossRef]
22. Yang, J.; Zhang, N.; Kang, C.; Xia, Q. A state-independent linear power flow model with accurate estimation of voltage magnitude. *IEEE Trans. Power Syst.* **2016**, *32*, 3607–3617. [CrossRef]
23. Wu, L. A tighter piecewise linear approximation of quadratic cost curves for unit commitment problems. *IEEE Trans. Power Syst.* **2011**, *26*, 2581–2583. [CrossRef]
24. Wang, M.F.; Au, F. Precise integration methods based on Lagrange piecewise interpolation polynomials. *Int. J. Numer. Methods Eng.* **2010**, *77*, 998–1014. [CrossRef]
25. Liu, B.; Zhang, Q.; Ge, X.; Yuan, Z. CVaR-based approximations of Wasserstein distributionally robust chance constraints with application to process scheduling. *Ind. Eng. Chem. Res.* **2020**, *59*, 9562–9574. [CrossRef]
26. Yan, J.; Li, K.; Bai, E.W.; Deng, J.; Foley, A.M. Hybrid probabilistic wind power forecasting using temporally local Gaussian process. *IEEE Trans. Sustain. Energy* **2015**, *7*, 87–95. [CrossRef]
27. Zhang, J.; Yan, J.; Infield, D.; Liu, Y.; Lien, F.S. Short-term forecasting and uncertainty analysis of wind turbine power based on long short-term memory network and Gaussian mixture model. *Appl. Energy* **2019**, *241*, 229–244. [CrossRef]
28. Yuan, T.I.A.N.; Keyou, W.A.N.G.; Guojie, L.; Weichun, G.; Huanhuan, L. Dynamic stochastic optimal power flow baesd on second-order cone programming considering wind power correlation. *Autom. Electr. Power Syst.* **2018**, *42*, 41–47.
29. Lee, D.; Baldick, R. Probabilistic wind power forecasting based on the laplace distribution and golden search. In Proceedings of the 2016 IEEE/PES Transmission and Distribution Conference and Exposition (T&D), Dallas, TX, USA, 2–4 May 2016; IEEE: Piscataway, NJ, USA, 2016.
30. Wang, Z.; Shen, C.; Liu, F.; Wu, X.; Liu, C.C.; Gao, F. Chance-constrained economic dispatch with non-Gaussian correlated wind power uncertainty. *IEEE Trans. Power Syst.* **2017**, *32*, 4880–4893. [CrossRef]

designs

MDPI

Article

End-to-End Deep Reinforcement Learning Control for HVAC Systems in Office Buildings

Xuyang Zhong [1,*], Zhiang Zhang [2,*], Ruijun Zhang [3] and Chenlu Zhang [4]

1 Department of Civil Engineering, Faculty of Engineering, Lishui University, Lishui 323000, China
2 Department of Architecture and Built Environment, University of Nottingham Ningbo China, 199 East Taikang Road, Ningbo 315100, China
3 Department of Architecture and Built Environment, The University of Nottingham, Nottingham NG7 2RD, UK; ruijun.zhang@nottingham.ac.uk
4 Center for Building Performance and Diagnostics, Carnegie Mellon University, 5000 Forbes Avenue, Pittsburgh, PA 15213, USA; zhangchenluzcl@gmail.com
* Correspondence: xuyangz7@hotmail.com (X.Z.); zhiang.zhang@nottingham.edu.cn (Z.Z.)

Abstract: The heating, ventilation, and air conditioning (HVAC) system is a major energy consumer in office buildings, and its operation is critical for indoor thermal comfort. While previous studies have indicated that reinforcement learning control can improve HVAC energy efficiency, they did not provide enough information about end-to-end control (i.e., from raw observations to ready-to-implement control signals) for centralized HVAC systems in multizone buildings due to the limitations of reinforcement learning methods or the test buildings being single zones with independent HVAC systems. This study developed a model-free end-to-end dynamic HVAC control method based on a recently proposed deep reinforcement learning framework to control the centralized HVAC system of a multizone office building. By using the deep neural network, the proposed control method could directly take measurable parameters, including weather and indoor environment conditions, as inputs and control indoor temperature setpoints at a supervisory level. In some test cases, the proposed control method could successfully learn a dynamic control policy to reduce HVAC energy consumption by 12.8% compared with the baseline case using conventional control methods, without compromising thermal comfort. However, an over-fitting problem was noted, indicating that future work should first focus on the generalization of deep reinforcement learning.

Keywords: HVAC control; deep reinforcement learning; thermal comfort; energy efficiency; A3C

Citation: Zhong, X.; Zhang, Z.; Zhang, R.; Zhang, C. End-to-End Deep Reinforcement Learning Control for HVAC Systems in Office Buildings. *Designs* **2022**, *6*, 52. https://doi.org/10.3390/designs6030052

Academic Editors: Zbigniew Leonowicz, Arsalan Najafi and Michał Jasiński

Received: 10 May 2022
Accepted: 30 May 2022
Published: 4 June 2022

1. Introduction

The proper control of heating, ventilation, and air conditioning (HVAC) systems is a crucial element for reducing the amount of energy used by buildings and improving occupants' thermal comfort [1,2]. The control of HVAC systems can usually be divided into the supervisory level and local level [3]. Supervisory-level control sets the setpoints or operation commands, whilst local-level control controls the HVAC actuators in response to supervisory-level control signals. This study focuses on supervisory-level control because of its generality. Different HVAC systems may have dramatically different local control structures due to differences in the system design, but they may share a similar supervisory control interface [3–5].

The most commonly found HVAC supervisory control strategy is static rule-based control, in which there is a set of if-then-else rules to determine supervisory-level setpoints or operation commands [6]. However, such simple control strategies may not achieve high HVAC energy efficiency and improved indoor thermal comfort because of the slow thermal response of buildings, dynamic weather conditions, and dynamic building internal loads [7]. Additionally, most static rule-based control strategies consider only indoor air temperature as the metric for thermal comfort, but thermal comfort is actually affected by a number of factors, including air temperature, radiant temperature, humidity, etc. [8].

1.1. Model Predictive Control

Model predictive control (MPC) has become popular over the past few years due to its potential for significant HVAC energy savings. MPC uses a building model to predict the future building performance, in which case the optimal control decisions for the current time step can be made. There have been a number of studies using MPC to control HVAC systems, such as controlling the supply air temperature setpoint for air handling units (AHUs) [9], controlling the on/off status of the HVAC system [10], controlling the ventilation airflow rates [11], and controlling zone air temperature setpoints [12], and most of them show significant energy savings.

While promising, MPC is still hard to implement in the real world because of the difficulties of HVAC modeling. The classic MPC requires low-dimensional and differentiable models; for example, the linear quadratic regulator needs a linear dynamics and quadratic cost function [13]. This is difficult for HVAC systems, especially for the supervisory control of centralized HVAC systems, not only because it has nonlinear dynamics but also because it involves a number of control logics that make it non-continuous. For example, the control logic for a single-speed direct-expansion (DX) coil may be "turn on the DX-coil if there is indoor air temperature setpoint not-met in more than five rooms". Such logic is hard to represent with a continuous mathematical model because of the if-then-else condition. Therefore, in most previous MPC studies, either the building had an independent air conditioner for each room rather than a centralized system (such as [14–16]), or the MPC was used to directly control the local actuators rather than to set supervisory-level commands (such as [17–19]). Neither way generalizes well for typical multizone office buildings, which usually have centralized HVAC systems and non-uniform HVAC design.

To address the modeling difficulties of MPC for HVAC systems, white-box building model (physical-based model) based predictive control was proposed in [9,14,20]. This method may significantly reduce the modeling difficulties of MPC, because the white-box building model generalizes well for different buildings, and there are a number of software tools available for modeling. However, white-box building models, such as EnergyPlus models, are usually high-dimensional and non-differentiable. Heuristic search must be implemented for MPC. Given the fact that the white-box building model can be slow in computation, the scalability and feasibility of this type of MPC in the real world are questionable.

1.2. Model-Free Reinforcement Learning HVAC Control

Since model-based optimal control, such as MPC, is hard to use for HVAC systems, model-free reinforcement learning control becomes a possible alternative. To the authors' knowledge, reinforcement learning control for HVAC systems has not yet been well studied. Either the reinforcement learning methods used are too simple to reveal their full potential, or the test buildings are too unrealistic. For example, Liu and Henze [21] applied very simple discrete tabular-setting Q-learning to a small multizone test building facility to control its global thermostat setpoint and thermal energy storage discharge rate for cost savings. Regardless of the limited real-life experiment showing 8.3% cost savings compared with rule-based control, the authors admitted that the "curse of dimensionality" of such a simple reinforcement learning method limited its scalability. In the following research by the same authors [22], a more advanced artificial neural network (ANN) was used to replace simple tabular-setting Q-learning; however, the results indicate that the use of ANN did not show clear advantages, probably due to the limited computation resources at that time.

The deep neural network (DNN) has become enormously popular lately in the machine learning community due to its strong representation capacity, automatic feature extraction, and automatic regularization [23–25]. Deep reinforcement learning methods take advantage of DNN to facilitate end-to-end control, which aims to use raw sensory data without complex feature engineering to generate optimal control signals that can be directly used to control a system. For example, Mnih et al. [26] proposed a deep Q-network that could directly take raw pixels from Atari game frames as inputs and play the game at a human level. More details about deep reinforcement learning can be found in Section 2.

Deep reinforcement learning methods have been widely studied not only by machine learning and robotics communities but also by the HVAC control community. Table 1 summarizes the HVAC control studies performed in recent years using deep reinforcement learning. Researchers have demonstrated via simulations and practical experiments that deep reinforcement learning can improve the energy efficiency for various types of HVAC systems. However, there are sparse data describing the implementation of end-to-end control for multizone buildings. On the one hand, the test buildings in several studies, including [27–31], were single zones with independent air conditioners. On the other hand, conventional deep reinforcement learning methods cannot effectively solve multizone control problems. Yuan et al. [32] showed that the direct application of deep Q-learning to a multizone control problem would make the training period too long. Ding et al. [33] proposed a multi-branching reinforcement learning method to solve this problem, but the method required a fairly complicated deep neural network architecture and therefore could not be scaled up for large multizone buildings. Based on deep reinforcement learning, Zhang et al. [4] proposed a control framework for a multizone office building with radiant heating systems. In this study, however, "reward engineering" (i.e., a complicated reward function of reinforcement learning) needed to be designed to help ensure that the reinforcement learning agent could learn efficiently, in which case end-to-end control could not be achieved.

Table 1. An overview of studies focusing on deep reinforcement learning methods for HVAC systems.

Reference	Reinforcement Learning Method	Test Building	HVAC System
[27]	Linear state-value function approximation	Two buildings, each having a single zone	Independent air conditioners
[28]	Continuous-action Q-learning	Multiple apartments, each having a single zone	Independent air conditioners
[29]	Continuous-action Q-learning	A hall with multiple single rooms	Independent air conditioners
[30]	Model-assisted fitted Q-learning	A lab room	An independent HVAC unit
[31]	Model-assisted fitted Q-learning	A single chamber	A heat pump system
[32]	Deep Q-learning	A multizone office building	A variable air volume system
[33]	Multi-branching reinforcement learning	A multizone office building	A variable air volume system
[4]	Policy gradient	A multizone office building	Radiant heating systems

1.3. Objectives

As discussed above, conventional rule-based supervisory HVAC control often results in unnecessary energy consumption and thermal discomfort. Better supervisory control methods should be found, but model-based optimal control, such as MPC, may not be practical for multizone office buildings. While previous studies have indicated that reinforcement learning control can be promising in terms of energy savings and thermal comfort, data from these studies did not provide enough information about the implementation of end-to-end control (i.e., from raw observations to the ready-to-implement control signals) for centralized HVAC systems in multizone buildings, mainly due to the limitations of reinforcement learning methods or the test buildings being single zones with independent HVAC systems. In this study, a supervisory-level HVAC control method was developed using the deep reinforcement learning framework in order to achieve end-to-end control for a typical multizone office building with a centralized HVAC system. The performance of the proposed control method, including both learning performance and building performance, were critically evaluated. The limitations of the proposed method are discussed, and the direction of future work is proposed.

2. Background of Reinforcement Learning

2.1. Markov Decision Process

According to [34], a standard reinforcement learning problem is that a learning agent interacts with the environment in a number of discrete steps to learn how to maximize the reward returned from the environment (Figure 1). Agent–environment interactions in one step can be expressed as a tuple (S_t, A_t, S_{t+1}, R_{t+1}), where S_t represents the state of the environment at time t, A_t is the action chosen by the agent to interact with the environment at time t, S_{t+1} is the resulting environmental state after the agent takes action, and R_{t+1} is the reward received by the agent from the environment. Ultimately, the goal of reinforcement learning control is to learn an optimal policy $\pi : S_t \rightarrow A_t$ that maximizes the accumulated future reward $\sum_t^{T\infty} R_t$.

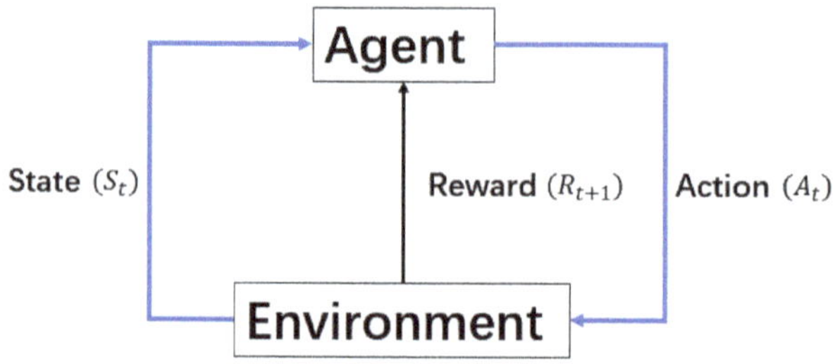

Figure 1. The Markov decision process framework.

The above-mentioned standard reinforcement learning problem is a Markov decision process (MDP) if it obeys the Markov property; that is, the environment's state of the next time step (S_{t+1}) only depends on the environment's state at this time step (S_t) and the action at this time step (A_t) and is not related to the state action history before this time step t. Most reinforcement learning algorithms implicitly assume that the environment is an MDP. However, empirically, many non-MDP problems can still be well solved by those reinforcement learning algorithms.

In reinforcement learning, there are three important concepts, including the state-value function, action-value function, and advantage function (as shown in Equations (1)–(3), where γ is the reward discount factor) [35]. Intuitively, the state-value function represents how much reward can be expected if the agent is at state s following policy π; the action-value function represents how much reward can be expected if the agent is at state s taking action a and then following policy π; and the advantage function, showing the difference between the action-value function and state-value function, basically indicates how good an action is with respect to the state.

$$v_\pi(s) = \mathbb{E}_\pi\left[\sum_{k=0}^{\infty} \gamma^k R_{t+k+1} | S_t = s\right] \tag{1}$$

$$q_\pi(s,a) = \mathbb{E}_\pi\left[\sum_{k=0}^{\infty} \gamma^k R_{t+k+1} | S_t = s, A_t = a\right] \tag{2}$$

$$a_\pi(s,a) = q_\pi(s,a) - v_\pi(s) \tag{3}$$

where $\mathbb{E}_{a\sim\pi(s)}[a_\pi(s,a)] = 0$. For the optimal policy π^*, there is

$$v_{\pi^*}(s) = \max_a q_{\pi^*}(s,a) = \max_a \mathbb{E}[R_{t+1} + \gamma v_{\pi^*}(S_{t+1}) | S_t = s, A_t = a] \tag{4}$$

2.2. Policy Gradient

Reinforcement learning problems are usually solved by learning an action-value function $q_\pi(s,a)$, and the resulting policy is $\pi'(s) = \underset{a}{\text{argmax}} q_\pi(s,a)$ if the greedy policy is used. In addition, there is another approach to reinforcement learning (known as the policy gradient) that learns the optimal policy directly without learning the action-value function. Compared with the greedy policy, the advantages of the policy gradient include better convergence properties, greater effectiveness in high-dimensional or continuous action spaces, and a better ability to learn stochastic policies [36]. The policy gradient was therefore used in this study.

The goal of the policy gradient is to learn the parameter θ in $\pi_\theta(s,a) = Pr(a|s,\theta)$ that maximizes average reward per time step $J(\theta)$, as shown in Equation (5):

$$J(\theta) = \sum_s d_{\pi_\theta}(s) \sum_a R_s^a \pi_\theta(s,a) \tag{5}$$

where $d_{\pi_\theta}(s)$ is the stationary distribution for state s of the Markov chain starting from s_0 and following policy π_θ, and R_s^a is the reward of the agent at state s taking action a. Gradient descent was used to maximize Equation (5). The gradient of $J(\theta)$ with respect to θ is shown in Equation (6):

$$\begin{aligned} \nabla_\theta J(\theta) &= \sum_s d_{\pi_\theta}(s) \sum_a R_s^a \pi_\theta(s,a) \frac{\nabla_\theta \pi_\theta(s,a)}{\pi_\theta(s,a)} \\ &= \sum_s d_{\pi_\theta}(s) \sum_a R_s^a \pi_\theta(s,a) \nabla_\theta \log \pi_\theta(s,a) \end{aligned} \tag{6}$$

According to the policy gradient theorem, Equation (6) can be rewritten as:

$$\nabla_\theta J(\theta) = \mathbb{E}_{\pi_\theta}[\nabla_\theta \log \pi_\theta(s,a) q_{\pi_\theta}(s,a)] \tag{7}$$

However, $q_{\pi_\theta}(s,a)$ usually has a large variance, which may harm the convergence of the policy gradient method. To solve this problem, a baseline function $B(s)$ can be subtracted from $q_{\pi_\theta}(s,a)$ in Equation (7). Because $B(s)$ is not a function of a, $\mathbb{E}_{\pi_\theta}[\nabla_\theta \log \pi_\theta(s,a)B(s)]$ equals zero. Therefore, subtracting a baseline function from $q_{\pi_\theta}(s,a)$ does not change the expected value of Equation (7) but reduces its variance. A good choice of $B(s)$ is $v_{\pi_\theta}(s)$. Then, the new policy gradient function is:

$$\begin{aligned} \nabla_\theta J(\theta) &= \mathbb{E}_{\pi_\theta}[\nabla_\theta \log \pi_\theta(s,a)(q_{\pi_\theta}(s,a) - v_{\pi_\theta}(s))] \\ &= E_{\pi_\theta}[\nabla_\theta \log \pi_\theta(s,a) a_{\pi_\theta}(s,a)] \end{aligned} \tag{8}$$

The policy gradient in the form of Equation (8) is called advantage actor critic (A2C), which is the main reinforcement learning method used in this study.

2.3. *Deep Reinforcement Learning*

The size of the state space of the reinforcement learning problem can easily be very large for real-life problems. Simple tabular settings, i.e., using a lookup table to store the state values and action values for every state and every action, cannot work for a large discrete state space or a continuous state space. Instead, the value functions and policy can be estimated using the function approximation, i.e., $v_\pi(s,\theta)$, $q_\pi(s,a,\theta)$, and $\pi(s,a,\theta)$, where state values, action values, and policy are a function with respect to θ. If a deep neural network is used as the function approximation, then it is called deep reinforcement learning.

The advantages of a deep neural network are its representation capacity, automatic feature extraction, and good generalization properties. Therefore, complicated feature engineering and results post-processing are no longer needed, making end-to-end control possible.

3. Methodology

Model-free deep reinforcement learning was used in this study, where the reinforcement learning agent interacted with the simulated building model offline to learn a good control policy and then controlled the real building online [21]. Since the building model was used as a simulator offline, slow computation and a non-differentiable model were no longer problems. EnergyPlus (Version 8.6 developed by the National Renewable Energy Laboratory, Golden, CO, USA) was used as the building simulation engine [37].

As shown in Figure 2, a multizone building simulator was used for offline model-free reinforcement learning (training), but only one zone was used as the training simulator. After learning, a control policy was obtained, and this control policy was used to control all zones in the testing simulator. Note that the testing simulator had perturbations to ensure the fairness of testing. The details of the simulators and perturbations can be found in Section 4.1.

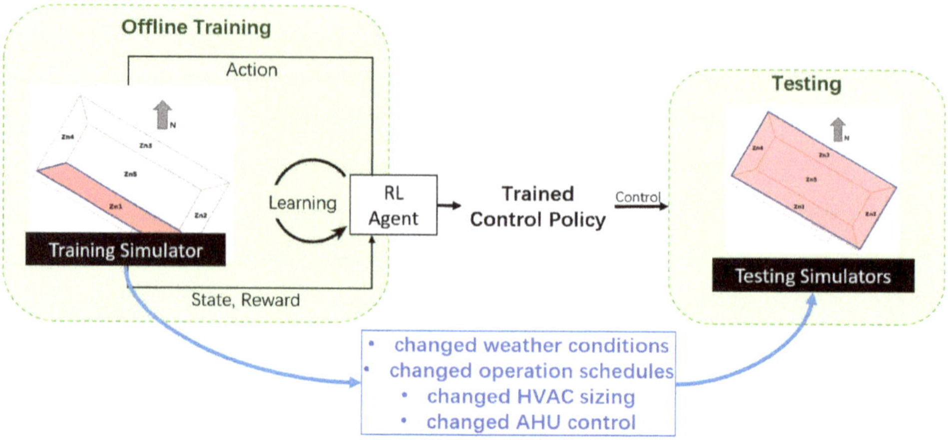

Figure 2. The schematic workflow of reinforcement learning control in this study.

3.1. *State, Action, and Reward Design*

For reinforcement learning, state, action, and reward design are critical for learning convergence (as described in Section 2.1). To take advantage of the deep reinforcement learning method, only raw observable or controllable parameters for our state, action, and reward design were used, with no extra data manipulation.

3.1.1. State

The state included the current time step's weather conditions, the environmental conditions of the controlled zone, and the HVAC power demand, which are summarized in Table 2.

Table 2. A description of the states selected for reinforcement learning.

State	Remarks
Weather conditions	
Site outdoor air dry-bulb temperature	Unit: °C
Site outdoor air relative humidity	Unit: %
Site wind speed	Unit: degree from north
Site diffuse solar radiation rate	Unit: W/m^2
Site direct solar radiation rate	Unit: W/m^2
Zone environmental conditions	
Zone thermostat heating setpoint temperature	Unit: °C
Zone thermostat cooling setpoint temperature	Unit: °C
Zone air temperature	Unit: °C
Zone air relative humidity	Unit: %
Occupancy-related zone average predicted percentage of dissatisfied using Fanger model	Unit: %, hereafter called OPPD. If the zone occupancy status is 0, OPPD is 0; otherwise, OPPD is the Fanger model predicted percentage of dissatisfied
Zone occupancy status	1 or 0
HVAC power demand	
Total HVAC power demand	Unit: W, hereafter called E_{HVAC}. E_{HVAC} is the total electric power demand for the whole HVAC system.

Each item in the state should be normalized to between 0 and 1 for the optimization purpose of the deep neural network. Min–max normalization was used (as shown in Equation (9)), with the parameter's physical limits or the parameter's expected bounds as the min–max values. For example, the min–max values for relative humidity (%) were 0 and 100, and the min–max values for zone air temperature (°C) were 15 and 30. The temperature range was selected based on data in the literature [9,38], which shows that the typical range of setpoint temperatures for office buildings in Pennsylvania, USA, is 15 °C to 30 °C.

$$x_{norm} = \frac{x - x_{min}}{x_{max} - x_{min}} \tag{9}$$

3.1.2. Action

The control action of the agent was discrete and was designed as the adjustment to the last time step's air temperature heating and cooling setpoints in the controlled zone. There are four basic action types, including:

1. [*0.0, 0.0*] means no change to the last time step's zone air temperature heating and cooling setpoints.
2. [*+deltaValue*, *+deltaValue*] means add *deltaValue* to both the zone air temperature heating and cooling setpoints of the last time step. This basically means that the zone needs more heating. Note that *deltaValue* should be greater than zero.
3. [*−deltaValue*, *−deltaValue*] means subtract *deltaValue* from both the zone air temperature heating and cooling setpoints of the last time step. This basically means that the zone needs more cooling. Note that *deltaValue* should be greater than zero.
4. [*−deltaValue*, *+deltaValue*] means subtract *deltaValue* from the zone air temperature heating setpoint of the last time step and add *deltaValue* to the zone air temperature cooling setpoint of the last time step. This basically means that the zone needs no air conditioning. Note that *deltaValue* should be greater than zero.

The value of *deltaValue* is a tunable parameter, and the action space can consist of the basic action types with different *deltaValue* simultaneously. In Section 4, different action spaces were tested based on the four basic action types. Note that the maximum setpoint value and the minimum setpoint value were enforced to be 30 °C and 15 °C, respectively.

3.1.3. Reward

The objective of the control method is to minimize the HVAC energy consumption and thermal discomfort. Therefore, a convex combination of the OPPD and EHVAC was used as the reward (both OPPD and EHVAC here are min–max-normalized scalars):

$$-(\lambda * a + (1-\lambda) * EHVAC),\ where\ a = \begin{cases} OPPD & OPPD \leq Lmt_{ppd} \\ 1.0 & OPPD > Lmt_{ppd} \end{cases} \tag{10}$$

λ is a tunable parameter representing the relative importance of HVAC energy efficiency and indoor thermal comfort, and $\lambda \in [0,1]$. Lmt_{ppd} is also a tunable parameter to penalize a large OPPD. Specifically, Lmt_{ppd} is a hyperparameter to control the penalty level for thermal discomfort. For example, if Lmt_{ppd} is 0.15, this means that the penalty for thermal discomfort will be amplified to the maximum if OPPD is larger than 0.15. Different values of λ and Lmt_{ppd} were tested, as described in Section 4.4.1, to evaluate the effects of λ and Lmt_{ppd} on control performance.

3.2. Asynchronous Advantage Actor Critic (A3C)

Policy gradient, as discussed in Section 2.2, was the main reinforcement learning training method used in this study. Specifically, a state-of-the-art deep reinforcement learning variation of A2C, asynchronous advantage actor critic (A3C) [39], was used. In the A3C method, rather than having only one agent to interact with the environment, a number of agents interact with copies of the same environment independently but update the same global action-value or policy function network asynchronously. Still asynchronously, the agents update their own action-value or policy function network to be the same as the global one in a certain frequency. The purpose of this method is to ensure that the tuples $(S_t, A_t, S_{t+1}, R_{t+1})$ used to train the global network are roughly independent. Compared with the non-asynchronous methods, A3C significantly reduces the memory usage and training time cost. Details of the algorithm can be seen in Algorithm S3 of [40].

To solve the reinforcement learning problem using the advantage actor critic method, we should have two deep neural networks: one is $\pi_\theta(s,a)$ to approximate the policy, and the other is $v_{\theta_v}(s)$ to approximate the state-value function. Therefore, according to Equations (4) and (8), θ can be learned by gradient descent, which is:

$$\theta \leftarrow \theta + \alpha \mathbb{E}_{\pi_\theta}[\nabla_\theta \log \pi_\theta(s,a)(q_{\pi_\theta}(s,a) - v_{\theta_v}(s))] = \theta + \alpha \mathbb{E}_{\pi_\theta}\left[\nabla_\theta \log \pi_\theta(s,a)\left(R' + \gamma v_{\theta_v}(s') - v_{\theta_v}(s)\right)\right] \quad (11)$$

θ_v can also be learned using stochastic gradient descent with the mean squared loss function, which is:

$$\theta_v \leftarrow \theta_v - \alpha \mathbb{E}_{\pi_\theta}\left[\partial(v_{true} - v\theta_v(s))^2 / \partial\theta_v\right] = \theta_v - \alpha \mathbb{E}_{\pi_\theta}\left[\partial\left(R' + \gamma v_{\theta_v}(s') - v\theta_v(s)\right)^2 / \partial\theta_v\right] \quad (12)$$

In Equations (11) and (12), α is the step size for gradient descent, R' is the actual reward at state s taking action a, and s' is the next state from state s taking action a.

4. Experiments and Results

4.1. Training and Testing Building Models

Experiments were carried out based on EnergyPlus (version 8.6, developed by the National Renewable Energy Laboratory, Golden, CO, USA) simulations. The target building in this study was selected based on the EnergyPlus v8.6 "5ZoneAutoDxVAV" example file, and Pennsylvania, USA, was selected as the location of the building due to better access to data about the environmental conditions of this site. The building was a single-level five-zone office building, the plan and dimensions of which can be seen in Figure 3. The types of building fabrics, along with their thermal properties, can be seen in Table 3. The building had four exterior zones and one interior zone. All zones were regularly occupied by office workers. Each zone had a 0.61 m high return plenum. Windows were installed on all four facades, and the south-facing facade was shaded by overhangs. The lighting load, office equipment load, and occupant density were 16.15 W/m^2, 10.76 W/m^2, and 1/9.29 m^2, respectively.

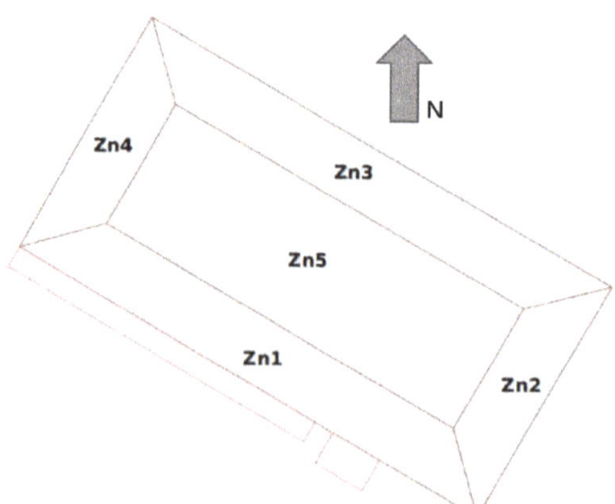

	Area (m2)	Height (m)
Zn1	100	2.4
Zn2	43	2.4
Zn3	100	2.4
Zn4	43	2.4
Zn5	186	2.4

Figure 3. The training and testing building model plan.

Table 3. The type of building fabrics, along with their thermal properties.

	Type	U-Value (W/m^2K)	Solar Heat Gain Coefficient
External walls	Mosaic tiles (0.005 m) + heavy concrete (0.203 m) + mass wall insulation (0.049 m) + gypsum plaster (0.013 m)	1.1	
Roof	Concrete tiles (0.025 m) + asphalt (0.020 m) + cement (0.050 m) + concrete (0.150 m) + insulation (0.100 m) + gypsum plaster (0.010 m)	0.42	
Ground floor	Floor tiles (0.005 m) + gypsum plaster (0.0127 m) + concrete (0.150 m)	4.25	
Windows	Clear glass (0.003 m) + air (0.013 m) + clear glass (0.003 m)	1.25	0.76

The HVAC system of the building model was a centralized variable air volume (VAV) system with terminal reheat. The cooling source in the AHU was a two-speed DX coil, and the heating source in the AHU was an electric heating coil. The terminal reheat was also an electric heating coil.

To ensure fair evaluation of the control method, two building models with several differences were developed, called the training model and the testing model. The deep reinforcement learning agent was trained using the training model. The two models shared the same geometry, envelope thermal properties, and HVAC systems. Differences between these two models are summarized in Table 4. To test the building model, the weather file was changed to a place that was about 200 km away, the occupant and equipment schedules were changed to be stochastic using the occupancy simulator [41], the HVAC equipment was more over-sized, and the AHU supply air temperature setpoint control strategy was changed to be simpler.

Table 4. Differences between the training model and the testing model.

Model	Weather File	Occupant and Equipment Schedule	HVAC Sizing	AHU Control
Training	Allegheny County, PA (TMY3)	Deterministic	Oversized by 20% from EnergyPlus auto-sizing	Dynamically change the AHU supply air temperature setpoint from 12 °C to 18 °C based on the warmest zone
Testing	State College, PA (TMY3)	Stochastic	Oversized by 44% from EnergyPlus auto-sizing	Set the AHU supply air temperature setpoint to 12 °C or 18 °C based on the outdoor air temperature

The EnergyPlus simulator was wrapped by the OpenAI Gym [42] for the convenience of the reinforcement learning implementation. The ExternalInterface function of EnergyPlus was used for data communication between the building model and the reinforcement learning agent during the run time.

For both training and testing, the run period of the EnergyPlus models was from Jan 1st to Mar 31st, which was the period for the whole winter season for Pennsylvania, USA. The simulation time step was 5 min. Therefore, for the discrete control of reinforcement learning control, the control time step was also 5 min.

4.2. A3C Model Setup

4.2.1. Policy and State-Value Function Network Architecture

As discussed in Section 3.2, the A3C method needs two function approximation neural networks, one for the policy and the other for the state-value function. Figure 4 shows the architecture of the networks. Rather than two separate networks, a shared multilayer feed-forward neural network was used. The output from the shared network was fed into a Softmax layer and a linear layer in parallel, where the Softmax layer outputs the policy and the linear layer outputs the state value. Note that the output of the Softmax layer was a vector with the length the same as the total number of discrete actions, and each entry in the vector corresponded to the probability of taking the action.

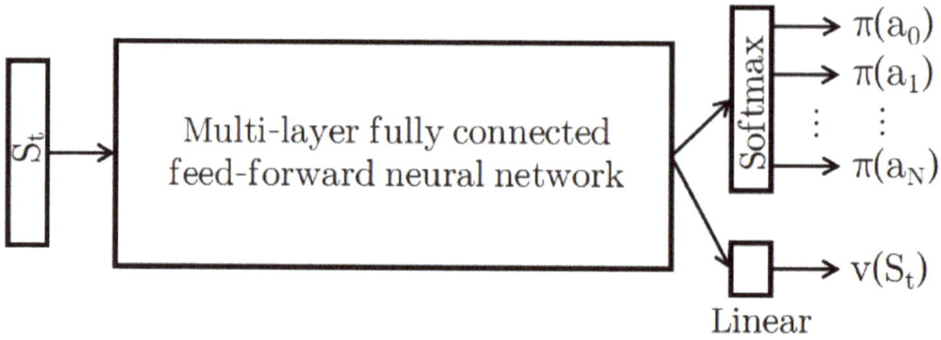

Figure 4. The policy and state-value function network architecture.

4.2.2. Hyperparameters

The shared network of Figure 4 has four hidden layers, each of which has 512 hidden units with rectifier nonlinearity. RMSProp [43] was used for optimization, and a single optimizer was shared across all agents in A3C. The learning rate was fixed to 0.0001, and the RMSProp decay factor was 0.99. To avoid too large a gradient in gradient descent, which would harm the convergence, all gradients were clipped so that their L2 norm was less than or equal to 5.0. The total number of interactions between the A3C agents and the environment was 20 million. The entropy of policy π was added to the policy gradient to regularize the optimization so that the agent would not overly commit to a deterministic policy in the training [44]. The weight for this regularization term was 0.01, as suggested by [40].

A building usually has slow dynamics, and the state observation of the current time step is not sufficient for the agent to make a good action choice. Recent n state observations can be stacked to be the effective state observation of the agent [26]. For example, rather than just observing the current zone indoor air temperature, the agent observes the zone indoor air temperatures of current and past $n-1$ time steps to make a decision. As suggested by [26], n was set to 24 in this study.

4.3. Baseline Control Strategies

The conventional fixed-schedule control strategy for indoor air heating and cooling temperature setpoints was used as the baseline. The values of the heating and cooling setpoints are usually determined by the facility manager based on experience. In this study, two sets of heating/cooling setpoints were selected, one representing the "colder" control case and the other representing the "warmer" control case.

1. B-21.1: the indoor air heating and cooling setpoints were 21.1 °C and 23.9 °C from 7:00 to 21:00 (training model) or from 7:00 to 18:00 (testing model) on weekdays and 15.0 °C and 30.0 °C at all other times;
2. B-23.9: the indoor air heating and cooling setpoints were 23.9 °C and 25.0 °C from 7:00 to 21:00 (training model) or from 7:00 to 18:00 (testing model) on weekdays and 15.0 °C and 30.0 °C at all other times.

It should be noted that the building model had default indoor air heating/cooling temperature setpoints, which were 21.1 °C/23.9 °C from 7:00 to 18:00 on weekdays and 7:00 to 13:00 on weekends and 12.8 °C/40.0 °C at all other times. The baseline control schedules B-21.1 and B-23.9 were only implemented when comparing them with deep reinforcement learning control. For other times, such as during the training period, the default control schedule was used. This is because the baseline schedules were manipulated to match the known building occupancy schedule, which might not be known in reality.

Such manipulation was to ensure fair comparison because the proposed reinforcement learning control method had an occupancy-related control feature.

4.4. Training

The reinforcement learning agent was trained using the training building model. An 8-core 3.5 GHz computer was used to carry out the training process. The period of the training process was 5 h. In the training, it controlled the indoor air temperature heating and cooling setpoints of Zn1 (see Figure 3) only and tried to minimize the thermal discomfort of Zn1 and the HVAC energy consumption of the whole building. Therefore, as discussed in Section 3.1.1, the agent's state observations were the weather conditions, environmental conditions of Zn1, and the whole building HVAC power demand. The reason for only controlling one zone during the training instead of controlling all five zones is to reduce the action space dimensions. The speed of convergence of deep reinforcement learning with a discrete action space relies on the action space dimension. In this study, the action space dimension increased exponentially with the increased number of controlled zones. Considering that all five zones were served by the same HVAC system and had similar thermal properties and functions, we chose to train the agent on one zone only and then applied the trained agent to all five zones to control the whole building.

4.4.1. Parameter Tuning

λ and Lmt_{ppd} in the reward function (see Section 3.1.3) and different combinations of *deltaValue* in the action space (see Section 3.1.2) were tuned. Two different values of λ (0.4 and 0.6) were studied; three different values of Lmt_{ppd} (0.15, 0.30, and 1.0) were studied; and two different *deltaValues* (1.0 and 0.5) were studied. This resulted in three action spaces:

1. act1 = Zip{(0.0, 1.0, −1.0, −1.0), (0.0, 1.0, −1.0, 1.0)};
2. act2 = Zip{(0.0, 1.0, −1.0, −1.0, 0.5, −0.5, −0.5), (0.0, 1.0, −1.0, 1.0, 0.5, −0.5, 0.5)};
3. act3 = Zip{(0.0, 1.0, −0.5, −0.5), (0.0, 0.5, −0.5, 0.5)}.

Therefore, in total, 18 cases with different hyperparameters were trained. Each value in parentheses represents an action choice for the heating setpoint and the cooling setpoint, respectively, and the zipped tuples of both parentheses are the final action space. For example, in act1, actions include (1) no change in either the heating or cooling setpoint; (2) increase both heating and cooling setpoints by 1 °C; (3) decrease both heating and cooling setpoints by 1 °C; and (4) decrease the heating setpoint by 1 °C and increase the cooling setpoint by 1 °C. The performance of each training case was evaluated by the mean and the standard deviation of the Zn1 OPPD of occupied time steps (hereafter called OPPD Mean and OPPD Std) and the total HVAC energy consumption of the run period from 1 January to 31 March (hereafter called E_{HVAC}). The hyperparameters of training cases are listed in Table 5.

4.4.2. Optimization Convergence

Reinforcement learning can be viewed as an optimization problem that looks for a control policy that maximizes the cumulative reward. Figure 5 shows the history of the cumulative reward for one simulation period (1 January to 31 March) for all cases in the training. Each subplot in the figure shows the reward history of cases with the same λ and Lmt_{ppd}. Note that different subplots in the figure have different scales of the y-axis because different training cases do not share the same reward function. For the convergence study, the relative value of the reward is more important than the absolute value of the reward. It can be found that all training cases had a fairly fast convergence speed, which usually converged between 5 million and 10 million training steps. In addition, a smaller value of Lmt_{ppd} had better convergence performance. This may be because a smaller Lmt_{ppd}, which leads to a more stringent requirement on thermal comfort, gives the agent a clearer signal about how good or how bad a state and an action are. Even though, in principle, a larger action space may take more time to converge, it is not clear in this study. It is interesting to find that the cases with $\lambda = 0.6$ (larger penalty on discomfort) had better convergence performance than the cases with $\lambda = 0.4$ (smaller penalty on discomfort). The reason for this difference is still not clear.

Table 5. The training results.

Case	Hyperparameters			Zn1 OPPD		Building
	λ	Lmt_{ppd}	Acts	Mean (%)	Std (%)	E_{HVAC} (kWh)
1	0.60	0.15	act1	6.70	2.97	5566
2	0.60	0.30	act1	8.45	6.25	5256
3	0.60	1.00	act1	9.74	6.11	5388
4	0.60	0.15	act2	6.94	2.89	5499
5	0.60	0.30	act2	7.38	4.23	5797
6	0.60	1.00	act2	7.08	4.77	5626
7	0.60	0.15	act3	9.30	6.27	5671
8	0.60	0.30	act3	7.34	4.05	7144
9	0.60	1.00	act3	8.84	4.80	5560
10	0.40	0.15	act1	7.60	3.07	5385
11	0.40	0.30	act1	10.11	6.05	5333
12	0.40	1.00	act1	10.27	7.84	5201
13	0.40	0.15	act2	7.25	3.76	5976
14	0.40	0.30	act2	8.02	4.38	5493
15	0.40	1.00	act2	8.07	4.26	5640
16	0.40	0.15	act3	6.64	3.09	5831
17	0.40	0.30	act3	9.34	5.96	5352
18	0.40	1.00	act3	9.84	6.81	5523
B-21.1	N/A	N/A	N/A	10.75	7.80	5003
B23.9	N/A	N/A	N/A	7.92	4.06	5436

Note: for all cases in this table, only Zn1 was controlled by the reinforcement learning agent or baseline control strategy; all four of the other zones were controlled using the model default control strategy.

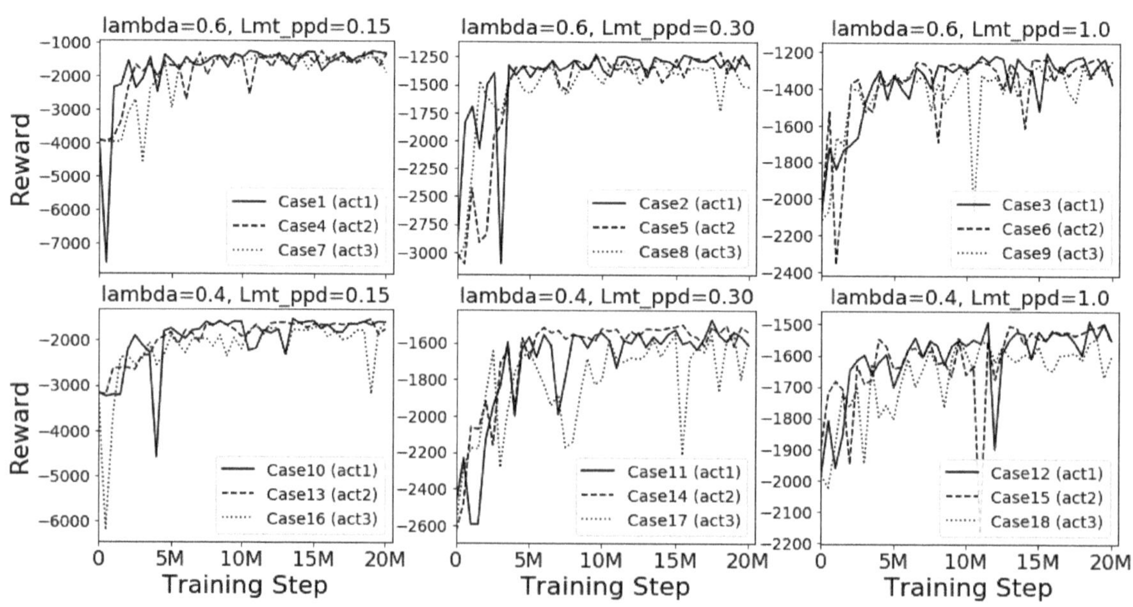

Figure 5. The history of one simulation period's cumulative reward for all cases in the training.

4.4.3. Performance Comparison

Table 5 shows the HVAC energy consumption and thermal comfort performance of all training cases and baseline cases. It shows that almost all training cases had less than 10% mean OPPD, and the standard deviation is fairly small. It is also found that smaller Lmt_{ppd} is favorable because it can increase the thermal comfort performance in most cases and does not necessarily increase the HVAC energy consumption. For different λ values, it is not expected that a smaller λ sometimes results in increased HVAC energy consumption. It may be because, in this study, optimizing the building's total HVAC energy consumption is difficult since the agent can only control one out of the five zones. Different action spaces were also studied, but there were no clear findings about the relationship between the action space and HVAC energy and thermal comfort performance.

Compared with the B-21.1 case, all reinforcement learning cases had better thermal comfort performance but higher HVAC energy consumption. This is as expected because the B-21.1 case had a low indoor air heating temperature setpoint. For the B-23.9 case, the comparison is more complex because some reinforcement learning cases had better thermal comfort performance and worse HVAC energy efficiency or vice versa. Among the 18 training cases, case 10 was selected as the best one compared with the B-23.9 case. Case 10 had slightly better thermal comfort performance in both the mean and standard deviation of OPPD, and it also had slightly lower HVAC energy consumption. Therefore, training case 10 was selected for the subsequent study.

To visually inspect the learned control policy of the agent, Figure 6 shows the control behavior snapshot of the case 10 agent on three days in winter. It can be found that the agent had learned to change the setpoints according to the occupancy. Additionally, the agent had learned to preheat the space before occupants arrived in the morning. In addition, the agent could decrease the heating setpoint when the zone internal heating gain (e.g., solar heating gain) was sufficient to keep the space warm at noon and in the afternoon of the day. However, the agent did not start to decrease the heating setpoint until the zone became unoccupied. The agent had to take nearly an hour to decrease the heating setpoint to the minimum value, which wasted HVAC energy. The OPPD of training case 10 in Figure 6 was kept lower than 10% most of the time. However, it is interesting to find that the OPPD reached above 15% in the afternoon of 01/09. The primary reason is the too high mean radiant temperature of the zone caused by strong afternoon solar radiation. The agent did decrease the cooling setpoint in response to this situation, but cooling was still not enough to offset the effect of the high mean radiant temperature. This shows that the agent is not well trained to deal with this type of situation. Compared with the B-23.9 case, the reinforcement learning agent tended to overheat the space in the morning and then let the indoor air temperature float, rather than keep the heating setpoint constant. The reason is probably that the agent wants to heat the space quickly in the morning in order to create a warm environment before occupancy. There are lots of small fluctuations in heating and cooling setpoints in the reinforcement learning case because the reinforcement learning agent gives a stochastic policy rather than a deterministic one. The stochastic policy is used because it helps the agent to explore unknown states. It is easy to change the stochastic policy to the deterministic one if needed.

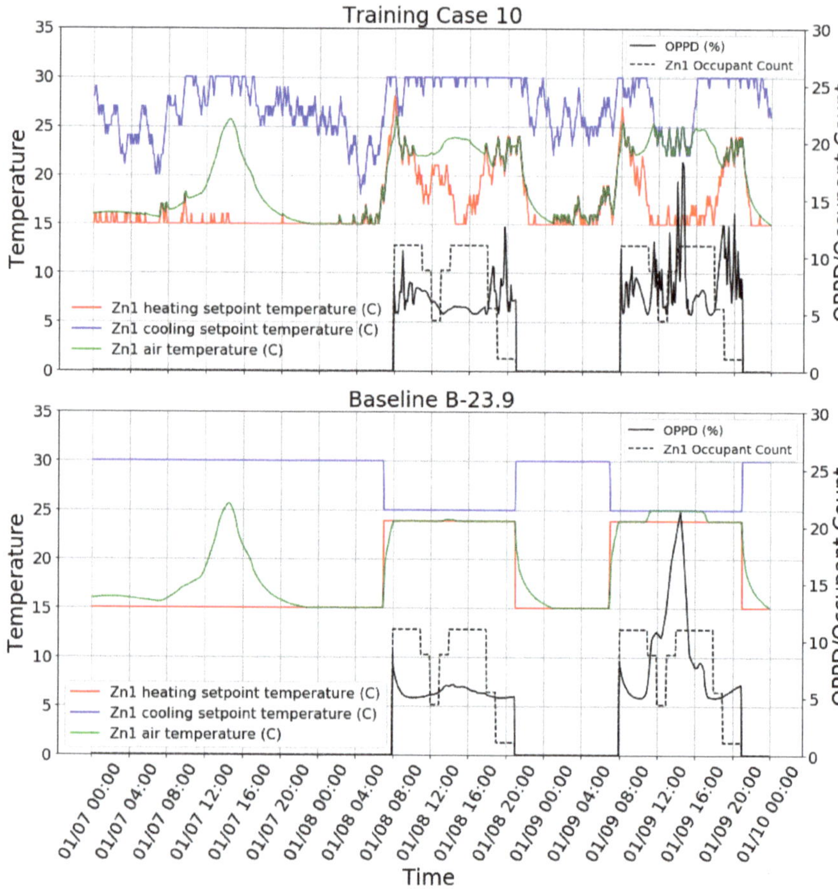

Figure 6. Training: control behavior snapshot of training case 10 (**top**) and the baseline case B-23.9 (**below**).

4.5. Testing

The trained reinforcement learning agent of training case 10 was tested in three scenarios, including single-zone testing in the testing building model, multizone testing in the training building model, and multizone testing in the testing building model. The trained agent's performance in the testing was also evaluated by OPPD Mean, OPPD Std, and E_{HVAC}.

4.5.1. Single-Zone Testing in the Testing Building Model

The trained agent in training case 10 was tested using the testing building model to control Zn1 of the building model, which was the same zone that the agent was trained on. All other zones still had setpoints with the fixed schedule. Table 6 shows the performance comparison between the reinforcement learning agent and baseline cases. It can be found that the reinforcement learning agent had a performance between the two baseline cases: its thermal comfort performance was worse than that of B-23.9, and its HVAC energy consumption was higher than that of B-21.1. The control behavior snapshot of the reinforcement learning agent and B-23.9 is shown in Figure 7. It can be found that the agent in this testing scenario still had a reasonable control policy but did not perform as well as in the training case. Firstly, the heating setpoint was sometimes too low during the occupied time even though the zone air temperature was not warm enough, e.g., at around noon on 01/09. Secondly, the cooling setpoint was sometimes too low during the unoccupied time, which triggered the cooling of the zone, e.g., on 01/07 from 8:00 to 16:00. An interesting finding is that there was a spark on OPPD in the B-23.9 case between 01/08 18:00 and 01/08 19:00 because the schedule set the heating setpoint to 15 °C while the zone was still occupied. This did not occur in the reinforcement learning case because it takes the occupancy as an input.

Table 6. The results of single-zone testing in the testing building model.

Case	Zn1 Mean OPPD (%)	Zn1 Std OPPD (%)	Building E_{HVAC} (kWh)
Trained agent 10	10.57	7.65	5803
B-21.1	13.25	9.52	5550
B-23.9	8.58	7.03	5942

Note: For all above cases, the control strategy of all zones except for Zn1 was the default control strategy of the building model.

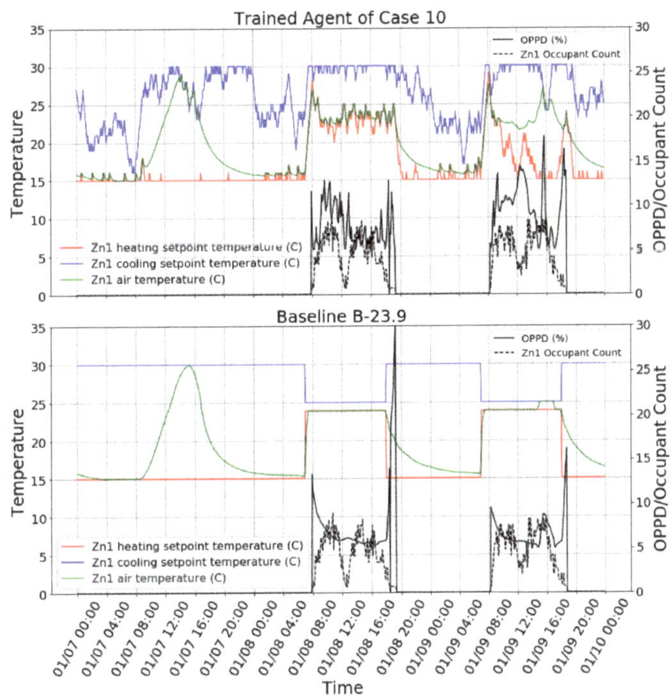

Figure 7. Single-zone testing in the testing building model: control behavior snapshot of the trained agent of case 10 (**top**) and the baseline case B-23.9 (**below**).

4.5.2. Multizone Testing in the Training Building Model

The trained reinforcement learning agent (case 10) was tested in the training building model to control all zones rather than just one. As shown in Table 7, case 10-0 achieved good thermal comfort for all zones but consumed much more energy than the baseline cases. The high HVAC energy consumption was primarily caused by the fact that the agent sometimes increased the heating setpoint during unoccupied times. This strange behavior of the trained agent is partially because the agent over-fitted to the HVAC energy consumption pattern in the training. Two additional tests were conducted to further analyze the agent's performance. One test used the trained agent along with a night setback rule: heating and cooling setpoints were set to 15 °C and 30 °C between 21:00 and 06:00 (case 10-1 in Table 6). The other test applied a mask to the state observation EHVAC: EHVAC was always zero in the testing (case 10-2 in Table 7). The results show that case 10-1 consumed 12.8% less HVAC energy than B-23.9 and achieved good thermal comfort performance, although not as good as B-23.9. Case 10-2 overcame the "unnecessary heating" problem of case 10-0, but it did not achieve as good a thermal comfort performance as case 10-1 because one state observation was masked. However, as expected, case 10-2 consumed even less HVAC energy consumption than case 10-1 at the price of worse thermal comfort.

Table 7. The results of multizone testing in the training building model.

Case	Zn1 OPPD Mean (%)	Zn1 OPPD Std (%)	Zn2 OPPD Mean (%)	Zn2 OPPD Std (%)	Zn3 OPPD Mean (%)	Zn3 OPPD Std (%)	Zn4 OPPD Mean (%)	Zn4 OPPD Std (%)	Zn5 OPPD Mean (%)	Zn5 OPPD Std (%)	Building E_{HVAC} (kWh)	Note
10-0	8.13	5.96	8.90	6.56	7.75	5.62	7.70	5.91	9.45	6.05	6983	As-is trained agent case 10
10-1	8.31	7.35	9.32	7.62	8.02	7.10	7.95	7.24	9.83	6.85	5328	Trained agent case 10 + night setback rule
10-2	10.46	11.11	9.86	9.08	9.67	10.47	9.97	11.09	10.37	8.12	4938	Trained agent case 10 + mask on state input EHVAC
B-21.1	10.79	7.86	10.33	7.44	9.89	7.73	10.25	7.88	7.29	4.16	4809	N/A
B-23.9	8.09	4.34	8.74	5.52	6.37	1.90	6.78	2.37	7.51	2.56	6109	N/A

4.5.3. Multizone Testing in the Testing Building Model

The trained reinforcement learning agent (case 10) was tested in the testing building model to control all zones. This is the most stringent test because both the building model and the control mode are different from the training. As shown in Table 8, the agent did not perform well in terms of either thermal comfort or HVAC energy efficiency. Firstly, the agent had worse thermal comfort performance than both B-21.1 and B-23.9; secondly, the agent consumed more energy than B-21.1. This means that using B-21.1 is better than using the trained agent in terms of both energy efficiency and thermal comfort. To find the reasons behind the agent's poor performance, the control behavior snapshot of Zn1 on three days in winter is plotted in Figure 8. It is clear that, for the reinforcement learning control case, high OPPD occurred in the morning because occupants arrived earlier and the agent started to increase the heating setpoint. We calculated the value count for the time that OPPD was higher than 20% for Zn1. It shows that about 70% of the larger-than-20% OPPD samples occurred between 06:00 and 10:00 (not included). This is partially because the trained agent over-fitted to the occupancy schedule of the training building model. In the training building model, occupants arrived at exactly 08:00 every workday, but in the testing building model, a stochastic occupancy schedule was used, in which there is some possibility that occupants arrive before 08:00. One observation in favor of the agent is that the B-23.9 case may have had high OPPD in the evening because the heating setpoint was set back to 15 °C while the zone was still occupied. The agent performed better regarding this problem because it takes occupancy as one of the inputs. For the whole building, the reinforcement learning case had 17% fewer larger-than-20% OPPD samples than B-23.9 for the time between 18:00 and 21:00 (not included) during the simulation period.

Table 8. The results of multizone testing in the testing building model.

Case	Zn1 OPPD Mean (%)	Zn1 OPPD Std (%)	Zn2 OPPD Mean (%)	Zn2 OPPD Std (%)	Zn3 OPPD Mean (%)	Zn3 OPPD Std (%)	Zn4 OPPD Mean (%)	Zn4 OPPD Std (%)	Zn5 OPPD Mean (%)	Zn5 OPPD Std (%)	Building E_{HVAC} (kWh)
10	13.84	14.31	12.12	11.87	12.66	13.53	13.45	13.55	9.40	10.37	5884
B-21.1	13.24	9.52	11.67	8.52	12.18	9.45	13.24	9.38	7.98	6.84	5752
B23.9	8.54	7.00	8.99	7.17	7.10	6.54	7.76	6.79	7.57	4.88	6965

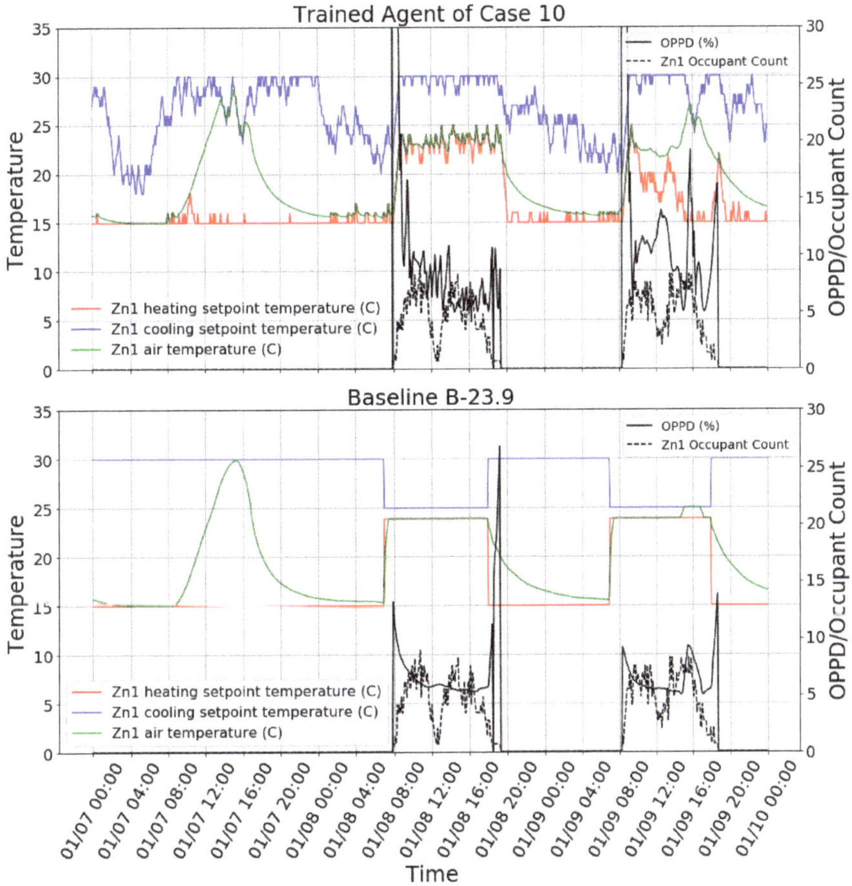

Figure 8. Multizone testing in the testing building model: control behavior snapshot of the trained agent of case 10 (**top**) and the baseline case B-23.9 (**below**).

4.6. Discussion

Optimization and generalization are two main problems in machine learning. Optimization is about how well the machine learning method can learn from the training data to minimize some loss functions. Generalization is about how well the trained machine learning model (or agent) performs with unseen data (or environments).

It was found in this study that the deep reinforcement learning control method had good convergence performance in the training, which usually converged long before the maximum learning step was reached. This finding is consistent with existing studies on deep reinforcement learning [4,45]. It was also found that all training cases could achieve good thermal comfort performance, and one training case was better than the B-23.9 baseline case in terms of both thermal comfort and HVAC energy efficiency. This shows that the proposed deep reinforcement learning control method could be well optimized.

Generalization performance is more difficult to evaluate for building control. Ideally, the trained agent's performance in controlling a real building is a good evaluation method. However, in this study, no real buildings were available. Therefore, the agent was evaluated in three testing scenarios. In the first testing scenario, the trained agent was used to control the same zone as in the training but with different weather conditions, operation schedules, etc. In this case, the agent could still perform reasonably, although not as well as in the training case. The agent might over-fit to the weather conditions of the training building model if it could not provide enough heating to the zone. In the second testing scenario, the trained agent was used to control different zones from the training, but the building model was exactly the same as in the training. This case clearly shows that the trained agent over-fitted to the HVAC energy profile in the training. When forcing a night setback rule for the agent, it achieved good thermal comfort performance in all zones and saved 12.8% HVAC energy consumption compared with the B-23.9 baseline case. Thirdly, the agent was used to control different zones from those in the training, and the building model was also different. In this case, the agent did not perform well. The trained agent might have over-fitted to the occupancy schedule of the training building model. Therefore, it can be concluded that the trained agent experienced the over-fitting problem. This problem was also reported in [46–48].

It must be admitted that there is a lack of a systematic method to diagnose the over-fitting problem of deep reinforcement learning control. All testing scenarios in this section can only conclude that the trained agent has an over-fitting problem, and there is no strong conclusion about where it over-fits. To the authors' knowledge, there is still no good theory behind the generalization of deep learning [49].

5. Conclusions and Future Work

Reinforcement learning control for HVAC systems has been thought to be promising in terms of achieving energy savings and maintaining indoor thermal comfort. However, previous studies did not provide enough information about end-to-end control for centralized HVAC systems in multizone buildings, mainly due to the limitations of reinforcement learning methods or the test buildings being single zones with independent HVAC systems. This study developed a supervisory HVAC control method using the advanced end-to-end deep reinforcement learning framework. Additionally, the control method was applied to a multizone building with a centralized HVAC system, which is not commonly seen in the existing literature. The control method directly took the measurable environmental parameters, including weather conditions and indoor environmental conditions, to control the indoor air heating and cooling setpoints of the HVAC system. A3C was used to train the deep reinforcement learning agent in a single-level five-zone office building model. During the training, the reinforcement learning agent only controlled one out of the five zones, with the goal of minimizing the controlled zone's thermal discomfort and the HVAC energy power demand of the whole building.

It was shown that the proposed deep reinforcement learning control method had good optimization convergence properties. In the training, it learned a reasonable control policy for the indoor air heating and cooling setpoints in response to occupancy, weather conditions, and internal heat gains. After hyperparameter tuning, a good training case was found, which achieved better thermal comfort and HVAC energy efficiency compared with the baseline case. It was also found that the penalty on large OPPD was beneficial to convergence.

By applying the trained agent to control all five zones of the training building model, 12.8% HVAC energy savings in comparison to one baseline case was achieved with good thermal comfort performance; however, a setpoint night setback rule must be enforced for the agent because of its over-fitting problem. The agent failed to achieve good performance in terms of both thermal comfort and HVAC energy efficiency if it was applied to control all five zones of the testing building model, also due to the over-fitting problem.

Future work should first focus on generalization techniques of deep learning. Dropout or batch normalization should be first considered to reduce over-fitting. The choice of the weather and occupancy schedule for training should be performed carefully to ensure that they are representative. Feature augmentation methods can be considered. Multi-task reinforcement learning is also a good candidate method to enhance deep reinforcement learning generalization performance. Secondly, multi-agent reinforcement learning or other methods that can be trained directly to provide a control policy for multiple zones should be studied. The current method in this study was trained for controlling one zone only, which may not be suitable for multizone control. Last but not least, the study was only tested using simulation models, rather than real buildings. The authors are now working on implementing the proposed control method in a real small office building.

Author Contributions: Conceptualization, X.Z., Z.Z. and R.Z.; methodology, X.Z. and Z.Z.; software, Z.Z.; validation, X.Z., Z.Z. and R.Z.; formal analysis, X.Z. and Z.Z.; investigation, X.Z., Z.Z. and C.Z.; resources, Z.Z. and C.Z.; data curation, Z.Z.; writing—original draft preparation, X.Z. and Z.Z.; writing—review and editing, X.Z., Z.Z. and R.Z.; visualization, X.Z. and Z.Z.; supervision, Z.Z.; project administration, Z.Z.; funding acquisition, Z.Z. All authors have read and agreed to the published version of the manuscript.

Funding: This research was funded by the University of Nottingham Ningbo China, grant number I01210100007.

Institutional Review Board Statement: Not applicable.

Informed Consent Statement: Not applicable.

Data Availability Statement: The data presented in this study are available on request from the corresponding author.

Acknowledgments: The authors would like to thank the Department of Architecture and Built Environment for providing materials used for experiments and simulations.

Conflicts of Interest: The authors declare no conflict of interest.

References

1. Szczepanik-Scislo, N.; Antonowicz, A.; Scislo, L. PIV measurement and CFD simulations of an air terminal device with a dynamically adapting geometry. *SN Appl. Sci.* **2019**, *1*, 370. [CrossRef]
2. Szczepanik-Scislo, N.; Schnotale, J. An Air Terminal Device with a Changing Geometry to Improve Indoor Air Quality for VAV Ventilation Systems. *Energies* **2020**, *13*, 4947. [CrossRef]
3. Bae, Y.; Bhattacharya, S.; Cui, B.; Lee, S.; Li, Y.; Zhang, L.; Im, P.; Adetola, V.; Vrabie, D.; Leach, M.; et al. Sensor impacts on building and HVAC controls: A critical review for building energy performance. *Adv. Appl. Energy* **2021**, *4*, 100068. [CrossRef]
4. Zhang, Z.; Chong, A.; Pan, Y.; Zhang, C.; Lam, K.P. Whole building energy model for HVAC optimal control: A practical framework based on deep reinforcement learning. *Energy Build.* **2019**, *199*, 472–490. [CrossRef]
5. Jiang, Z.; Risbeck, M.J.; Ramamurti, V.; Murugesan, S.; Amores, J.; Zhang, C.; Lee, Y.M.; Drees, K.H. Building HVAC control with reinforcement learning for reduction of energy cost and demand charge. *Energy Build.* **2021**, *239*, 110833. [CrossRef]
6. Esrafilian-Najafabadi, M.; Haghighat, F. Occupancy-based HVAC control systems in buildings: A state-of-the-art review. *Build. Environ.* **2021**, *197*, 107810. [CrossRef]
7. Chen, Y.; Chen, Z.; Yuan, X.; Su, L.; Li, K. Optimal Control Strategies for Demand Response in Buildings under Penetration of Renewable Energy. *Buildings* **2022**, *12*, 371. [CrossRef]
8. Tardioli, G.; Filho, R.; Bernaud, P.; Ntimos, D. An Innovative Modelling Approach Based on Building Physics and Machine Learning for the Prediction of Indoor Thermal Comfort in an Office Building. *Buildings* **2022**, *12*, 475. [CrossRef]
9. Zhao, J.; Lam, K.P.; Ydstie, B.E.; Loftness, V. Occupant-oriented mixed-mode EnergyPlus predictive control simulation. *Energy Build.* **2016**, *117*, 362–371. [CrossRef]
10. Zhao, T.; Wang, J.; Xu, M.; Li, K. An online predictive control method with the temperature based multivariable linear regression model for a typical chiller plant system. *Build. Simul.* **2020**, *13*, 335–348. [CrossRef]
11. Talib, R.; Nassif, N. "Demand Control" an Innovative Way of Reducing the HVAC System's Energy Consumption. *Buildings* **2021**, *11*, 488. [CrossRef]
12. Dong, B.; Lam, K.P. A real-time model predictive control for building heating and cooling systems based on the occupancy behavior pattern detection and local weather forecasting. *Build. Simul.* **2014**, *7*, 89–106. [CrossRef]
13. Ma, X.; Bao, H.; Zhang, N. A New Approach to Off-Line Robust Model Predictive Control for Polytopic Uncertain Models. *Designs* **2018**, *2*, 31. [CrossRef]
14. Ascione, F.; Bianco, N.; De Stasio, C.; Mauro, G.M.; Vanoli, G.P. Simulation-based model predictive control by the multi-objective optimization of building energy performance and thermal comfort. *Energy Build.* **2016**, *111*, 131–144. [CrossRef]
15. Garnier, A.; Eynard, J.; Caussanel, M.; Grieu, S. Predictive control of multizone heating, ventilation and air-conditioning systems in non-residential buildings. *Appl. Soft Comput.* **2015**, *37*, 847–862. [CrossRef]
16. Wang, C.; Wang, B.; Cui, M.; Wei, F. Cooling seasonal performance of inverter air conditioner using model prediction control for demand response. *Energy Build.* **2022**, *256*, 111708. [CrossRef]
17. Váňa, Z.; Cigler, J.; Široký, J.; Žáčeková, E.; Ferkl, L. Model-based energy efficient control applied to an office building. *J. Process Control* **2014**, *24*, 790–797. [CrossRef]

18. Kumar, R.; Wenzel, M.J.; ElBsat, M.N.; Risbeck, M.J.; Drees, K.H.; Zavala, V.M. Stochastic model predictive control for central HVAC plants. *J. Process Control* **2020**, *90*, 1–17. [CrossRef]
19. Toub, M.; Reddy, C.R.; Razmara, M.; Shahbakhti, M.; Robinett, R.D.; Aniba, G. Model-based predictive control for optimal MicroCSP operation integrated with building HVAC systems. *Energy Convers. Manag.* **2019**, *199*, 111924. [CrossRef]
20. Kwak, Y.; Huh, J.-H.; Jang, C. Development of a model predictive control framework through real-time building energy management system data. *Appl. Energy* **2015**, *155*, 1–13. [CrossRef]
21. Liu, S.; Henze, G.P. Experimental analysis of simulated reinforcement learning control for active and passive building thermal storage inventory: Part 2: Results and analysis. *Energy Build.* **2006**, *38*, 148–161. [CrossRef]
22. Liu, S.; Henze, G.P. Evaluation of Reinforcement Learning for Optimal Control of Building Active and Passive Thermal Storage Inventory. *J. Sol. Energy Eng.* **2006**, *129*, 215–225. [CrossRef]
23. Jayalaxmi, P.L.S.; Saha, R.; Kumar, G.; Kim, T.-H. Machine and deep learning amalgamation for feature extraction in Industrial Internet-of-Things. *Comput. Electr. Eng.* **2022**, *97*, 107610. [CrossRef]
24. Chen, Y.; Tong, Z.; Zheng, Y.; Samuelson, H.; Norford, L. Transfer learning with deep neural networks for model predictive control of HVAC and natural ventilation in smart buildings. *J. Clean. Prod.* **2020**, *254*, 119866. [CrossRef]
25. Othman, K. Deep Neural Network Models for the Prediction of the Aggregate Base Course Compaction Parameters. *Designs* **2021**, *5*, 78. [CrossRef]
26. Mnih, V.; Kavukcuoglu, K.; Silver, D.; Graves, A.; Antonoglou, I.; Wierstra, D.; Riedmiller, M. Playing Atari with Deep Reinforcement Learning. *arXiv* **2013**, arXiv:1312.5602. [CrossRef]
27. Dalamagkidis, K.; Kolokotsa, D.; Kalaitzakis, K.; Stavrakakis, G.S. Reinforcement learning for energy conservation and comfort in buildings. *Build. Environ.* **2007**, *42*, 2686–2698. [CrossRef]
28. Fazenda, P.; Veeramachaneni, K.; Lima, P.; O'Reilly, U.-M. Using reinforcement learning to optimize occupant comfort and energy usage in HVAC systems. *J. Ambient. Intell. Smart Environ.* **2014**, *6*, 675–690. [CrossRef]
29. Capozzoli, A.; Piscitelli, M.S.; Gorrino, A.; Ballarini, I.; Corrado, V. Data analytics for occupancy pattern learning to reduce the energy consumption of HVAC systems in office buildings. *Sustain. Cities Soc.* **2017**, *35*, 191–208. [CrossRef]
30. Costanzo, G.T.; Iacovella, S.; Ruelens, F.; Leurs, T.; Claessens, B.J. Experimental analysis of data-driven control for a building heating system. *Sustain. Energy Grids Netw.* **2016**, *6*, 81–90. [CrossRef]
31. Fang, J.; Feng, Z.; Cao, S.-J.; Deng, Y. The impact of ventilation parameters on thermal comfort and energy-efficient control of the ground-source heat pump system. *Energy Build.* **2018**, *179*, 324–332. [CrossRef]
32. Yuan, X.; Pan, Y.; Yang, J.; Wang, W.; Huang, Z. Study on the application of reinforcement learning in the operation optimization of HVAC system. *Build. Simul.* **2021**, *14*, 75–87. [CrossRef]
33. Ding, X.; Du, W.; Cerpa, A. OCTOPUS: Deep Reinforcement Learning for Holistic Smart Building Control. In Proceedings of the 6th ACM International Conference on Systems for Energy-Efficient Buildings, Cities, and Transportation (BuildSys '19), New York, NY, USA, 13–14 November 2019.
34. Morales, E.F.; Escalante, H.J. Chapter 6—A brief introduction to supervised, unsupervised, and reinforcement learning. In *Biosignal Processing and Classification Using Computational Learning and Intelligence*; Torres-García, A.A., Reyes-García, C.A., Villaseñor-Pineda, L., Mendoza-Montoya, O., Eds.; Academic Press: Cambridge, MA, USA, 2022; pp. 111–129. [CrossRef]
35. Sun, G.; Ayepah-Mensah, D.; Xu, R.; Boateng, G.O.; Liu, G. End-to-end CNN-based dueling deep Q-Network for autonomous cell activation in Cloud-RANs. *J. Netw. Comput. Appl.* **2020**, *169*, 102757. [CrossRef]
36. Bommisetty, L.; Venkatesh, T.G. Resource Allocation in Time Slotted Channel Hopping (TSCH) networks based on phasic policy gradient reinforcement learning. *Internet Things* **2022**, *19*, 100522. [CrossRef]
37. Crawley, D.B.; Lawrie, L.K.; Winkelmann, F.C.; Buhl, W.F.; Huang, Y.J.; Pedersen, C.O.; Strand, R.K.; Liesen, R.J.; Fisher, D.E.; Witte, M.J.; et al. EnergyPlus: Creating a new-generation building energy simulation program. *Energy Build.* **2001**, *33*, 319–331. [CrossRef]
38. Aliaga, L.G.; Williams, E. Co-alignment of comfort and energy saving objectives for U.S. office buildings and restaurants. *Sustain. Cities Soc.* **2016**, *27*, 32–41. [CrossRef]
39. Zhou, J.; Xue, Y.; Xu, D.; Li, C.; Zhao, W. Self-learning energy management strategy for hybrid electric vehicle via curiosity-inspired asynchronous deep reinforcement learning. *Energy* **2022**, *242*, 122548. [CrossRef]
40. Mnih, V.; Badia, A.; Mirza, M.; Graves, A.; Lillicrap, T.; Harley, T.; Silver, D.; Kavukcuoglu, K. Asynchronous Methods for Deep Reinforcement Learning. *arXiv* **2016**, arXiv:1602.01783. [CrossRef]
41. Luo, X.; Lam, K.P.; Chen, Y.; Hong, T. Performance evaluation of an agent-based occupancy simulation model. *Build. Environ.* **2017**, *115*, 42–53. [CrossRef]
42. Brockman, G.; Cheung, V.; Pettersson, L.; Schneider, J.; Schulman, J.; Tang, J.; Zaremba, W. OpenAI Gym. *arXiv* **2016**, arXiv:1606.01540. [CrossRef]
43. Tieleman, T.; Hinton, G. Lecture 6.5-rmsprop: Divide the Gradient by a Running Average of its Recent Magnitude. *Neural Netw. Mach. Learn.* **2012**, *4*, 26–30.
44. Williams, R.J.; Peng, J. Function Optimization using Connectionist Reinforcement Learning Algorithms. *Connect. Sci.* **1991**, *3*, 241–268. [CrossRef]
45. Fang, X.; Gong, G.; Li, G.; Chun, L.; Peng, P.; Li, W.; Shi, X.; Chen, X. Deep reinforcement learning optimal control strategy for temperature setpoint real-time reset in multi-zone building HVAC system. *Appl. Therm. Eng.* **2022**, *212*, 118552. [CrossRef]

46. Homod, R.Z.; Togun, H.; Hussein, A.K.; Al-Mousawi, F.N.; Yaseen, Z.M.; Al-Kouz, W.; Abd, H.J.; Alawi, O.A.; Goodarzi, M.; Hussein, O.A. Dynamics analysis of a novel hybrid deep clustering for unsupervised learning by reinforcement of multi-agent to energy saving in intelligent buildings. *Appl. Energy* **2022**, *313*, 118863. [CrossRef]
47. Ala'raj, M.; Radi, M.; Abbod, M.F.; Majdalawieh, M.; Parodi, M. Data-driven based HVAC optimisation approaches: A Systematic Literature Review. *J. Build. Eng.* **2022**, *46*, 103678. [CrossRef]
48. Yang, L.; Nagy, Z.; Goffin, P.; Schlueter, A. Reinforcement learning for optimal control of low exergy buildings. *Appl. Energy* **2015**, *156*, 577–586. [CrossRef]
49. Zhang, C.; Bengio, S.; Hardt, M.; Recht, B.; Vinyals, O. Understanding deep learning requires rethinking generalization. *arXiv* **2017**, arXiv:1611.03530. [CrossRef]

mathematics

MDPI

Article

Enhanced Teaching Learning-Based Algorithm for Fuel Costs and Losses Minimization in AC-DC Systems

Shahenda Sarhan [1,2], Abdullah M. Shaheen [3], Ragab A. El-Sehiemy [4,*] and Mona Gafar [5,6]

1 Faculty of Computing and Information Technology, King Abdulaziz University, Jeddah 21589, Saudi Arabia; sssarhan@kau.edu.sa
2 Faculty of Computers and Information Sciences, Mansoura University, Mansoura 35516, Egypt
3 Electrical Engineering Department, Faculty of Engineering, Suez University, Suez 41522, Egypt; abdullahshaheen2015@gmail.com
4 Electrical Engineering Department, Faculty of Engineering, Kafrelsheikh University, Kafrelsheikh 33516, Egypt
5 Department of Computer Science, College of Science and Humanities in Al-Sulail, Prince Sattam Bin Abdulaziz University, Kharj 16278, Saudi Arabia; mona2004egypt@yahoo.com
6 Machine Learning and Information Retrieval Department, Artificial Intelligence, Kafrelsheikh University, Kafrelsheikh 33516, Egypt
* Correspondence: elsehiemy@eng.kfs.edu.eg

Abstract: The Teaching Learning-Based Algorithm (TLBA) is a powerful and effective optimization approach. TLBA mimics the teaching-learning process in a classroom, where TLBA's iterative computing process is separated into two phases, unlike standard evolutionary algorithms and swarm intelligence algorithms, and each phase conducts an iterative learning operation. Advanced technologies of Voltage Source Converters (VSCs) enable greater active and reactive power regulation in these networks. Various objectives are addressed for optimal energy management, with the goal of attaining economic and technical advantages by decreasing overall production fuel costs and transmission power losses in AC-DC transmission networks. In this paper, the TLBA is applied for various sorts of nonlinear and multimodal functioning of hybrid alternating current (AC) and multi-terminal direct current (DC) power grids. The proposed TLBA is evaluated on modified IEEE 30-bus and IEEE 57-bus AC-DC networks and compared to other published methods in the literature. Numerical results demonstrate that the proposed TLBA has great effectiveness and robustness indices over the others. Economically, the reduction percentages of 13.84 and 21.94% are achieved for the IEEE 30-bus and IEEE 57-bus test systems when the fuel costs are minimized. Technically, significant improvement in the transmission power losses with reduction 28.01% and 69.83%, are found for the IEEE 30-bus and IEEE 57-bus test system compared to the initial case. Nevertheless, TLBA has faster convergence, higher quality for the final optimal solution, and more power for escaping from convergence to local optima compared to other published methods in the literature.

Keywords: teaching-learning-based algorithm; multi-terminal HVDC grids; economic power flow; valve point loading effect

MSC: 68T20

Citation: Sarhan, S.; Shaheen, A.M.; El-Sehiemy, R.A.; Gafar, M. Enhanced Teaching Learning-Based Algorithm for Fuel Costs and Losses Minimization in AC-DC Systems. *Mathematics* **2022**, *10*, 2337. https://doi.org/10.3390/math10132337

Academic Editors: Zbigniew Leonowicz, Arsalan Najafi and Michał Jasiński

Received: 22 April 2022
Accepted: 18 June 2022
Published: 4 July 2022

1. Introduction

To satisfy the ever-increasing household and industrial loads, the development of electric power networks has become a must-do operation. As power systems expand, power losses grow, resulting in a waste of huge amounts of money annually. Furthermore, the proper functioning of electrical networks takes into account a variety of factors such as fuel cost reduction, environmental pollution, network losses, security, quality, and stability [1]. As a consequence, for the effective supply of electricity, the operational condition is separate to the main objective functions such as reducing power losses, seeking to avoid voltage irregularities, and growing system security while complying to numerous equality and inequality restrictions.

Optimal power flow and economic dispatch (ED) seem to be crucial minimization aspects in power systems that necessitate efficient generator interoperability, strategic planning, and scheduling [2]. In [3], a slime mould technique driven by customizable weight vector to control the series among positive and negative propagation waves was utilized for the minimized ED problem. In [4], a bi-stage self-adaptive differential evolution (DE) approach of k-nearest neighbours relying computation system has been designed to address numerous metaheuristic issues, and it was suggested that the ED problem be addressed in the upcoming years. A Manta ray forage optimizer with non-dominated sorting approach was developed in [5] to solve the multi-objective load flow encompassing solar, wind and small-hydro energy production. In [6], a social network searching algorithm was used to schedule the power network outputs with non-dominated electrical losses and fuel costs. A multi-verse algorithm for minimizing the dynamic ED management issue in electricity frameworks utilizing valve point effects was presented in [7].

In most countries, high voltage alternating current (HVAC) technology is used in conjunction with electric power components and the incorporation of significant alternative electricity sources [8]. However, shortcomings caused by excessive system losses, expenditures, and reactive power compensation requirements as the length of transmission circuits rises drawing HVAC technologies inappropriate for linking bulk systems or faraway renewable energy production companies [9]. Based on voltage source converters (VSCs), the transmission technology of HV direct current (HVDC) has arisen as an appealing option. It has outstanding features to control the voltage in the AC system by appropriate management of the reactive power injection and absorption. Regardless of the DC transferring power, the VSC scheme can govern respectively real and reactive power throughout its station at the same time [10,11]. An AC-DC load flow procedure was introduced for managing the VSCs-HVDC in power systems by de-coupling the AC network from the DC power network together with the VSC transformer stations; however, its relevance was demonstrated using simple frameworks of 5 and 14-bus test systems [12]. In [13], a sequential algorithm to perform a load flow assessment in hybridized AC-DC networks incorporating all their operational types in the steady-state model. In [14], a sequential method relying on Gauss-Seidel and modified Gauss techniques was used to handle the operation in an AC-DC system, with DC sides controlled by injecting current into the linking stations. Nevertheless, their implementation was executed on a small IEEE 9-bus system because the utilized AC-DC formulations have been solved independently [14]. A quasi-AC alternative centred on relaxing the semi-definite coding framework was also addressed [15]; however, the test data have only been conducted on a basic IEEE 30-bus network. Additionally, in Ref. [16], a sequential method to solve the AC-DC system equations separately for each repeat, employing the interface variables projected from the AC load flow until the solution convergence was achieved. Even though it is simple to construct, it may confront convergence issues under certain instances. Owing to the significance of environmental and economic power operations in HVAC systems hybridized with HVDC systems, an OPF optimization model has been developed [17].

VSCs system developments can be extensively simulated, keeping in view their converter station, transformer, phase reactor, and filter parts [18]. Power losses in the parts of VSCs system were typically represented by a quadratic relationship of the converter current [18–20]. In [21], a hybrid AC-DC distribution system was presented considering the integration of distributed generators with AC and DC soft open. However, the presented methodology in [21] has been dedicated for minimizing the system power losses as a single target. In [22], an analysis based on the invested costs and the gained benefit for HVDC and AC options for integrating the offshore wind turbines or bulk power has been handled. Notwithstanding, the investigation of HVDC frameworks was limited to a two-terminal configuration. Otherwise, integrating a linked DC system within an established AC system complicates the coordinated control consideration of these structures [23]. However, despite this, the DC load flow calculations in [24] were overlooked, as were specific AC-DC system characteristics in [25,26] were neglected. Also, A second-order cone programme solver has also been applied for hybrid power networks to investigate VSC-DC mechanisms in an optimisation problem [27]. There is also a primal-dual interior point method merged with upgraded Jacobian and Hessian matrices [28]. The impact of tap changer situations and VAr variation in the AC configuration was ignored in these reviews, and some applied methodologies were dependent on the initial estimate in certain cases, based on finite assumptions that restricted the required precision.

Despite advances in artificial intelligence-based metaheuristic solvers, including crow search optimizer [29] and manta ray forage technique [30], little attention has been paid to solving the OPF challenging problem in hybrid AC systems. In [31], genetic metaheuristic method has been performed to optimise the OPF for minimizing the power losses in hybrid AC-DC power systems. Ref. [32] used the DE algorithm to solve the OPF issue in hybridized AC-DC systems as a minimization goal. In addition, techniques focusing on material equilibrium state [33] and marine predators' simulation [34] were established to address multi-objective OPF modelling in AC-DC systems.

The teaching–learning–based Algorithm (TLBA) is a population-based intelligent algorithm that mimics the teaching–learning process in a classroom [35]. TLBA's iterative computing process is separated into two phases, unlike standard evolutionary algorithms and swarm intelligence algorithms, and each phase conducts an iterative learning operation. Since its debut by Rao and colleagues, TLBA has garnered the attention of an increasing number of academics because to several of its merits, including its simple idea, lack of algorithm-specific parameters, quick convergence, and ease of implementation while being effective [36,37]. The TLBA has fairly recently been used effectively to solve numerous engineering design issues such as parameter identification of Photovoltaic (PV) panels [38,39], operation assessment of integrated PV and batteries with power system [40], harmonic elimination inside inverters [41], robots manipulator calibration [42], condition prediction in water supply pipes [43], welding processes [44], optimal design of electrical filters [45], expansion planning of power generation in electrical networks [46], tsallis-entropy-based feature selection [47], service restoration problem in delivery networks [48] and reactive power management in power grids [49]. The TLBA's strengths and effective implementations in a broad range of engineering design problems are the prime motivations for its utilization in this study. TLBA is used to various sorts of nonlinear and multimodal functioning of hybrid alternating current (AC) and multi-terminal direct current (DC) power grids. The proposed TLBA is evaluated on modified IEEE 30-bus AC-DC networks and compared to other published methods in the literature.

This paper's main contribution could be explained simply:

- Various objective targets for optimized energy management in AC-DC transmission systems are handled, by achieving technical and economic benefits.
- TLBA is developed with the capability to handle the operational optimization OPF problem by reducing the power losses and the total power generation costs in AC-DC transmission networks with successful application on two IEEE systems comprising 30 and 57 buses
- For IEEE 30 and 57 bus systems, very high success rates are demonstrated accompanying to the proposed TLBA
- Moreover, when compared to other reported techniques in the literature, TLBA possesses higher convergence rate, greater quality for the ultimate best solution, and more strength for attempting to escape from convergence to optimum.

2. Problem Formulation

In high voltage AC-DC systems, the main operation target is technical and economical by determining the optimal decision variables for attaining a variety of defined aims in AC-DC networks that are subject to different equality and inequity constraints.

2.1. Problem Objectives

Primarily, the total fuel costs (TFCs), in \$/h, are the sum of the fuel costs of each generator. Therefore, TFCs minimization objective function is the first one (M_1) that can be mathematically modelled in (1) [50,51]:

$$M_1 = \sum_{i=1}^{Ng} cc_i + bb_i P_{Gi} + aa_i {P_{Gi}}^2 \quad (1)$$

where P_{Gi} indicates the real output power in MW of generator i, and aa_i, bb_i, and cc_i are the related cost coefficients.

On the other side, the TFCs may be formulated considering numerous ripples due to the loading impacts of the valve point. Therefore, it could be mathematically modelled as additional rectified terms in sinusoids forms to the cost model in (1) [52] as follows:

$$M_1 = \sum_{i=1}^{Ng} cc_i + bb_i P_{Gi} + aa_i {P_{\mathrm{Gi}}}^2 + |ee_i(\sin ff_i(P_{\mathrm{Gi}}^{\min} - P_{Gi}))| \quad (2)$$

where, ee_i and ff_i refer to the additional cost coefficients of the valve point loadings [53]:

Secondly, the entire transmission losses (*ETLs*) (M_2) is handled with three parts in such systems, as described in (3), including the losses in AC system ($Loss_{AC}$) that described in (4), the losses in DC system ($Loss_{DC}$) that described in (5), and the losses in VSC stations ($Loss_{VSC}$) that are described in (6) [54]:

$$M_2 = Loss_{AC} + Loss_{DC} + Loss_{VSC} \quad (3)$$

$$Loss_{AC} = \sum_{i=1}^{Nb} \sum_{j=1}^{Nb} G_{ij}(-2(V_i V_j cos\,\theta_{ij} + V_i^2 + {V_j}^2)) \quad (4)$$

$$Loss_{DC} = \sum_{i,j \in N_{bDC}} R_{ij} I_{ij}^2 \quad (5)$$

$$Loss_{VSC} = \sum_{i=1}^{N_V} A_i {Ic_i}^2 + B_i Ic_i + C_i \quad (6)$$

where G_{ij} is to the conductance of the line connected between buses i and j: N_b indicate the buses number; V and θ are the voltage and phase angle; R_{ij} refers to the resistance of the DC link between buses i and j: N_{bDC} indicates the DC buses number; I_{ij} indicates the DC Ampere flow over the link between buses i and j; A, B and C are the factors of the losses due to each VSC (i).

2.2. Control and Dependent Variables in AC-DC Network

The control variables in AC-DC systems are changed to involve the following variables that corresponding to the DC side. So, the implementation is extended to sense the variables added to the AC variables in the main grid. Also, current and voltage sensors are needed at different lines and buses to check several operational constraints.

Firstly, related to the AC network, the control variables are:

(a) (Vg_1: Vg_{Ng}) Voltages at generators
(b) (Pg_1: Pg_{Ng}) Real output power of each generator
(c) (Tap_1: Tap_{Nt}) tap setting of transformers
(d) (Qc_1: Qc_{Nq}) Reactive output power of VAr devices

where, N_g, N_t, and N_q refer to, accordingly the generators number, the number of on-load tap transformers, and the number of VAr devices [55].

Secondly, related to the VSC type, the control variables are [56]:

(a) (V_{dc}-Q_c) Constant voltage and reactive power, respectively, at DC and AC buses.

(b) (V_{dc}-V_c) Constant voltages at DC and AC buses.
(c) (P_{dc}-Q_c) Constant real power and constant reactive power, respectively, trough the DC line and at AC bus
(d) (P_{dc}-V_c) Constant active power and voltage, respectively, trough the DC line and at the AC bus

Similarly, at first, some dependent variables are related to the AC network which are

(a) (SF_1: SF_{Nf}) Apparent flow over AC branches
(b) (VL_1: VL_{NPQ}) Voltages at load buses
(c) (Qg_1: Qg_{Ng}) Reactive powers of each generator

where, Nf, and NPQ are, respectively the number of the branches, and the number of load buses.

Secondly, related to the VSC type, dependent variables are

(a) Real Power flow over DC links.
(b) Voltages at DC buses

2.3. Equality Constraints

There are two forms of equality restrictions which are belonging to the balanced real and reactive powers flow in the AC system as defined in Equations (7) and (8), the balanced power flow in the DC system as defined in Equation (9).

$$-P_{Li}+P_{Gi} - V_i\sum_{j=1}^{Nb} V_j(B_{ij}\ sin\ \theta_{ij} + G_{ij}cos\ \theta_{ij}) = 0,\ i = 1, \ldots Nb \tag{7}$$

$$-Q_{Li}+Q_{Gi} - V_i\sum_{j=1}^{Nb} V_j(-B_{ij}cos\theta_{ij}+G_{ij}sin\theta_{ij}) = 0,\ i = 1, \ldots\ Nb \tag{8}$$

$$S_{kj} = [Vs_k]\left[I_{kj}\right]^* = [Vs_k]\left[\frac{Vs_k - Vc_j}{\left[R_{jk}\right] + j\left[X_{jk}\right]}\right]^* = \left[P_{kj}\right] + j\left[Q_{kj}\right],\ k = 1 : N_A,\ j = 1 : N_V \tag{9}$$

where P_L and P_G are the real powers of loads and generators; B_{ij} is the line susceptance connected between buses i and j; Q_L and Q_G are the reactive powers of load and generators; S_{kj} is the injected MVA from AC system to the VSCs; R_{dc} refers to the DC resistance of the link; P and Q are, correspondingly, the injected powers of active and reactive type. Vc_j indicates the VSC voltage; R_{jk} + jX_{jk} is the equivalent impedance of the VSC accessories; Vs_k is the voltage at the AC connected bus. N_V and N_A are, accordingly, the VSCs number and the AC buses whereas Ik_j symbolizes the injected current.

2.4. Inequality Constraints

Also, the operating limitations I the AC-DC system should be maintained within the permissible bounds which can be mathematically described as follows:

$$Pg_G^{\max} \geq Pg_G \geq Pg_G^{\min}, G = 1\ :\ Ng \tag{10}$$

$$Qg_G^{\max} \geq Qg_G \geq Qg_G^{\min}, G = 1\ :\ Ng \tag{11}$$

$$Vg_G^{\max} \geq Vg_G \geq Vg_G^{\min}, G = 1\ :\ Ng \tag{12}$$

$$Qc_q^{\max} \geq Qc_q \geq Qc_q^{\min}, q = 1\ :\ Nq \tag{13}$$

$$Tap_T^{\max} \geq Tap_T \geq Tap_T^{\min}, T = 1\ :\ Nt \tag{14}$$

$$|SF_{line}| \leq SF_{line}^{\max},\ \text{line} = 1 : Nf \tag{15}$$

$$V_{L_k}^{\max} \geq V_{L_k} \geq V_{L_k}^{\min}, k = 1\ :\ NPQ \tag{16}$$

$$Ps_j^{\max} \geq Ps_j \geq Ps_j^{\min}, j = 1 : N_V \tag{17}$$

$$Qs_j^{\max} \geq Qs_j \geq Qs_j^{\min}, j = 1 : N_V \tag{18}$$

$$Vc_j^{\max} \geq Vc_j \geq Vc_j^{\min}, j = 1 : N_V \tag{19}$$

$$V_{d_c,j}^{\max} \geq V_{d_c,j} \geq V_{d_c,j}^{\min}, j = 1 : N_{bDC} \tag{20}$$

$$\frac{d_j^{\max}}{2} \geq \sqrt{(Ps_j - P_o)^2 - (Qs_j - Q_o)^2} \geq \frac{d_j^{\min}}{2}, j = 1 : N_V \tag{21}$$

where, (P_o, Q_o) indicated the circle's centre connected to the VSC's PQ-capacity and d is its diameter. The superscripts "min" and "max" denote the linked variable's lowest and highest bounds.

3. Proposed TLBA for OPF Problem in AC-DC Grids

3.1. TLBA Concept

TLBA seems to be a population adaptive technique that simulates the teaching–learning procedure in a classroom [35]. Unlike basic evolutionary algorithms and swarm intelligent computational methods, the iterative computing method of TLBA is divided into two stages, with each stage performing an adaptive learning procedure. First and foremost,

$$Y_j = Y_{\min} + rand(0,1).[Y_{\max} - Y_{\min}] \quad j = 1,2, \ldots\ldots. N_s \tag{22}$$

where, $Y_{\max}$ and $Y_{\min}$ represent the maximum and minimum bounds due to the decision variables and N_s is the students' number in a population.

The fundamental TLBA has been split into two stages: teaching and learning.

Initially, during the teaching stage of development, the teacher is regarded as the person with the deepest expertise, understanding, and skillset (the best student with minimum objective). In this stage, the teacher (Yt) continues to strive to increase the classroom mean (Ym). As a result, the jth student new knowledge (Y_{new}) following the teaching stage of development is as follows:

$$Y_{new} = Y_j + rand(0,1).[Yt - (FT.Ym)] \quad j = 1,2, \ldots\ldots. N_s \tag{23}$$

$$\text{where, } FT = round[1 + rand(0,1)] \tag{24}$$

where, Y_j is the jth student in the classroom and round is an integer approximated value which is randomly generated; FT indicates the factor of learning variation.

Conversely, through peer engagement, students gain experience and skill across the learning stage. Consequently, the jth student (Y_j) continues to strive to improve his/her investigative information and knowledge in the classroom by learning from another randomly selected person involved (Y_k), where k and j are different.

$$Y_{new} = \begin{cases} Y_j + rand(0,1).[Y_j - Y_k] & if\ F(Y_j) \leq F(Y_k) \\ Y_j + rand(0,1).[Y_k - Y_j] & if\ F(Y_j) > F(Y_k) \end{cases} \tag{25}$$

where, $F(Y_k)$ and $F(Y_j)$ are, respectively the objective values related to the students k and j.

As illustrated, based on Y_j and Y_k, two outcomes are possible: if Y_j is preferable than Y_k, Y_k is shifted towards Y_j. Alternatively, it is pushed away from Y_j.

A pseudocode of the TLBA is described in Algorithm 1.

Algorithm 1. A pseudocode of the TLBA.
Input: Number of students (N_s), lower limits ($Y_{\min}$) and upper limits ($Y_{\max}$), Maximum number of iterations
Output: Minimum fitness solution
1: procedure TLBA
2: Set $It = 1$
3: Initialize the population of students (Y_j), $Y_j = Y_{\min}$ + rand*($Y_{\max} - Y_{\min}$)
4: Evaluate the fitness functions of each student j as ($F(Y_j)$)
5: while (It < Maximum number of iterations) do
6: Evaluate the learning changing factor (FT), FT=round[1+rand(0,1)]
7: Select the instructor with the best solution obtained in all population (Y_t)
8: Extract the classroom mean (Y_m)
9: Apply the teaching phase to update the position of the member (Y_{new}) based on Eq. (23)
10: Evaluate the fitness functions as ($F(Y_{new})$)
11: Compare the new and current members and accept the one with better fitness value.
12: Randomly select a member (Y_k)
13: Apply the learning phase to update the position of the member (Y_{new}) based on Eq. (25)
14: Evaluate the fitness functions as ($F(Y_{new})$)
15: Compare the new and current members and accept the one with better fitness value.
16: End while
17: Find the best solution with the minimum fitness
18: End procedure

3.2. Proposed TLBA for OPF Problem in AC-DC Grids

This sub-section illustrates the developed teaching-learning studying-based algorithm for OPF Problem in AC-DC grids. For handling the OPF problem in AC-DC grids, the proposed TLBA is enhanced. The new solutions of infeasible dimensions must always be treated appropriately in order to determine whether one student is superior to the other. Therefore, each new solution is checked for each dimension as follows:

$$Y_{new,d} = \begin{cases} Y_{\max,d} & if\ Y_{new,d} > Y_{\max,d} \\ Y_{\min,d} & if\ Y_{new,d} < Y_{\min,d} \\ Y_{new,d} & Else \end{cases} \tag{26}$$

Also, the balancing equations in AC-DC grids, which express the equality restrictions, are assured fundamentally for dealing with the discussed problem using the consecutive Load flow approach [57]. The Newtonian algorithm typically finds a solution if the load flow in the AC-DC grid is met.

Additionally, the operating boundaries of the independent variables are begun fulfilling their boundaries, and if any of them are breached through the iterations, they are set at the nearest limits, as illustrated in Equation (26). In the investigated objectives, the operational limitations of the dependent variables in AC-DC grid are checked as well and any violation in the regarding constraints are penalized and added to the fitness function. The proposed TLBA is dedicated for solving the OPF problem in AC-DC grids, as described in Figure 1.

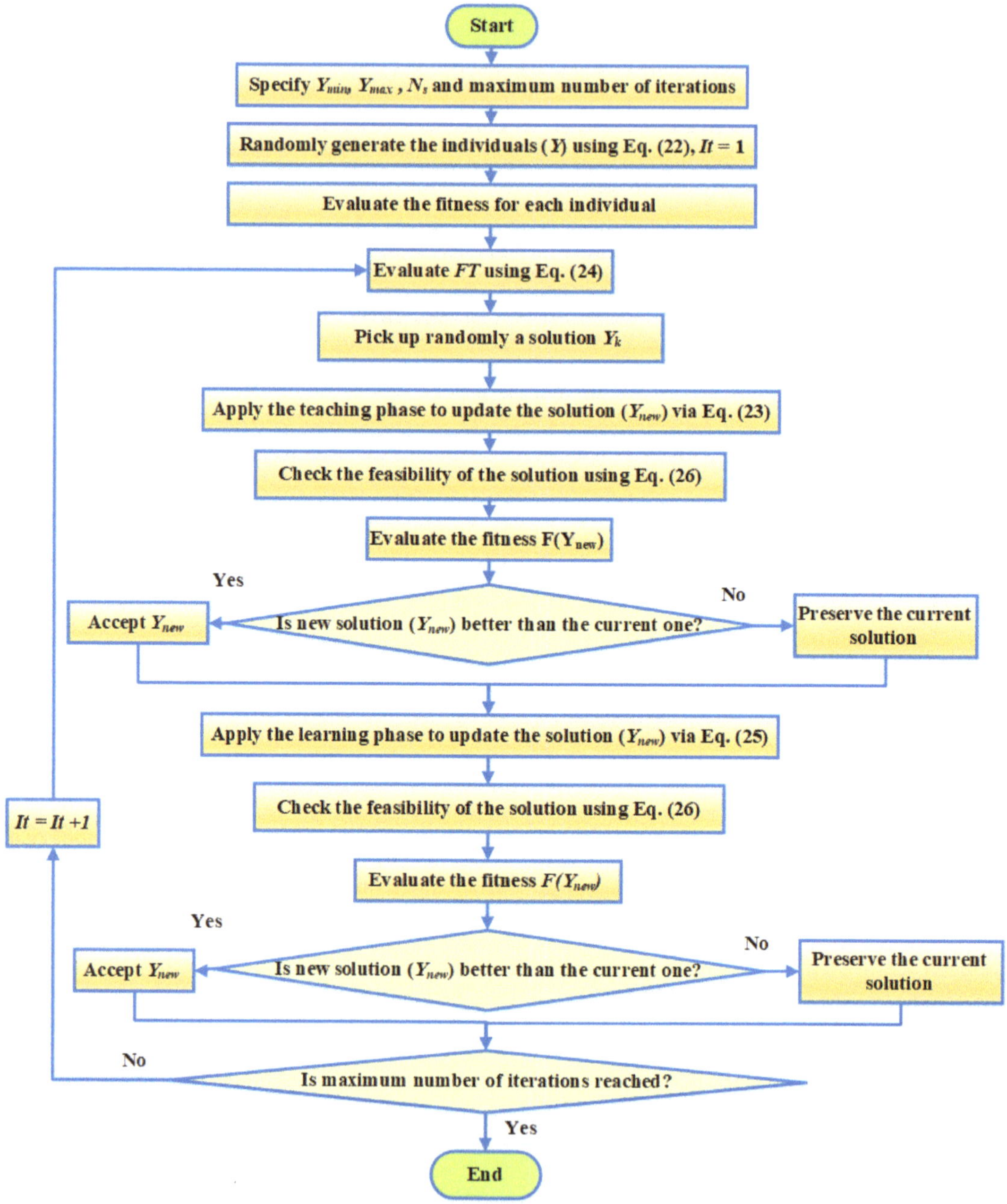

Figure 1. Proposed TLBA for solving the OPF problem in AC-DC grids.

4. Simulation Results

The proposed TLBA is applied in MATLAB and included in this section to solve the economic technical OPF issue in AC systems using modified IEEE 30 and 57 bus schemes. The population of students is 50 and 100 for the two examined networks, while the number of iterations is 300. The suggested TLBA is repeated 15 times and compared to some of the many other methods published in the literature.

4.1. *Results of the IEEE 30-Bus Network*

The initial IEEE 30-bus test system consists of 6 generators, 30 buses, 41 transmission branches, 4 on-load tap transformer, and 9 VAr sources. Its buses and branch data are derived from [58]. The cost parameters are derived from [59]. The modified system consists of two DC grid systems. The generators voltage has highest and lowest ranges of, respectively, 1.1 and 0.95 per-unit (pu). For the tap changing transformer, the permissible range is [0.9–1.1] pu. The highest and lowest voltage values for the load buses are assumed to be, respectively, 1.05 and 0.95 p.u. VSC 1 in the first DC system is under V_{dc}-Q_c controlling, whereas VSCs 2 and 3 are under P_{dc}-V_c controlling. VSC 4 is in V_{dc}-Q_c operating setting in the second DC system, whereas VSCs 5 and 6 are under P_{dc}-V_c controlling. The highest and lowest voltage values for the VSC stations and DC buses are 1.1 and 0.9 pu, correspondingly, and the conversion power for the VSC stations is 100 MVA. Two instances are analyzed where the goal of minimizing the TFCs is considered first, and the minimization of the ETLs is taken into account second.

4.1.1. Minimization of the TFCs of the IEEE 30-Bus Network

In the first instance, the TFCs minimization is considered in its quadratic form with additional sinusoid terms. The proposed TLBA is run, and the optimal results are shown in Table 1. As shown, the TLBA minimizes the TFCs value from 975.64 of 840.6166 $/h which indicates to a huge reduction percentage of 13.84%. Also, the convergence characteristics related to the proposed TLBA for this instance are shown in Figure 2.

Table 1. Simulation results of TLBA for the minimization of the TFCs of the IEEE 30-bus network.

Variables	Initial	TLBA	Variables	Initial	TLBA
Vg_1	1.05	1.057230475	Pg_2	80	44.23301348
Vg_2	1.04	1.038516219	Pg_5	50	20.24856
Vg_5	1.01	1.015932627	Pg_8	20	10.00046
Vg_8	1.01	1.023835298	Pg_{11}	20	10
Vg_{11}	1.05	1.046280787	Pg_{13}	20	12.00001
Vg_{13}	1.05	1.036847246	Qs_1	17.31	3.538584
$T_{6–9}$	1.078	0.900472851	Qs_4	−17.45	0.391427
$T_{6–10}$	1.069	1.083855873	Vc_2	1	1.024533
$T_{4–12}$	1.032	1.009042356	Vc_3	1	0.97523
$T_{28–27}$	1.068	0.986765581	Vc_5	1	1.053674
Qc_{10}	19	0.176292937	Vc_6	1	1.051539
Qc_{12}	0	29.16232583	Ps_2	25.74	14.35307
Qc_{15}	0	0.308572782	Ps_3	52.53	33.43576
Qc_{17}	0	2.737558014	Ps_5	40.44	41.01778
Qc_{20}	0	4.153172252	Ps_6	18.45	17.98847
Qc_{21}	0	9.478169587	$V_{dc,1}$	1.06	1.099989
Qc_{23}	0	0.825606456	$V_{dc,4}$	1.06	1.09993
Qc_{24}	4.3	9.310278385	ECFs ($/h)	975.64	840.6166
Qc_{29}	0	0.002124205	ETLs (MW)	11.9236	13.32178
Pg_1	105.32	199.935112			

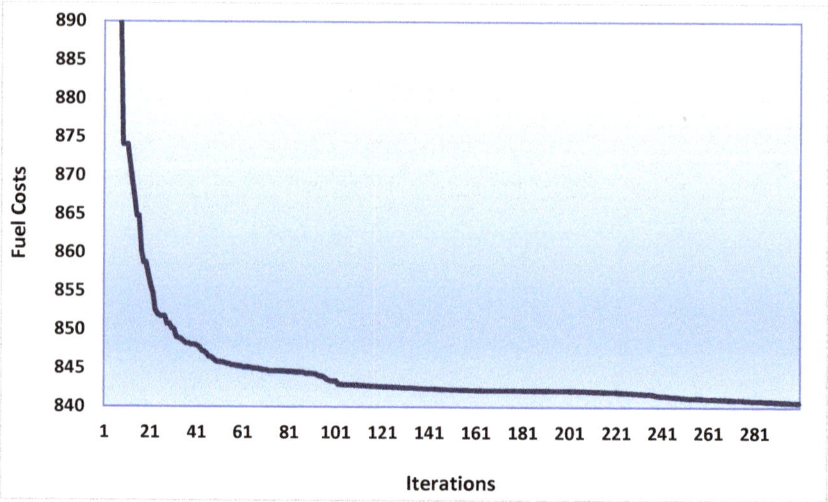

Figure 2. Convergence curves of TLBA for the minimization of the TFCs of the IEEE 30-bus network.

As illustrated, the proposed TLBA has significant convergent performance in avoiding local minima since it provides successive decreasing in the obtained objective.

Otherwise, Table 2 tabulates comparative results with other reported techniques of GWO [29], CSA [29], PSO [29] and ICSA [29]. In Appendix A, Table A1 identifies the settings of the control parameters for the methods established and reported in the comparisons. Table 2 deduces the great superiority of the proposed TLBA in finding the least TFCs of 840.6166 $/h where GWO, CSA, PSO and ICSA obtains TFCs of 854.43, 848.93, 846.25 and 842.34 $/h. Then, the TLBA achieves the most economical solution compared with the competitive algorithms.

Table 2. Comparative results of the IEEE 30-bus network for the minimization of the TFCs.

Method	ECFs ($/h)
Proposed TLBA	840.6166
ICSA [29]	842.34
PSO [29]	846.25
CSA [29]	848.93
GWO [29]	854.43

To investigate the analysis of the proposed TLBA in terms of average success rate and convergence characteristics, the minimizing of the TFCs for IEEE 30-bus system is considered. Table 3 tabulates the related absolute difference between the best and worst, its percentage and the regarded success rate. The absolute difference between the best and worst, its percentage and the success rate are computed at different percentage of the convergence including 50, 66.67, 83.33 and 100%. The proposed TLBA provides higher exploitation ability which is increased with the increasing of the convergence level. It always achieves small difference percentage, which is less than 0.5% at all the convergence levels. It always achieves high success rate which is increased with the increasing of tolerance level. At 83.33%, it provides more than 90% success rate at tolerances of 0.5 and 0.25%. Also, it provides more 86.67% and 46.67% success rates are achieved at tolerances of 0.1 and 0.05%, respectively. At 100% convergence, the proposed TLBA achieves 100% success rate at all tolerance levels. Decreasing the tolerance rates leads to decrease the success rates at different progress stages. An increased success rate is achieved for increasing the iteration number for all tolerance levels.

Table 3. Success rate of the proposed TLBA for the minimization of the TFCs of the IEEE 30-bus network.

		At 50% Convergence Iterations = 150	At 66.67% Convergence Iterations = 200	At 83.33% Convergence Iterations = 250	At 100% Convergence Iterations = 300
|Best-worst| ($/h)		2.4629	2.5588	3.0342	3.9384
|Best-worst| (%)		0.2927	0.3042	0.3610	0.4691
Success rate	Tolerance of 0.5%	100%	100%	100%	100%
	Tolerance of 0.25%	93.33%	93.33%	93.33%	100%
	Tolerance of 0.1%	26.67%	46.67%	86.67%	100%
	Tolerance of 0.05%	6.67%	26.67%	46.67%	100%

4.1.2. Minimization of the ETLs of the IEEE 30-Bus Network

In this case, the minimization of the ETLs is considered. The proposed TLBA is run, and the optimal simulation results that obtained by the proposed TLBA are reported in Table 4 compared with the initial operating condition. As shown, the proposed TLBA minimizes the ETLs values from 11.9236 MW to 8.582753 MW which indicates to a significant reduction percentage of 28.01%. Then, more technical improvement is noticed. However, the associated fuel costs are increased from 975.64 to 1044.197 $/h.

Table 4. Simulation results of TLBA for the minimization of the ETLs of the IEEE 30-bus network.

Variables	Initial	TLBA	Variables	Initial	TLBA
Vg_1	1.05	0.988792	Pg_2	80	80
Vg_2	1.04	0.986402	Pg_5	50	49.99999
Vg_5	1.01	0.975554	Pg_8	20	34.98962
Vg_8	1.01	0.981124	Pg_{11}	20	29.99579
Vg_{11}	1.05	1.043618	Pg_{13}	20	39.99997
Vg_{13}	1.05	1.099908	Qs_1	17.31	6.661652
$T_{6–9}$	1.078	0.930013	Qs_4	−17.45	5.704899
$T_{6–10}$	1.069	0.900445	Vc_2	1	0.981841
$T_{4–12}$	1.032	0.900137	Vc_3	1	0.90435
$T_{28–27}$	1.068	0.900415	Vc_5	1	1.087169
Qc_{10}	19	16.29461	Vc_6	1	1.08373
Qc_{12}	0	0.287574	Ps_2	25.74	6.568036
Qc_{15}	0	0.13763	Ps_3	52.53	20.83097
Qc_{17}	0	7.530908	Ps_5	40.44	15.98558
Qc_{20}	0	3.03858	Ps_6	18.45	12.34799
Qc_{21}	0	9.415924	$V_{dc,1}$	1.06	1.099742
Qc_{23}	0	1.233828	$V_{dc,4}$	1.06	1.099997
Qc_{24}	4.3	7.06592	ECFs ($/h)	975.64	1044.197
Qc_{29}	0	0.160811	ETLs (MW)	11.9236	8.582753
Pg_1	105.32	56.99738243			

Also, the convergence characteristics related to the proposed TLBA for this instance is shown in Figure 3. As illustrated, the proposed TLBA has better convergent performance in avoiding local minima.

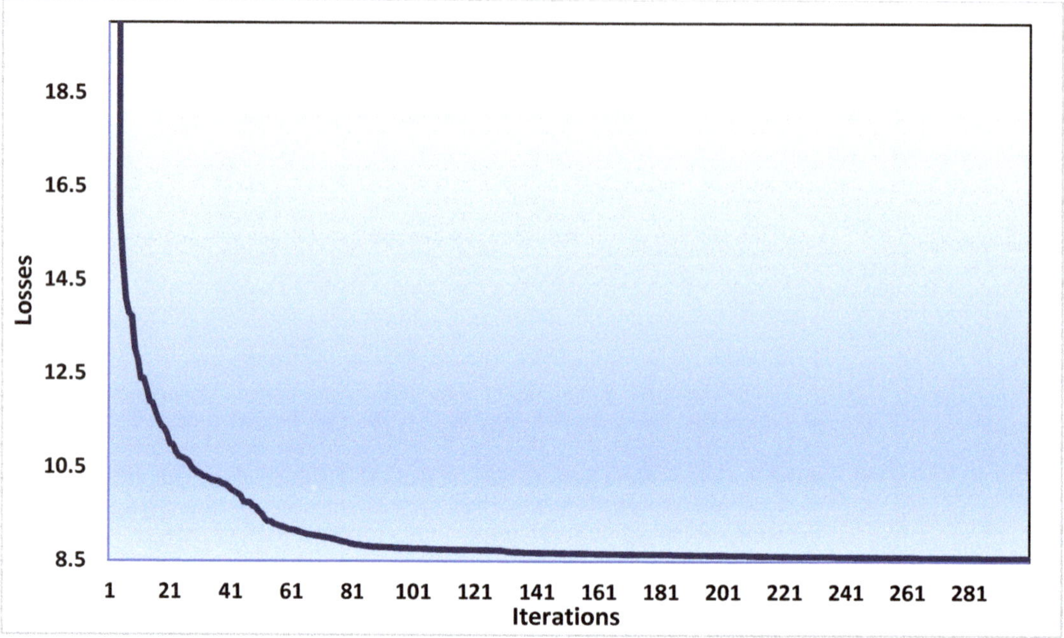

Figure 3. Convergence curves of TLBA for the minimization of the ETLs of the IEEE 30-bus network.

Also, for the minimization of the ETLs, Table 5 tabulates comparative results with various reported techniques of CSA [30], PSO [56], MVO [34], MPO [56], IMPO [34] and MRFO [30]. This table deduces the great superiority of the proposed TLBA in finding the least ETLs of 8.5827 MW where CSA, PSO, MVO, MPO, IMPO and MRFO [30] obtains ETLs of 9.57, 9.078, 9.005, 8.75, 8.66 and 8.5704 MW. The accepted level of accuracy, in terms of the technical merits, in ETLs is noted and compared with other methods in the literature.

Table 5. Comparative results of the IEEE 30-bus network for the minimization of the ETLs.

Method	ETLs (MW)
Proposed TLBA	8.5827
MRFO [30]	8.5704
IMPO [34]	8.66
MPO [56]	8.75
MVO [34]	9.005
PSO [56]	9.078
CSA [30]	9.57

To investigate the analysis of the proposed TLBA in terms of average success rate and convergence characteristics, Table 6 tabulates the related absolute difference between the best and worst, its percentage and the regarded success rate. Minimizing the ETLs for a IEEE 30-bus system is considered in Table 6. As shown, the proposed TLBA provides higher exploitation ability, which is increased with the increasing of the convergence level. It always achieves a small difference percentage which is less than 0.5% at all the convergence levels. It always achieves high success rate, which is increased with the increasing of tolerance level. At 83.33%, it provides more than 90% success rate at tolerances of 1, 0.75 and 0.5%. At 100% convergence, the proposed TLBA achieves 100% success rate at all tolerance levels. From the tabulated success rates, it is possible to state that:

- Decreasing the tolerance rates leads to decrease the success rates
- An increased success rate is achieved for increasing the iteration number

Table 6. Success rate of the proposed TLBA for the minimization of the ETLs of the IEEE 30-bus network.

		At 50% Convergence Iterations = 150	**At 66.67% Convergence Iterations = 200**	**At 83.33% Convergence Iterations = 250**	**At 100% Convergence Iterations = 300**
|Best-worst| (MW)		0.2274	0.1865	0.1967	0.1880
|Best-worst| (%)		2.6357	2.1670	2.2892	2.1910
Success rate	Tolerance of 1%	53.33%	100%	100%	100%
	Tolerance of 0.75%	33.33%	86.67%	100%	100%
	Tolerance of 0.5%	6.67%	53.33%	100%	100%
	Tolerance of 0.25%	0%	6.67%	60%	100%

4.2. Results of the IEEE 57-Bus Network

The original IEEE 57-bus testing network includes 57 buses, 8 generators, 8 lines, 17 on-load tapping transformers, and 3 reactive sources. Its branch and bus data is based on [60]. As illustrated in Figure 4, the modified system consists of one DC grid system with five VSCs and four DC connected lines. The generators and loads voltage have highest and lowest ranges of, respectively, 1.06 and 0.94 pu. For the tap changing transformer, the permissible range is [0.9–1.1] pu. The VSCs may be found on buses 26–29 and 52, accordingly. VSC 1 is under V_{dc}-Q_c controlling, whereas VSCs 2–5 are under P_{dc}-V_c controlling. The highest and lowest voltage values for the VSC stations and DC buses are 1.1 and 0.9 pu, correspondingly, and the conversion power for the VSC stations is 100 MVA. For this system, three instances are analysed, each with a distinct aim in mind. The first introduces the goal of minimizing the TFCs in its quadratic form, while the second one takes the minimization of the ETLs into account.

4.2.1. Minimization of the TFCs of the IEEE 57-Bus Network

In the first instance, the minimization of the TFCs is considered in its quadratic form. The proposed TLBA is run, and the optimal results are shown in Table 7 where their convergence characteristics are described in Figure 5. As shown, based on the proposed TLBA, the TFCs are reduced from 53,673.1 to 41,894.89 $/h compared with the initial case. This indicates to a huge reduction percentage of 21.94%.

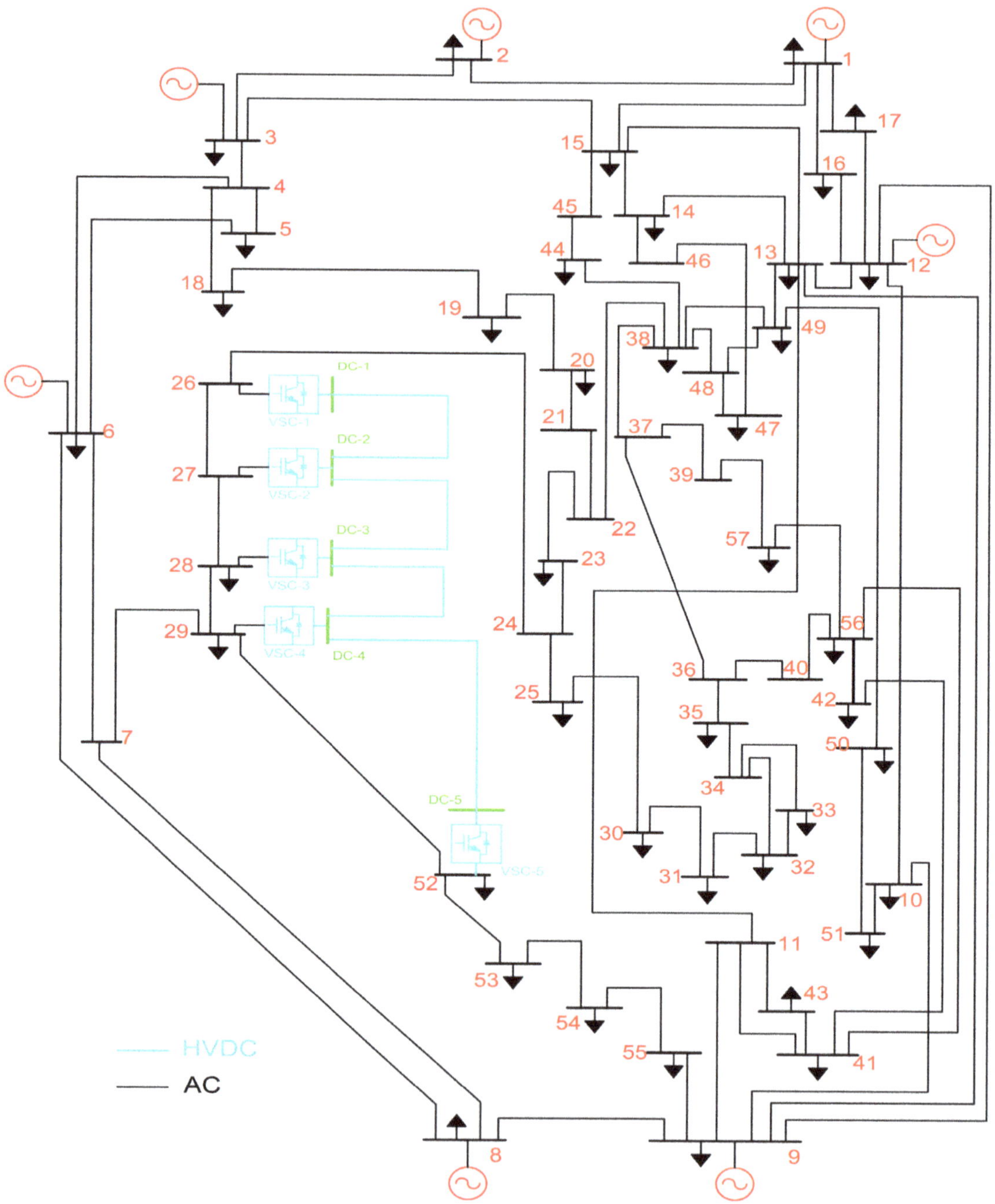

Figure 4. IEEE 57-bus with AC-DC network.

Table 7. Simulation results of TLBA for the minimization of the TFCs of the IEEE 57-bus network.

Variables	Initial	TLBA	Variables	Initial	TLBA
Vg1	1.01	0.987674	Tap 9–55	0.94	1.01373
Vg2	1.01	0.986699	Qc18	10	12.9268
Vg3	1.01	0.988601	Qc25	5.9	7.510958
Vg6	1.01	1.006765	Qc53	6.3	12.06029
Vg8	1.01	1.027091	Pg1	502.84	141.9835
Vg9	1.01	0.992088	Pg2	0	85.46196
Vg12	1.01	0.982979	Pg3	40	45.14887
Tap 4–18	0.97	0.90711	Pg6	0	80.09542
Tap 4–18	0.978	0.915543	Pg8	450	455.9096
Tap 21–20	1.043	0.968473	Pg9	0	99.83689
Tap 24–25	1	0.950536	Pg12	310	363.1258
Tap 24–25	1	0.947538	Qs1	17.31	5.636424
Tap 24–26	1.043	1.1	Vc2	1	0.96259
Tap 7–29	0.967	1.040925	Vc3	1	0.967439
Tap 34–32	0.975	0.943351	Vc4	1	0.975098
Tap 11–41	0.955	0.90019	Vc5	1	0.964904
Tap 15–45	0.955	0.91526	Ps2	25.47	4.062106
Tap 14–46	0.9	0.912562	Ps3	52.53	−5.37369
Tap 10–51	0.93	0.92415	Ps4	−59.91	−43.1476
Tap 13–49	0.895	0.900441	Ps5	−59.91	16.62977
Tap 11–43	0.958	0.900242	Vdc,1	1	1.081541
Tap 40–56	0.958	0.973081	ECFs ($/h)	53673.1	41894.89
Tap 39–57	0.98	0.952192	ETLs (MW)	52.04	20.00579

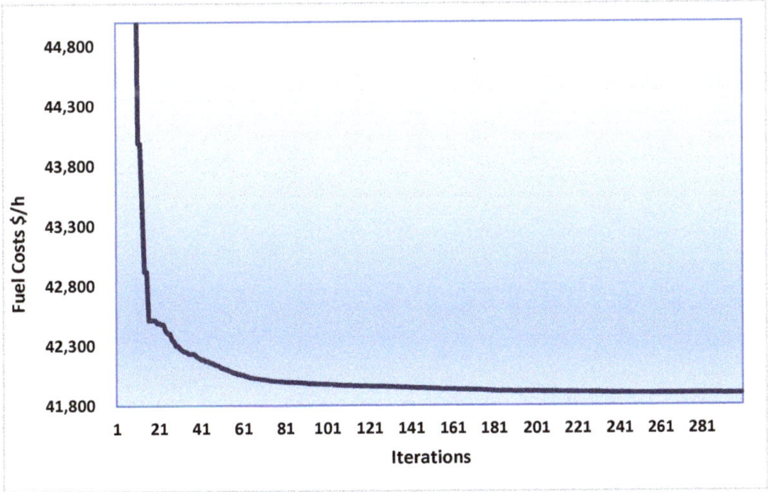

Figure 5. Convergence curves of TLBA for the minimization of the TFCs of the IEEE 57-bus network.

For this instance, Table 8 tabulates comparative results with other reported techniques in CSA [30], MVO [17], PSO [30], MPO [17], MRFO [30] and IMPO [17]. This table deduces the great superiority of the proposed TLBA in finding the least TFCs of 41,888.86 $/h where

CSA, MVO, PSO, MPO, MRFO and IMPO obtains TFCs of 42,050.2, 43,628.05, 41,932.8, 41,987.91, 41,923.6 and 41,920.67 $/h.

Table 8. Comparative results for the minimization of the TFCs of the IEEE 57-bus network.

Method	ECFs ($/h)
Proposed TLBA	41,894.89
IMPO [17]	41,920.67
MRFO [30]	41,923.6
MPO [17]	41,987.91
PSO [30]	41,932.8
MVO [17]	43,628.05
CSA [30]	42,050.2

4.2.2. Minimization of the ETLs of the IEEE 57-Bus Network

In the second instance, the minimization of the ETLs is considered. The proposed TLBA is run, and the optimal results are shown in Table 9. As shown, the proposed TLBA obtains lower ETLs value from 52.04 of 15.67 MW which indicates to a huge reduction percentage of 69.83%. The voltage level at generation buses are close to 1.0 p.u and therefore enhanced voltage profile. Also, the convergence characteristics related to the proposed TLBA are for this instance is shown in Figure 6. As illustrated, the proposed TLBA has better convergent performance in avoiding local minima. Also, for minimizing the power losses, Table 10 tabulates comparative results with various reported techniques of CSA [30], PSO [30], MPO [17], MRFO [30] and IMPO [17]. This table deduces the great superiority of the proposed TLBA in finding the least ETLs of 15.6711 MW where CSA, PSO, MPO, MRFO and IMPO obtains ETLs of 18.635, 17.337, 16.20859, 16.82 and 16.10132 MW. Then, more technical benefits are achieved using TLBA.

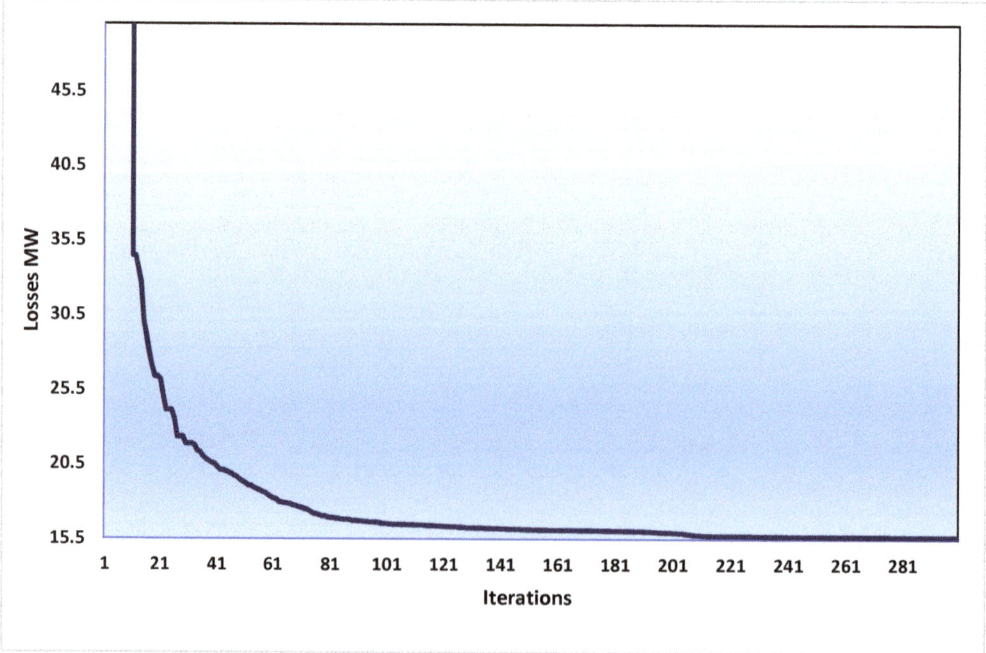

Figure 6. Convergence curves of TLBA for the minimization of the ETLs of the IEEE 57-bus network.

Table 9. Simulation results of TLBA for the minimization of the ETLs of the IEEE 57-bus network.

Variables	Initial	TLBA	Variables	Initial	TLBA
Vg1	1.01	0.977438	Tap 9–55	0.94	0.994384
Vg2	1.01	0.971793	Qc18	10	11.94515
Vg3	1.01	0.979725	Qc25	5.9	15.53356
Vg6	1.01	0.978621	Qc53	6.3	13.19981
Vg8	1.01	0.988716	Pg1	502.84	185.6365
Vg9	1.01	0.966121	Pg2	0	0.829644
Vg12	1.01	0.971125	Pg3	40	139.896
Tap 4–18	0.97	0.904931	Pg6	0	99.99999
Tap 4–18	0.978	0.910205	Pg8	450	322.511
Tap 21–20	1.043	0.983688	Pg9	0	99.99899
Tap 24–25	1	1.037482	Pg12	310	409.9903
Tap 24–25	1	0.936694	Qs1	17.31	−2.29511
Tap 24–26	1.043	1.090701	Vc2	1	0.955538
Tap 7–29	0.967	1.002518	Vc3	1	0.962063
Tap 34–32	0.975	0.926401	Vc4	1	0.969298
Tap 11–41	0.955	0.900318	Vc5	1	0.964268
Tap 15–45	0.955	0.905755	Ps2	25.47	4.210106
Tap 14–46	0.9	0.905533	Ps3	52.53	−3.32897
Tap 10–51	0.93	0.912182	Ps4	−59.91	−41.8902
Tap 13–49	0.895	0.900007	Ps5	−59.91	17.64526
Tap 11–43	0.958	0.900157	Vdc,1	1	1.084429
Tap 40–56	0.958	0.998685	ECFs ($/h)	53673.1	44894.97
Tap 39–57	0.98	0.977515	ETLs (MW)	52.04	15.6711

Table 10. Comparative results for the minimization of the ETLs of the IEEE 57-bus network.

Method	ETLs (MW)
Proposed TLBA	15.6711
IMPO [17]	16.10132
MRFO [30]	16.82
MPO [17]	16.20859
PSO [30]	17.337
CSA [30]	18.635

4.3. Statistical Analysis of the Proposed TLBA in Solving the OPF Problem

For the modified IEEE 30-bus AC-DC network, a statistical analysis is conducted by displaying the minimum, mean and maximum of the objectives obtained for the proposed TLBA as shown in Figure 7. As shown, the proposed TLBA has superior performance. For minimizing the TFCs, the proposed TLBA obtains small values of the indices of minimum, mean and maximum of the TFCs of 840.616, 841.838 and 843.433, respectively. For minimizing the ETLs, the proposed TLBA obtains small values of the indices of minimum, mean and maximum of the ETLs of 8.58, 8.635 and 8.771, respectively. For both studied cases, the proposed TLBA declares very small standard deviation of 0.8475 $/h and 0.04993 MW. This suggests significant robust performance of the enhanced TLBA.

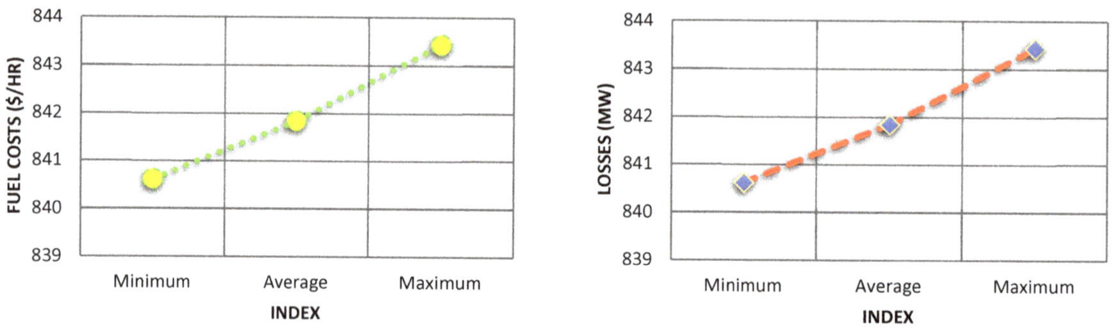

(a) (b)

Figure 7. Statistical analysis curves of TLBA for the first system. (**a**) Instance 1; (**b**) Instance 2.

For the modified IEEE 57-bus AC-DC network, the minimum, mean and maximum of the objectives obtained for the proposed TLBA are described in Figure 8. As shown, the proposed TLBA has superior performance. For minimizing the TFCs, the proposed TLBA obtains small values of the indices of minimum, mean and maximum of the TFCs of 41,894.89, 41,929.594 and 41,981.34, respectively. For minimizing the ETLs, the proposed TLBA obtains small values of the indices of minimum, mean and maximum of the ETLs of 15.6711, 16.0413 and 16.637, respectively. For both studied cases, the proposed TLBA showed very small standard deviation of 24.99 $/h and 0.3182 MW. This suggests a significantly robust performance of the enhanced TLBA.

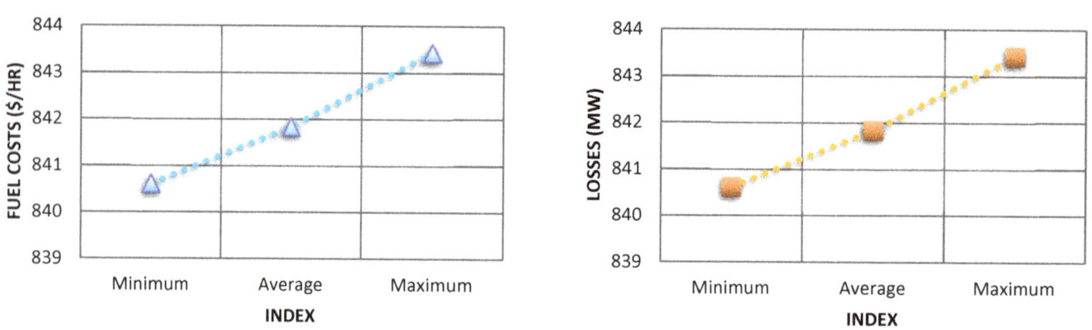

(a) (b)

Figure 8. Statistical analysis curves of TLBA for the second system. (**a**) Instance 1; (**b**) Instance 2.

5. Conclusions and Discussion

The TLBA algorithm is a powerful and effective optimization approach. TLBA mimics the teaching–learning process in a classroom, where TLBA's iterative computing process is separated into two phases, unlike standard evolutionary algorithms and swarm intelligence algorithms, and each phase conducts an iterative learning operation. TLBA is used to various sorts of nonlinear and multimodal functioning of hybrid alternating current and multi-terminal direct current power grids. Advanced technologies of Voltage source converter enable greater active and reactive power regulation in these networks. Various goals for optimal energy management are presented, with the aim of achieving economic and technical benefits. The proposed TLBA is evaluated on modified IEEE 30-bus and 57-bus AC-DC networks and compared to other published methods in the literature. For the IEEE 30 bus system, huge reduction percentages of 13.84 and 28.01% in the overall fuel costs and transmission power losses are achieved utilizing the proposed TLBA compared to the initial case. The proposed TLBA obtains huge reductions in the costs and losses with 21.94 and 69.83%, also for the IEEE 57 bus system, compared to the initial case. For both systems, very high success rates are demonstrated for the proposed TLBA. Therefore, these numerical results demonstrate that the proposed TLBA has great effectiveness and robustness indices over the others. Nevertheless, TLBA has faster convergence, higher quality for the final optimal solution, and more power for escaping from convergence to local optima. Different success rates are achieved that correspond with two criteria the progress of iteration number and lowering the tolerance rates.

In this study, significant technical and economical improvements are achieved for different test systems. However, some limitations are important to consider, since the application of presented TLBA is dependent on the control parameter settings. Therefore, the main limitation of this study is the pre-specification requirement of the control parameter settings, which are the number of students and the maximum number of iterations. In order to appropriately adapt them for any test system, a parametric analysis should be assessed in order to optimally extract the control parameter settings. Also, the comparison is executed with different recent techniques, which are reported in the literature. However, there are several new algorithms which are created monthly with different characteristics. Therefore, applications of modern optimization algorithms such as equilibrium, slime mould [61] and tunicate [62] optimizers, etc., can also be considered as another trend of future studies, especially for the high number of objectives and constraints. Even with the previous benefits, the modelling of various renewable energy resources must be included in future work to show the merits of this study because the penetration of renewable energy resources can be considered as the need for hybrid AC-DC networks.

Author Contributions: Conceptualization, A.M.S.; Data curation, S.S.; Formal analysis, R.A.E.-S. and M.G.; Funding acquisition, S.S.; Investigation, M.G.; Methodology, A.M.S.; Resources, R.A.E.-S. and M.G.; Software, A.M.S.; Writing—review & editing, S.S., R.A.E.-S. and M.G. All authors have read and agreed to the published version of the manuscript.

Funding: This research received no external funding.

Institutional Review Board Statement: Not applicable.

Informed Consent Statement: Not applicable.

Data Availability Statement: All required data are involved in the article.

Conflicts of Interest: The authors declare no conflict of interest.

Abbreviations

AC	Alternating Current
CSA	Crow Search Algorithm
DC	Direct Current
ELD	Economic Load Dispatch
ETL	Entire Transmission Loss
GA	Genetic Approach
GWO	Grey Wolf Optimizer
HVAC	High Voltage Alternating Current
HVDC	High Voltage Direct Current
ICSA	Improved Crow Search Algorithm
IMPO	Improved Marine Predators Algorithm
MPO	Marine Predators Algorithm
MRFO	Manta Ray Foraging Optimizer
MVO	Multi-Verse Optimizer
NSMRFO	Non-Dominated Sorting Manta Ray Foraging Optimization
OPF	Optimal Power Flow
PSO	Particle Swarm Optimizer

PV	Photovoltaic
TFC	Total Fuel Cost
TLBA	Teaching Learning-Based Algorithm
VSC	Voltage Source Converter

Appendix A

Table A1 tabulates the control parameter values used for TLBA, ICSA, PSO, CSA and GWO which are the methods established and reported in the comparisons.

Table A1. Parameter settings for the methods that were used.

Algorithm	Settings of the Control Parameters
TLBA	Number of students is 50 and 100 for IEEE 30 and 57 bus systems Maximum number of iterations is 300. Learning changing factor (FT) = randi([1 2])
ICSA	Number of crows is 50 and 100 for IEEE 30 and 57 bus systems Maximum number of iterations is 300. Awareness probability = 0.3 Flight length = 2
PSO	Number of particles is 50 and 100 for IEEE 30 and 57 bus systems Maximum number of iterations is 300. cognitive coefficient (c_1) = 2; social coefficient (c_2) = 2 Constriction coefficient (chi) = 1
CSA	Number of crows is 50 and 100 for IEEE 30 and 57 bus systems Maximum number of iterations is 300. Awareness probability = 0.3 Flight length = 2
GWO	Number of grey wolves is 50 and 100 for IEEE 30 and 57 bus systems Maximum number of iterations is 300. An adaptive parameter (a) which is decreased linearly from 2 to 0 as follows: a = 2-iter × ((2)/MaxIt)

References

1. Ali, M.H.; Kamel, S.; Hassan, M.H.; Tostado-Véliz, M.; Zawbaa, H.M. An improved wild horse optimization algorithm for reliability based optimal DG planning of radial distribution networks. *Energy Rep.* **2022**, *8*, 582–604. [CrossRef]
2. Khoa, T.H.; Vasant, P.M.; Singh, M.S.B.; Dieu, V.N. Swarm based mean-variance mapping optimization for convex and non-convex economic dispatch problems. *Memetic Comput.* **2017**, *9*, 91–108. [CrossRef]
3. Kamboj, V.K.; Kumari, C.L.; Bath, S.K.; Prashar, D.; Rashid, M.; Alshamrani, S.S.; AlGhamdi, A.S. A Cost-Effective Solution for Non-Convex Economic Load Dispatch Problems in Power Systems Using Slime Mould Algorithm. *Sustainability* **2022**, *14*, 2586. [CrossRef]
4. Wu, L.; Liu, Z.; Wei, H.L.; Wang, R. An efficient bilevel differential evolution algorithm with adaptation of lower level population size and search radius. *Memetic Comput.* **2021**, *13*, 227–247. [CrossRef]
5. Daqaq, F.; Kamel, S.; Ouassaid, M.; Ellaia, R.; Agwa, A.M.; Sa, A.M.A. Non-Dominated Sorting Manta Ray Foraging Optimization for Multi-Objective Optimal Power Flow with Wind/Solar/Small-Hydro Energy Sources. *Fractal Fract.* **2022**, *6*, 194. [CrossRef]
6. El-Sehiemy, R.; Elsayed, A.; Shaheen, A.; Elattar, E.; Ginidi, A. Scheduling of Generation Stations, OLTC Substation Transformers and VAR Sources for Sustainable Power System Operation Using SNS Optimizer. *Sustainability* **2021**, *13*, 11947. [CrossRef]
7. Acharya, S.; Ganesan, S.; Kumar, D.V.; Subramanian, S. A multi-objective multi-verse optimization algorithm for dynamic load dispatch problems. *Knowl. Based Syst.* **2021**, *231*, 107411. [CrossRef]
8. Korompili, A.; Wu, Q.; Zhao, H. Review of VSC HVDC connection for offshore wind power integration. *Renew. Sustain. Energy Rev.* **2016**, *59*, 1405–1414. [CrossRef]
9. Renedo, J.; Ibrahim, A.A.; Kazemtabrizi, B.; Garcia-Cerrada, A.; Rouco, L.; Zhao, Q.; Garcia-Gonzalez, J. A simplified algorithm to solve optimal power flows in hybrid VSC-based AC/DC systems. *Int. J. Electr. Power Energy Syst.* **2019**, *110*, 781–794. [CrossRef]
10. Sau-Bassols, J.; Zhao, Q.; García-González, J.; Prieto-Araujo, E.; Gomis-Bellmunt, O. Optimal power flow operation of an interline current flow controller in an hybrid AC/DC meshed grid. *Electr. Power Syst. Res.* **2019**, *177*, 105935. [CrossRef]
11. Ma, Q.; Wei, W.; Chai, W.; Mei, S. Solvability region of AC–DC power systems with volatile renewable energy sources. *Energy Rep.* **2022**, *8*, 1463–1472. [CrossRef]
12. El-Hawary, M.E.; Ibrahim, S.T. A new approach to AC-DC load flow analysis. *Electr. Power Syst. Res.* **1995**, *33*, 193–200. [CrossRef]

13. Beerten, J.; Belmans, R. Development of an open source power flow software for high voltage direct current grids and hybrid AC/DC systems: MATACDC. *IET Gener. Transm. Distrib.* **2015**, *9*, 966–974. [CrossRef]
14. Messalti, S.; Belkhiat, S.; Saadate, S.; Flieller, D. A new approach for load flow analysis of integrated AC-DC power systems using sequential modified Gauss-Seidel methods. *Eur. Trans. Electr. Power* **2012**, *22*, 421–432. [CrossRef]
15. Shi, C.; Tang, A.; Yang, H.; Yan, H.; Lu, Z. Quasi-AC Optimal Power Flow for VSC-MTDC Systems. In Proceedings of the 2020 IEEE 4th Information Technology, Networking, Electronic and Automation Control Conference, ITNEC 2020, Chongqing, China, 12–14 June 2020. [CrossRef]
16. Zhao, Y.; Liu, Y. A practical AC-DC load flow program based on sequential solution method. In Proceedings of the Proceedings. International Conference on Power System Technology, Kunming, China, 13–17 October 2002; Volume 2. [CrossRef]
17. al Harthi, M.; Ghoneim, S.; Elsayed, A.; El-Sehiemy, R.; Shaheen, A.; Ginidi, A. A Multi-Objective Marine Predator Optimizer for Optimal Techno-Economic Operation of AC/DC Grids. *Stud. Inf. Control* **2021**, *30*, 89–99. [CrossRef]
18. Zhao, Q.; García-González, J.; Gomis-Bellmunt, O.; Prieto-Araujo, E.; Echavarren, F.M. Impact of converter losses on the optimal power flow solution of hybrid networks based on VSC-MTDC. *Electr. Power Syst. Res.* **2017**, *151*, 395–403. [CrossRef]
19. Beerten, J.; Cole, S.; Belmans, R. Modeling of multi-terminal vsc hvdc systems with distributed dc voltage control. *IEEE Trans. Power Syst.* **2014**, *29*, 34–42. [CrossRef]
20. Baradar, M.; Ghandhari, M. A multi-option unified power flow approach for hybrid AC/DC grids incorporating multi-terminal VSC-HVDC. *IEEE Trans. Power Syst.* **2013**, *28*, 2376–2383. [CrossRef]
21. Khan, M.O.; Wadood, A.; Abid, M.I.; Khurshaid, T.; Rhee, S.B. Minimization of network power losses in the ac-dc hybrid distribution network through network reconfiguration using soft open point. *Electronics* **2021**, *10*, 326. [CrossRef]
22. Wang, S.; Zhu, J.; Trinh, L.; Pan, J. Economic assessment of HVDC project in deregulated energy markets. In Proceedings of the 3rd International Conference on Deregulation and Restructuring and Power Technologies, DRPT 2008, Nanjing, China, 6–9 April 2008. [CrossRef]
23. Feng, W.; Tjernberg, L.B.; Mannikoff, A.; Bergman, A. A new approach for benefit evaluation of multiterminal VSC-HVDC using a proposed mixed AC/DC optimal power flow. *IEEE Trans. Power Deliv.* **2014**, *29*, 432–443. [CrossRef]
24. Lotfjou, A.; Fu, Y.; Shahidehpour, M. Hybrid AC/DC transmission expansion planning. *IEEE Trans. Power Deliv.* **2012**, *27*, 579–591. [CrossRef]
25. Hotz, M.; Member, S.; Utschick, W.; Member, S. *hynet*: An Optimal Power Flow Framework for Hybrid AC/DC Power Systems. *IEEE Trans. Power Syst.* **2020**, *35*, 1036–1047. [CrossRef]
26. Maulik, A.; Das, D. Optimal power dispatch considering load and renewable generation uncertainties in an AC–DC hybrid microgrid. *IET Gener. Transm. Distrib.* **2019**, *13*, 1164–1176. [CrossRef]
27. Baradar, M.; Hesamzadeh, M.R.; Ghandhari, M. Second-order cone programming for optimal power flow in VSC-type AC-DC grids. *IEEE Trans. Power Syst.* **2013**, *28*, 4282–4291. [CrossRef]
28. Cao, J.; Du, W.; Wang, H.F.; Bu, S.Q. Minimization of transmission loss in meshed AC/DC grids with VSC-MTDC networks. *IEEE Trans. Power Syst.* **2013**, *28*, 3047–3055. [CrossRef]
29. Shaheen, A.M.; Elsayed, A.M.; El-Sehiemy, R.A. Optimal Economic–Environmental Operation for AC-MTDC Grids by Improved Crow Search Algorithm. *IEEE Syst. J.* **2022**, *16*, 1270–1277. [CrossRef]
30. Elattar, E.E.; Shaheen, A.M.; Elsayed, A.M.; El-Sehiemy, R.A. Optimal power flow with emerged technologies of voltage source converter stations in meshed power systems. *IEEE Access* **2020**, *8*, 166963–166979. [CrossRef]
31. Pinto, R.T.; Rodrigues, S.F.; Wiggelinkhuizen, E.; Scherrer, R.; Bauer, P.; Pierik, J. Operation and power flow control of multi-terminal DC networks for grid integration of offshore wind farms using genetic algorithms. *Energies* **2013**, *6*, 1. [CrossRef]
32. Sayah, S. Modified differential evolution approach for practical optimal reactive power dispatch of hybrid AC–DC power systems. *Appl. Soft Comput. J.* **2018**, *73*, 591–606. [CrossRef]
33. Abdul-hamied, D.T.; Shaheen, A.M.; Salem, W.A.; Gabr, W.I.; El-sehiemy, R.A. Equilibrium optimizer based multi dimensions operation of hybrid AC/DC grids. *Alexandria Eng. J.* **2020**, *59*, 4787–4803. [CrossRef]
34. Elsayed, A.M.; Shaheen, A.M.; Alharthi, M.M.; Ghoneim, S.S.M.; El-Sehiemy, R.A. Adequate operation of hybrid AC/MT-HVDC power systems using an improved multi- objective marine predators optimizer. *IEEE Access* **2021**, *9*, 51065–51087. [CrossRef]
35. Rao, R.; Savsani, V.; Vakharia, D. Teaching–Learning-Based Optimization: An optimization method for continuous non-linear large scale problems. *Inf. Sci.* **2012**, *183*, 1–15. [CrossRef]
36. Zou, F.; Chen, D.; Xu, Q. A survey of teaching–learning-based optimization. *Neurocomputing* **2018**, *335*, 366–383. [CrossRef]
37. Xue, R.; Wu, Z. A Survey of Application and Classification on Teaching-Learning-Based Optimization Algorithm. *IEEE Access* **2019**, *8*, 1062–1079. [CrossRef]
38. Mi, X.; Liao, Z.; Li, S.; Gu, Q. Adaptive teaching–learning-based optimization with experience learning to identify photovoltaic cell parameters. *Energy Rep.* **2021**, *7*, 4114–4125. [CrossRef]
39. Li, L.; Xiong, G.; Yuan, X.; Zhang, J.; Chen, J. Parameter Extraction of Photovoltaic Models Using a Dynamic Self-Adaptive and Mutual- Comparison Teaching-Learning-Based Optimization. *IEEE Access* **2021**, *9*, 52425–52441. [CrossRef]
40. Ashtiani, M.N.; Toopshekan, A.; Astaraei, F.R.; Yousefi, H.; Maleki, A. Techno-economic analysis of a grid-connected PV/battery system using the teaching-learning-based optimization algorithm. *Sol. Energy* **2020**, *203*, 69–82. [CrossRef]
41. Yaqoob, M.T.; Rahmat, M.K.; Maharum, S.M.M. Modified teaching learning based optimization for selective harmonic elimination in multilevel inverters. *Ain Shams Eng. J.* **2022**, *13*, 101714. [CrossRef]

42. Le, P.-N.; Kang, H.-J. Robot Manipulator Calibration Using a Model Based Identification Technique and a Neural Network With the Teaching Learning-Based Optimization. *IEEE Access* **2020**, *8*, 105447–105454. [CrossRef]
43. Elshaboury, N.; Abdelkader, E.M.; Al-Sakkaf, A.; Alfalah, G. Teaching-Learning-Based Optimization of Neural Networks for Water Supply Pipe Condition Prediction. *Water* **2021**, *13*, 3546. [CrossRef]
44. Safari, M.; de Sousa, R.J.A.; Rabiee, A.H.; Tahmasbi, V. Investigation of Dissimilar Resistance Spot Welding Process of AISI 304 and AISI 1060 Steels with TLBO-ANFIS and Sensitivity Analysis. *Metals* **2021**, *11*, 1324. [CrossRef]
45. Yang, N.-C.; Liu, S.-W. Multi-Objective Teaching–Learning-Based Optimization with Pareto Front for Optimal Design of Passive Power Filters. *Energies* **2021**, *14*, 6408. [CrossRef]
46. Jalalzad, S.H.; Yektamoghadam, H.; Haghighi, R.; Dehghani, M.; Nikoofard, A.; Khosravy, M.; Senjyu, T. A Game Theory Approach Using the TLBO Algorithm for Generation Expansion Planning by Applying Carbon Curtailment Policy. *Energies* **2022**, *15*, 1172. [CrossRef]
47. Wu, D.; Jia, H.; Abualigah, L.; Xing, Z.; Zheng, R.; Wang, H.; Altalhi, M. Enhance Teaching-Learning-Based Optimization for Tsallis-Entropy-Based Feature Selection Classification Approach. *Processes* **2022**, *10*, 360. [CrossRef]
48. Ayalew, M.; Khan, B.; Alaas, Z.M. Optimal Service Restoration Scheme for Radial Distribution Network Using Teaching Learning Based Optimization. *Energies* **2022**, *15*, 2505. [CrossRef]
49. Alghamdi, A.S. A New Self-Adaptive Teaching–Learning-Based Optimization with Different Distributions for Optimal Reactive Power Control in Power Networks. *Energies* **2022**, *15*, 2759. [CrossRef]
50. Bouchekara, H.; Chaib, A.; Abido, M.; El-Sehiemy, R. Optimal power flow using an Improved Colliding Bodies Optimization algorithm. *Appl. Soft Comput.* **2016**, *42*, 119–131. [CrossRef]
51. Gupta, S.; Kumar, N.; Srivastava, L. Bat Search Algorithm for Solving Multi-objective Optimal Power Flow Problem. *Autom. Wirel. Syst. Electr. Eng.* **2019**, *553*, 347–362. [CrossRef]
52. Shaheen, A.M.; Ginidi, A.R.; El-Sehiemy, R.A.; Elattar, E.E. Optimal economic power and heat dispatch in Cogeneration Systems including wind power. *Energy* **2021**, *225*, 120263. [CrossRef]
53. Bentouati, B.; Khelifi, A.; Shaheen, A.M.; El-Sehiemy, R.A. An enhanced moth-swarm algorithm for efficient energy management based multi dimensions OPF problem. *J. Ambient Intell. Humaniz. Comput.* **2020**, *12*, 9499–9519. [CrossRef]
54. El Sehiemy, R.A.; El Ela, A.A.A.; Shaheen, A. A Multi-Objective Fuzzy-Based Procedure for Reactive Power-Based Preventive Emergency Strategy. *Int. J. Eng. Res. Afr.* **2014**, *13*, 91–102. [CrossRef]
55. Shaheen, A.M.; El-Sehiemy, R.A.; Alharthi, M.M.; Ghoneim, S.S.; Ginidi, A.R. Multi-objective jellyfish search optimizer for efficient power system operation based on multi-dimensional OPF framework. *Energy* **2021**, *237*, 121478. [CrossRef]
56. Shaheen, A.M.; El-Sehiemy, R.A.; Elsayed, A.M.; Elattar, E.E. Multi-objective manta ray foraging algorithm for efficient operation of hybrid AC/DC power grids with emission minimisation. *IET Gener. Transm. Distrib.* **2020**, *15*, 1314–1336. [CrossRef]
57. Beerten, J.; Cole, S.; Belmans, R. A sequential AC/DC power flow algorithm for networks containing multi-terminal VSC HVDC systems. In Proceedings of the IEEE PES General Meeting, PES 2010, Minneapolis, MN, USA, 25–29 July 2010. [CrossRef]
58. Shaheen, A.M.; El-Sehiemy, R.A. Application of multi-verse optimizer for transmission network expansion planning in power systems. In Proceedings of the International Conference on Innovative Trends in Computer Engineering ITCE 2019, Aswan, Egypt, 2–4 February 2019; pp. 371–376. [CrossRef]
59. Shaheen, A.M.; El-Sehiemy, R.A.; Elattar, E.E.; Abd-Elrazek, A.S. A Modified Crow Search Optimizer for Solving Non-Linear OPF Problem With Emissions. *IEEE Access* **2021**, *9*, 43107–43120. [CrossRef]
60. Zimmerman, M.-S.C.R.D. Matpower [Software]. Available online: https://matpower.org (accessed on 1 August 2021).
61. Sarhan, S.; Shaheen, A.M.; El-Sehiemy, R.A.; Gafar, M. An Enhanced Slime Mould Optimizer That Uses Chaotic Behavior and an Elitist Group for Solving Engineering Problems. *Mathematics* **2022**, *10*, 1991. [CrossRef]
62. Kaur, S.; Awasthi, L.K.; Sangal, A.; Dhiman, G. Tunicate Swarm Algorithm: A new bio-inspired based metaheuristic paradigm for global optimization. *Eng. Appl. Artif. Intell.* **2020**, *90*, 103541. [CrossRef]

MDPI
St. Alban-Anlage 66
4052 Basel
Switzerland
www.mdpi.com

MDPI Books Editorial Office
E-mail: books@mdpi.com
www.mdpi.com/books

www.ingramcontent.com/pod-product-compliance
Lightning Source LLC
Chambersburg PA
CBHW061500220926
51850CB00004B/138
9783036594262